Reference Series in Phytochemistry

Series Editors
Jean-Michel Mérillon
Faculty of Pharmaceutical Sciences
University of Bordeaux
Bordeaux, France

Kishan Gopal Ramawat
Botany Department
M.L. Sukhadia University
Udaipur, India

This reference works series provides a platform for all information on plant metabolites and phytochemicals, their chemistry, properties, applications, and methods. By the strictest definition, phytochemicals are chemicals derived from plants. However, the term is often used to describe the large number of secondary metabolic compounds found in and derived from plants. These metabolites exhibit a number of nutritional and protective functions for human welfare such as colorants, fragrances and flavorings, amino acids, pharmaceuticals, hormones, vitamins and agrochemicals. Besides food, fibers, fuel, cloth and shelter, a vast number of wild plants can hence provide important sources for medicines, especially in developing countries for their traditional health systems. Natural products have inspired and provided the foundation to the bulk of FDA-approved compounds and there is tremendous increase in natural products and natural products derived compounds that have been registered against many prevailing diseases. Natural product industry has shown tremendous growth and is expected to continue to do so in the near future. The present series compiles reference information on various topics and aspects about phytochemicals, including their potential as natural medicine, their role as chemo-preventers, in plant defense, their ecological role, their role in plants as well as for pathogen adaptation, and disease resistance. Volumes in the series also contain information on methods such as metabolomics, genetic engineering of pathways, molecular farming, and obtaining metabolites from lower organisms and marine organisms besides higher plants. The books in the series are hence of relevance in various fields, from chemistry, biology, biotechnology, to pharmacognosy, pharmacology, botany, or medicine. Each volume is edited by leading experts and contains authoritative contributions by renowned authors.

More information about this series at http://www.springer.com/series/13872

Sumita Jha
Editor

Transgenesis and Secondary Metabolism

With 108 Figures and 33 Tables

Springer

Editor
Sumita Jha
Department of Botany
Centre of Advanced Study
University of Calcutta
Kolkata, India

ISBN 978-3-319-28668-6 ISBN 978-3-319-28669-3 (eBook)
ISBN 978-3-319-28670-9 (print and electronic bundle)
DOI 10.1007/978-3-319-28669-3

Library of Congress Control Number: 2017933076

Printed on acid-free paper

This Springer imprint is published by Springer Nature
The registered company is Springer International Publishing AG
The registered company address is: Gewerbestrasse 11, 6330 Cham, Switzerland

Preface

Bioactive molecules available in plants are largely inaccessible to modern synthesis methods. Most of the commercially valuable substances are still extracted directly from plants, which are under pressure from human harvesting. The ultimate goal of this book is to provide an insight into the various biotechnological approaches currently undertaken globally to find alternative methods for production of natural secondary metabolites to ensure that global biodiversity is preserved so that the resources that could considerably benefit mankind are not lost. Plant genetic manipulations in relation to synthesis, accumulation, and production of secondary metabolites have been well investigated in large number of species around the world. The chapters thus focus on case studies as well as original research reviews in genetic and biotechnological approaches to enhance the production of given metabolite or a group of related compounds. The book also includes chapters on indirect genetic approaches to improve metabolite production including *Agrobacterium* mediated transformations as well as metabolic engineering of the relevant biosynthetic pathways, overexpression of genes encoding key enzymes, metabolic phytochemistry, and molecular pharming. We hope that the book will be useful for researchers in academia and industry.

The book is intended to serve the needs of graduate students, scholars, and researchers in the field of botany, agriculture, pharmacy, biotechnology, and phytochemistry; industrial scientists; and those involved in marketing phytochemicals and their extracts.

Finally, I would like to acknowledge all our contributors who have made immense efforts to ensure the scientific quality of this book. I am grateful to Dr. David Alan Tepfer for agreeing to write a chapter based on his lifetime experiences and sharing them with us. I thank Dr. K.G. Ramawat for constant encouragement. We thank all our colleagues at Springer for excellent support.

Sumita Jha

Contents

About the Editors

Prof. Dr. Sumita Jha (nee Mukherjee) received her M.Sc. (1975) and Ph.D. (1981) from the University of Calcutta, Kolkata, in India. She joined the same University as UGC (New Delhi) Research Scientist in 1985 and as a faculty member in Botany in 1990. She was appointed Associate Professor in 1993 and became Professor in 2001. She served as Program Coordinator of UGC Centre of Advanced Study in Botany (2004–2015), as Head Department of Botany (2007–2009), and as Head Department of Genetics (2009–2011; 2013–2014), Calcutta University. She has been involved in developing teaching on plant biology, plant genetics, and biotechnology.

Professor Jha's group has developed transgenic cell and organ cultures in a number of rare, endangered indigenous medicinal plants for production of high value pharmaceuticals. Her research is supported by funding from Department of Science and Technology and Department of Biotechnology, Government of India.

Prof. Jha under an Indo-French Project funded by IFCPAR/CEFIPRA (2002–2005) developed a strong collaboration with Dr. David Tepfer, INRA, Versailles, France, leading to the development and exchange of novel methods and information on secondary metabolism in transformed plant cell and organ cultures and proposing that natural transformation operates as an adaptive resource in evolution, particularly in plant–microorganism interactions. She also visited AFRC Institute of Food Research (Plant Biotechnology Group), Norwich, UK (1991), and Plant Science Division, Nottingham University, Nottingham, UK (1999), as Visiting Scientist under Royal Society–Indian National Science Academy scientist exchange program.

Prof. Jha has published more than 130 research papers in internationally recognized journals and 20 book chapters. To her credit, Professor Jha has mentored 20 students for their Ph.D. degree and five are working for Ph.D. She has served on various advisory committees and has garnered prestigious awards for her contributions to academic excellence. Notable among these are the INSA Science Academy Medal for Young Scientist (1983), the Prof. Hiralal Chakravarty Award by ISCA (1989), the UGC Career Award for young teachers (1994–1997), Fellow of the National Academy of Sciences, India (2008), and Fellow of the West Bengal Academy of Science and Technology (2015).

Contributors

Hafiz A. Abdelgadir Agricultural Research Council, Vegetable and Ornamental Plants (VOP), Pretoria, South Africa

University of Limpopo, Sovenga, South Africa

Eduardo Bezerra Almeida Jr. Graduate Program in Health Sciences, Biological and Health Sciences Center, Federal University of Maranhão, São Luís, MA, Brazil

Laboratory of Botanical Studies, Department of Biology, Biological and Health Sciences Center, Federal University of Maranhão, São Luís, MA, Brazil

Sara Adrián L. Andrade Department of Plant Biology, Institute of Biology, State University of Campinas, Campinas, SP, Brazil

Randolph R. J. Arroo Leicester School of Pharmacy, De Montfort University, Leicester, UK

Suchitra Banerjee Department of Plant Biotechnology, Central Institute of Medicinal and Aromatic Plants (CSIR-CIMAP), Lucknow, Uttar Pradesh, India

Mercedes Bonfill Plant Physiology Laboratory, Faculty of Pharmacy, University of Barcelona, Barcelona, Spain

Priscilla Rodrigues Valadares Campana Serviço de Fitoquímica e Prospecção Farmacêutica, Diretoria de Pesquisa e Desenvolvimento, Fundação Ezequiel Dias, Belo Horizonte, MG, Brazil

Sheela Chandra Department of Bio-Engineering, Birla Institute of Technology, Mesra, Ranchi, Jharkhand, India

Vera Lúcia de Almeida Serviço de Fitoquímica e Prospecção Farmacêutica, Diretoria de Pesquisa e Desenvolvimento, Fundação Ezequiel Dias, Belo Horizonte, MG, Brazil

Antonio Marcus de Andrade Paes Graduate Program in Health Sciences, Biological and Health Sciences Center, Federal University of Maranhão, São Luís, MA, Brazil

Laboratory of Experimental Physiology, Department of Physiological Sciences, Biological and Health Sciences Center, Federal University of Maranhão, São Luís, MA, Brazil

Flavia Maria Mendonça do Amaral Department of Pharmacy, Biological and Health Sciences Center, Federal University of Maranhão, São Luís, MA, Brazil

Ian du Plooy Agricultural Research Council, Vegetable and Ornamental Plants (VOP), Pretoria, South Africa

Zakheleni P. Dube Green Technologies Research Centre, University of Limpopo, Sovenga, South Africa

Biswajit Ghosh Plant Biotechnology Laboratory, Department of Botany, Ramakrishna Mission Vivekananda Centenary College, Kolkata, India

Mihir Halder Department of Botany, University of Calcutta, Kolkata, India

Kathleen Hefferon Department of Food Sciences, Cornell University, Ithaca, NY, USA

Sumita Jha Department of Botany, Centre of Advanced Study, University of Calcutta, Kolkata, India

Raj Kumar Joshi Centre of Biotechnology, Siksha O Anusandhan University, Bhubaneswar, Odisha, India

Sana Khan Plant Biotechnology Division, CSIR-Central Institute of Medicinal and Aromatic Plants, Lucknow, Uttar Pradesh, India

Tomotsugu Koyama Bioorganic Research Institute, Suntory Foundation for Life Sciences, Seika, Kyoto, Japan

Dijana Krstić-Milošević Institute for Biological Research "Siniša Stanković", University of Belgrade, Belgrade, Serbia

Madhumita Kumari Department of Bio-Engineering, Birla Institute of Technology, Mesra, Ranchi, Jharkhand, India

Subrata Kundu Plant Biotechnology Laboratory, Department of Botany, Ramakrishna Mission Vivekananda Centenary College, Kolkata, India

Daniel Lipiński Department of Biochemistry and Biotechnology, Poznan University of Life Sciences, Poznan, Poland

Aleksandra Luwańska Institute of Natural Fibres and Medicinal Plants, Poznan, Poland

Abdullah Makhzoum Department of Biology, The University of Western Ontario, London, ON, Canada

Sonia Malik Graduate Program in Health Sciences, Biological and Health Sciences Center, Federal University of Maranhão, São Luís, MA, Brazil

Phatu W. Mashela Green Technologies Research Centre, University of Limpopo, Sovenga, South Africa

Erika Matsumoto Bioorganic Research Institute, Suntory Foundation for Life Sciences, Seika, Kyoto, Japan

Paulo Mazzafera Department of Plant Biology, Institute of Biology, State University of Campinas, Campinas, SP, Brazil

Shakti Mehrotra Plant Biotechnology Division, Central Institute of Medicinal and Aromatic Plants, Lucknow, Uttar Pradesh, India

Mohammad Hossein Mirjalili Department of Agriculture, Medicinal Plants and Drugs Research Institute, Shahid Beheshti University, G.C., Tehran, Iran

Manoj K. Mishra Department of Botany, Lucknow University, Lucknow, Uttar Pradesh, India

Rukmini Mishra Centre of Biotechnology, Siksha O Anusandhan University, Bhubaneswar, Odisha, India

Sonal Mishra National Institute of Plant Genome Research, New Delhi, India

Pratibha Misra Genetics and Plant Molecular Biology Division, Council of Scientific and Industrial Research, National Botanical Research Institute, Lucknow, Uttar Pradesh, India

Adinpunya Mitra Natural Product Biotechnology Group, Agricultural and Food Engineering Department, Indian Institute of Technology Kharagpur, Kharagpur, West Bengal, India

Jatindra Nath Mohanty Centre of Biotechnology, Siksha O Anusandhan University, Bhubaneswar, Odisha, India

Denise Fernandes Coutinho Moraes Department of Pharmacy, Biological and Health Sciences Center, Federal University of Maranhão, São Luís, MA, Brazil

Kinuyo Morimoto Bioorganic Research Institute, Suntory Foundation for Life Sciences, Seika, Kyoto, Japan

Chiranjit Mukherjee Natural Product Biotechnology Group, Agricultural and Food Engineering Department, Indian Institute of Technology Kharagpur, Kharagpur, West Bengal, India

Department of Botany, Chandernagore College, Chandernagore, West Bengal, India

Swapna Mukherjee Department of Microbiology, Dinabandhu Andrews College, Garia, Kolkata, India

Jun Murata Bioorganic Research Institute, Suntory Foundation for Life Sciences, Seika, Kyoto, Japan

Satyabrata Nanda Centre of Biotechnology, Siksha O Anusandhan University, Bhubaneswar, Odisha, India

Bhekumthetho Ncube Research Centre for Plant Growth and Development, School of Life Sciences, University of KwaZulu-Natal Pietermaritzburg, Scottsville, South Africa

Ashwell R. Ndhlala Division Crop Science, Agricultural Research Council, Pretoria, South Africa

Green Technologies Research Centre, University of Limpopo, Sovenga, South Africa

Eiichiro Ono Research Institute, Suntory Global Innovation Center (SIC) Ltd., Seika, Kyoto, Japan

Pallavi Pandey Department of Plant Biotechnology, Central Institute of Medicinal and Aromatic Plants (CSIR-CIMAP), Lucknow, Uttar Pradesh, India

Vibha Pandey Department of Plant Molecular Biology, University of Delhi South Campus, New Delhi, India

Adrian J. Parr Institute of Food Research, Norwich Research Park, Colney, Norwich, UK

Christie A. M. Peebles Chemical and Biological Engineering Department, Colorado State University, Fort Collins, CO, USA

Milica Pešić Institute for Biological Research "Siniša Stanković", University of Belgrade, Belgrade, Serbia

Ana Podolski-Renić Institute for Biological Research "Siniša Stanković", University of Belgrade, Belgrade, Serbia

Kgabo M. Pofu Green Technologies Research Centre, University of Limpopo, Sovenga, South Africa

Division Crop Protection, Agricultural Research Council, Pretoria, South Africa

Laiq ur Rahman Plant Biotechnology Division, CSIR-Central Institute of Medicinal and Aromatic Plants, Lucknow, Uttar Pradesh, India

Kaliyaperumal Rekha Department of Applied Bioscience, College of Life and Environmental Sciences, Konkuk University, Seoul, Republic of Korea

Maria Nilce Sousa Ribeiro Department of Pharmacy, Biological and Health Sciences Center, Federal University of Maranhão, São Luís, MA, Brazil

Marina Pereira Rocha Serviço de Biotecnologia Vegetal, Diretoria de Pesquisa e Desenvolvimento, Fundação Ezequiel Dias, Belo Horizonte, MG, Brazil

Dipasree Roychowdhury Department of Botany, University of Calcutta, Kolkata, India

Syed Saema Department of Biotechnology, Integral University, Lucknow, Uttar Pradesh, India

Honoo Satake Bioorganic Research Institute, Suntory Foundation for Life Sciences, Seika, Kyoto, Japan

Katarzyna Seidler-Łożykowska Institute of Natural Fibres and Medicinal Plants, Poznan, Poland

Cláudia Gontijo Silva Serviço de Biotecnologia Vegetal, Diretoria de Pesquisa e Desenvolvimento, Fundação Ezequiel Dias, Belo Horizonte, MG, Brazil

Sanjeewa Singhabahu Human Genetics Unit, Faculty of Medicine, University of Colombo, Colombo, Sri Lanka

Gaurav Singh Genetics and Plant Molecular Biology Division, Council of Scientific and Industrial Research, National Botanical Research Institute, Lucknow, Uttar Pradesh, India

Sailendra Singh Department of Plant Biotechnology, Central Institute of Medicinal and Aromatic Plants (CSIR-CIMAP), Lucknow, Uttar Pradesh, India

Debabrata Sircar Natural Product Biotechnology Group, Agricultural and Food Engineering Department, Indian Institute of Technology Kharagpur, Kharagpur, West Bengal, India

Department of Biotechnology, Indian Institute of Technology Roorkee, Roorkee, Uttarakhand, India

Ryszard Słomski Department of Biochemistry and Biotechnology, Poznan University of Life Sciences, Poznan, Poland

Vikas Srivastava National Institute of Plant Genome Research, New Delhi, India

Department of Plant Sciences, Central University of Jammu, Samba, Jammu (J & K), India

Angelina Subotić Institute for Biological Research "Siniša Stanković", University of Belgrade, Belgrade, Serbia

Jiayi Sun Chemical and Biological Engineering Department, Colorado State University, Fort Collins, CO, USA

David Tepfer Institut National de la Recherche Agronomique, PESSAC, Versailles, France

Muthu Thiruvengadam Department of Applied Bioscience, College of Life and Environmental Sciences, Konkuk University, Seoul, Republic of Korea

Snežana Trifunović Faculty of Chemistry, University of Belgrade, Belgrade, Serbia

Milana Trifunović-Momčilov Institute for Biological Research "Siniša Stanković", University of Belgrade, Belgrade, Serbia

Johannes Van Staden Research Centre for Plant Growth and Development, School of Life Sciences, University of KwaZulu-Natal Pietermaritzburg, Scottsville, South Africa

Karolina Wielgus Institute of Natural Fibres and Medicinal Plants, Poznan, Poland

Part I

General Aspects of Transgenesis and Secondary Metabolism

DNA Transfer to Plants by *Agrobacterium rhizogenes*: A Model for Genetic Communication Between Species and Biospheres

1

David Tepfer

> *So what molecular biology has done you see, is to prove beyond any doubt but in a totally new way the complete independence of the genetic information from events occurring outside or even inside the cell, to prove by the very structure of the genetic code and the way it is transcribed that no information from outside of any kind can ever penetrate the inheritable genetic message.*
> *Jacques Monod one of the founders of molecular biology, quoted in the Eighth day of Creation, by Horace Freeland Judson, Simon and Schuster 1979.*

Abstract

Agrobacterium rhizogenes genetically transforms dicotyledonous plants, producing a transformed phenotype caused by the Ri TL-DNA (root-inducing, left hand, transferred DNA). Phenotypic changes include wrinkled leaves, reduced apical dominance, shortened internodes, changes in flowering, including a switch from biennialism to annualism, and altered secondary metabolite production, including increases in alkaloids. The transformed phenotype is correlated with a reduction in the accumulation of polyamines; it is mimicked using an inhibitor of polyamine synthesis. Roots transformed by *A. rhizogenes* grow in axenic culture, permitting the production of secondary metabolites in bioreactors, the modeling of the rhizosphere, and the propagation of arbuscular micorrhizal fungi for biofertilization.

A general view of parasexual DNA transfer postulates the exchange of genetic information among genetically distant plant genomes, with *A. rhizogenes* acting as an intermediary, thanks to its wide host spectrum for DNA transfer to plant,

D. Tepfer (✉)
Institut National de la Recherche Agronomique, PESSAC, Versailles, France
e-mail: david.tepfer@gmail.com

S. Jha (ed.), *Transgenesis and Secondary Metabolism*, Reference Series in Phytochemistry, DOI 10.1007/978-3-319-28669-3_19

fungal, and animal cells and to exchange with other bacteria, including *Acinetobacter baylyi*, which uses homologous recombination to incorporate plant DNA into its genome. Marker exchange served to document DNA transfer from leaves and roots to *A. baylyi*. Transferred functions in this hypothetical system connecting phylogenetically distant genomes included genes encoding antibiotic resistance, nutritional mediators of plant/microorganism interactions (calystegins and betaines), and an elicitor of plant host defense responses (β-cryptogein), whose expression in tobacco resulted in increased resistance to *Phytophthora*. Thus, DNA encoding a trait of adaptive significance in a plant could be acquired by soil bacteria and eventually transferred into multiple plant species, thanks to the presence on the Ri TL-DNA of genes that increase developmental plasticity (organ formation) in the host plant, ensuring the sexual transmission of the foreign DNA. The image of genetic football is invoked to convey the multiple facets of this largely theoretical system of this parasexual DNA transfer.

The plausibility of a role for DNA transfer in the origin and future of our biosphere was tested by attaching unprotected DNA and seeds of *Arabidopsis thaliana* and tobacco to the outside of the International Space Station to simulate an interplanetary transfer of life. Seeds and fragments of DNA survived 18 months of exposure, indicating that DNA transfer could play a role in biosphere formation and evolution, particularly when protected from short wavelength UV by flavonoids in the seed coat.

Keywords

Agrobacterium rhizogenes • *Rhizobium* • *Phytophthora* • Arbuscular mycorrhizal fungi • Ri TL-DNA • Horizontal gene transfer • Rhizosphere • Polyamines • Panspermia • Evolution

Abbreviations

crypt	Gene encoding β-cryptogein
DFMO	DL-Difluoromethylornithine
DNA	Deoxyribonucleic acid
HGT	Horizontal gene transfer
nptII	Gene encoding kanamycin resistance
PCR	Polymerase chain reaction
Ri TL-DNA	Root-inducing left hand, transferred DNA
RNA	Ribonucleic acid
rolA,B...	Root locus A B... from the Ri TL-DNA

Contents

1 Introduction

The soil bacterium, *Agrobacterium rhizogenes*, naturally transfers DNA to dicotyledonous plant cells [1]. It uses the same DNA transfer mechanism as *Agrobacterium tumefaciens* [2], but in *A. rhizogenes* the transferred DNA (Ri T-DNA) is root inducing, while in *A. tumefaciens* the transferred DNA (Ti T-DNA) is tumor inducing (Fig. 1). (A right-hand, TR, DNA may also be transferred.) Tumor formation by *A. tumefaciens* results from the expression of genes carried by Ti T-DNA that encode the production of the plant hormones, auxin and cytokinin, in the transformed cells [3]. In contrast, the roots induced by *A. rhizogenes* regenerate shoots that carry Ri T-DNA into whole plants and their progeny [4]. These genetically transformed roots and shoots express the Ri T-DNA, and they are phenotypically altered in a similar fashion in different species [4]. Because the physiological basis for the transformed phenotype is still only partially understood, Ri T-DNA remains a source of information about how genotype leads to phenotype through conserved biochemical mechanisms. Some of the characteristics of the transformed phenotype have led to practical applications. See Sect. 9. This chapter describes the effects of Ri T-DNA on plant morphology, development, secondary metabolism, and plant/microorganism interactions. It will also consider *A. rhizogenes* in an ecological and evolutionary context, including in the coevolution of biospheres.

A comprehensive review would be too vast for this chapter, so in the name of brevity and simplicity no attempt is made to be exhaustive, and the reader is asked to search the literature for omissions of similar results that were published in parallel by other authors. Instead of writing a traditional review, I have taken the unusual step of recounting a personal history from an end-of-career vantage point, illustrating with

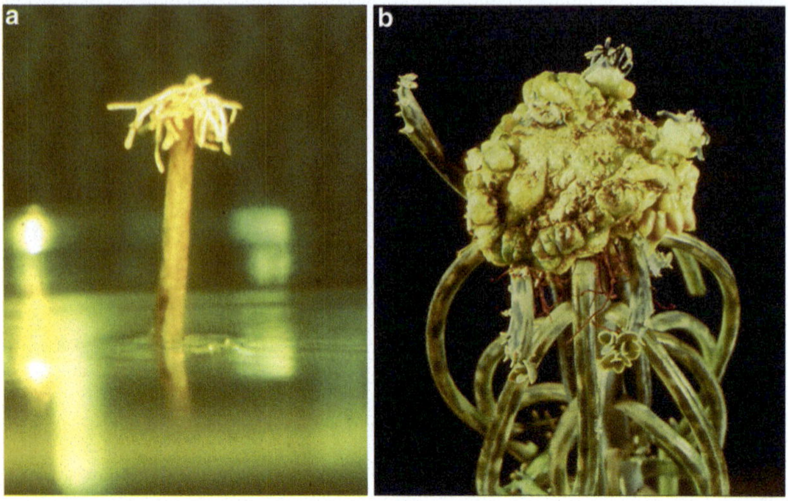

Fig. 1 Morphological responses at the site of inoculation. (**a**) root induction by *A. rhizogenes* on a morning glory (*Calystegia sepium*) stem segment; (**b**) tumor induction by *A. tumefaciens* on a decapitated kalanchoë plant (Photos D. Tepfer)

examples the ramifications that grew out of a simple initial observation and explaining how the human context molded the research.

1.1 Early Encounters

Seen in retrospect, lives and work are structured by early encounters. My father, Sanford (Sandy) Tepfer, was an evolutionary and developmental plant biologist at the University of Oregon (USA). My formative years were peopled by his colleagues, including Jacob (Jake) Strauss, who introduced me to plant molecular biology, and Howard (Howie) Bonnet, who taught me in vitro root culture, photobiology, and the scientific method. Family friends and neighbors included chemists, e.g., Virgil Boekelheide and Richard (Dick) Noyes; molecular biologists, e.g., Aron Novick, George Streisinger, and Sydney Bernhard; and microbial and plant ecologists, Richard (Dick) Castenholtz and Stanley (Stan) Cook; and botanists J. Strauss, H. Bonnet, and S. Tepfer. I remember Barbara McClintock and on another occasion the evolutionary biologist, E.B. Ford at our dining table – and meeting James Watson in a dark alley in Eugene, Oregon (coming out of a bar, arm in arm with Sydney Bernhard) when I was a teenager. There were frequent seminars in our living room and tea and cookies for students almost every evening. These encounters primed later interactions (see below) with botanists, chemists, microbiologists, molecular biologists, ecologists, and evolutionary biologists.

The idea of applying the just-emerging molecular techniques to plant biology came from my father's colleague, J. Strauss, who died young, shortly after I watched him unpack a Sorvall centrifuge, a Cary spectrophotometer, and a fraction collector. These exotic treasures were attainable because the Russians orbited Sputnick around the Earth in 1957 (when I was 11 years old), causing an injection of Federal grant money into American science. My father's lab was quickly filled with graduate students and stacks of plastic Petri dishes. My presence there was tolerated, on the condition that I kept my hands in my pockets. The desire to get my hands on things in labs was thus born, and it has not waned nearly 60 years later.

In 1964, my family went to France for my father's first sabbatical leave, and I met Georges Morel and his graduate student, Arlette Ménagé (later Goldman), at the Institut National de la Recherche Agronomique (INRA) in Versailles. A. Goldman was working on the opines, which are exotic amino acid derivatives in axenic plant tissue cultures derived from tumors induced by *Agrobacterium tumefaciens* [5]. She was to become my principal collaborator in Versailles 15 years later.

Well before this first visit to INRA, Armin Braun had proposed that crown gall tumors could be caused, among other things, by genetic transformation [6], which induced the production of auxins and cytokinins in the transformed tissue [7] (Braun was influenced by his Rockefeller Institute colleagues, Oswald Avery et al., who first demonstrated genetic transformation in *Streptococcus pneumoniae* [8]). Specificity between opine synthesis in the transformed plant and opine catabolism in the bacteria was determined by the bacterial strain, and it was later put forward as evidence supporting DNA transfer in crown gall [9–11]. J. Tempé and Annik Petit later showed that opines induced the conjugation of the plasmid that encoded their catabolism [12], and opine synthesis genes were found in *Agrobacterium* T-DNAs. See reference [13]. Opine-like substances were also involved in the nitrogen-fixing relationship between *Rhizobium* and legumes [14].

My undergraduate research and Master's degree under H. Bonnet at the University of Oregon concerned the control by light of geotropism in morning glory roots grown in vitro [15]. The root culture and photobiology were to prove useful for producing axenic cultures of the roots induced by *A. rhizogenes* and in exposing plant seeds to space travel. Starting in the early 1970s, I was a graduate student at the University of California at Irvine (UCI), working under Donald (Don) Fosket, who encouraged me to follow-up on J. Strauss's ambition to do molecular biology in plants. Don had recently purchased a Sorvall and a spectrophotometer, and ultracentrifuges were available. My project stalled on the lack of methods for isolating RNA from plants but was saved by a timely publication [16]. I thus managed to do some pre-cloning molecular biology, concerning the control of protein synthesis by cytokinins [17, 18].

Howard Schneiderman was an important mentor at UCI. He had a Master's degree in botany but had devoted his Ph.D. and career to *Drosophila* developmental biology. He and Peter Bryant put Susan Germeraad, a fellow graduate student, on a project to genetically transform fruit flies by injecting DNA into their eggs. I was introduced through them to DNA transfer in its pre-embryonic stages, and it was several times the subject of departmental seminars, including one by Clarence Kado on *A. tumefaciens*, cytokinins, and crown gall. A few years later, H. Schneiderman launched genetic engineering in plants by convincing Monsanto Company that agrochemicals should be replaced by transferred genes. I attended the first plant molecular biology conference in 1976 in Strasbourg, France, and after hearing Jozef (Jeff) Schell's seminar, I was convinced that crop plants would soon be fixing nitrogen. After UCI, I was a postdoc at the Molecular Biology Institute at UCLA, where I learned to manipulate DNA and to use in vitro translation systems. DNA cloning was new, and Gilbert and Maxim sequencing was being done with photocopied protocols, prior to publication. Biologists made their own electrophoresis equipment and enzymes. It was an exciting time to be a biologist.

A decisive move was precipitated by my father's proposal that we return to Paris for his third sabbatical. He suggested that I contact Jacques Tempé, who was working on crown gall, a field that was coming alive with the first Southern hybridization data showing bacterial DNA in axenic plant cell cultures induced by *A. tumefaciens* [19]. I had recently skied in Park City, Utah, with Ed Southern, inventor of the famous hybridization method, so everything seemed to point toward *Agrobacterium* research in France. I concocted a molecular biology project to work with J. Tempé in Versailles on the expression in plant cells of Ti T-DNA, and it was funded by NATO. I showed up, dripping with sweat, at INRA (Institut National de la Recherche Agronomique) in Versailles on a hot day in the fall of 1978, having roller-skated through the park behind the Louis XIV's palatial abode.

I was already a confirmed francophile and francophone (from my father's first sabbatical in 1964), but nothing prepared me for the shock of trying to do what I thought of as science at INRA. A French agricultural research station was very different from an American university. On the bright side, there were natural product chemists, including J. Tempé and A. Goldmann, excellent bacterial geneticists, including Jean Dénarié, Pierre Boistard, and Charles Rosenberg, working on *Rhizobium*, and a fine protein chemist, Jean-Claude Pernollet, working on a fungal elicitor. On the dark side was a vast array of often dysfunctional infrastructure and staff. (Even the electrical supply was unreliable.) Worse was the constant interpersonal conflict, often degenerating into sabotage. Aside from defective infrastructure and difficult human relations, there was no molecular biology (nor funding for it) and most of the necessary biochemical supplies had to be ordered from Sigma in the USA. I slowly became aware that in spite of the dysfunction and continuous infighting, novel collaborations with competent scientists could be forged in the unfamiliar subjects of plant breeding, pathology, soil science, nitrogen fixation, and natural product chemistry. I also slowly realized that the molecular project I had picked was not right for the environment at INRA.

A year of effort was wasted before arriving at the end of my fellowship and admitting defeat, but I was miraculously saved by a *deus ex machina* in the form of an unsolicited letter automatically extending my fellowship for another year. I was relieved to have a job but anxious to find a new project that was more in tune with local conditions, and I had only a year to obtain results. J. Tempé was on sabbatical leave in Australia. I needed to find a project that could be done quickly and that did not involve sophisticated molecular biology. For once, I felt free to explore, unencumbered by commitments to a pre-established research program.

2 *A. rhizogenes*, Root Cultures, the Transformed Phenotype

I saw Pierre Guyon washing fresh carrots in the lab sink. Later I saw him walk by with a tray of Petri dishes containing carrot disks on solid medium. A couple of weeks later I saw him carrying a tray of Petri dishes containing carrot disks covered with roots. That was my introduction to *A. rhizogenes*. He explained that he and J. Tempé were trying to find opines in the roots induced by *A. rhizogenes*. A paper had just appeared indicating that there were large plasmids in *A. rhizogenes* [20], and Tempé was hoping to use opines in the roots as evidence for DNA transfer.

I had used cultured morning glory roots to study the effects of light on geotropism under H. Bonnett [15], and I knew that excised carrot roots could be grown but with difficulty. I set up a series of carrot disks on agar + water medium, and I inoculated stem segments from morning glory plants growing behind the lab. Ten days later I had roots, which I excised, passed through multiple rinses in sterile water, and plated on White's root growth medium [21]. I directed the roots into the agar to keep their tips from sliding over the surface and carrying along contaminating *A. rhizogenes*, and I marked the apex of each root on the underside of the Petri dish. A career-changing surprise was waiting for me the next day. Many of the roots had elongated as much as a centimeter, which was completely unexpected based on my experience with morning glory. That evening I told my wife that I could not be leaving on the vacation in Corsica that we had planned with her parents.

The roots elongated as fast as a millimeter per hour and produced profuse laterals. Controls taken from germinated seeds grew slowly. I was quickly filling Petri dishes with roots, which I maintained as clones of each original root excised from the point of inoculation (Fig. 2). I was in root heaven. No bacteria grew when subcultured roots were crushed and plated on bacterial media. While passing through the medium, the roots had quickly outgrown the bacteria that had induced them. To my amusement, when the Petri dishes were full, the roots forced up the lids and grew across the shelf. The wild type transformed morning glory roots I had worked with in Oregon [22] had been slow growing in comparison, and they had rarely produced lateral branches. The only disappointment was that the morning glory roots did not regenerate shoots, while those in Oregon had produced shoots in response to light. I queried H. Bonnet, who surmised that I was working with *Calystegia sepium*, which does not regenerate. I inoculated *C. arvensis* stem segments, and a month later the resulting roots had produced formed shoots. I transferred the regenerants to soil in the green house, and to my surprise they had wrinkled leaves, and they were highly branched, like the roots they had come from. I used somatic embryogenesis to regenerate the carrot roots produced using *A. rhizogenes* A4, but the carrot roots induced by strain 8196 produced embryos directly from roots without hormonal treatment. Transformed tobacco roots regenerated without intervention, and inoculation of leaves sometimes produced plantlets directly [4]. All of the regenerants had wrinkled leaves and reduced stature and apical dominance (Fig. 3). The carrot plants had converted from biennial to annual flowering (see also Fig. 14). There was clearly

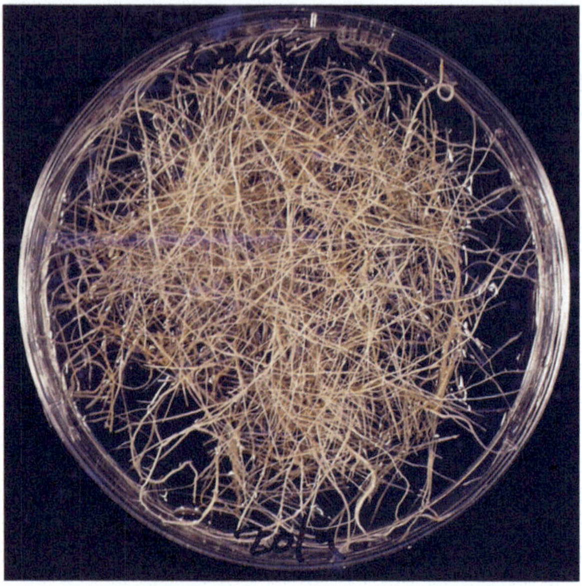

Fig. 2 Carrot root clone, genetically transformed by *A. rhizogenes* (Photo D. Tepfer)

Fig. 3 Changes in stature in tobacco, variety Mammoth, (regenerants from roots induced by *A. rhizogenes* strain A4), showing reduced apical dominance and wrinkled leaves. *Left* to *right*, wild type control, transformed plant of T phenotype, transformed plant of T' phenotype, transformed plant of an extreme T' phenotype. Ly-Yan Sun, the graduate student who made the plants, is on the far *left* (See also Fig. 5) (Photo D. Tepfer)

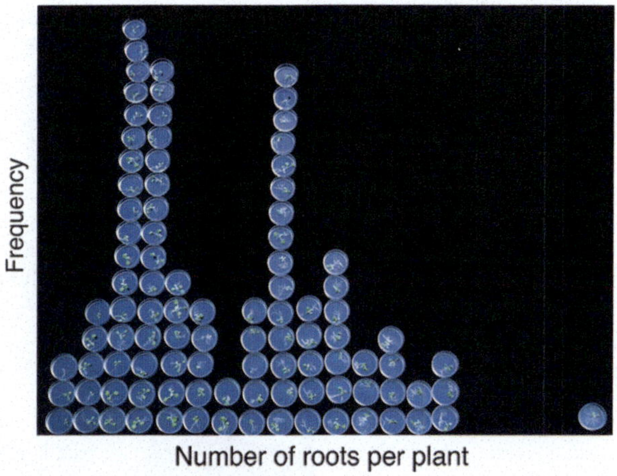

Fig. 4 Progeny of a transformed tobacco plant, variety Xanthi, resulting from crossing a wild type (female) with a primary transformant (male). Seeds were sewn in Petri dishes, which were sorted by root number per plant, starting with one root per plant on the *left*. The root number phenotype segregated as a 1:1 Mendelian dominant, with an outlier phenotype on the far *right* (Photo D. Tepfer)

a phenotype in both roots and regenerated plants that was similar in different plant species and similarly induced by the two bacterial strains I had used (A4 and 8196). This transformed phenotype included a variety of morphological changes and a surprising increase in regeneration capacity. It remained to show that roots and whole plants were genetically transformed.

J. Tempé returned from sabbatical leave in Australia to find his lab filled with roots. He had learned how to use high voltage paper electrophoresis to assay for two new opines, agropine and mannopine, whose synthesis is encoded by T-DNA and whose catabolism is encoded by the bacterial plasmid that carries the T-DNA. My *A. rhizogenes* A4 roots contained agropine in surprisingly high amounts and those induced by *A. rhizogenes* 8196 contained mannopine [23], providing evidence for genetic transformation. I went to Mary Dell Chilton's lab in Saint Louis (USA) to do Southern hybridizations with root and plant DNA that I had extracted in Versailles. I had to return to Versailles before the films were developed, and when they finally were, the results were positive but grossly over exposed.

Just as I had finished setting up to repeat the Southern blots in Versailles, J. Tempé expelled me from his lab to make way for a visit from M.D. Chilton to repeat the Southern blots with DNA from roots produced under my tutelage by a student in a neighboring lab. I set up my Southern hybridization experiments in another lab in Versailles, using DNA from roots and regenerated plants. All of the results came in within a few days of each other. There was T-DNA from *A. rhizogenes* in the roots that M.D. Chilton analyzed [24] and in all of the roots and plants that I had produced, including their progeny [4, 25–27]. The transformed genotype was inherited as a Mendelian dominant in tobacco (Fig. 4) [4, 25–28], and it co-segregated with the transformed phenotype in morning glory, carrot, and tobacco. In tobacco, it came in two general intensities, T and T', with the more intense, T' phenotype reverting to the T phenotype in lateral shoots (Fig. 5). The plants of T phenotype were homozygous for the T-DNA insertion and the T' plants were heterozygous [29]. See references [30–35]. In tobacco, the ability to grow under reduced gas exchange,

Fig. 5 Reversion of the T' phenotype to T phenotype in tobacco, variety Xanthi (Reprinted from Ref. [28]) (Photo D. Tepfer)

attributed to wild type Ri T-DNA [4], was later explained by a reduction in ethylene production [36].

In the meantime, a colleague in Versailles, Jean-Pierre Bourgin, had alerted me to a paper in German, describing earlier experiments in which tobacco roots induced by *A. rhizogenes* had been converted to callus, which had regenerated into plants, showing the altered morphology I had described [37]. Thus, transgenic plants had been produced using *A. rhizogenes* before mine. The authors' conclusion that they were genetically transformed was substantiated by sexual transmission of the phenotype. I cited their work in papers and seminars, and I invited the first author, Claudius Ackerman, to do Southerns and Northerns in Versailles, which proved that the descendants of the plants he had produced in 1973 [37] were genetically transformed. He had produced transgenic plants well before mine, but it turned out that nature had done it before all of us.

Clearly, no human intervention should be necessary for gene transfer from *A. rhizogenes* to morning glory, since root production at a wound site in the soil would produce a secondary root system, which would regenerate transformed plants that would pass the foreign genes on to their progeny. In keeping with this thought experiment, positive hybridization signals were reported with Ri plasmid probes in wild type *Nicotiana* [38] and *C. arvensis* plants [25]. Ancient Ri TL and TR-DNAs were recently described in domesticated sweet potato (*Ipomoea*) [39], a member of the *Convolvulaceae* and a relative of *C. arvensis,* and we would later see hybridization signals in apple [40]. See [39] for recent literature. Later, on a visit to Madison, Jerry Slightom and I made a phage Lambda library from *C. arvensis* DNA that I took to Versailles with the intention of purifying and mapping positive clones for sequencing.

This never happened. I became overwhelmed by the complexities of doing research in Versailles. I had been obliged to obtain and maintain my own infrastructure, including a prefab building to use as a lab. The director of INRA, Jacques Poly, had been encouraging, but INRA had not accepted a grant from Agrigenetics to pay the costs of the research. In spite of promises to the contrary, INRA did not provide adequate funding. The grant game in France was fraught with cronyism, entrenched territoriality, and corrupt administrators. I was the American bull in the china shop of French science. Things got even worse with the transfer of French national science funding resources to Europe, increasing paper work and politics. When J. Poly retired, I lost the only moral and financial support that I had at INRA. I was exhausting myself with trivia and attacked on all fronts. I realized that if I wanted to do science, France was not the place for me. The local powers clearly wanted me to leave, and they certainly did not want me working on A. rhizogenes and Ri T-DNA. It seemed unlikely that my ambition to work out how the genes carried by the Ri TL-DNA altered development could be fulfilled in Versailles.

Nevertheless, I decided to ignore the realities and to stay. The reasons were in part personal (an attachment to living in France), but more importantly, at INRA I was free to self-direct my scientific activities. Elsewhere, I would not be able to so hands-on science myself, which I craved. I would have to teach undergraduates and build up a grant-dependent factory of graduate students, postdocs, and technicians. In Versailles, I could have a minimum of internal support in the form of some operating expenses and personnel, allowing me to pursue, albeit on a small scale, subjects of my choosing. I had to lower my ambitions and depend more on collaborations, which turned out to be a source of diversity and inspiration, I tried to keep this collaborative approach organized under the general subject of rhizosphere biology. I decided to leave the Lambda clones in the refrigerator and to let E. Nester's very competent lab in Seattle pursue the hybridization signals in wild type plants. They described the remnants of an ancient Ri T-DNA, which they named the cellular T-DNA (cT-DNA) in *Nicotiana glauca* [35, 38]. I was free to diversify into rhizosphere subjects that were related to the evolution and developmental biology of the Ri T-DNA and parasexual DNA transfer in general, but this required knowing more about the Ri TL-DNA.

3 Ri TL-DNA Structure and Function

A decisive encounter was meeting J. Slightom in Madison, Wisconsin. He proposed to sequence the Ri TL-DNA that was causing the transformed phenotype, using our clones (produced by Lise Jouanin and Françesca Leach) from the Ri plasmid (which contained a second T-DNA, the TR-DNA, which was not causing the transformed phenotype). In the mid-1980s, sequencing on this scale was heroic. In all, J. Slightom sequenced approximately 100,000 base pairs, using the Gilbert and

Maxim technique. To my knowledge, nobody has detected an error in the final 21,126 bp sequence.

My objective was to correlate the DNA sequence with insertional mutagenesis, transcriptional, biochemical, and developmental studies. I obtained everything necessary to do insertional mutagenesis on the Ri TL-DNA from the Fredrick (Fred) Asubel lab in Boston, and I recruited a bacteriologist in Versailles to do the job. The Northern analysis of transcription [28] was done thanks to a collaboration with Mylène Durand-Tardif and Richard Broglie in Nam Chua's lab in New York. See also reference [41]. The missing link was the mutagenesis, which for trivial reasons did not happen in Versailles. Fortunately, it was done in Eugene Nester's lab in Seattle [42]. The key for me was J. Slightom's sequence, which revealed the locations of potential genes (open reading frames or ORFs) and identified protein sequences and transcriptional signals.

An observation made by a visiting student, Hervé Levesque, suggested that there was interesting evolutionary biology hidden in the Ri TL-DNA. He spotted structural similarities among some of the putative proteins encoded by these putative genes (ORFs), allowing their alignment into a gene family. I named them *plast* genes for their effects on developmental plasticity (the propensity to form organs) in the roots and regenerated plants carrying Ri T-DNA. The *plast* genes came in acidic and basic forms, with the acidic on the right and the basic on the left [43]. Interspersed with the *plast* genes were other functions that had seemingly fallen among them. For example, ORF 10 (called *rolA*) encoded a short, basic protein, suggesting a nucleic acid binding protein. ORF 8 was a homologue of the Ti T-DNA gene, IAAM, the first of two genes necessary for auxin synthesis in soil bacteria; intriguingly, it was fused to a *plast* gene. With J. Slightom's further collaboration, the border sequences defining the ends of the Ri TL-DNA were found to be the same as those in the Ti T-DNAs, indicating the same transfer mechanism for the Ti and Ri plasmids [2]. A comparison between the Ri and Ti T-DNAs allowed us to make a case for a common ancestor [43]. These structural studies provided a framework for the ideas I had nurtured since the original observation of the growth of transformed roots.

4 Ecological and Evolutionary Hypotheses

The ecological and evolutionary implications of natural gene transfer by *A. rhizogenes* were difficult to ignore. During the 1980s, I published symposium volume papers and a review outlining some of these hypotheses [1, 25–27], which I attempted to test over the following years in both laboratory and thought experiments. The subject also found its way into the discussion sections of peer-reviewed papers. See for example [4, 43, 44]. The concept of species and their genomes as impervious to outside genetic influence was ingrained in biological thinking (see quote from J. Monod at the beginning of this chapter). DNA transfer from *A. rhizogenes* to a plant and its progeny went against the dictum that acquired characters are not inherited and that the genome is an impenetrable fortress. In contrast, I saw the transformed phenotype as a source of heritable variability that functioned in diverse species. Under stress, a community could share genetic information of adaptive significance. The T-DNA transfer mechanism and the *plast* genes insured the passage of bacterial T-DNA into fertile plants and their progeny. The structure of the Ri TL-DNA suggested that other unrelated functions were carried along with the diverged *plast* genes, insuring that they would penetrate the genomes and the species of diverse plants. If the direction of transfer were reversed, and plant DNA found its way into the Ri T-DNA, *A. rhizogenes* would provide a genetic bridge between plant species, thanks to the broad spectrum of its interactions. Many seminars to this effect produced positive reactions but a few (particularly in France) elicited anger. Parasexual genetic exchange was taken to be Lamarckian, as opposed to Darwinian. Acquired characteristics, encoded in a T-DNA, were inherited, raising the specter of Lysenko and the dark years of eugenics in the Soviet Union. My seminar attendees were right to feel threatened. The textbook dogma about acquired characters not being heritable was out of date, as was the concept of species. Nature was the original genetic engineer [27].

4.1 DNA Transfer from Plants to Soil Bacteria, Genetic Football

The origin of the genes that *A. rhizogenes* was sending into plants is a fundamental question. I had long thought that they could come from diverse sources, including plants themselves. For instance, the genes in the Ri TL-DNA found in wild type plants could be the original source of the bacterial versions. Necrotic plant biomass enters the soil environment, where it is digested by bacteria. As roots grow, cap cells slough off, liberating their contents into the rhizosphere. DNA persists in the environment [45–47], and bacteria import it; thus, some of this liberated plant DNA could be taken up by bacteria, become incorporated into a T-DNA, and transferred back into plants. Since T-DNA transfer to plants has a wide host range, soil bacteria could provide a genetic link between plants of different species (Fig. 6). I thought of plant and microbial communities as engaged in genetic football [1]. Attempts to demonstrate DNA transfer from plants to bacteria proved to be technically complex, in part due to the presence of DNA encoding marker genes as contaminants in laboratories and as natural constituent of the soil metagenome.

Ultimately, Johann de Vries and Wilfried Wackernagel used a simple and sensitive method for detecting the uptake and incorporation of a DNA from an antibiotic resistance gene in the soil bacterium, *Acinetobacter baylyi* strain BD413 (Fig. 6). The method was marker exchange, in which an intact *nptII* (kanamycin resistance gene) from the source DNA replaces a defective *nptII* in the target DNA through homologous recombination [48]. We used this system (kindly provided by J. de Vries and W. Wackernagel) to documented DNA transfer from six species of plants into *A. baylyi*. (Results were first presented at the European Society for Evolutionary Biology, Barcelona, Spain, 1999.) Sources included crushed leaves, intact leaves, and intact plants in vitro. Transfer was dose dependent and DNase sensitive; the problem of contaminating DNA was resolved by introducing silent mutations in the

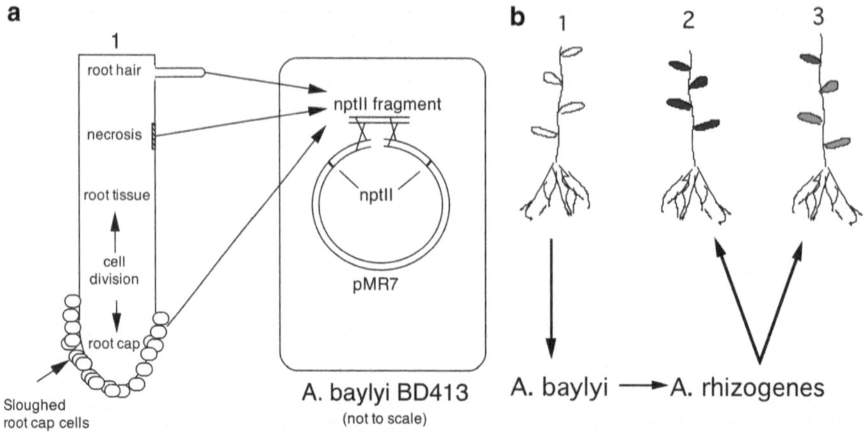

Fig. 6 Parasexual DNA transfer between plants and soil bacteria. (**a**) DNA, released from roots by cap cells, necrotic wounds, and root hairs, is assimilated by *A.* baylyi, where it recombines with homologous sequences through marker exchange [48]. In this laboratory example, fragments of an *nptII* antibiotic resistance gene from the root rescue a deletion in the *nptII* gene carried by plasmid MR7, rendering *A. baylyi* resistant to kanamycin [49]. (**b**) DNA from plant species 1 is transferred (via *A. baylyi* and *A. rhizogenes*) into species 2 and 3 [25, 26, 50]

source plant *nptII,* which were detected in the bacterial target [49]. See references [45, 51, 52].

Once in *A. baylyi*, it was easy to imagine how conjugation could move a DNA sequence into other bacteria, e.g., *Agrobaterium* having a DNA transfer system for plants. However, DNA incorporation in *A. baylyi* depended on homology between the incoming foreign DNA and sequences carried by *A. baylyi*. Homology could be provided by the Ri T-DNA already in the plant host. (The Ri plasmid is maintained in *A. baylyi*, Message and Tepfer, unpublished results.) Thus, DNA transfer in either direction would provide the homology for shuttling other sequences between plant species via *A. rhizogenes*, e.g., those interspersed among the *plast* genes. The direction of the initial transfer could not be known, and it was not important, once the homologous sequences were in place. But what sorts of genes would make this journey back and forth? The opine synthesis genes were transferred to plants from bacteria, but suppose they had come from plants via a pre-established homology? These questions inspired a fresh look at nutritional relationships between plants and bacteria in the rhizosphere.

5 DNA Transfer, Plant-Microorganism Interactions, Opines, Calystegins, and Betaines

The microorganisms that live around roots are nourished by root exudates, root necrosis, and the contents of the cap cells that are sloughed off during root growth [53]. Bacterial catabolism of plant secondary metabolites is adaptive and specific to plant species, whether the metabolites are encoded in a T-DNA or not. I thought of these generalized opines or nutritional mediators as the currency in the economic system of the rhizosphere. *A. rhizogenes* was closely related to *Rhizobium*, capable of nodulating legume roots and fixing nitrogen in the nodules. J. Tempé identified opine-like substances in the relationship between *Rhizobia* and their legume hosts [14, 54]. Most conveniently, there was a first rate *Rhizobium* lab on the INRA campus in Versailles.

While using high voltage paper electrophoresis and silver nitrate staining to assay for agropine in *C. sepium* transformed by *A. rhizogenes*, I was intrigued by a pair of substances in the roots of both transformed and control plants. They reacted with silver nitrate, like agropine and manopine, and they accumulated preferentially in the underground organs of both control and transformed plants in quantities similar to agropine, so I thought they could be exclusive sources of nutrition for plant-associated soil bacteria. I screened for metabolism of these calystegins, as I named them, in 42 laboratory strains of soil bacteria, and discovered that they were only metabolized by *Rhizobium meliloti* strain 41, but not by the other soil bacteria [55]. This strain was serendipitously in my collection thanks to interactions with the Versailles *Rhizobium* group (J. Denarié, P. Boistard et al.). They had been looking for the genes for nodulation and nitrogen fixation on a large plasmid, pRme41a, so they had marked it with a transposon insertion, and they had cured the host *Rhizobium* of the plasmid, showing that the symbiotic genes were not on pRme41a. I tested these derivatives for calystegin catabolism: the host *Rhizobium* lacking the plasmid did not catabolize calystegins – nor did the transposon insertion [55], which turned out to be in the calystegin catabolism region. The odds of this happening were like hitting a jackpot in the slot machine of laboratory research.

These simple experiments led to the characterization of the calystegin catabolic region on pRme41 [56]. Screening in the subterranean organs of 105 species of plants revealed calystegins in only three species: *C. sepium*, *C. arvensis*, and *Atropa belladonna* [57]. A. Goldman (who had discovered nopaline and octopine) purified the calystegins, and their structures [57] showed that they were novel tropane derivatives that resembled glycosidase inhibitors. Glycosidases are important in cell to cell recognition, among other functions, and they are thus poisons. Thanks to a collaboration with Russell Molyneux and Alan Elbein, the glycosidase prediction proved to be true, and they inhibited seed germination and root elongation. Furthermore, the roots of plants producing calystegins carried bacteria that catabolized them and nonproducing plant roots did not [58]. See [59] for different result.

An ecological interpretation seemed obvious. Plants liberate carbon and nitrogen into the soil in the form of exotic molecules that are catabolized only by soil bacteria that carry special catabolic genes. But calystegins are not just exclusive carbon and nitrogen sources; they are also allelopathic glycosidases that kill animals and plants in the rhizosphere. Thus, the bacterial catabolic genes would not only provide exclusive nutrition to soil bacteria but they would protect the host plant by detoxifying the soil, and (conveniently for spreading them among bacteria) they are carried on a self-transmissible plasmid. The opines and T-DNA transfer were just one facet of a general network of nutritional relationships. Similarities with the digestive tube in animals were easy to imagine – so were the possibilities for genetic football. But how did nitrogen fixation fit into this picture? Calystegin catabolism was not necessary for nodulation and nitrogen fixation. *R. meliloti* (pRme41) was probably isolated from soil where morning glory had grown recently. (Rhizobia have a saprophytic existence, aside from their nodulation and nitrogen-fixing activities.)

We set out to find another nutritional mediators that might be more directly related to nodulation and nitrogen fixation. We focused on the betaines present in alfalfa seeds in large quantities, thinking that the germinating seeds would liberate betaines and initiate interactions with *Rhizobium* through nutritional selection. We localized betaine catabolism to the *Rhizobium* symbiotic plasmid (pSym), surrounded by genes involved in nitrogen fixation [60]. Mutagenesis showed that in keeping with the functions of neighboring genes, catabolism of the betaine, stachydrine, was required for efficient nodulation [61]. Thanks to the collaboration of Michael Burnet and J. Slightom, the catabolic region was dissected and sequenced, revealing similarities to bacterial genes involved in the detoxification of xenobiotics [62].

We concluded that nutritional relationships are important in nodulation, just as they are in DNA transfer, but that allelopathy can also be involved. Layers of complexity had been added during the coevolution of plants and bacteria. Genes encoding anabolism and catabolism of nutritional mediators were potential hitch-hikers on the basic *plast* gene T-DNA vehicle. There was no evidence for DNA transfer in *Rhizobium* symbiosis, but there was no need for it, since that was amply carried out by a sister bacterium, *A. rhizogenes*. The origin of the catabolic and anabolic gene pairs was thus not important, since they could be kicked back and forth in genetic football. More importantly, the work on nutritional mediators provided evidence for the principle [53, 63] that plant/microorganism relations are modulated by nutritional relationships. The anabolic genes (carried by Ri and Ti T-DNAs) are clearly the objects of genetic football in the case of the opines, and other nutritional mediators (e.g., calystegins and betaines) were candidates as well. What other functions could serve as genetic footballs?

6 DNA Transfer, *Phytophthora* Resistance

A colleague in Versailles, J.C. Pernollet, was working on a family of defense-inducing, small proteins, called cryptogeins, in *Phytophthora*, the fungus responsible for the late blight that caused the Irish potato famine in the mid-nineteenth century. They are secreted in large amounts to scavenge sterols. It was thought that plants use these elicitins as indicators of the presence of *Phytophthora cryptogea* and as a signal to turn on their defenses, including the production of antimicrobial metabolites. J.C. Pernollet et al. had determined the amino acid sequence of β-cryptogein from the purified protein [64]. A genetic football experiment was obvious: if the gene for β-cryptogein was expressed in a host plant, it could set off a defense response and protect the host. A postdoctoral fellow, Helène Gousseau, synthesized a DNA sequence that encoded the β-cryptogein protein described by J.C. Pernollet et al. The resulting gene was also mutated to replace a key amino acid,

Fig. 7 Tobacco plants decapitated and inoculated with *Phythophtora* at points shown by *arrows*. *Left*, plant carrying the *crypt* gene encoding β-cryptogein, a fungal elicitor of plant defense responses. Necrosis was stopped, and the plant made lateral branches. *Right*, a wild type control. (Reprinted from Ref. [45]) (Photo D. Tepfer)

Lysine-13 by a Valine, and it was expressed in yeast by another postdoctoral fellow, Michael O'Donohue [65]. When injected into tobacco leaves, the wild type *β*-cryptogein from the synthetic gene produced the same necrotic lesions as the purified natural protein, while (as predicted from earlier work) the mutant induced relatively little necrosis [66]. Expression of the synthetic cryptogein gene and the mutant under a constitutive promoter in tobacco by a visiting student, Catherine Boutteaux, was correlated with varying degrees of *Phytophthora* resistance [44]. The necrosis induced by inoculation of decapitated plants with mycelium or by infection of root systems with zoospores was reduced in the transformants, including the K13V mutant (Fig. 7) [44]. See also [67]. The degree of resistance increased when the foreign gene was in the hemizygous state, which was reminiscent of the increased penetration of the T' phenotype that was associated with hemizygosity in tobacco transformed by native Ri T-DNA (see above).

The mechanism of increased resistance was not constitutive systematic acquired resistance (SAR), because levels of salicyclic acid and PR proteins stayed at basal levels in control and transgenic plants until they were challenged by *Phytophthora* [44]. A working hypothesis was that increased antifungal secondary metabolites were interfering with *Phytophthora* infection. In collaboration with Sumita Ja, we tested this indirectly using other metabolites by expressing the β-cryptogein gene in other plants (See Sect. 9, below.)

We had made a genetic football out of β-crytogein, but could nature have done likewise? Could *Agrobacterium,* coupled with *A. baylyi,* mediate gene transfer between a fungal pathogen and the plant host? Necrosis of *Phytopthora* would liberate fungal genes for elicitins like β-cryptogein into the soil, but they would have to somehow be incorporated into a T-DNA to be transferred to plants, which would require DNA homology between a soil bacterium, such as *A. baylyi*, and *Phytophthora*. Some bacteria, including *Agrobacterium,* can genetically transform fungi [68, 69], so T-DNAs could be present in the bacterial and fungal genomes that would provide the homologies required for the transfer of genes like β-cryptogein) into *A. baylyi*, where Ri plasmids are be maintained (Message and Tepfer, unpublished results). Conceptually at least, these experiments brought fungi into the genetic football game, leaving moot the question of the origin of the transferred DNAs. Such transfers would take place on an evolutionary time scale, so human intervention was necessary to demonstrate their feasibility in nature, but the elements necessary for the existence of a multidirectional network of genetic links between plants, fungi and bacteria were in identified, and the transfer from a fungus to a plant (with a little help from soil bacteria and laboratory scientists) resulted in increased resistance to the fungal pathogen, showing it could have adaptive significance in nature.

7 DNA Transfer, Plant Development, Polyamines

In the meantime, I had not lost interest in untangling the biochemistry connecting the Ri TL-DNA to the transformed phenotype. One approach was electronic: as data banks filled with DNA sequences and, by extension, amino acid sequences and functions for enzymes and structural proteins, I periodically searched for protein sequence homologues for Ri T-DNA genes, but with no obvious success. A classical chemical approach proved more fruitful. A quick look at the Merck index suggested that more information had accumulated about the biochemicals than about the enzymes that produced them, and chemical structural similarities had led us to the identification of the calystegins as glycosidase inhibitors. Biochemicals are conserved among species, and diverse but related structures and functions have evolved from simple precursors. The transformed phenotype was conserved in numerous dicot species. Logically, the Ri T-DNA genes would function through conserved biochemistry.

I met a polyamine biochemist, Josette Martin-Tanguy (INRA, Dijon, France) at a NATO conference in Copenhagen. Since polyamines are implicated in plant development (e.g., in flowering), and they are ubiquitous in living organisms, we decided to look for possible correlations between changes in polyamine accumulation and the transformed phenotype. A series of papers was the result [29, 36, 70–74], and some of these are discussed below.

In Versailles, we produced T and T' tobacco plants expressing wild type Ri T-DNA or just *rolA* or *rolC* from the Ri TL-DNA. We also provided tobacco plants singly expressing these genes under the constitutive control of the 35S-CAMV (Caluliflower Mosaic Virus) promoter. In Dijon, J. Martin-Tanguy and her student, Daniel Burtin, showed that tobacco plants displaying the transformed phenotype had reduced free polyamines and polyamine hydroxycinnamic acid conjugates, and that this reduction occurred in direct proportion to the severity of the phenotype:

e.g., there was greater reduction of polyamine and polyamine conjugate titers in the T' phenotype than in the T phenotype [74]. Feeding polyamines to plants of intense T' phenotype caused attenuation to the T phenotype [29]. An early step in polyamine synthesis is inhibited by α-DL-Difluoromethylornithine (DFMO), which produced a phenotype in tobacco that resembled the transformed phenotype caused by the Ri TL-DNA, confirming that the inhibition of polyamine accumulation via the ornithine pathway is an essential step in the chain of biochemical events that lead to the transformed phenotype [73].

Another series of experiments concerned plants expressing *rolA,* driven by the wild type promoter. In accord with references [75, 76], we concluded that *rolA* was the major determinant in the T' phenotype. Furthermore, inhibition of the polyamine conjugates, rather than the polyamines themselves, was the biochemical correlate. 35S CAMV *rolA* plants had an extreme T' phenotype, and they were unable to flower [29]. This extreme phenotype could not be attenuated by watering 35S CAMV *rolA* plants with polyamines, because they were deficient in the polyamine conjugates whose accumulation in the top of the shoot occurs prior to flowering in wild type plants. Conjugated polyamines were not available for watering experiments, so P_{35S}-*rolA* shoots were grafted by J. Martin-Tanguy et al. onto wild type plants that had been induced to flower and had accumulated conjugated polyamines in their shoots. The P_{35S}-*rolA* scion flowered, but the flowers aborted, falling off the plant [29] (Fig. 8). This last defect was corrected by watering with putrescine alone. A control wild type rootstock that had not been induced to flower did not restore flowering. A final series of attenuation experiments performed by a graduate student, Li-Yan Sun, using plants expressing *rolA*

Fig. 8 Attenuation of extreme phenotypes in Xanthi tobacco due to Ri TL-DNA and its *rolA* (ORF 10) gene driven by the 35S CAMV promoter. *Left*, extreme T' phenotype in a plant transformed by wild type Ri T-DNA, with reversion to the attenuated T phenotype in a lateral shoot (see also Fig. 5). *Center*, phenotype due to 35S *rolA* in a shoot grafted onto wild type tobacco that was not induced to flower and had not accumulated polyamine conjugates. *Right*, same genotype as middle plant but grafted onto a wild type plant that had been induced to flower and had accumulated the polyamine conjugates (Reprinted from Refs. [29, 73])

from wild type regulatory sequences, showed a correlation between phenotypic attenuation, reduction in *rolA* transcripts and methylation of sites in the 3' regulatory region [72]. The phenotypic changes associated with transformation by the Ri TL-DNA and *rolA* alone were attenuated by polyamines and polyamine conjugates.

To be transmitted through meiosis, the Ri T-DNA encodes increased plasticity, i.e., root and shoot formation, but increased plasticity is not adaptive if it interferes with fertility. Attenuation is accomplished through homozygosity [29] through segregation of truncated T-DNAs [77], through physiological compensation (by making more polyamines and their conjugates) [29] and by decreasing transcription, e.g., through methylation of the 3' regulatory region [29]. The foreign DNA is thus regulated to insure sexual transmission. Attenuation through homozygosity would select for progeny from self-fertilization and counter select those from out crossing. It should thus drive speciation.

8 DNA Transfer, Biospheres, Flavonoids

Sydney Leach, a physicist friend, suggested that we work together on the origin of life, which seemed like a complex and distant subject until I realized that for the sake of simplicity, life could be defined as DNA and that the origin of life was a problem in DNA transfer. Life was present early in the history of the Earth and all life on Earth that has been examined uses essentially the same genetic code, strongly pointing to a common origin for the life forms we know. These two facts are compatible with, but do not prove panspermia, the ancient hypothesis that life is everywhere. The European Space Agency (ESA) called for proposals to test the plausibility of the dispersal of life through space by attaching life forms known to survive in extreme environments (extremophiles) to the outside of the International Space Station (ISS). While orbiting the Earth, extremophiles would be exposed to conditions encountered in space, the equivalent of an interplanetary voyage in our solar system (albeit with reduced cosmic radiation, due to partial protection from Earth's magnetic field in low orbit). This was the chance to do the ultimate genetic football experiment (Fig. 9), kicking the genetic foot ball into orbit, thanks to ESA and NASA.

I proposed exposing plant seeds in space, because they resist desiccation, radiation, and they have protective seed coats containing flavonoids that stop destructive UV light. We (including Andreja Zalar and S. Leach) prepared for two exposures in space to determine the survival of both unprotected DNA and seeds and to thus test the plausibility of panspermia. Using the UV beam from the synchrotron at the University of Aarhaus, Denmark, we measured short wavelength UV absorption in known and potential UV shields, finding that flavonoids had UV absorption characteristics similar to those of DNA [79, 80]. Preparatory experiments at the DLR (German Aerospace Center) in Cologne, Germany with Arabidopsis (*A. thaliana*) mutants lacking sunscreens showed that flavonoids (and sinapate esters to a lesser extent) were important in protecting seeds from part of the UV spectrum that is

Fig. 9 EXPOSE and SEEDS on the Columbus module of the ISS (Photo courtesy NASA). EXPOSE (*small red box*) is on the EuTEF platform. SEEDS is covered by a protective shutter (closed in this photo, taken prior to the start of exposure). Columbus, the European module, is to the *left* of EXPOSE. Proximal sources of UV and solar wind shadows include Columbus, the other experiments surrounding EXPOSE on the EuTEF platform and the protective shutter, which was perpendicular to the surface of SEEDS during exposure but closed during transport. *Insert*, (*large red box*), position of SEEDS on EXPOSE, with the shutters (Sh) open with seeds on the *lower right*. *White arrow*, direction of flight, except during shuttle docking, when it was inverted 180°. The external dimensions of Expose were 440 mm × 380 mm × 250 mm (Reprinted from Ref. [78])

deleterious to unprotected organisms in space travel [81]. Exposures to simulated space conditions also showed that Arabidopsis and tobacco seeds were many orders of magnitude more resistant than other potential space travelers, including UV-resistant bacterial spores [81].

The first exposure on the outside of the ISS lasted 18 months; 23% of the exposed Arabidopsis and tobacco seeds germinated and produced fertile plants after return to Earth [78]. We concluded that an unprotected Arabidopsis seed could theoretically survive a direct transfer from Mars to Earth, and that resistance was largely due to flavonoids. Furthermore, the survival of a 110 bp fragment of unprotected DNA was detected by PCR. A second experiment on the ISS established an end point for resistance of Arabidopsis seeds to space travel and localized the part of the UV spectrum that was most deleterious (Tepfer and Leach, submitted). Further laboratory experiments on Earth showed that morning glory seeds, which have thick seed coats and are long-lived in nature, were much more resistant to $UV_{245\ nm}$ than Arabidopsis and tobacco seeds, suggesting that larger seeds with more protective coats should survive much longer exposures (Tepfer and Leach, submitted). The lack of morphological mutants in the plants that grew from exposed seeds led us to measure the structure and function of a *nptII* marker gene carried by the plastids in exposed tobacco seeds. We used both quantitative PCR and marker exchange in *A. baylyi* to show that DNA damage was being circumvented and repaired, and that other targets were more limiting to seed survival, e.g., ribosomes, membranes, and proteins (Tepfer and Leach, submitted).

These experiments were a thinly disguised excuse for doing another genetic football experiment. The use of marker exchange to measure DNA integrity in exposed seeds was a proxy for DNA coming to Earth and entering the biosphere through homologous recombination in a naturally transformable bacterium, such as *A. baylyi*. Marker exchange showed how extraterrestrial DNA sequences (exoDNA) would influence Darwinian evolution on Earth. Fragments of the *nptII* gene survived unprotected exposure to full space conditions; thus, naked or protected DNA from Earth's biosphere could be carried by updrafts into the stratosphere, returning to Earth at a later time and reentering our biosphere through homologous recombination in bacteria such as *A. baylyi*. Mutations acquired in space would thus provide variability for Darwinian evolution. The forces needed to eject seeds from Earth via meteorite impact were too great for seeds to survive, but fragments of seeds did survive in simulations of those ejections [82].

To be efficient, natural transformation of *A. baylyi* requires at least a few hundred base pairs of one sided homology between the incoming and recipient DNA [83], limiting the inward transfer of extraterrestrial DNA to sequences originating on Earth and suggesting that life on Earth is insulated from exoDNA. On the other hand, if DNA transfer into the biosphere were not homology independent, partially degraded DNA and bacteria carried by micrometeorites or defunct extraterrestrial organisms, resembling seeds, could enter Earth's biosphere. Among the mostly unknown microbes, there could be species capable (e.g., when under stress) of incorporating heterologous DNA into their genomes. Alternatively, DNA sequence evolution could be open to exoDNA via universal, homologous sequences like the Ri TL-DNA that would genetically connect distant biospheres. This discussion is relevant to proposals to send humans to Mars (see below).

Nucleic acid transfer through space is a plausible way to explain the past origin of the life we know on Earth. However, the future of our life will be limited by increased heating from the Sun. For life is to survive, it will have to be exported, e.g., in a vector like a plant seed, to exohabitats. Ironically, dispersal away from Earth has already occurred through the contamination of unmanned space probes with UV-resistant bacterial spores [84, 85]. Exospermia (as opposed to endospermia) is thus happening through our space explorations, in a reversal of the directed panspermia hypothesis evoked to explain the origin of life on Earth [86].

If we are directing panspermia, why not do it as best as we can? What are the most resistant life forms that can be sent to exohabitats, and how can their resistance be improved? In the case of seeds, they could be chosen for inherent resistance (e.g., morning glory) and also genetically modified, e.g., to accumulate more of the flavonoid sunscreens [87, 88]. They could also be coated with UV screens and loaded with free-living, beneficial bacteria (e.g., nitrogen-fixing *Rhizobium*). Genetic redundancy in nuclear DNA can be increased in plants by increasing ploidy levels [89]. The redundancy of vulnerable components, necessary for recovery, such as ribosomes, chloroplasts and mitochondria, might also be improved. The latter two might be genetically modified with high performance bacterial DNA repair systems, like those in from *Deinococcus radiodurans* [90].

Humans are poorly suited for space travel, but they would be excellent vectors for microbial life. A human death on Mars could liberate roughly 10^{14} microbial cells [91]. Seeds and selected microorganisms could be sent to specific exohabitats [78, 85, 92] in a vanguard for human colonization. While seeds might survive space travel and germinate, what would happen to the plants they produce? The deleterious effects of UV on mature plants have been described only on Earth, where atmospheric ozone filters out $UV_{<300\ nm}$. DNA absorbs strongly at wavelengths below 300 nm. Thus, sunlight for photosynthesis must be filtered via external shading, increased inherent filtering, e.g., through increased flavonoids, and through an atmospheric filter, like oxygen and derived ozone. An ideal exohabitat would include enough atmospheric oxygen to protect against UV and to supply mitochondrial respiration.

Voyager I is over 20 billion km away from the sun. Humans are currently engaged in genetic football on a cosmic scale, but time is running out for life on Earth. In roughly one billion years the Earth will be too hot for even microbial life to survive; thus, only 20% of life's tenure on Earth remains for finding another habitat, and at our present pace of self-destruction, humans will be extant and healthy enough to accomplish exospermia for a much shorter period. In the meantime, better UV and radiation resistance in plants could be useful in the event of damage to the ozone layer, increased radiation from nuclear wars or during magnetic pole reversal.

9 Biotechnological Applications

Soon after the initial culture of transformed roots *in vitro*, numerous biotechnical applications became obvious or were proposed by collaborators. It seemed important, while doing the basic research outlined above, to stay connected to the real world, so applications were pursued both in-house and through collaborations. The most obvious of these, the use of the Ri plasmid as a genetic transformation vector for crop plants, was aggressively developed by several competent labs, so my participation was not needed, and my proposal that the transformed phenotype was a better marker than antibiotic resistance was not accepted (Vectors were soon made with Ti T-DNA in which the plant hormone genes were removed.). Applications using the transformed phenotype (e.g., transformed roots and altered root and shoot architectures) were more interesting than genetic engineering, because they led me into unfamiliar subjects. Thus, on a much smaller scale, through seminars and the dissemination of *A. rhizogenes* bacterial strains, we encouraged the use of wild type *A. rhizogenes* and the transformed phenotype as a means of genetically improving plants without resorting to DNA manipulation [27].

Since transformation by *A. rhizogenes* occurs in nature, it is often considered to be natural and not to involve genetic manipulation. Thus, a cottage industry has developed that uses wild type *A. rhizogenes* in clever and inexpensive biotechnological applications. I find that low budget research is often more inventive and adventurous than the well financed version, because it requires more imagination. Also, it can be done in developing countries, where innovation has a bright future [93]. I also believe that inventions generated in the public sector should belong to the public domain, and INRA is a governmental research institute. Nevertheless, five patents were awarded to us, but not pursued by INRA. All intellectual property described here is now in the public domain. To my knowledge, the only research currently in commercial application is the production of an AM fungal inoculum on transformed roots. See Sect. 9.3.

9.1 Chimeric Plants Through Genetic Grafting

In nature, *A. rhizogenes* presumably produces a transformed, secondary root system. Claudi Lambert, a graduate student, showed that *A. rhizogenes* could root recalcitrant apple root stocks and that the opines were translocated into the wild type plant parts of chimeric apple trees from a genetically transformed root system [94, 95]. See also references [96–98]. We envisioned using genetic grafting to improve edible plant parts, e.g., apples, using genes transferred to the root system, but not to the aerial parts. A transgenic apple tree showed increased root and shoot production, as expected with augmented plasticity.

9.2 Use of Transformed Roots to Determine Cadmium Availability in Contaminated Soils

A Versailles colleague, René Prost, suggested that we use transformed root cultures to measure the availability of cadmium and other toxic heavy metals in soils. Heavy metals are released by mining and manufacturing, from sewage sludges and batteries (to name just a few sources). They are spread through water, wind, human activity and by animals and plants that absorb and concentrate them. Once released, they stay in the environment, from which they are readily taken up by both animals and plants. However, heavy metals can become complexed with soil constituents and thus taken out of biological circulation.

While it is relatively simple to measure heavy metals in the soil, there was no inexpensive and reliable way measure their availability to living organisms (bioavailability), which is a crucial parameter in predicting toxicity. We thus set out with a postdoctoral fellow, Lionel Metzger, to use transformed roots as a living assay for bioavailability. The result was a feasibility study that showed that transformed tobacco and morning glory roots could be used to bioassay Cd availability in contaminated sewage sludges [99]. We also noted that transformed roots in culture take up little space, that they are inexpensive to produce and that they avoid the variable and complex influences of the aerial plant parts, providing simplicity and reproducibility.

The next project was to use what we had learned about Cd uptake and toxicity to stabilize and detoxify extremely polluted waste dumps, where no plants could grow. I coordinated a proposal to the European Union to attempt this reclamation using morning glory, genetically modified to specifically nourish a heavy metal sequestering soil bacterium. The objective was to first stabilize the site against erosion, then to harvest the morning glory as it removed Cd from the soil. (This was before the concept of phytoremediation was invented.) The proposal was not accepted.

9.3 Use of Transformed Roots to Propagate Arbuscular Mycorrhizal Fungi In Vitro

Jacques Mugnier, a colleague at a nearby Rhône-Poulenc lab, used my transformed morning glory roots to grow an obligate, rhizosphere symbiont, an arbuscular mycorrhizal (AM) fungus [100] (Fig. 10). See also [101]. AM fungi are essential for the growth of about 80% of the vascular plants. Hyphae invade root cortical cells, where they exchange minerals (e.g., phosphorus) for energy-rich nutrients from the plant. Hyphae form extensive networks in the soil, complement the host's root system and physiologically connect plants of different species. Attempts had been made to produce their spores in pot-grown plants, but they were stopped for fear of propagating pathogens. Axenic root cultures solved this problem. An inoculum consisting of spores and transformed roots is currently used to produce biological fertilizer for soil remediation in India, thanks to research by Alok Adholeya and his coworkers at TERI University in New Delhi, India [102]. We have used transformed roots to explore similarities between AM fungal and Rhizobial mutualisms [102] in a collaboration with Boovaraghan Balaji and Yves Piché (Université Laval, Québec).

I was intrigued by the possibility that plants of different species could be connected through a common network of AM fungi. I concocted a two compartment model that relied on staining and autoradiography for following ^{23}P transport and uptake into the roots by the fungus. The basal parts of the roots were nourished by a rich, solid medium (simulating the plant) in a small Petri dish inserted in a large Petri

Fig. 10 Root segment, oriented vertically, from a culture transformed by *A. rhizogenes* and propagating hyphae of an arbuscular mycorrhizal fungus, *Glomus mosseae*, running horizontally (Co-cultivation J. Mugnier; Photo D. Tepfer)

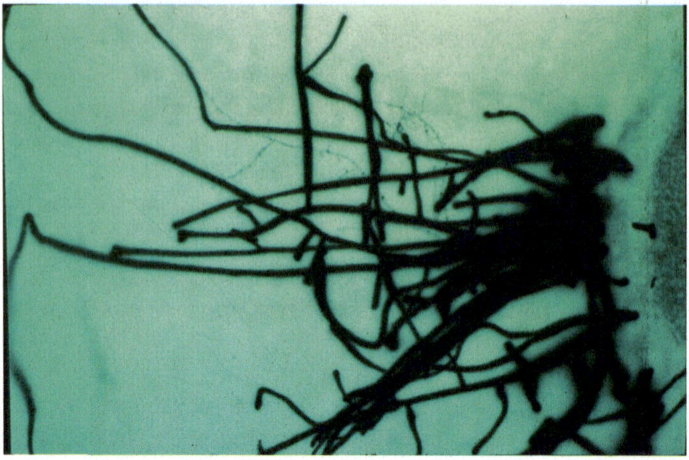

Fig. 11 Autoradiograph of transformed *C. sepium* roots and AM fungal hyphae (*Gigaspora margarita*), labeled with ^{32}P, taken up from the medium in a large Petri dish filled with agar + water. The roots were nourished through their proximal parts by a rich medium in a small Petri dish set inside the large Petri dish. The small dish (formerly on the *right*) was removed to allow for autoradiography, after cutting the roots at its edge. The *thick lines* are roots, Photo and the *thin lines* are hyphae (Co-culture, J. Mugnier. Autoradiography and experimental design, Photo D. Tepfer.)

dish, where the fungal/root interaction took place in a basic agar + water medium. We could monitor the movement of ^{32}P from a point source in the large compartment (simulating the soil) into hyphae and the roots by removing the outer compartment and drying the medium onto filter paper like an electrophoresis gel. Staining of the hyphae and autoradiography revealed the spatial and functional relationships between the roots, the hyphae and ^{32}P (Fig. 11). Transformed roots in vitro could be used to model other rhizosphere relationships, free from the complexity of the soil [16]. The experiment I never did was to connect the roots of two species using two small Petri dishes by AM fungi in a single large Petri dish and to look for chemical connections between them, including DNA transfer (genetic football), perhaps mediated by bacteria harbored by the fungus.

9.4 Use of Genetic Transformation to Alter Secondary Metabolite Production

Early in the adventure of growing transformed roots, I was confronted with a problem of over-supply, which was solved by eating the rapidly accumulating carrot roots (Fig. 2). Compared to normal roots in culture, they had an intense, spicy, carrot flavor with an unexpected hint of pepper. This taste test was the first indication that secondary metabolite production was altered and enhanced in plant tissues transformed by *A. rhizogenes*.

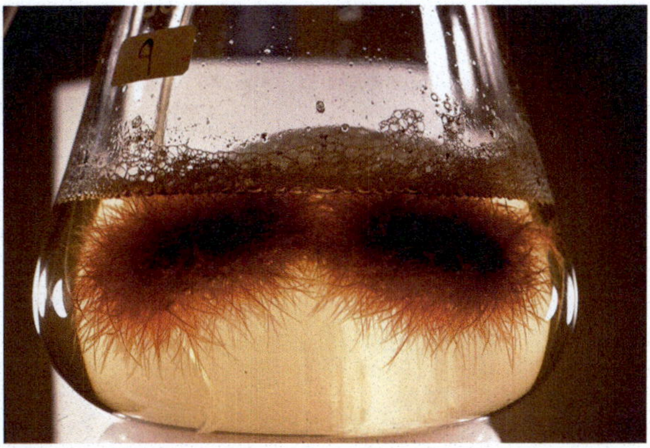

Fig. 12 Red beet roots transformed by wild type *A. rhizogenes* in an early stage of culture with aeration by sterile air pumped through an aquarium stone. Bioreactors are not needed to grow transformed roots. Cultures can be aerated with simple aquarium equipment plus a filter to sterilize incoming air, and the pressure produced is sufficient to remove medium for extraction and renewal. In one such culture, morning glory root density was so high that they had to be sawed to remove them from the culture vessel and the aquarium stone was never found, apparently digested by the roots (Photo D. Tepfer)

Plant cells grown in bioreactors have been used in attempts to produce high value pharmaceuticals. Transformed roots are better candidates for such endeavors because they grow fast, and (unlike cell cultures) they are genetically stable. These advantages became evident when a colleague, Gerard Jung at a nearby Rhône-Poulenc lab grew my *C. sepium* roots in a 30 l yeast fermenter. He produced two kilos (fresh weight) in two weeks, and he showed that they accumulated increased amounts (10x the titers in wild type leaves) of the tropane alkaloid, cuscohigrine [103]. See a simple bioreactor, Fig.12) In general, transformed roots accumulate the metabolites found in normal roots; e.g., red beet roots are a bright red (Fig. 12), but quantitative and qualitative changes take place in transformed roots and plants. An example is given below.

Scented lemon geraniums (*Pelargonium*) are the source of most of the natural rose fragrance in expensive cosmetics. They are also popular houseplants, but they suffer from having few branches and long internodes, a defect corrected by transformation with wild type Ri T-DNA (Fig. 13) They were attractive subjects for exploring the world of fragrance secondary metabolites and shoot system architecture, and when transformed by wild type Ri T-DNA, they produced quantitatively more and qualitatively different monoterpenes, i.e., fragrances [104]. Geraniol increased 2.0–4.4 fold; cineole increased 3.3–13 fold. Other oxygenated monoterpenes decreased. Overall essential oil production increased. The differences were obvious to human volunteers in blind tests (unpublished results). Leaf production increased by a factor of 3–4 times, compared to wild type controls, giving an overall increased production of geraniol of about tenfold [104]. Furthermore, the transformed plants had improved architecture and

Fig. 13 Transgenic scented geranium plants. *Left*, transformed by wild type *A. rhizogenes*. *Right*, wild type control. Internode distance and apical dominance were decreased, improving stature for use as a house plant. Fragrance production was increased (Reprinted from Ref. [104]). See Fig. 16 for the drought stress phenotypes in the same plants (Photo D. Tepfer)

drought resistance (Figs. 13 and 16). I imagined a houseplant perfume mini-factory, tailored by genetic transformation to the user's fragrance preferences. A collaboration with John Sanford extended our results to other scented geraniums, having very different fragrances, but to my knowledge there is no current commercial application.

Secondary metabolites took center stage thanks to visit from S. Ja (University of Calcutta, India). After showing that transformation by *A. rhizogenes* could increase secondary metabolites *Tylophora indica* [105, 106] and *Withania somnifera* [107], we received a grant from CEFIPRA to attempt to go beyond the improvements associated with transformation by Ri T-DNA. Our approach was to use genetic elicitation with the β-cryptogen gene to trick the host plant into perceiving a fungal attack coming from within its cells (see above, Sect. 6) and thus increase secondary metabolite production. We transformed *C. sepium, C. arvensis, T. indica, W. somnifera* and Arabidopsis with Ri T-DNA plus β-cryptogein and obtained secondary metabolite increases beyond those due to the Ri T-DNA (Fig. 14). In Arabidopsis, transformation with the *crypt* gene produced increases in the titers of certain flavonoids [87] (This research is described in more detail elsewhere in this volume.).

Transgenic mimicry of pathogen attack was thus correlated with increased growth and secondary metabolite accumulation in five species for four classes of medicinal substances. These results again suggested that the resistance to *Phytophthora* associated with the β-cryptogein gene in tobacco [44, 67] was due at least in part to increased antifungal metabolites, and they provided a further example of the possible adaptive significance of genetic football. But how does it work? One simple hypothesis is suggested by the role of polyamines in producing the transformed phenotype. Polyamine accumulation is reduced by Ri TL-DNA, which could shift carbon allocation into other anabolic pathways, e.g., for antifungal and medicinal substances.

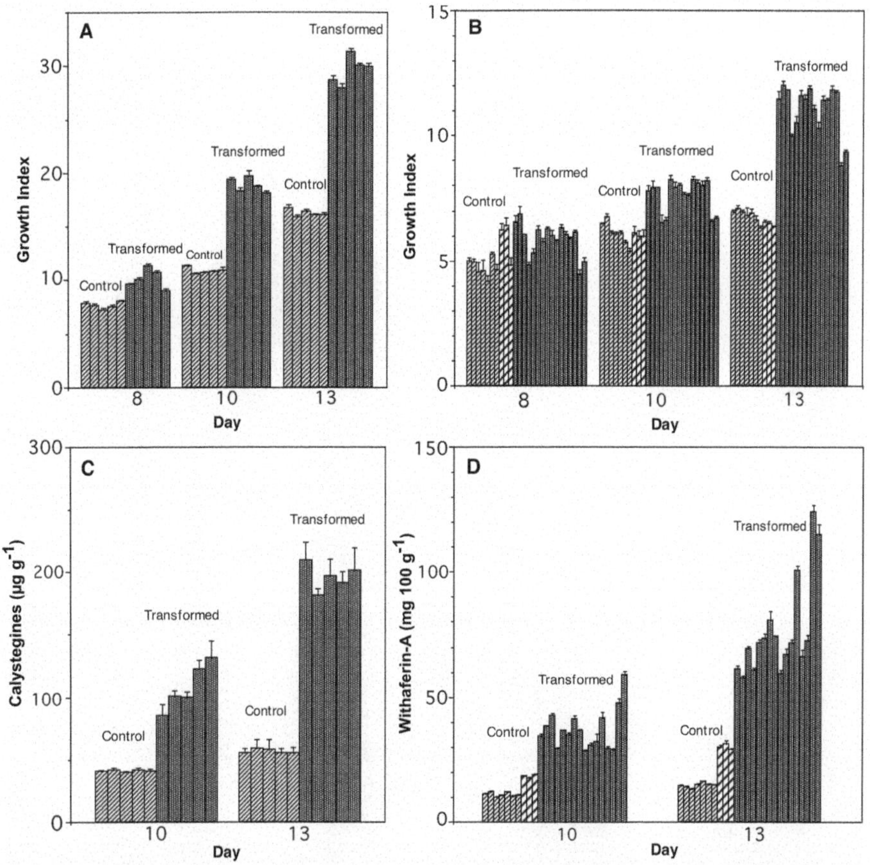

Fig. 14 Effects of genetic transformation with the *crypt* gene on growth and secondary metabolite production. The *crypt* gene was introduced into *C. sepium* (**a**, **c**) and *W. somnifera* (**b**, **d**) using *A. rhizogenes* LBA9402. Growth (**a**, **b**) and secondary metabolite accumulation (**c**, **d**) were recorded over time. Bars represent individual root clones from independent transformation events: dark hatching, roots containing *crypt* and Ri T-DNA; bold hatching, roots containing just Ri T-DNA produced by wild type *A rhizogenes*, carrying the *crypt* gene construct, but which did not receive the *crypt* gene from the pBin19 binary vector; light hatching, roots produced with wild type *A. rhizogenes* (Reprinted from Ref. [87])

9.5 Changing Root and Shoot Architecture

The effects of Ri TL-DNA genes on growth and development have been the subject of numerous studies. A comprehensive review is outside the scope of this chapter, but to give a few examples: in our hands the promotion of annualism and reduced apical dominance was due to *rolC* in carrot (Fig. 15) and Belgian endive (*Cichorium intybus*) [77]; extreme dwarfing was associated with *rolD* in carrot [77]; extreme wrinkling and dwarfing were attributed to *rolA* in tobacco

Fig. 15 Changing phenotype in carrot with wild type Ri T-DNA and its *rolC* gene. (**a**) *Left*, plants transformed by wild type Ri T-DNA, showing reduced apical dominance, flowered as annuals and produced diffuse root systems, instead of carrots; *right*, wild type controls, regenerated from wild type root cultures were biennials, and they produced normal carrots. The plants are of approximately the same age. The difference in size is probably because the products of photosynthesis were stored in carrots in the controls, but reinvested in making leaves in the transformed plants. (**b**) Root system of a plant transformed by *rolC* (ORF 12) from the Ri TL-DNA, showing reduced apical dominance. (**c**) Shoot system of the same plant, showing reduced apical dominance and annual flowering (Photos D. Tepfer)

(Fig. 8) [72], and wild type Ri T-DNA caused reduced apical dominance and internode shortening in numerous species (Figs. 3, 5, 8, 13, and 14).

A grant from the Rockefeller Foundation revived a long-standing, personal concern about climate change and drought resistance. Global warming was upon us, and improving root system architecture was one approach to adapting crop plants to changed access to water. Nonirrigated, upland, rice was the chosen target because of the vulnerability of subsistence farmers in developing countries to even small climate changes. Experiments using scented geraniums grown in the greenhouse in columns of sand, with water and nutrients injected either from the top or from part way down the side, showed that plants transformed by wild type Ri T-DNA responded better to water limitations than the wild type by making use of the expected increased plasticity of their root development when the source of moisture was lowed in the column of sand (Fig. 16). The experiments were done in a large, asymmetrical greenhouse that had evenly distributed

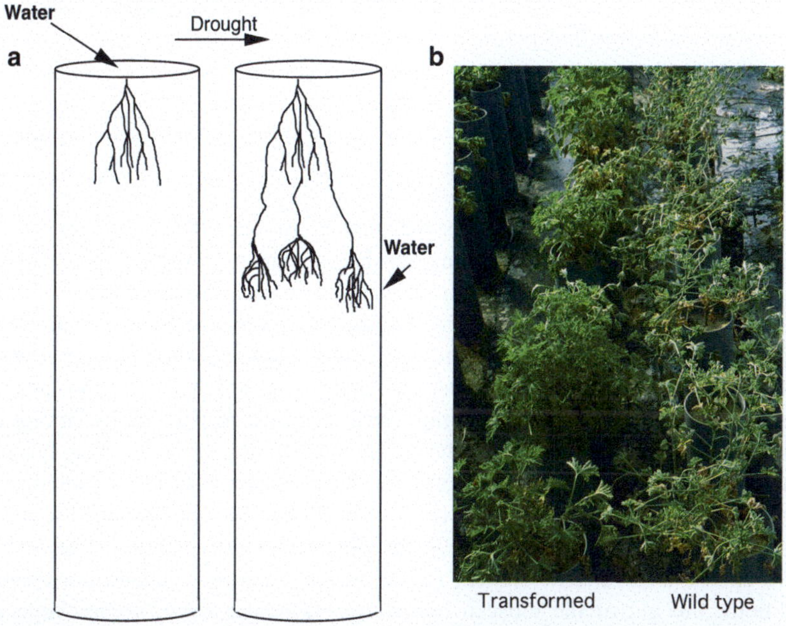

Fig. 16 Resistance to simulated drought in scented geranium transformed by wild type Ri T-DNA. (**a**) System for simulating drought in a column of sand in PVC drainage pipe (unpublished results). *Left*, prior to drought, water plus mineral nutrients were periodically provided through a capillary at the base of the plant. (Aerial parts are not shown.) To simulate drought, the capillary was inserted *lower down* in the column. Root growth and architecture were revealed by lifting the PVC pipe, which liberated the root system intact. Plants responded to drought by sending down deep roots that developed profuse branches at the lowered source of moisture. This adaptation was faster and more efficient in the transformed plants. (**b**) Experiment in the greenhouse part way through drought simulation, with the transformed plants on the *left* and the wild type on the *right*. The transformed plants adapted their root systems to the drought conditions and stayed green. The wild type plants did not adapt and had turned brown by the end of the experiment (Photo D. Tepfer)

light exposure. Unfortunately, a windstorm damaged the greenhouse, which was not repaired because INRA had other plans for it. We were given a much smaller space with uneven lighting, and the experiment could thus not be repeated and published.

However, the real goal was to produce phenotypic changes in rice root systems, and for this we were fortunate to collaborate with Michael Davey et al., who showed that our 35S CAMV *rolA* construct was capable of changing phenotype in rice. The phenotype was similar to that described in tobacco transformed with the same gene [29], with extreme leaf wrinkling and severe alterations of growth and architecture in both shoots and roots. Unfortunately, the grant was not renewed, but we showed that at least one of the Ri TL-DNA genes produced a phenotype in rice that recalled to the one we had described for dicot species.

As I write this, 15 years later, I am discouraged by the lack of progress in reducing the human activities that contribute to climate change, and I continue to lament the potentially beneficial scientific and biotechnological opportunities that were missed due to petty, self-interested political controversies over genetically modified plants. It was thus pertinent to evaluate the importance of genetic football in biosafety.

9.6 Biosafety

The genetic football model predicts the transfer of plant DNA to bacteria. We observed this in the laboratory, but unpublished attempts to evaluate its occurrence in hydroponics and in the soil were thwarted by the presence of antibiotic resistance markers, e.g., *nptII,* in wild bacteria. We therefore made three silent mutations (creating a new restriction site) in the part of the incoming *nptII* that replaces the deletion in the resident *nptII* in *A. baylyi*, restoring *nptII* activity. It was thus possible to unequivocally demonstrate DNA transfer from plants to bacteria by cutting and sequencing a PCR product from the rescued *nptII*, thus proving that bacterial sequences had come from the plant and not from a bacterial or DNA contaminant. We used this system when we showed DNA transfer from the roots and leaves of transgenic plants in the laboratory [49]. See Sect. 4.1.

We choose a novel environment for doing the experiment under more natural conditions: the gut of the tobacco hornworm, *Manduca sexta*, a very large and beautiful insect larva that consumes tobacco plants with gusto and produces copious feces. The marker exchange assay seemed suited for detecting DNA transfer, since it does not rely on intact DNA sequences, and we used (among others) the multi-copy DNA source employed in the seed experiments in space: a tobacco chloroplast insertion of *nptII*, present at about 5,000 copies per leaf cell, allowing flooding the horn worm gut with *nptII*. The hornworms consumed transgenic tobacco plants that had been sprayed with *A. baylyi* containing the deleted *nptII.* We were unable to detect DNA transfer, probably due to high DNAse activity in the hornworm intestine [108]. The follow-up experiments were to take place in the soil and hydroponic microcosms, but personnel and funding difficulties prevented further research.

We nevertheless showed that tobacco hornworms are not obvious intermediates in the game of genetic football.

9.7 HIV Vaccine

The most ambitious biotechnical application of DNA transfer was an attempt to express a viral antigen in plants in the hope of eventually producing an oral AIDS vaccine in collaboration with Anna Kostrazk (a shared graduate student), Simon Wain-Hobson and Monica Sala (Pasteur Institute), and with Tomasz Pniewski (Polish Academy of Sciences). Development of an AIDS vaccine and vaccines based on the expression of antigens in plants are both distant goals. We reasoned that while immediate success was unlikely, in the name of future advances it was important to make a preliminary attempt. Oral tolerance is an inherent problem in vaccines produced in plants, largely because plant secondary metabolites can both elicit oral tolerance and serve as adjuvants. We used the Hepatitis B virus as a model and expressed HBsAg in tobacco. Dried leaves were fed to mice to assay for immune responses and oral tolerance. We concluded that secondary metabolites in plants can act as adjuvants, boosting the immune response, but that their levels needed to be kept low to avoid oral tolerance [109].

The significance of these experiments for genetic football recalled our experiments with the β-cryptogein gene: elicitation of a host defense response was obtained by expressing a gene, encoded in the pathogen, that serves as a signal to turn on the host's defenses. In this case the pathogen was a virus, rather than a fungus, and the genetic host was a plant, which was eaten by a mammalian viral host. These experiments illustrate another level of possible complexity in the genetic football game, and they make me wonder if genes encoding antigens might be subjects for genetic football.

10 Conclusions

It seems like a long way from natural genetic transformation by *A. rhizogenes* to sending seeds into space to find a new home for life, but looking back, the long and circuitous path can be explained by the cast of chemists, physicists, and biologists who visited my parents' house when I was growing up. Starting with a list of interesting characters, the scenario wrote itself, aided by serendipity, setbacks, and desperate attempts to find funding that forced me to take new directions and collaborations. The admonition that I keep my hands in my pockets when visiting my father's lab set off a desire to do hands-on research, which is still unquenched. The frustrations of managing a lab in a hostile environment and the distractions of writing grant proposals, reports, reviewing papers, etc. only contributed to the frustration. Fortunately, the last two years of the space project were spent working blissfully alone, using my accumulated equipment and supplies. I happily did all of the experimental work after the seeds came back from space [78] (Tepfer and Leach, submitted), confirming my belief that the hands nourish the mind.

My great luck was having eclectic collaborations and a small, faithful core of collaborators in Versailles. The result is a view of genetics, ecology, and evolution that pokes holes in the concept of species, and fits into the general discussion of horizontal gene transfer (HGT), which is coming into general acceptance with discoveries of its occurrence in the animal kingdom [110]. Genetic football is based on a few facts gleaned from experiments constrained by lack of tools and knowledge. Some experiments were done in thought only, and the DNA transfers described here were facilitated by human intervention, but they show that genetic footballs can be passed across species and kingdom boundaries and even kicked long distances through space. Parasexual DNA transfer does occur in nature, and the basic protagonists might not be genes but rather protein functional domains, encoded in nucleic acids, which is why the term horizontal gene transfer should be changed to horizontal DNA transfer. It is still too soon to use amino acid sequences and protein domain functions to reliably reconstruct past parasexual DNA transfers, but as computer modeling and protein biochemistry improve, such an approach could become viable. In the meantime, there are many laboratory experiments to do in real time, e.g., using transformed roots and bacteria growing in bioreactors and in multi-compartmental rhizosphere models to follow markers encoded in DNA and exchanged on a human time scale.

10.1 A General Model for the Role of DNA Transfer in Ecology and Evolution

Bacteria send DNA into plants, but they also receive DNA, e.g., from plants, from their environment. Plant to plant DNA transfer is accomplished through bacterial intermediates, but for the transferred DNA to have evolutionary significance, it must enter a germ line, which in plants means forming a fertile shoot from the cell that received the foreign DNA. This developmental regeneration is accomplished through *plast* genes, carried by the Ri TL-DNA, and these genes have diverged to be functional in many different species. Part of their action is through changes in polyamine accumulation. Plant DNA enters the microbial metagenome through soil bacteria, e.g., *A. baylyi*, which are adept at incorporating foreign DNA, and then it spreads through the bacterial community, e.g., via conjugation, eventually finding its way into a T-DNA, at which point it can be transferred back into the plant community by *A. rhizogenes*. Fungi [111, 112] and even animals [110] participate in this horizontal DNA transfer. There are interesting rules and constraints in this genetic football game [1, 26], and nutrition is an important driving force. Plants are the primary producers of biological energy. They synthesize complex secondary metabolites that can specifically nourish rhizosphere bacteria carrying genes to catabolize them. These nutritional mediators are the currency in the economics of parasexual

DNA transfer, and it takes a catabolic key to unlock their energy. The production of some of them is encoded by genes for opine synthesis, found in soil bacteria and transferred to plants via *Agrobacterium*; others are currently encoded in the plant genome (e.g., calystegins and betaines), but who knows – maybe they were or will become the objects of DNA transfer?

Besides nutritional mediators, other candidates for transfer include DNA sequences that alter development (branching, flowering, root geotropism, etc.) and genes that confer resistance to pathogens. The gene is not necessarily the fundamental functional unit, since through recombination (e.g., in *A. baylyi*) functional domains in homologous protein-encoding DNA sequences can be switched. In a multidirectional genetic exchange system, there is no way to know the origin of a DNA segment, since the genetic code is essentially the same in all of our life forms and DNA spreads horizontally, perhaps explaining the use of a similar genetic code in all organisms tested so far. The biosphere thus resembles a mosaic, with species lines blurred through the sharing of genetic information from all life forms.

And beyond our biosphere? Of the extremophiles tested so far, seeds are impressively adapted to conserving DNA integrity. Something like a seed might have brought life to Earth – perhaps including multiple life forms, explaining the appearance of the Archea, Bacteria, and Eukaryota at the root of the tree of life [92]. Looking into the future, seeds could serve as vectors for disseminating multiple life forms away from Earth. Only about 20% of life's tenure on Earth remains, and humans will be present and capable of dispersing life through space for only a small part of that. In the meantime, biological communities will adapt to environmental stresses by sharing DNA across species and kingdom boundaries.

Acknowledgments The core research was performed at the Institut National de la Recherche Agronomique by Tepfer, D., Goldmann, A., Message, B., Pamboukdjian, N., and Maille, M. I am indebted to my collaborators for their hard work, dedication, and tolerance: Ackerman, C. Ba, A.M., Balaji, B., Bandyopadhyay, M., Ben-Hayyim, G., Blackhalla, N., Boivin, C., Bonnett, H., Bouchez, D., Broglie, R., Burchell, M., Burnet, M., Burtin, D., Casse-Delbart, F., Charbonnier, C., Charpin, I., Chaudhuri, K., Chevalier, D., Chilton, M.D., Chioccioli, M., Cocking, E., Damon, J.P., Das, S., Davey, M., David, C., Delepelaire, P., Dénarié, J., Deni, J., Devendra, B.N., Drong, R., Ducrot, P. H., Durand-Tardif, M., El Amrani, A., Elbein, A.D., Fosket, D., Garcia-Gonzales, R., Ghosh, B., Goldmann, A., Gonzaleza, M.C., Guetarda, D., Guyon, P., Heisler, L., Helgeson, J., Hoffmann, S. V., Corbineau, F., Jerling, A., Jha, S., Jouanin, L., Jung, G., Kenney, J., Kollmann, A., Kostrzak, A., Lallemand, J.Y., Lambert, C., Larue, T.A., Leach F., Leach, S., Lee, S.-H., Lépingle, A., Levesque, H., Limami A., Maille, M., Mansouri, H., Martin, C., Martin-Tanguy, J., Message, B., Metzger, L., Michael, A., Milat, M.L., Molyneux, R., Monneuse, M.O., Mugnier, J., Pamboukdjian, N., Pan, Y. T., Paynot, M., Pellegrineschi, A., Perica, M.C., Petit, A., Piche, Y., Pniewski, T., Power, B., Prost, R., Rosenberg, C., Rouzé, P., Sala M., Seruga, M., Slightom, J., Sun, L.Y., Tempé, J., Touraud, G., Wain-Hobson, S., Yacoub, A. For materials, moral support, and advice I am indebted to (among others) the following benefactors: Adholeya, A., Bevan, M., Bourgin, J.P., Chua, N.-H., Cambell, R., Crespin, M., de Vries, J., Descoins, C., Field, D., Grunstein, M., Helgeson, J., Horneck, G., Kemp, J., Maliga, P., Ohkawa, H., Rabbow, E., Schneiderman, H., Shields, R., Tempé, J., Tepfer, M., Tepfer, R., Tobin, E., Torrey, J., and Wackernagel, W. Support (partial list) came from CEFIPRA, the Rockefeller Foundation, the Universities of Oregon, Aarhaus, Kobe, Singapore and Zagreb, from INRA, the CNRS, ESA, the CNES, TERI, the DLR, and NATO.

References

1. Tepfer D (1989) Ri T-DNA from *Agrobacterium rhizogenes:* a source of genes having applications in rhizosphere biology and plant development, ecology, and evolution. In: Kosuge T, Nester E (eds) Plant-microbe interactions, vol 3. McGraw Hill, New York, pp 294–342
2. Slightom JL, Jouanin L, Leach F, Drong RF, Tepfer D (1985) Isolation and identification of TL-DNA/plant junctions in *Convolvulus arvensis* transformed by *Agrobacterium rhizogenes* strain A4. Embo J 4(12):3069–3077
3. Nester EW, Gordon MP, Amasino RM, Yanofsky MF (1984) Crown gall: a molecular and physiological analysis. Annu Rev Plant Physiol Plant Mol Biol 35:387–413
4. Tepfer D (1984) Transformation of several species of higher plants by *Agrobacterium rhizogenes:* sexual transmission of the transformed genotype and phenotype. Cell 37(3): 959–967
5. Menage A, Morel G (1964) Sur la presence d'un acide amine nouveau dans le tissu de crown-gall. Physiol Veg 2:1–8
6. Braun A (1947) Thermal studies on the factors responsible for tumor induction in crown gall. Am J Bot 34:234–240
7. Braun AC (1958) A physiological basis for autonomous growth of the crown-gall tmor cell. Proc Natl Acad Sci U S A 44:344–349
8. Avery O, Mac Leod C, Mc Carty M (1944) Studies on the chemical nature of the substance inducing transformation of pneumococcal types. J Exp Med 79:137–158
9. Petit A, Delhaye S, Tempé J, Morel G (1970) Recherches sur les guanidines des tissus de crown-gall. Mise en evidence d'une relation biochimique specifique entre les souches *d'Agrobacterium tumefaciens* et les tumeurs qu'elles induisent. Physiol Veg 8:205–209
10. Goldman A, Tempe J, Morel G (1968) Quelques particularités de diverses souches d'*Agrobacterium tumefaciens*. CR Seances Soc Biol Ses Fil 162:630–631
11. Tempé J, Goldmann A (1982) Occurence and biosynthesis of opines. In: Kahl G, Schell J (eds) Molecular biology of plant tumors. Academic, New York, pp 451–459
12. Tempe J, Estrade C, Petit A (1978) The biological significance of opines: II. The conjugative activity of the Ti-plasmids of *Agrobacterium tumefaciens.* In: Sta. Path. Veg. Phytobact A, France (ed) The IVth international conference on plant pathogenic bacteria, Angers, pp 153–160
13. Nester EW, Gordon MP (1991) Molecular strategies in the interaction between *Agrobacterium* and its hosts. In: Hennecke H, Verma DPS (eds) Advances in molecular genetics of plant – microbe interactions, vol I, Current plant science and biotechnology in agriculture. Kluwer, Dordrecht, pp 3–9
14. Tempe J, Petit A, Bannerot H (1982) Presence de substances semblables a des opines dans des nodosites de luzerne. C R Acad Sci 295(5):413–416
15. Tepfer D, Bonnett H (1972) The role of phytochrome in the geotropic behavior of roots of *Convolvulus arvensis*. Planta 106(4):311–324
16. Tepfer D, Yacoub A (1986) Modèle de rhizosphère utile notamment pour l'étude des parasites des plantes. France Pat 2575759 Patent
17. Tepfer D, Fosket D (1975) Phosphorylation of ribosomal protein in soybean. Phytochemistry 14:1161–1165
18. Tepfer D, Fosket D (1978) Hormone-mediated translational control of protein synthesis in cultured cells of Glycine max. Dev Biol 62:486–497
19. Chilton MD, Drummond MH, Merlo DJ, Sciaky D (1978) Highly conseverd DNA of Ti plasmids overlaps T_DNA maintained in plant tumours. Nature 275(5676):147–149
20. Moore L, Warren G, Strobel G (1979) Involvement of a plasmid in the hairy root disease of plants caused by *Agrobacterium rhizogenes*. Plasmid 2:617–626
21. White P (1938) Cultivation of excised roots of dicotyledonous plants. Am J Bot 25:348–356

22. Torrey J (1958) Endogenous bud and root formation by isolated roots of *Convolvulus* grown in vitro. Plant Physiol 33(4):258–263
23. Tepfer D, Tempé J (1981) Production d'agropine par des racines formées sous l'action d'*Agrobacterium rhizogenes*, souche A4. C R Acad Sci 292(série III):153–156
24. Chilton MD, Tepfer D, Petit A, David C, Casse-Delbart F, Tempe J (1982) *Agrobacterium rhizogenes* Inserts T-DNA into the genomes of the host plant root cells. Nature 295(5848):432–434
25. Tepfer D (1982) La transformation génétique de plantes supérieures par *Agrobacterium rhizogenes*. In: 2e Colloque sur les Recherches Fruitières. Centre Technique Interprofessionnel des Fruits et Légumes, Bordeaux, pp 47–59
26. Tepfer D (1983) The biology of genetic transformation of higher plants by *Agrobacterium rhizogenes*. In: Puhler A (ed) Molecular genetics of the bacteria-plant interaction. Springer, Berlin, pp 248–258
27. Tepfer D (1983) The potential uses of *Agrobacterium rhizogenes* in the genetic engineering of higher plants: nature got there first. In: Lurquin P, Kleinhofs A (eds) Genetic engineering in eukaryotes. Plenum Press, New York, pp 153–164
28. Durand-Tardif M, Broglie R, Slightom J, Tepfer D (1985) Structure and expression of Ri T-DNA from *Agrobacterium rhizogenes* in *Nicotiana tabacum*. Organ and phenotypic specificity. J Mol Biol 186(3):557–564
29. Martin-Tanguy J, Sun LY, Burtin D, Vernoy R, Rossin N, Tepfer D (1996) Attenuation of the phenotype caused by the root-inducing, left-hand, transferred DNA and Its rolA gene (Correlations with changes in polyamine metabolism and DNA methylation). Plant Physiol 111(1):259–267
30. David C, Chilton MD, Tempé J (1984) Conservation of T-DNA in plants regenerated from hairy root cultures. Biotechnology 2:73–76
31. White F, Ghidossi G, Gordon M, Nester E (1982) Tumor induction by *Agrobacterium rhizogenes* involves the transfer of plasmid DNA to the plant genome. Proc Natl Acad Sci U S A 79:3193–3197
32. Costantino P, Spano L, Pomponi M, Benvenuto E, Ancora G (1984) The T-DNA of *Agrobacterium rhizogenes* is transmitted through meiosis to the progeny of hairy root plants. J Mol Appl Genet 2:465–470
33. Spano L, Pomponi M, Costantino P, Van Slogteren G, Tempe J (1982) Identification of T-DNA in the root-inducing plasmid of the agropine type *Agrobacterium rhizogenes* 1855. Plant Mol Bio 1:291–300
34. Spano L, Costantino P (1982) Regeneration of plants from callus cultures of roots induced by *Agrobacterium rhizogenes* on tobacco. Z Pflanzenphysiol 106:87–92
35. Furner I, Huffman G, Amasino R, Garfinkel D, Gordon M, Nester E (1986) An *Agrobacterium* transformation in the evolution of the genus *Nicotiana*. Nature 319:422–427
36. Martin-Tanguy J, Corbineau F, Burtin D, Ben-Hayyim G, Tepfer D (1993) Genetic transformation with a derivative of *rolC* from *Agrobacterium rhizogenes* and treatment with a-Aminoisobutyric acid produce similar phenotypes and reduce ethylene production and the accumulation of water-insoluble polyamine-hydroxycinnamic acid conjugates in tobacco flowers. Plant Sci 93(1-2):63–76
37. Ackermann C (1977) Pflanzen aus *Agrobacterium rhizogenes*-tumoren aus *Nicotiana tabacum*. Plant Sci Lett 8:23–30
38. White F, Garfinkel D, Huffman G, Gordon M, Nester E (1983) Sequences homologous to *Agrobacterium rhizogenes* T-DNA in the genomes of uninfected plants. Nature 301:348–350
39. Kyndt T, Quispe D, Zhai H, Jarret R, Ghislain M, Liu Q, Gheysen G (2015) The genome of cultivated sweet potato contains *Agrobacterium* T-DNAs with expressed genes: an example of a naturally transgenic food crop. Proc Natl Acad Sci U S A 112(18):5844–5849
40. Lambert C, Tepfer D (1992) Use of *Agrobacterium rhizogenes* to create transgenic apple trees having an altered organogenic response to hormones. Theor Appl Genet 85(1):105–109

41. Taylor BR, White FF, Nester EW, Gordon MP (1985) Transcription of *Agrobacterium rhizogenes* A4 T-DNA. Mol Gen Genet 201:546–553
42. White F, Taylor B, Huffman G, Gordon M, Nester E (1985) Molecular and genetic analysis of the transferred DNA regions of the root inducing plasmid of *Agrobacterium rhizogenes*. J Bacteriol 164:33–44
43. Levesque H, Delepelaire P, Rouzé P, Slightom J, Tepfer D (1988) Common evolutionary origin of the central portions of the Ri TL-DNA of *Agrobacterium rhizogenes* and the Ti T-DNAs of *Agrobacterium tumefaciens*. Plant Mol Biol 11:731–744
44. Tepfer D, Boutteaux C, Vigon C, Aymes S, Perez V, O'Donohue MJ, Huet J, Pernollet J (1998) *Phytophthora* resistance through production of a fungal protein elicitor (β-cryptogein) in tobacco. Mol Plant Microbe Interact 11–1:64–67
45. de Vries J, Heine M, Harms K, Wackernagel W (2003) Spread of recombinant DNA by roots and pollen of transgenic potato plants, identified by highly specific biomonitoring using natural transformation of an *Acinetobacter* sp. Appl Environ Microbiol 69(8):4455–4462
46. Paget E, Lebrun M, Freyssinet G, Simonet P (1998) The fate of recombinant plant DNA in soil. Eur J Soil Biol 34:81–88
47. Pote J, Mavingui P, Navarro E, Rosselli W, Wildi W, Simonet P, Vogel TM (2009) Extracellular plant DNA in Geneva groundwater and traditional artesian drinking water fountains. Chemosphere 75(4):498–504
48. de Vries J, Wackernagel W (1998) Detection of *nptll* (kanamycin resistance) genes in genomes of transgenic plants by marker-rescue transformation. Mol Gen Genet 257(6):606–613
49. Tepfer D, Garcia-Gonzales R, Mansouri H, Seruga M, Message B, Leach F, Perica MC (2003) Homology-dependent DNA transfer from plants to a soil bacterium under laboratory conditions: implications in evolution and horizontal gene transfer. Transgenic Res 12 (4):425–437
50. Tepfer D, Metzger L, Prost R (1989) Use of roots transformed by *Agrobacterium rhizogenes* in rhizosphere research: applications in studies of cadmium assimilation from sewage sludges. Plant Mol Biol 13(3):295–302
51. Ceccherini M, Pote J, Kay E, Van VT, Marechal J, Pietramellara G, Nannipieri P, Vogel TM, Simonet P (2003) Degradation and transformability of DNA from transgenic leaves. Appl Environ Microbiol 69(1):673–678
52. Bertolla F, Frostegard A, Brito B, Nesme X, Simonet P (1999) During infection of its host, the plant pathogen *Ralstonia solanacearum* develops a state of competence and exchanges genetic material. Mol Plant Microbe Interact 12:1–6
53. Nutman PS (1965) The relation between nodule bacteria and the legume host in the rhizosphere and in the process of infection. In: Baker KF, Synder WC (eds) Ecology of soil-borne pathogens. University of California Press, Berkeley, pp 231–247
54. Murphy PJ, Heycke N, Banfalvi Z, Tate ME, de Bruijn F, Kondorosi A, Tempé J, Schell J (1987) Genes for the catabolism and synthesis of an opine-like compound in *Rhizobium meliloti* are closely linked and on the Sym plasmid. Proc Natl Acad Sci U S A 84:493–497
55. Tepfer D, Goldmann A, Pamboukdjian N, Maille M, Lépingle A, Chevalier D, Dénarié J, Rosenberg C (1988) A plasmid of *Rhizobium meliloti* 41 encodes catabolism of two compounds from root exudate of *Calystegia sepium*. J Bacteriol 170(3):1153–1161
56. Boivin C, Malpica C, Rosenberg C, Dénarié J, Goldman A, Fleury V, Maille M, Message B, Tepfer D (1989) Metabolic signals in the rhizosphere: catabolism of calystegins and trigonelline by *Rhizobium meliloti*. In: Lugtenberg B (ed) Signal molecules in plant and plant-microbe interactions, Nato ASI Series. Springer, Berlin, pp 401–407
57. Goldmann A, Milat ML, Ducrot PH, Lallemand JY, Maille M, Lepingle A, Charpin I, Tepfer D (1990) Tropane derivatives from *Calystegia sepium*. Phytochemistry 29(7):2125–2127
58. Goldmann A, Message B, Tepfer D, Molyneux RJ, Duclos O, Boyer FD, Pan YT, Elbein AD (1996) Biological activities of the nortropane alkaloid, calystegine B2, and analogs: structure-function relationships. J Nat Prod 59(12):1137–1142

59. Guntli D, Burgos S, Moënne-Loccoz Y, Défago G (1999) Calystegine degradation capacities of microbial rhizosphere communities of *Zea mays* (calystegine-negative) and *Calystegia sepium* (calystegine-positive). FEMS Microbiol Ecol 28:75–84
60. Goldmann A, Boivin C, Fleury V, Message B, Lecoeur L, Maille M, Tepfer D (1991) Betaine use by rhizosphere bacteria: genes essential for trigonelline, stachydrine, and carnitine catabolism in *Rhizobium meliloti* are located on pSym in the symbiotic region. Mol Plant Microbe Interact 4(6):571–578
61. Goldmann A, Lecoeur L, Message B, Delarue M, Schoonejans E, Tepfer D (1994) Symbiotic plasmid genes essential to the catabolism of proline betaine, or stachydrine, are also required for efficient nodulation by *Rhizobium meliloti*. FEMS Microbiol Lett 115:305–312
62. Burnet M, Goldmann A, Message B, Drong R, El Amrani A, Loreau O, Slightom J, Tepfer D (2000) The stachydrine catabolism region in *Sinorhizobium meliloti* encodes a multi-enzyme complex similar to the xenobiotic degrading systems in other bacteria. Gene 244(1-2):151–161
63. Tempe J, Guyon P, Tepfer D, Petit A (1979) The role of opines in the ecology of the Ti plasmids of *Agrobacterium*. In: Timmis KN, Pühler A (eds) Plasmids of medical, environmental and commercial importance. Elsevier/North-Holland Biomedical Press, pp 353–363
64. Huet JC, Pernollet JC (1993) Sequences of acidic and basic elicitin isoforms secreted by *Phytophthora megasperma*. Phytochemistry 33:797–805
65. O'Donohue MJ, Gousseau H, Huet JC, Tepfer D, Pernollet JC (1995) Chemical synthesis, expression and mutagenesis of a gene encoding β-cryptogein, an elicitin produced by *Phytophthora cryptogea*. Plant Mol Biol 27(3):577–586
66. O'Donohue MJ, Boissy G, Huet JC, Nespoulous C, Brunie S, Pernollet JC (1996) Overexpression in *Pichia pastoris* and crystallization of an elicitor protein secreted by the phytopathogenic fungus, *Phytophthora cryptogea*. Protein Expr Purif 8(2):254–261
67. Keller H, Pamboukdjian N, Ponchet M, Poupet A, Delon R, Verrier J, Roby D, Ricci P (1999) Pathogen-induced elicitin production in transgenic tobacco generates a hypersensitive response and nonspecific disease resistance. Plant Cell 11:223–235
68. Bundock P, Den Dulk-Ras A, Beijersbergen A, Hooykaas PJJ (1995) Trans-kingdon T-DNA transfer from *Agrobacterium tumefaciens* to *Saccharomyces cerevisiae*. EMBO J 14(13): 3206–3214
69. Hanif M, Guillermo Pardo A, Gorfer M, Raudaskoski M (2002) T-DNA transfer and integration in the ectomycorrhizal fungus *Suillus bovinus* using hygromycin B as a selectable marker. Curr Genet 41:183–188
70. Ben-Hayyim G, Martin-Tanguy J, Tepfer D (1996) Changing root and shoot architecture with the *rolA* gene from *Agrobacterium rhizogenes*: interactions with gibberellic acid and polyamine metabolism. Physologia Plantarum 96:237–243
71. Ben-Hayyim G, Damon JP, Martin-Tanguy J, Tepfer D (1994) Changing root system architecture through inhibition of putrescine and feruloyl putrescine accumulation. FEBS Lett 342(2):145–148
72. Sun LY, Monneuse MO, Martin-Tanguy J, Tepfer D (1991) Changes in flowering and accumulation of polyamines and hydroxycinnamic acid-polyamine conjugates in tobacco plants transformed by the *rolA* locus from the Ri TL-DNA of *Agrobacterium rhizogenes*. Plant Sci 80:145–156
73. Burtin D, Martin-Tanguy J, Tepfer D (1991) α-DL-Difluoromethylornithine, a specific, irreversible inhibitor of putrescine biosynthesis, induces a phenotype in tobacco similar to that ascribed to the root-Inducing, left-hand transferred DNA of *Agrobacterium rhizogenes*. Plant Physiol 95(2):461–468
74. Martin-Tanguy J, Tepfer D, Paynot M, Burtin D, Heisler L, Martin C (1990) Inverse relationship between polyamine levels and the degree of phenotypic alteration induced by the root-inducing, left-hand transferred DNA from *Agrobacterium rhizogenes*. Plant Physiol 92(4):912–918
75. Sinkar V, Pythoud F, White F, Nester E, Gordon M (1988) *rol A* locus of the Ri plasmid directs developmental abnormalities in transgenic plants. Genes Dev 2:688–698

76. Sinkar V, White F, Furner I, Abrahamsen M, Pythoud F, Gordon M (1988) Reversion of aberrant plants transformed with *Agrobacterium rhizogenes* is associated with the transcriptional inactivation of the TL-DNA genes. Plant Physiol 86:584–590
77. Limami MA, Sun LY, Douat C, Helgeson J, Tepfer D (1998) Natural genetic transformation by *Agrobacterium rhizogenes*. Annual flowering in two biennials, belgian endive and carrot. Plant Physiol 118(2):543–550
78. Tepfer D, Zalar A, Leach S (2012) Survival of plant seeds, their UV screens, and *nptII* DNA for 18 months outside the International Space Station. Astrobiology 12(5):517–528
79. Zalar A, Tepfer D, Hoffmann SV, Kollmann A, Leach S (2007) Directed exospermia: II. VUV-UV spectroscopy of specialized UV screens, including plant flavonoids, suggests using metabolic engineering to improve survival in space. I J Astrobiol 6(04):291–301
80. Zalar A, Tepfer D, Hoffmann SV, Kenney JM, Leach S (2007) Directed exospermia: I. Biological modes of resistance to UV light are implied through absorption spectroscopy of DNA and potential UV screens. I J Astrobiol 6(03):229–240
81. Zalar A (2010) Resistance of *Arabidopsis thaliana* seeds exposed to monochromatic and simulated solare polychromatic UV radiation: Preparation for the EXPOSE space missions to the International Space Station (ISS). Université de Versailles Saint-Quentin-en-Yvelines
82. Jerling A, Burchell M, Tepfer D (2008) Survival of seeds in hypervelocity impacts. I J Astrobiol 7:217–222
83. de Vries J, Wackernagel W (2002) Integration of foreign DNA during natural transformation of *Acinetobacter* sp. by homology-facilitated illegitimate recombination. Proc Natl Acad Sci U S A 99(4):2094–2099
84. Link L, Sawyer J, Venkateswaran K, Nicholson W (2004) Extreme spore UV resistance of Bacillus pumilus isolates obtained from an ultraclean spacecraft assembly facility. Microb Ecol 47(2):159–163
85. Tepfer D (2008) The origin of life, panspermia and a proposal to seed the Universe. Plant Sci 175(6):756–760
86. Crick FHC, Orgel LE (1973) Directed panspermia. Icarus 19:341–346
87. Chaudhuri K, Das S, Bandyopadhyay M, Zalar A, Kollmann A, Jha S, Tepfer D (2009) Transgenic mimicry of pathogen attack stimulates growth and secondary metabolite accumulation. Transgenic Res 18(1):121–134
88. Yu O, Shi J, Hession AO, Maxwell CA, McGonigle B, Odell JT (2003) Metabolic engineering to increase isoflavone biosynthesis in soybean seed. Phytochemistry 63(7):753–763
89. Blakeslee AF, Avery AG (1937) Methods of inducing doubling of chromosomes in plants by treatment with colchicine. J Hered 28(12):393–411
90. Mattimore V, Battista JR (1996) Radioresistance of *Deinococcus radiodurans*: functions necessary to survive ionizing radiation are also necessary to survive prolonged desiccation. J Bacteriol 178(3):633–637
91. Savage DC (1977) Microbial ecology of the gastrointestinal tract. Annu Rev Microbiol 31(1):107–133
92. Tepfer D, Leach S (2006) Plant seeds as model vectors for the transfer of life through space. Astrophys Space Sci 306(1-2):69–75
93. Tepfer D, Smeekens S, Lepri O (2000) Expect genetic engineering of plants in unexpected places. International Herald Tribune, 8 Feb, p 8
94. Lambert C, Thomas G, Leger D, Pamboukdjian N, Tepfer D (1988) Utilisation de la transformation génétique par *Agrobacterium rhizogenes* pour améliorer la rhizogénèse d'arbres fruitiers. In: 8e Colloque sur les recherches fruitières. INRA/CTIFL, Bordeaux, pp 176–183
95. Lambert C, Tepfer D (1991) Use of *Agrobacterium rhizogenes* to create chimeric apple trees through genetic grafting. Bio/Technology 9(1):80–83
96. Strobel GA, Nachmias A (1985) *Agrobacterium rhizogenes* promotes the initial growth of bare root stock almond. J Gen Microbiol 131:1245–1249
97. Strobel G (1986) "Hairy root" bacterium shows horticultural promise. Am Nurseryman 163:77–85

98. Strobel GA, Nachmias A, Satouri S, Hess WH (1988) Improvements in the growth and yield of live trees by transformation with the Ri plasmid of *A. rhizogenes*. Can J Bot 66:2581–2585
99. Metzger L, Fouchault I, Glad C, Prost R, Tepfer D (1992) Estimation of cadmium availability using transformed roots. Plant and Soil 143:249–257
100. Mugnier J, Mosse B (1987) Vesicular-arbuscular mycorrhizal infection in transformed root-inducing T-DNA roots grown axenically. Phytopathology 77:1045–1050
101. Becard G, Piche Y (1989) New aspects on the acquisition of biotrophic status by a vesicular-arbuscular mycorrhizal fungus, *Gigaspora margarita*. New Phytol 112(1):77–83
102. Balaji B, Ba AM, Larue TA, Tepfer D, Piché Y (1994) *Pisum sativum* mutants insensitive to nodulation are also insensitive to invasion in vitro by the mycorrhizal fungus, *Gigaspora margarita*. Plant Sci 102(2):195–203
103. Jung G, Tepfer D (1987) Use of genetic transformation by the Ri T-DNA of *Agrobacterium rhizogenes* to stimulate biomass and tropane alkaloid production in *Atropa belladonna* and *Calystegia sepium* roots grown in vitro. Plant Sci 50:145–151
104. Pellegrineschi A, Damon JP, Valtorta N, Paillard N, Tepfer D (1994) Improvement of ornamental characters and fragrance production in lemon-scented geranium through genetic transformation by *Agrobacterium rhizogenes*. Biotechnology 12(1):64–68
105. Chaudhuri KN, Ghosh B, Tepfer D, Jha S (2005) Genetic transformation of *Tylophora indica* with *Agrobacterium rhizogenes* A4: growth and tylophorine productivity in different transformed root clones. Plant Cell Rep 24(1):25–35
106. Chaudhuri KN, Ghosh B, Tepfer D, Jha S (2006) Spontaneous plant regeneration in transformed roots and calli from *Tylophora indica*: changes in morphological phenotype and tylophorine accumulation associated with transformation by *Agrobacterium rhizogenes*. Plant Cell Rep 25(10):1059–1066
107. Bandyopadhyay M, Jha S, Tepfer D (2007) Changes in morphological phenotypes and withanolide composition of Ri-transformed roots of *Withania somnifera*. Plant Cell Rep 26:599–609
108. Deni J, Message B, Chioccioli M, Tepfer D (2005) Unsuccessful search for DNA transfer from transgenic plants to bacteria in the intestine of the tobacco horn worm, *Manduca sexta*. Transgenic Res 14(2):207–215
109. Kostrzak A, Cervantes Gonzalez M, Guetard D, Nagaraju DB, Wain-Hobson S, Tepfer D, Pniewski T, Sala M (2009) Oral administration of low doses of plant-based HBsAg induced antigen-specific IgAs and IgGs in mice, without increasing levels of regulatory T cells. Vaccine 27(35):4798–4807
110. Dunning Hotopp J (2011) Horizontal gene transfer between bacteria and animals. Trends Genet 27(4):157–163
111. Heinemann JA, Sprague GF Jr (1989) Bacterial conjugative plasmids mobilize DNA transfer between bacteria and yeast. Nature 340(6230):205–209
112. Moran N, Jarvik T (2010) Lateral transfer of genes from fungi underlies carotenoid production in aphids. Science 317(5845):624–627

Secondary Metabolism and the Rationale for Systems Manipulation

2

Bhekumthetho Ncube, Ashwell R. Ndhlala, and
Johannes Van Staden

Abstract

Plant-derived secondary metabolites provide mankind with a number of economic products that range from pharmaceutical drugs, fragrances, insecticides to flavours and dyes. The incomprehensible amount of chemically and functionally diverse products of secondary metabolism are synthesized through a complex network of enzymatically catalyzed metabolic pathways. The pool of enzymes employed in this biocatalytic landscape includes an assortment of substrate-, stereo-, and regio-specific types. The enzyme-driven reactivity and regio- and stereo-chemistry during the multistep conversion of substrates into diverse products offers lucrative manipulative points of exploitation in metabolic engineering. Parallel to the rich pool and flexibility of enzymes are the numerous genes and other regulatory mechanisms of metabolism that equally offer limitless opportunities for further manipulation to produce novel compounds. These characteristic features of secondary metabolism forms the rationale for its exploitation in producing fine products for human benefit. However, for effective strategies in metabolic engineering, the basic understanding of pathways, gene regulations, and enzymes involved as well as factors affecting the metabolism is required.

B. Ncube (✉) • J. Van Staden (✉)
Research Centre for Plant Growth and Development, School of Life Sciences, University of KwaZulu-Natal Pietermaritzburg, Scottsville, South Africa
e-mail: Ncubeb@ukzn.ac.za; bbkncube@yahoo.com; rcpgd@ukzn.ac.za

A.R. Ndhlala (✉)
Division Crop Science, Agricultural Research Council, Pretoria, South Africa

Green Technologies Research Centre, University of Limpopo, Sovenga, South Africa
e-mail: NdhlalaA@arc.agric.za; ashwelln@yahoo.com

S. Jha (ed.), *Transgenesis and Secondary Metabolism*, Reference Series in Phytochemistry, DOI 10.1007/978-3-319-28669-3_23

Keywords
Biosynthesis • Enzymes • Genes • Metabolic engineering • Secondary metabolism • Transcriptional factors

Abbreviations

CHS	Chalcone synthase
DMAPP	Dimethylallyl pyrophosphate
DOXP	1-deoxy-D-xylulose-5-phosphate
EPSP	Enolpyruvylshikimate-3-phosphate
ETI	Effector-triggered immunity
FDA	Food and Drug Administration
FPP	Farnesyl diphosphate
GAP	Glyceraldehyde 3-phosphate
GGPP	Geranylgeranyl diphosphate
GPP	Geranyl diphosphate
HMG-CoA	3-hydroxy- 3-methylglutaryl-CoA
IPP	Isopentenyl pyrophosphate
MAMPS	Microbe-associated molecular patterns
MAP	Mitogen-activated protein
MEP	Methyl-D-erythritol-4-phosphate
MVA	Mevalonate
MVAPP	Mevalonate 5-diphosphate
NLR	NOD-like receptor
PAL	phenylalanine ammonia lyase
PAMP	Pathogen-associated molecular patterns
PEP	Phosphoenolpyruvate
PKS	Polyketide synthases
PRR	Pattern recognition receptors
PTI	Pattern-triggered immunity
RLK	Receptor-like kinases
RLP	Receptor-like proteins
TCA	Tricarboxylic acid
TF	Transcription factors
UV	Ultra violet

Contents

1 Introduction

With a plethora of biotic and abiotic stress factors that constantly confront plants in their natural habitat, their adaptation demands a great degree of phenotypic plasticity. Apart from the preformed physical barriers, plants have evolved an innate immune system to cope with a diverse range of stressful factors. Being organisms devoid of mobility, sensing of stress signals and their subsequent transduction into appropriate responses are crucial components for plant adaptation and survival in ecosystems. However, under natural conditions, plants are potentially exposed to multitudes of concurrent stresses, which can have complex effects on the immune system response. The efficiency and effectiveness of plant defence responses would thus depend on the ability to trigger a rapid defence reaction following the recognition of the stress stimuli. Because plants, unlike animals, lack circulating defender cells and a somatic adaptive immune system, every plant cell must, in principle, recognize external stimuli and be able to effect systemic signals [1–3]. At the cellular level, the perception of external threats and the efficient integration of such information for an appropriate response often require robust and adaptable molecular systems [1, 4].

At the helm of plant survival within ecosystems, is their interaction with pathogenic microbes. The signaling events that lead to cellular and physiological responses within the plant in the plant-microbe interaction have been studied extensively. Studies have identified a two-tier perception system employed by plant innate immunity, distinct in their cellular localization. At the first line of innate immune response, plant pattern-triggered immunity (PTI) is initiated upon perception of evolutionarily conserved microbial signatures, called pathogen- or microbe-associated molecular patterns (PAMPs/MAMPs) [1, 4–7]. Perception of these molecular patterns is mediated through cell surface-resident pattern recognition receptors (PRRs), which are often encoded by receptor-like kinases (RLKs) or receptor-like proteins (RLPs) in plants [4, 6, 7]. Upon specific recognition of PAMP by the cognate PRR, the host plant elicits a series of complex cellular responses and physiological changes, among which include the activation of an evolutionarily conserved mitogen-activated protein (MAP) kinase cascade and profound gene transcriptional reprogramming, all of which collectively lead to a resistance response [8–10]. The second layer involves intracellular immune receptors, most often of the NOD-like receptor (NLR) type, which directly or indirectly recognize virulence effectors secreted within host cells by pathogens, thereby inducing effector-triggered immunity (ETI) [4]. In both PTI and ETI, a variety of host proteins ranging from immune receptors and signaling components to regulatory and antimicrobial proteins are engaged to orchestrate stimulus perception, defence signal activation and transduction, and execution of defence responses [1, 8].

The activation of a range of overlapping and/or specific intracellular immune responses is achieved by modulation of the activity of multiple transcriptional regulators, both DNA-binding TFs and their regulatory proteins which are able to reprogram transcription towards the activation of defence signaling [11]. Further downstream, in metabolic perspective, among the diverse, although interconnected network of defence metabolic responses, plant secondary metabolism holds key to the plant's adaptive mechanism [12]. Secondary metabolism generates a rich repertoire of structural and functionally diverse metabolites through a coordinated system of inherent metabolic network of biosynthetic pathways. For this reason, plants have become the most prolific factories for low molecular weight molecules that benefit mankind. To the producing plants, the phytochemical derivatives of secondary metabolism confer a multitude of adaptive, growth, and evolutionary advantages. The diverse molecular changes that are associated with the metabolism are understood to be preserved genetically, functionally, and structurally to confer selective and adaptive advantage for the host [13]. Combination of factors such as gene duplication, neofunctionalization, and positive selection are some of the suggested mechanisms behind the evolution of this diversity [13, 14]. The biosynthesis of secondary metabolites is controlled at the level of expression of the biosynthetic genes by developmental and tissue-specific factors as well as by external signals [15, 16]. Furthermore, synthesis of some metabolites is induced by some phytohormones produced in response to stress, such as (methyl)jasmonate [(Me)JA].

Owing to a host of economic, therapeutic, and various other uses of the products and intermediary metabolites of secondary metabolism that benefit humankind, secondary metabolism is thus an interesting target for molecular plant breeding. Genetic engineering is, in this respect, one of the promising approaches. Other approaches include systems manipulation through diversion of the flux into a competitive pathway or an increase in the catabolism of the target metabolite. Despite the detailed mechanisms that remain to be elucidated in some metabolic pathways, the framework underlying perception, signaling, and responses are being uncovered. One major constraint has been the poor characterization of plant secondary metabolic pathways at the level of biosynthetic intermediates and enzymes. This consequently has led to very few genes known from plant secondary metabolism. The understanding of the complex network of metabolic pathways holds significant promise to providing genetic resources and tools for systems manipulation and thus provides a broad-spectrum of economically important compounds.

2 Plant Secondary Metabolism in Perspective

Secondary metabolism comprises a coordinate series of coupled enzymatic conversions that utilizes limited products of primary/central metabolism as substrates. The link between metabolic fluxes of central metabolism and secondary metabolism implies a demand for resources and the existence of coordinated gene expression networks at the interface of the two metabolisms [17]. The metabolic architecture of secondary metabolism is highly organized into systematic mechanisms that integrate into developmental, morphological, and biochemical regulatory patterns of the entire plant metabolic network. For example, regulation of glucosinolate biosynthesis in *Arabidopsis thaliana* is not restricted to the metabolic space surrounding its biosynthesis but is tightly linked to more distal metabolic networks of primary metabolism [18]. Transgenic *A. thaliana* overexpress two clades of genes, *ATR1*-like and *MYB28*-like genes, that regulate the aliphatic and indole glucosinolate biosynthetic pathways. Transgenic *A. thaliana* concomitantly induced genes that are involved in the sulfur assimilation pathways as well as in the formation of precursor molecules for the biosynthesis of both glucosinolates. In the same system, all genes responsible for the enzymatic reaction steps of the TCA cycle from oxaloacetate to methionine were induced concurrently with genes encoding enzymes that regulate the committed steps in aliphatic glucosinolate biosynthesis in plants overexpressing the TFs. These changes were accompanied with changes in the levels of the affected central metabolites. The relatively broad view of transcripts and metabolites altered in transgenic plants overexpressing the different factors (*ATR1* and *MYB28*) [18, 19] underlined novel links of glucosinolate metabolism to additional metabolic pathways, including those of jasmonic acid, folate, benzoic acid, and various phenylpropanoids. While there is no evidence that these sets of TFs bind directly to the upstream regions of genes belonging to the central metabolic pathways, the findings pointed to a shared and coordinated transcriptional regulation of primary and secondary pathways [18–20].

The two most characteristic inherent lineaments of secondary metabolism are the structural diversity and high intraspecific and interspecific variability of its products. The high degree of plasticity of secondary metabolism which, in contrast to primary metabolism, allows for structural and chemical modifications of its products is identified as the basis for the generation of chemical diversity. Cumulative evidence suggests that many transcriptional factors (TFs) coordinate the activation of secondary metabolism genes concurrently with the expression of genes in upstream pathways of central metabolism [17, 19, 21]. The basic biochemistry and molecular biology of the biosynthetic pathways of secondary metabolism elucidated to date suggests the diversification of secondary metabolites to be originating from the elaboration of a few central intermediates [12, 13, 22]. Furthermore, the vestigial structural and mechanistic traits that characterize the biosynthetic enzymatic pathways during the diversification of substrate to product specificities is conserved via a few but complex biogenetic routes. The catalytic landscape adapts all possible lineages as the enzyme family use simple transformations in order to utilize new substrates and ensure product selectivity.

Based on their biosynthetic origins, secondary metabolites can be broadly categorized into three main groups: nitrogen or sulfur containing compounds such as alkaloids and glucosinolates, phenolic compounds, and terpenoid/isoprenoids, respectively. In plants, the three main biosynthetic sources of secondary metabolites are: the shikimate pathway, the isoprenoid pathway, and the polyketide pathway. The basic skeletons resulting from these general secondary metabolite pathways are further modified in various specific ways in each plant species. The modifications encompasses, among others the introduction of further substituents, new functional groups, isomerisation, ring opening followed by new ring closures, and coupling of different secondary metabolites. Introduction of functional groups is usually due to oxidative reactions which may include the introduction of hydroxy, methoxy, epoxy, aldehyde, and carboxyl groups. The enzymes involved are often cytochrome P450 enzymes, dioxygenases, and peroxidases. Peroxidases and dioxygenases occur widely in plants and are encoded by multigene families and generally possess broad substrate specificity. Generally, the P450 enzymes are bound to the endoplasmic reticulum membranes. They are among others involved in the oxidative phenol coupling, which plays an important role in the isoquinoline alkaloid biosynthesis of aromatic hydroxylations in various phenylpropanoid pathways [23].

2.1 Shikimate Pathway

The shikimate pathway plays a pivotal role in the biosynthesis of precursors for aromatic compounds in plants and microorganisms [24–27]. The pathway links metabolism of carbohydrates to biosynthesis of aromatic compounds. The metabolic sequence converts the primary metabolites phosphoenolpyruvate (PEP) and erythrose-4-phosphate to chorismate, the last common precursor for the three aromatic amino acids: phenylalanine, tyrosine, and tryptophan. The main trunk of the shikimate pathway consists of seven enzyme-catalyzed reactions with all pathway intermediates considered to be branch point compounds that may serve as substrates for other metabolic pathways.

The pathway is initiated with the condensation of erythrose-4-phosphate and PEP. A cyclic compound, 3-dehydroquinate is obtained in a series of downstream reactions. The remaining five steps serve to introduce a side chain and two of the three double bonds that convert this cyclohexane into the benzene ring, the hallmark of aromatic compounds. In two further downstream steps, 3-dehydroquinate yields shikimate, which after phosphorylation is coupled by the enzyme EPSP synthase with phosphoenolpyruvate to give 5-enolpyruvylshikimate-3-phosphate (EPSP). The last step of the shikimate pathway is the concerted 1,4-*trans* elimination of phosphate from EPSP to yield chorismate [24, 28]. Although conversion of substrate into product does not involve any overall oxidation or reduction, the enzyme that catalyzes this reaction, chorismate synthase, surprisingly requires a reduced flavin cofactor for this activity [28, 29]. According to Garnem [30], the reaction mechanism of this enzyme is stereochemically ambiguous. It is from chorismate that the pathway diverges into two major branches, leading to the synthesis of phenylalanine/tyrosine and tryptophan, respectively.

In plants, the aromatic amino acids are the precursors for a large variety of secondary metabolites with aromatic ring structures that often make up a substantial amount of the total dry weight of a plant [30]. In terms of carbon flux, the phenylpropanoid pathway is one of the most important metabolic pathways [26, 30, 31], with the enzyme chorismate mutase being the major regulatory point. Among the many aromatic secondary metabolites produced through this pathway are flavonoids, anthocyanins, phenolic compounds, many phytoalexins, indole acetate, alkaloids such as morphine, UV light protectants, lignin, and lignans [26]. Fundamental to the biosynthesis of these metabolites is the enzyme phenylalanine ammonia lyase (PAL), which catalyses the conversion of phenylalanine into *trans-cinnamic* acid, the first committed step of the pathway. Some plants have several iso-enzymes of this enzyme and may be under different regulation, such as inducible after wounding or by UV-light. Apart from the phenylpropanoid pathway, carbon fluxes also emanates from some minor branches from the shikimate pathway leading to isochorismate, 4-aminobenzoic acid, and 4-hydroxybenzoic acid, from which series of different secondary metabolites are derived [23].

All enzymes of the shikimate pathway have been isolated and elucidated and their respective DNAs characterized from numerous organisms, including plants. The cDNAs of plants encode proteins with amino terminal signal sequences for plastid import, suggesting that plastids are the exclusive locale for chorismate biosynthesis. The lack of physiological feedback inhibition mechanisms for this pathway in plants further suggest that pathway regulation may occur exclusively at the genetic level [24]. Several of the pathway enzymes occur in isoenzymic forms whose expression varies with changing environmental conditions, between and within species, and from organ to organ in specific subcellular compartments and at specific times [30]. The regulation of the differential biosynthesis of diverse and sometimes complex molecular structures may involve regulation of the supply of the precursors influencing the rate-limiting step for carbon flow through the shikimate pathway [32, 33].

2.2 Isoprenoid/Terpenoid Pathway

Isoprenoids constitute one of the most structurally and functionally diverse families of natural compounds that can be divided into classes of both primary and secondary metabolites. As primary metabolites, they have functions in photosynthesis, respiration, membrane fluidity, and regulation of growth and development [34]. The diverse chemical properties of isoprenoids serve important biological functions such as electron transport, photoprotection, hormone-based signaling, regulation of transcription and post-translational processes, meiosis, apoptosis, glycoprotein biosynthesis, and protein degradation. They are also structural components of cell and organelle membranes [34–36]. The isoprenoids classified as secondary metabolites include monoterpenes, sesquiterpenes, and diterpenes, terpenoid indole alkaloids, flavonoids, and isoflavonoids [35, 37]. Compartmentation is an important aspect of terpenoid biosynthesis.

Biosynthetically, all isoprenoids are assembled from two universal 5-C precursors, isopentenyl pyrophosphate (IPP) and its isomer dimethylallyl pyrophosphate (DMAPP) which, in plants, is synthesized via two different pathways. The isomerization of IPP into DMAPP is catalyzed by the enzyme IPP isomerase. The highly reactive DMAPP forms the starter molecule of terpenoid biosynthesis. Two biosynthetic routes to IPP and DMAPP exist for plants and other organisms. The cytosolic mevalonate (MVA) pathway, the first to be identified, utilizes seven enzymes to supply the precursors in most eukaryotes (all mammals), archaea, a few eubacteria, the cytosol and mitochondria of plants and fungi [35, 38]. The pathway is initiated with the synthesis of acetoacetyl-CoA from the condensation of two acetyl-CoA molecules in a reaction catalyzed by acetoacetyl-CoA thiolase. A third acetyl-CoA is then condensed with acetoacetyl-CoA to form 3-hydroxy-3-methylglutaryl-CoA (HMG-CoA) by HMG-CoA synthase. The NADPH-dependent HMG-CoA reductase then converts the CoA derivative to MVA. Through ATP-dependent steps, two consecutive phosphorylation reactions catalyzed by mevalonate kinase and phosphomevalonate kinase then convert (MVA to mevalonate 5-diphosphate (MVAPP). The diphosphate is subsequently decarboxylated by mevalonate diphosphate decarboxylase to yield a pool of IPP. An IPP isomerase then converts some of the IPP to DMAPP [35, 39].

The plastidial non-mevalonate route, alternatively known as the 1-deoxy-D-xylulose-5-phosphate (DOXP) or 2C-methyl-D-erythritol-4-phosphate (MEP) has only been fully elucidated very recently [40, 41]. The pathway consists of eight reactions catalyzed by nine enzymes and proceeds in the stroma of plastids. In the first reaction of the pathway, glyceraldehyde 3-phosphate (GAP) is condensed with pyruvate by the enzyme DOXP synthase to yield DOXP. The pathway's characteristic intermediate, MEP, is generated through the rearrangement of DOXP and side chain synthesis by the enzyme 1-deoxy-D-xylulose 5-phosphate reducto-isomerase. Several steps downstream of these reactions yield D-l-deoxyxylulose-5-phosphate, which upon rearrangement, results in the branched skeleton of IPP. The IPP isomerase maintains the supply of DMAPP [35].

In energy terms, when comparing the two pathways for DMAPP/IPP synthesis from glucose, the MVA pathway is more energy efficient, with a net gain in NAD(P)H reducing equivalents. The MEP pathway on the other hand is more carbon efficient, with only two GAP molecules required towards DMAPP/IPP synthesis as opposed to three in the MVA pathway [42].

From the elaboration of these precursors to further synthesize the myriad of different isoprenoid molecules, several of reactions are involved. Detailed description of these reactions has already been described in numerous reviews [43–45]. From the C_5 isoprene building block, a family of specific skeletons is assembled from basically three chemical reactions. Firstly, elongation reactions based on "head-to-tail" condensation of an allylic substrate with IPP lead to isoprenes of increasing size, which represent the scaffold of isoprenoid biosynthesis [45]. Through either condensation by a so-called "head-to-head" condensation of isoprene units, or cyclization of linear precursors, the different terpenes are assembled (mono to polyisoprenoids) [39]. In this way, C_{10} monoterpenes are synthesized starting from geranyl diphosphate (GPP), C_{15} sesquiterpenes from farnesyl diphosphate (FPP), C_{20} diterpenes from geranylgeranyl diphosphate (GGPP), C_{30} triperpenes from the condensation of two FPP, and finally C_{40} tetraterpenes from the condensation of two GGPP units. The elongation of linear hydrocarbon phosphates (C_{10} up to C_{40} or more exceptionally up to a thousand units) is catalyzed by prenyltransferases while terpene cyclases converts these simple molecules into a diverse array of chiral, carbocyclic skeletons. A number of compounds, referred to as meroterpenes, are of mixed biosynthetic origin. Further oxidation reactions and rearrangements result in an almost limitless number of conceivable structures [39, 44]. The enormous diversity in mono-, sesqui-, and diterpene skeletons is largely due to selective terpenoid synthases and cyclases, a large family of enzymes which catalyze the cyclization of differently folded GPP-, FPP-, or GGPP-molecules. From the various basic skeletons, further skeletal diversity can be introduced in subsequent biosynthetic steps in which among others various cytochrome P-450 enzymes play a key role [44, 46, 47].

A multitude of diverse secondary metabolites are derived from all the mentioned groups of terpenoids. All MEP-derived isoprenoids are considered as typical and exclusive plastidial isoprenoids. In general, besides gibberellic acids, strigolactones, abscisic acid, and some major cytokinins representing vital phytohormones, as well as molecules which function in photosynthesis like carotenoids, α-tocopherol, and plastoquinone, respectively, other MEP-derived compounds play mainly a role in secondary metabolism [43]. Despite a strict organellar compartmentalization and fundamentally distinct metabolic chemistry, the MVA and MEP pathways does not appear to function independently from the MEP pathway but rather interacts through metabolic cross-talk [39, 48]. The co-operation of both pathways entails that various carbon sources can be utilized to satisfy a special and rapid need for specific end-products. From a metabolic engineering perspective, the ability to control such variabilities as carbon flux could help facilitate increased production of specific products of commercial value.

2.3 Polyketide Pathway

The fatty acids are the basis for various secondary metabolites synthesized through the polyketide pathway, but the pathway also leads directly to secondary metabolites. In plants, polyketides are synthesized by type III polyketide synthases (PKSs) through condensation of acetyl (ketide) units with a CoA- linked starter molecule [49]. The structural diversity of the plant polyketides results from a range of starter substrates that can be used by the PKSs and from subsequent modifications via regiospecific condensation, cyclization, aromatization, hydroxylation, glycosylation, acylation, prenylation, sulfation, and methylation reactions [6]. The PKS enzyme reactions involve the loading of a starter molecule, the extension of the polyketide chain, and cyclization of the linear intermediate [50]. The enzymes can utilize a variety of CoA-linked starter substrates, a few of which include: acetyl-CoA, malonyl-CoA, methylmalonyl-CoA, p-coumaroyl-CoA, cinnamoyl-CoA, N-methylanthraniloyl-CoA, n-hexanoyl-CoA, isobutyryl-CoA, isovaleryl-CoA, and 3-hydroxybenzoyl-CoA [49, 50]. Chalcone synthase (CHS) is the archetypal plant-specific type III PKSs that catalyzes sequential condensation of p-coumaroyl-CoA with three molecules of malonyl-CoA to yield naringenin chalcone [27]. The enzyme stilbene synthase utilizes the same substrates to produce stilbenes such as resveratrol. From naringenin chalcone, a diverse range of flavonoids are created by the combined actions of functionalizing enzymes that include, isomerases, reductases, hydroxylases, glycosyltransferases, acyltransferases, methyltransferases, and prenyltransferases [51–54]. Naringenin chalcone and resveratrol are not the only plant polyketides that can result from the condensation of p-coumaroyl-CoA and malonyl-CoA. The condensation of coumaryl-CoA with one malonyl-CoA leads to benzalacetones, for example [51]. Different polyketides can be synthesized by varying the number of malonyl-CoA and p-coumaroyl-CoA molecules used in the elongation step. The type of cyclization or its absence also influences the resulting product [50]. Other plant PKSs catalyze the condensation of three malonyl-CoA units while employing varying starter substrates. The range of compound backbones generated by these PKSs includes chalcones, stilbenes, phloroglucinols, resorcinols, benzophenones, biphenyls, bibenzyls, chromones, acridones, pyrones, and curcuminoids [50, 54]. Other examples of plant secondary metabolites derived from the polyketide pathway are coniine, plumbagin, and anthraquinones.

2.4 Secondary Metabolites and Their Benefits

The plant-environment interaction is inevitable and unavoidable interaction occurring in plant ecosystems. To the producing plant, most secondary metabolites are understood to play important ecologically adaptive roles that enable the host to interact defensively with its environment. Apart from serving defensive and adaptive roles, some secondary metabolites serve as attractants for pollinators or seed dispersers, as signal compounds and as allelochemicals against competitors. Alkaloids, for example, are thought to confer a survival benefit through their ability to bind to cellular targets in antagonistic organisms [12, 55]. The toxic nature of most alkaloids gives them a general defensive mechanism for the producing organism. Caffeine is one such example of alkaloids which has been demonstrated to act as a natural insecticide in plants. When genes involved in caffeine synthesis were overexpressed in *Nicotiana tabacum*, the resulting increase in caffeine production improved the tolerance of the plants to certain insect pests [56].

The successful evolutionary adaptation of plants to land can largely be attributed to the massive ubiquitous synthesis of phenolic compounds. Polyphenols form an integral part of the cell wall structure in plants, mainly in the form of polymeric materials such as lignins, which serve as mechanical support and barriers against microbial invasion. Apart from cell wall structural roles, a number of adaptive, defensive, and certain species-specific distinguishing roles exists in some plants. Flavonoids are constituents of a variety of plant parts, including, leaves, fruits, seeds, flowers, and roots, with over 4000 different variants identified [57]. Flavonoids are water-soluble pigments which often account for pink, red, orange, scarlet, purple, blue, blue-black, and some yellow colors. Structural and chemical diversity of flavonoids is related to their diverse properties and roles in plants. In the producing plants, flavonoids provide protection against ultraviolet radiation, invading pathogens, and herbivores [58]. Tannins, another group of polyphenols, are characteristic of the chemical defence of plants and act as quantitative-dosage dependent-barriers to predators that may feed on them [59]. Condensed tannins may act as feeding deterrents in reproductive tissues and developing fruit and also impart astringency to fresh fruit, fruit juices, and wine [59]. The relevant physiological effects of tannins upon predation are assumed to be derived from their ability to complex with proteinaceous materials. Isoprenoids on the other hand serve numerous biochemical functions that include: electron transport chains, as components of membranes (sterols), in subcellular targeting and regulation (prenylation of proteins), as photosynthetic pigments (carotenoids, side chain of chlorophyll), as hormones (gibberellins, brassinosteroids, abscisic acid, cytokinins), and as plant defence compounds as well as attractants for pollinators [12, 59].

2.4.1 Economic Importance

A reservoir of more than 200,000 known secondary metabolites, with many more that continue to be discovered, provides mankind with a biogenic resource to exploit for economically important products such as pharmaceuticals, dyes, flavors, and fragrances. Of the various economic products derived from plant secondary metabolites, pharmaceutical drugs and/or medicinal uses are among the most prominent. Literature is replete with excellent reviews that discuss the various applications of these phytoproducts [12, 60–66].

In the alkaloid family of secondary metabolites, the tropane class of alkaloids serve an important class of plant-derived anticholinergic compounds, such as hyoscyamine and scopolamine and the narcotic tropical anesthetic cocaine, that occur mainly in Solanaceae. The central nervous system stimulant, cocaine, is a tropane which is found outside Solanaceae (in *Erythmxylon coca*). Scopolamine is commonly used in modern medicine in the form of a transdermal patch to combat motion sickness. Cocaine, which served as a "lead," and a starting structure that medicinal chemists modified for the development of an optimized drug, synthetic topical anesthetics, is also one of the tropane alkaloid examples. *Datura* leaves are usually smoked for the hallucinogenic effects of scopolamine [62, 66]. Cocaine is illicitly applied to mucus membranes for its addictive stimulatory effects. The pharmaceutical potential of the Amaryllidaceae alkaloids began after the commercialization of galanthamine as an Alzheimer's drug due to its potent and selective inhibitory activity against the enzyme acetylcholinesterase [65]. The lycorane derivatives pancratistatin and narciclasine hold promising chemotherapeutic potential due to their potent selective anticancer properties, although the specificity of the compounds to cancer cells and their mechanism of action remain unknown [67, 68]. Narciclasine has been shown to disrupt organization of the actin skeleton in cancer cells at very minute concentrations as well as increasing survival in preclinical models of human glioblastoma multiforme by markedly decreasing mitotic rates without inducing apoptosis [65]. Many of the terpenoid indole alkaloids are physiologically active in mammals. Camptothecin possesses antitumor activity which is due to its ability to inhibit DNA topoisomerase. Through inhibition of Tat-mediated transcription [69], the compound also inhibits anti-retroviruses such as HIV and the equine infectious anemia virus [69, 70]. The three antitumor and chemotherapeutic agents, vincristine, vinblastine, and ajmalicine, isolated from *Catharanthus roseus* are some examples of the terpenoid indole alkaloids with therapeutic benefits. Quinine from *Cinchona officinalis* is an antimalarial drug while strychnine is a rat poison and homeopathic drug from *Strychnos nuxvomica.* Topotecan and irinotecan are to date some of the US Food and Drug Administration (FDA) approved agents for the treatment of ovarian and colon cancer [71]. Caffeine is known to have sensory and stimulatory effects when consumed as a psychostimulant drug in coffee (*Coffea arabica* and *Coffea canephora*) and tea (*Camellia sinensis*) [72]. Xanthines produce numerous physiological effects that include positive inotropic and chronotropic effects on the heart, decreased airway resistance in the lungs, and respiratory stimulation [73].

Polyphenols exhibit a considerable free radical scavenging activity, determined largely by their reactivity as hydrogen- or electron- donating agents and the stability of the resulting antioxidant-derived radical that prevents the oxidation of various food ingredients, particularly fatty acids and oils [74]. The compounds of this group represent some of the most potent ingredients of nutraceuticals and functional foods currently available in the market and among these components are the anthocyanins, proanthocyanidins, flavanones, resveratrol, isoflavones, lignans coumarins, ellagic acid, and ellagitannins [75]. The ability of flavonoids to inhibit spore germination of pathogens has been exploited in the treatment of human pathogenic diseases. Numerous flavonoids have been characterized as antifungal, antibacterial, antiviral, anti-inflammatory, antioxidant, antitumor, antihepatotoxic, antilipolytic, vasodilator, immunostimulant, and antiallergic agents [63, 76, 77]. Consistent with most polyphenolic antioxidants, both the configuration and total number of hydroxyl groups, in addition to the flavan backbone, substantially influence several mechanisms of antioxidant activity in flavonoids. Tannins (hydrolyzable and condensed) are rich in highly reactive hydroxyl groups which emanate from each of the benzene ring constituents. These functional groups form complexes with proteins, including enzymes and polymers such as cellulose and hemicellulose [63, 78, 79]. Most of the medicinal beneficial effects exerted by tannins as constituents of drugs, food, and herbal remedies are largely due to their interaction with enzymes (proteins) within cell systems.

Artemisinin, an antimalarial sesquiterpenoid isolated from *Artemisia annua,* and taxol, a high-value diterpenoid-derived anticancer drug from the bark of *Taxus brevifolia,* are some of the examples of pharmaceuticals from a family of terpenes. Artemisinin, in the form of combination therapies, is to date the only effective treatment for multidrug-resistant strains of the malarial parasite *Plasmodium falciparum*. Digitoxin, the glycone digitoxigenin extracted from foxglove (*Digitalis*), is used widely in carefully prescribed doses for treatment of congestive heart disease. Azadirachtin A is a powerful insect antifeedant terpenoid compound isolated from *Azadirachta indica* [47, 80, 81].

The other major group of economically important natural products is that of flavors and fragrances. This group comprises both pure chemical entities and mixtures of compounds (e.g., various essential oils). Anthocyanins, flavonoids, and carotenoids are some of the well-known examples. Apart from the actual economical value derived from secondary metabolites to date, an enormous potential value exists for new drugs and other product development.

3 Genetic and Enzymatic Regulation Secondary Metabolic Systems

Diversity in chemical structures is paralleled by the presence of large gene families whose products carry out enzymatic reactions in plant secondary metabolism. Linking the overwhelming amount of both structural and quantitative variation in plant secondary metabolism to the underlying genes has been fundamental to technological breakthroughs in natural product production. The fact that secondary metabolites are derived from several different precursors and that the intersection of their metabolism with other metabolic pathways produces key metabolic intermediate compounds suggests complex regulation of their synthesis. Either coordinately or separately, such regulatory networks are expected to modulate levels of each metabolite to optimise fitness, as signalled by developmental and/or environmental cues [81]. Each plant produces a broad spectrum of secondary metabolites, all with different functions. On the basis that a diverse range of secondary metabolites serve different functions in the producing plant, the expression of the pathways involved would thus be different. Some pathways are expressed constitutively in specific tissues while others are induced at gene level after perception of external signals such as wounding or pathogenic infection, elicitors, or endogenous signal molecules such as salicylic acid. The regulation of certain biosynthetic pathways at the gene level also forms part of the developmental differentiation process of plant cells [82]. Ectopic expression of specific transcription factors can redirect the metabolic differentiation of plant cells by acting simultaneously and coordinately on different events, including the regulation of the expression of genes that encode biosynthetic enzymes and proteins necessary for metabolite storage and differentiation of appropriate subcellular compartments.

Transcription factors, a diverse group of proteins that recognize specific DNA sequences in the promoters of the genes, negotiate the regulation of gene expression at the level of transcription. TFs mediate the assembly of the basal transcription machinery resulting in the activation of RNA polymerase II and mRNA synthesis. The control of specific sets of genes within the metabolic network is accomplished by the combinatorial interaction among TFs, between TFs and non-DNA-binding proteins, and between TFs and *cis*-regulatory elements in an organized hierarchical gene regulatory networks TF [83, 84]. The organ-specific and tissue-specific regulation of anthocyanin biosynthesis, for example, is controlled by specific transcription factors, which are structurally and functionally well conserved between species [85]. In maize, anthocyanin biosynthesis is regulated by a combination of two transcription factor species that are encoded by two families of regulatory genes, *R/B* and *C1/Pl*. The *R/B* family encodes transcription factors that share homology with the basic helix–loop–helix (bHLH) protein encoded by the proto-oncogene *c-MYC*, whereas the *C1/Pl* genes encode proteins that have homology to the proto-oncogene *c-MYB* product. R and C1 interact to regulate anthocyanin biosynthesis in the maize kernel [86]. Homologous regulatory genes regulate anthocyanin synthesis in other parts of maize and other plant species [86].

Studies on the transcriptional regulation that alters accumulation of defensive glucosinolate metabolites in *A. thaliana* across environmental stresses have identified hundreds of unique potential regulatory interactions with a nearly complete complement of 21 promoters for the aliphatic glucosinolate pathway [19]. Phenotypic validation demonstrated that more than 75% of the tested TF mutants significantly altered the accumulations of glucosinolates and that these were conditional upon the environment and tissue type. The conclusion derived is that defence chemistry within the plant has a highly intricate transcriptional regulatory system that may allow for the optimization of defence metabolite accumulation corresponding to specific environmental cues. Similarly, the phenylpropanoid metabolism, in which phenolic compounds are derived, is regulated by coordinate changes of gene expression accompanied by changes in the expression of genes that encode enzymes in primary metabolism [87, 88]. Most of the genes encoding the enzymes of phenylpropanoid metabolism contain, within their promoters, well-conserved motifs that conform to the motifs recognized by plant MYB TFs and some of which are able to transactivate genes encoding phenylalanine ammonia-lyase (PAL) in primary metabolism [89, 90]. Transketolase activity was identified as an important determinant of photosynthetic and phenylpropanoid metabolism and that the provision of precursors from central metabolism co-limits flux into the shikimate pathway and phenylpropanoid metabolism [87]. In terpenoid indole alkaloid biosynthesis in *Catharanthus roseus*, overexpression of *ORCA3*, a jasmonate-responsive AP2/ERF-domain family TF, led to an induction of genes encoding two enzymes (anthranilate synthase and D-1-deoxyxylulose 5-phosphate synthase) involved in central metabolism [15, 16]. *ORCA3* overexpression resulted in enhanced expression of several metabolite biosynthetic genes and consequently in increased accumulation of terpenoid indole alkaloids [16]. Studies indicate that *ORCA3* is a central regulator of terpenoid indole alkaloid biosynthesis that acts pleiotropically on several steps of the terpenoid indole alkaloid pathway and activates the biosynthesis of terpenoid indole alkaloid precursors [91].

Regulation of secondary metabolic pathways also occurs at the level of biosynthetic enzymes. Posttranslational modifications, enzyme turnover, feedback inhibition or activation, allosteric interactions, co-factor availability, and phosporylation are some of the few examples of possible regulatory mechanisms at this level. Anthranilate synthase, the first enzyme in the tryptophan branch of the aromatic amino acid biosynthesis is strongly inhibited by its end product, tryptophan. This serves as a typical example of feedback inhibition regulation mechanism. Further downstream, in the same metabolic pathway, in the branch leading to phenylalanine, the first enzyme chorismate mutase is induced by tryptophan while inhibited by phenylalanine and tyrosine [92, 93]. Similar genes encoding enzymes of certain secondary metabolite pathways are differently regulated by having different promoters.

3.1 Compartmentation as a Regulation Mechanism

Compartmentation forms another important aspect of metabolic regulation, both at cellular and subcellular levels. Alkaloid biosynthesis and accumulation, for example, are associated with a variety of cell types in different plants, including epidermis, endodermis, pericycle, phloem parenchyma, phloem sieve elements and companion cells, specialized mesophyll, and laticifers. A common paradigm is the involvement of multiple cell types and the implied transport of pathway intermediates and/or products [94]. The complex intracellular compartmentation of alkaloid biosynthesis is thought to have occurred as a consequence of adapting compartmented reactions of primary metabolism to participate in alkaloid biosynthesis [95]. The subcellular trafficking of pathway intermediates also creates an important level of metabolic regulation that could not occur if enzymes and substrates diffused freely in the cytosol.

4 Metabolic Systems Manipulation: the Rationale

The heritable and adaptable nature of secondary metabolism and subsequent chemical diversity presents a unique biogenic resource that can be rationally engineered using the unique perspectives of evolution, genomics, and structural biology to create novel economic products. Metabolic engineering exploits an integrated, systems-level approach for optimizing a desired cellular phenotype. A multitude of factors, the complex integrated regulatory mechanisms and coordinated networks of metabolic routes leading to the synthesis of specific metabolites, shape the profiles and fluxes of plant secondary metabolites. For this reason, exploiting the plant biosynthetic capacity presents limitless exciting opportunities although with equally complex challenges. Understanding the basic network of metabolic intermediates and enzymes forms the fundamental basis for manipulating the secondary metabolism. Beyond this level, knowledge of the spatial and temporal regulatory architectures of secondary metabolic pathways and the ways in which they are integrated into broader metabolic networks is essential. It forms the focal point in the exploitation of TFs for predictive plant metabolic engineering.

Much of this rich chemical diversity arises from a limited pool of chemical scaffolds which are subsequently modified though specific chemical substitutions as necessitated by substrate and/or regio-specific enzymes. The enzyme-driven reactivity during the multistep conversion of substrates into precise products in the biocatalytic landscape of secondary metabolism is one of the lucrative key points of exploitation. The exploitation of enzymes, particularly those that exhibit strict stereospecificity, is an interesting aspect in the production of novel compounds. Protein engineering which alters the substrate specificity of biosynthetic enzymes also offers an opportunity for the biosynthesis of novel metabolites. Recruitment of heterologous proteins enables extension of existing pathways to obtain new chemical products, alter posttranslational protein processing, and degrade recalcitrant wastes [96, 97]. Transgenic plants with altered enzyme activities are also some of the powerful tools to study the metabolic control architecture of secondary metabolites.

The enormous number of enzyme-catalyzed reactions in secondary metabolism provides unprecedented opportunities for the selection of suitable enzymes for use in metabolic engineering. Metabolic perturbation of the known biochemical network, based on the knowledge of the metabolic system of interest, forms the basis for metabolic engineering. The unique architectural network and regulatory systems that characterize secondary metabolism in terms of enzymes, genes, substrates, and intermediates form the rationale for systems manipulation. Against this background, Sauer [98] suggest that the functional behavior of a network emerges through the nonlinear gene, protein, and metabolite interactions across multiple metabolic and regulatory layers. Intracellular reaction rates are thus the functional end points of these interactions in metabolic networks, hence are highly relevant for systems biology. The fact that secondary metabolites production are influenced by environmental factors further adds to the multiple dimensions of the metabolic manipulation level points. Following this logic, varied manipulation of varied environmental factors can thus trigger positive abrupt activation of qualitative and quantitative changes in plant secondary metabolite accumulation [12, 99, 100]. Other approaches entail diverting the carbon flux into a competitive pathway or an increase in the catabolism of the target compound. The biosynthesis of certain metabolites are under strict developmental regulation in plants, a characteristic attribute due to which exploitation of cultured cells for the production of certain valuable metabolites has been severely restricted [12].

Due to their regulatory role in secondary metabolism, TFs add to the range of techniques for plant metabolic engineering to increase the production of valuable compounds. Parallel to this, artificial TFs can also be used as valuable tools for plant metabolic engineering. The use of specific transcription factors would avoid the time-consuming step of acquiring knowledge about all enzymatic steps of the often poorly characterized biosynthetic pathways. Transcriptional factors can also be used to drive flux through a pathway and the expression of an enzyme used to divert pathway intermediates to the desired final product. Because isopentenyl and dimethylallyl pyrophosphates are the universal precursors to all isoprenoids including microorganisms, methods can be developed for expressing complex multigene pathways for the biosynthesis of plant secondary metabolites in microorganisms. Strains developed in this study can serve as platform hosts for the production of any terpenoid compound for which a terpene synthase gene is available. Gain-of-function mutagenesis with a strong constitutive promoter that is carried on an insertion element such as *Agrobacterium tumefaciens* T-DNA is one of the tools that can be exploited to engineer secondary metabolism. Another interesting approach to exploit secondary metabolism for the production of new compounds is through combinatorial biochemistry by introducing enzymes with a different substrate specificity [101].

The general plasticity, chemodiversity of the products, and the limitless regulatory mechanisms and enzymes involved in secondary metabolism offer enormous options for metabolic alteration approaches. In addition to the genetic, enzymatic, and transcriptome metabolic alteration techniques, numerous other approaches can be exploited that range from the use of elicitors (abiotic and biotic) in cell cultures to the manipulation of environmental factors including plant growth regulators at cellular, organ, and whole plant level.

5 Conclusions

The chemodiversity of secondary metabolites produced by plants is a result of the complex metabolic pathways involved in their biosynthesis as mediated by specific enzymes, which add new structural elements to the basic skeletons. However, one compound may derive from completely different pathways in different plants. Elicitation of the biosynthesis of secondary metabolites can be effected by factors such as cellular development as well as induced by other exogenous or external signals. The pathways are regulated at numerous levels within the biocatalytic landscape, and these include at the level of genes, enzymes, compartmentation, and thus transport. This, therefore, positions plants as natural chemical factories able to carry out combinatorial chemistry that mankind exploits to their benefit. The pleiotropic action of central transcription factors on a wide array of genes involved in metabolic differentiation of plant cells enable the development of strategies to engineer complex metabolic pathways to yield useful products. The rationale for metabolic engineering is rooted on the general plasticity of this metabolism. However, at the helm of metabolic engineering is the consideration of the entire metabolic network to redirect central metabolites into secondary metabolites without compromising plant fitness. Understanding the entire network of pathways involved, including their regulation, genes, and enzymes, is key to successful metabolic engineering.

Acknowledgments The National Research Foundation of South Africa is gratefully acknowledged for financial support.

References

1. Jones JD, Dangl JL (2006) The plant immune system. Nature 444:323–329
2. Nishimura MT, Dangl JL (2014) Paired plant immune receptors. Science 344:267
3. Williams SJ, Sohn KH, Wan L, Bernoux M, Sarris PF, Segonzac C et al (2014) Structural basis for assembly and function of a heterodimeric plant immune receptor. Science 344:299–303
4. Macho AP, Zipfel C (2014) Plant PRRs and the activation of innate immune signaling. Mol Cell 54:263–272
5. Dodds PN, Rathjen JP (2010) Plant immunity: towards an integrated view of plant–pathogen interactions. Nat Rev Genet 11:539–548
6. Wu S, Shan L, He P (2014) Microbial signature-triggered plant defense responses and early signaling mechanisms. Plant Sci 228:118–126
7. Schwessinger B, Ronald PC (2012) Plant innate immunity: perception of conserved microbial signatures. Plant Biol 63:451–482

8. Boller T, Felix G (2009) A renaissance of elicitors: perception of microbe-associated molecular patterns and danger signals by pattern-recognition receptors. Annu Rev Plant Biol 60:379–406
9. Ma Y, Zhao Y, Walker RK, Berkowitz GA (2013) Molecular steps in the immune signaling pathway evoked by plant elicitor peptides (Peps): CPKs, NO and ROS are downstream from the early Ca^{2+} signal. Plant Physiol 163:1459–1471
10. Meng X, Zhang S (2013) MAPK cascades in plant disease resistance signaling. Annu Rev Phytopathol 51:245–266
11. Buscaill P, Rivas S (2014) Transcriptional control of plant defence responses. Curr Opin Plant Biol 20:35–46
12. Ncube B, Van Staden J (2015) Tilting plant metabolism for improved metabolite biosynthesis and enhanced human benefit. Molecules 20:12698–12731
13. Wink M (2003) Evolution of secondary metabolites from an ecological and molecular phylogenetic perspective. Phytochemistry 64:3–19
14. Benderoth M, Textor S, Windsor AJ, Mitchell-Olds T, Gershenzon J, Kroymann J (2006) Positive selection driving diversification in plant secondary metabolism. Proc Natl Acad Sci U S A 103:9118–9123
15. Mol J, Grotewold E, Koes R (1998) How genes paint flowers and seeds. Trends Plant Sci 3:212–217
16. van der Fits L, Memelink J (2000) ORCA3, a jasmonate-responsive transcriptional regulator of plant primary and secondary metabolism. Science 289:295–297
17. Aharoni A, Galili G (2011) Metabolic engineering of the plant primary–secondary metabolism interface. Curr Opin Biotechnol 22:239–244
18. Malitsky S, Blum E, Less H, Venger I, Elbaz M, Morin S et al (2008) The transcript and metabolite networks affected by the two clades of Arabidopsis glucosinolate biosynthesis regulators. Plant Physiol 184:2021–2049
19. Li B, Gaudinier A, Tang M, Taylor-Teeples M, Nham NT, Ghaffari C et al (2014) Promoter-based integration in plant defense regulation. Plant Physiol 166:1803–1820
20. Grubb CD, Abel S (2006) Glucosinolate metabolism and its control. Trends Plant Sci 11:89–100
21. Soltis NE, Kliebenstein DJ (2015) Natural variation of plant metabolism: genetic mechanisms, interpretive caveats, evolutionary and mechanistic insights. Plant Physiol doi:10.1104/pp.15.01108
22. Pichersky E, Gang DR (2000) Genetics and biochemistry of secondary metabolites in plants: an evolutionary perspective. Trends Plant Sci 5:439–445
23. Verpoorte R (2000) Metabolic engineering of plant secondary metabolism. In: Verpoorte R, Alfermann AW (eds) Metabolic engineering of plant secondary metabolism. Kluwer, London
24. Herrmann KM, Weaver LM (1999) The shikimate pathway. Annu Rev Plant Biol 50:473–503
25. Jensen RA (1986) The shikimate/arogenate pathway: link between carbohydrate metabolism and secondary metabolism. Physiol Plant 66:164–168
26. Bentley R, Haslam E (1990) The shikimate pathway-a metabolic tree with many branche. Crit Rev Biochem Mol Biol 25:307–384
27. Luckner M (2013) Secondary metabolism in microorganisms, plants and animals. Springer Science and Business Media, Berlin
28. Hawkes T, Lewis T, Coggins J, Mousdale D, Lowe D, Thorneley R (1990) Chorismate synthase. Pre-steady-state kinetics of phosphate release from 5-enolpyruvylshikimate 3-phosphate. Biochem J 265:899–902
29. White PJ, Millar G, Coggins JR (1998) The overexpression, purification and complete amino acid sequence of chorismate synthase from *Escherichia coli* K12 and its comparison with the enzyme from *Neurospora crassa*. Biochem J 251:313–322
30. Herrmann KM (1995) The shikimate pathway: early steps in the biosynthesis of aromatic compounds. Plant Cell 7:907–919
31. Haslam E (1993) Shikimic acid: metabolism and metabolites. Wiley, Chichester

32. Görlach J, Schmid J, Amrhein N (1994) Abundance of transcripts specific for genes encoding enzymes of the prechorismate pathway in different organs of tomato (*Lycopersicon esculentum* L.) plants. Planta 193:216–223
33. Jones J, Henstrand J, Handa A, Herrmann K, Weller S (1995) Impaired wound induction of DAHP synthase and altered stem development in transgenic potato plants expressing a DAHP synthase antisense construct. Plant Physiol 108:1413–1421
34. Chappell J (1995) The biochemistry and molecular biology of isoprenoid metabolism. Plant Physiol 107:1–6
35. Hunter WN (2007) The non-mevalonate pathway of isoprenoid precursor biosynthesis. J Biol Chem 282:21573–21577
36. Sacchettini JC, Poulter CD (1997) Creating isoprenoid diversity. Science 277:1788–1789
37. Ibrahim RK, Anzellotti D (2003) The enzymatic basis of flavonoid biodiversity. Recent Adv Phytochem 37:1–36
38. Goldstein JL, Brown MS (1990) Regulation of the mevalonate pathway. Nature 343:425–430
39. Hemmerlin A, Harwood JL, Bach TJ (2012) A raison d'être for two distinct pathways in the early steps of plant isoprenoid biosynthesis? Prog Lipid Res 51:95–148
40. Eisenreich W, Bacher A, Arigoni D, Rohdich F (2004) Biosynthesis of isoprenoids via the non-mevalonate pathway. Cell Mol Life Sci 61:1401–1426
41. Rohmer M (1999) The discovery of a mevalonate-independent pathway for isoprenoid biosynthesis in bacteria, algae and higher plants. Nat Prod Rep 16:565–574
42. Gruchattka E, Hädicke O, Klamt S, Schütz V, Kayser O (2013) In silico profiling of *Escherichia coli* and *Saccharomyces cerevisiae* as terpenoid factories. Microb Cell Fact 12:1–18
43. Bouvier F, Rahier A, Camara B (2005) Biogenesis, molecular regulation and function of plant isoprenoids. Prog Lipid Res 44:357–429
44. Kuzuyama T, Seto H (2003) Diversity of the biosynthesis of the isoprene units. Nat Prod Rep 20:171–183
45. Kellogg BA, Poulter CD (1997) Chain elongation in the isoprenoid biosynthetic pathway. Curr Opin Chem Biol 1:570–578
46. Mihaliak CA, Karp F, Croteau R (1993) 10 Cytochrome P-450 Terpene Hydroxylases. Methods Plant Biochem 9:261
47. McGarvey DJ, Croteau R (1995) Terpenoid metabolism. Plant Cell 7:1015–1026
48. Liao P, Hemmerlin A, Bach TJ, Chye M-L (2016) The potential of the mevalonate pathway for enhanced isoprenoid production. Biotechnol Adv. doi:10.1016/j.biotechadv.2016.03.005
49. Austin MB, Noel JP (2003) The chalcone synthase superfamily of type III polyketide synthases. Nat Prod Rep 20:79–110
50. Abe I, Morita H (2010) Structure and function of the chalcone synthase superfamily of plant type III polyketide synthases. Nat Prod Rep 27:809–838
51. Borejsza-Wysocki W, Hrazdina G (1996) Aromatic polyketide synthases (purification, characterization, and antibody development to benzalacetone synthase from raspberry fruits). Plant Physiol 110:791–799
52. Sasaki Y, Konishi T, Nagano Y et al (1995) Compartmentalization of two forms of acetyl-CoA carboxylase and plant tolerance towards herbicides. In: Kader JC, Mazliak P (eds) Plant lipid metabolism. Kluwer, Dordrecht
53. Ferrer J-L, Austin M, Stewart C, Noel J (2008) Structure and function of enzymes involved in the biosynthesis of phenylpropanoids. Plant Physiol Biochem 46:356–370
54. Lussier F-X, Colatriano D, Wiltshire Z, Page JE, Martin VJ (2012) Engineering microbes for plant polyketide biosynthesis. Comput Struct Biotechnol J 3:1–11
55. López-Meyer M, Nessler CL (1997) Tryptophan decarboxylase is encoded by two autonomously regulated genes in *Camptotheca acuminata* which are differentially expressed during development and stress. Plant J 11:1167–1175
56. Cauli O, Morelli M (2005) Caffeine and the dopaminergic system. Behav Pharmacol 16:63–77

57. Dixon RA, Paiva NL (1995) Stress-induced phenylpropanoid metabolism. Plant Cell 7:1085–1097
58. Harborne JB, Williams CA (2000) Advances in flavonoid research since 1992. Phytochemistry 55:481–504
59. Bartram S, Jux A, Gleixner G, Boland W (2006) Dynamic pathway allocation in early terpenoid biosynthesis of stress-induced lima bean leaves. Phytochemistry 67:1661–1672
60. Gurib-Fakim A (2006) Medicinal plants: traditions of yesterday and drugs of tomorrow. Mol Aspects Med 27:1–93
61. Chotani G, Dodge T, Hsu A, Kumar M, LaDuca R, Trimbur D et al (2000) The commercial production of chemicals using pathway engineering. Biochim Biophys Acta (BBA) 1543:434–455
62. Croteau R, Kutchan TM, Lewis NG (2000) Natural products (secondary metabolites). Biochem Mol Biol Plant 24:1250–1319
63. Burda S, Oleszek W (2001) Antioxidant and antiradical activities of flavonoids. J Agric Food Chem 49:2774–2779
64. Fredholm BB, Bättig K, Holmén J, Nehlig A, Zvartau EE (1999) Actions of caffeine in the brain with special reference to factors that contribute to its widespread use. Pharmacol Rev 51:83–133
65. Houghton PJ, Ren Y, Howes M-J (2006) Acetylcholinesterase inhibitors from plants and fungi. Nat Prod Rep 23:181–199
66. Facchini PJ (1999) Plant secondary metabolism: out of the evolutionary abyss. Trends Plant Sci 4:382–384
67. Facchini PJ, De Luca V (1995) Phloem-specific expression of tyrosine/dopa decarboxylase genes and the biosynthesis of isoquinoline alkaloids in *Opium poppy*. Plant Cell 7:1811–1821
68. Unver N (2007) New skeletons and new concepts in Amaryllidaceae alkaloids. Phytochem Rev 6:125–135
69. Pandey S, Kekre N, Naderi J, McNulty J (2005) Induction of apoptotic cell death specifically in rat and human cancer cells by pancratistatin. Artif Cells Blood Substit Immobil Biotechnol 33:279–295
70. McLachlan A, Kekre N, McNulty J, Pandey S (2005) Pancratistatin: a natural anti-cancer compound that targets mitochondria specifically in cancer cells to induce apoptosis. Apoptosis 10:619–630
71. McNulty J, Thorat A, Vurgun N, Nair JJ, Makaji E, Crankshaw DJ et al (2010) Human cytochrome P450 liability studies of trans-dihydronarciclasine: a readily available, potent, and selective cancer cell growth inhibitor. J Nat Prod 74:106–108
72. Kato M, Kanehara T, Shimizu H, Suzuki T, Gillies FM, Crozier A et al (1996) Caffeine biosynthesis in young leaves of *Camellia sinensis*: *in vitro* studies on N-methyltransferase activity involved in the conversion of xanthosine to caffeine. Physiol Plant 98:629–636
73. Porkka-Heiskanen T, Strecker RE, Thakkar M, Bjørkum AA, Greene RW, McCarley RW (1997) Adenosine: a mediator of the sleep-inducing effects of prolonged wakefulness. Science 276:1265–1268
74. Crozier A, Jaganath IB, Clifford MN (2006) Phenols, polyphenols and tannins: an overview. In: Crozier A, Clifford MN, Ashihara H (eds) Plant secondary metabolites-occurrence, structure and role in the diet. Blackwell, Oxford
75. Espín JC, García-Conesa MT, Tomás-Barberán FA (2007) Nutraceuticals: facts and fiction. Phytochemistry 68:2986–3008
76. Waterman PG, Mole S (1994) Analysis of phenolic plant metabolites. Blackwell Scientific, London
77. Williams CA, Harborne JB, Geiger H, Hoult JRS (1999) The flavonoids of *Tanacetum parthenium* and *T. vulgare* and their anti-inflammatory properties. Phytochemistry 51:417–423
78. Wallace G, Fry SC (1994) Phenolic components of the plant cell wall. Int Rev Cytol 151:229–268

79. Schofield P, Mbugua D, Pell A (2001) Analysis of condensed tannins: a review. Anim Feed Sci Technol 91:21–40
80. Zulak KG, Bohlmann J (2010) Terpenoid biosynthesis and specialized vascular cells of conifer defense. J Integr Plant Biol 52:86–97
81. Celenza JL, Quiel JA, Smolen GA, Merrikh H, Silvestro AR, Normanly J et al (2005) The Arabidopsis ATR1 Myb transcription factor controls indolic glucosinolate homeostasis. Plant Physiol 137:253–262
82. Pierpoint W (1994) Salicylic acid and its derivatives in plants: medicines, metabolites and messenger molecules. Adv Bot Res 53:412
83. Grotewold E (2008) Transcription factors for predictive plant metabolic engineering: are we there yet? Curr Opin Biotechnol 19:138–144
84. Dixon RA (2005) Engineering of plant natural product pathways. Curr Opin Plant Biol 8:329–336
85. Mol J, Jenkins G, Schäfer E, Weiss D, Walbot V (1996) Signal perception, transduction, and gene expression involved in anthocyanin biosynthesis. Crit Rev Plant Sci 15:525–557
86. Grotewold E, Sainz MB, Tagliani L, Hernandez JM, Bowen B, Chandler VL (2000) Identification of the residues in the Myb domain of maize C1 that specify the interaction with the bHLH cofactor R. Proc Natl Acad Sci U S A 97:13579–13584
87. Henkes S, Sonnewald U, Badur R, Flachmann R, Stitt M (2001) A small decrease of plastid transketolase activity in antisense tobacco transformants has dramatic effects on photosynthesis and phenylpropanoid metabolism. Plant Cell 13:535–551
88. Tamagnone L, Merida A, Parr A, Mackay S, Culianez-Macia FA, Roberts K et al (1998) The AmMYB308 and AmMYB330 transcription factors from *Antirrhinum* regulate phenylpropanoid and lignin biosynthesis in transgenic tobacco. Plant Cell 10:135–154
89. Shufflebottom D, Edwards K, Schuch W, Bevan M (1993) Transcription of two members of a gene family encoding phenylalanine ammonia-lyase leads to remarkably different cell specificities and induction patterns. Plant J 3:835–845
90. Weisshaar B, Jenkins GI (1998) Phenylpropanoid biosynthesis and its regulation. Curr Opin Plant Biol 1:251–257
91. Gantet P, Memelink J (2002) Transcription factors: tools to engineer the production of pharmacologically active plant metabolites. Trends Pharmacol Sci 23:563–569
92. Schmid J, Amrhein N (1995) Molecular organization of the shikimate pathway in higher plants. Phytochemistry 39:737–749
93. Poulsen C, Verpoorte R (1991) Roles of chorismate mutase, isochorismate synthase and anthranilate synthase in plants. Phytochemistry 30:377–386
94. Ziegler J, Facchini PJ (2008) Alkaloid biosynthesis: metabolism and trafficking. Annu Rev Plant Biol 59:735–769
95. De Luca V, St Pierre B (2000) The cell and developmental biology of alkaloid biosynthesis. Trends Plant Sci 5:168–173
96. Bailey JE, Sburlati A, Hatzimanikatis V, Lee K, Renner WA, Tsai PS (1996) Inverse metabolic engineering: a strategy for directed genetic engineering of useful phenotypes. Biotechnol Bioeng 52:109–121
97. Bailey JE (1991) Toward a science of metabolic engineering. Science 252:1668–1675
98. Sauer U (2006) Metabolic networks in motion: 13C-based flux analysis. Mol Syst Biol 2:1–10
99. Ncube B, Finnie J, Van Staden J (2012) Quality from the field: The impact of environmental factors as quality determinants in medicinal plants. S Afr J Bot 82:11–20
100. Ncube B, Finnie JF, Van Staden J (2014) Carbon–nitrogen ratio and *in vitro* assimilate partitioning patterns in *Cyrtanthus guthrieae* L. Plant Physiol Biochem 74:246–254
101. Sato F, Hashimoto T, Hachiya A, Tamura K-i, Choi K-B, Morishige T et al (2001) Metabolic engineering of plant alkaloid biosynthesis. Proc Natl Acad Sci U S A 98:367–372

Plant Cell Cultures as Producers of Secondary Metabolites: *Podophyllum* Lignans as a Model

3

Cláudia Gontijo Silva, Vera Lúcia de Almeida, Priscilla Rodrigues Valadares Campana, and Marina Pereira Rocha

Abstract

Podophyllums have been used extensively as medicinal plants and are part of the folklore in Asian and American cultures. Their use dates back to the Chinese culture of 2,000 years ago where they were used as an antitumor drug. The resin podophyllin was first recommended as an antiviral agent and provided clues for new applications. The most important secondary metabolite isolated from the rhizomes and roots of the *Podophyllum* species is podophyllotoxin and its related lignans. This lignan is the precursor of the semisynthetic drugs etoposide, teniposide, and etopophos, which are clinically used in the therapeutic treatment of cancer. Moreover, other derivatives have shown different types of biological activity. With an increasing worldwide market for anticancer drugs, supplies of podophyllotoxin for the pharmaceutical industry are under great pressure. As the chemical synthesis of podophyllotoxin is not economic on a commercial scale, supplies are still obtained from wild populations of *Podophyllum*. Concern has been expressed about the shortage of *Podophyllum* which is now an endangered species due to overexploitation and a lack of cultivation. Attempts to increase plant yields have improved through in vitro technology while the production of podophyllotoxin will require further studies. This chapter provides an overview of the *Podophyllum* species and its lignans. *Podophyllum* biotechnology still presents challenges to be overcome and some of these are, in part, discussed in this chapter.

C.G. Silva (✉) • M.P. Rocha (✉)
Serviço de Biotecnologia Vegetal, Diretoria de Pesquisa e Desenvolvimento, Fundação Ezequiel Dias, Belo Horizonte, MG, Brazil
e-mail: claudia.gontijo@funed.mg.gov.br; marina.rocha@funed.mg.gov.br

V.L. de Almeida (✉) • P.R.V. Campana (✉)
Serviço de Fitoquímica e Prospecção Farmacêutica, Diretoria de Pesquisa e Desenvolvimento, Fundação Ezequiel Dias, Belo Horizonte, MG, Brazil
e-mail: vera.almeida@funed.mg.gov.br; priscilla.campana@funed.mg.gov.br

S. Jha (ed.), *Transgenesis and Secondary Metabolism*, Reference Series in Phytochemistry, DOI 10.1007/978-3-319-28669-3_3

Keywords
Plant cell cultures • *Podophyllum* species • Lignans • Podophyllotoxin • Biosynthetic pathway • Analytical methods • Biological activities • Biotechnological approaches • Tissue culture

Abbreviations

2,4-D	2,4-Dichlorophenoxyacetic acid
AChE	Acetylcholinesterase enzyme
B5	Gamborg medium (1968)
BAP	6-Benzylaminopurine
CAD	Cinnamyl alcohol-dehydrogenase
CCC	Countercurrent chromatography
cv	Cultivar
CYP	Cytochrome P450 enzymes
d. wt	Dried weight
DEPBG	4′-Demethylepipodophyllotoxin-D-benzylidene glucoside
DIR	Dirigent protein
gusA	*β*-Glucuronidase reporter gene
HPTLC	High-performance TLC
HSV-I	Herpes simplex virus type-1
HSV-II	Herpes simplex virus type-2
IAA	Indole-3-acetic acid
IUCN	International Union for Conservation of Nature
LD50	Lethal dose 50
MCMV	Murine cytomegalovirus
MS	Murashige and Skoog medium (1962)
NAA	1-Naphthaleneacetic acid
NAD	Nicotinamide adenine dinucleotide
NP-CC	Normal-phase column chromatography
npt II	Neomycin phosphotransferase gene
PAL	Phenylalanine ammonium liase
PDA	Photodiode array
PLR	Pinoresinol–lariciresinol reductase
PTLC	Preparative thin-layer chromatography
RP-HPLC	Reversed-phase high-performance liquid chromatography
SAR	Structure–activity relationship
SIRD	Secoisolariciresinol dehydrogenase
TLC	Thin-layer chromatography
UDP	Uridine diphosphate
UPLC	Ultra-performance liquid chromatography
UV–vis	Ultraviolet and visible spectrophotometry
VM26	Teniposide
VP16	Etoposide

Contents

1 Introduction

The term *Podophyllum* is derived from the Greek words podos (foot) and phyllon (leaf) and refers to the leaf's shape which looks like a duck's foot [1]. Some of the species are known as mayapple because they bloom in May. The use of *Podophyllum* dates back to the Chinese culture where it was used over 2,000 years ago as an antitumor drug [2]. Podophyllums have been claimed as medicinal plants with an extensive folklore use in Asian and American cultures for the treatment of skin cancers and warts [3]. Likewise, its extracts were indicated as antidotes against poisonous and suicide agents. The species *Podophyllum hexandrum* and *Podophyllum peltatum* were included in the Pharmacopoeia of India [4], while the former is also in the Ayurvedic Pharmacopoeia [5]. Its rhizomes and roots yield a resinous mixture known as podophyllin that contains lignans.

In 1820, the resin was listed in the US Pharmacopeia as a purgative and cholagogue [6], but its use declined due to its toxicity [7]. The situation changed when Kaplan in 1942 recommended a preparation containing an alcoholic extract of podophyllin in mineral oil for the treatment of venereal warts [8]. From this event, there emerged a renewed interest in the resin and a search for new applications. The therapeutic properties exhibited by the resin are attributed to podophyllotoxin and its related compounds.

The *Podophyllum* species have been extensively investigated for its remarkable anticancer lignans related to podophyllotoxin. Because the total synthesis of podophyllotoxin is uneconomic, the supply is still obtained from wild *Podophyllum* populations. The latter are considered to be "endangered" or "vulnerable" because they grow in only a few areas and their yield is small, this, together with a high demand and unskilled overexploitation [9–11]. There is a need for careful conservation of these genetic resources, otherwise many plants may be lost forever and become extinct.

With an increasing worldwide market for anticancer drugs, such as etoposide phosphate, teniposide, and other derivatives used for the treatment of several cancers [10], supplies of podophyllotoxin for the pharmaceutical industry are under great pressure. Commercially there are few plant sources for this lignan, a relatively rare natural product, and thus the availability is limited. It is still obtained by extraction of the rhizomes and roots of *P. hexandrum* and *P. peltatum* [12]. In India, plant material comes from wild populations of *P. hexandrum* growing in the Himalayan region; these have become an endangered species due to overexploitation and a lack of organized cultivation [13]. Concern has been expressed about the shortage of raw material even though the supply of wild-collected rhizomes from China has been already reported [14]. To overcome this situation, bioprospecting studies for podophyllotoxin in other genera have been reported [12, 15]. On the other hand, alternative sources for its production could be solved by using a biotechnological approach.

Considerable interest has been centered on the *Podophyllum* species. Therefore, in this chapter we describe briefly some of the botanical aspects, the biosynthesis pathway, and medicinal uses as well as the methods for the extraction and analysis of lignans. In addition, a few derivatives of podophyllotoxin and its biological properties will be presented. We have attempted to describe a few biotechnological approaches applied to the *Podophyllum* species in order to obtain organ and cell cultures as well as some strategies including precursor feeding, elicitation, immobilization, and transgenic and endophytic cultures for the production of podophyllotoxin.

2 Botanical Features of *Podophyllum*

The genus *Podophyllum* consists of hardy, herbaceous perennials, rhizomatous herbs with annual aerial stems. Early shoots show an umbrella shape giving rise to large leaves. There is a considerable variation in the leaf shape in most Asian species. Podophyllums spread by using a root system originating from underground short and thick rhizomes including *P. hexandrum* and *P. pleianthum*. In other species such as *P. peltatum*, it occurs as a long and thin creeping rhizome [1]. The flowers bloom as solitary or in clusters and are white, pink, or red-purple with some producing an unpleasant odor. The large fruit is a berry of yellow, red, or purple color, spherical or elliptical, and with many seeds embedded in the pulp. The fleshy fruit has been reported to be edible when ripe while the other parts are toxic [1]. The morphology of many species has been described in detail by Shaw [16].

Podophyllums are not popular garden plants although the ornamental foliage of Asian species can be an attractive to growers. There some accounts referring to cultivation of clones and inbreeding lines of *P. hexandrum* as ornamental perennials [17].

The placement of this small genus within the family Berberidaceae was the subject of investigations by a few authors, and different classifications have been proposed [18–20]. Despite of these studies, the phylogenetic relationship between the genera within the Berberidaceae needs to be further investigated [21, 22]. Meanwhile, some issues regarding the botanical classification of *Podophyllum* remain to be resolved. Besides that, a few taxonomic revisions of the genus have been reported by Shaw [16, 17, 23, 24]. Therefore, in the present chapter, the authors decided to follow the classification of *Podophyllum* species according to Shaw [16].

3 Ethnobotanical Uses

There are many ethnobotanical records based on the healing properties of *Podophyllum* species; however, this section focuses on the uses of *P. hexandrum*, *P. peltatum*, *P. pleianthum*, and *P. sikkimensis*.

P. hexandrum is included in the Hartwell's survey which is a compilation of over 3,000 species and their ethnomedical uses against cancer [25]. The Indian *Podophyllum* has been used in traditional medicine due to its purgative, emetic, cytotoxicity, and antitumor activity [26]. To date, it has been described as an antileukemic crude plant drug by the "vaidyas" who are the medical practitioners of Ayurveda in India [27].

The podophyllin resin has a number of applications in the treatment of several disorders by traditional systems of medicine including Ayurveda, Unani, Siddha, and the Tibetan medicines [7]. According to an early account, it was administered in small doses for chronic constipation while in an overdose acted as a poison [28]. The powdered of roots were indicated as a purgative while the decoction was employed in the treatment of diarrhea and liver problems [29] The rhizomes, roots, and fruits are the main parts of the plant used for medicinal purposes such as the treatment of ulcers, hepatic disorders, wounds, tuberculosis, and constipation [27]. In addition, the ripe fruit is eaten in some parts of the Indian Himalayas, where it has been used to promote conception [29]. It was also used in the treatment of fever in Indian traditional medicine. Furthermore, the rhizomes were prescribed for the treatment of gynecological disorders by the Tibetan traditional medicine [7].

The use of *P. peltatum* goes back to the folklore medicine of the Maine Penobscot Indians, who applied the resin to treat poisonous snake bites, and its roots were used as a suicide agent and poison [30]. The resin was also employed to treat cancer [31] and as a vermifuge [32]. Among the Cherokee Indians, the rhizome and whole plant were used as a slow-acting purgative and an anti-helminthic, while the juice was indicated for ear ailments and for dermatological dressing for ulcers and sores as well as an antirheumatic [33]. The Native Americans used to treat some skin disorders and tumors with an extract obtained from the roots [34].

P. pleianthum is one of the oldest traditional Chinese herbs. Preparations using its rhizomes either in water or alcohol have long been used in Taiwan and China [35]. In the traditional Chinese medicine, some of the main uses include the treatment of snake bite, lymphadenopathy, tumors, weakness, and other disorders [16, 35]. Due to its toxic side effects, its clinical use as an anticancer drug has been limited. However, it is still used in Taiwan by health-care professionals for the recovery from a postpartum condition, chronic headache, hepatoma, lumbago, and others. On the other hand, cases of intoxication have already been reported, and some have led to the death of humans and animals [35].

Regarding the species *P. sikkimensis*, only a limited amount of data relating to their use are available in the literature. The latex obtained from its unripe fruits is applied to treat tumors, while the roots are used as a blood purifier, vermifuge, purgative, and cardiac tonic in small doses. It is also used to treat peristalsis, allergy, and skin inflammations as well as a hepatic stimulant [36].

4 The Biosynthetic Pathway of Lignans

Lignans are a group of phenolic secondary metabolites found in plants and, to a lesser extent, in other organisms [37–39]. They are classified in eight groups based on the oxygenation and cyclization patterns: furofuran, furan, dibenzylbutane, dibenzylbutyrolactone, aryltetralin, arylnaphthalene, dibenzocyclooctadiene, and dibenzylbutyrolactol [39].

The biosynthesis of lignans occurs from the shikimate pathway via the amino acid L-phenylalanine or L-tyrosine, the major phenylpropane (C_6–C_3) building blocks (Scheme 1). Elimination of ammonia from the C_3 side chain of these amino acids generates the *E*-cinnamic acid (**1**) by phenylalanine ammonium lyase (PAL) [37].

From *E*-cinnamic acid (**2**), C_6–C_3 alcohols, called monolignols, are synthesized, e.g., *E*-*p*-coumaryl alcohol (**3**), *E*-coniferyl alcohol (**4**), and *E*-sinapyl alcohol (**5**) (Scheme 2). Phenolic oxidative coupling of two of these C_6–C_3 monomers generates the lignans. There are some variations as to how the monomeric units can be coupled. The term lignan is often restricted to molecules in which the two monomeric units are coupled at the *beta*-carbon of the side chain, e.g., (+)-pinoresinol (**6**). Compounds containing other types of coupling are referred to as neolignans [37, 39, 40]. Lignan and neolignan are normally synthesized enantiomerically pure since the coupling reaction is stereochemically controlled. Cyclization and other modifications generate significant structural diversity among this class of secondary metabolites.

There has been a widespread interest in lignan research since the discovery that these metabolites possess important medicinal applications. In this scenario, *Podophyllum*-derived lignans, especially (−)-podophyllotoxin (**1**), are of particular interest mostly because of their antitumor activity. Podophyllotoxin (**1**), an aryltetralin lactone lignan, is the precursor of the semisynthetic antitumor drugs etoposide, etopophos®, and teniposide. The biological activities of *Podophyllum*-derived lignans are detailed in Sect. 6 of this chapter.

A significant amount of resources have been dedicated to the elucidation of the biosynthesis pathway of these medicinally important lignans, including the

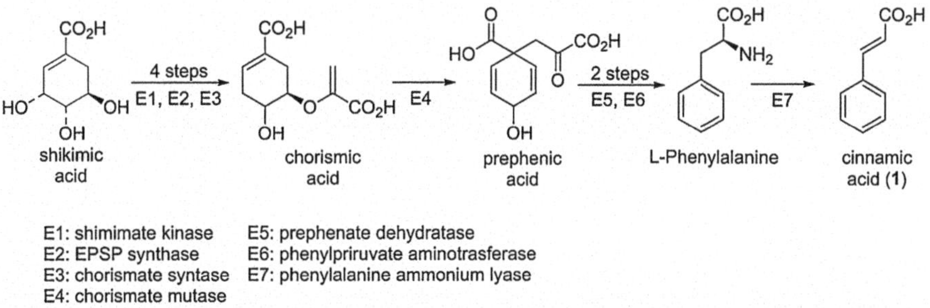

Scheme 1 Biosynthesis pathway of cinnamic acid: intermediates and enzymes involved

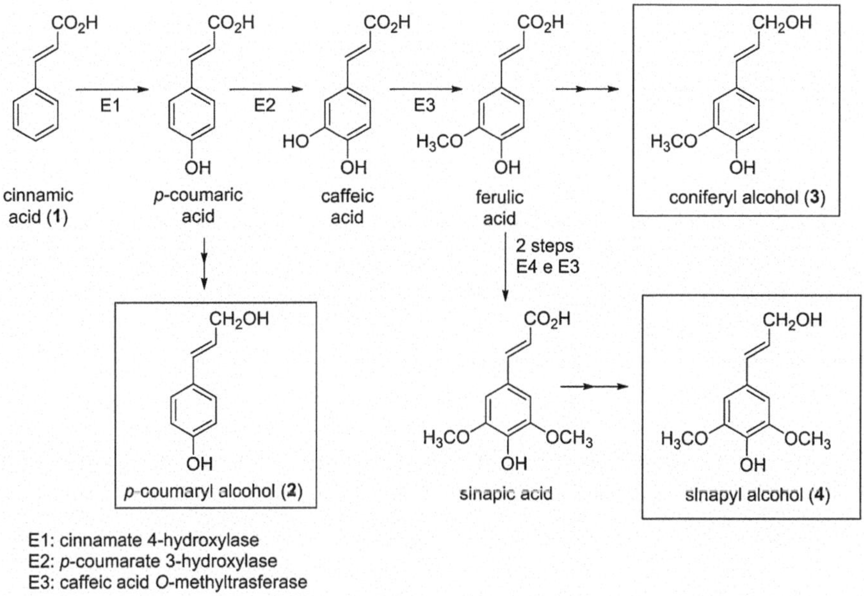

Scheme 2 Biosynthesis pathway of monolignols: intermediates and enzymes involved

molecular characterization of the enzymes involved [40, 41]. This knowledge is essential to the development of technology that allows these compounds to be obtained in vitro. This technology will be further discussed in Sects. 8 and 9 of this chapter.

The biosynthetic pathway of podophyllotoxin and its related lignans (Scheme 3) has yet to be fully elucidated. After monolignol formation, the first step is the synthesis of (+)-pinoresinol (**6**) via stereoselective oxidative coupling of two *E*-coniferyl alcohol-derived radicals [37, 40]. The enzyme involved in this step has yet to be identified. However, Davin et al. [42] demonstrated that a dirigent protein (DIR) was required for the stereospecific dimerization of *E*-coniferyl alcohol.

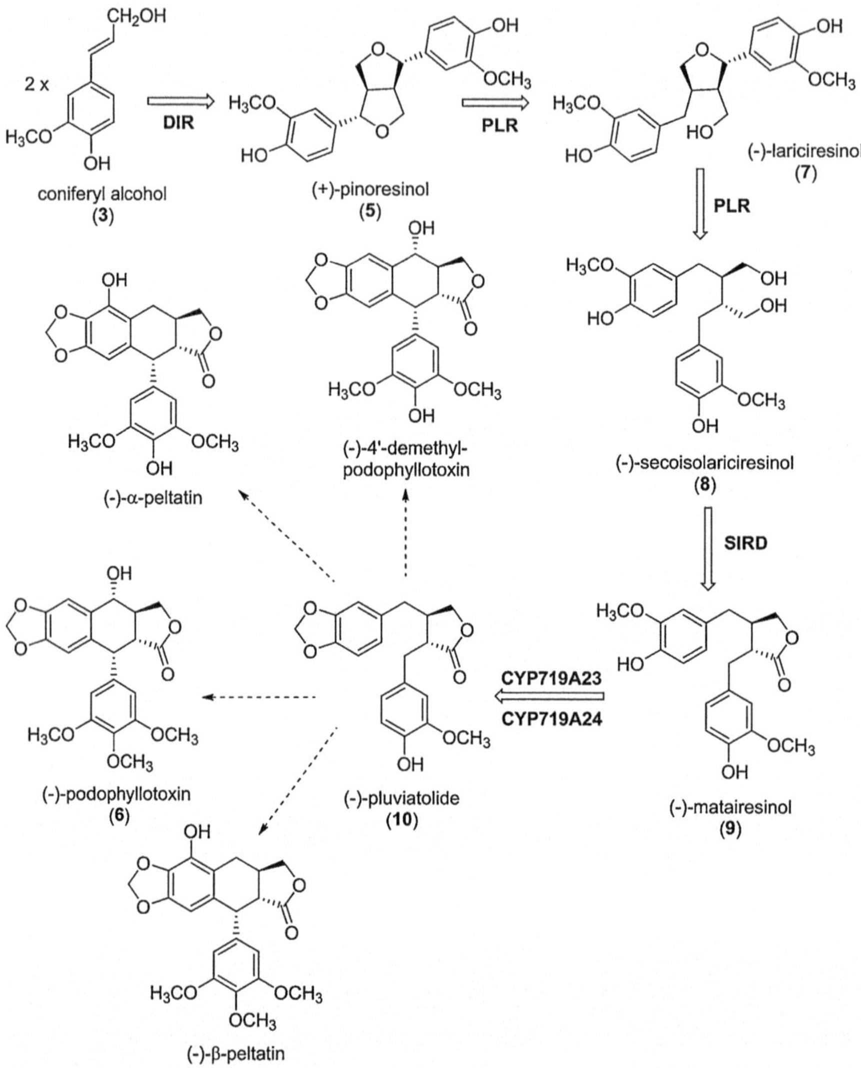

Scheme 3 Partial biosynthesis pathway of (−)podophyllotoxin (**6**) and related lignans. Broken lines represent unidentified steps. *DIR* dirigent protein, *PLR* pinoresinol–lariciresinol reductase, *SIRD* secoisolariciresinol dehydrogenase

Following this step, (+)-pinoresinol (**6**) is enantiospecifically reduced to (−)-lariciresinol (**7**) and then to (−)-secoisolariciresinol (**8**) by pinoresinol–lariciresinol reductase (PLR) [43]. Pinoresinol can also undergo a glycosylation by a UDP-glucose-dependent glucosyltransferase UGT71A18. In fact, PLR-catalyzed metabolism and UGT71A18-directed glucosylation are competitive pathways that participate in the regulation of lignan biosynthesis via pinoresinol metabolism [40].

Next, a stereospecific dehydrogenation occurs, converting (−)-secoisolariciresinol (**8**) into (−)-matairesinol (**9**) by a NAD-dependent secoisolariciresinol dehydrogenase (SIRD) identified in *Forsythia intermedia* and *P. hexandrum* [44].

The use of next-generation sequencing of *P. hexandrum* and *P. peltatum* transcriptomes followed by bioinformatic analysis allowed the identification of biosynthetic genes for the homologous enzymes, CYP719A23 (from *P. hexandrum*) and CYP719A24 (from *P. peltatum*), responsible for the conversion of (−)-matairesinol (**9**) into (−)-pluviatolide (**10**) [41]. Until now, the podophyllotoxin biosynthesis pathway is elucidated up to this intermediate compound.

Possible intermediates of (−)-podophyllotoxin (**1**) biosynthesis from (−)-pluviatolide (**10**) have been proposed recently based on studies using crude enzymatic assays. However, the steps leading to these intermediates have not been validated, since neither their chemical structures have been confirmed nor the enzymes/genes involved have been identified [40, 41, 45, 46].

It is expected that in the near future, the molecular characterization of the remaining key enzymes involved in lignan biosynthesis will be revealed by the use of advanced biotechnological tools, such as next-generation sequencing, which will allow the biosynthesis pathway of **1** to be fully elucidated.

5 *Podophyllum* Species and Its Lignans

Lignans are widespread in the plant kingdom; however, the aryltetralin group has been found in only a few botanical families. The genus *Podophyllum* is a rich source of 1-aryltetralin-type lignans (Fig. 1) with antitumor activity [47]. Within the genus *Podophyllum,* the species *P. peltatum* is the most intensively studied followed by *P. hexandrum*, although others such as *P. pleianthum*, *P. versipelle*, *P. delavayi*, and *P. sikkimensis* were also subjected to investigation in the search for lignans related to podophyllotoxin.

Podophyllum hexandrum Royle, syn. *P. emodi* Wall, *Sinopodophyllum hexandrum* (Royle) Ying, is quite often named as the Himalayan mayapple or Indian mayapple, and it is distributed along the Indian Himalayas to Bhutan, Nepal, Tibet and Northern Yunnan (China), Afghanistan, Pakistan, and India [7]. Before over-collection for commercial supply, it was common in the Alpine Himalayas, growing at an altitude of 3,000–4,000 m [48]. This species has long been a rich source of many lignans such as podophyllotoxin (**1**), 4′-demethylpodophyllotoxin (**11**), podophyllotoxin glucoside (**12**), deoxypodophyllotoxin (**13**), 4′-demethylpodophyllotoxin glucoside (**14**), 4′-demethylisopicropodophyllone (**15**), podophyllotoxone (**16**), 4′-demethylpodophyllotoxone (**17**), picropodophyllotoxin (**18**), isopicropodophyllone (**19**), 4′-demethyldeoxypodophyllotoxin (**20**), α-peltatin (**21**), and β-peltatin (**22**) [26, 49].

Podophyllum peltatum L., syn. *Anapodophyllum peltatum* Moench, is commonly known as the American *Podophyllum*, wild mandrake, mayapple root, mayweed, wild lemon, or devil's apple and is widespread and common in Eastern North America [6]. The lignans isolated from its underground parts are

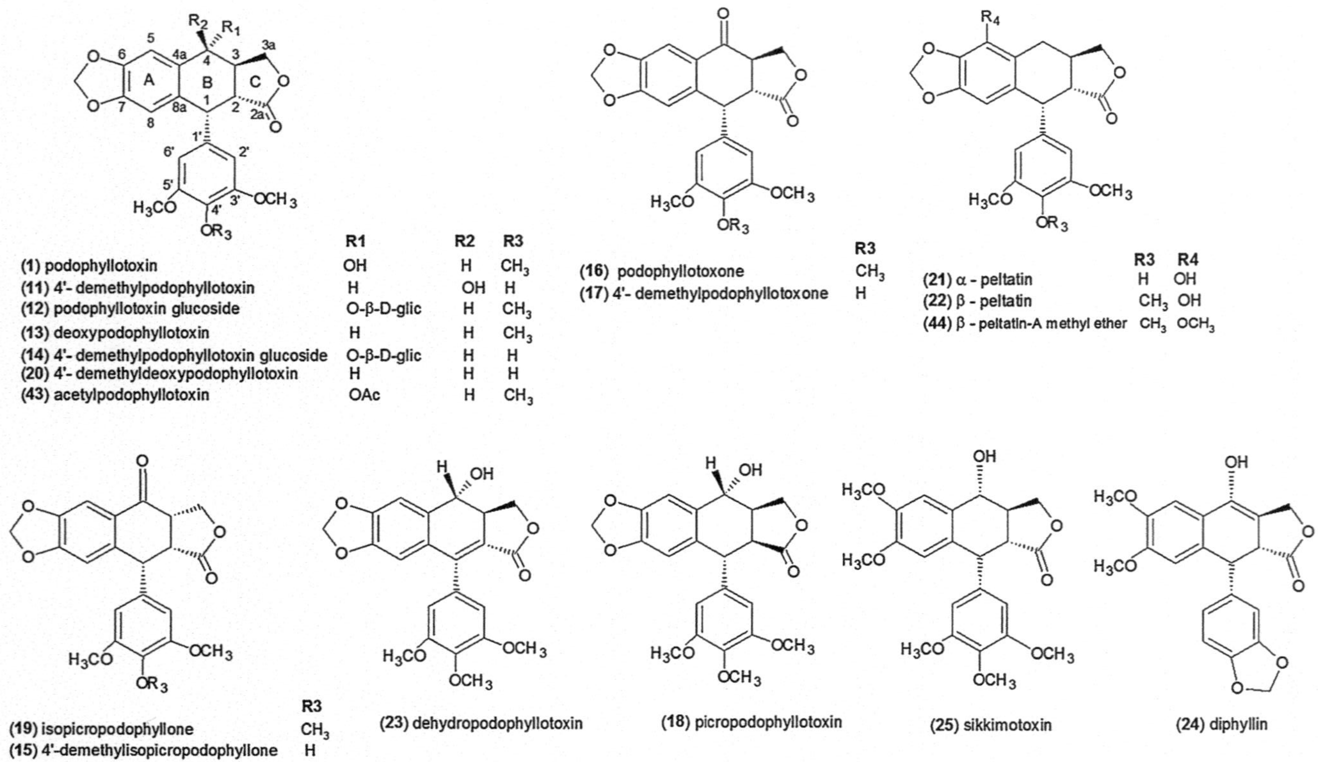

Fig. 1 Aryltetralin lignans structurally related to podophyllotoxin

podophyllotoxin (**1**), 4′-demethylpodophyllotoxin (**11**), deoxypodophyllotoxin (**13**), 4′-demethyldeoxypodophyllotoxin (**20**), β-peltatin (**22**), α-peltatin (**21**), podophyllotoxone (**16**), 4′-demethylpodophyllotoxone (**17**), isopicropodophyllone (**19**), 4′-demethylisopicropodophyllone (**15**), and dehydropodophyllotoxin (**23**) [49–51].

Podophyllum pleianthum Hance, syn. *Dysosma pleiantha* (Hance) Woodson, named as Bajiaolian in Chinese is widely used in Taiwan. It is the only species of the genus found growing at an altitude of 1,000–2,500 m in the central and northern highlands in Taiwan [35]. This species has been become threatened in China and Taiwan due to overharvesting for local herbal medicine [1, 52]. Podophyllotoxin (**1**) is the main lignan found in its rhizomes and roots although others were reported as, for example, 4′-demethylpodophyllotoxin glucoside (**14**), podophyllotoxin glucoside (**12**), 4′-demethylpodophyllotoxin (**11**), 4′-demethyldeoxypodophyllotoxin (**20**), dehydropodophyllotoxin (**23**), diphyllin (**24**), and podophyllotoxone (**16**) [35].

Podophyllum versipelle Hance, syn. *Dysosma versipellis* (Hance) M. Cheng ex Ying, is a species that grows in restricted areas in the southeastern China and Taiwan [52]. According to the IUCN database (2013), it has been considered as one of the four endangered or vulnerable *Dysosma* species [9]. Its roots yielded podophyllotoxin (**1**), α-peltatin (**21**), β-peltatin (**22**), podophyllotoxin (**1**), podophyllotoxin glucoside (**2**), 4′-demethylpodophyllotoxin (**11**), deoxypodophyllotoxin (**13**), 4′-demethyldeoxypodophyllotoxin (**20**), diphyllin (**24**), and podophyllotoxone (**9**) [53–55].

Podophyllum sikkimensis Chatt & Mukh. This species has the vernacular name Homochari in the Sikkin Himalaya (India) where it is used for medicinal purposes by local ethnic groups. An analysis of the resin from the rhizomes of *P. sikkimensis* led to the isolation of different components from those of *P. peltatum* and *P. hexandrum* [56]. The resin contains some flavonoids and a lactone known as sikkimotoxin (**25**). The latter has properties analogous to the podophyllotoxin [56, 57].

Podophyllum delavayi Franch, syn. *Dysosma delavayi*, is another Chinese mayapple native to Western China and distributed in the provinces of Sichuan and Yunnan [13] (Fig. 1).

6 Biological Activity of Podophyllotoxin and Related Lignans

The most significant therapeutic applications of podophyllotoxin (**1**) and its derivatives are as antineoplastic and antiviral agents. A widespread variety of other biological activities, such as insecticidal, antimalarial, and fungicidal, have been ascribed to *Podophyllum* spp. and its 1-aryltetralin-type lignans [58, 59].

According to Li et al. [60], *Podophyllum* spp. are considered to be pharmaceutical crops which are cultivated species used for the extraction or preparation of therapeutic substances by the pharmaceutical industry.

Regarding the pharmacological importance of *Podophyllum* lignans, a wide range of podophyllotoxin derivatives have been synthesized (Fig. 2) aiming to obtain compounds with fewer side effects, high potency, and improved pharmacokinetics characteristics [59, 61, 62].

Excellent reviews of podophyllotoxin derivatives from an historical point of view, biological activity, synthesis, and structure–activity relationship (SAR) studies are available in the literature [62–65]. In this chapter, we will highlight some of the biological activities reported for these compounds.

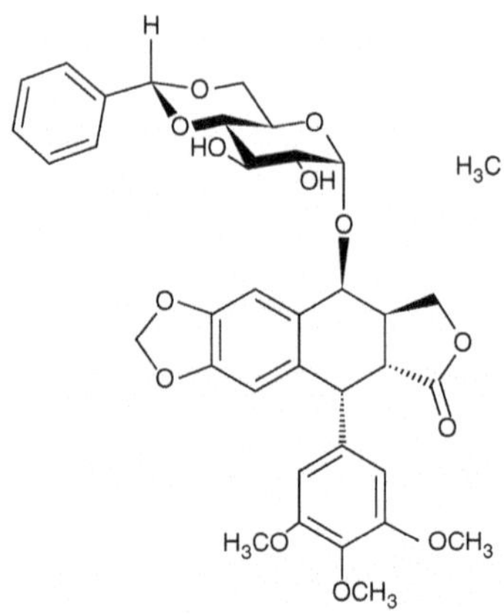

(26) DEPBG
4′-demethylepipodophylltoxin-β-D-benzylidene glucoside

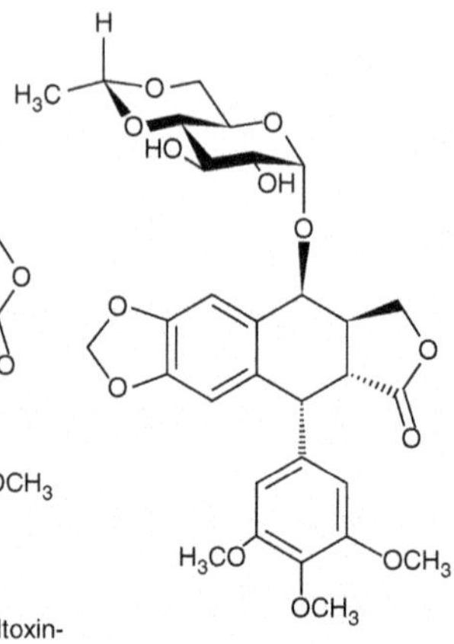

(27) VP16
etoposide

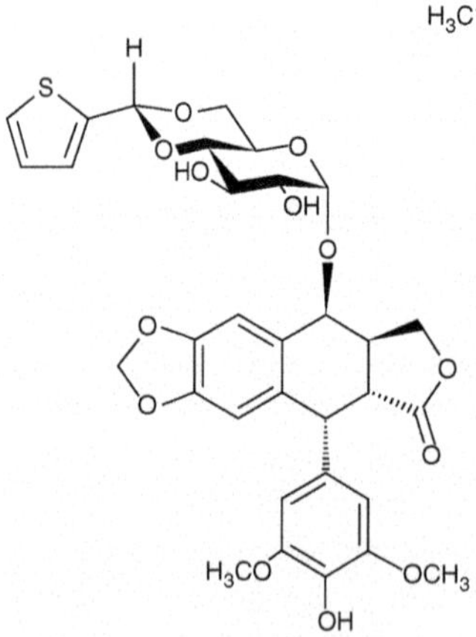

(28) VM 26
teniposide

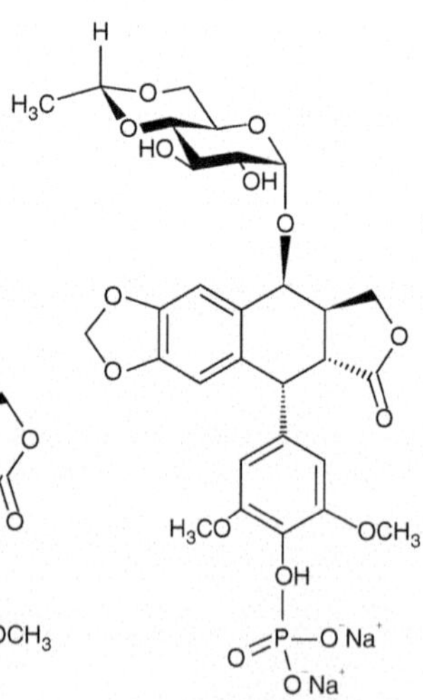

(29) Etopophos
ethoposide photsphate

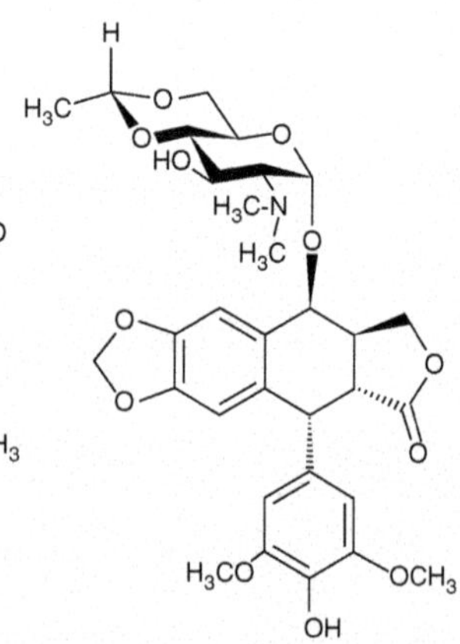

(30) NK611

Fig. 2 (continued)

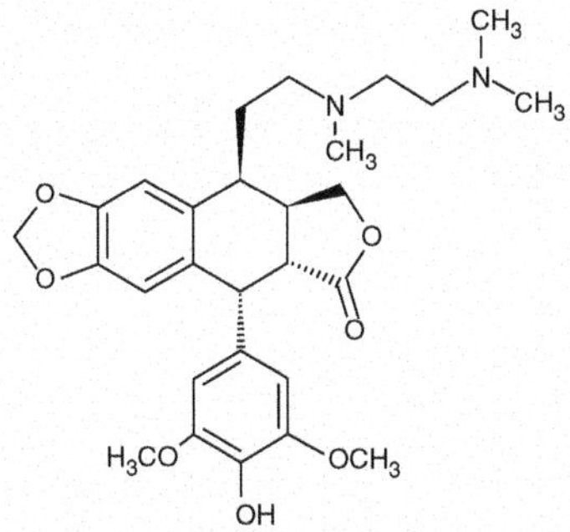

(31) TOP-53

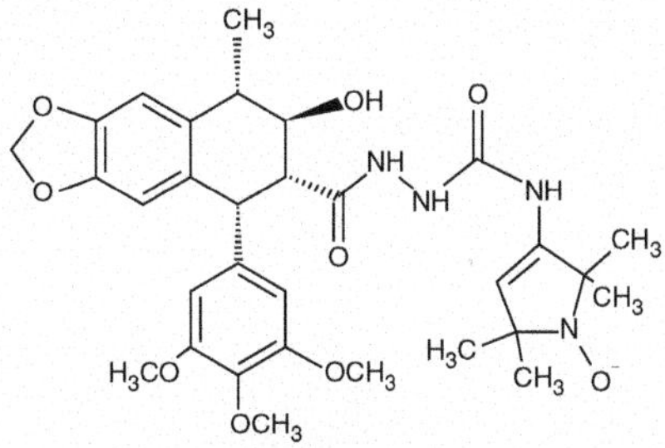

(32) GP-11
N′-podophyllic acid-N-
(3-(2,2,5,5-tetramethylprrolinrnyloxy))
semicarbazide

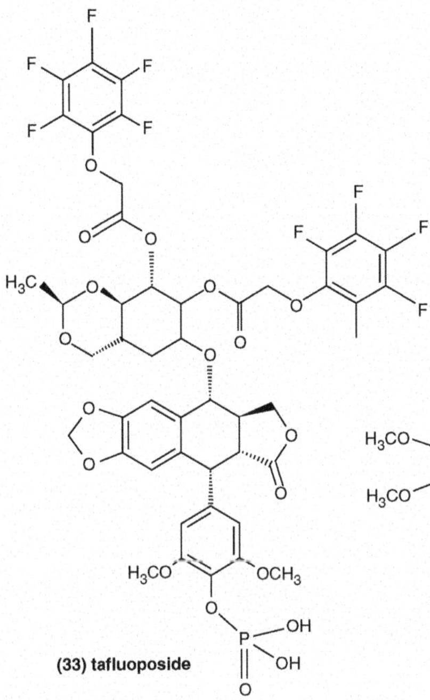

(33) tafluoposide

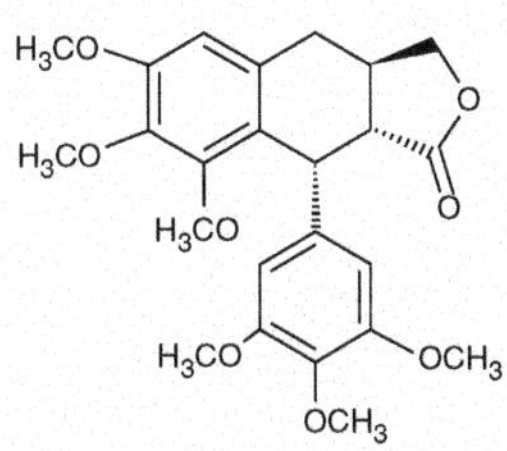

(34)

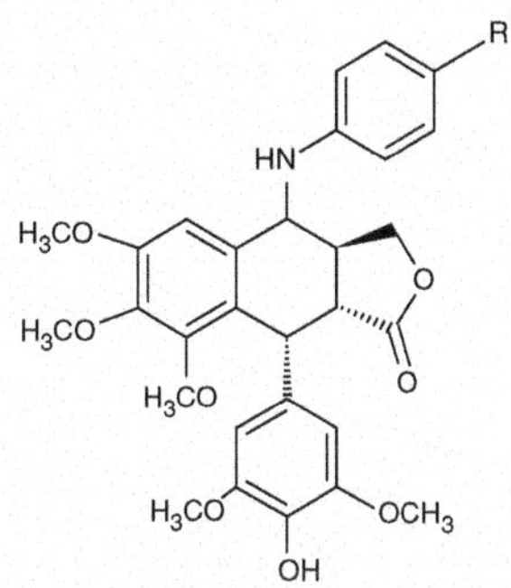

(35) R = CN
(36) R = CO_2Et
(37) R = NO2
(38) R = F

Fig. 2 (continued)

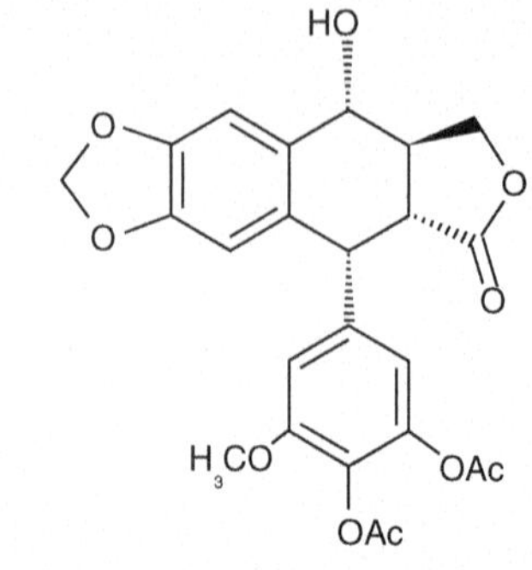

(39)

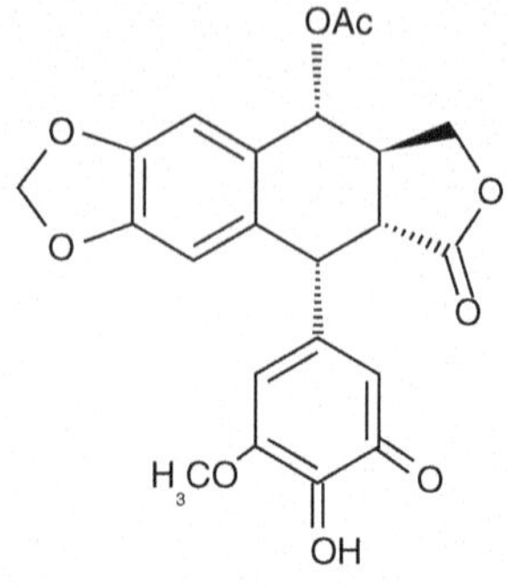

(40)

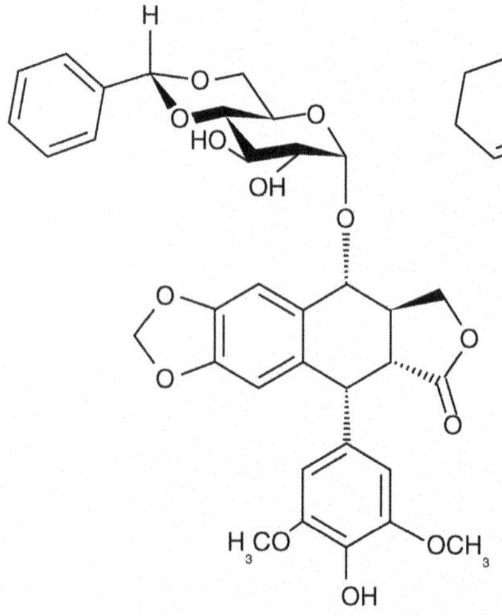

(41) AS 3739

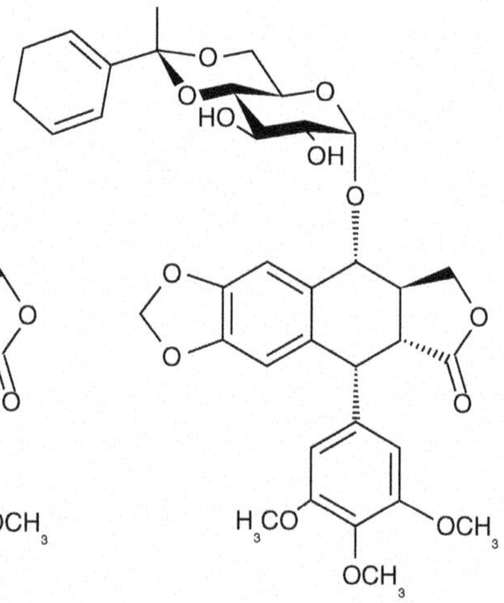

(42) AS 3738

Fig. 2 Examples of podophyllotoxin derivatives with biological activity

6.1 Antitumor Activity

Kaplan used podophyllin to treat condylomata acuminata with good results [8], and the studies on its mechanism of action realized by King and Sullivan [66] and Sullivan and Wechester [67] stimulated the interest in podophyllin and its components. However, the idea to use it clinically as an antitumor agent was abandoned because of its side effects. In the literature, there are many reports about the discovery and development of drugs from podophyllotoxin derivatives [59, 61, 68].

The investigation of semisynthetic glycoconjugates led to the discovery of the antitumor compounds 4′-demethylepipodophyllotoxin-D-benzylidene glucoside (DEPBG, **26**), etoposide (VP16, **27**), and teniposide (VM26, **28**), as well as etopophos® (**29**). The clinical success of VP16, VM26, and etopophos® stimulated the interest in developing derivatives with better antitumor activity. Examples of some important analogs are NK-611 (**30**), Top 53 (**31**), GL-11 (**32**), and tafluposide (**33**) [61].

The proposed mechanism of antitumor activity of podophyllotoxin derivatives includes the inhibition of tubulin polymerization for podophyllotoxin-like compounds or inhibiting DNA topoisomerase II for etoposide-like compounds [68].

Podophyllotoxin reversibly bind to tubulin, leading to inhibition of microtubules formation in the metaphase of mitosis, at the same binding site of colchicine [58]. The acetal products of podophyllotoxin glucosides and peltatins have the same mechanism of action [69].

Etoposide (**21**) and teniposide (**22**) stabilize the covalent DNA–enzyme ternary complex inhibiting the catalytic activity of topoisomerase II irreversibly. These compounds act in late S or G2 phases of the cell cycle, preventing the DNA repair by topoisomerase II [61, 62, 69, 70]. In addition to these two mechanisms of action, a third mechanism has also been proposed. The metabolic activation of the dimethoxyphenol ring (E ring) of podophyllotoxin analogs may produce metabolites that can deactivate the DNA by forming chemical adducts. It has been shown that the 3′,4′-catechol derivatives of etoposide can be formed and oxidized to 3′,4′-ortho-quinone in the presence of cytochrome P-450 and oxygen, horseradish peroxidase, or prostaglandin E synthase. The compounds formed bind at DNA and this may contribute to the activity of these compounds [62]. Reviews of podophyllotoxin derivatives and the proposed mechanism of action are available in the literature [59, 62, 64, 69, 71].

The development of synthetic and semisynthetic derivatives of podophyllotoxin, the structure–activity relationship studies, and understanding the mechanism of podophyllotoxin antitumor activity have aided to propose analogs with improved properties. The structural features important to the antitumor activity of podophyllotoxin can be summarized in some aspects (Fig. 3): (a) *Z* configuration of the A/B ring, (b) *E*-lactone with 2α and 3β in ring C is essential to the activity, (c) 4-β configuration is essential with various substitutions, (d) the free 4′-hydroxy is crucial, (e) the dioxolane A ring is optimal, and (f) the free rotation of ring E is required [62, 71].

6.2 Antiviral Activity

The investigation of antiviral activity of lignans began with the use of an alcoholic extract against *condyloma acuminate,* a disease caused by the papillomavirus [69].

Several mechanisms of the antiviral action of lignans have been proposed [64, 65]. Bedows and Hatfield [72] showed that podophyllotoxin (**1**), deoxypodophyllotoxin (**13**), and β-peltatin (**22**) were active against the measles virus (RNA virus) and herpes simplex virus type I (HSV-I, DNA virus). They proposed that this activity is likely due to the disruption of cellular microtubules interfering with viral replication.

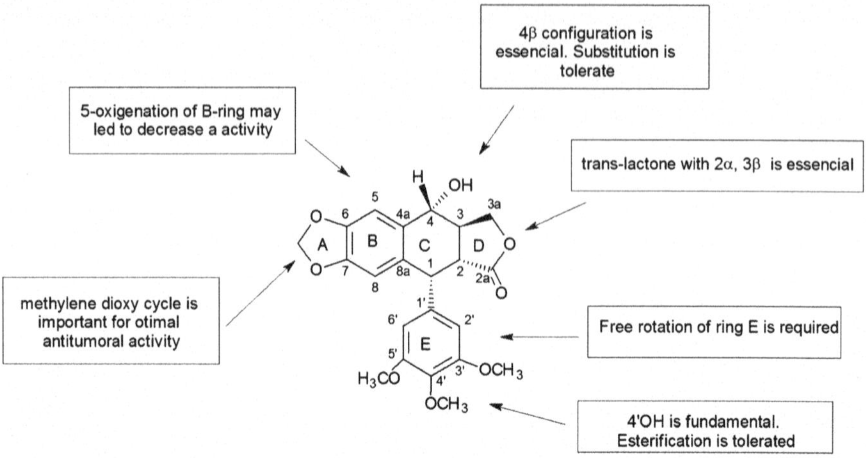

Fig. 3 Structure–activity relationship of podophyllotoxin analogs for the Topo II inhibition (Source: Adapted from [62])

In addition, some podophyllotoxin derivatives (**34–38**, Fig. 2) showed an inhibition of HIV replication by the inhibition of the reverse transcriptase viral [63, 69, 73]. Macrae et al. [74] evaluated the antiviral activity podophyllotoxin (**1**), α-peltatin (**21**), and diphyllin (**25**) against murine cytomegalovirus (MCMV, DNA virus) and the Sindbis virus (RNA virus). Podophyllotoxin (241 nM) and α-peltatin (250 nM) showed good inhibition against the MCMV virus (74 and 85%, respectively) and low inhibition against the Sindbis virus (3 and 5%, respectively).

Castro et al. [59] synthesized some podophyllotoxin derivatives modified in the E ring and evaluated against herpes simplex virus type II (HSV-II). The compounds (**39**) and (**40**) were active at 23 and 25 μg/mL, respectively.

6.3 Other Activities

Due to a significant immunosuppressive activity shown by etoposide and other podophyllotoxin analogs [75], researches led to the development of CPH 82 (Reumacon®) which is a mixture of two similar benzylidenated podophyllotoxin glucosides, AS 3739 (**41**) and AS 3738 (**42**) (Fig. 2). CPH 82 is a nonsteroid antirheumatic drug used in the therapy of rheumatoid arthritis.

The insecticidal activity of podophyllotoxin and its derivatives, as well as others lignans, have been reported [76, 77]. Miyazawa et al. [78], in the search for new insecticidal compounds from plants, described the insecticidal activity of podophyllotoxin (**1**) and acetylpodophyllotoxin (**43**, Fig. 1.), isolated by bioassay-guided fractionation from dichloromethane extract of *P. hexandrum* roots active against larvae of *Drosophila melanogaster*. Podophyllotoxin (**1**) showed LC_{50} 0.24 μmol/L diet against larvae and LD_{50} 22 μmol/adults against adults. Deoxypodophyllotoxin (**2**) showed LC_{50} 0.64 μmol/L diet against larvae and LD_{50} 80 μmol/adults against adults. In studies with podophyllotoxin (1) and deoxypodophyllotoxin (**2**), Inamori et al. [79] showed insecticidal activity of (**1**) and (**2**) against larvae of *Epilachna sparsa orientalis* (95% of mortality at 20 ppm and 85% mortality at 500 ppm, respectively) and insecticidal activity of (**2**) against *Culex pipiens molestus* (90% of mortality at 20 ppm). In the work by Russell et al. [80], the assay with β-peltatin-A-methyl ether (**44**, Fig. 1) resulted in a 98% mortality of housefly (*Musca domestica*) larvae at 100 ppm in a chemically defined diet.

Ethanolic extract of rhizomes and roots of *P. hexandrum* (4 mg/mL) showed AChE inhibition in vitro and antioxidant activity (IC_{50} 21.56 μg/mL) in β-carotene/linoleic acid system [81]. These activities combined indicate a potential application of this species for the treatment of neurodegenerative diseases.

7 Analytical Methods of Analysis

Many methods have been used in the extraction of lignans from plant material, such as Soxhlet extractor, accelerated solvent extraction, and percolation and digestion with hot solvents [82]. The aryltetralin lignans can be obtained from the rhizomes, roots, and leaves of the *Podophyllum* species. The common methods to obtain lignans from *Podophyllum* involve extraction of the dried plant material with alcohol using warm methods such as heat reflux and Soxhlet extraction [83, 84]. Other methods described consist of an initial extraction with other less polar organic solvents, such as petroleum ether, n-hexane, and halogenated hydrocarbon, aiming to remove the lipids [85, 86].

Purification of crude methanol extracts is quite time-consuming and laborious, so usually to obtain enriched fraction, the crude extracts are partitioned with ethyl acetate or other halogenated hydrocarbon solvent [84, 86]. Another classical method described by Chatterjee and Chakravarti [56] is the precipitation of lignans from ethanolic extracts using hydrochloric acid solution followed by cooling.

Even though the techniques of analysis and isolation of natural products have developed greatly in recent years, the classical thin-layer chromatography (TLC) still remains a useful analytical method due to its easy performance, low cost, celerity, and efficiency [87]. Furthermore, this method allows the simultaneous analysis of multiple samples [88]. TLC techniques have been used in the analysis of plant cell and tissue cultures for a prompt characterization of lignans [51, 82, 86, 89]. Aryltetralin lignans structurally related to podophyllotoxin can be detected on TLC at 254 nm using plates with a fluorescent indicator. The characterization of the main lignans can be achieved by spraying coloring reagents such as sulfuric acid in ethanol followed by heating and using fast blue salt B [86, 90]. Jackson and Dewick [49] described the detection *of Podophyllum* lignans with nitric acid-acetic acid reagent (10:3) and heating. The lignan bands show red or brown coloration due to the

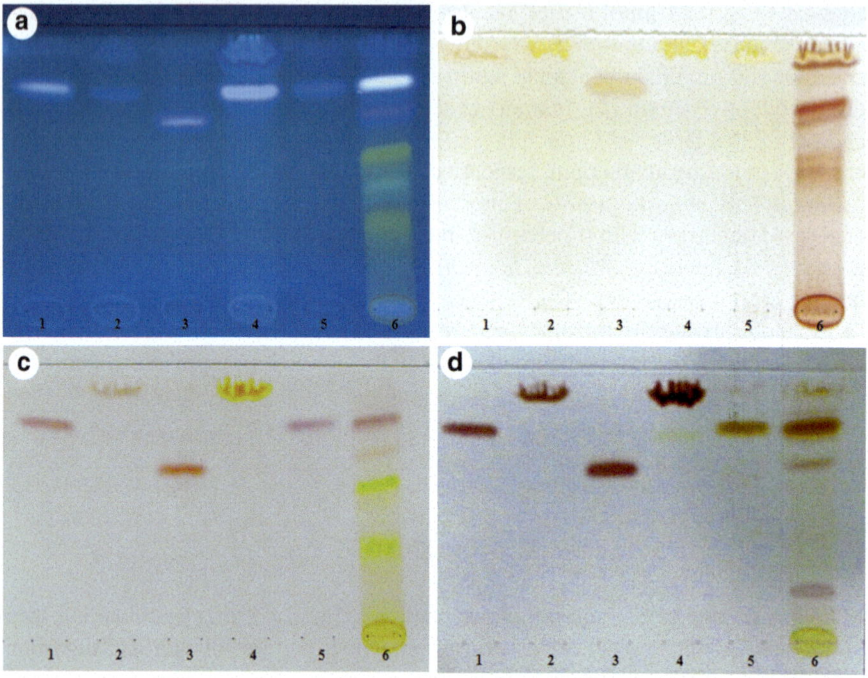

Fig. 4 TLC profile of *Podophyllum*-based lignans: *1*. podophyllotoxin, *2*. deoxypodophyllotoxin, *3*. 4′-demethyllpodophyllotoxin, *4*. podophyllotoxone, *5*. picropodophyllotoxin, *6*. EtOH extract of *Podophyllum hexandrum*. Eluent: Chloroform to MeOH (9:1). Detection (*A*). $UV_{365\ nm}$, untreated (*B*). Fast blue salt B and NaOH 10% in EtOH (*C*). H_2SO_4 50% in EtOH, heated (*D*). Nitric acid: acetic acid (30:9)

formation of quinone derivatives via oxidation or demethylation reactions [12, 49] (Fig. 4).

Isolation of podophyllotoxin has been carried out using a preparative TLC (PTLC) [12, 51] and other chromatographic methods, e.g., normal-phase column chromatography (NP-CC) [54], reversed-phase high-performance liquid chromatography (RP-HPLC) [91, 92], and countercurrent chromatography (CCC) [55, 93].

Precipitation using an aqueous solution of Na_2CO_3 has been reported [26]. However, lactones and ketolactones of podophyllotoxin group can undergo epimerization in C2 with basic solutions (Fig. 5). In the presence of mild base, podophyllotoxin (**1**) can epimerizes slowly to picropodophyllotoxin (**18**) [94], leading to artifact formation [51].

The quantitative analysis of lignans has been reported using several analytical methods, for instance, capillary zone electrophoresis [95], UV–vis spectrophotometry [12], HPLC [54, 96, 97], and UPLC [98]. Mishra et al. [99] used HPLC-PDA and HPTLC densitometry to quantify podophyllotoxin (**1**) in dry cell mass obtained from tissue cultures of P. *hexandrum*.

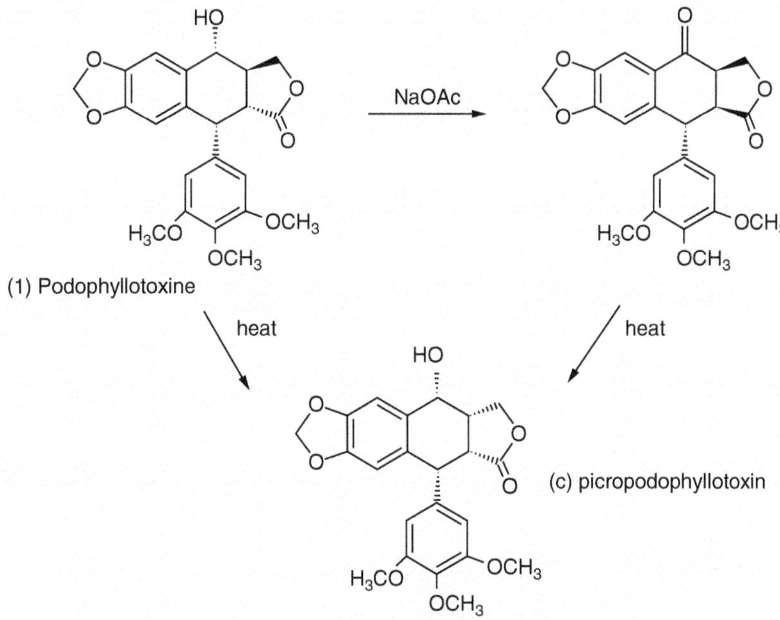

Fig. 5 Podophyllotoxin epimerization to picropodophyllotoxin

The quantities and relative proportions of lignans may vary according to the *Podophyllum* species (Table 1), agronomic aspects [100], and genetic variants in the population [11, 101].

Jackson and Dewick [49] reported the lignan content in different species of *Podophyllum*. The quantification of lignans was performed using UV–vis spectrophotometry. According to the results reported (Table 1), podophyllotoxin represents the main lignan of *P. hexandrum* and *P pleianthum*. Although *P. peltatum* is one of the species used to commercially obtain podophyllotoxin [102], β-peltatin is the major constituent of this species.

8 Background on Cultivation and Propagation of *Podophyllum* spp.

Regarding the history of cultivation of Podophyllums, in this chapter only data on *P. hexandrum* and *P. peltatum* will be reported since both species can supply podophyllotoxin to the pharmaceutical industry.

P. hexandrum has been found in the wild since 1820. It has been found growing naturally in some Asian countries, but there was no cultivation on a commercial scale [103, 104]. In the early 1900s, the exportation of wild-collected rhizomes from India as a source of income resulted in overexploitation [57]. In addition, plant populations in accessible areas were reduced due to the unregulated collection of

Table 1 Lignan content (mg/g dried root/rhizome) of *Podophyllum* species

Lignan	*Podophyllum Species*			
	P. hexandrum	*P. peltatum*	*P. pleianthum*	*P. versipelle*
Podophyllotoxin (**1**)	42.7	2.5	1.35	3.2
4′-Demethylpodophyllotoxin (**11**)	4.5	0.07	0.41	0.14
Deoxypodophyllotoxin (**13**)	0.17	0.23	0.10	tr
4′-Demethyldeoxypodophyllotoxin (**20**)	0.10	0.07	0.03	tr
β-peltatin (**22**)	0.10	3.3	–	13.5
α-peltatin (**21**)	0.07	2.5	–	1.2
Podophyllotoxone (**16**)	0.57	0.20	0.37	0.11
4′-Demethylpodophyllotoxone (**17**)	0.13	0.07	0.13	tr
Isopicropodophyllone (**19**)	0.33	0.07	0.20	–
4′-Demethylisopicropodophyllone (**15**)	0.07	0.03	0.05	–

tr - traces Data from Ref. [49]

rhizomes and roots by untrained people. Earlier work reported that *P. hexandrum* can be propagated easily either from rhizomes or seeds [105]. However, plants raised from rhizome cuttings needed at least 12 years to produce a marketable rhizome [106]. Efforts were made to preserve and cultivate the species on a commercial scale in India as a means to reduce the pressure of harvesting. There are a number of reports referring to vegetative propagation through rhizome cuttings and seeds with limited results [106–108]. The most economical time period for the harvesting of roots and rhizomes was 5 years for plants raised by vegetative propagation and 6 years when they were raised from seeds [14]. Moreover, it was reported that rhizomes of cultivated plants in a subalpine garden exhibited a lignan content lower than that of the wild plants [109].

P. peltatum was one of the species collected by the locals in the Appalachian mountains as a source of income rather than being cultivated for medicinal use [110]. The interest in cultivation was stimulated by the demand for the podophyllin resin in the drug trade from dried rhizomes [6]. It has been found growing in large colonies in moist woodland areas in eastern North America [6] although it is absent from Florida [16].

Like other species within the Berberidaceae, it is also noteworthy that some Podophyllums have been cultivated as ornamental plants [20] and on some occasions for inducing mutations in other plants. In the UK and France, *P. peltatum* was introduced in 1664 for cultivation in woodland gardens [16, 32] whereas *P. hexandrum* was recorded as an alien garden species [111]. Moreover, *P. hexandrum* is easy to grow in the garden, and it has been cultivated in North America and Europe as an ornamental plant [16]. Other Asian counterparts can be found in cultivation as foliage plants but require some protection against the frost. In general, Podophyllums are resistant to low temperatures, but young leaves can be damaged by late frosts.

Podophyllum species are perennial herbaceous plants with a short annual growth period [112]. As has been previously described, conventional propagation can be achieved by seeds and cuttings of rhizomes and roots. Rhizomes cuttings were considered useful for propagating elite clones of high-yielding plants [113]. Moreover, dormancy of seeds has been considered to be a major constraint in some species of *Podophyllum* [10]. Propagation and conservation of *P. hexandrum* have been reviewed by Nadeem et al. [114].

Experiments on cultivation of Podophyllums were carried out in the Pharmacy garden at the University of Nottingham (UK). Fieldwork with *P. hexandrum* lines has shown that plants cultivated at Nottingham are long-lived and produced large clonal colonies [111]. The rhizomes were kept growing for about 6 years after which the growth rate declined. Vegetative propagation through rhizome cuttings has been found to be inefficient and a slow method compared with seedlings.

P. hexandrum cv. Majus plants were originally obtained from the Bressingham Gardens (Bressingham, Norfolk, UK) and cultivated individually in pots containing a mixture of Levington M3 soilless and John Innes No. 2 compost and maintained in a cold frame and then planted out (Fig. 6a). Seeds of self-pollinations were used to increase stocks. In a population of approximately 20 self-pollinated individuals, seeds were obtained from the ripe fruits (Fig. 6b) and used for tissue culture studies [51].

Fig. 6 Plants, ripe fruits, and germination in vitro of seeds of *Podophyllum hexandrum* (**a**). *P. hexandrum* cv. Majus plants (8 years old) cultivated at the University of Nottingham showing ripe fruits. (×0.10) (**b**). Ripe fruits, each containing approximately 50 seeds, harvested from a population of self-pollinated seed-derived plants. (×0.75) (**c**). Seeds germinating on moist filter paper after 35 days in the dark. (×0.93) (**d**). An axenic seedling cultured on full-strength MS semi-solidified agar medium lacking growth regulators after 35 days in the light. (×1.25) (**e**). Dormant buds, rhizomes, and roots of *P. hexandrum* cv. Majus used in the vernalization experiments. (×1.0) (**f**). Plant (approx. 5–6 years old) maintained in the glasshouse at 20 ± 2 °C (night) and 22 ± 2 °C (day) with natural daylight supplemented by a 16 h photoperiod (151 μmol $m^{-2}s^{-1}$) provided by "daylight" fluorescent tubes. (×0.40)

Overall, propagation of Podophyllums easily achieved from seeds although this method is prone to some constraints. It was reported that seed desiccation is a problem particularly to *P. peltatum* because its seeds cannot survive for long if stored in paper packets [1]. Research at Nottingham has shown that with *P. peltatum* and *P. hexandrum*, if clean seeds were stored in moist conditions at 4 °C, their full viability is retained for at least 2 years [16]. Several workers have reported that the presence of inhibitors in the fruit pulp, a post-harvesting ripening period, hard seed coat, low seed viability, and poor seed germination can be a problem in propagation from seeds [106, 108, 114, 115]. Other factors such as seed and hypocotyl dormancy also affect the establishment of seedlings [113]. In order to break seed dormancy and improve germination, some treatments were applied using gibberellic acid, KNO_3, chilling, acid scarification, storage in the dark, and others [113, 116, 117]. In a study, the radicle emerged after approximately 35 d on a moisture filter paper (Fig. 6c) for both stored and fresh seeds, indicating that a post-harvesting ripening period was required for in vitro seed germination [51]. On the other hand, high germination was achieved (60–90%) when freshly collected seeds from different locations of *P. hexandrum* were transferred to sterile sand without any treatment [113].

The species *P. hexandrum*, *P. peltatum*, and *P. pleianthum* showed a seasonal availability which can be responsible for the limited supply of plant material. For example, to overcome the limited supply of *P. hexandrum* for the tissue culture studies, axenic cultures were successfully established on full-strength MS semi-solidified agar medium lacking growth regulators (Fig. 6d). In a study, plants were exposed to a cold chilling (vernalization) to see if they could break a predetermined period of dormancy. Dormant plants of *P. hexandrum* with buds, rhizomes, and roots (Fig. 6e) maintained in the glasshouse were transferred to an illuminated vernalization chamber. After 8 weeks, plants were returned to glasshouse conditions, but their underground parts remained dormant, suggesting that this species may require a longer period of vernalization [51]. In addition, the maintenance of plants (Fig. 6f) during the winter in the glasshouse at 20 ± 2 °C (night) and 22 ± 2 °C (day) with natural daylight supplemented by a 16 h photoperiod (151 μmol $m^{-2}s^{-1}$) provided by "daylight" fluorescent tubes did not break dormancy [51].

Alternatively, *Podophyllum* species can be propagated by tissue culture. This approach is being used for commercial-scale plant multiplication in several species, including medicinal, that are the reservoir of useful secondary metabolites. In additional, endangered, threatened, and rare species have successfully been grown and conserved by micropropagation.

Podophyllum species have been investigated by researchers either working in governmental institutions or private companies which explored different strategies on micropropagation for over two decades. Micropropagation is more efficient than propagation via rhizome cuttings because it can supply uniform and consistent plant material for extraction of lignans. A number of studies have been undertaken to preserve and multiply germplasm as well as to establish cell and organ cultures from lignan-accumulating *Podophyllum* plants. As a result the species *P. hexandrum* [89, 114, 118, 119] and *P. peltatum* [47, 120, 121] had received much attention probably due to its content of lignans. The others, *P. pleianthum* and *P. versipellis*, have also being studied [53, 122]; however, these species contain low yields of podophyllotoxin.

In *P. hexandrum*, the growing awareness of decreasing plant resource has helped to focus attention on alternative sources for mass propagation. In this context, plant regeneration has been achieved through root-derived callus [123] via somatic embryogenesis in callus and cells cultures [89, 114] and organogenesis based on root cultures in liquid MS medium [124] as well as by multiple shoot formation cultured on solidified MS medium [119]. Additionally, somatic embryos were obtained from embryogenic cell cultures from a root-derived callus under dark conditions [51], and further details of this are described in the next section.

The need for a reliable supply of podophyllotoxin has also stimulated other research strategies which will be discussed in Sect. 9 of this chapter.

9 In Vitro Studies for the Production of Podophyllotoxin

9.1 Organ and Cell Cultures as an Alternative Source

Organ and cell suspension cultures have been used as an alternative source for the production of podophyllotoxin and were summarized by some workers [125, 126].

It has been reported that undifferentiated callus tissues from different explants of *P. peltatum* were able to produce podophyllotoxin. The content of lignan in callus from rhizomes was higher than that found in leaves, roots, and stem cultures [127, 128]. The production of podophyllotoxin was affected by light quality and intensity as well as by the composition of the culture medium used for callus initiation. In this regard, kinetin played an important role besides NAA (1-naphthaleneacetic acid) for callus induction and with 2,4-D for maximum podophyllotoxin production. This study has been reported as the first successful production of podophyllotoxin by in vitro culture.

Uden and collaborators [129] have shown that cultures of dark-grown callus of *P. hexandrum* had a higher content of podophyllotoxin (ca. 0.3% d. wt.) compared with light-grown cultures. In another study, root-derived callus from in vitro cultured seedlings showed positive results in terms of lignan production [130]. This study has also shown that for some cell lines, high levels of podophyllotoxin are related to the tissue differentiation status elicited by BAP. Moreover, the presence of auxins, in particular 2,4-D, was necessary for callus induction.

It has been suggested that the initiation and establishment of callus cultures of *P. hexandrum* represent a difficult task [131]. For example, an initial callus response from root explants has been shown to be very slow. Cell suspensions were established in the dark from root-derived callus cultured in a liquid MS medium containing 2,4-D and kinetin, respectively. It was interesting to note that such cultures became embryogenic, producing somatic embryos at different stages of development as well as small cell aggregates in the culture medium [51]. It will be, therefore, worthwhile to investigate whether the embryogenic cell culture system offers possibilities for the in vitro propagation and/or genetic manipulation of *P. hexandrum*. Both systems, callus and cell cultures, are quite often time-consuming and require more expertise into tissue culture techniques. In another study, callus cultures were obtained for roots and rhizomes from in vitro plants, and 1 year later all the surviving callus lines accumulated podophyllotoxin [132].

In a study with *P. peltatum*, different types of buds were induced from a terminal bud originated by rhizome tip cultures from wild plants [120]. The lignan content from in vitro rooted buds and plantlets was found to be at comparable concentrations to those of plants collected from wild populations [83]. Moreover, the accumulation of podophyllotoxin was higher in adventitious root cultures than those in embryogenic cells clumps [133].

The production of podophyllotoxin has been reported in callus and cultured rhizomes of *P. versipellis* [53]. The lignan content was higher in the cultured rhizomes (3.20 mg g^{-1}) than in the calli (1.17 mg g^{-1}). This study has also shown that rhizomes from wild plants exhibited a slightly higher content of podophyllotoxin (6.01 mg g^{-1}) than those from plants grown in the garden (5.89 mg g^{-1}). These results agree with those of Liu et al. [134] showing that plants from tissue culture accumulated less of this lignan compared with their wild counterparts.

10 Biotechnological Approaches to Improve Podophyllotoxin Production

10.1 Optimizing Culture Conditions and Manipulation of Nutrient Levels

In general, the composition of the medium and nutritional elements significantly affects cultural growth and the synthesis of secondary metabolites. In cell cultures of *P. hexandrum*, the pH of the medium and the concentration of IAA were related to the accumulation of podophyllotoxin [135]. Moreover, both B5 and MS media were reported to be useful for cell growth and the synthesis of this lignan, respectively. In addition, the amount of nitrogen (60 mM) and phosphate (1.25 mM) with a combination of ammonium salts and nitrate at the ratio of 1:2 showed optimum results for the lignan production, and glucose was indicated as a source of carbon [132, 136].

In a recent study, the effects of the micronutrients Na^{+}, K^{+}, Fe^{+2}, and Mn^{+2} were correlated in the production of podophyllotoxin in vitro seedlings of *P. hexandrum* [100]. The increase in the levels of Fe^{+2} and Mn^{+2} in the culture medium enhanced the yield of lignan by 60% and 34%, respectively. Another interesting finding was the enhancement of the activity of the enzymes phenylalanine ammonia lyase (PAL) and cinnamyl alcohol-dehydrogenase (CAD) by Fe^{+2} and Mn^{+2}. The highest activity levels of both enzymes were more correlated with Fe^{+2} leading to an increase of 2.66-fold for PAL and 1.76-fold for CAD. Based on these results, the authors suggest that Fe^{+2} has a role in the formation of coniferyl alcohol, a key precursor in the biosynthetic pathway of podophyllotoxin, as discussed in Sect. 4 [137].

10.2 Precursor Feeding

An exogenous supply of a biosynthetic precursor or intermediate compounds to the culture medium may increase the yield of the desired product. Phenylpropanoid precursors, such as phenylalanine, tyrosine, cinnamic acid, caffeic acid, coumaric acid, ferulic acid, and coniferin, as well as the compound 3,4-methylenedioxycinnamic acid were used for the improvement of podophyllotoxin in root-derived-cells from *P. hexandrum* [138]. Among them, only coniferin induced an increase by 12.8-fold in the content of this lignan. In another study, the content of podophyllotoxin also increased after feeding the cell cultures with coniferyl alcohol complexed with β-cyclodextrin [139].

Coniferin is not commercially available but it has been found in the roots of *Linum flavum*. The hairy roots of *L. flavum* were cocultured with *P. hexandrum* cell cultures in order to offer an alternative source of coniferin [140]. The latter was produced by the hairy roots and used by the cell cultures of *Podophyllum* for the production of podophyllotoxin.

Feeding experiments showed that the addition of tryptophan (250 mg L^{-1}), an indirect precursor in the biosynthesis of podophyllotoxin, resulted in the accumulation of 2.7 times more podophyllotoxin than that control from of untreated cells [141]; however, cell growth decreased, though biomass yield did not improve with other concentrations of tryptophan.

10.3 Elicitation

Elicitors are signals triggering the formation of secondary metabolites. The treatment with elicitors can stimulate the production of these metabolites in plant cell cultures as well as activate the genes involved in their biosynthesis. In the cell suspension cultures of *P. hexandrum*, the reduction in cell viability and biomass as well as in podophyllotoxin content was associated with the browning of the medium and clumping of the cells [142]. Such browning may be attributed to the presence of quinones derived from the oxidation of phenolic compounds, which are produced or result from existing biosynthetic pathways, wounding of the tissues, and medium composition deficiencies [143]. These problems could be avoided by the addition of polyvinylpyrrolidone (PVP) and pectinase to the MS culture medium, supplemented with IAA, which inhibited the staining of the medium and allowed the cells to grow into the suspensions as smaller cell aggregates and single cells. Although pectinase has been reported to act as an elicitor to the production of ursene derivatives, in the cell suspensions of *P. hexandrum*, it was used only to initiate the cultures and then removed without affecting the production of podophyllotoxin.

The production of podophyllotoxin was reported using embryogenic cells and adventitious root cultures of *P. peltatum* [133]. Both systems were subjected to a treatment with methyl jasmonate (20 μM) which resulted in the enhancement of the lignan content. In another study, salicylic acid and methyl jasmonate were found to increase the production of podophyllotoxin in the cell cultures of *P. hexandrum*, whereas the former resulted in a significant increase [132].

In addition, elicitation with methyl jasmonate (100 μM) was used to investigate whether *P. hexandrum* cell cultures showed changes in the proteome profile [144]. The treatment led to an increase in podophyllotoxin accumulation as well as offered a model system to profile modulations on proteins. This model could be used, in the future, to study the proteins closely related to phenylpropanoid/monolignol biosynthesis and other defense responses.

10.4 Immobilized Cells

Immobilization of cells by polymers is another approach used to improve the production of secondary metabolites [145]. It was attempted with *P. hexandrum* cell cultures using calcium alginate combined with L-phenylalanine and L-tyrosine [138]. However, there was no increase on the podophyllotoxin in the cultures.

10.5 Transgenic Cultures

Agrobacterium-mediated transformation has been used to improve methods of inserting foreign and novel genes into the plants. In addition, it offers an alternative system for the production of secondary metabolites. Assessment of the genetic transformation of *Podophyllum* has already been reported. According to an early record [146], some members of Berberidaceae including *Podophyllum* were not susceptible to infection with *A. tumefaciens* strain ChrIIb. The inoculated explants failed to produce crown galls. There has been limited research on genetic engineering of *Podophyllum* with only a few reports of transgenic hairy roots and plants.

Different strains of *Agrobacterium rhizogenes* (LBA 9402, R 1601, and 8196) and *A. tumefaciens* (C 58, T 37, and 1065) were used to infect organs and tissues of *P. hexandrum* in order to evaluate whether *Agrobacterium* per se could be used for gene transfer in this species [51, 147]. The explants were harvested from cultivated seed-derived *P. hexandrum* cv. Majus plants maintained at the University of Nottingham (UK). The enhancement of virulence of *A. rhizogenes* strain R 1601 was attempted by the addition of acetosyringone (100 μM) to the bacterial cell culture. Acetosyringone (4-acetyl-2,6-dimethoxyphenol) increased *Agrobacterium*-mediated transformation in many plants including some recalcitrant species. The production of hairy roots was not observed on the explants inoculated with *A. rhizogenes* strain R 1601 alone or in combination with acetosyringone. Moreover, the expression of the *gus* reporter gene was not detected on the explants from in vitro seedlings and inoculated with any of the evaluated strains of *A. tumefaciens*. The non-responsiveness of *P. hexandrum* to *Agrobacteria* inoculation may be caused, in part, by the presence of lignans in this species, leading to an inactivation of the *Agrobacteria*. It was speculated that the cytotoxicity of the aryltetralin lignans present in *Podophyllum* species may have blocked bacterial infection

[51]. Furthermore, failure to transform roots from axenic cultures of *P. hexandrum* with *A. rhizogenes* strain LBA 9402 had previously been reported [148]. Interestingly, Giri et al. [149] reported that different strains of *A. rhizogenes* (A4, 15834, and K 599) were not only effective in transforming *P. hexandrum*, but the transgenic callus synthesis podophyllotoxin showed a threefold increase compared to the controls.

It has been also reported that *A. tumefaciens*-mediated transformation of *P. peltatum* [150]. In this study, embryogenic callus was co-cultivated with *A. tumefaciens* harboring a binary vector pBI 121 carrying the genes *gus* and *npt II*. It was reported that transformation efficiency was enhanced by the addition of acetosyringone (50 μM) during the infection by the *Agrobacterium* used. Molecular work confirmed that both genes, *gus* and *npt II*, were integrated into transgenic *P. peltatum* plants.

Recently, Rajesh and collaborators [151] reported the successful regeneration of transgenic *P. hexandrum* plants using an efficient *Agrobacterium*-mediated transformation method. In this study, embryogenic cells were infected with *A. tumefaciens* strain EHA 105 with binary vector pCAMBIA 2301 (CAMBIA, Canberra, Australia) containing the genes *npt II* and *gusA*. According to the authors, this method could be used in the future to genetically engineer *P. hexandrum* to enhance the podophyllotoxin content via the transfer and expression of the genes pinoresinol synthase and secoisolariciresinol dehydrogenase which are involved in the biosynthesis of podophyllotoxin.

Some parameters such as the type of plant and the state of tissue that will be infected, genotype, the *Agrobacterium* strain, the conditions of co-cultivation, and selection and regeneration of transformed tissues are reported to affect the transformation of many plants including *Podophyllum* species [51, 151]. Moreover, the concentration of acetosyringone and type of antibiotic are other parameters that can contribute to the efficiency on transformation of *P. hexandrum* [151].

10.6 Endophytic Fungi as Producers of Podophyllotoxin

The production of podophyllotoxin by endophytes is a strategy that could be considered for commercial application, aiming to meet the ever-increasing demand for this natural product, and, in turn, it would help to preserve wild populations of *Podophyllum*. Further studies will be needed in this area in order to evaluate if the endophytes from *Podophyllum* spp. would be able to produce high yields of podophyllotoxin. Additionally, one might expect that investigation into other species would perhaps allow the isolation and identification of new endophytes.

As far as we know, there are only two reports regarding to the production of podophyllotoxin by fungal endophytes from *Podophyllum* species. These findings are interesting, and it suggested a possible gene transfer between the plant and the endophytic fungus [37]. It has been reported that two strains of *Phialocephala fortinii* were isolated from the rhizomes of wild *P. peltatum* plants [152]. Their cultures produced podophyllotoxin at low yields in the broth culture and the process was reported in a US patent [153]. Furthermore, this lignan was found to be produced by an endophyte of *P. hexandrum* identified as *Trametes hirsuta* [154].

11 Conclusions

Podophyllum species and its lignans, especially podophyllotoxin, are important natural products with a diverse spectrum of biological activities. Podophyllotoxin was the precursor for the development of semisynthetic drugs that are successfully used clinically use for cancer therapy. To date, this lignan remains as a starting material for the production of these anticancer drugs as well as a lead compound for the development of new drugs. The ever-increasing demand for this lignan by the pharmaceutical industry has endangered the Asian *Podophyllum* spp. Podophyllotoxin has already been produced by plant cell cultures of *P. hexandrum*, *P. peltatum*, and *P. versipellis* although their use as a commercial source for the supply of this lignan requires much more effort. For this reason, these species are worth exploring further using different strategies in plant biotechnology in order to improve their lignan content. To improve yields metabolic engineering offers promising perspectives but requires the understanding of the regulation of the secondary metabolite pathways on the levels of product, enzymes, and genes, including aspects of transport and compartmentation. Moreover, a better understanding of the regulatory process would help to influence the in vitro production of the lignans by this approach. Furthermore, conservation strategies should be improved considering the commercial value of the *Podophyllum* species as well as their ethnobotanical uses.

Acknowledgments CG Silva is grateful to Dr PM Dewick who first introduced her to the research on *Podophyllum* lignans and tissue culture of Berberidaceae species at the University of Nottingham (UK). PRV Campana is grateful to FAPEMIG for the fellowship (BIP-00224-15). VL Almeida is grateful to CNPq for the fellowship (249299/2013-5). The authors are grateful to AWA Linghorn for revising the manuscript.

References

1. Dewick PM, Shaw JMH (1988) Podophyllums. Garden 113(5):233–238
2. Dewick PM (1996) Tumour inhibitors from plants. In: Evans WC, Trease GE, Evans WC (eds) Pharmacognosy, 14th edn. WB Saunders, London
3. Cordell GA, Farnsworth NR, Beecher CWW, Soejarto DD, Kinghorn AD, Pezzuto JM, Wall ME, Wani MC, Brown DM, O'Neill MJ, Lewis JA, Tait RM, Harris TJR (1993) Novel strategies for the discovery of plant-derived anticancer agents. In: Kinghorn AD, Balandrin MF (eds) Human medicinal agents from plants. American Chemical Society, Washington, DC
4. The Pharmacopoeia of India (1955) Government of India Press, Calcutta, The Pharmacopoeia of India
5. Karnick CR (1994) Pharmacopoeial standards of herbal plants, vol I. Sri Satguru Publications, New Delhi

6. Meijer W (1974) *Podophyllum peltatum* – may apple a potential new cash-crop plant of Eastern North America. Econ Bot 28:68–72
7. Chaurasia OP, Ballabh B, Tayade A, Kumar R, Kumar GP, Singh SB (2012) *Podophyllum* L.: an endangered and anticancerous medicinal plant – an overview. Indian J Tradit Knowl 11(2):234–241
8. Kaplan IW (1942) Condylomata acuminata. New Orleans Med Surg J 94:388–390
9. Guo R, Mao YR, Cai JR, Wang JY, Wu J, Qiu YX (2014) Characterization and cross-species transferability of EST-SSR markers developed from the transcriptome of *Dysosma versipellis* (Berberidaceae) and their application to population genetic studies. Mol Breed 34:1733–1746
10. Ali M, Sharma V (2013) An urgent need for conservation of *Podophyllum hexandrum* (Himalayan Mayapple): an economically important and threatened plant of cold desert of Ladakh, India. J Bio Chem Res 30(2):741–747
11. Prakash V, Bisht H (2010) Genetic polymorphism in inter population variation of *Podophyllum hexandrum* Royle: an endangered medicinal plant of Himalaya, India. Rep Opin 2(7):55–58
12. Broomhead AJ, Dewick PM (1990) Tumour-inhibitory aryltetralin lignans in *Podophyllum versipelle*, *Diphylleia cymosa* and *Diphylleia grayi*. Phytochemistry 29(12):3831–3837
13. Choudhary DK, Kaul BL, Khan S (1998) Cultivation and conservation of *Podophyllum hexandrum* – an overview. J Med Aromat Plant Sci 20:1071–1073
14. Konuklugil B, Shaw JMH (1996) Alpines and medicines: a look at *Podophyllum* and *Linum*. Q Bull Alp Gard Soc 64:334–340
15. Bedir E, Khan I, Moraes RM (2002) Bioprospecting for podophyllotoxin. In: Janick J, Whipkey A (eds) Trends in new crops and new uses. ASHS Press, Alexandria
16. Shaw JMH (2002) The genus *Podophyllum*. Part III. In: Stearn WT, Green PS, Mathew B (eds) The genus *Epimedium* and other herbaceous Berberidaceae. Royal Botanic Gardens, Kew
17. Shaw JMH (1999) Variation in *Podophyllum hexandrum* and its nomenclatural consequences. In: Andrews S, Leslie AC, Alexander C (eds) Taxonomy of cultivated plants, third international symposium. Royal Botanic Gardens, Kew
18. Terabayashi S (1982) Systematic consideration of the Berberidaceae. Acta Phytotaxonom et Geobot 33:355–370
19. Loconte H, Estes JR (1989) Phylogenetic systematics of Berberidaceae and Ranunculales (Magnoliidae). Syst Bot 14(4):565–579
20. Stearn WT (1993) Berberidaceae. In: Heywood VH (ed) Flowering plants of the world. BT Batsford, London
21. Kim Y, Kim S, Kim CH, Jansen RK (2004) Phylogeny of Berberidaceae based on sequences of the chloroplast gene *ndhF*. Biochem Syst Ecol 32:291–301
22. Maqbol M (2011) Mayapple: a review of the literature from a horticultural perspective. J Med Plants Res 5(7):1037–1045
23. Shaw JMH (1996) A taxonomic revision of *Podophyllum* L. (Berberidaceae). MPhil. thesis, University of Nottingham
24. Shaw JMH (1999) New taxa, combinations and taxonomic notes on *Podophyllum* L. New Plantsman 6(3):158–165
25. Hartwell JL (1971) Plants used against cancer: a survey. Lloydia 34(4):431–435
26. Mulik MB, Laddha KS (2015) Isolation and characterization of aryltetraline type lignan from roots of *Podophyllum emodi*. Int J Health Sci Res 5(8):537–540
27. Sharma LK, Kumar A (2007) Traditional medicinal practices of Rajasthan. Indian J Tradit Knowl 6(3):531–533
28. Dymock W, Warden CJH, Hooper D (1890) Pharmacographica indica. Keegan Paul, London
29. Kapahi BK (1990) Ethno-botanical investigation in Lahaul (Himachal Pradesh). J Econ Tax Bot 14(1):49–55
30. Kelly MG, Hartwell JL (1954) The biological effects and the chemical composition of podophyllin: a review. J Nat Cancer Inst 14(4):967–1010
31. Hartwell JL (1967) Plant used against cancer: a survey. Lloydia 30(4):379–436
32. Sanecki KN (1996) The book of herbs. Selectabook, London

33. Hamel PB, Chiltoskey MU (1975) Cherokee plants. Herald Publishing Co., Sylva
34. Simpson BB, Ogorzaly M (1995) Medicinal plants. In: Prancan KM, Barter PW, Lurhs M (eds) Economic botany: plants in our world, 2nd edn. McGraw-Hill, New York
35. Karuppaiya P, Tsay HS (2015) Therapeutic values, chemical constituents and toxicity of Taiwanese *Dysosma pleiantha*: a review. Toxicol Lett 236:90–97
36. Singh HB, Prasad P, Rai L (2002) Folk medicinal plants in Sikkim Himalayas of India. Asian Folk Stud 61:295–310
37. Dewick PM (2009) Medicinal natural products: a biosynthetic approach. Wiley, West Sussex
38. Kusari S, Lamsho M, Spiteller M (2009) *Aspergillus fumigatus* Fresenius, an endophytic fungus from *Juniperus communis* L. Horstmann as a novel source of the anticancer pro-drug deoxypodophyllotoxin. J Appl Microbiol 107:1019–1030
39. Suzuki S, Umezawa T (2007) Biosynthesis of lignans and norlignans. J Wood Sci 53:273–284
40. Satake H, Koyama T, Bahabadi SE, Matsumoto E, Ono E, Murata J (2015) Essences in metabolic engineering of lignan biosynthesis. Metabolites 5:270–290
41. Marques JV, Kim KW, Lee C, Costa MA, May GD, Crow JA, Davin LB, Lewis NG (2013) Next generation sequencing in predicting gene function in podophyllotoxin biosynthesis. J Biol Chem 288:466–479
42. Davin LB, Lewis NG (2003) A historical perspective on lignan biosynthesis: monolignol, allylphenol and hydroxycinnamic acid coupling and downstream metabolism. Phytochem Rev 2:257–288
43. Nakatsubo T, Mizutani M, Suzuki S, Hattori T, Umezawa T (2008) Characterization of *Arabidopsis thaliana* pinoresinol reductase, a new type of enzyme involved in lignan biosynthesis. J Biol Chem 283:15550–15557
44. Xia ZQ, Costa MA, Pélissier HC, Davin LB, Lewis NG (2001) Secoisolariciresinol dehydrogenase purification, cloning, and functional expression. J Biol Chem 276:12614–12623
45. Wankhede DP, Biswas DK, Rajkumar S, Sinha AK (2013) Expressed sequence tags and molecular cloning and characterization of gene encoding pinoresinol/lariciresinol reductase from *Podophyllum hexandrum*. Protoplasma 250(6):1239–1249
46. Kumar P, Pal T, Sharma N, Kumar V, Sood H, Chauhan RS (2015) Expression analysis of biosynthetic pathway genes vis-à-vis podophyllotoxin content in *Podophyllum hexandrum* Royle. Protoplasma 252(5):1253–1262
47. Lata H, Moraes RM, Douglas A, Scheffler BE (2002) Assessment of genetic diversity in *Podophyllum peltatum* by molecular markers. In: Janick J, Whipkey A (eds) Trends in new crops and new uses. ASHS Press, Alexandria
48. Rao RR, Hajra PK (1993) Podophyllaceae. In: Sharma BD, Balakrishnan NP (eds) Flora of India, vol 1. Botanical Survey of India, Calcutta
49. Jackson DE, Dewick PM (1985) Tumour-inhibitory aryltetralin lignans from *Podophyllum pleianthum*. Phytochemistry 24:2407–2409
50. Kofod H, Jørgensen C (1953) Dehydropodophyllotoxin, a new compound isolated from *Podophyllum peltatum* L. Acta Chem Scand 8:1296–1297
51. Silva CG (2000) Tissue culture and phytochemical studies of *Podophyllum*, *Diphylleia* and *Passiflora* species. PhD thesis, University of Nottingham
52. Guan BC, Gong X, Zhou SL (2011) Development and characterization of polymorphic microsatellite markers in *Dysosma pleiantha* (Berberidaceae). Am J Bot 9:210–212
53. Jiang W, Chen L, Pan Q, Qiu Y, Shen Y, Fu C (2012) An efficient regeneration system via direct and indirect organogenesis for the medicinal plant *Dysosma versipellis* (Hance) M. Chen and its potential as a podophyllotoxin source. Acta Physiol Plant 34:631–639
54. Jiang R, Zhou J, Hon P, Li S, Zhou Y, Li L, Ye W, Xu H, Shaw P, But PP (2007) Lignans from *Dysosma* versipellis with inhibitory effects on prostate cancer cell lines. J Nat Prod 70:283–286
55. Yang Z, Wu Y, Wu S (2016) A combination strategy for extraction and isolation of multi-component natural products by systematic two-phase solventextraction-13C nuclear magnetic resonance pattern recognition and following conical counter-current chromatography

separation: podophyllotoxins and flavonoids from *Dysosma versipellis* (Hance) as examples. J Chromatogr A 1431:184–196
56. Chatterjee R, Chakravarti SC (1952) Resin *sikkimensis*. I Sikkimotoxin: a lactone from *Podophyllum sikkimensis* R Chatterjee et Mukerjee. J Am Pharm Assoc 41(8):415–419
57. Chatterjee R (1952) Indian *Podophyllum*. Econ Bot 6(4):342–354
58. Butt TI, Amjad MS (2015) Ethnopharmacology, phytochemistry and cytotoxicity of emerging biotechnological tool: Mayapple (*Podophyllum hexandrum* L.: Berberidaceae). J Coast Life Med 3(8):652–657
59. Castro MA, del Corral JMM, Gordaliza MM, Zurita AG, de la Puente ML, Betancur-Galvis LA, Sierra J, San Feliciano A (2003) Synthesis, cytotoxicity and antiviral activity of podophyllotoxin analogues modified in the E-ring. Eur J Med Chem 38:899–911
60. Li S, Yuan W, Yang P, Antoun MD, Balick MJ, Cragg GM (2010) Pharmaceutical crops: an overview. Pharm Crops 1:1–17
61. Liu L, Tian J, Qian K, Zhao X, Morris-Natschke SL, Yang L, Nan X, Tian X, Lee K-H (2015) Recent progress on C-4-modified podophyllotoxin analogs as potent antitumor agents. Med Res Rev 35(1):1–62
62. Liu YQ, Yang L, Tian X (2007) Podophyllotoxin: current perspectives. Curr Bioact Compd 3:37–66
63. Botta B, Monache GD, Misiti D, Vitali A, Zappia G (2001) Aryltetralin ligans: chemistry, pharmacology and biotransformations. Curr Med Chem 8:1363–1381
64. Apers S, Vlietink A, Pieters L (2003) Lignans and neolignans as lead compounds. Phytochem Rev 2:201–207
65. Charlton JL (1988) Antiviral activity of lignans. J Nat Prod 61:1447–1451
66. King ML, Sullivan MM (1946) The similarity of the effect of podophyllin and colchicine and their use in the treatment of Condylomata acuminata. Science 104:244–245
67. Sullivan BJ, Wechsler HJ (1947) The cytological effects of podophyllin. Science 105:433
68. Lv M, Xu H (2011) Recent advances in semisynthesis, biosynthesis, biological activities, mode of action, and structure-activity relationship of podophyllotoxins: an update (2008–2010). Mini Rev Med Chem 11:901–909
69. Lee K, Xiao Z (2003) Lignans in treatment of cancer and others diseases. Phytochem Rev 2:341–362
70. Brandão HN, David JP, Couto RD, Nascimento JAP (2010) Química e farmacologia de quimioterápicos antineoplásicos derivados de plantas. Quim Nova 33(6):1359–1369
71. Lee KH, Xiao Z (2005) Podophyllotoxins and analogs. In: Cragg GM, Kingston DGI, Newman DJ (eds) Anticancer agents from natural products, 1st edn. CRC Press, Boca Raton
72. Bedows E, Hatfield GM (1982) An investigation of the antiviral activity of *Podophyllum peltatum*. J Nat Prod 45(6):725–729
73. Lee CT, Lin VC, Zhang S, Zhu X, VanVliet D, Hu H, Beers SA, Wang ZL, Cosentino M, Morris-Natschke SL, Lee K (1997) Anti-AIDS agents. 291. Anti-HIV activity of modified podophyllotoxin derivatives. Bioorg Med Chem Lett 7(22):2897–2902
74. MacRae WD, Hudson JB, Towers GH (1989) The antiviral action of lignans. Planta Med 55 (6):531–535
75. Stahelin H, von Wartburg A (1991) The chemical and biological route from podophyllotoxin glucoside to etoposide: Ninth Cain Memorial Award Lecture. Cancer Res 51:1–15
76. Harmatha J, Dinan L (2003) Biological activities of lignans and stilbenoids associated with plant–insect chemical interactions. Phytochem Rev 2:321–330
77. MacRae WR, Towers GHN (1984) Biological activities of lignans. Phytochemistry 23(6): 1207–1220
78. Miyazawa M, Fukuyama M, Yoshio K, Kato T, Ishikawa Y (1999) Biologically active components against *Drosophila melanogaster* from *Podophyllum hexandrum*. J Agric Food Chem 47:51–085110
79. Inamori Y, Kubo M, Tsujibo H, Ogawa M, Baba K, Kozawa M, Fujita E (1986) The biological activities of podophyllotoxin compounds. Chem Pharm Bull 34(9):3928–3932

80. Russell GB, Singh P, Fenemore PG (1976) Insect control chemicals from plants III. Toxic lignans from *Libocedrus bidwillii*. Aust J Biol Sci 29:99–103
81. Rocha MP, Campana PRV, Almeida VL, Silva CG (2015) Berberidaceae species as source of inhibitors of acetylcholinesterase and antioxidants. Planta Med 82:PC21
82. Ayres DC, Loike JD (1990) Lignans: chemical, biological and clinical properties. Cambridge University Press, Cambridge
83. Bastos JK, Kopycki WJ, Burandt CL Jr, Nanayakkara NP, McChesney JD (1995) Quantitative determination of podophyllotoxin and related compounds in *Podophyllum* species by reverse phase high performance liquid chromatography. Phytochem Anal 6(2):101–105
84. Jackson DE, Dewick PM (1984) Aryltetralin lignans from *Podophyllum hexandrum* and *Podophyllum peltatum*. Phytochemistry 23:1147–1152
85. Sharma E, Arora BS (2015) Identification of aryltetralin lignans from *Podophyllum hexandrum* using hyphenated techniques. Int J Pharm Sci Drug Res 7(1):83–88
86. Opletal L, Sovová H (2008) TLC of lignans. In: Waksmundzka-Hajnos M, Sherma J, Kowalska T (eds) Thin layer chromatography in phytochemistry. CRC Press, Boca Raton
87. Kumar J (2015) Studies on in vitro production of podophyllotoxin from *Podophyllum hexandrum* using callus and cell cultures. PhD thesis, University of Biotechnology and Management Sciences
88. Srinivasan V, Roberts SC, Shuler ML (1997) Combined use of six-well polystyrene plates and thin layer chromatography for rapid development of optimal plant cell culture processes: application to taxane production by *Taxus* sp. Plant Cell Rep 16(9):600–604
89. Arumugan N, Bhojwani SS (1990) Somatic embryogenesis in tissue cultures of *Podophyllum hexandrum*. Can J Bot 68(3):487–491
90. Wagner H, Bladt S (2001) Plant drug analysis: a thin layer chromatography atlas, 2nd edn. Springer, Berlin
91. Medrado HHS, David JM, David JP, Brandão HN (2015) Distribuição, atividade biológica, síntese e métodos de purificação de podofilotoxina e seus derivados. Quim Nova 28(2):243–258
92. Zhao L, Liu ZL, Fan PC, Zhang ZW, Liu X, Zhan YJ, Tian X (2009) HPLC-DAD and HPLC-ESI-MS separation, determination and identification of the spin-labeled diastereoisomers of podophyllotoxin. J Sep Sci 32(9):1323–1332
93. Zhu Q, Liub F, Xub M, Linb X, Wang X (2012) Ultrahigh pressure extraction of lignan compounds from *Dysosma versipellis* and purification by high-speed counter-current chromatography. J Chromatogr B 905:145–149
94. Cassady JM, Douros JD (1980) Antitumor agents based on natural product models. Academic, New York
95. Ding Z, Deng S, Zhao X, Li F, Cao Q, Zhu H (2005) Determination of podophyllotoxin and 4′-demethylepipodophyllotoxin by capillary electrophoresis. Chromatographia 62:209–213
96. Zhang J, Shi S, Peng M, Juan Pan J, Wan H, Zhou C (2014) Simultaneous determination of five active compounds from *Dysosma* difformis roots by HPLC. J Liq Chromatogr Relat Technol 37:1226–1236
97. Pandey H, Nandi SK, Palni LMS (2013) Podophyllotoxin content in leaves and stems of *Podophyllum hexandrum* Royle from Indian Himalayan region. J Med Plant Res 7:3237–3241
98. Zhen L, Guangyi Y, Minghui L, Ling Z (2013) Determination of Podophllotoxin in *Dysosma versipellis* by UPLC. China Pharmacist 16(4):548–549
99. Mishra N, Acharya R, Gupta AP, Singh B, Kaul VK, Ahuja OS (2005) A simple microanalytical technique for determination of podophyllotoxin in *Podophyllum hexandrum* roots by quantitative RP-HPLC and RP-HPTLC. Curr Sci 18(9):1372–1373
100. Li MF, Li W, Yang DL, Zhou LL, Li TT, Su XM (2013) Relationship between podophyllotoxin accumulation and soil nutrients and the influence of Fe and Mn on podophyllotoxin biosynthesis in *Podophyllum hexandrum* tissue culture. Plant Physiol Biochem 71:96–102
101. Pandey H, Kumar A, Palni LMS, Nandi SK (2015) Podophyllotoxin content in rhizome and root samples of *Podophyllum hexandrum* Royle populations from Indian Himalayan region. J Med Plants Res 9(9):320–325

102. Och M, Och A, Ciesla T, Kubrak T, Pecio T, Stochmal A, Kock J, Bogucka-Kocka A (2016) Study of cytotoxic activity, podophyllotoxin and deoxypodophyllotoxin content in selected *Juniperus* species cultivated in Poland. Pharm Biol 53(6):831–837
103. Chaudhri II (1956) Medicinal plants from West Pakistan: *Podophyllum emodi* L. Pakistan J Sci 8(5):230–233
104. Hooper D (1913) Drug culture in British India. Pharm J 90:552–554
105. Troup RS (1915) A note on the cultivation of *Podophyllum emodi*. Indian Forester 41:361–365
106. Badhwar RL, Sharma BK (1963) A note on the germination of *Podophyllum* seeds. Indian Forester 89:445–447
107. Krishnamurthy T, Karira GV, Sharma BK, Bhatia K (1965) Cultivation and exploitation of *Podophyllum hexandrum* Royle. Indian Forester 91(7):470–475
108. Nautiyal MC, Rawat AS, Bhadula SK, Purohit AN (1987) Seed germination in *Podophyllum hexandrum*. Seed Res 15(2):206–209
109. Prasad P (2001) Impact of cultivation on active constituents of the medicinal plants *Podophyllum hexandrum* and *Aconitum heterophyllum* in Sikkin. Plant Genet Resour Newsl 124:33–35
110. Krochmal A (1968) Medicinal plants and Appalachia. Econ Bot 22:332–337
111. Shaw JMH (1996) Notes on *Podophyllum* (Berberidaceae) as potential British aliens. BSBI News 73:43–45
112. Rahman MMA (1988) Studies on the biosynthesis of lignans and their production in plant cell cultures. PhD thesis, University of Nottingham
113. Kharkwal AC, Kushwaha R, Prakash O, Ogra RK, Bhattacharya A, Nagar PK, Ahuja PS (2008) An efficient method of propagation of *Podophyllum hexandrum*: an endangered medicinal plant of the Western Himalayas under ex situ conditions. J Nat Med 62:211–216
114. Nadeem M, Palni LMS, Purohit AN, Pandey H, Nandi SK (2000) Propagation and conservation of *Podophyllum hexandrum* Royle: an important medicinal herb. Biol Conserv 92:121–129
115. Nautiyal MC (1982) Germination studies on some high altitude medicinal plant species. In: Kaushik P (ed) Indigenous medicinal plants symposium. Today and Tomorrow's Printers and Publishers, New Delhi, pp 107–112
116. Choudhary DK, Kaul BL, Khan S (1996) Breaking seed dormancy of *Podophyllum hexandrum* Royle. Ex. Camb. (syn. *P. emodi* Wall Ex. Honigberger). J Non-Timber Prod 3(1/2):10–12
117. Sharma R, Sharma S, Sharma SS (2006) Seed germination behavior of some medicinal plants of Lahaul and Spiti cold desert (Himachal Pradesh): implications for conservation and cultivation. Curr Sci 90(8):1113–1118
118. Heyenga AG (1989) Tissue culture of *Podophyllum hexandrum* and production of anticancer lignans. PhD thesis, University of Nottingham
119. Chakraborty A, Bhattacharya D, Ghanta S, Chattopadhya S (2010) An endangered protocol for in vitro regeneration of *Podophyllum hexandrum*, a critically endangered medicinal plant. Indian J Biotechnol 9:217–220
120. Moraes-Cerdeira RM, Burandt CL, Bastos JK, Nanayakkara NPD, McChesney JD (1998) In vitro propagation of *Podophyllum peltatum*. Planta Med 64(1):42–45
121. Kim YS, Lim S, Choi YE, Anbazhagan VR (2007) High frequency plant regeneration via somatic embryogenesis in *Podophyllum peltatum* L., an important source of anticancer drug. Curr Sci 92(5):662–666
122. Chuang MJ, Chang WC (1987) Embryoid formation and plant regeneration in callus cultures derived from vegetative tissues of *Dysosma pleiantha*. J Plant Physiol 128:279–283
123. Sultan P, Shawl AS, Ramteke PW, Jan A, Chisti N, Jabeen N, Shabir S (2006) In vitro propagation for mass multiplication of *Podophyllum hexandrum*: a high value medicinal herb. Asian J Plant Sci 5(2):179–184
124. Silva CG, Power JB, Lowe KC, Davey MR (1998) Plant regeneration from root explants of the medicinal plant *Podophyllum hexandrum* cultured in liquid medium. J Exp Bot 49:88–89

125. Petersen M, Alfermann AW (2001) The production of cytotoxic lignans by plant cell cultures. Appl Microbiol Biotechnol 55:135–142
126. Arro RRJ, Alfermann AW, Medarde M, Petersen M, Pras N, Woolley JG (2002) Plant cell factories as a source for anti-cancer lignans. Phytochem Rev 1:27–35
127. Kadkade PG (1981) Formation of podophyllotoxins by *Podophyllum peltatum* tissue cultures. Naturwissenschaften 68:481–482
128. Kadkade PG (1982) Growth and podophyllotoxin production in callus tissues of *Podophyllum peltatum*. Plant Sci Lett 25:107–115
129. van Uden W, Pras N, Visser JF, Malingré TM (1989) Detection and identification of podophylltoxin produced by cell cultures derived from *Podophyllum hexandrum* Royle. Plant Cell Rep 8:165–168
130. Heyenga AG, Lucas JA, Dewick PM (1990) Production of tumor-inhibitory lignans in callus cultures of *Podophyllum hexandrum*. Plant Cell Rep 9:382–385
131. Giri A, Narasu ML (2000) Production of podophyllotoxin from *Podophyllum hexandrum*: a potential natural product for clinically useful anticancer drugs. Cytotechnology 34:17–26
132. Majumder A, Jha S (2009) Biotechnological approaches for the production of potential anticancer leads podophyllotoxin and paclitaxel: an overview. J Biol Sci 1(1):46–69
133. Anbazhagan VR, Ahn CH, Harada E, Kim YS, Choi YE (2008) Podophyllotoxin production via cell and adventitious root cultures of *Podophyllum peltatum*. In Vitro Cell Dev Biol – Plant 44:494–501
134. Liu L, Gao ZP, Jiang PF (1997) Comparison of chemical composition between the tissue culture plants and the wild plants. Chin J Chin Mater Med 22:593–594
135. Chattopadhyay S, Srivastava AK, Bhojwani SS, Bisaria VS (2001) Development of suspension culture of *Podophyllum hexandrum* for production of podophyllotoxin. Biotechnol Lett 23:2063–2066
136. Chattopadhyay S, Mehra RS, Srivastava AK, Bhojwani SS, Bisaria VS (2003) Effect of major nutrients on podophyllotoxin production in *Podophyllum hexandrum* suspension culture. Appl Microbiol Biotechnol 60:541–546
137. Jackson DE, Dewick PM (1984) Biosynthesis of *Podophyllum* lignans-I. Cinnamic acid precursors of podophyllotoxin in *Podophyllum hexandrum*. Phytochemistry 3:1029–1035
138. van Uden W, Pras N, Malingré TM (1990) On the improvement of the podophyllotoxin production by phenylpropanoid precursor feeding to cell cultures of *Podophyllum hexandrum* Royle. Plant Cell Tiss Org Cult 23:217–224
139. Woerdenbag HJ, van Uden W, Frijlink HW, Lerk CF, Pras N, Malingré TM (1990) Increased podophyllotoxin production in *Podophyllum hexandrum* cell suspension cultures after feeding coniferyl alcohol as a β-cyclodextrin complex. Plant Cell Rep 9:97–100
140. Lin H, Kian HK, Doran PM (2003) Production of podophyllotoxin using cross-species coculture of *Linum* flavum hairy roots and *Podophyllum hexandrum* cell suspensions. Biotechnol Prog 19:1417–1426
141. Majumder A (2012) Influence of an indirect precursor on podophyllotoxin accumulation in cell suspension cultures of *Podophyllum hexandrum*. Plant Tissue Cult Biotechnol 22(2): 171–177
142. Chattopadhyay S, Srivastava AK, Bisaria VS (2002) Optimization of culture parameters for production of podophyllotoxin in suspension culture of *Podophyllum hexandrum*. Appl Microbiol Biotechnol 102(103):381–393
143. Preece JE, Compton ME (1991) Problems with explant exudation in micropropagation. In: Bajaj YPS (ed) Biotechnology in agriculture and forestry, vol 17. Springer, Berlin
144. Bhattacharyya D, Sinha R, Ghanta S, Chakraborty A, Hazra S, Chattopadhyay S (2012) Proteins differentially expressed in elicited cell suspension culture of *Podophyllum hexandrum* with enhanced podophyllotoxin content. Proteome Sci 10:34
145. Verpoorte R, Contin A, Memelink J (2002) Biotechnology for the production of plant secondary metabolites. Phytochem Rev 1:13–25
146. de Cleene M, de Ley J (1976) The host range of crown gall. Bot Rev 42(4):389–466

147. Silva CG, Davey MR, Power JB (2001) Evaluation of *Agrobacterium* for gene transfer in *Podophyllum hexandrum*. In: Kreis W (ed) Traits, tracks and traces. Druckerei S. Kriegelstein, Erlangen
148. Oostdam A, Mol JNM, van der Plas LHW (1993) Establishment of hairy root cultures of *Linum* flavum producing the lignan 5-methoxypodophyllotoxin. Plant Cell Rep 12:474–477
149. Giri A, Giri CC, Dhingra V, Narasu ML (2001) Enhanced podophyllotoxin production from *Agrobacterium rhizogenes* transformed cultures of *Podophyllum hexandrum*. Nat Prod Lett 15:229–235
150. Anbazhagan VR, Kim YS, Choi YE (2009) Production of transgenic *Podophyllum peltatum* via *Agrobacterium tumefaciens*-mediated transformation. Biol Plant 53:637–642
151. Rajesh M, Jeyaraj M, Sivanandhan G, Subramanyam K, Mariashibu TS, Mayavan S, Dev GK, Anbashagan VR, Manickavasagam M, Ganapathi A (2013) *Agrobacterium*-mediated transformation of the medicinal plant *Podophyllum hexandrum* Royle (syn. *P. emodi* Wall. Ex Hook.f & Thomas). Plant Cell Tiss Org Cult 114:71–82
152. Eyberger AL, Dondapati R, Porter JR (2006) Endophyte fungal isolates from *Podophyllum peltatum* produce podophyllotoxin. J Nat Prod 69(8):1121–1124
153. Porter JL, Eyberger AL (2004) Endophytes for the production of podophyllotoxin. US Patent Appl. 10/455,844
154. Puri SC, Nazir A, Chawla R, Arora R, Riyaz-ul-Hasan S, Anna T, Ahmed B, Verma V, Singh S, Sagar R, Sharma A, Kumar R, Sharma RK, Zazi GN (2006) The endophytic fun*gus Trametes hirsuta* as a novel alternative source of podophyllotoxin and related aryltetralin lignans. J Biotechnol 122(4):494–510

Secondary Metabolite Production in Transformed Cultures

4

Stevioside Glycosides Production from *Stevia rebaudiana* Hairy Root Cultures

Madhumita Kumari and Sheela Chandra

Abstract

The study elucidates the production of hairy roots through *Agrobacterium rhizogenes* mediated transformation of leaf and stem explants of *S. rebaudiana* for secondary metabolite production. Hairy root culture was established successfully using leaf and stem explants of *S. rebaudiana*. Hairy roots grown in Murashige and Skoog (MS) media at temperature 25 °C kept in 16 h photoperiod proved best for stevioside glycosides production. Among different strength of MS and B5 media, full strength MS media produced stevioside 0.412 ± 0.008 mg mL^{-1} and Reb A 0.100 ± 0.008 mg mL^{-1}, whereas hairy roots on B5 media produced stevioside 0.383 ± 0.002 mg mL^{-1} and Reb A 0.098 ± 0.005 mg mL^{-1}. Growth of hairy root was found to be maximal at 25 ± 2 °C and with the increase of temperature, growth of hairy roots decreases. Secondary metabolite production was not much affected by increase in temperature up to 31 °C with approximately similar stevioside content detectable at 25 °C (0.425 ± 0.08 mg mL^{-1}), 28 °C (0.415 ± 0.08 mg mL^{-1}), and at 31 °C (0.386 ± 0.02 mg mL^{-1}). At 35 °C, stevioside content decreases rapidly (0.090 ± 0.02 mg mL^{-1}) and Reb A was completely not detected. Results were validated through ultra high performance liquid chromatography-electrospray ionization mass spectrometry (LC-ESI/MS) studies.

Keywords

Bioproduction • *Stevia rebaudiana* • In vitro cultures • Steviosides glycosides • Biosynthetic pathway

M. Kumari • S. Chandra (✉)
Department of Bio-Engineering, Birla Institute of Technology, Mesra, Ranchi, Jharkhand, India
e-mail: schandra@bitmesra.ac.in

S. Jha (ed.), *Transgenesis and Secondary Metabolism*, Reference Series in Phytochemistry, DOI 10.1007/978-3-319-28669-3_1

Abbreviations

°C	Degree Celsius
%	Percent
2,4-D	2,4-Dichlorophenoxy acetic acid
Ads	Adenine sulphate
B5	Gamborg B5 medium
BAP	6-Benzyladenine
HPLC	High performance liquid chromatography
h	Hour
IAA	Indole-3-acetic acid
IBA	Indole-3-butyric acid
Kn	Kinetin
min	Minutes
MS	Murashige and Skoog's medium
PGRs	Plant growth regulators
rpm	Rotation per minute
SD	Standard deviation
μL	Microlitre

Contents

1 Introduction

Stevia rebaudiana (Bertoni) is a small perennial herb, belonging to the asteraceae family. It is native to certain regions of Paraguay and Brazil in South America. The plant is also known as a honey leaf and candy leaf [1]. *S.rebaudiana* contains diterpene glycosides viz. stevioside and rebaudioside A, which are responsible for its sweet taste with zero calories [2] and are estimated to be 250–300 times sweeter than sucrose. These glycosides possess a number of therapeutics properties in addition to their sweetness value. They regulate the blood glucose level by stimulating insulin secretion so that they can be used as an alternative sweetener by hyperglycemic patients [3, 4]. Steviol glycosides can also be used as an antihypertensive [5], antitumor [6], vasodilator [7], antimicrobial, [8] and neuroprotective drugs [9]. They are heat- and pH-stable with a good shelf life and can be added in cooking, baking, or in beverages. In a number of countries, Stevia was approved as dietary supplements. It might be a source of a number of pharmaceutical drugs. Since Stevia is highly versatile as an additive, it has gained a great boostin popularity in the past few years and is progressively becoming the focal point of attention amongst food and beverage producers. Thus, various therapeutics and sweetening properties are the most important attributes of Stevia, which makes it a commercially important plant.

For commercial cultivation, homogenous range of improved plants is required and plants germinated from seeds show a degree of variability. Also in field conditions, a wide range of variation occurs due to external environmental conditions such as plant pathogen, temperature, drought, and water logging, which leads to variation in composition and sweetening levels [10]. Seeds of Stevia have very poor germination potential [11–13]. In today's world, production of stress tolerant varieties, enhanced plant biomass, and production of medicinally important secondary metabolites are considerable issues. Earlier, vegetative propagation was generally used for cultivation of Stevia. Although this technique is limited, numbers of explants were obtained from a single plant and that may raise possibilities of pathogen accumulation in tissues. These in vitro tissue culture techniques might prove an alternative tool to conventional methods for comparatively rapid multiplication of elite medicinal plantlets, which gives disease-free, resistant plant with high biomass and secondary metabolites.

In plant cell cultures, secondary metabolite augmentation has limiting culture parameters and therefore requirement of knowledge and introduction of new techniques are necessary. Many plant secondary metabolites get accumulated in hairy roots. Therefore, in the recent past hairy root culture has received a lot of attention in research field. The need of taking the research form laboratory to industry has led to the development of this new technology. Genetic engineering of hairy roots has played a key role and has provided new direction to hairy root research. Hairy root cultures are preferred over other methods of transformation as they have fast, hormone-independent growth, are highly branched, lack geotropism, shows lateral branching, are genetic stability and due to the ease of elicitor treatment [14].

Table 1 Secondary metabolites production in hairy root cultures of different plants

Plant species	Secondary metabolite	References
Atropa belladonna	Scopolamine	[16]
Artemisia annua	Artemisinin	[17]
Camptothecaacuminata	Camptothecin	[18]
Datura innoxia	Hysocyamine and scopolamine	[19]
D. quercifolia	Scopolamine and hysocyamine	[20]
D. candida	Scopolamine and hysocyamine	[21]
Duboisialeichhardtii	Scopolamine	[22]
Droseraburmanii	Plumbagin	[23]
Echinacea purpurea	Cichoric acid	[24]
Ginkgo biloba	Ginkgolides	[25]
Glycyrrhizaglabra	Glycyrhizin	[26]
Hyoscyamusniger	Hysocyamine and scopolamine	[19]
Hyoscyamusniger	Scopolamine	[27]
Papaver somniferum	Morphine, codeine	[28]
Panax ginseng	Ginsenosides	[29]
Psoraleacorylifolia	Isoflavones	[30]
Przewalskiatangutica	Tropane alkaloids	[31]
PodophyllumhexandrumRoyle	Podophyllotoxin	[32]
Plumbagorosea	Plumbagin	[33]
Rauvolfiamicrantha	Ajmalicine, Ajmaline	[34]
Rubiatinctoria	Anthraquinone	[35]
Rubiacordifolia	Anthraquinones	[36]
Solanum khasianum	Solasodine	[37]
Stevia rebaudiana	Chlorogenic acid	[38]
Tylophoraindica	Tylophorine	[39]
Withaniasomnifera	Withanolide A	[40]

In vitro culture of plant cells is now a mature technology with successful applications in crop improvement. The major limitation to wide industrial application of plant cell culture is the maintenance of stable cell lines [15]. Hairy roots can synthesize more than a single metabolite and so prove economical for commercial production. A number of secondary metabolites have been reported to be produced from hairy root cultures (Table 1).

Over the last decade, transformed hairy roots have been developed in number of important medicinal plants [41]. Different types of explants like hypocotyls, cotyledons, petioles, and young leaves are most frequently used for *Agrobacterium*-mediated transformation [42, 43]. The extent of secondary metabolite release in hairy root cultures varies between different plant species. As a consequence, much effort has been put into the use of in vitro cultures as one attractive biotechnological strategy for producing this natural compound of commercial interest.

2 Distribution and Localization of Diterpene Glycosides in *S. rebaudiana*

The sweet diterpene glycosides of *Stevia* have been the subject of a number of reviews. The leaves of *S. rebaudiana* contain at least eight diterpene glycosides, viz., stevioside and rebaudiosides. In 1931, isolation of stevioside was done by Bridel and Lavieille [2]. In 1952, the chemical structure of stevioside (Fig. 1) was established and described as an aglycon, steviol with glycoside of three glucose molecules. During the 1970s, other compounds were isolated, including rebaudioside A (Fig. 2), with sweetness potency even higher than stevioside.

Fig. 1 Stevioside

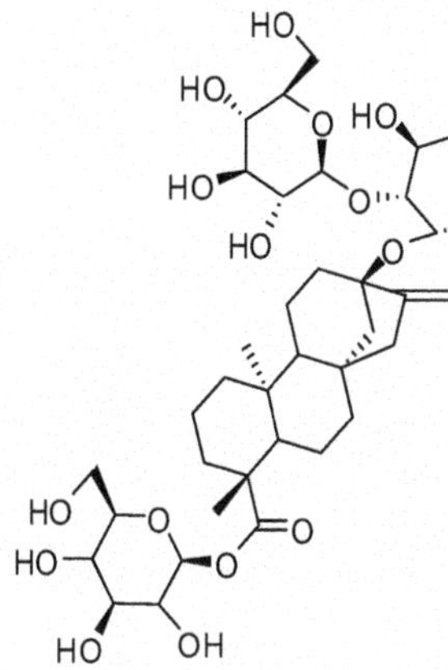

Fig. 2 Rebaudioside A

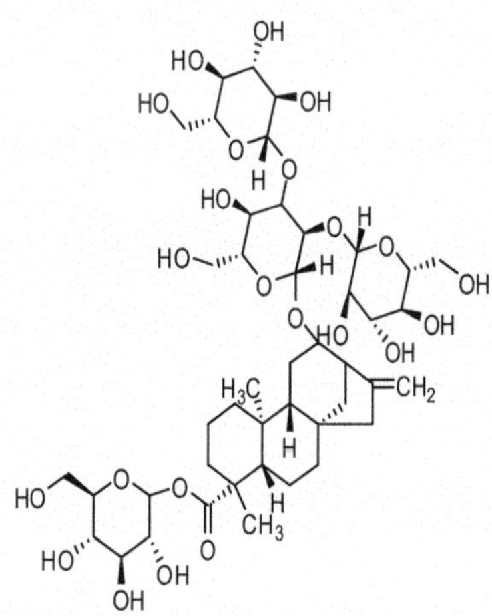

3 Biosynthetic Pathway of Steviol Glycosides

Steviol glycosides biosynthesis pathway shares some common steps with GA (Gibberellic acid) biosynthesis. Steviol glycoside biosynthesis occurs in leaves and transported to different parts. In vivo labeling with [1-13C] glucose and NMR spectroscopy showed that main precursor steviol is synthesized via the plastid localized methylerythritol 4-phosphate (MEP) pathway (Fig. 3) [44].

Aglyconesteviol is glycosylated by various glucosyltransferases present in the cytoplasm. Steviol has two hydroxyl groups, one present at C-19 of C-4 carboxyl

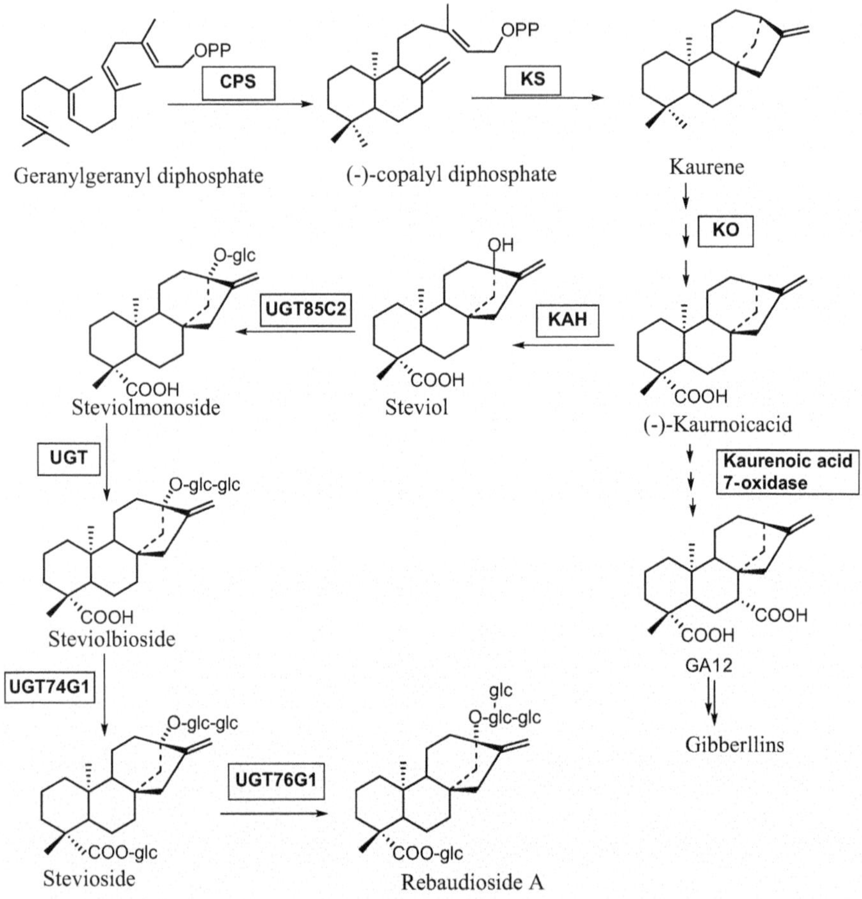

Fig. 3 Biosynthetic Pathway of Steviol Glycosides (Redrawn from Brandle and Telmer [44]). (Abbreviations: copalyl diphosphate synthase (*CPS*), kaurene synthase (*KS*), kaurene oxidase (*KO*), kaurenoic acid 13-hydroxylase (*KAH*))

and other at C-13. Glycosylation starts at C-13 by UGT85C2 which produces steviolmonoside. Shibata et al. [45] used 13-O- and 19-O-methylsteviol as substrates for crude *Stevia* leaf enzyme extracts to determine which active group is glucosylated first. They found that only 19-O-steviol could serve as a substrate and concluded that synthesis of SGs starts with the glucosylation of the 13-hydroxyl of steviol. Steviolmonoside is then glycosylated to produce steviolbioside. UGT (uridine diphosphate-dependent glycosyltransferase) of this step is not yet identified. Finally, stevioside is produced by UGT74G1 by glucosylation at C-19 position (Fig. 3). Rebaudioside A is synthesized by glucosylation of steviosideat C-13 by UGT76G1 [45].

3.1 Cellular Components Involved in Biosynthesis

The precursor of diterpenoids, kaurene is synthesized in the chloroplast by terpene cyclases (Fig. 4). Like all diterpenes, steviol is synthesized from GGDP (geranyl

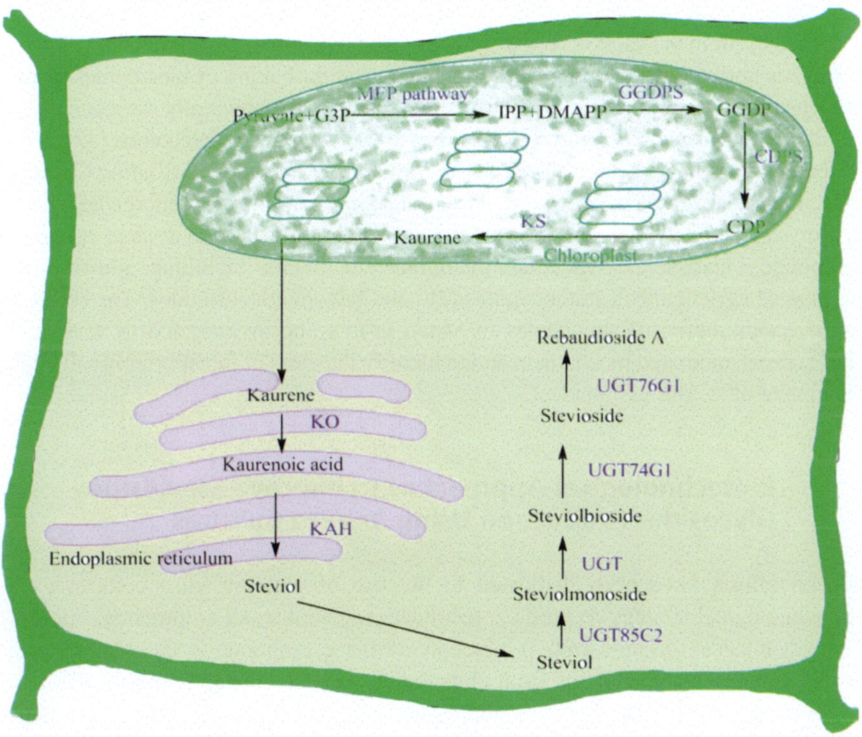

Fig. 4 Cellular components involved in biosynthesis of steviol glycosides

geranyl diphosphate), first by protonation-initiated cyclization to copalyl diphosphate (CDP) by CDP synthase (CPS). Next, kaurene is produced from CDP by an ionization-dependant cyclization catalysed by KS (kaurene synthase).

Kaurene is then converted to steviol by the activity of enzymes present at the membrane of endoplasmic reticulum. It is oxidized in a three-step reaction to kaurenoic acid, by kaurene oxidase (KO), a P450 mono-oxygenase that also functions in GA biosynthesis. Steviol biosynthesis diverges from GA biosynthesis with the hydroxylation of kaurenoic acid by KAH (*ent*-kaurenoic acid 13-hydroxylase). This is the first committed step and the enzyme is of significant interest for use in biotechnology. Steviol synthesizes various steviol glycosides in the cytosol that ultimately accumulate in the vacuole. The aglyconesteviol has two hydroxyl groups, one attached to the C-19 of the C-4 carboxyl and the other attached to the C-13, both of which can be glycosylated.

3.2 Transport of Steviol Glycosides to Vacuole

The final phase of glycoside accumulation is the translocation of glycosylated steviol out of the cytosol and accumulation in the vacuole. In *Stevia*, steviol glycosides are known to occur in the vacuole, but the mechanism by which they are trafficked into the vacuole is not yet understood [46]. Vesicle-mediated trafficking of metabolites seems like a possible scenario for biosynthetic pathways which are physically associated with the ER (endoplasmic reticulum) or organized in ER-associated metabolons [47].

Recent work has revealed an equally important role for the ATP Binding Cassette (ABC) superfamily of transporters. ABC transporters are directly energized by the hydrolysis of ATP [48] and have been shown to transport a diverse array of compounds across the vacuolar membrane in plants including glutathione-conjugated agrichemicals anthocyanins [49] and flavone glucuronides. The energetics of accumulation of steviosides in *Stevia* need to be investigated to confirm a direct carrier-mediated mechanism and to identify the class of transporter involved in the uptake of steviol glycosides.

4 Biotechnological Approaches to Improve Steviosides Glycosides Production Using *In Vitro* Cultures

Several efforts have been dedicated to the use of plant in vitro cultures as a biotechnological strategy to produce secondary metabolites of commercial interest. The advantages for industrial production of these compounds include uniform product quality, independence from climate and seasonal changes, supply stability, and a closer relationship between supply and demand. Several studies have been done in *Stevia* in vitro cultures for glycosides production.

Recent studies of Kumari and Chandra [50] revealed micropropagation in *S. rebaudiana* from leaf and nodal explants and production of high-value secondary metabolites. A combination of PGRs (plant growth regulators) proved better than single for both callusing and direct-shoot multiplication from leaf explants.

Treatment of Kn and IAA (1.5 mg L^{-1}) each showed best callusing response (85.5 ± 0.33%). For shoot proliferation from callus, Kn (2.5 mg L^{-1}) with IAA (1.5 mg L^{-1}) showed maximum number of shoots (5.3 ± 0.3) proliferating from callus with longest set of 9.03 ± 0.14 cm. Direct organogenesis from leaf explant and Kn (1.5 mg L^{-1}) with BAP (2.5 mg L^{-1}) gave maximum number (8.6 ± 0.33) of shoots from leaf explant with longest shoot length (5 ± 0.11) cm. HPLC studies showed that both the secondary metabolites (stevioside, 0.451 ± 0.001 mg g^{-1}, and Reb A, 0.131 ± 0.005 mg g^{-1}) are higher in in vitro shoots developed through organogenesis from callus cultures [50].

5 Biotechnological Approaches Involving Genetic Transformation Studies in *S. rebaudiana*

5.1 Establishment of Hairy Root Cultures Through Infection with *Agrobacterium rhizogenes* on Aseptic Explants of *S. rebaudiana*

S. rebaudiana is an important medicinal plant and only few reports are available on hairy root culture of this plant. No study has been done on improvement of steviol glycoside production through hairy root culture. The following experiments were performed to establish a transformation system so that in future transgenic *S. rebaudiana* could be developed for various economic and medicinal purposes.

Agrobacterium rhizogenes strain ATCC 15834 was used for the study. The bacterial cultures were grown in 50 ml of yeast extract broth (YEB) supplemented with 50 mg L^{-1} rifampicin. Plant material of *S. rebaudiana* was collected from Indigenous Medicinal Plant Garden of Birla Institute of Technology, Mesra campus, and then stem and shoot tips were selected for surface sterilization in laminar airflow. A culture of bacteria was inoculated in 50 mL YEB media and incubated in a rotary shaker for 24 h at 28 °C at 100 rpm in the dark. After attaining optimum growth (O.D 0.9), bacterial suspension was then centrifuged at 4000 rpm for 10 min and pellet was collected. The resulting pellet from 50 ml suspension was suspended in 250 ml MS liquid media and was acclimatized for 4 h at 28 °C in dark. This was further used for co-culture. Surface sterilized explants were wounded and co-cultured in MS liquid media containing bacterial culture for 30 min. Infected explants were washed with autoclaved distilled water for 1–2 times and excess water was removed by blotting paper. All the explants were then inoculated on hormone-free MS agar medium and cultured in the dark at 25 ± 2 °C. After 2 days of co-cultivation of plant tissues and bacterial cells, the explants were transferred onto hormone-free MS agar medium with 500 mg L^{-1} cefotaxime, a bacteriostatic agent, and cultured for 7 days. One week later, the explants were retransferred onto hormone-free MS agar medium with 250 mg L^{-1} cefotaxime. This procedure was repeated for 2 or 3 times with an interval of 1 week, and gradually the concentration of cefotaxime was reduced. The number of responsive explants and number of hairy roots per explants were recorded after 15, 30, and 45 days.

Table 2 Primers designed for amplification of *rol* C gene

S. No.	Primer name	Primer sequence
1	*Rol* F	5'-TGTGACAAGCAGCGATGAGC-3'
2	*Rol* R	5'-GATTGCAAACTTGCACTCGC-3'

The hairy roots so obtained from the above transformation procedure were subjected to molecular biology studies to confirm integration of Ri plasmid into the host genome. DNA isolation from hairy roots was done using DNA isolation kit (QIAGEN DNeasy plant mini-kit) following the manufacturer instructions.

Polymerase chain reaction (PCR) identification of the rooting locus gene *rol* C was performed using DNA from the hairy roots as template and the non-transformed roots as control, respectively. Primers (*rol* C F and *rol* C R, details shown in Table 2) and 100 ng of genomic DNA isolated from hairy roots were used for amplification.

Amplifications were performed using a GeneAmp PCR System 9700 Thermocycler (Applied Biosystems, Foster City, CA, USA) that was programmed for an initial denaturation step of 3 min at 94 °C and 35 cycles (each consisting of 1 min at 94 °C, 1 min at 53.5 °C, and 1 min at 72 °C), followed by a final extension at 72 °C for 6 min. Amplified products were resolved on 1.4% agarose gel by electrophoresis.

5.2 Steviol Glycosides Production in Hairy Root Cultures of *S. rebaudiana*

Purified hairy roots were cut by sterilized scalpel under laminar airflow and inoculated on hormone-free liquid MS medium. After 3 weeks, regenerated hairy roots were transferred to different media according to test conditions. Some physical parameters like effect of culture media, temperature, and light on steviol glycoside production was also observed.

5.3 Extraction and Analysis of Stevioside Glycosides Through HPLC

Hairy roots from different sets of flasks were collected and dried to a constant weight at 40 °C and grounded to make fine powder. For extraction of stevioside glycosides, dried powder was dipped in methanol/water in the 4:1. The mixture was left at room temperature for overnight. Extract was then filtered next day using Whatmann number 1 filter paper and dried. These sample extract were redissolved in distilled water and filtered through a 0.2 μm Millipore filter for analysis by HPLC. Stevioside glycosides were separated using a Waters HPLC system (WatersCorporation, Milford, MA) in dC18 column (Waters Atlantis, 4.6 mm × 150 mm; 4 μm particle size). 20 μL of extracts were injected in a HPLC system. The solvents optimized for isocratic elution consisted of acetonitrile and MilliQ water with 0.1% orthophosphoric acid in the ratio of 20:30 with flow rate of 0.5 mL min^{-1}. The detector was set at 210 nm. The compounds from samples were identified by comparing the retention time with the corresponding retention time of standards. Quantification of compounds was done using standard curves. All experiments were repeated at least three times. The results were presented as mg mL^{-1} of extracts.

5.4 Statistical Analysis

Each treatment consisted of five Erlenmeyer flasks, each containing five explants ($n = 25$). All experiments were repeated three times. Data obtained from all experiments were presented as the mean ± standard error of three replications.

6 Hairy Root Induction and Stevioside Glycoside Production

Hairy root research for the production of valuable secondary metabolites has received a lot of attention to overcome the limiting parameters to improve the technique for enhanced production of metabolites [14]. Many plant secondary metabolites of interest are accumulated in roots. Harvesting roots is destructive for the plants and hence there has been increasing interest in developing hairy root cultures from several medicinal plant species.

The agropine type of *A. Rhizogenes* ATCC 15834 strain was used in the present study, successfully provided consistent hairy roots from leaves and stems of *S. rebaudiana*. Root initials were observed at the infection sites, within 15–20 days of infection (Fig. 5). Whereas, it took 2 weeks in *Arachis hypogea* [51], 3 weeks in *Ocimum basilicum* [52] and *Azadirachta indica* [53], 4 weeks in *Ipomoea batatas* [54] and 2–5 weeks in *Saussurea involucrata* [55]. As different strains of *A. rhizogenes* have differential susceptibility to infect different plants it might be the key factor for such variations [51].

Leaves of *S. Rebaudiana* were more susceptible to infection than stem explants. Initially, hairy roots were mostly unbranched and densely covered with root hairs giving a furry appearance. The hairy roots had an average root thickness of 2 mm. After 10–15 days of growth, it starts branching and becomes highly branched with number of lateral branches (Fig. 5).

During initial experiments, leaf explants were infected with *A. rhizogenes* ATCC 15834. However, no hairy roots were obtained after 1 month of inoculation. Consequently, further experiments were performed with stem and leaf explants with or without petioles. When leaf explants with petioles were inoculated with freshly grown *A. rhizogenes* suspensions in MS medium, hairy roots were induced from micro-calli of leaf petioles or directly from the cut edges of leaf petiole, but the amount of root induction was higher from microcalli than directly from the cut edges of leaf petioles. Independent hairy root lines were developed for *S. rebaudiana*on MS media containing cefotaxime to eliminate excess *Agrobacterium*. To assess the genetic status of the hairy roots, PCR-based analysis targeted towards *A. rhizogenes rol* C gene was performed. As this gene is located on independent TLDNAs of the Ri plasmid pRi15834, its presence confirms the integration of T-DNA to the host genome. The *rol* C gene was amplified (Fig. 6) from all the hairy root lines selected

Fig. 5 Hairy root initiation in *S. rebaudiana* leaf explants after co-cultivation with *A. Rhizogenes* ATCC 15834 (**a**) co-cultivated explants after 7 days (**b**) initiation of hairy roots from leaf explants after 15 days inoculated on MS basal media. (**c**) subcultured hairy roots after 20 days (**d**) highly branched hairy roots after 45 days inoculated on MS basal media [bar = 10 mm]

on the basis of vigorous growth in hormone-free MS media indicates successful integration of TL-DNA in the *S. Rebaudiana* genomes.

7 Effect of Different Conditions on Hairy Rootsgrowth and Secondary Metabolites Production

7.1 Effect of Murashige and Skoog (MS) and Gamborg's B5 Media on Biomass and Secondary Metabolite Content of *S. rebaudiana* Hairy Roots

Three different strengths (Full, ½, and ¼) of MS and Gamborg's B5 media were used. 100–200 mg of hairy roots was inoculated on each 50 mL of media containing in 250 mL Erlenmeyer flasks. Cultures were kept on shaker with 120 rpm, temperature 25 ± 2 °C, and 16 h photoperiod. After 21 days, not much difference was found on secondary metabolite content of hairy roots grown on different strengths of MS and B5 media. Hairy roots on full strength MS media possess stevioside 0.412 ± 0.008 mg mL^{-1} and Reb A 0.100 ± 0.008 mg mL^{-1}whereas hairy roots

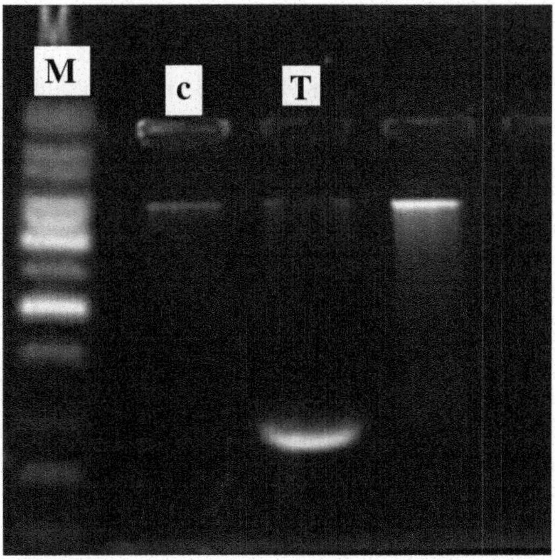

Fig. 6 PCR Image of amplified *rol* C gene where *T* indicates for test sample and *C* indicates control which was not amplified

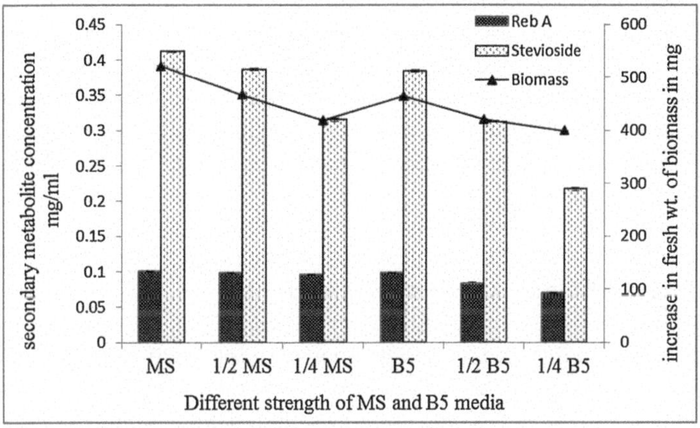

Fig. 7 Effect of MS and B5 media on secondary metabolite content and biomass of *S. rebaudiana* hairy roots after 21 days

on B5 media possess stevioside 0.383 ± 0.002 mg mL^{-1} and Reb A 0.098 ± 0.005 mg mL^{-1} (Fig. 7).

Biomass of hairy roots decreases with the decreasing strength of both MS and B5 media. On MS media, it decreased from 520 ± 1.33 mg (full MS) to 419 ± 0.52 mg (1/4MS), and on B5 media, from full strength to 1/4th strength it decreased from 464.6 ± 0.6 mg to 398.6 ± 0.8 mg. Full strength for both the media proved better for biomass and secondary metabolite content; however, on MS media, biomass and secondary metabolite content of hairy roots was better, so for further study full strength MS media was used. In *Valeriana officinalis* hairy roots B5 and ½ B5 basal media were the best for growth of hairy roots. MS medium showed highest growth rates when supplemented with a 20:20 ratio (mM) of NH_4+ to NO_3− [56]. In similar study, manipulation of MS and B5 medium salts was done for growth and solasodine production in hairy root cultures of *Solanum khasianum* Clarke [57]. The study showed that growth phase of roots was dependent on composition of different salts of B5 and MS and solasodine production increased with a decrease in total nitrogen. Hence, it was possible to enhance the secondary metabolite production and growth of hairy root cultures by formulating the medium. Different concentration of MS and B5 media affects the induction of hairy roots after co-cultivation period [58]. Maximum numbers of hairy roots were obtained by 72 h co-cultivation, and proliferation of hairy roots increased in 1/2MS and B5 media during co-cultivation period compared to the MS full strength and 1/4MS medium.

The diversity of plant materials adaptable to culture in hairy root as well as callus cultures has recently been reviewed [59]. *Agrobacterium* can be used to introduce new gene into the cells. This approach was applied in *Atropa belladonna* hairy roots in which the enzyme that converts hyoscyamine into scopolamine was overexpressed [60] and in improving productivity in *Cinchona* hairy roots [61]. Some of the biosynthetic pathways are not expressed in roots but in leaves and shoots. However, hairy root cultures have been shown to accumulate those metabolites also. For example, artemisinin was thought to accumulate only in the aerial part of the *Artemisia annua* but several reports have shown that hairy roots can produce artemisinin [62]. Normal root cultures can also produce secondary metabolites but reports suggest that the transformed root cultures exhibit far better growth kinetics in terms of biomass production, fast growth, more lateral branching, and higher yield of secondary metabolites [63].

7.2 Effect of Light and Dark Conditions

Hairy root cultures after 15 days turn green when exposed to 16 h photoperiod and showed the production of high amount of stevioside. Stevioside production in green hairy roots (0.49 ± 0.43 mg mL^{-1}) was found comparable to in vivo leaves (0.52 ± 0.14 mg mL^{-1}). Further optimization of different parameters may enhance the production. Recent studies on hairy root cultures of *S. rebaudiana* showed production of chlorogenic acid and its derivatives (CADs) by Fu et al. [38]. They are present in low amounts in *S. rebaudiana* in comparison to stevioside glycosides. Three major CADs (3-caffeoylquinic acid, 3-CQA), 3, 5-dicaffeoylquinic acid (3, 5-CQA), and 4, 5-dicaffeoylquinic acid (4, 5-CQA) were detected through HPLC-MS studies. Optimum production of these secondary metabolites reached 105.58 mg g^{-1} and total yield of these three compounds was 234.40 mg/100 mL.

7.3 Effect of Temperature on Biomass and Secondary Metabolite Content of Hairy Root

Current study showed that optimum temperature for growth of hairy root was 25 ± 2 °C (Fig. 8), and with the increase of temperature, growth of hairy roots decreases. Highest increase in biomass (527.33 ± 0.35 mg) was observed at 25 °C. Secondary metabolite production was not much affected by increase in temperature up to 31 °C with approximately similar stevioside content at 25 °C (0.425 ± 0.08 mg mL^{-1}), 28 °C (0.415 ± 0.08 mg mL^{-1}), and at 31 °C (0.386 ± 0.02 mg mL^{-1}). At 35 °C, stevioside content decreases rapidly (0.090 ± 0.02 mg mL^{-1}) and Reb A was completely not detected.

The extent of secondary product release in hairy root cultures varies among different plant species. In hairy root cultures of *Nicotiana rustica,* 10–50% of the nicotine was excreted [64], and in hairy root cultures of *L.erythro rhizon*, 25% of shikonin was excreted [19], whereas, in hairy root cultures of *Catharanthus roseus*, for example, only 3–5% of total indole alkaloids were found in the media [65]. Rhodes et al. [66] found differences in product release between different root clones of the same species of *Datura stramonium*. However, the excretion of the product was related to the onset of senescence in roots maintained for long period in the stationary phase, with most of the alkaloids remaining intracellular during active growth phase [66].

First report on examination of stevioside production through hairy root culture came earlier in 1991 by Yamazaki et al. [67]. Steviol glycosides were not produced by the hairy root cultures in the dark conditions. Addition of cytokinins (Kn and BAP) to solid medium in the light resulted in morphological changes of hairy root cultures and turned them into callus. Stevioside was not detected in the hairy root cultures. However, stevioside was detected in the primary shoot tip cultures from the

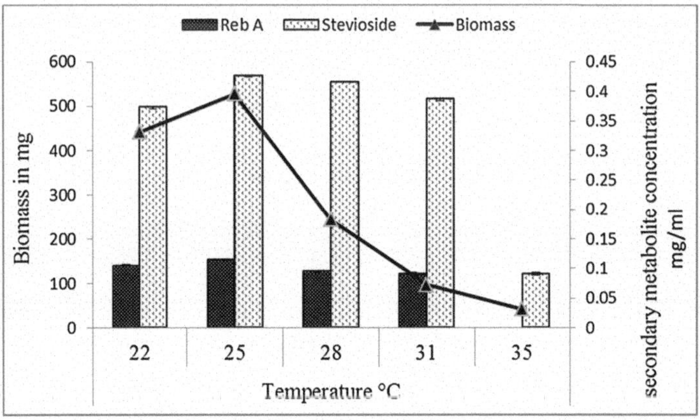

Fig. 8 Effect of temperature on growth of hairy roots and secondary metabolite content

in vitro cultured plantlets produced in liquid medium in presence of light. As per their studies, results suggested that the organ for the synthesis site of steviol glycosides was leaves, not roots in the whole plants. In contradiction to this, the present study reveals the production of stevioside glycosides in hairy roots of *S. rebaudiana*. Further, Reis et al. [68] established adventitious root culture of *S. rebaudiana* Bertoni in a roller bottle system and screened the root extract for the presence of steviol glycosides through HPLC [68]. They showed similar results with Yamazaki et al. [67]. In the present study, light conditions favored the production of steviosides glycosides in hairy root cultures of *S. rebaudiana*.

8 Conclusions

Hairy root culture flourished well in liquid MS media compared to B5 and high branching was observed. Secondary metabolite assessment revealed that it produced both stevioside and Reb A. Colour of hairy roots changed to green when kept in 16 h photoperiod and produced higher amount of stevioside than that of hairy roots kept in dark conditions. The present study gives an insight for an alternate method for stevioside glycosides production and opens a path for scale up studies in bioreactor.

References

1. Carakostas MC, Curry LL, Boileau AC, Brusick DJ (2008) Overview: the history, technical function and safety of rebaudioside A, a naturally occurring steviol glycoside, for use in food and beverages. Food Chem Toxicol 46:1–10
2. Bridel M, Lavielle R (1931) Sur le principesucre des feuilles de kaa-he-e (*Stevia rebaundiana* B). Acad Sci Paris Comptes Rendus 192:1123–1125
3. Kighom AD, Soejarto DD (1985) Current status of stevioside as a sweetening agent for human use. In: Wagner H, Hikino H, Farnsworth R (eds) Economics and medicinal plant research. Academic Press, London, pp 1–52
4. Mizutani K, Tanaka O (2002) Use of *Stevia rebaudiana* sweeteners in Japan. In: Kinghorn AD (ed) Stevia, the genus Stevia. Medicinal and aromatic plants industrial profile, vol 19. Taylor and Francis, London, pp 178–195
5. Ferri LA, Alves-Do-Prado W, Yamada SS, Gazola S, Batista MR, Bazotte RB (2006) Investigation of the antihypertensive effect of oral crude stevioside in patients with mild essential hypertension. Phytother Res 20:732–736
6. Yasukawa K, Kitanaka S, Seo S (2002) Inhibitory effect of stevioside on tumor promotion by 12-O-tetradecanoylphorbol-13-acetate in two-stage carcinogenesis in mouse skin. Biol Pharm Bull 25:1488–1490
7. Bornia ECS, Amaral VD, Bazotte RB, Alves-do-prado W (2008) The reduction of arterial tension produced by stevioside is dependent on nitric oxide synthase activity when the endothelium is intact. J Smooth Muscle Res 44:1–8
8. Jayaraman S, Manonharan MS, Illanchezian S (2008) *In-vitro* antimicrobial and antitumor activities of *Stevia rebaudiana* (Asteraceae) leaf extracts. Trop J Pharm Res 7:1143–1149

9. Xu D, Du W, Zhao L, Davey AK, Wang J (2008) The neuroprotective effects of isosteviol against focal cerebral ischemia injury induced by middle cerebral artery occlusion in rats. Planta Medica 74:816–821
10. Nakamura S, Tamura Y (1985) Variation in the main glycosides of Stevia (*Stevia rebaudiana* Bertoni). Jpn J Trop Agric 29:109–116
11. Goettemoeller J, Ching A (1999) Seed germination in *Stevia rebaudiana*. In: Janick J (ed) Perspectives on new crops and new uses. ASHS Press, Alexandria, VA, pp 510–511
12. Macchia M, Andolfi L, Ceccarini L, Angelini LG (2007) Effects of temperature, light and pre-chilling on seed germination of *Stevia rebaudiana* (Bertoni) Bertoni Accessions. Ital J Agron/Riv Agron 1:55–62
13. Abdullateef RA, Osman M (2011) Effects of visible light wavelengths on seed germinability in *Stevia rebaudiana* Bertoni. Int J Biol 3:83–91
14. Guillon S, Guiller JT, Pati PK, Rideau M, Gantet P (2006) Harnessing the potential of hairy roots: dawn of a new era. Trends Biotechnol 24:403–409
15. Kurz WGW (1988) Bottlenecks towards commercialization of product synthesis by plant cell cultures. In: Kurz WGW (ed) Primary and secondary metabolism of plant cell cultures ZI. Cambridge University Press, Cambridge, pp 83–92
16. Bonhomme V, Laurain-Mattar D, Lacoux J, Fliniaux M, Jacquin-Dubreuil A (2000) Tropane alkaloid production by hairy roots of *Atropa belladonna* obtained after transformation with *Agrobacterium rhizogenes* 15834 and *Agrobacterium tumefaciens* containing *rol* A, B, C genes only. J Biotechnol 81:151–158
17. Wang JW, Zheng LP, Tan RX (2006) The preparation of an elicitor from a fungal endophyte to enhance artemisinin production in hairy root cultures of *Artemisia annua*L. Sheng Wu Gong Cheng Xue Bao 22:829–834
18. Lorence A, Medina-Bolivar F, Nessler CL (2004) Camptothecin and 10 hydroxycamptothecin from *Camptotheca acuminate* hairy roots. Plant Cell Rep 22:437–441
19. Shimomura K, Sauerwein M, Ishimaru K (1991) Tropane alkaloids in the adventitious and hairy root cultures of solanaceous plants. Phytochemistry 30:2275–2278
20. Dupraz JM, Christen P, Kapetanidis I (1994) Tropane alkaloids in transformed roots of *Datura quercifolia*. Planta Med 60:158–162
21. Christen P, Robert MF, Phillipson JD, Evans WC (1991) Alkaloids of hairy root cultures of a *Datura candida* hybrid. Plant Cell Rep 9:101–104
22. Mano Y, Ohkawa H, Yamada Y (1989) Production of tropane alkaloids by hairy root cultures of *Duboisia Leichhardtii* transformed by *Agrobacterium rhizogenes*. Plant Sci 59:191–201
23. Putalun W, Udomsin O, Yusaku G et al (2010) Enhanced plumbagin production from in vitro cultures of *Drosera burmanii* using elicitation. Biotechnol Lett 32:721–724
24. Abdoli M, Moieni A, NaghdiBadi H (2013) Influence of KNO_3, $CaCl_2$ and $MgSO_4$ concentrations on growth and cichoric acid accumulation in hairy root culture of purple coneflower (*Echinacea purpurea*L.). J Med Plant 12:45
25. Ayadi R, Tremouillaux-Guiller J (2003) Root formation from transgenic calli of *Ginkgo biloba*. Tree Physiol 23:713–718
26. Mehrotra S, Kukreja AK, Khanuja SPS, Mishra BN (2008) Genetic transformation studies and scale up of hairy root culture of *Glycyrrhiza glabra* in bioreactor. Elect J Biotechnol 11:1–7
27. Zhang L, Ding R, Chai Y, Bonfill M, Moyano E, Oksman-Caldentey KM, Xu T, Pi Y, Wang Z, Zhang H, Kai G, Liao Z, Sun X, Tang K (2004) Engineering tropane biosynthetic pathway in *Hyoscyamus niger* hairy root cultures. PNAS 101:6786–6791
28. Bonhomme VL, Laurain-Mattar D, Fliniaux MA (2004) Hairy root induction of *Papaver somniferum* var. album, a difficult- to-transform plant, by *A. rhizogenes* LBA 9402. Planta 218:890–893
29. Kunshi M, Shimomura K, Takida M, Kitanaka S (1998) Growth and ginsenoside production of adventitious and hairy root cultures in an interspecific hybrid ginseng (*Panaxginseng_P. quinquefolium*). Nat Med 52:1–4

30. Shinde AN, Malpathak N, Fulzele PD (2009) Enhanced production of phytoestrogeniciso-flavones from hairy root cultures of *Psoralea corylifolia* L. using elicitation and precursor feeding. Biotechnol Bioprocess Eng 14:288–294
31. Lan X, Quan H (2010) Hairy root culture of *Przewalskia tangutica* for enhanced production of pharmaceutical tropane alkaloids. J Med Plant Res 4:1477–1481
32. Li W, Li MF, Yang DL, Xu R, Zhang R (2009) Production of podophyllotoxin by root culture of *Podophyllum hexandrum* Royle. Elect J Biol 5:34–39
33. Satheeshkumar K, Jose B, Sonia EV, Seeni S (2009) Isolation of morphovariants through plant regeneration in *A. rhizogenes* induced hairy root cultures of *Plumbago rosea* L. Indian J Biotechnol 8:435–441
34. Sudha CG, Obul Reddy B, Ravishankar GA, Seeni S (2003) Production of ajmalicine and ajmaline in hairy root cultures of *Rauvolfia micrantha* Hook f., a rare and endemic medicinal plant. Biotechnol Lett 25:631–636
35. Saito K, Yamazaki T, Okuyama E, Yoshihira K, Shimomura K (1991) Anthraquinone production by transformed root cultures of *Rubia ticntorum:* influence of phytohormone and sucrose concentration. Phytochem 30:2977–2980
36. Shin S, Kim Y (1996) Production of anthraquinone derivatives by hairy roots of *Rubia cordifolia* varpratensis. Saengyak Hakhoechi 27:301–308
37. Jacob A, Malpathak N (2004) Green hairy root cultures of *Solanum khasianum* Clarke – a new route to in vitro solasodine production. Curr Sci 87:1442–1447
38. Fu X, Yin ZP, Chen JG, Shangguan XC, Wang X, Zhang QF, Peng DY (2015) Production of chlorogenic acid and its derivatives in hairy root cultures of *Stevia rebaudiana*. J Agric Food Chem 63:262–268
39. Chaudhuri KN, Ghosh B, Tepfer D, Jha S (2006) Spontaneous plant regeneration in transformed roots and calli from *Tylophora indica*: changes in morphological phenotype and tylophorine accumulation associated with transformation by *Agrobacterium rhizogenes*. Plant Cell Rep 25:1059–1066
40. Murthy HN, Dijkstra C, Anthony P, White DA, Davey MR, Power JB, Hahn EJ, Paek KY (2008) Establishment of *Withania somnifera* hairy root cultures for the production of withanolide A. J Int Plant Biol 50:975–981
41. Giovannini A (2006) Tissue culture, cell culture and genetic transformation by wild type *Agrobacterium rhizogenes* in Mediterranean Helichrysum. In: Teixeira da Silva JA (ed) Floriculture, ornamental and plant biotechnology advances and tropical issues II. Global Science Books, Ikenobe
42. Sevón N, Oksman-Caldentey KM (2002) *Agrobacterium rhizogenes*-mediated transformation: root cultures as a source of alkaloids. Planta Med 68:859–868
43. Bensaddek L, Villarreal ML, Fliniaux M (2008) Induction and growth of hairy roots for the production of medicinal compounds. Electron J Integr Biosci 3:2–9
44. Brandle BE, Telmer PG (2007) Steviol glycoside biosynthesis. Phytochemistry 68:1855–1863
45. Shibata H, Sonoke S, Ochiai H, Nishihashi H, Yamada M (1991) Glucosylation of steviol and steviol-glucosides in extracts from *Stevia rebaudiana* Bertoni. Plant Physiol 95:152–156
46. Mu-zuan C, Wei-lan H, Shao-pang C, De-yao W (1983) Observations of the leaf cell vacuole of *Stevia rebaudiana* Bertoni under the electron microscope. Acta Bot Sinica 25:426–432
47. Jorgensen K, Rasmussen AV, Morant M, Nielsen AH, Bjarnholt N, Zagrobelny M, Bak S, Moller BL (2005) Metabolon formation and metabolic channeling in the biosynthesis of plant natural products. Curr Opin Plant Biol 8:280–291
48. Schmitt L, Tampe R (2002) Structure and mechanism of ABC transporters. Curr Opin Struct Biol 12:754–760
49. Goodman CD, Casati P, Walbot V (2004) A multidrug resistance associated protein involved in anthocyanin transport in Zea mays. Plant Cell 16:1812–1826
50. Kumari M, Chandra S (2015) Stevioside glycosides from in vitro cultures of *Stevia rebaudiana* and antimicrobial assay. Braz J Bot. doi:10.1007/s40415-015-0193-3

51. Medina-Bolivar F, Condori J, Rimando AM, Hubstenberger J, Shelton K, O'Keefe SF, Bennett S, Dolan MC (2007) Production and secretion of resveratrol in hairy root cultures of peanut. Phytochemistry 68:1992–2003
52. Tada H, Murakami Y, Omoto T, Shimomura K, Ishimaru K (1995) Rosamarinic acid and related phenolics in hairy root cultures of *Ocimum basilicum*. Pergamon 42:431–434
53. Satdive RK, Fulzele DP, Eapen S (2007) Enhanced production of azadirachtin by hairy root cultures of *Azadirachta indicia*. Juss by elicitation and media optimization. J Biotechnol 128:281–289
54. Nishiyama Y, Yamakawa T (2004) Effect of medium composition on the production of anthocyanins by hairy root cultures of *Ipomoea batatas*. Plant Biotechnol 21:411–414
55. Fu CX, Zhao DX, Xue XF, Jin ZP, Ma FS (2005) Transformation of *Saussurea involucrata* by *Agrobacterium rhizogenes*: hairy root induction and syringin production. Process Biochem 40:3789–3794
56. Parizi AP, Farsi M, Nematzadeh GA, Mirshamsi A (2014) Impact of different culture media on hairy roots growth of *Valeriana officinalis* L. Acta Agric Slov 103:299–305
57. Jacob A, Malpathak N (2005) Manipulation of MS and B5 components for enhancement of growth and solasodine production in hairy root cultures of *Solanum khasianum* Clarke. Plant Cell Tiss Org Cult 80:247–257
58. Bivadi V, Zakaria RA, Zare N, Yazdani B (2014) Effects of different tissue culture conditions in hairy roots induction in *Hypericum perforatum* L. Intl Res J Appl Basic Sci 8:597–604
59. Karuppusamy S (2009) A review on trends in production of secondary metabolites from higher plants by in vitro tissue, organ and cell cultures. J Med Plant Res 3:1222–1239
60. Hashimoto T, Yun DZ, Yamada Y (1993) Production of tropane alkaloids in genetically engineered root cultures. Phytochemistry 32:713–718
61. Hallard D, Van der Heijden R, Verpoorte R, Lopez-Cardoso I, Pasquali G, Memelink J, Hoge JHC (1997) Suspension cultured transgenic cells of *Nicotiana tabacum* expressing tryptophan decarboxylase and strictosidine synthase cDNAs from *Catharanthus roseus* produce strictosidine upon feeding of secologanin. Plant Cell Rep 17:50–54
62. Weathers PJ, Cheetham RD, Follansbee E, Teoh T (1994) Artemisinin production by transformed roots of *Artemisia annua*. Biotechnol Lett 16:1281–1286
63. Mukundan U, Hjortso MA (1990) Effect of fungal elicitor on thiophene production in hairy root cultures of *Tagetes patula*. Appl Microbiol Biotechnol 33:145–147
64. Hamill JD, Parr AJ, Robins RJ, Rhodes MJC (1986) Secondary product formation by cultures of *Beta vulgaris* and *Nicotiana rustica* transformed with *Agrobacterium rhizogenes*. Plant Cell Rep 5:111–114
65. Toivonen L, Balsevich J, Kurz WGW (1989) Indole alkaloid production by hairy root cultures of *Catharanthus roseus*. Plant Cell Tiss Organ Cult 18:79–93
66. Rhodes MJC, Robins RJ, Hamill JD, Parr AJ, Walton NJ (1987) Secondary product formation using *Agrobacterium rhizogenes* transformed hairy root cultures. IAPTC Newsl 53:2–15
67. Yamazaki T, Flores HE, Shimomura K, Yoshihira K (1991) Examination of steviol glucosides production by hairy root and shoot cultures of *Stevia rebaudiana*. J Nat Prod 54:986–992
68. Reis RV, Borges APPL, Chierrito TPC, Souto ERS, Souza LM, Iacomini M, Oliveira AJB, Gonçalves RAC (2011) Establishment of adventitious root culture of *Stevia rebaudiana* Bertoni in a roller bottle system. Plant Cell Tiss Organ Cult 106:329–335

Antimalarial Compound Synthesis from Transgenic Cultures

5

Subrata Kundu, Swapna Mukherjee, and Biswajit Ghosh

Abstract

Malaria is a devastating disease that affects the enormous health and economic costs throughout the world. About 214 million people in Africa, India, Southeast Asia, and South America are infected with the endemic malaria, and about four and a half lakh deaths, including children below the age of five, have been registered in the year 2015. In order to reduce the global burden of malaria, effective antimalarial drugs are continually being investigated, especially against extremely mutable parasite *Plasmodium falciparum* that has the ability to develop multidrug resistance characteristics. Artemisinin, a sesquiterpene endoperoxide lactone, isolated from the aerial parts of *Artemisia annua* L. plant is highly effective, harmless, and the best therapeutic agent against drug-resistant strains of *Plasmodium* sp. The World Health Organization (WHO) has implemented the use of artemisinin-based combination therapies (ACTs) with preparations containing an artemisinin derivative for the first-line treatment of malaria. In spite of its immense potential, the low concentration will be (0.1–1.2%) of artemisinin in plant is a serious limitation for its commercial and pharmaceutical production. Nevertheless, de novo synthesis of artemisinin is not economically feasible. Therefore a plethora of approach has been developed to optimize and scale up the cost-effective production of artemisinin to make an important contribution in global malaria eradication programs. In this chapter, we have discussed different features of transgenic approaches, including genetic manipulation of plants as well as microbes to increase antimalarial compound production.

S. Kundu • B. Ghosh (✉)
Plant Biotechnology Laboratory, Department of Botany, Ramakrishna Mission Vivekananda Centenary College, Kolkata, India
e-mail: ghosh_b2000@yahoo.co.in

S. Mukherjee
Department of Microbiology, Dinabandhu Andrews College, Garia, Kolkata, India

S. Jha (ed.), *Transgenesis and Secondary Metabolism*, Reference Series in Phytochemistry, DOI 10.1007/978-3-319-28669-3_22

Keywords
Malaria • Antimalarial drug • Transformed culture • Metabolic engineering • Hairy root culture • *Artemisia annua* • Artemisinin • Artemisinin-based combined therapy

Abbreviations

AaAOS	Allene oxide synthase gene
ACT	Artemisinin-based combination therapy
ADS	Amorpha-411-diene synthase enzyme
ads	Amorpha-411-diene synthase gene
ALDH1	Aldehyde dehydrogenase 1 enzyme
cpr	Cytochrome P450 reductase gene
cps	β-Caryophyllene synthase gene
CYP71AV1	Cytochrome P450 monooxygenase enzyme
cyp71av1	Cytochrome P450 monooxygenase gene
DBR2	Artemisinic aldehyde Δ11 (13) reductase enzyme
dbr2	Artemisinic aldehyde Δ11 (13) reductase gene
dw	Dry weight
dxr	1-Deoxy-D-xylulose 5-phosphate reductase gene
FPP	Farnesyl diphosphate
fps	Farnesyl pyrophosphate synthase gene
hmgr	Hydroxymethylglutaryl coenzyme A gene
HMGR	3-hydroxy-3-methylglutaryl-CoA reductase enzyme
ipt	Isopentenyl transferase gene
sqs	Squalene synthase gene
WHO	World Health Organization

Contents

1 Introduction

Malaria is a mosquito-borne disease and is the deadliest infectious disease that is responsible for the ravaging effect throughout the world especially in the tropical zones. The severity of this ailment was accentuated in the World Health Organization report (2015), representing that 214 million people were infected with malaria and about Four and a half lakh patients died of the disease, most of which are in the African Region followed by Southeast Asian Region [1]. Regardless of the marvelous efforts for the control of malaria, the global mortality rate has not decreased significantly over the decade. Moreover malaria-infected people are highly vulnerable to other infectious diseases, including AIDS, measles, anemia, respiratory tract infections, and diarrhea [2–4]. During pregnancy, it increases the chances of miscarriages and delivery of infants with low birth weight and mortality [5]. The major control of this devastating parasitic disease is to restrict the multiplication of the mosquito vector as well as the effective treatment with antimalarial drugs. Since the discovery of the disease, unprecedented efforts have been made toward the discovery of novel antimalarial drugs to control the spread of this ailment, and a series of natural and synthetic drugs, including quinine, chloroquine, primaquine, mefloquine, and sulfadoxine–pyrimethamine, have been applied [6, 7]. The great significance of plant-derived drugs for the treatment of the disease is highlighted by quinine (derived from *Cinchona* tree). Unfortunately, due to the high mutable nature of the malaria parasite, the spread of drug resistance has played an important role in the occurrence and severity of this epidemic disease in the world. Resistance to the major and effective drugs like quinine, chloroquine, primaquine, and mefloquine has been reported in many prevalent regions of the world [8]. The highly adaptive nature of the malaria parasite also accentuates the obstacles in obtaining an antimalarial vaccine. Under such circumstances WHO has recommended artemisinin-based combination therapies (ACTs) as a first-line treatment for malaria [9–11]. ACTs comprise semisynthetic artemisinin derivatives paired with distinct chemical classes of longer-acting drugs. It is the most favored treatment process for cerebral as well as for chloroquine-resistant malaria due to its high effectiveness, quick action, and nonexistence of severe side effects [10].

It has been reported that the plants of the genus *Artemisia* are the only natural source of artemisinin and *Artemisia annua* is the major source [12, 13]. Unfortunately, the supply of artemisinin is limited because of the low content of artemisinin (0.1–1.2% dry weight), ultimately resulting in the hike of the cost of ACTs – too high for the population of malarial victims below poverty lines [14–16]. Therefore, there is an earnest need of improving the artemisinin yield *in planta* either by developing high-yielding varieties of *A. annua* or finding alternative sources. Efforts are being adopted worldwide to improve its production by employing various approaches such as conventional breeding and biochemical as well as physiological methods [2, 17–23]. Although these approaches exhibited possibilities but so far, they are unable to meet the global requirements. The chemical synthesis of artemisinin is complicated with low yields and high cost. Therefore the enhanced production of artemisinin in cell/tissue culture or whole plants is the major way to explore it at industrial level. There is a plethora of boundaries to obtain plant-derived compounds because they might be formed only during a particular developmental stage or under precise seasonal or nutrient availability situations. In such circumstances metabolic engineering offers prospects to procure such compounds through the modern molecular biological techniques including cloning, recombinant DNA technology, and modulation of biosynthetic pathways.

Genetic modification of the artemisinin biosynthetic pathway is a promising approach to increase artemisinin production in transgenic *A. annua*. Several efforts have been exploited in the last decade to accomplish higher synthesis of artemisinin through bioengineering of *A. annua* L. plants and other heterologous organisms. In this chapter, we summarize the progress made in the overproduction of artemisinin and other antimalarial compound by biotechnological approaches. We have highlighted the efforts made by several groups of researchers in the production of artemisinin by biotechnological approaches, including in vitro transformed culture, metabolic regulation of biosynthesis, genetic engineering, and bioreactor technology.

2 Malaria: World's Most Devastating Parasitic Ailment

Malaria has caused greater suffering and death over the course of human history compared to any other single disease. Malaria-like febrile illnesses (with names like "the ague" or "paludism") have been described since the time of Hippocrates as fevers that were periodic and associated with marshes and swamps. The word "malaria" signified "bad air" (mal'aria) according to Italian. It was not until the 1880s and 1890s that Alphonse Laveran, Ronald Ross, Battista Grassi, and others were able to identify the malaria parasite and link the transmission of malaria to mosquitoes. Malaria is a life-threatening disease caused by parasites that are transmitted to people through the bites of infected female mosquitoes. About 3.2 billion people – almost half of the world's population – are at risk of malaria. There are currently over 100 countries and territories mainly in many tropical and subtropical areas.

2.1 Causative Agent and Mode of Transmission

Five species of the parasite belonging to the genus *Plasmodium* are the causative agents of malaria. *Plasmodium* belongs to the phylum Apicomplexa, a large group of eukaryotic microorganisms possessing unique organelle termed as apicoplast as well as apical complex that is responsible for causing malaria in human. Among them four species (*P. falciparum*, *P. vivax*, *P. malariae,* and *P. ovale*) are primarily responsible for malaria in human [24]. It has been reported that the fifth species *P. knowlesi* also causes malaria in certain forested areas of Southeast Asia [25]. Malaria triggered by *P. falciparum* and *P. vivax* poses many devastating health problems. *P. falciparum* is confined on the African continent and is responsible for maximum deaths, whereas *P. vivax* has broad geographic distribution compared to

P. falciparum as it can develop in the vector at lower temperatures and can survive at higher altitudes. Female mosquitoes of the genus *Anopheles* spread malaria from person to person. Among the 400 different species of *Anopheles* mosquitoes, only 30 are major vectors for malarial transmission [2]. The intensity of disease transmission primarily depends on factors including the type of vector, the host, as well as the environment. Disease spread is more penetrating in places where the lifespan of mosquito is relatively longer. Nearly about 90% of the world's malaria cases are found in Africa due to the long lifespan and robust human-biting habit of the African vector species [1]. The transmission of disease is also found to be seasonal with ultimate during the rainy season. The disease becomes epidemic when climate and other conditions become favorable for transmission in areas where people have little or no immunity against malaria [2].

2.2 Disease Etiology and Severity

It has been reported that the number of malaria cases globally has been declined to 18% from an estimated 262 million in 2000 to 214 million in 2015. Most of the infections have occurred in the African Region (88%), followed by the Southeast Asian Region (10%), and the Eastern Mediterranean Region (2%). The number of malaria deaths globally fell from an estimated 839 000 in 2000 to 438 000 in 2015 with the decline of 48%. Malaria infection in humans is started when sporozoites are injected with the saliva of a female *Anopheles* spp. feeding mosquito. Then the sporozoites migrate to the liver through the bloodstream of the infected person and invade hepatocytes and then asymptomatically divide into merozoites. The asexual blood-stage life cycle of the parasite begins when the liver-released merozoites invade the erythrocytes. Then the merozoites develop through ring, trophozoite, and schizont stages to produce about 16–32 daughter merozoites. The new merozoites invade other erythrocytes to continue the intraerythrocytic cycle. This asexual blood-stage cycle of infection is primarily responsible for the clinical pathologies of malaria. Some intraerythrocytic stages develop into gametocytes that can be ingested by a mosquito during feeding. The mosquito sexual reproduction and further development of the parasites lead to the generation of new sporozoites that are ready to infect a new host.

3 Prevention of the Disease

The control of the vector is the main way to prevent and reduce malaria transmission. The insecticide-treated mosquito nets (ITNs) and indoor residual spraying (IRS) with insecticides are the major guidelines by WHO for reducing the control of the vector. IRS is the application of stable formulations of insecticides to the inside of houses to kill resting adult female mosquitoes. The primary contributions of IRS in reducing malaria transmission are reducing the life span of female mosquitoes, so that they can no longer transmit malaria parasites; thus the density of the vector mosquitoes decreases.

3.1 Role of Plant Natural Products as Antimalarial Agent

The evolution of human civilization and modernization and the importance of natural products for medicine and health have been enormous. We have long used naturally occurring substances for medical purposes. Especially, plants have played a leading medical role in most cultures. The study of the use of plants by indigenous people (ethnobotany), followed by phytochemical, preclinical, and clinical studies, is an important approach to the discovery and development of traditional medicines [26]. After plant species are identified as potentially useful for the pharmaceutical industry, their active principles and properties are thoroughly studied. In some cases the crude plant extract or the pure active principle is directly employed as a medicine. In the last centuries, the pharmaceutical industry has profited from the huge varieties of plant bioactive molecules to produce medicines and has synthesized molecules of similar biological activities. In Western countries, about 25% of the molecules used are of natural plant origin.

Since the recognition of malaria, a huge number of plant species with antimalarial properties have been identified by various research groups. It has been reported that more than 1,277 plants belonging to 160 families were used traditionally for the treatment of malaria [27–40]. The crude extracts of these medicinal plants either have in vitro or in vivo anti-plasmodial activity directed to the erythrocytic stage of malaria parasites. Traditional medicines have been used to treat malaria for thousands of years and are the source of the two main groups (artemisinin and quinine derivatives) of modern antimalarial drugs.

The discovery of quinine from *Cinchona* sp. (family Rubiaceae comprise 23 species of plants) and its subsequent development as a dependable antimalarial drug represented a milestone in the history of modern medicine for malaria. The most popular species are *C. officinalis*, *C. succirubra*, and *C. ledgeriana*, with a good yield of quinine. The quinine content are varies from species to species in Cinchona species, e.g., *C. calisaya* (0–4%), *C. pubescens* (1–3%), *C. officinalis* (2–7.5%), *C. ledgeriana* (3–13%), *C. succirubra* (4–14%), etc. Quinine, as a component of the bark of the *Cinchona* (quina-quina) tree, was used to treat malaria from as early as the 1600s, when it was referred to as the "Jesuits' bark," "cardinal's bark," or "sacred bark." In 1820 two scientists, Pelletier and Caventou, isolated an alkaloid chemical in the bark which provided the highest antimalarial effect and named it *quinine*. Before 1820, the bark of the *Cinchona* tree was first dried, ground to a fine powder, and then mixed into a liquid (commonly wine) before being drunk. *Cinchona* bark, in the form of a decoction, infusion, or maceration, is still used for the treatment of malaria today in several countries. The final end products of alkaloid biosynthesis in *Cinchona* can be divided into three groups: the quinoline alkaloids (quinine, cinchonine, etc.), the indole alkaloids quinamine and cinchonamine, and the quasi-dimeric cinchophyllines, which are principally found in the leaves of *Cinchona* plants. Quinine is a *Cinchona* alkaloid that belongs to the aryl amino alcohol group of drugs. Quinine continues to play a significant role in the management of malaria, but in the 1960s, drug-resistant malaria parasites developed and spread rapidly in Southeast Asia and Africa; therefore, existing antimalarial drugs, such as quinine, chloroquine, and sulfadoxine–pyrimethamine, became less efficient.

The species *Artemisia annua* L. (Asteraceae) is native to China. Its ancient Chinese name, qinghao, literally means "green herb." *Artemisia annua* – sweet wormwood or qinghao – was used by Chinese herbal medicine practitioners for at least 2000 years. In 1596, Li Shizhen, a famous herbalist, recommended this herb for fever and specified that the extract be prepared in cold water. In 1967, the government of the People's Republic of China established a program to screen traditional plants for drug activities in an effort to professionalize traditional medicine. Qinghao was tested in this program and found to have potent antimalarial activity. In the 1970s, Chinese scientists first isolated colorless needle-like crystals with antimalarial activity from *A. annua* and named artemisinin. The chemical structure of artemisinin was confirmed as a sesquiterpene lactone with an endoperoxide bridge of melting point 156–157 °C and molecular formula $C_{15}H_{22}O_5$. Later research showed that artemisinin and its derivatives had good effects on cerebral malaria, chloroquine-resistant falciparum malaria, and chloroquine-sensitive malaria. Artemisinin-based combination therapies (ACTs) were recommended by the World Health Organization to control the spread of malaria. Youyou Tu, a Chinese scientist, discovered low-temperature extraction of sweet wormwood (*A. annua*) that provides the most effective preparation against malaria parasites. It is noteworthy that Youyou Tu is the first Chinese scientist awarded with Nobel Prize in Physiology or Medicine in 2015. However, the low content of artemisinin in *A. annua* (0.1–1.2%) makes artemisinin-based antimalarial drug expensive, while malaria endemic areas are mainly concentrated in some economically poor areas in Africa, Asia, and Latin America. Artemisinin is commercially produced by extraction from sweet wormwood (*Artemisia annua*) at a cost of $400–1100/kg. So the price of the drug has become the key obstacle to popularizing ACTs.

3.2 Antimalarial Drugs

The principal and effective antimalarial drugs against human malaria appeared approximately in 1770 and contained infusions of bark from the *Cinchona* spp. Quinine, an aminoquinoline alkaloid isolated from the bark of *Cinchona* species (*Rubiaceae*), is one of the primitive and important antimalarial drugs. This alkaloid was found to be the sole active compound effective against *P. falciparum* for almost three centuries. But after World War II, it was replaced by a series of synthetic drugs including 8-aminoquinolines (e.g., primaquine), 4-aminoquinolines (e.g., chloroquine, amodiaquine), and folic acid synthesis inhibitors (e.g., proguanil, pyrimethamine). These synthetic drugs were expected to eradicate malaria worldwide, but the unfortunate appearance of drug-resistant *P. falciparum* strains in the 1960s had complicated the treatment of malaria worldwide [31, 32]. By the early 1980s, several strains of *P. falciparum* had become multidrug resistant, and today chloroquine resistance is widespread all over world [33, 34].

Table 1 The combination of artemisinin derivative with the copartner drug used in ACTs

Artemisinin derivatives	Copartner drug
Artesunate	Mefloquine
Artesunate	Sulfadoxine–pyrimethamine
Artemether	Lumefantrine
Artesunate	Amodiaquine
Artesunate	Pyronaridine
Dihydro artesunate	Piperaquine

The resistance of the highly mutable parasites toward antimalarial drugs is a major threat in achieving efficient malaria control and ultimate eradication. The resistance to chloroquine and sulfadoxine–pyrimethamine has led to the implementation of artemisinin-based combination therapies (ACTs) as the first line of treatment against malaria [9, 11]. ACTs comprise semisynthetic artemisinin derivatives paired with distinct chemical classes of longer-acting drugs (Table 1). These derivatized artemisinins are remarkably effective against the pathogenic asexual blood stages of *Plasmodium* parasites and also act on the transmissible sexual stages. These combinations therapy increased clinical and parasitological curing rate and decreased the selection pressure for the occurrence of resistance. The list of antimalarial drugs and their chemical structures along with their mode of action is shown in Fig. 1 and Table 2.

Artemisinin (qinghaosu) is a highly oxygenated sesquiterpene lactone containing unique 1-, 2-, and 4-trioxane ring structure that is responsible for the antimalarial activity. It was first isolated from the aerial parts of traditional Chinese medicinal herb *A. annua* L. plants by the Chinese scientists (Klayman 1985). *A. annua* L, an aromatic annual herbaceous medical plant belonging to the Asteraceae (Compositae) family (Fig. 2). Although it has been reported that artemisinin biosynthesis was exclusive to *A. annua* plant, recent studies have shown that artemisinin is also produced in some other species of the genus *Artemisia* including, *A. absinthium*, *A. bushriences*, *A. cina*, *A. dracunculus*, *A. dubia*, *A. indica*, *A. japonica*, *A. moorcroftiana*, *A. parviflora*, *A. roxburghiana*, *A. sieberi*, *and A. vulgaris* [35–38]. But unfortunately the artemisinin content of these species is significantly low compared to *A. annua* (0.5–1.2% dry weight). Artemisinin has been detected in leaves, buds, flowers, and seeds of different *Artemisia* species [39–43]. It has been reported that artemisinin is not found in roots of field-grown plants or in pollen [12, 44]. The highest concentration of artemisinin is found in the glandular trichomes that are present in leaves, stems, and flowers of the plant [45, 46].

4 Strategies for Enhanced Production of Antimalarial Compound

The synthesis of plant-derived secondary compounds is very complicated due to several reasons including the species- or genus-specific expression as well as developmental stage-related restriction of synthesis. Nevertheless the synthesis of

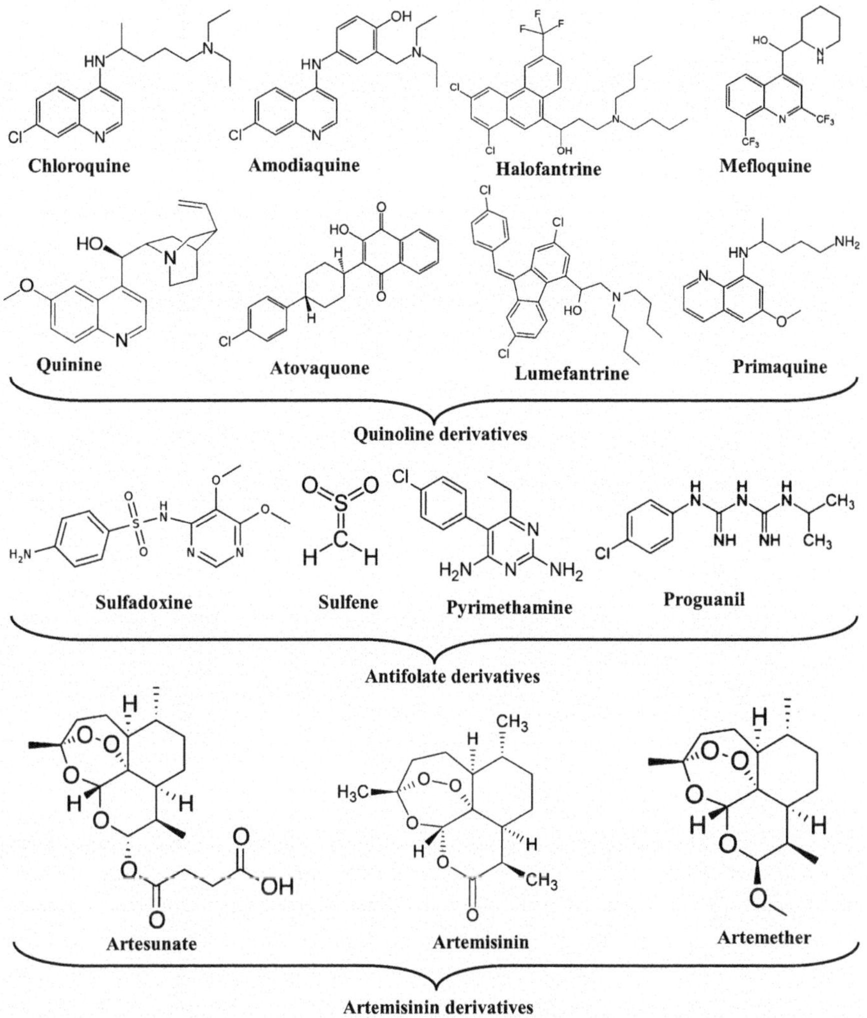

Fig. 1 The chemical structure of different types of major antimalarial drugs

several important metabolites also depends on season, biotic or abiotic stress, or accessibility of nutrients [47, 48]. The chemical syntheses of these compounds are also hampered due to chemical complexity, stereospecificity, and economic feasibility. In such circumstances metabolic engineering through modern molecular biology techniques including recombinant DNA technology and strategies for manipulation of information of the biosynthetic pathways offers prospects to overcome the deficiency of availability of such compounds at industrial levels.

The cultivated *A. annua* L. plants are the major natural source of the antimalarial drug worldwide [12, 13]. Nevertheless low yield of artemisinin in *A. annua* L. is a major drawback for the commercialization of this important medicinal plant. The

Table 2 Classification of antimalarial drugs with their mode of action

Antimalarial derivatives	Chemical family	Drug name	Mode of action
Quinoline derivatives	4-Aminoquinolines	Chloroquine	Inhibition of heme detoxification
		Amodiaquine	Inhibition of heme detoxification
	Amino alcohols	Quinine	Inhibition of heme detoxification
		Mefloquine	Inhibition of heme detoxification
		Halofantrine	Not precisely known
		Lumefantrine	Not precisely known
	8-Aminoquinolines	Primaquine	Not precisely known
	Naphthoquine	Atovaquone	Inhibits the respiratory reaction of parasite
Antifolate derivatives	Sulfa drugs	Sulfadoxine, sulfene	Inhibition of folate biosynthesis
	Pyrimethamine, proguanil	Pyrimethamine, proguanil	Inhibits the dihydrofolate reductase activity
Artemisinin derivatives	Endoperoxidases	Artesunate	Induces DNA double-strand breaks in *P. falciparum* with simultaneous increase in intercellular ROS level
		Artemisinin	Free radical mediated oxidative stress, alkylation of heme, and other protein, inhibition of PfATP6, interfere with plasmodial mitochondrial function
		Artemether	Not precisely known

cost of chemical synthesis of artemisinin is not affordable to most of the developing countries for the treatment of this devastating ailment [49, 50]. Several efforts have been made by the researchers throughout the world to enhance artemisinin content *in planta*. A significant progress has been achieved in the elucidation of detailed artemisinin biosynthetic pathway and characterization of the rate-limiting genes in order to modulate artemisinin contents. As a result the biosynthetic pathway of artemisinin is almost completely elucidated. Briefly, the first committed step of this pathway is the conversion of farnesyl diphosphate (FPP) to amorpha-4,11-diene that is catalyzed by amorpha-4,11-diene synthase (ADS) enzyme. Consequently, amorpha-4,11-diene is hydroxylated to artemisinic alcohol and ultimately oxidized to artemisinic aldehyde and artemisinic acid by a multifunctional enzyme cytochrome P450 monooxygenase (CYP71AV1). It has been reported that dihydroartemisinic acid is the precursor molecule of artemisinin in *A. annua*. The artemisinic aldehyde Δ11 (13) reductase (DBR2) enzyme catalyzes the conversion of artemisinic aldehyde to dihydroartemisinic aldehyde that oxidized to dihydroartemisinic acid by aldehyde dehydrogenase 1 (ALDH1). Consequently, the yield of artemisinin is positively correlated with the level of FPP and on the activity of other enzymes utilizing this substrate. These information aided the precise genetic manipulation of *A. annua* L. to increase the

Fig. 2 *A. annua* plant (**a**) at vegetative stage (**b**) at flowering stage (**c**). Hairy roots cultured on solid basal media and (**d**) hairy roots cultured in liquid basal media

artemisinin content by overexpression of endogenous and heterologous genes as well as suppression of genes that encode the rate-limiting enzymes of the competing branched pathways. The transgenic approaches for the synthesis of artemisinin as well as its precursor molecules *in planta* and heterologous system including engineered microbes and yeasts are summarized in Table 3. The approaches and major and salient findings of the studies of different investigators are summarized in Fig. 3 and discussed below.

Table 3 Transgenic approaches for enhancement of artemisinin and its precursor molecule production in *A. annua* as well as in heterologous hosts

Genes	Mode	Production system	Products	References
ipt	Overexpressed	*A. annua* L.	Artemisinin	[50]
fps	Overexpressed	*A. annua* L.	Artemisinin	[51–54]
hmgr	Overexpressed	*A. annua* L.	Artemisinin	[55, 56]
ads	Overexpressed	*A. annua* L.	Artemisinin	[57]
hmgr and *fps*	Co-overexpressed	*A. annua* L.	Artemisinin	[58]
Cyp71av1 and *cpr*	Co-overexpressed	*A. annua* L.	Artemisinin	[59]
Cyp71av1, *cpr*, and *dxr*	Co-overexpressed	*A. annua* L.	Artemisinin	[61]
ads, *cyp71av1*, and *cpr*	Co-overexpressed	*A. annua* L.	Artemisinin	[63]
AaWRKY1	Overexpressed	*A. annua* L.	Artemisinin	[62]
ads and *hmgr*	Co-overexpressed	*A. annua* L.	Artemisinin	[64, 65]
rol B and *rol C*	Co-overexpressed	*A. annua* L.	Artemisinin	[16]
β-Glucosidase	Overexpressed	*A. annua* L.	Artemisinin	[67]
AabZIP1	Overexpressed	*A. annua* L.	Artemisinin	[68]
thmgr, *cpr*, *ads*, *cyp71av1*, and *dbr2*	Co-overexpressed	*N. tabaccum*	Artemisinin	[71]
CRY-1	Overexpressed	*A. annua* L.	Artemisinin	[72]
AaERF1 and *AaERF2*	Co-overexpressed	*A. annua* L.	Artemisinin	[73]
AaPYL9	Overexpressed	*A. annua* L.	Artemisinin	[74]
fps, *cyp71av1*, *and cpr*	Co-overexpressed	*A. annua* L.	Artemisinin	[66]
sqs	Suppressed	*A. annua* L.	Artemisinin	[82, 83, 85, 86]
cps	Suppressed	*A. annua* L.	Artemisinin	[84]
ads	Overexpressed	*Escherichia coli*	Amorphadiene	[87]
hmgr and *cyp71av1*	Overexpressed	*Saccharomyces cerevisiae*	Artemisinic acid	[88]
hmgs and *hmgr*	Overexpressed	*Escherichia coli*	Amorpha-4,11-diene	[89]
Genes from MVA pathway and ERG-20	Overexpressed	Yeast CEN.PK2	Amorpha-4,11-diene	[90]
Plant dehydrogenase and second cytochrome	Overexpressed	Yeast S228C	Artemisinic acid	[91]
AaAOS	Overexpressed	*A. annua* L.	Artemisinin	[70]

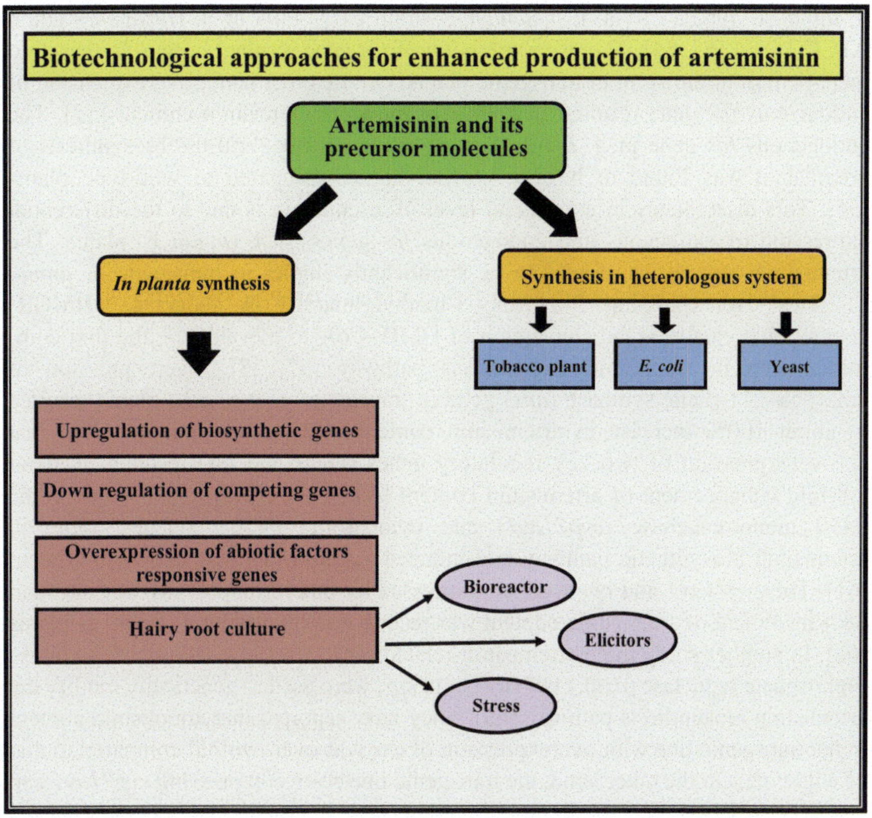

Fig. 3 Schematic representation of different types of biotechnological approaches for enhanced synthesis of artemisinin

4.1 Upregulation of Exogenous and Artemisinin Biosynthetic Pathway Genes

In 1996 first transgenic *A. annua* plant was developed in order to commercialize the production of artemisinin [51]. Since then, plenty of transgenic plants, with overexpression of enzymes in the early steps of terpene biosynthesis pathway, have been established for higher production of artemisinin. The impact of overexpression of isopentenyl transferase (*ipt*) gene on physiological and biochemical parameters of transgenic *A. annua* L. plants was assessed [52]. The transgenic plants were found to accumulate higher amount of cytokinins, chlorophylls, as well as artemisinin (30–70%) compared to those of non-transgenic plants. The *fps* gene of cotton tagged under CaMV 35S promoter was transferred and overexpressed into

A. annua L. plants L. via *A. tumefaciens* strain LBA 4404 or *A. rhizogenes* strain ATCC 15834 [53, 54]. The overexpression of *fps* gene resulted in two- to threefold increment of artemisinin in transgenic plants. On the other hand, overexpression of endogenous *fps* gene resulted in 34.4% increase in artemisinin content [55]. The endogenous *fps* gene in *A. annua* L. was overexpressed, and the biosynthesis of artemisinin was found to be 2.5–3.6-fold higher compared to wild-type plants [56]. This discrepancy in expression level of artemisinin is due to the differential expression of exogenous and endogenous *fps* genes in *A. annua* L. plants. The artemisinin content was found to be significantly higher in transgenic *A. annua* L. lines overexpressing 3-hydroxy-3-methylglutaryl-CoA reductase (HMGR) enzyme that catalyzes the conversion of HMG–CoA to mevalonate, the first committed step in isoprenoid biosynthesis pathway [57, 58]. Overexpression of amorpha-4,11-diene synthase (*ads*) gene in transgenic *A. annua* L. plants resulted in about 100% increase in artemisinin content [59]. It has been reported that co-overexpression of two key regulatory genes (*hmgr* and *fps*) resulted in about 1.8-fold enhancement of artemisinin content [60]. Overexpression of cytochrome P450 monooxygenase (*cyp71av1*) and cytochrome P450 reductase (*cpr*) in artemisinin biosynthetic pathway had induced 2.4-fold higher artemisinin content [61]. The *cyp71av1* and *cpr* were co-expressed by other group of investigator, and 38% increment of artemisinin content was recorded in transgenic *A. annua* L. plants [62]. In another study three artemisinin-related genes, namely, 1-deoxy-D-xylulose 5-phosphate reductase (*dxr*), *cyp71av1*, and *cpr*, were used to genetically modify the artemisinin biosynthesis pathway [63]. They have reported that artemisinin content in the transgenic line with overexpression of *dxr* was over twofold compared to that of wild type. On the other hand, the transgenic line co-overexpressing *cyp71av1* and *cpr* gene resulted in over 3 times increment of artemisinin production. In the subsequent study, 22.3 mg g^{-1} dw artemisinin was found in transgenic plants overexpressing artemisinic aldehyde Δ11 (13) reductase (*dbr2*) gene [64]. *A. tumefaciens*-mediated transformation of *A. annua* plants with *ads*, *cyp71av1*, and *cpr* genes resulted in stable transgenic lines in which the content of artemisinin was found to be 2.4-fold higher than that of control plants [65]. In another study, *hmgr* gene from *Catharanthus roseus* and *ads* gene from *A. annua* L. were co-overexpressed in transgenic plants [66]. They have reported 7.65-fold enhancement of artemisinin content in transgenic plant compared to that of non-transgenic plant. This group of researchers have also established transgenic *A. annua* L. plants overexpressing both *hmgr* and *ads* genes of *A. annua* to observe modulation of artemisinin content. Significant enhancement of artemisinin content (over 1.5-fold) was recorded in transgenic lines as compared to that of non-transgenic plants [16]. Co-overexpression of three key genes of artemisinin biosynthetic pathway, namely, *fps*, *cyp71av1*, and *cpr* genes in *A. annua* through *Agrobacterium*-mediated transformation, resulted in the successful establishment of transgenic line with about 3.6-fold higher (2.9 mg g^{-1} FW) artemisinin than that of the control [67]. Transformation of *A. annua* with *rol B* and *rol C* genes results in the enhancement of its secondary metabolites including artemisinin and its derivatives over fourfold [68]. The expression of β-glucosidase gene in *A. annua* through *Agrobacterium*-mediated transformation resulted in 66% increase in trichome density of flower. The 2.56% increase in artemisinin content was found in transgenic line [69]. Overexpression of basic leucine zipper family transcription factor (AabZIP1) in *A. annua* resulted in significantly increased accumulation of artemisinin [70]. Recently, *dbr2* gene was introduced to engineer the biosynthetic pathway of artemisinin in transgenic *A. annua* plants, and it was recorded that artemisinin and its direct precursor, dihydroartemisinic acid, were remarkably increased in the transgenic plants [71]. The overexpression of allene oxide synthase (*AaAOS*) gene involved in jasmonate biosynthesis pathway was found to be strongly correlated with the artemisinin content upon treatment with MeJ, ABA, and ethylene [72]. Several efforts have also been implemented in numerous studies to synthesize artemisinin in other plants. The *N. tabacum* plant was transformed with MEGA vector carrying *thmgr* of yeast and *cpr, ads, cyp71av1*, and *dbr2* from *A. annua* L. [73]. These transgenic tobacco lines overexpressing *ads* gene in mitochondria had been shown to produce higher levels of artemisinin (5.0–6.8 μg g^{-1} dw).

4.2 Overexpression of Abiotic Factors Responsive Genes

The level of terpenoids can also be enhanced by modulation of carbon flux through the regulation of *hmgr* genes upon numerous abiotic factors including light, phytohormones, as well as nutrient stress. It has been reported that overexpression of blue light receptor, *CRY1* of *Arabidopsis thaliana* in *A. annua* L. plants, resulted in the enhancement of 30–40% artemisinin content in transgenic plant compared to the non-transgenic plants [74]. Overexpression of jasmonic acid-responsive transcription factors AaERF1 and AaERF2 strongly induces the expression levels of ADS and CYP71AV1 level in transgenic plant [75]. In this transgenic plant, the artemisinin and artemisinic acids were also elevated drastically ranging from 19 to 67 and 11 to 76%, respectively, in AaERF1 while 24–51 and 17–121%, respectively, in AaERF2-transgenic lines. In another study, overexpression of an ABA-responsive gene, AaPYL9, has directed to the increment of artemisinin content (74–95%) upon ABA treatment [76].

4.3 Downregulation of Endogenous Genes Involved in Artemisinin-Competing Pathways

The artemisinin biosynthetic pathway and its regulation have been unrevealed by several reports [77–81]. It has been well established that FPP plays a central role as an intermediate precursor for the synthesis of various isoprenoid products such as essential oils, carotenoids, artemisinin, and sterols. Therefore diversion of FPP toward other metabolic pathways ultimately reduces artemisinin biosynthesis in *A. annua* L. plants. It was reported that inhibition of sterol biosynthesis resulted in higher accumulation of artemisinin in *A. annua* L. plants [82, 83]. The study was further supported by enhanced artemisinin biosynthesis in transgenic *A. annua* L. plants by suppressing sterol biosynthesis through hairpin–RNA-mediated RNAi (RNA interference) technology [84]. They have reported that suppressing the expression of *sqs* (squalene synthase) gene, a key enzyme of sterol pathway, resulted in about 3.14-fold increase in artemisinin content compared to that of untransformed control plants. In another study antisense *SS* cDNA (*asSS*) was introduced into the genome of *A. annua* with elevated expression of *ads* gene. A 3.7-fold increase in artemisinin content was recorded in transgenic plant [85]. The suppression of β-caryophyllene synthase (*cps*) gene, a sesquiterpene synthase, competing as a precursor of artemisinin by antisense technology, resulted in significant reduction of endogenous CPS expression, whereas artemisinin content was found to be up to 54.9% higher compared to that of wild-type control [86]. The level of *ads*, *cyp71av1*, and *cpr* gene as well as artemisinin content was significantly induced in transgenic *A. annua* plants expressing the genomic integrated antisense squalene synthase gene (asSS) [87]. Recently the expression levels of four branch-pathway genes, *cps*, *bfs*, germacrene A synthase, and *sqs*, were inhibited by antisense in *A. annua*. Interestingly significant increase in artemisinin content was reported in all the transgenic plants [88].

4.4 Metabolic Engineering to Produce Antimalarial Precursor Molecules in Microbial System

Metabolic engineering of the biosynthetic pathways of terpenoids in microbial hosts including *Escherichia coli* and *Saccharomyces cerevisiae* is an alternative source for their large-scale production. Modulation in the MVA and MEP pathways along with the incorporation of enzymes including ADS and CYP71AV1 in the host is essential for the synthesis of artemisinin or its precursor molecules in microbial system. Heterologous production of amorphadiene (0.5 g L^{-1}), an artemisinin precursor, was successfully achieved using two partitioning bioreactors using the *E. coli* system engineered with the mevalonate pathway (using truncated *thmgr*) from *S. cerevisiae* and ads gene from *A. annua* plants [89]. It has been reported that genetic engineering of *S. cerevisiae* with amorphadiene synthase and *CYP71AV1* genes from *A. annua* resulted in the production of artemisinic acid to a high titer (up to 100 mg L^{-1}) [90]. Nevertheless synthesized artemisinic acid is secreted out and retained on the outside of the engineered yeast that resulted in a cheap purification process to obtain the desired product. The substitution of yeast genes for HMG–CoA synthase and HMG–CoA reductase with equivalent genes from *Staphylococcus aureus* in the recombinant *E. coli* and optimization of fed-batch fermentation process resulted in a consistent increase in amorphadiene content [91]. The modified strain of *S. cerevisiae*, overexpressing all the genes from MVA pathway including ERG20 (*fps*) gene, leads to over 250-fold increment of dihydroartemisinic acid content [92]. Synthetic biological approach was employed to develop strains of *S. cerevisiae* with high-yielding capacity of artemisinic acid [93]. They have demonstrated complete biosynthetic pathway in Baker's yeast containing the plant dehydrogenase and a second cytochrome that provided an efficient biosynthetic production of artemisinic acid to 25 g L^{-1}. The engineered yeast is capable of producing artemisinic acid at significantly higher amount compared to that of *A. annua* plant. This artemisinic acid can also be reduced to dihydroartemisinic acid that ultimately auto-oxidized to artemisinin. Transgenic microbes enable the consecutive partial synthesis of artemisinin that ultimately provides smart alternative source for the commercial production of artemisinin. But yield optimization and industrial scale-up is very much mandatory to raise the production to such a level that ultimately resulted in the reduction of cost of ACTs, significantly below their present cost.

4.5 In Vitro Transformed Culture for Synthesis of Antimalarial Compound

Normally, medicinal plants are harvested from the wild, which causes habitat destruction leading to depletion of the irreplaceable valuable medicinal plant genetic diversity. Most often systematic cultivation of medicinal plants has been proposed and adopted as a substitute to collection from the wild condition. However, sometimes the spectra of compounds produced by medicinal plants under cultivation are quite different from those of wild populations. The fluctuations in quantities of phytochemicals are due to unfavorable environmental conditions, infestation, and diseases. In vitro plant tissue culture methodology, an attractive alternative system for the uniform production of phytochemicals, can continually provide high-value medical natural products. Some of the exclusive advantages that in vitro culture-based production technology provides include the following:

(a) Steady production supply independent of the geographical position, constraint of seasonal fluctuations of plant and environmental variations.
(b) Uniform growth under defined environment.
(c) Ease with which cultivation conditions can be altered and optimized to upregulate or downregulate the pathway expression.
(d) Biosynthetic steps expressed at low level or for very limited time in intact plants can be prolonged.
(e) No use of harmful pesticides and herbicides that are generally used during field.

Agrobacterium tumefaciens and *A. rhizogenes* plant pathogens contain Ti and Ri plasmids, which induce crown galls and hairy roots, respectively, during plant–*Agrobacterium* interactions. T-DNA genes of these plasmids have been found to influence tumor or hairy root development.

Over the past few years, hairy root cultures, produced via *A. rhizogene-m*ediated transformation, have emerged as ideal biotechnological sources of valuable natural products because of their genetic stability and sizable biomass production without the external application of phytohormones. A major characteristic of hairy root is their general ability to produce the desired phytochemicals concomitantly with growth. This is relevant because, unlike the production being repressed during the growth phase of dedifferentiated cell culture, it is possible to get constant and standardized production of the desired phytochemicals from growing hairy roots.

Transformed root cultures of *Cinchona ledgeriana* have been generated by infecting shoots cultured in vitro with *Agrobacterium rhizogenes* LBA 9402 [94]. A wide range of alkaloids including quinine, cinchonidine and quinidine was synthesized by transformed culture. The maximum production of alkaloids was found to be about 50 μg g–1 fresh weight after 45 days of culture.

Presently, *Artemisia annua* field growing plant resources cannot meet the increasing worldwide demands, while chemical synthesis is difficult and expensive because of its endoperoxide bridge. Therefore, biotechnology approach for increasing the accumulation of artemisinin level via metabolic engineering in transgenic *A. annua* plants and in genetically modified microbes is sought as a novel means for large-scale production and cost-effective commercialization of artemisinin.

Agrobacterium tumefaciens-mediated transformation in *A. annua* reported different strains and vectors. Most of the reports are restricted on the efficiency of transformation, but very few were concerned with artemisinin synthesis [95–97].

Hairy root cultures have been reported to accumulate those metabolites that normally synthesize in the aerial parts of the plant. Although it has been assumed that artemisinin accumulates only in the aerial part of the plants [98], numerous reports have established that hairy roots can also produce artemisinin [99–106]. Several efforts were implemented in the enhancement of artemisinin content through the establishment of hairy roots of transgenic strains of *A. annua* L. plants by *Agrobacterium rhizogene*-mediated genetic transformation [107–109]. The rate-limiting genes, including *fps* [110], *ads* [98, 111–113], and *hmgr* [57, 58, 60, 66, 97], involved in artemisinin biosynthetic pathway have been transformed and overexpressed. Hairy root cultures of *A. annua* L. have also been established with transformed *fps* gene from *G. arboreum*, and artemisinin content of the regenerated transgenic plants was significantly higher compared to that of non-transgenic plant [114]. The improvement of artemisinin production in hairy root culture by optimizing chemical and physical environmental factors has been extensively analyzed and implemented successfully. It has been reported that artemisinin accumulation increased significantly by improving the conditions of light irradiation in hairy root cultures of *A. annua* L. [115]. It was also reported that red light at 660 nm induces artemisinin content (31 mg g^{-1} dry cells) that was 67% higher than those obtained under white light [116]. Several elicitors including chitosan, methyl jasmonate (MeJ), salicylic acid (SA), and yeast extract (YE) can act as stimulants in artemisinin production in hairy root culture. It has been reported that artemisinin production by hairy roots of *A. annua* L. was increased up to sixfold with the supplementation of 150 mg L^{-1} chitosan alone with MeJ and 1.5-fold with the addition of chitosan and YE [117]. The synergistic effect of cerebrosides and nitric oxide on stimulation of artemisinin synthesis was studied. It has been reported that 2.3-fold increase of artemisinin was achieved on *A. annua* L. hairy root culture by application of cerebrosides and nitric oxide [118]. The ratio of NO_3/NH_4 and total initial nitrogen concentration also modulate the yield of artemisinin in hairy roots. It has been reported that in the ratio of NO_3/NH_4 at 5:1 (w/w) with the optimum concentration of total nitrogen increase, artemisinin production was 57% compared to that in standard MS medium [119]. The effect of gibberellic acid (GA_3) on the growth and secondary metabolite production of hairy roots of *A. annua* was investigated, and it was noticed that 28.9 μM GA_3 produces maximum amount of artemisinin in culture medium [120]. In another study, tetraploid clones of *A. annua* L. were generated from the YUT16 hairy root clone using colchicine. It has been reported that tetraploid clones produced up to six times more artemisinin compared to that of the parent [121]. The treatment of mycelial extracts from the endophytic fungus *Colletotrichum* sp. increased artemisinin content up to 44% over the control [116]. 549.1 mg l^{-1} artemisinin production was achieved by application of 0.3 mg total sugar mL^{-1} medium, elicitor derived from mycelial extracts of *Penicillium chrysogenum* 3446 for 3 days [122]. The effects of media sterilization methods and effect of type of sugar on growth and artemisinin accumulation of *A. annua* hairy roots were also extensively analyzed. It was reported that artemisinin accumulation from filter-sterilized medium was less than that from autoclaved medium [109]. The addition of (22S, 23S) homobrassinolide at 1 μg L^{-1} to hairy root cultures resulted in an increment of 57% over the control [119]. The effect of temperature on growth and artemisinin biosynthesis in hairy roots was investigated at a range of 15 °C–35 °C, and it was observed that the highest artemisinin content was achieved at 30 °C [123]. The stimulation of artemisinin biosynthesis in *A. annua* hairy root culture upon elicitation with oligogalacturonides was also reported [124]. The hairy roots of another species of *Artemisia* (*A. dubia* and *A. indica*), infected with *Agrobacterium rhizogenes* strains LBA 9402, produced maximum artemisinin [125]. Hairy root culture of *A. dubia* was established by another group of researchers, and 79 ± 3 μg g^{-1} of dry weight artemisinin was recorded upon exposure of 13.8 mg L^{-1} of SA [126]. The enhancement of artemisinin content up to threefold in hairy root culture of *A. annua* was reported by application of synthetic cytokinin (2-chloro-4-pyridyl)-N-phenylurea to the culture media [127]. The elicitor-induced modulation of artemisin biosynthesis was analyzed in hairy root of *A. annua*. The combination of NO donor sodium nitroprusside and oligosaccharide elicitor from *Fusarium oxysporum* mycelium induces artemisinin synthesis up to twofold compared to that of the control [128]. The hairy root system was developed using the *A. rhizogene* LBA 9402 strain to enhance artemisinin content in *A. annua* [129]. In a separate study, Baldi and Dixit have reported enhanced artemisinin production by the cell suspension culture of *A. annua* upon elicitation of MeJ in the culture medium [130].

4.6 Scale-Up of Hairy Root Cultures Through Bioreactor

The implementation of bioreactor is very much required for the production of hairy root-based secondary metabolite at an industrial level [131–133]. Usually gas- and liquid-phase bioreactors are used for hairy root culture in order to produce secondary metabolites. The liquid-phase reactor is user-friendly, and the growth and production kinetics can be easily monitored compared to gas-phase bioreactors [134]. To scale up the production of artemisinin, using hairy root culture in bioreactor, optimization of its growth and production ability in bioreactors with different bioreactor configuration is also necessary [131, 132, 135]. The cultivation of hairy roots was done in a modified 3-L stirred tank bioreactor, and 18 g biomass L^{-1} (on dry weight basis) and 4.63 mg L^{-1} of artemisinin were produced after 28 days [136]. They have reported that the artemisinin content enhanced up to 2.2-fold upon elicitation with methyl jasmonate. They have also established fed-batch cultivation that resulted in 0.32 mg g^{-1} accumulation of artemisinin after 16 days of cultivation [137]. A few reports are available in the literature, which focus on in vitro production of artemisinin in liquid-phase bioreactors using hairy root culture [79, 137–139]. Hairy roots in nutrient mist bioreactor (NMB) combined with novel cultivation strategies have been successfully established for scale-up of artemisinin production [140]. They have reported highest artemisinin content of 25.78 mg L^{-1} in the gas-phase NMB cultivation. Table 4 summarized the amount of artemisinin accumulation using different types of bioreactors.

Table 4 The different types of bioreactors used for artemisinin production using hairy root culture with artemisinin content

Sl no.	Type	Artemisinin content	References
1	Nutrient mist bioreactor	0.07–0.29 μg g^{-1}DW	[129]
2	Modified bubble column reactor	0.025 mg L-1	[136]
3	Modified nutrient mist bioreactor	0.031 mg L-1	[136]
4	Modified stirred tank bioreactor (STR)	4.63 mg L-1	[134]
5	Modified STR with elicitor	10.33 mg L-1	[134]
6	Model based fed-batch cultivation in modified STR	13.68 mg L-1	[135]
7	Modified bubble column reactor	1.50 mg L-1	[139]
8	Nutrient mist reactor	1.87 mg L-1	[139]
9	Combination bioreactor (modified NMB)	25.78 mg L-1	[139]

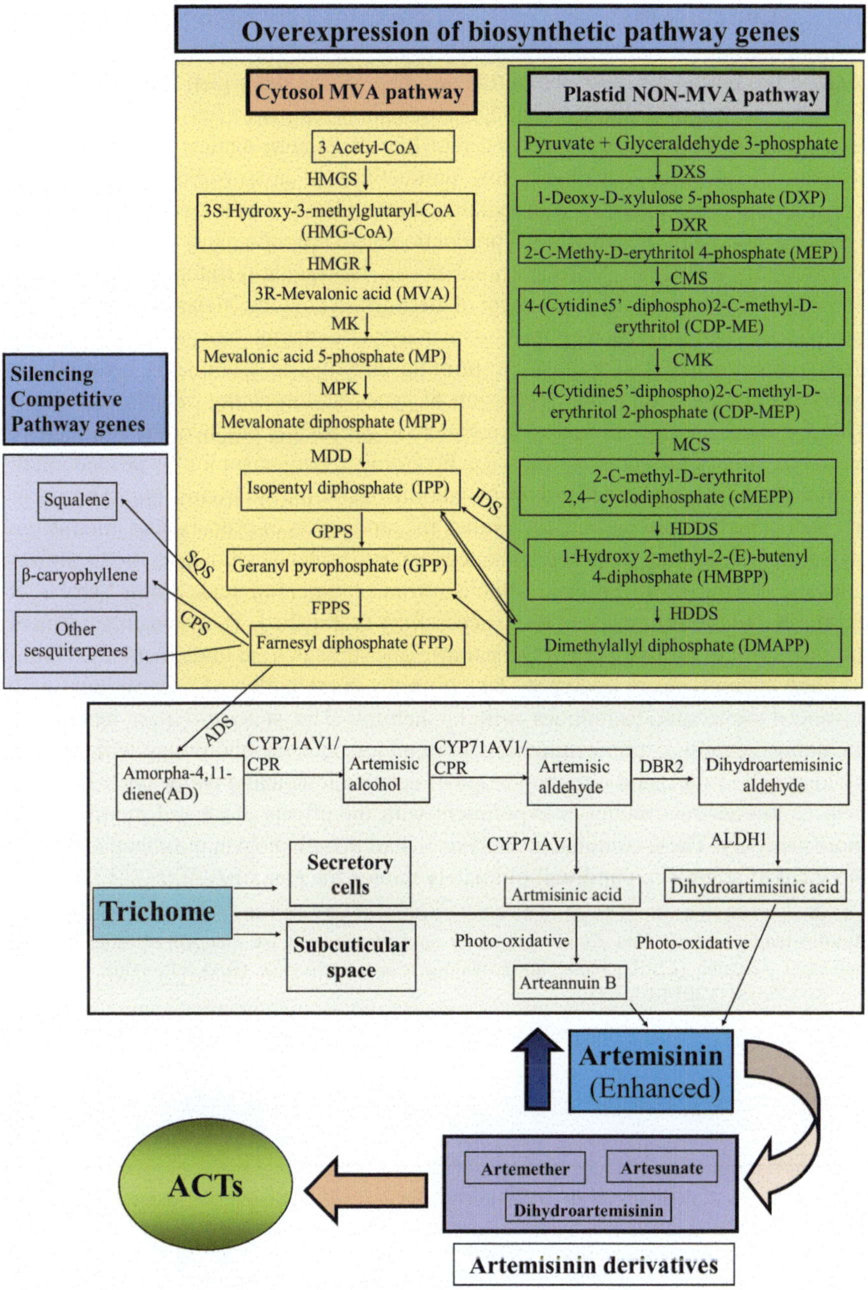

Fig. 4 Transgenic approaches for enhanced artemisinin biosynthesis in *A. annua* by upregulation artemisinin biosynthetic pathway genes and downregulation of competing pathway genes. [*HMGS* 3-hydroxy-3-methyl-CoA synthase, *HMGR* 3-hydroxy-3-methyl glutaryl coA reductase, *MK* mevalonate kinase, *MPK* mevalonate-5-phosphate kinase, *MDD* mevalonate diphosphate

5 Conclusions

Malaria is a global health problem that severely affects more than four lakh people annually, including children. Although artemisinin resistance has been reported in few countries, still the ACTs are the first line of defense against this devastating ailment. Unfortunately, relatively low productivity of artemisinin from *A. annua* L. plants and nonavailability of alternative economically feasible synthetic or semi-synthetic process of its commercial production have been the major blockage for the industrial production. A long number of transgenic strategies have been implemented for in vivo enhancement of artemisinin content. Metabolic engineering exploits the favorable perceptions to increase artemisinin biosynthesis by either overexpressing genes encoding rate-limiting enzymes in artemisinin biosynthetic pathways or suppressing the expression of genes of competing pathways (Fig. 4). This approach revealed an innovative system to genetically engineer the complete or partial secondary metabolic pathways in heterologous organism for the production of economically important secondary metabolites particularly artemisinin. The introduction of genetically modified microbes for production of precursor of artemisinin also added an advantage in this enhancement program. Nevertheless the optimization and scaling-up to produce artemisinin at commercial scale using hairy root culture-based bioreactors were also successfully established. The yield enhancement approach will be more fruitful if these transgenic strategies are implemented on high-yielding chemotypes or ecotypes. Therefore the combination of conventional with advanced molecular techniques will be helpful. The state-of-the-art techniques including genomics, transcriptomics, proteomics, and metabolomics with system biology will also expand our knowledge of detailed interacting pathways that would help us to construct model to experiment with the effects of altered modifications more precisely. These cumulative analysis will ultimately help in the enhancement of the yield of artemisinin and will ultimately reduce the cost of ACTs.

Acknowledgments Subrata Kundu gratefully acknowledges to the Council of Scientific and Industrial Research (CSIR), India, for providing research associate (RA) fellowship (Sanction No. 08/525(0004)/2014-EMR1).

Fig. 4 (continued) decarboxylase, *GPPS* geranyl diphosphate synthase, *FPPS* farnesyl pyrophosphate synthase, *ADS* amorpha-4,11-diene synthase, *SQS* squalene synthase, *CPS* β caryophyllene synthase, *DXS* 1-deoxy-D-xylulose-5-phosphate synthase, *DXR* 1-deoxy-D-xylulose-5-phosphate reductoisomerase, *CMS* 4-diphosphocytidyl-2C-methyl-D-erythritol 4-phosphate synthase, *CMK* 4-diphosphocytidyl-2C-methyl-D-erythritol kinase, *MCS* 2C-methyl-D-erythritol 2,4-cyclodiphosphate synthase, *HDDS* 1-hydroxy-2-methyl-2-(E)-butenyl 4-diphosphate synthase, *IDS* isopentyl diphosphate synthase, *CYP71AV1* cytochrome p450 monooxygenase, *CPR* cytochrome p450 reductase, *DBR2* artemisinic aldehyde Δ11(13) reductase, *ALDH1* aldehyde dehydrogenase 1, *ACTs* artemisinin-based combination therapy]

References

1. World Malaria report (2015) http://www.who.int/malaria/publications/world-malaria-report-2015/en/
2. Graham IA, Besser K, Blumer S, Branigan CA, Czechowski T, Elias L, Guterman I, Harvey D, Isaac PG, Khan AM et al (2010) The genetic map of *Artemisia annua* L. identifies loci affecting yield of the antimalarial drug artemisinin. Science 327:328–331
3. Malaney P, Spielman A, Sachs J (2004) The malaria gap. Am J Trop Med Hyg 71:141–146
4. Kotepui M, Piwkham D, PhunPhuech B, Phiwklam N, Chupeerach C, Duangmano S (2015) Effects of malaria parasite density on blood cell parameters. PLoS One 10:e0121057. doi:10.1371/journal.pone.0121057
5. Ward SA, Sevene EJ, Hastings IM, Nosten F, McGready R (2007) Antimalarial drugs and pregnancy: safety, pharmacokinetics, and pharmacovigilance. Lancet Infect Dis 7:136–144
6. Fidock DA, Rosenthal PJ, Croft SL, Brun R, Nwaka S (2004) Antimalarial drug discovery: efficacy models for compound screening. Nat Rev Drug Discov 3:509–520. doi:10.1038/nrd1416
7. Buchholz K, Burke TA, Williamson KC, Wiegand RC, Wirth DF, Marti M (2011) A high-throughput screen targeting malaria transmission stages opens new avenues for drug development. J Infect Dis 203:1445–1453. doi:10.1093/infdis/jir037
8. Schlitzer M (2007) Malaria chemotherapeutics part I: history of antimalarial drug development, currently used therapeutics and drugs in clinical development. Chem Med Chem 2:944–986
9. World Malaria Report (2005) http://www.who.int/malaria/publications/atoz/9241593199/en/
10. White NJ (2008) Qinghaosu (Artemisinin) the price of success. Science 320:331–334. doi:10.1126/science.1155165
11. Duffy PE, Mutabingwa TK (2006) Artemisinin combination therapies. Lancet 367:2037–2039
12. Klayman DL (1985) Qinghaosu (Artemisinin): an antimalarial drug from China. Science 228:1049–1055
13. Liu J, Ni M, Fan J, Tu Y, Wu Z, Wu Y, Zhou W (1979) Structure and reaction of Qinghaosu (in Chinese). Acta Chim Sin 37:129–143
14. Weathers P, Elkholy S, Wobbe KK (2006) Invited review: artemisinin: the biosynthetic pathway and its regulation in *Artemisia annua*, a terpenoid-rich species. In Vitro Cell Dev Biol Plant 42:309–317
15. Nair P, Misra A, Singh A, Shukla AK, Gupta MM et al (2013) Differentially expressed genes during contrasting growth stages of *Artemisia annua* for artemisinin content. PLoS One 8(4):e60375. doi:10.1371/journal.pone.0060375
16. Alam P, Kamaluddin, Khan MA, Mohammad A, Khan R, Abdin MZ (2014) Enhanced artemisinin accumulation and metabolic profiling of transgenic *Artemisia annua* L. plants over-expressing by rate limiting enzymes from isoprenoid pathway. J Plant Interact 9:655–665. doi:10.1080/17429145.2014.893030
17. Delabays N, Simonnet X, Gaudin M (2001) The genetics of artemisinin content in *Artemisia annua* L. and the breeding of high yielding cultivars. Curr Med Chem 8:1795–1801
18. Delabays N, Darbellay C, Galland N (2002) Variation and heritability of artemisinin content in *Artemisia annua* L. In: Wright CW (ed) Artemisia. Taylor & Francis, London, pp 197404–209999
19. Jain DC, Mathur AK, Gupta MM, Singh AK, Verma RK, Gupta AP, Kumar S (1996) Isolation of high artemisinin-yielding clones of *Artemisia annua*. Phytochemistry 6:993–1001
20. Townsend T, Segura V, Chigeza G, Penfield T, Rae A, Harvey D, Bowles D, Graham IA (2013) The use of combining ability analysis to identify elite parents for *Artemisia annua* F1 hybrid production. PLoS One 8(4):e61989. doi:10.1371/journal.pone.0061989
21. Paul S, Khanuja SP, Shasany AK, Gupta MM, Darokar MP, Saikia D, Gupta AK (2010) Enhancement of artemisinin content through four cycles of recurrent selection with relation to

heritability, correlation and molecular marker in *Artemisia annua* L. Planta Med 76:1468–1472
22. Pandey N, Pandey-Rai S (2014) Short term UV-B radiation-mediated transcriptional responses and altered secondary metabolism of in vitro propagated plantlets of *Artemisia annua* L. Plant Cell Tiss Org Cult 116:371–385. doi:10.1007/s11240-013-0413-0
23. Pandey N, Pandey-Rai S (2014) Modulations of physiological responses and possible involvement of defense-related secondary metabolites in acclimation of *Artemisia annua* L. against short-term UV-B radiation. Planta 240:611–627. doi:10.1007/s00425-014-2114-2
24. Muerhoff AS, Birkenmeyer LG, Coffey R, Dille BJ, Barnwell JW, Collins WE, Sullivan JS, Dawson GJ, Desai SM (2010) Detection of *Plasmodium falciparum*, *P. vivax*, *P. ovale*, and *P. malariae* merozoite surface protein 1-p19 antibodies in human malaria patients and experimentally infected nonhuman primates. Clin Vaccine Immunol 17:1631–1638. doi:10.1128/CVI.00196-10
25. Kantele A, Jokiranta TS (2011) Review of cases with the emerging fifth human malaria parasite, *Plasmodium knowlesi*. Clin Infect Dis 52:1356–1362. doi:10.1093/cid/cir180
26. Graz B, Kitua AY, Malebo HM (2011) To what extent can traditional medicine contribute a complementary or alternative solution to malaria control programmes? Malar J 10 (Suppl 1):S6
27. Willcox M, Burford G, Bodeker G (2004) An overview of ethnobotanical studies on plants used for the treatment of malaria. In: Willcox M, Bodeker G, Rasoanaivo P (eds) Traditional medicinal plants and malaria. CRC Press, Boca Raton, pp 187–197
28. Batista R, Silva ADJ Jr, de Oliveira AB (2009) Plant-derived antimalarial agents: new leads and efficient phytomedicines. Part II Non-alkaloidal natural products. Molecules 14:3037–3072
29. Kaur K, Jain M, Kaur T, Jain R (2009) Antimalarials from nature. Bioorg Med Chem 17:3229–3256
30. Oliveira AB, Dolabela MF, Braga FC, Jacome RLRP, Varotti FP, Povoa MM (2009) Plant-derived antimalarial agents: new leads and efficient phytomedicines. Part I alkaloids. An Acad Bras Cienc 81:715–740
31. Phillipson JD, O'Neill MJ (1987) Antimalarial and amoebicidal natural products. In: Hostettmann K, Lea PJ (eds) Biologically active natural products. Clarendon, Oxford, pp 49–64
32. Winter RW, Kelly JX, Smilkstein MJ, Dodean R, Bagby GC, Rathbun RK, Levin JI, Hinrichs D, Riscoe MK (2006) Evaluation and lead optimization of anti-malarial acridones. Exp Parasitol 114:47–56
33. Maguire JD, Sumawinata IW, Masbar S, Laksana B, Prodjodipuro P, Susanti I et al (2002) Chloroquine-resistant *Plasmodium malariae* in South Sumatra, Indonesia. Lancet 360:58–60
34. World Malaria Report (2010) http://www.who.int/malaria/world_malaria_report_2010/en/
35. Arab HA, Rahbari S, Rassouli A, Moslemi MH, Khosravirad F (2006) Determination of artemisinin in Artemisia sieberi and anticoccidial effects of the plant extract in broiler chickens. Trop Anim Health Prod 38:497–503
36. Hsu E (2006) The history of qinghao in the Chinese material medica. Trans R Soc Trop Med Hyg 100:505–508
37. Mannan A, Ahmed I, Arshad W, Asim MF, Qureshi RA, Hussain I, Mirz B (2010) Survey of artemisinin production by diverse Artemisia species in northern Pakistan. Malaria J 9:310
38. Mannan A, Ahmed I, Arshad W, Hussain I, Mirza B (2011) Effects of vegetative and flowering stages on the biosynthesis of artemisinin in Artemisia species. Arch Pharm Res 34:1657–1661
39. Acton N, Klayman DL, Rollman IJ (1985) Reductive electrochemical HPLC assay for artemisinin (qinghaosu). Planta Med 51:445–446
40. Liersch R, Soicke H, Stehr C, Tallner H-U (1986) Formation of artemisinin in *Artemisia annua* during one vegetation period. Planta Med 52:387–390
41. Martinez BC, Staba J (1988) The production of artemisinin in *Artemisia annua* L. tissue cultures. Adv Cell Cult 6:69–87

42. Singh A, Kaul VK, Mahajan VP, Singh A, Misra LN, Thakur RS, Husain A (1988) Introduction of *Artemisia annua* in India and isolation of artemisinin, a promising antimalarial drug. Indian J Pharm Sci 48:137–138
43. Madhusudanan K (1989) Mass spectral studies on artemisinin, dehydroartemisinin, and arteether. Indian J Chem 28B:751–754
44. Ferreira JFS, Janick J (1995) Floral morphology of *Artemisia annua* with special reference to trichomes. Int J Plant Sci 156:807–815
45. Duke MV, Paul RN, Elsohly HN, Sturtz G, Duke SO (1994) Localization of artemisinin and artemisitene in foliar tissues of glanded and glandless biotypes of *Artemisia annua* L. Int J Plant Sci 155:365–372
46. Ferreira JFS, Janick J (1997) *Artemisia annua*: botany, horticulture, pharmacology. In: Janick J (ed) Horticultural reviews, vol 19. Wiley, New York, pp 319–371
47. Kala CP (2006) Medicinal plants of the high altitude cold desert in India: diversity, distribution and traditional uses. Int J Biodivers Sci Manag 2:1–14
48. Verpoorte R, Contin A, Memelink J (2002) Biotechnology for the production of plant secondary metabolites. Phytochem Rev 1:13–25
49. Haynes RK (2006) From artemisinin to new artemisinin antimalarials: biosynthesis, extraction, old and new derivatives, stereochemistry and medicinal chemistry requirements. Curr Top Med Chem 6:509–537
50. Zhu C, Cook SP (2012) A concise synthesis of (?) – artemisinin. J Am Chem Soc 134:13577–13579
51. Vergauwe A, Cammaert R, Vandenberghe D, Genetello C, Van Montagu M, Vanden Eeckhout E (1996) *Agrobacterium tumefaciens*-mediated transformation of *Artemisia annua* L. and regeneration of transgenic plant. Plant Cell Rep 15:929–937
52. Sa G, Mi M, He-Chun Y, Ben-Ye L, Guo-feng L, Kang C (2001) Effects of ipt gene expression on the physiological and chemical characteristics of *Artemisia annua* L. Plant Sci 160:691–698
53. Chen DH, Liu CJ, Ye HC, Li GF, Liu BY, Meng YL, Chen XY (1999) Ri-mediated transformation of *Artemisia annua* with a recombinant farnesyl diphosphate synthase gene for artemisinin production. Plant Cell Tissue Org Cult 57:157–162
54. Chen D, Ye H, Li G (2000) Expression of a chimeric farnesyl diphosphate synthase gene in *Artemisia annua* L. transgenic plants via Agrobacterium tumefaciens-mediated transformation. Plant Sci 155:179–185
55. Han JL, Liu BY, Ye HC, Wang H, Li ZQ, Li GF (2006) Effects of over expression of the endogenous farnesyl diphosphate synthase on the artemisinin content in *Artemisia annua* L. J Integr Plant Biol 48:482–487
56. Banyai W, Kirdmanee C, Mii M, Supaibulwatana K (2010) Overexpression of farnesyl pyrophosphate synthase (FPS) gene affected artemisinin content and growth of *Artemisia annua* L. Plant Cell Tissue Org 103:255–265
57. Aquil S, Husaini AM, Abdin MZ, Rather GM (2009) Overexpression of the HMG-CoA reductase gene leads to enhanced Artemisinin biosynthesis in transgenic *Artemisia annua* plants. Planta Med 75:1–6
58. Nafis T, Akmal M, Ram M, Alam P, Ahlawat S, Mohammad A, Abdin MZ (2011) Enhancement of artemisinin content by constitutive expression of HMG CoA reductase gene in high yielding strain of *Artemisia annua* L. Plant Biotechnol Rep 5:53–60
59. Ma C, Wang HH, Lu X, Wang H, Xu G, Liu B (2009) Terpenoid metabolic profiling analysis of transgenic *Artemisia annua* L. by comprehensive two-dimensional gas chromatography time-of flight mass spectrometry. Metabolomics 5:497–506
60. Wang YY, Jing FY, Yu SY, Chen YF, Wang T, Liu P, Wang GF, Sun XF, Tang KX (2011) Co-overexpression of the HMGR and FPS genes enhances artemisinin content in *Artemisia annua* L. J Med Plants Res 5:3396–3403
61. Jing FY, Zhang L, Li MY, Tang KX (2008) Over-expressing CYP7AV1 and CPR genes enhances Artemisinin content in *Artemisia annua* L. J Agric Sci Tech Iran 10(3):64–70

62. Shen Q, Chen YF, Wang T, Wu SY, Lu X, Zhang L, Zhang FY, Jiang WM, Wang GF, Tang KX (2012) Overexpression of the cytochrome P450 monooxygenase (cyp71av1) and cytochrome P450 reductase (cpr) genes increased artemisinin content in *Artemisia annua* (Asteraceae). Genet Mol Res 11:3298–3309
63. Xiang LE, Zeng LX, Yuan Y, Chen M, Wang F, Liu XQ, Zeng LJ, Lan XZ, Liao ZH (2012) Enhancement of artemisinin biosynthesis by overexpressing dxr, cyp71av1 and cpr in the plants of *Artemisia annua* L. Plant Omics J 5:503–507
64. Tang KX, Shen Q, Chen YF, Wang T, Wu SY, Lu X (2012) Overexpression DBR2 gene increased artemisinin content in *Artemisia annua* L. Shanghai Jiao Tong University. Patent CN201210014227.2
65. Lu X, Shen Q, Zhang L, Zhang F, Jiang W, Lv Z, Yan T, Fu X, Wang G, Tang K (2013) Promotion of artemisinin biosynthesis in transgenic *Artemisia annua* by overexpressing *ADS*, *CYP71AV1* and *CPR* genes. Ind Crop Prod 49:380–385
66. Alam P, Abdin MZ (2011) Over-expression of HMG-CoA reductase and amorpha-4,11-diene synthase genes in *Artemisia annua* L. and its influence on artemisinin content. Plant Cell Rep 10:1919–1928
67. Chen Y, Shen Q, Wang Y, Wang T, Wu S, Zhang L et al (2013) The stacked over-expression of *FPS*, *CYP71AV1* and *CPR* genes leads to the increase of artemisinin level in *Artemisia annua* L. Plant Biotechnol Rep 7:287–295
68. Dilshad E, Cusido RM, Palazon J, Estrada KR, Mand B, Mirza B (2015) Enhanced artemisinin yield by expression of rol genes in *Artemisia annua*. Malar J 14:424. doi:10.1186/s12936-015-0951-5
69. Singh ND, Kumar S, Daniell H (2016) Expression of β-glucosidase increases trichome density and artemisinin content in transgenic *Artemisia annua* plants. Plant Biotechnol J 14:1034–1045. doi:10.1111/pbi.12476
70. Zhang F, Fu X, Lv Z, Lu X, Shen Q, Zhang L, Zhu M, Wang G, Sun X, Liao Z, Tang K (2015) A basic leucine zipper transcription factor, AabZIP1, connects abscisic acid signaling with artemisinin biosynthesis in *Artemisia annua*. Mol Plant 8:163–175. doi:10.1016/j.molp.2014.12.004
71. Yuan Y, Liu W, Zhang Q, Xiang L, Liu X, Chen M, Lin Z, Wang Q, Liao Z (2015) Overexpression of artemisinic aldehyde Δ11 (13) reductase gene-enhanced artemisinin and its relative metabolite biosynthesis in transgenic *Artemisia annua* L. Biotechnol Appl Biochem 62:17–23. doi:10.1002/bab.1234
72. Lu X, Zhang F, Shen Q, Jiang W, Pan Q, Lv Z, Yan T, Fu X, Wang Y, Qian H, Tang K (2014) Overexpression of allene oxide cyclase improves the biosynthesis of Artemisinin in *Artemisia annua* L. PLoS One 9(3):e91741. doi:10.1371/journal.pone.0091741
73. Farhi M, Marhevka E, Ben-Ari J, Algamas-Dimantov A, Liang Z, Zeevi V, Edelbaum O, Spitzer-Rimon B, Abeliovich H, Schwartz B, Tzfira T, Vainstein A (2011) Generation of the potent anti-malarial drug artemisinin in tobacco. Nat Biotechnol 29:1072–1074. doi:10.1038/nbt.2054
74. Hong GJ, Hu WL, Li JX, Chen XY, Wang LJ (2009) Increased accumulation of artemisinin anthocyanins in *Artemisia annua* expressing the Arabidopsis blue light receptor CRY1. Plant Mol Biol Rep 27:334–341. doi:10.1007/S11105-008-0088-6
75. Yu ZX, Li JX, Yang CQ, Hu WL, Wang LJ, Chen XY (2012) The jasmonate-responsive AP2/ERF transcription factors AaERF1 and AaERF2 positively regulate artemisinin biosynthesis in *Artemisia annua* L. Mol Plant 5:353–365
76. Zhang F, Lu X, Lv Z, Zhang L, Zhu M, Jiang W, Wang G, Sun X, Tang K (2013) Overexpression of the Artemisia orthologue of ABA receptor, AaPYL9, enhances ABA sensitivity and improves artemisinin content in *Artemisia annua* L. PLoS One 8(2): e56697
77. MaujiRam, Khan MA, Jha P, Khan S, Kiran U, Ahmad MM, Javed S, Abdin MZ (2010) HMG-CoA reductase limits artemisinin biosynthesis and accumulation in *Artemisia annua* L. plants. Acta Physiol Plant 32:859–866

78. Towler MJ, Weathers PJ (2007) Evidence of artemisinin production from IPP stemming from both the mevalonate and the nonmevalonate pathways. Plant Cell Rep 26:2129–2136
79. Liu C, Zhao Y, Wang Y (2006) Artemisinin: current state and perspectives for biotechnological production of an antimalarial drug. Appl Microbiol Biotechnol 72:11–20
80. Bertea CM, Freije JR, Van der Woude H, Verstappen FW, Perk L, Marquez V, De Kraker JW, Posthumus MA, Jansen BJM, De Groot AE, Franssen MCR, Bouwmeester HJ (2005) Identification of intermediates and enzymes involved in the early steps of artemisinin biosynthesis in *Artemisia annua*. Planta Med 71:40–47
81. Abdin MZ, Israr M, Rehman RU, Jain SK (2003) Artemisinin, a novel antimalarial drug: biochemical and molecular approaches for enhanced production. Planta Med 69:289–299
82. Woerdenbag HJ, Jos FJ, van Uden W, Pras N, Malingre TM, Alfermann AW (1993) Production of the new antimalarial drug artemisinin in shoot cultures of *Artemisia annua* L. Plant Cell Tissue Org Cult 32:247–257
83. Kudakasseril GJ, Lam L, Staba EJ (1987) Effect of sterol inhibitors on the incorporation of 14C-isopentenyl pyrophosphate into artemisinin by a cell-free system from *Artemisia annua* tissue cultures and plants. Planta Med 53:280–284
84. Zhang L, Jing F, Li F, Li M, Wang Y, Wang G, Sun X, Tang K (2009) Development of transgenic *Artemisia annua* (Chinese wormwood) plants with an enhanced content of artemisinin, an effective anti-malarial drug, by hairpin-RNA mediated gene silencing. Biotechnol Appl Biochem 52:199–207
85. Feng L, Yang R, Yang X, Zeng X, Lu W, Zeng Q (2009) Synergistic re-channeling of mevalonate pathway for enhanced artemisinin production in transgenic *Artemisia annua*. Plant Sci 177:57–67. doi:10.1016/j.plantsci.2009.03.014
86. Chen JL, Fang HM, Ji YP, Pu GB, Guo YW, Huang LL, Du ZG, Liu BY, Ye HC, Li GF, Wang H (2011) Artemisinin biosynthesis enhancement in transgenic *Artemisia annua* plants by downregulation of the b-caryophyllene synthase gene. Planta Med 77:1759–1765. doi:10.1055/s-0030-1271038
87. Yang R, Feng L, Yang X, Yin L, Xu X, Zeng Q (2008) Quantitative transcript profiling reveals down-regulation of a sterol pathway relevant gene and overexpression of Artemisinin biogenetic genes in transgenic *Artemisia annua* plants. Planta Med 74:1510–1516. doi:10.1055/s-2008-1081333
88. Lv Z, Zhang F, Pan Q, Fu X, Jiang W, Shen Q, Yan T, Shi P, Lu X, Sun X, Tang K (2016) Branch pathway blocking in *Artemisia annua* is a useful method for obtaining high yield Artemisinin. Plant Cell Physiol 57:588–602. doi:10.1093/pcp/pcw014
89. Martin VJJ, Pitera DJ, Withers ST, Newman JD, Keasling JD (2003) Engineering a mevalonate pathway in *Escherichia coli* for production of terpenoids. Nat Biotechnol 7:796–802
90. Ro DK, Paradise EM, Ouellet M, Fisher KJ, Newman KL, Ndungu JM, Ho KA, Eachus RA, Ham TS, Kirby J, Chang MC, Withers ST, Shiba Y, Sarpong R, Keasling JD (2006) Production of the antimalarial drug precursor artemisinic acid in engineered yeast. Nature 440:940–943
91. Tsuruta H, Paddon CJ, Eng D, Lenihan JR, Horning T, Anthony LC, Regentin R, Keasling JD, Renninger NS, Newman JD (2009) High-level production of amorpha-4,11-diene, a precursor of the antimalarial agent artemisinin, in *Escherichia coli*. PLoS One 4:e4489
92. Westfall PJ, Pitera DJ, Lenihan JR, Eng D, Woolard FX, Regentin R, Horning T, Tsuruta H, Melis DJ, Owens A, Fickes S, Diola D, Benjamin KR, Keasling JD, Leavell MD, McPhee DJ, Renninger NS, Newman JD, Paddon CJ (2012) Production of amorphadiene in yeast, and its conversion to dihydroartemisinic acid, precursor to the antimalarial agent artemisinin. Proc Natl Acad Sci 109.E111–E118
93. Paddon CJ, Westfall PJ, Pitera DJ, Benjamin K, Fisher K, McPhee D, Leavell MD, Tai A, Main A, Eng D, Polichuk DR et al (2013) High-level semisynthetic production of the potent antimalarial artemisinin. Nature 496:528–532
94. Hamill JD, Robins RJ, Rhodes MJ (1989) Alkaloid production by transformed root cultures of *Cinchona ledgeriana*. Planta Med 55:354–357

95. Ghosh B, Mukherjee S, Jha S (1997) Genetic transformation of *Artemisia annua* by *Agrobacterium tumefaciens* and artemisinin synthesis in transformed cultures. Plant Sci 122:193–199
96. Vergauwe A, Van Geldre E, Inze D, Van Montagu M, Van Den Eeckhout E (1998) Factors influencing *A. tumefaciens* mediated transformation of *Artemisia annua* L. Plant Cell Rep 8:105–110
97. Han JL, Wang H, Ye HC, Liu Y, Li ZQ, Zhang L et al (2005) High efficiency of genetic transformation and regeneration of *Artemisia annua* L. via *Agrobacterium tumefaciens*-mediated procedure. Plant Sci 168:73–80
98. Wallaart TE, Bouwmeester HJ, Hille J, Poppinga L, Maijers NCA (2001) Amorpha-4,11-diene synthase: cloning and functional expression of a key enzyme in the biosynthetic pathway of the novel antimalarial drug artemisinin. Planta 212:460–465
99. Weathers PJ, Cheetham RD, Follansbee E, Tesh K (1994) Artemisinin production by transformed roots of *Artemisia annua*. Biotechnol Lett 16:1281–1286
100. Jaziri M, Shimomura K, Yoshimatsu K et al (1995) Establishment of normal and transformed root cultures of *Artemisia annua* L. for artemisinin production. J Plant Physiol 145:175–177
101. Liu CZ, Wang YC, Ouyang F, Ye HC, Li GF (1999) Improvement of artemisinin accumulation in hairy root culture of *Artemisia annua* L. by fungal elicitor. Bioprocess Eng 20:161–164
102. Giri A, Ravindra ST, Dhingra V, Narasu ML (2000) Influence of different strains of *Agrobacterium rhizogenes* on induction of hairy roots and artemisinin production in *Artemisia annua*. Curr Sci 81:378–382
103. Liu CZ, Zhao Y, Wang YC (2006) Artemisinin: current state and perspectives for biotechnological production of an antimalarial drug. Appl Microbiol Biotechnol 71:11–20
104. Covello PS (2008) Making artemisinin. Phytochemistry 69:2881–2885
105. Rao SR, Ravishankar GA (2002) Plant cell cultures: chemical factories of secondary metabolites. Biotechnol Adv 20:1001–1153
106. Guillon S, Tre'mouillaux-Guiller J, Pati PK, Rideau M, Gantet P (2006) Hairy root research: recent scenario and exciting prospects. Curr Opin Plant Biol 9(3):341–346. doi:10.1016/j.pbi.2006.03.008
107. He XZM, Li G, Liang Z (1983) Callus induction and regeneration of plantlets from *Artemisia annua* and changes in qinghaosu contents. Acta Bot Sin 25:87–90
108. Mukherjee S, Ghosh B, Jha S (1995) Genetic transformation of *Artemisia annua* by *Agrobacterium rhizogenes*. Indian J Exp Biol 33:868–871
109. Weathers PJ, De Jesus-Gonzalez L, Kim YJ et al (2004) Alteration of biomass and artemisinin production in *Artemisia annua* hairy roots by media sterilization method and sugars. Plant Cell Rep 23:414–418
110. Matsushita Y, Kang W, Charlwood BV (1996) Cloning and analysis of a cDNA encoding farnesyl diphosphate synthase from *Artemisia annua*. Gene 172:207–209
111. Mercke P, Bengtsson M, Bouwmeester HJ et al (2000) Molecular cloning, expression, and characterization of amorpha-4,11-diene synthase, a key enzyme of artemisinin biosynthesis in *Artemisia annua* L. Arch Biochem Biophys 381:173–180
112. Chang YJ, Song SH, Park SH, Kim SU (2000) Amorpha-4, 11-diene synthase of *Artemisia annua*: cDNA isolation and bacterial expression of a terpene synthase involved in artemisinin biosynthesis. Arch Biochem Biophys 383:178–184
113. Picaud S, Olofsson L, Brodelius M, Brodelius PE (2005) Expression, purification, and characterization of recombinant amorpha-4,11-diene synthase from *Artemisia annua* L. Arch Biochem Biophys 436:215–226
114. Wu S, Schalk M, Clark A et al (2006) Redirection of cytosolic or plastidic isoprenoid precursors elevates terpene production in plants. Nat Biotechnol 24:1441–1447
115. Liu CZ, Guo C, Wang YC, Ouyang F (2002) Effect of light irradiation on hairy root growth and artemisinin biosynthesis of *Artemisia annua* L. Process Biochem 38:581–585
116. Wang JW, Zhang Z, Tan RX (2001) Stimulation of artemisinin production in *Artemisia annua* hairy roots by the elicitor from the endophytic *Colletotrichum* sp. Biotechnol Lett 23:857–860

117. Putalun W, Luealon W, De- Eknamkul W, Tanaka H, Shoyama Y (2007) Improvement of artemisinin production by chitosan in hairy root cultures of *Artemisia annua* L. Biotechnol Lett 29:1143–1146
118. Wang JW, Zheng LP, Zhang B, Zou T (2009) Stimulation of artemisinin synthesis by combined cerebrosides and nitric oxide elicitation in *Artemisia annua* hairy roots. Appl Microbiol Biotechnol 85:285–292
119. Wang JW, Tan RX (2002) Artemisinin production in *Artemisia annua* hairy root cultures with improved growth by altering the nitrogen source of the medium. Biotechnol Lett 24:1153–1156
120. Smith TC, Weathers PJ, Cheetham RC (1997) Effect of gibberellic acid on hair root cultures of *Artemisia annua* growth and artemisinin production. In Vitro Cell Dev Biol Plant 33:75–79
121. De Jesus-Gonzalez L, Weathers PJ (2003) Tetraploid *Artemisia annua* hairy roots produce more artemisinin than diploids. Plant Cell Rep 21:809–813
122. Liu C, Xu W, Ouyang F, Ye H, Li G (1999) Improvement of artemisinin accumulation in hairy root cultures of *Artemisia annua* L by fungal elicitor. Bioprocess Eng 20:161–164
123. Guo C, Liu CZ, Ye HC, Li GF (2004) Effect of temperature on growth and artemisinin biosynthesis in hairy root cultures of *Artemisia annua*. Acta Botan Boreali-Occident Sin 24:1828–1831
124. Zhang B, Zou T, Lu Y, Wang J (2010) Stimulation of artemisinin biosynthesis in *Artemisia annua* hairy roots by oligogalacturonides. Afr J Biotechnol 9:3437–3442
125. Mannan A, Shaheen N, Arshad W, Qureshi RA, Zia M, Mirza B (2008) Hairy roots induction and artemisinin analysis in *Artemisia dubia* and *Artemisia indica*. Afr J Biotechnol 7:3288–3292
126. Ali M, Kiani BH, Mannan A, Ismail T, Mirza B (2012) Enhanced production of artemisinin by hairy root cultures of *Artemisia dubia*. J Med Plants Res 6:1619–1622
127. Oba EG, Inthima P, Supaibulwatana K (2012) Effect of (2-chloro-4-pyridyl)-N-phenylurea (CPPU) on artemisinin production in hairy root culture of *Artemisia annua* L. Afr J Biotechnol 11:13992–13997
128. Zheng LP, Guo YT, Wang JW, Tan RX (2008) Nitric oxide potentiates oligosaccharide-induced artemisinin production in *Artemisia annua* hairy roots. J Integr Plant Biol 50:49–55. doi:10.1111/j.1744-7909.2007.00589.x
129. Ahlawat S, Saxena P, Alam P, Wajid S, Abdin MZ (2014) Modulation of artemisinin biosynthesis by elicitors, inhibitor, and precursor in hairy root cultures of *Artemisia annua* L. J Plant Interact 9:811–824
130. Baldi A, Dixit VK (2008) Enhanced artemisinin production by cell cultures of *Artemisia annua*. Curr Trends Biotechnol Pharm 2:341–348
131. Kim Y, Wyslouzil BE, Weathers PJ (2001) A comparative study of mist and bubble column reactors in the in vitro production of artemisinin. Plant Cell Rep 20:451–455
132. Kim YJ, Weathers PJ, Wyslouzil BE (2002) Growth of *Artemisia annua* hairy roots in liquid- and gas-phase reactors. Biotechnol Bioeng 80:454–464
133. Patra N, Srivastava AK, Sharma S (2013) Study of various factors for enhancement of artemisinin in *Artemisia annua* hairy roots. Int J Chem Eng Appl 4:157–160
134. Weathers PJ, Towler MJ, Xu J (2010) Bench to batch: advances in plant cell culture for producing useful products. Appl Microbiol Biotechnol 85:2354–2364
135. Liu C, Wang Y, Guo C, Ouyang F, Ye H, Li G (1998) Enhanced production of artemisinin by *Artemisia annua* L. hairy root cultures in a modified inner-loop airlift bioreactor. Bioprocess Eng 19:389–392
136. Patra N, Srivastava AK (2014) Enhanced production of artemisinin by hairy root cultivation of *Artemisia annua* in a modified stirred tank reactor. Appl Biochem Biotechnol 174:2209–2222
137. Patra N, Srivastava AK (2015) Use of model-based nutrient feeding for improved production of artemisinin by hairy roots of *Artemisia annua* in a modified stirred tank bioreactor. Appl Biochem Biotechnol 177:373–388

138. Kim YJ, Weathers PJ, Wyslouzil BE (2003) Growth dynamics of *Artemisia annua* hairy roots in three culture systems. Biotechnol Bioeng 83:428–443
139. Liu CZ, Wang YC, Zhao B, Guo C, Ouyang F, Ye HC, Li GF (1999) Development of a nutrient mist bioreactor for growth of hairy roots. In Vitro Cell Dev-Pl 35:271–274
140. Patra N, Srivastava AK (2016) Artemisinin production by plant hairy root cultures in gas- and liquid-phase bioreactors. Plant Cell Rep 35:143–153

Evaluation of *Agrobacterium tumefaciens* Usefulness for the Transformation of Sage (*Salvia officinalis L.*)

6

Aleksandra Luwańska, Karolina Wielgus, Katarzyna Seidler-Łożykowska, Daniel Lipiński, and Ryszard Słomski

Abstract

Herbal plants have aroused people's interest for centuries as possible remedies for a number of conditions, in particular chronic and incurable diseases. Today, by combining genetic engineering methods with in vitro culture techniques, these plants can be modified for biomedical purposes as alternative systems for the production of biopharmaceuticals. The present research has examined possibilities of obtaining transgenic sage and aimed to establish a basis for preparing an oral, plant-produced recombinant vaccine against dental decay. Sage was chosen based on the fact that it is being used for treating oral cavity inflammations and in prophylaxis. The strain LBA4404 of *Agrobacterium tumefaciens*, most frequently used in the transformation of dicotyledons, was chosen for the transgenesis. The authors assessed the usefulness of this bacteria strain for sage transformation.

High antibacterial activity of sage against a wide range of microorganisms, including *A. tumefaciens*, may prevent or limit genetic modification in plants obtained by vector-based methods for transgenesis. Therefore, nonvector methods should be recommended to obtain the successful transformation of sage due to its strong antibacterial effectiveness.

Keywords

Sage • In vitro cultures • Transgenesis • *Agrobacterium tumefaciens* • Antibacterial activity • Secondary metabolites

---

A. Luwańska (✉) • K. Wielgus (✉) • K. Seidler-Łożykowska (✉)
Institute of Natural Fibres and Medicinal Plants, Poznan, Poland
e-mail: aleksandra.luwanska@iwnirz.pl; karolina.wielgus@iwnirz.pl; katarzyna.lozykowska@iwnirz.pl

D. Lipiński (✉) • R. Słomski (✉)
Department of Biochemistry and Biotechnology, Poznan University of Life Sciences, Poznan, Poland
e-mail: lipinskidaniel71@gmail.com; slomski@up.poznan.pl

S. Jha (ed.), *Transgenesis and Secondary Metabolism*, Reference Series in Phytochemistry, DOI 10.1007/978-3-319-28669-3_2

Contents

1 Introduction

1.1 Botanical Description of Sage

Salvia genus belongs to the Nepetoideae subfamily in the Lamiaceae family. About 900 species of the plant have been recognized [1]. Sage (*Salvia officinalis* L.), a small evergreen plant, originating from the Mediterranean and Asia Minor, has been used for ages as a medicinal plant [2]. Sage is a perennial dwarf shrub. It grows up to 40–80 cm in height and has silvery-green lance-shaped leaves covered with tomentose. The plant blooms in the second year of cultivation, and its violet flowers are attractive to insects. The fruit is a schizocarp which splits into four black smooth-surfaced spherical mericarps [3]. The weight of 1000 seeds is approximately 8 g. Seeds are used to establish sage plantations. Sage needs rich, permeable, moist soil. The plants are resistant to drought. In Poland, however, they can be sensitive to frost, particularly during snowless winters. A perfect site for growing sage should be warm, isolated, and protected against winds. In crop rotation, sage can be cultivated after rape, root crops, or legumes. Sage is not a good forecrop as it leaves the field dried and weedy. Sage is cultivated in Europe, Russia, North America, and Africa. In Poland it is cultivated on an area of 600 ha. The Polish sage cultivar "Bona" produces the yield of 3 $t \cdot ha^{-1}$ of dry herb, containing 1.4% of essential oils. As 95% of flowers are male fertile, the seed yield is approximately 0.5 $t \cdot ha^{-1}$.

1.2 Sage Secondary Metabolites and Their Activity

Sage leaves (*Salviae folium*) and herb (*Salviae herba*) are raw materials harvested before flowering. Sage raw material contains essential oil, tannins, flavonoids, saponins, resins, and vitamins. The major components of sage oil are monoterpenes: α- and β-thujone, camphor, borneol, and 1,8-cineole [4–7]. The precursors of all sage

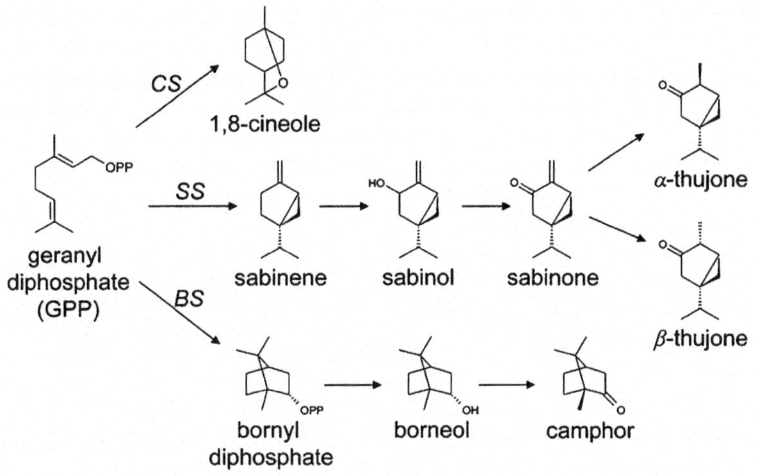

Fig. 1 Three terpene synthases and their products: 1,8-cineole synthase (*CS*) for the one-step formation of 1,8-cineole, (+)-sabinene synthase (*SS*) as a first step to α- and β-thujone, and (+)-bornyl diphosphate synthase (BS) en route to borneol and camphor (based on the data of Grausgruber-Gröger et al. [7])

monoterpenes are geranyl diphosphate (GGPP) and neryl diphosphate formed in the plastids via the 1-deoxy-D-xylulose-5-phosphate pathway [7]. Three distinct monoterpene synthases are responsible for the first steps in the formation of the major monoterpenes (Fig. 1). The (+)-sabinene synthase (SS) catalyzes the production of sabinene, which undergoes further rearrangements leading to the two major monoterpenes α- and β-thujone. The 1,8-cineole synthase (CS) produces in one step 1,8-cineole. Finally, (+)-bornyl diphosphate synthase (BS) produces bornyl diphosphate, which is subsequently hydrolyzed to borneol and then oxidized to camphor.

Monoterpenes are produced and accumulated mainly in epidermal glands. After their formation in the metabolically highly active secretory cells, they are transported through the cytoplasm and the cell wall and stored in subcuticular oil storage cavities. Grausgruber-Gröger et al. [7] investigated the seasonal influence on the formation of the main monoterpenes in young leaves of the field-grown sage plants in two cultivars at the level of mRNA expression, analyzed by qRT-PCR, and at the level of end products, analyzed by gas chromatography. All monoterpene synthases and monoterpenes were significantly influenced by cultivar and season. 1,8-Cineole synthase and its end product 1,8-cineole remained constant until August and then slightly decreased. The thujones increased steadily during the vegetative period. The transcript level of their corresponding terpene synthase, however, showed its maximum in the middle of the vegetative period and declined afterwards. Camphor remained constant until August and then declined, exactly correlated with the mRNA level of the corresponding terpene synthase.

Sage raw material also contains catechin; tannins; triterpenes; flavonoids; bitters (carnosol); organic acids; vitamins B1, C, and PP; and carotenes [8–11]. In many studies, sage is mentioned as a source of polyphenol compounds. Rosmarinic acid, an ester of caffeic acid and 3,4-dihydroxyphenyl lactic acid, was identified as one of the active components with high biological activity. The phenylpropanoid and tyrosine-derived pathways are both involved in the biosynthesis of rosmarinic acid in plants (Fig. 2). L-phenylalanine (L-Phe) is transformed to 4-coumaroyl-CoA by the enzymes of the phenylpropanoid pathway (phenylalanine ammonia lyase, cinnamic acid 4-hydroxylase, and hydroxycinnamic acid: coenzyme A ligase). On the other side, tyrosine can be transaminated to 4-hydroxyphenyllactate (pHPL) by two enzymes: tyrosine aminotransferase and hydroxyphenylpyruvate reductase. The hydroxycinnamoyl moiety of hydroxycinnamoyl-CoA is then transferred to the aliphatic hydroxyl group of a hydroxyphenyllactate by hydroxycinnamoyl-CoA: hydroxyphenyllactate hydroxycinnamoyltransferase ("rosmarinic acid synthase," RAS) which was shown to prefer the monohydroxylated substrates, 4-coumaroyl-CoA and pHPL. The hydroxyl groups in positions 3 and 3′ of the aromatic rings are finally introduced by cytochrome P450 monooxygenases [11].

Sage is known for its multiple pharmacological effects including antibacterial, antiviral, antioxidant, anti-inflammatory, antidiabetic, and antitumor properties, which are related to its active compounds [7, 11–18]. Sage essential oil has a strong inhibiting effect against the multiplication of numerous species of gram-positive and gram-negative bacteria, including those resistant to antibiotics. The oil also neutralizes bacterial toxins and has an antioxidant effect [19, 20]. The antiseptic effect of the oil is used as an alternative method for controlling microorganism multiplication and for food preservation [21, 22].

Sage extracts are used in treating gastrointestinal inflammations, bleeding, and ulcers [23]. Sage is also recommended for rinsing inflamed mucous membranes in the oral cavity and throat [24]. It is also used in excessive intestinal fermentation, flatulence, gastritis, and diarrhea as it regulates the functioning of the digestive system.

Sage raw material is an ingredient of numerous medications and medicinal preparations. What is more, in vitro research has demonstrated that sage is highly active against the Herpes simplex viruses HSV1 and HSV2 [24]. The chemical compounds in the raw materials carnosol, epirosemanol, and methyl carnosol play an important role in the prophylaxis of numerous diseases of affluence, such as atherosclerosis, diabetes, cataract, liver diseases, Parkinson disease, and Alzheimer disease [25–27]. Sage extracts protect cells from DNA damage, stimulate its repair [28], and inhibit the development of cancer cells [29]. In recent years it has been discovered that diterpenes isolated from sage activate PPARs, which influences their antidiabetic effect [30]. Sage preparations are also used in treating menopause symptoms [31]. Sage is also used as herbal tea and spice and in cosmetics and perfumery [32].

High thujone content in the essential oil can produce negative effect of the raw material. The EU Scientific Committee on Food [33] indicates that α-thujone acts as a GABA blocker and is three times more potent than β-thujone and that thujone

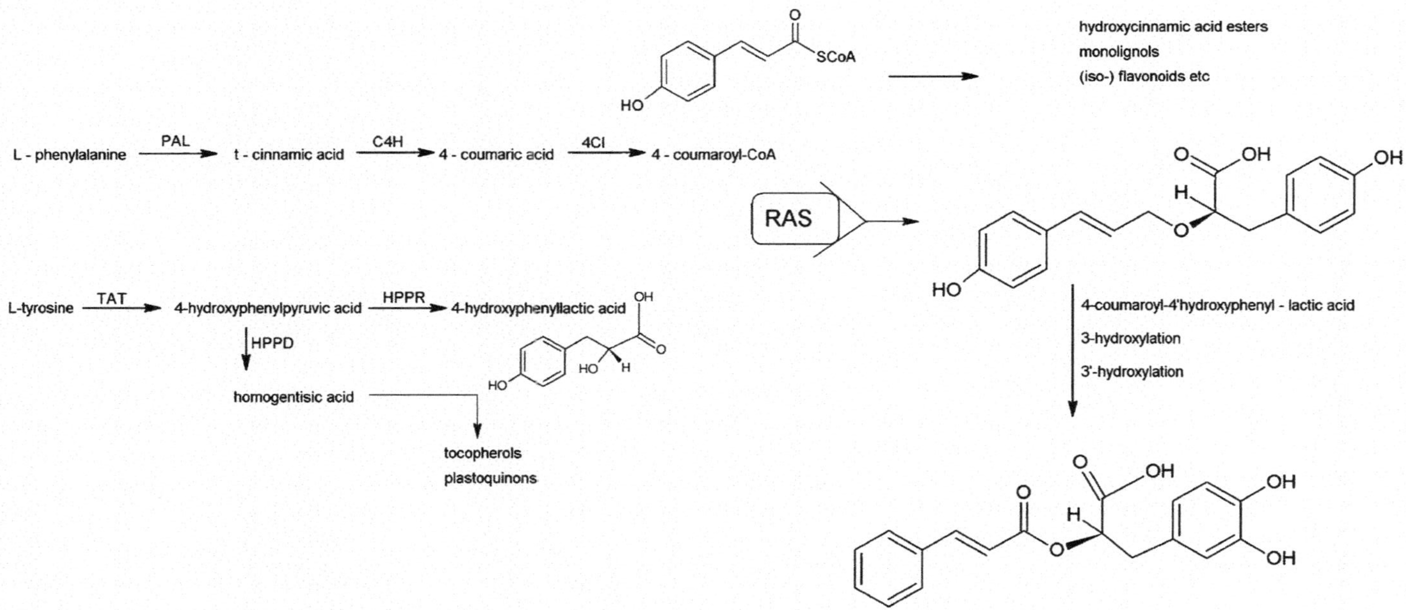

Fig. 2 The biosynthetic pathway leading to rosmarinic acid with some side reactions. PAL phenylalanine ammonia lyase, C4H cinnamic acid 4-hydroxylase, 4CL 4-coumarate:CoAligase, TAT tyrosine aminotransferase, HPPR hydroxyphenylpyruvate reductase, HPPD hydroxyphenylpyruvate dioxygenase, and RAS rosmarinic acid synthase (based on data of Ejtahed et al. [11])

causes convulsions and epileptiform seizures. The maximum sufficiently protective daily dose of thujone for humans is 7 mg [33]. What is more, there have been reports on toxic effects of sage on newborns and infants [34].

1.3 In Vitro Production of Sage Secondary Metabolites

Herbal raw material derived from in vitro cultures is gaining popularity as an alternative method of obtaining secondary metabolites for multipurpose use. The fact that in vitro cultures allow for free control of the process may have a significant effect on the increase in secondary metabolite synthesis [35]. The development of biotechnological methods for producing standardized valuable phytochemicals (e.g., food additives, pharmaceuticals, and pesticides) based on plant in vitro cultures offers the advantages of continuous production under controlled conditions, independently of environmental factors (geographical latitude, climatic change, and seasonal variation) [36]. The potential of in vitro *Salvia* systems as tools for controlled production of valuable secondary metabolites has been investigated [18]. In particular, considerable attention has been paid to *S. miltiorrhiza*, although the content of *S. officinalis* active substances from in vitro system was also reported (Table 1). The more recent findings confirmed usefulness of cell suspensions and hair root cultures as the potential of terpenoids and polyphenols accumulation.

Only a few reports have described cell suspensions and hairy roots system from sage as a source of secondary metabolites. In a comparative study of various sage in vitro systems with different levels of differentiation, it was found that hairy roots and cell suspensions were the best systems for rosmarinic acid production, whereas the shoot cultures were the only system that produced carnosic acid and carnosol [19]. Bolta et al. [39] observed that the biosynthesis of ursolic acid by sage cell culture was adversely affected by cell differentiation: the single-cell fraction of sage cell suspension accumulated almost 50-fold higher amount of ursolic acid than highly aggregated suspension culture. However, the tested sage cell suspensions exhibited negligible biomass accumulation when forced to grow as single cells [39]. Hairy root cultures are the most effective in vitro systems for the production of valuable secondary metabolites whose biosynthesis occurs primarily in the roots of mature plants [60]. The modified metabolism of sage hairy roots may also result in production of novel biologically active compounds that are not typically found in the field-growing plants [18]. Genetic engineering, elicitation, [51, 54, 59] and even bioreactor designs may improve the yields of produced secondary metabolites of sage hairy roots. Grzegorczyk and Wysokińska [70] reported that sage hairy roots accumulated up to 1.6-fold more biomass when cultivated in nutrient sprinkle bioreactor in comparison with shake-flasks cultivation. They also found that the amount of accumulated rosmarinic acid from this bioreactor system was up to fivefold higher than that detected in mature sage plants. Also Grzegorczyk et al. [19] found that hairy roots of sage accumulated higher amount of rosmarinic acid (31 mg rosmarinic acid/g DW) than mature plant roots, in vitro shoots, or cell suspension cultures, and therefore their extracts exhibited higher antioxidant activity.

Table 1 Recently reported bioactive terpenoids and polyphenols from *Salvia* in vitro systems (based on the data of Marchev et al. [18])

Secondary metabolite	Species/in vitro system	References
Triterpenes	*S. tomentosa*/callus; *S. scabiosifolia*/callus; *S. officinalis*/cell suspension; *S. officinalis*/shoot culture; *S. sclarea*/hairy roots; *S. austriaca*/hairy roots	Georgiev et al. [37]; Marchev et al. [38]; Bolta et al. [39]; Grzegorczyk et al. [19]; Kuźma et al. [40]; Kuźma et al. [41]; Kuźma et al. [42]
Tanshinones	*S. miltiorrhiza*/cell suspension; *S. miltiorrhiza*/hairy roots	Wu and Shi [43]; Yuan et al. [44]; Zhao et al. [45], (2011) [46]; Gupta et al. [47]; Yan et al. [48]; Yang et al. [49], [50], [51]; Kai et al. [52], [53]; Gu et al. [54]; Liang et al. [55]
Hydrophilic phenolic compounds	*S. miltiorrhiza*/hairy roots; *S. miltiorrhiza*/cell suspension	Yuan et al. [44]; Xiao et al. [56]; Dong et al. [57]; Gu et al. [54]; Hao et al. [58]
Phenolic acids	*S. officinalis*/hairy roots; *S. officinalis*/cell suspension; *S. miltiorrhiza*/hairy roots *S. miltiorrhiza*/cell suspension *S. tomentosa*/hairy roots	Grzegorczyk et al. [19]; Xiao et al. [56]; Dong et al. [57]; Zhao et al. [46]; Xiao et al. [59]; Gu et al. [54]

Sage cell suspension cultures currently seem to be more useful as model systems for biochemical, biosynthetic, and genetic studies. However, it should be considered as a potential method for the commercial production of sage secondary metabolites. Sage is also an attractive subject for hairy roots induction, mainly because of its potential to produce a wide range of biologically active compounds that accumulate in the roots [18].

2 Estimation of Sage Secondary Metabolites in In Vitro Cultures

In vitro culture conditions (the type and concentration of plant growth regulators) are different for every species. Micropropagation conditions for species of the *Salvia* genus, such as *Salvia sclarea* [61], *Salvia fruticosa* [62], and *Salvia guaranitica* [63], have been established. In the Institute of Natural Fibres and Medicinal Plants (INF&MP), an effective protocol for the micropropagation of sage cultivar "Bona," which ensures a rapid growth of the plants, has been developed [64]. Effective micropropagation and successful regeneration of the plants were achieved by using cytokinins: 6-benzylaminopurine (BA) or meta-topolin (mT) in the concentration of 0.3 mg/L in apical and axillary bud cultures.

In the 2-week cyclic culture period, the multiplication rate (MR) was similar in all combinations, except for the MS BA + NAA combination, where it was significantly lower (Table 2). Cultures grown on media containing BA or mT were characterized by a good condition of explants (creating buds and growing) during cyclic micropropagation, as well as high multiplication rate. Other combinations of growth regulators decreased the multiplication rate, and the explants were characterized by a worse physiological condition (vitrificated and necrosed). In her research, Gostin used BA during a 6-week sage culture, obtaining an effective regeneration of plants with good morphological parameters [65]. The author also obtained regenerated morphologically normal sage plants by using kinetin (KIN) together with 1-naphthaleneacetic acid (NAA). The usefulness of BA was documented together with other growth regulators: 2,4-dichlorophenoxyacetic acid (2,4-D) [20] or indole-3-acetic acid (IAA) [66]. The innovative use of mT, which had not been used in sage micropropagation before, had a positive effect on the proliferation of buds and may become an alternative for the widely used BA cytokinin.

As far as biosynthetic properties are concerned, callus and suspension cultures are particularly useful as a possible valuable source of secondary metabolites for industrial use. However, in the case of unorganized cultures, there may be differences in the composition of secondary metabolites as opposed to organized cultures. Analysis of sage chemical composition showed that the production of diterpenoid compounds (carnosic acid and carnosol) is closely linked to the differentiation of organs. As regards the production of rosmarinic acid, it was similar in callus and suspension cultures and in in vitro and in vivo grown plants [67]. Callus cultures and

Table 2 Selected plant growth regulators used in micropropagation of apical and axillary buds

Combination		Multiplication rate (MR)	Explants [%]			
			Creating buds	Growing	Vitrificated	Necrosed
Medium	MS 0.3 BA	**2.952a**	**91.1a**	**77.5a**	**5.5c**	**3.19bc**
	MS 0.3 mT	**2.732a**	**98.6a**	**93.9a**	**4.8c**	**1.14c**
	MS BA + IAA	2.550a	96.3a	86.8a	12.4bc	1.69c
	MS BA + NAA	1.847b	59.8c	36.7b	24.8b	38.48a
	MS 0.3 BA	3.095a	83.2b	33.7b	54.8a	11.58b
$L.S.D_{0,05}$		0.6843	15.28	22.14	14.83	8.564

suspension cultures [68, 69] as well as liquid bud cultures of sage may be a valuable source of phenolic acids used for controlled production of antioxidant substances in bioreactors [70].

The composition of certain active ingredients of the raw material obtained from apical and axillary bud cultures may vary due to somaclonal variation in plants multiplied in in vitro cultures [71–73].

In the present study, phytochemical analysis was conducted to compare sage raw material derived from field cultivation (in vivo) with in vitro cultures (Table 3).

The HPLC analysis and spectrophotometrical measurement showed an increase in polyphenolic compounds and polyphenol acids in plant material derived from in vitro culture and a decrease in flavonoids compared to the raw material derived from field cultivation. The increase in polyphenolic compounds and the decrease in flavonoids may be related to the physiological reaction of the plants grown in in vitro cultures caused by infection- and damage-related stress.

Correlation analysis of features showed an inversely proportional relation between the flavonoid content per quercetin and the content of polyphenolic compounds. Negative effect of drying on the flavonoid content per hyperoside was also noted in in vitro and in vivo cultures (Table 4).

Negative correlation between the content of flavonoids and polyphenolic compounds in the raw material derived from in vitro cultures may result from metabolic changes caused by increased stress. Lower intensity of light used in in vitro cultures as compared to in vivo cultures was an important stress factor for the plants. It might have limited the biosynthesis of flavonoids, increasing the activity of other phenol compounds with antioxidant properties [74].

The antioxidant properties of sage are determined by the synergistic effect of various phenolic compounds [75] or the presence of rosmarinic and carnosic acids [76]. The observed decrease in the amount of flavonoids and increased biosynthesis of other polyphenolic compounds (rosmarinic acid, other polyphenol acids) may have stemmed from triggering certain metabolic pathways at the cost of others under stress conditions in in vitro cultures. As regards polyphenol acids, which participate in the synthesis of lignins responsible for the stiffening of cell walls [77], their increased content limited plant growth in in vitro cultures as compared to field-cultivated plants. The amounts of tannins, which inactivate enzyme proteins in necrotic tissues, remained unchanged regardless of growing conditions.

3 Sage Genetic Transformation

The wide range of possibilities offered by secondary metabolites obtained from sage in vitro cultures triggered the search for new methods of improving *Salvia officinalis* through genetic transformation. The micropropagation protocol for sage cultivar "Bona" enabled for obtaining and storing massive amounts of selected plants with the best medicinal parameters for genetic modification. Few publications are to be found in literature regarding the transformation of sage. Researchers have investigated the use of *Agrobacterium rhizogenes* for obtaining transformed sage roots for

Table 3 Secondary metabolites content in sage leaves

Growing conditions	Content [%]					Loss of weight after drying [%]
	Flavonoids per		Tannins per pyrogallol	Polyphenolic compounds per rosmarinic acid	Polyphenol acids per chlorogenic acid	
	Perquercetin	Perhyperoside				
In vivo	0.52	0.74	1.52	4.07	4.33	6.53
In vitro	0.37	0.53	1.45	6.37	7.38	6.46

Table 4 Correlation coefficient of sage secondary metabolites obtained in in vitro and in vivo conditions

Feature	Flavonoids per quercetin	Flavonoids per hyperoside	Tannins	Polyphenolic compounds	Loss of weight after drying
Flavonoids per quercetin	–				
Flavonoids per hyperoside	0.5992	–			
Tannins	−0.3301	0.2925	–		
Polyphenolic compounds	−0.9815***	−0.5274	0.3474	–	
Loss of weight after drying	−0.4806	−0.9139*	−0.2883	0.3782	–

Correlation significance * $p < 0.05$; *** $p < 0.001$

the production of rosmarinic acid [78] and other species of sage: *S. sclarea*, *S. miltiorrhiza*, and *S. przewalskii* [79]. There is no data, however, on developing sage transformation procedures using *A. tumefaciens* or nonvector-based methods.

The aim of presented research was to investigate the possibilities for obtaining a transgenic variety of the sage and creating a basis for an oral, plant-produced recombinant vaccine against dental decay. The main research goal was to obtain transgenic sage containing a gene coding for the surface protein antigen I/II in *Streptococcus mutans*, the main factor responsible for the development of dental decay in humans. The antigen could be a valuable component of a vaccine against dental decay as it causes a relevant immunological response. Dental decay is considered a socially determined disease, with *Streptococcus mutans* as the primary pathogen in the etiology of the disease. Obtaining a transgenic sage mutation would be a good basis for developing a preparation based on the medicinal properties of the active ingredients found in the plant and the immunogenic properties of certain *S. mutans* proteins, without the need for their costly extraction and purification.

Prior to commencing the process of transformation, the concentration of the selection agent (kanamycin antibiotic), lethal to non-transgenic plants, was established. The transformed cells containing genes coding for the enzyme which decomposed the antibiotic survived the selection. Performing the selection was the easiest way to control the transformation process. However, antibiotic selection is currently avoided with plants grown for consumption [73]. Selectable markers determining the identification of transformants remain in the plant genome even after they have fulfilled their function and may affect microorganisms living in a given environment. Moreover, the presence of selectable markers may be an obstacle for subsequent transformations. The idea of the research excluded the danger of horizontal gene transfer in the cultivation, allowing for this kind of selection. Recent developments open new directions in research, aiming at eliminating or silencing selectable markers in genetically modified plants [80].

3.1 Methods of Transformant Selection

In order to establish the appropriate concentration of kanamycin, experiments on apical and axillary buds obtained both from seedlings and multiplants were conducted. The MS medium with 0.3 mg/L BA was used. Selectable concentrations for seedlings and multiplants were established by using six variants of kanamycin concentrations (0, 25, 50, 100, 150, 200 mg/L), with 100 explants in each combination in two replicates. The analyzed kanamycin concentrations affected the longevity of explants in a variety of ways (Fig. 3). Lethal kanamycin concentration for sage explants both from seedlings and multiplants was established at 200 mg/L (Fig.3f).

3.2 Methods of Plant Transformation

Prior to begin the process of sage transformation, *A. tumefaciens* LBA4404 strain was prepared according to the transformation protocol in the following stages:

(a) Frozen bacterial cells containing CRSA gene construct suspended in glycerol were cultured in Petri dishes with YEB_{KAN50} medium [81] and grown for 2 days at a temperature of 28 °C.
(b) The bacteria were cultured on minimal medium AB_{KAN50} [82] and grown for 1–2 days at 28 °C.
(c) A single colony was inoculated into 10 ml of liquid YEB_{KAN50} medium in 100 ml Erlenmeyer flasks and grown on vortex mixer at 250 rpm at 28 °C for 16–20 h.
(d) Fresh mg/L $_{KAN50}$ [83] medium was inoculated with *A. tumefaciens* suspension at a ratio of 1:400, and culture was continued at 28 °C to obtain late log phase of bacterial growth with an OD_{550} of 1.
(e) The culture was vortexed for 10 min at 10,000 rpm at 4 °C. Supernatant was discarded, and the bacteria in the deposit were resuspended in a fresh $MSGA_{KAN50}$ medium (based on Murashige and Skoog medium [84] with 1 μM of acetosyringone and 1 M of glucose). The ready suspension was poured onto Petri dishes which were used for inoculation.

A range of different standard media was used for culture of *A. tumefaciens* containing pBI-CRSA gene construct. The prepared media were sterilized in an autoclave for 20 min at 121 °C. The filtered selective antibiotic kanamycin was added to the sterile media in a concentration of 50 mg/L.

3.3 The Transgenesis of Sage Apical and Axillary Buds Was Conducted in Two Stages

Firstly, the initial transformation was carried out on 120 apical and axillary bud explants derived from 3-week- and 2-week-old multiplants and seedlings. Explants

Fig. 3 Establishing lethal kanamycin concentration for sage multiplant explants. (**a**) Positive control without kanamycin, (**b**) 25 mg/L, (**c**) 50 mg/L, (**d**) 100 mg/L, (**e**) 150 mg/L, (**f**) 200 mg/L lethal concentration

were described in detail in order to establish the influence of the explants' shape and size on its susceptibility to transformation. Secondly, on the basis of the results obtained in the first stage, 400 explants from apical and axillary buds derived from 3-week-old multiplants were selected for transformation.

At each stage a negative control (non-transformed explants on a selection medium) and a positive control (non-transformed explants on a medium without a selecting agent) were prepared for each kind of explants.

The explants that underwent transformation were placed in *A. tumefaciens* cell suspension for 10–15 min. Following the inoculation period, the buds were blotted dry on sterile tissue paper and placed on MS medium with 0.3 mg/L BA. A coculture was then grown for 2–3 days at 28 °C in semi-shade. In the next stage, excessive bacteria were removed by rinsing the explants in sterile water. The explants were then blotted dry on tissue paper and placed on $MSKAN_{200}$:T_{250} selection medium with BA phytohormone and kanamycin and Timentin®. In both stages of the experiment, the selective culture was monitored in 2-week intervals before subsequent passaging. The observations involved assessing the state of the explants (growing, vitrificated, necrosed). With longer culture time, multiplication rate of the plants and the presence of *A. tumefaciens* were determined. *A. tumefaciens* did not cause infection in the subsequent subcultures, which enabled for using a medium without Timentin®. Moreover, in the second stage of the experiment, after the eighth passage, it was decided to grow the culture without a selecting agent. In both stages of the experiment, the rooting of explants growing on the selection medium was tested. In the first stage, rooting was conducted on one-half MS medium containing 0.3 mg/L IAA with a selecting agent, while in the second stage, the rooting medium did not contain any selecting agent.

3.4 The Transgenesis of Sage Meristems of Apical and Axillary Buds

In the next stage of the research, isolated apical and axillary sage meristems were also subjected to transformation. Explants of less than 1 mm in size consisted almost exclusively of meristematic tissue so that bacteria would reach meristematic cells more easily, thus increasing the probability of their transformation. Transformation of isolated meristems was performed by directly injecting a needle inserted in *A. tumefaciens* suspension into the meristems under a binocular microscope. Thus prepared explants were placed on MS medium with 0.3 mg/L BA. A coculture was then grown for 2–3 days at 28 °C in semi-shade. In the next stage, excessive bacteria were removed by rinsing the explants in sterile water. The explants were then blotted dry on tissue paper and placed on two kinds of mediums: $MSKAN_{200}$: T_{250} regeneration selection medium (0.3 mg/L BA) containing Timentin® and kanamycin (200 mg/L) and regeneration medium MST250 (0.3 mg/L BA) containing only Timentin® in the concentration of 250 mg/L. Due to the small size of the isolated meristems, regeneration medium without a selecting agent was used. Selecting agent was introduced when explants began to expand into shoots. Meristems were derived both from seedlings and 3-week-old multiplants. Meristems obtained from seedlings (40 explants each) were inoculated with *A. tumefaciens* in the concentration of 10^8 and $<10^8$, additionally in an injected/non-injected combination (160 explants). There were three controls performed: two positive (MS 0.3 mg/L BA and MST250 0.3 mg/L BA) and one negative (MSK_{200}:T_{250} 0.3 mg/L BA). Each control consisted of 20 explants, non-injected and injected with the preparation needle (120 explants). The first examination of the explants was performed 2 weeks after rinsing the coculture explants and then after another 4 weeks since the first observation. After 5 rounds of selection explants subcultures, antibiotics were no longer administered.

Three explants growing on a selection medium were selected after the transformation. DNA was isolated from these samples following modified Gawal and Jarret (1991) method [85], and then transgene integration analysis was conducted using PCR. However, despite the fact that the plants grew on selection media, our research did not confirm the presence of transgene in the plant cells. It was suspected that temporary transfection took place, and the introduced DNA did not integrate with the sage genome, producing only temporary resistance to the selective antibiotic. Thus, the introduced gene might have been lost during cell division in plant multiplication. This may explain the gradual decrease in the vitality of the explants in selection cultures until their complete decay. Explants were selected during the 3rd–5th subculturing following the transformation. Further passages, due to the small size of the plants, provided material both for DNA isolation (6th–11th subcultures) and assessment of the rhizogenesis process, which indicates normal development of a plant.

The lack of implemented fragments of the *S. mutans* gene in the genome of the explants analyzed might have been caused by the instability of transformant strains or loss of the transgene during passages. A stable transgenic line is produced only when the transgene integrates with the genome in the right location. Loss of the transgene in subsequent passages may be also caused by the activation of processes leading to the identification and repair of DNA changes in meristematic cells. Explant rhizogenesis after transformation was not consistent with the described sage regeneration methodology, proving that genetic homeostasis was disturbed and explants did not develop normally.

The type of explant selected determined the process of transformation; apical and axillary buds ensured quick and effective regeneration. However, as the explants contained both meristems and leaf primordia, it was difficult for *A. tumefaciens* to reach the meristematic tissue. After a detailed analysis of all explants, transformation of isolated meristems was performed by directly injecting a needle inserted in *A. tumefaciens* suspension into the meristems. The aim was to ensure that the bacteria could reach the meristematic cells more easily and increase the probability of their transformation. Although several variants of the experiment were conducted, the stable transgene integration in producing plants was not obtained. However, valuable observations were made in control explant cultures regarding high development potential of isolated meristems. This type of culture may be used in the future to produce transgenic plants using nonvector methods.

Secondary metabolite content was measured in the raw material obtained in subsequent transformations of sage (Agro 1, Agro 2, Agro 3) and compared with the material grown in in vitro cultures (Table 5).

Biochemical analysis showed no differences in the content of flavonoids, polyphenolic compounds, polyphenol acids, and tannins in the material obtained after the transformation and that from in vitro cultures.

Table 5 Secondary metabolite content in sage leaves

	Content [%]					
	Flavonoids per		Tannins per pyrogallol	Polyphenolic compounds per rosmarinic acid	Polyphenol acids per chlorogenic acid	Loss of weight after drying
	Perquercetin	Perhyperoside				
In vitro	0.37	0.53	1.45	6.37	7.38	6.46
Agro 1	0.38	0.54	1.38	6.38	8.10	5.30
Agro 2	0.27	0.39	1.33	6.75	7.21	5.94
Agro 3	0.24	0.35	1.52	5.97	6.89	6.33

Yet it is worth noting that the material obtained after the first transformation (Agro 1) contained the highest amount of quercetin, hyperoside, and chlorogenic acid compared to the raw material after the second and third transformations and a similar amount compared to the material from in vitro cultures. As regards the raw material after the second (Agro 2) and third (Agro 3) transformations, it contained the highest amount of rosmarinic acid and tannins, respectively.

Transformation of explants without producing transformed plants did not affect the content of the sage active compounds analyzed as compared to the material derived from in vitro cultures.

The main goal of research studies whose aim is to improve medicinal plants by the use of biotechnological methods is to increase the synthesis of secondary metabolites [86]. However, in the presented research with the use of *A. tumefaciens*, the stable integration of the *Streptococcus mutans* gene fragments in sage genome was not achieved. Paradoxically, in the case of sage, the biosynthesis of active ingredients may inhibit *A. tumefaciens* growth and thus become an obstacle for producing genetically modified plants by the use of vector-based transgenesis systems. For that reason the influence of sage on the growth and development of *A. tumefaciens* colonies was examined.

4 The Influence of Sage on *Agrobacterium tumefaciens*

Sage displays high antibacterial activity against numerous microorganisms. It has been demonstrated that the compounds found in sage essential oil inhibit the multiplication of pathogenic bacteria, both gram-positive, i.e., *Staphylococcus aureus* and *Bacillus subtilis*, and gram-negative, i.e., *Escherichia coli* and *Pseudomonas aeruginosa* [9]. Moreover, they have a virucidal effect. Stanojevic et al. described the synergistic activity of antibacterial sage aqueous extracts and preservatives against selected bacteria, including *A. tumefaciens* [22]. Their research showed that the minimum inhibitory concentration (MIC) for the growth of *A. tumefaciens* was 20 mg/mL, which resulted in medium antibacterial activity compared with *B. subtilis* (MIC 10 mg/mL) and *E. coli* (MIC 40 mg/mL), but did not guarantee bacterial resistance. The research showed synergistic action of sage extracts with preservatives; sage extract in a concentration of 1 mg/mL with sodium nitrite effectively inhibited the growth of *A. tumefaciens*. For this reason, in the presented experiment, the influence of sage explants placed in synthetic media on bacterial growth during inoculation and coculture was determined. The experiment was performed on MS medium with 0.3 mg/L BA (6-benzylaminopurine) in three variants (Fig. 4):

1. Experimental coculture of *A. tumefaciens* with sage explants, which involved culturing the bacteria in a straight line and placing sage explants next to it on one side. Apical buds from 3-week-old multiplants were used as sage explants, as was done during the transformation. Seedlings and leaves with petioles were used as

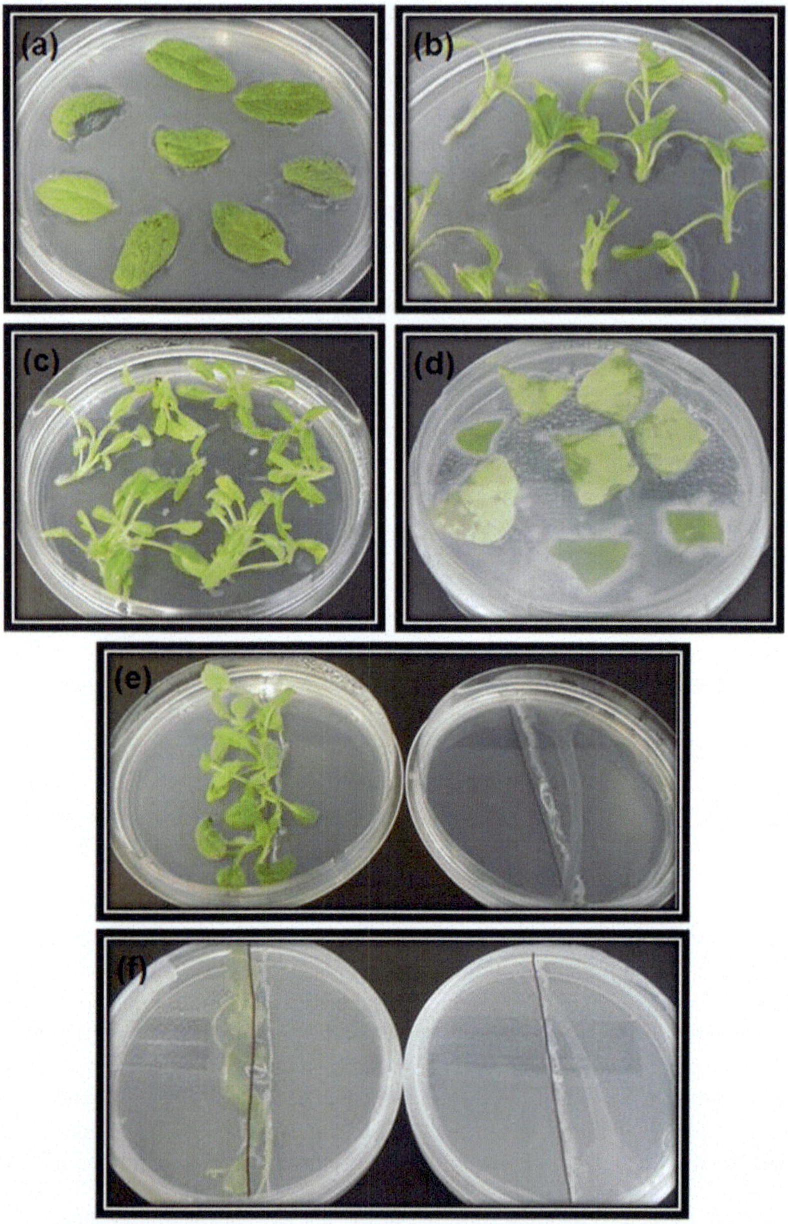

Fig. 4 The effect of sage explants on *A. tumefaciens* growth on the third day of coculture. (**a**) The inhibition of bacterial growth around sage leaves (second variant); (**b**) the inhibition of bacterial growth around apical buds (second variant); (**c**) the inhibition of bacterial growth on apical sage buds previously placed in *Agrobacterium* suspension (third variant) visible lack of bacterial growth on sage explants; (**d**) bacterial growth on tobacco explants previously placed in *Agrobacterium* suspension; (**e**) growth inhibition of bacteria inoculated linearly beside explants from sage seedlings as compared to the control (first variant); (**f**) bacterial growth inhibition in the presence of sage leaves as compared to the control (first variant)

well. The experiment was conducted in three replicates with 8 explants of each type per replicate.

2. Bacterial coculture cultured on the entire surface of the dish, with apical buds or leaves placed on it. The experiment was conducted in two replicates with 8 explants of each type per replicate.
3. Coculture in which apical buds were soaked in bacterial suspension for 15 min in order to establish the scale of *A. tumefaciens* growth on explants. In addition, a coculture of tobacco leaf fragments soaked in bacterial suspension was established in order to compare the influence of the two plant species on the growth and development of *A. tumefaciens*.

LBA4404 *A. tumefaciens* strain inoculated into the medium was in the same log growth phase (OD ~ 1) as the bacteria used in the transformation. The same strain of bacteria inoculated without sage explants was used as a control. Cultures were grown in semi-shade at 26 °C. The growth and development of the bacteria and explants were examined four times in the coculture of sage explant with *A. tumefaciens*: after the first, third, seventh, and fourteenth days of the experiment. Observations following the first and third days after the inoculation were particularly important because that was the time of inoculation in the previously conducted transformation. In the first variant, following the first day of the coculture, both the bacteria cultured in a straight line and the sage explants placed next to them developed normally. In the second variant, inhibition in the growth and development of bacteria placed beside the leaves and a 75% growth inhibition of sage apical buds was observed. In the variant with sage apical buds, which had been placed in *A. tumefaciens* suspension, a limited bacterial growth was observed.

The second coculture examination conducted on the third day of the culture followed the protocol of sage transformation, which had been carried out using *A. tumefaciens* in the transgenesis process (Fig. 4). Both in the first and second variants, bacterial growth inhibition was observed with various intensity compared to the control. In the first variant, apical buds, seedlings, and leaves limited the growth of the linearly inoculated bacteria (Fig. 4e–f). Bacterial growth inhibition around the leaves and complete lack of bacteria under explants were observed for the bacteria inoculated in the entire medium (Fig. 4a). Apical buds inhibited bacterial growth mainly in those places where the explant was in contact with the medium (Fig. 4b). In the third variant, there was a slight growth of bacteria, with no growth on explants (Fig. 4c). All the sage explants grew normally with no stress symptoms. The state of the coculture did not change on the seventh day of the culture (third examination). After 14 days, the sage explants gradually turned brown and decayed due to the shortage of nutrients in the medium. At the same time, an increased growth of *A. tumefaciens* was observed.

The conducted analysis leads to the conclusion that sage explants had an inhibiting effect on bacterial growth and development. This might have decreased the probability of a successful introduction of a fragment of bacterial plasmid (T-DNA) into the plant genome. The similar observation was given by Marchev et al. [87], who noticed difficulties during *A. tumefaciens* transformation of

S. tomentosa plants due to the extensive release of phenolic compounds that had strong antimicrobial and allelopathic activities [87]. This negative effect was postponed by using a temporary immersion system in combination with a two-phase cultivation protocol that incorporates Amberlite XAD-4 resin. This procedure enabled to obtain 100% transformation efficiency using *S. tomentosa* mature leaves.

5 Conclusions

Progress in plant biotechnology can also be observed in the context of medicinal plant species. Development of effective regeneration methods for plants which are crucial in phytotherapy, such as sage, creates a multitude of possibilities for their better use: exploring metabolic pathways and increasing production of secondary metabolites, micropropagation of uniform genotypes and their long-term storage, genetic transformation, and the production of biopharmaceuticals. The present research points to considerable difficulties in sage transgenesis using vector-based methods with *A. tumefaciens*. The problem of achieving stable transgene integration in sage is related to its high antibacterial activity against a wide range of microorganisms. The observed negative effect of active ingredients contained in sage explants on bacterial growth and development is a major obstacle for conducting an effective sage transformation using *A. tumefaciens*. Nonvector methods should be recommended to obtain the successful transformation of sage due to its strong antibacterial effectiveness.

Acknowledgments The presented work was financially supported by The Ministry of Science and Higher Education in Poland, grant number MNiSzW IT-1 – 13910.

References

1. Pederson JA (2000) Distribution and taxonomic implications of some phenolic in the family *Lamiaceae* determined by ESR spectroscopy. Biochem Syst Ecol 28:229–253
2. Kholmünzer S (2012) Pharmacognosy. PZWL, Warsaw (in Polish)
3. Kołodziej B (2010) Medicinal plant cultivation, handbook for farmers. PWRiL, Poznan, pp 413–416 (in Polish)
4. Putievsky BM, Vemin RC (1992) A contribution to quality and composition assay of (*Salvia officinalis* L.). Acta Pharm Jugosl 35:121–125
5. Khalil MY, Kandil MAM, Swaefy Hend MF (2008) Effect of three different compost levels on *Fennel* and *Salvia* growth character and their essential oils. Res J Agric Biol Sci 4:34–39
6. Gonceariuc M, Balmush Z, Kulcitki V, Gonceariuc N, Romanciuc G, Sirbu T (2012) Essential oil content and composition different of *Salvia officinalis* L. genotypes cultivated in Moldova. Muzeul Olteniei Craiova 28:7–10
7. Grausgruber-Gröger S, Schmiderer C, Steinborn R, Novak J (2012) Seasonal influence on gene expression of monoterpene synthases in *Salvia officinalis* (*Lamiaceae*). J Plant Physiol 169:353–359
8. Miura K, Kikuzaki H, Nakatani N (2001) Apianane terpenoids from *Salvia officinalis*. Phytochemistry 58:1171–1175

9. Radulescu V, Chiliment S, Oprea E (2004) Capillary gas chromatography–mass spectrometry of volatile and semi-volatile compounds of *Salvia officinalis*. J Chromatogr A 1027:121–126
10. Lu Y, Foo LY (2000) Flavonoid and phenolic glycosides from *Salvia officinalis*. Phytochemistry 55:263–267
11. Ejtahed RS, Radjabian T, Tafreshi SAH (2015) Expression analysis of phenylalanine ammonia lyase gene and rosmarinic acid production in *Salvia officinalis* and *Salvia virgata* shoots under salicylic acid elicitation. Appl Biochem Biotechnol 176:1846–1858
12. Baricevic DS, Sosa S, Della Loggia R, Tubaro A, Simonovska B, Krasana A, Zupancic A (2001) Topical anti-inflammatory activity of *Salvia officinalis* L. leaves: the relevance of ursolic acid. J Ethnopharmacol 75:125–132
13. Delamare APL, Moschen-Pistorello IT, Artico L, Atti-Serafini L, Echeverrigaray S (2007) Antibacterial activity of the essential oils of *Salvia officinalis* L. and *Salvia triloba* L. cultivated in South Brazil. Food Chem 100:603–608
14. Loizzo MR, Tundis R, Menichini F, Saab AM, Statti GA (2007) Cytotoxic activity of essential oils from *Labiatae* and *Lauraceae* families against in vitro human tumor models. Anticancer Res 27:3293–3299
15. Kamatou GP, Makunga NP, Ramogola WP, Viljoen AM (2008) South African *Salvia* species: a review of biological activities and phytochemistry. J Ethnopharmacol 119:664–672
16. Kelen M, Tepe B (2008) Chemical composition, antioxidant and antimicrobial properties of essential oils of three *Salvia* species from Turkish flora. Bioresour Technol 99:4096–4104
17. Loizzo MR, Saab AM, Tundis R, Menichini F, Bonesi M, Piccolo V, Statti GA, de Cindio B, Houghton PJ (2008) In vitro inhibitory activities of plant used in Lebanon traditional medicine against angiotensin converting enzyme (ACE) and digestive enzymes related to diabetes. J Ethnopharmacol 119:109–116
18. Marchev A, Haas C, Schulz S, Georgiev V, Steingroewer J, Bley T, Pavlov A (2014) Sage in vitro cultures: a promising tool for the production of bioactive terpenes and phenolic substances. Biotechnol Lett 36:211–221
19. Grzegorczyk I, Matkowski A, Wysokińska H (2007) Antioxidant activity of extracts from in vitro cultures of *Salvia officinalis* L. Food Chem 104:536–541
20. Santos-Gomes PC, Seabra RM, Andrade PB, Fernandes-Ferreira M (2002) Phenolic antioxidant compounds produced by in vitro shoots of sage (*Salvia officinalis* L.). Plant Sci 162:981–987
21. Viuda-Martos M, Ruiz-Navajas Y, Fernández-López J, Pérez-Álvarez JA (2008) Antibacterial activity of different essential oils obtained from spices widely used in Mediterranean diet. Int J Food Sci Technol 43:526–531
22. Stanojevic D, Comic L, Stefanovic O, Solujic-Sukdolak S (2010) In vitro synergistic antibacterial activity of *Salvia officinalis* L. and some preservatives. Arch Biol Sci 62(1): 167–173
23. Seidler-Łożykowska K (2008) Medicinal plant cultivars. Publisher by Polish Institute of Medicinal Plants, pp 1–24 (in Polish)
24. Schnitzler P, Nolkemper S, Stintzing FC, Reichling J (2008) Comparative in vitro study on the anti-herpetic effect of phytochemically characterized aqueous and ethanolic extracts of *Salvia officinalis* grown at two different locations. Phytomedicine 15:62–70
25. Szajdek A, Borowska J (2004) Antioxidant properties of plant-based food products. Żywn Nauka Technol Jakość 4(41):5–28 (in Polish)
26. Lima CF, Andrade PB, Seabra RM, Fernandes-Ferreira M, Pereira-Wilson C (2005) The drinking of a *Salvia officinalis* infusion improves liver antioxidant status in mice and rats. J Ethnopharmacol 97:383–389
27. Akhondzadeh S, Noroozian M, Mohammadi Mahammad R, Ohadinia S, Jamshidi Amir H, Khani M (2003) *Salvia officinalis* extract in the treatment of patients with mild to moderate Alzheimer's disease: a double blind, randomized and placebo-controlled trial. J Clin Pharm Ther 28:53–59

28. Ramos AA, Azqueta A, Pereira-Wilson C, Collins AR (2010) Polyphenolic compounds from *Salvia* species protect cellular DNA from oxidation and stimulate DNA repair in cultured human cells. J Agric Food Chem 58(12):7465–7471
29. El Hadri A, Gómez del Río MÁ, Sanz J, González Coloma A, Idaomar M, Ribas Ozonas B, Benedí González J, Sánchez Reus MI (2010) Cytotoxic activity of α-humulene and transcaryophyllene from *Salvia officinalis* in animal and human tumor cells. An R Acad Nac Farm 76(3):343–356
30. Christensen KB, Jørgensen M, Kotowska D, Petersen RK, Kristiansen K, Christensen LP (2010) Activation of the nuclear receptor PPARγ by metabolites isolated from sage (*Salvia officinalis* L.). J Ethnopharmacol 132(1):127–133
31. Bommer S, Klein P, Suter A (2011) First time proof of sage's tolerability and efficacy in menopausal women with hot flushes. Adv Ther 28(6):490–500
32. Ardalan A, Majid S (2012) Essential oil composition, phenolic content, antioxidant and antimicrobial activity in *Salvia officinalis* L. cultivated in Iran. Adv Environ Biol 6:221–226
33. European Medicines Agency (2016) Assessment report on *Salvia officinalis* L., *folium* and *Salvia officinalis* L., *aetheroleum*. EMA/HMPC/150801/2015, 2 Feb 2016. www.ema.europa.eu
34. Halicioglu O, Astarcioglu G, Yaprak I (2011) Toxicity of *Salvia officinalis* in a newborn and a child: an alarming report. Pediatr Neurol 45:259–260
35. Szpitter A, Królicka A (2005) Stimulating the effect of biotic elicitors on the production of pharmacologically active secondary metabolites in plant in vitro cultures. Biotechnologia 4(71):82–108 (in Polish)
36. Steingroewer J, Bley T, Georgiev V, Ivanov I, Lenk F, Marchev A, Pavlov A (2013) Bioprocessing of differentiated plant in vitro systems. Eng Life Sci 13:26–38
37. Georgiev V, Marchev A, Haas C, Weber J, Nikolova M, Pavlov A (2011) Production of oleanolic and ursolic acids by callus cultures of *Salvia tomentosa* Mill. Biotechnol Biotechnol Equip 25:34–38
38. Marchev A, Georgiev V, Badjakov I, Kondakova V, Nikolova M, Pavlov A (2011) Triterpenes production by rhizogenic callus of *Salvia scabiosifolia* Lam. obtained *via Agrobacterium rhizogenes* mediated genetic transformation. Biotechnol Biotechnol Equip 25:30–33
39. Bolta Ž, Baričevič D, Raspor P (2003) Biomass segregation in sage cell suspension culture. Biotechnol Lett 25:61–65
40. Kuźma Ł, Bruchajzerb E, Wysokińska H (2008) Diterpenoid production in hairy root culture of *Salvia sclarea* L. Z Naturforsch 63c:621–624
41. Kuźma Ł, Bruchajzer E, Wysokińska H (2009) Methyl jasmonate effect on diterpenoid accumulation in *Salvia sclarea* hairy root culture in shake flasks and sprinkle bioreactor. Enzym Microb Technol 44:406–410
42. Kuźma Ł, Kisiel W, Królicka M, Wysokińska H (2011) Genetic transformation of *Salvia austriaca* by *Agrobacterium rhizogenes* and diterpenoid isolation. Pharmazie 66:904–907
43. Wu J, Shi M (2008) Ultra high diterpenoid tanshinone production through repeated osmotic stress and elicitor stimulation in fed-batch culture of *Salvia miltiorrhiza* hairy roots. Appl Microbiol Biotechnol 78:441–448
44. Yuan Y, Liu Y, Lu D, Huang L, Liang R, Yang Z, Chen S (2009) Genetic stability, active constituent, and pharmacoactivity of *Salvia miltiorrhiza* hairy roots and wild plant. Z Naturforsch 64c:557–563
45. Zhao J-L, Zhou L-G, Wu J-Y (2010) Effects of biotic and abiotic elicitors on cell growth and tanshinone accumulation in *Salvia miltiorrhiza* cell cultures. Appl Microbiol Biotechnol 87:137–144
46. Zhao J, Lou J, Mou Y, Li P, Wu J, Zhou L (2011) Diterpenoidtanshinones and phenolic acids from cultured hairy roots of *Salvia miltiorrhiza* Bunge and their antimicrobial activities. Molecules 16:2259–2267
47. Gupta S, Liu R, Liaw S, Chan H, Tsay H (2011) Enhanced tanshinone production in hairy roots of '*Salvia miltiorrhiza* Bunge' under the influence of plant growth regulators in liquid culture. Bot Stud 52:435–443

48. Yan Q, Wu J, Liu R (2011) Modeling of tanshinone synthesis and phase distribution under the combined effect of elicitation and in situ adsorption in *Salvia miltiorrhiza* hairy root cultures. Biotechnol Lett 33:813–819
49. Yang D, Du X, Liang X, Han R, Liang Z, Liu Y, Liu F, Zhao J (2012) Different roles of the mevalonate and methylerythritol phosphate pathways in cell growth and tanshinone production of *Salvia miltiorrhiza* hairy roots. PLoS One 7:e46797. doi:10.1371/journal.pone.0046797
50. Yang D, Ma P, Liang X, Wei Z, Liang Z, Liu Y, Liu F (2012) PEG and ABA trigger methyl jasmonate accumulation to induce the MEP pathway and increase tanshinone production in *Salvia miltiorrhiza* hairy roots. Physiol Plant 146(173–1):b83
51. Yang D, Sheng D, Duan Q, Liang X, Wei Z, Liang Z, Liu Y (2012) PEG and ABA trigger the burst of reactive oxygen species to increase tanshinone production in *Salvia miltiorrhiza* hairy roots. J Plant Growth Regul 31:579–587
52. Kai G, Xu H, Zhou C, Liao P, Xiao J, Luo X, You L, Zhang L (2011) Metabolic engineering tanshinone biosynthetic pathway in *Salvia miltiorrhiza*. Metab Eng 13:319–327
53. Kai G, Liao P, Xu H, Wang J, Zhou C, Zhou W, Qi Y, Xiao J, Wang Y, Zhang L (2012) Molecular mechanism of elicitor-induced tanshinone accumulation in *Salvia miltiorrhiza* hairy root cultures. Acta Physiol Plant 34:1421–1433
54. Gu X, Chen J, Xiao Y, Di P, Xuan H, Zhou X, Zhang L, Chen W (2012) Overexpression of allene oxide cyclase promoted tanshinone/phenolic acid production in *Salvia miltiorrhiza*. Plant Cell Rep 31:2247–225
55. Liang Z, Yang D, Liang X, Zhang Y, Liu Y, Liu F (2012) Roles of reactive oxygen species in methyl jasmonate and nitric oxide-induced tanshinone production in *Salvia miltiorrhiza* hairy roots. Plant Cell Rep 31:873–888
56. Xiao Y, Gao S, Di P, Chen J, Chen W, Zhang L (2009) Methyl jasmonate dramatically enhances the accumulation of phenolic acids in *Salvia miltiorrhiza* hairy root cultures. Physiol Plant 137:1–9
57. Dong J, Wan G, Liang Z (2010) Accumulation of salicylic acid-induced phenolic compounds and raised activities of secondary metabolic and antioxidative enzymes in *Salvia miltiorrhiza* cell culture. J Biotechnol 148:99–104
58. Hao G, Ji H, Li Y, Shi R, Wang J, Feng L, Huang L (2012) Exogenous ABA and polyamines enhanced salvianolic acids contents in hairy root cultures of *Salvia miltiorrhiza* Bge. f. alba. Plant Omics J 5:446–452
59. Xiao Y, Zhang L, Gao S, Saechao S, Di P, Chen J, Chen W (2011) The c4h, tat, hppr and hppd genes prompted engineering of rosmarinic acid biosynthetic pathway in *Salvia miltiorrhiza* hairy root cultures. PLoS One 6:e29713. doi:10.1371/journal.pone.0029713
60. Dai Z, Cui G, Zhou S-F, Zhang X, Huang L (2011) Cloning and characterization of a novel 3-hydroxy-3-methylglutaryl coenzyme A reductase gene from *Salvia miltiorrhiza* involved in diterpenoid tanshinone accumulation. J Plant Physiol 168:148–157
61. Liu W, Chilcott CE, Reich RC, Hellmann GM (2000) Regeneration of *Salvia Sclarea* via organogenesis. In Vitro Cell Dev Biol Plant 36:201–206
62. Arikat NA, Jawad FM, Karama NS, Shibli RA (2004) Micropropagation and accumulation of essential oils in wild sage (*Salvia fruticosa* Mill.). Sci Hortic 100:193–202
63. Echeverrigaray S, Postingher Carrer R, Bavaresco Andrade L (2010) Micropropagation of *Salvia guaranitica* Benth. through axillary shoot proliferation. Braz Arch Biol Technol 53(4):883–888
64. Luwańska A, Wielgus K, Szalata M, Lipiński D, Zeyland J, Słomski R (2013) Optimisation of the conditions for in vitro regeneration of sage. Bull Plant Breed Acclimatization Inst 270:133–142 (In Polish)
65. Gostin I (2008) Effects of different plant hormones on *Salvia officinalis* cultivated in vitro. Int J Bot 4(4):430–436
66. Grzegorczyk I, Wysokińska H (2008) Liquid shoot culture of *Salvia officinalis* L. for micropropagation and production of antioxidant compounds; effect of triacontanol. Acta Soc Bot Pol 77(2):99–104

67. Grzegorczyk I, Bilichowski I, Mikiciuk-Olasik E, Wysokińska H (2005) In vitro cultures of *Salvia officinalis* L. as a source of antioxidant compounds. Acta Soc Bot Pol 74:17–21
68. Santos-Gomes PC, Seabra RM, Andrade PB, Fernandes-Ferreira M (2003) Determination of phenolic antioxidant compounds produced by calli and cell suspensions of sage (*Salvia officinalis* L.). J Plant Physiol 160:1025–1032
69. Ruffoni B, Raffi D, Rizzo A, Oleszek W, Giardi MT, Bertoli A, Pistelli L (2009) Establishment of in vitro *Salvia* cell biomass for the controlled production of antioxidant metabolites. Acta Hortic 829:423–427
70. Grzegorczyk I, Wysokińska H (2010) Antioxidant compounds in *Salvia officinalis* L. shoot and hairy root cultures in the nutrient sprinkle bioreactor. Acta Soc Bot Pol 79(1):7–10
71. Avato P, Morone Fortunato I, Ruta C, D'Elia R (2005) Glandular hairs and essential oils in micropropagated plants of *Salvia officinalis* L. Plant Sci 169:29–36
72. Randriamampionona D, Diallo B, Rakotoniriana F, Rabemanantsoa C, Cheuk K, Corbisier A-M, Mahillon J, Ratsimamanga S, El Jaziri M (2007) Comparative analysis of active constituents in *Centella asiatica* samples from Madagascar: application for ex situ conservation and clonal propagation. Fitoterapia 78:482–489
73. Michalik B (2009) Plant breeding based on genetic and biotechnology. PWRiL, Poznan (in Polish)
74. Buchanan B, Gruissem W, Jones R (2006) Biochemistry & molecular biology of plants. American Society of Prout Biologists (ASPB)
75. Spiridon I, Bodirlau R, Teaca C-A (2011) Total phenolic content and antioxidant activity of plants used in traditional Romanian herbal medicine. Cent Eur J Biol 6(3):388–396
76. Başkan S, Öztekin N, Erim FB (2007) Determination of carnosic acid and rosmarinic acid in sage by capillary electrophoresis. Food Chem 101:1748–1752
77. Wójcik-Wojtkowiak D, Politycka B, Weyman-Kaczmarkowa W (1998) Allelopathy. Publisher by August Cieszkowski Agricultural University of Poznan (in Polish)
78. Grzegorczyk I, Królicka A, Wysokińska H (2006) Establishment of *Salvia officinalis* L. hairy root cultures for the production of rosmarinic acid. Z Naturforsch 61c:351–356
79. Wysokińska H, Chmiel A (2006) Secondary metabolites production in cultures of transformed plant organs. Biotechnologia 4(75):124–135 (in Polish)
80. Tuteja N, Verma S, Kumar Sahoo R, Raveendar S, Lingeshwara Reddy IB (2012) Recent advances in development of marker-free transgenic plants: regulation and biosafety concern. J Biosci 37(1):167–197
81. Vervliet G, Holsters M, Teuchy H, Van Montagu M, Schell J (1975) Characterization of different plaque-forming and defective temperate phages in *Agrobacterium* strains. J Gen Virol 26:33–48
82. Ooms G, Hooykaas PJJ, Van Veen RJM, Van Beelen P, Regensburg-Tuïnk TJG, Schilperoort RA (1982) Octopine Ti-plasmid deletion mutants of *Agrobacterium tumefaciens* with emphasis on the right side of the T-region. Plasmid 7(1):15–29
83. Garfinkel DJ, Nester EW (1980) *Agrobacterium tumefaciens* mutants affected in crown gall tumorigenesis and octopine catabolism. J Bacteriol 144:732–743
84. Murashige T, Skoog F (1962) A revised medium for rapid growth and bio assays with tobacco tissue cultures. Physiol Plant 15:473–497
85. Gawal NJ, Jarret RL (1991) A modified CTAB DNA extraction procedure for Musa and Ipome. Plant Mol Biol Report 9:262–266
86. Kayser O, Quax WJ (2007) Medicinal Plant Biotechnology. Wiley-Vch Verlag GmbH & Co KGaA, Weinheim
87. Marchev A, Georgiev V, Ivanov I, Badjakov I, Pavlov A (2011) Two-phase temporary immersion system for *Agrobacterium rhizogenes* genetic transformation of sage (*Salvia tomentosa* Mill.). Biotechnol Lett 33:1873–1878

Ruta graveolens: Phytochemistry, Pharmacology, and Biotechnology

7

Sonia Malik, Denise Fernandes Coutinho Moraes, Flavia Maria Mendonça do Amaral, and Maria Nilce Sousa Ribeiro

Abstract

Ruta graveolens L., popularly known as rue, is a multipurpose herb belonging to family Rutaceae. It is a rich source of secondary metabolites mainly: coumarins, alkaloids, volatile oils, flavonoids, and phenolic acids. It has been used abundantly worldwide due to its diverse medicinal properties. Extract and essential oil obtained from this plant species have been shown to possess various pharmacological activities including contraceptive, anti-inflammatory, antimicrobial, antipyretic, antioxidant, analgesic, antihyperglycemic, free radical scavenging, hypotensive, antiviral, and antiplasmodial effects. In vitro assays performed with human cell lines have indicated the anticancer potential of furanoacridones and acridone alkaloids isolated from *R. graveolens*. In vitro approaches have been carried out for rapid clonal multiplication of *R. graveolens*. Application of hairy root culture has effectively observed to be beneficial for enhanced production of bioactive compounds from this plant species. A review of literature suggests that it is an interesting plant species to pharmaceutical industry due to its potential to produce several pharmacological effects.

Keywords

Bioactive compounds • Hairy root culture • In vitro culture • Medicinal plant • Ruta • Rutaceae • Pharmacology • Toxicology

S. Malik (✉)
Graduate Program in Health Sciences, Biological and Health Sciences Center, Federal University of Maranhão, São Luís, MA, Brazil
e-mail: sonia.unicamp@gmail.com; 777soniamalik@gmail.com

D.F.C. Moraes • F.M.M. do Amaral • M.N.S. Ribeiro
Department of Pharmacy, Biological and Health Sciences Center, Federal University of Maranhão, São Luís, MA, Brazil
e-mail: deniseufma2013@gmail.com; flavia.amaral@ufma.br; mnribeiro@ufma.br

S. Jha (ed.), *Transgenesis and Secondary Metabolism*, Reference Series in Phytochemistry, DOI 10.1007/978-3-319-28669-3_4

Contents

1 Introduction

Plants produce several variety and number of secondary metabolites that play a major role in adaptation with the environment. These compounds also represent important sources of pharmaceutical drugs due to their pharmacological activities. The metabolism from plants is quite diverse with long and complicated chemical reactions which, in many cases, is impossible to be carried out in laboratories. Thus many active raw materials used in pharmaceutical industries are not synthesized by the human and are being obtained just from plant source.

Biotechnology is being used as a tool to obtain bioactive compounds from plants which are unlikely to be synthesized in laboratories. It is also possible to improve the production of secondary metabolites from plants by using various biotechnological means [1–3]. Several strategies have been used to obtain the enhanced and improved production of compounds of interest, which include plant cell culture as well as modification in the gene expression to improve the enzymatic reactions of the metabolism [4, 5].

Among the interesting species to the pharmaceutical area, *Ruta graveolens* has gained considerable importance because of its medicinal purpose, and it is also a source of several secondary metabolites that have demonstrated different biological activities.

2 Botanical Characteristics

R. graveolens L. (Fig. 1) belongs to family Rutaceae, which comprises approximately 160 genera and 2070 species, distributed in tropical and temperate regions of the world, mainly America, Africa, and Australia [6, 7]. According to Engler [8, 9], the Rutaceae Jussieu family belongs to Rutales order. Takhtajan [10, 11] and Thorne

Fig. 1 Photograph of *Ruta graveolens* L.

[12] also placed this family into Rutales although Cronquist [13] and APG III [6] classified it as Sapindales.

Engler [8, 9] divided Rutaceae family into seven subfamilies: Rhabdodendroideae, Aurantioideae, Flindersioideae, Spathelioideae, Dictyolomatoideae, Rutoideae, and Toddalioideae. Thorne [12] has combined Toddalioideae to Rutoideae and suppressed Rhabdodendroideae, thus making five in total. Takhtajan [10, 11] classified it into six subfamilies, excluding also Rhabdodendroideae and the APG III recognized four subfamilies (Cneoroideae, Amyridoideae, Rutoideae, Aurantioideae) [6]. Cronquist [13] did not mention subfamilies. In all these classifications, *Ruta* genus is placed into Rutoideae subfamily.

Some of the species from Rutaceae family, which are having economic importance, include *Ruta*, *Citrus*, and *Pilocarpus*. This family includes species with edible fruit such as *Citrus aurantium* L. (sour orange), *C. sinensis* (L.) Osbeck (sweet orange), *Citrus limon* (L.) Osbeck (lemon), and *Citrus reticulate* Blanco (tangerine) [14]. Some plant species have been used as ornamental plants due to their attractive flowers such as *Hortia, Correa, Boronia, Choisya*, and *Clausena* [15, 16]. Species from *Ruta*, *Zanthoxylum*, and *Casimiroa* are known for their medicinal purposes. Genus *Pilocarpus* has species native from Brazil and Paraguay that produce pilocarpine, an alkaloid used to manufacture medicine to treat glaucoma [17].

Ruta graveolens L. is popularly known as rue, herb of grace (English), ruda (Spanish), raute (German), and arruda (Portuguese), and it has *Ruta hortensis* Mill. as scientific synonym [18]. It is a herb or sub-shrub with branches grown in clumps up to 60 m high. The plant is covered by trichomes, and flowers are small and yellow in color. Plant has scent, alternate, petiolated, and compound leaves that can reach up to 15 cm length. Each leaf has 2–5 leaflets that are fleshy, sessile, and with color varying from light green to bluish-green [19–21].

There are reports concerning difficulty in identifying *R. graveolens*, mainly to differentiate it from *R. chalepensis*. In this regards, Kanan and Babu [22] performed pharmacognostic studies in stem and leaves of *R. graveolens* and recognized microscopic characteristics that can be used to identify this plant species. The stem of *R. graveolens* has single layer epidermis followed by hypodermis. The cortex is divided into two layers: one of chlorenchyma and the other parenchyma. The chlorenchyma layer has a lot of air spaces between the cells, characterized by an aerenchyma and the parenchyma has normal intercellular spaces. The pericycle is made of fibers with wide and clearly visible lumen. The xylem and phloem vessels have usual elements. In the center of the stem, there is large pith formed by undifferentiated parenchyma. Starch grains and calcium oxalates are found in the stem. The lamina of the leaves has a single layer epidermis and dorsiventral mesophyll. Druses of calcium oxalate are abundant in leaves.

In our study about microscopic features of leaves in *R. graveolens*, we observed a single layer of epidermis in both sides (in cross section), mesophyll dorsiventral, numerous crystals of calcium oxalate as described by Kanan and Babu [22], and also frequent secretory cavities that store essential oil (Fig. 2). These characteristics can be used to identify *R. graveolens*, avoiding misidentification when it is used to produce interesting materials for human being.

3 Distribution and Medicinal Importance

R. graveolens is known for its medicinal properties since a quite long time ago. This plant species is native from Mediterranean region. Nowadays, it can be found in many different countries, including Brazil, Peru, Italy, India, South Africa, and others. There are many reports about the medicinal usage of this species by the ancient Greeks and Romans [23]. In several traditional medicine systems, *R. graveolens* is used as tonic, anthelmintic, emmenagogue, sudorific, antifertility, and also for respiratory disease, headache, heart problem, gastrointestinal disorders, neck pains, rheumatism, gout, intestinal cramps, convulsions, diabetes, fever, worms, kidney problem, earache, bladder and sinus [23–29]. Rue is also a plant used in magic rituals to treat supernatural folk illness as well as against evil eye or bad spirits influence [30, 31].

In Amazonic region, *R. graveolens* leaves have been used against headache, dizziness, brain weakness, flu with cough, fever, hoarse voice, stroke, toothache, numbness after bug sting, and intestinal pain. It is also used for personal protection as a magic herb against evil energies [32, 33]. Thring and Weitz [34] performed an ethnobotanical study in South Africa and described *R. graveolens* as the second most medicinal plant as reported by the questionnaire. Above description shows how important is *R. graveolens* as a medicinal plant worldwide.

Besides the medicinal purpose of this species, there are some reports on its toxic effects in liver and kidney and as abortive. According to Prabhu et al. [24], *R. graveolens* may damage important organs of the body, when taken in high dosage. The leaves of this species can cause chemical irritation in the skin. Some dermatitis has been reported mainly in children [30, 35].

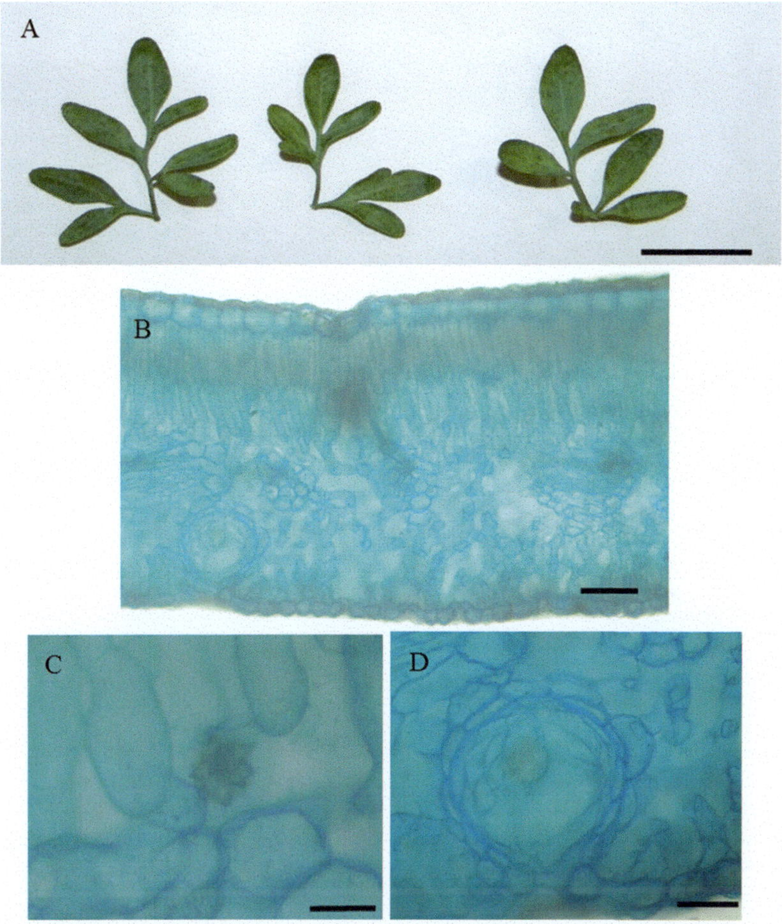

Fig. 2 *Ruta graveolens* L. (**a**). Leaves, bar 2 cm; (**b**). Cross section of leaves, bar 150 μm; (**c**). Detail of druse, bar 20 μm; (**d**). Detail of secretory cavity, scale 50 μm

4 Phytochemistry

4.1 Secondary Metabolites in *R. graveolens*

R. graveolens is a rich source of secondary metabolites mainly: coumarins, alkaloids, volatile oils, flavonoids, and phenolic acids [36–38]. The bioactive compounds of this species have been widely studied not only because of their interest in the chemistry of natural products but also because of the several biologically active compounds which provide a base for the use of *R. graveolens* in folk medicine and in the search of more biologically active compounds. Figure 3 depicts the

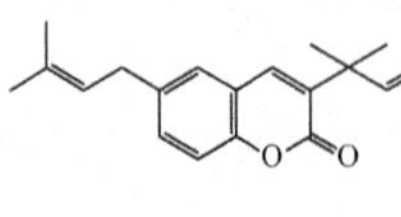

1

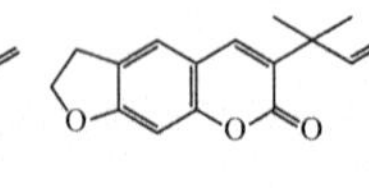

2

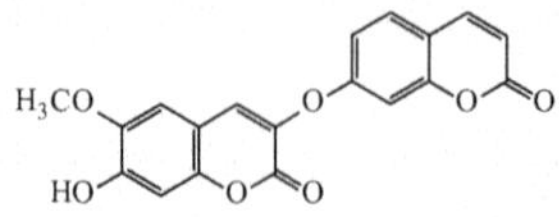

3

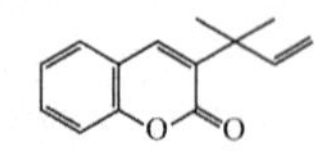

4

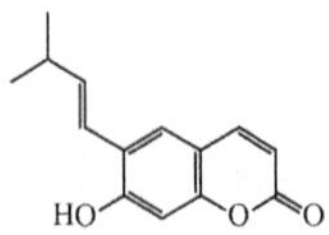

5

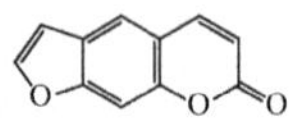

6

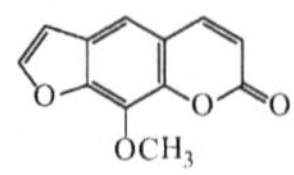

7

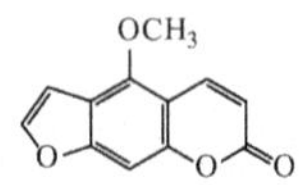

8

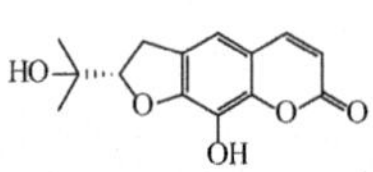

9

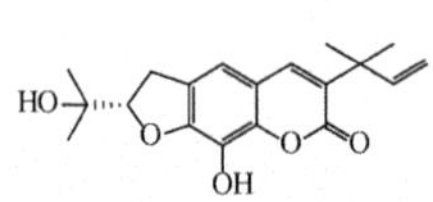

10

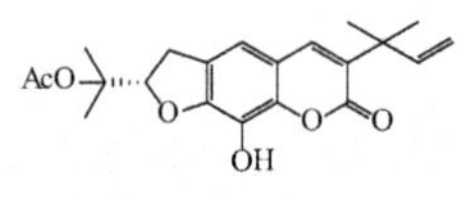

11

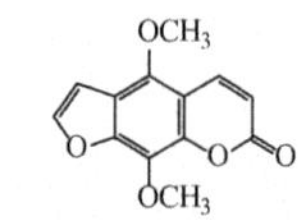

12

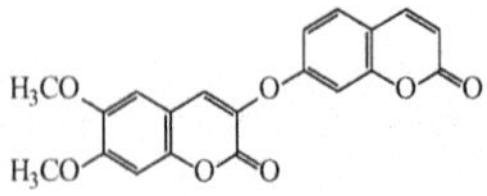

13

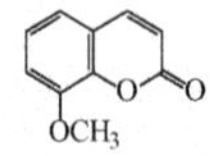

14

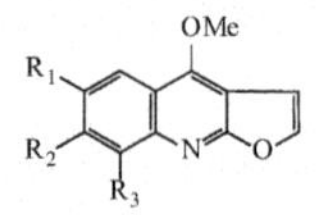

	R1	R2	R3
15	H	H	H
16	OMe	OMe	H
17	H	H	OH
18	H	OMe	OH
19	H	H	OMe

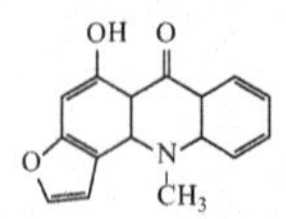

20

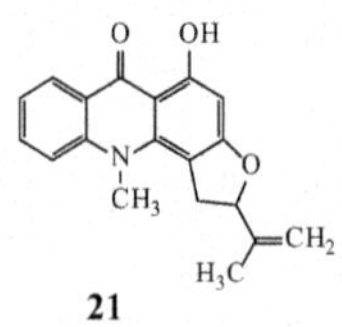

21

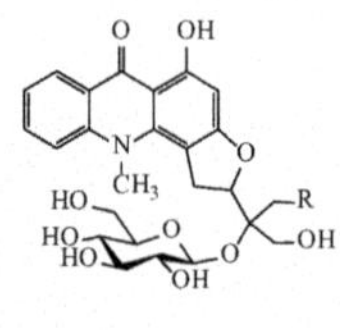

27 R= H
28 R= OH

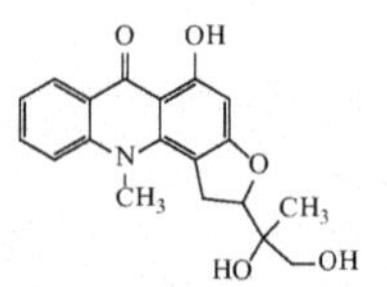

24

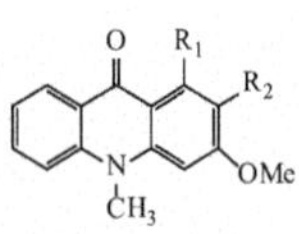

29 R1 = OH, R_2 = H
30 R1 = OH, R_2 = OMe

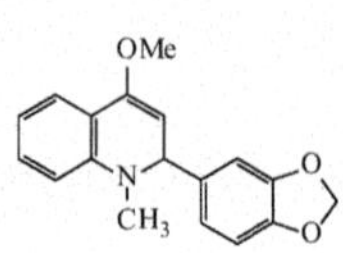

31

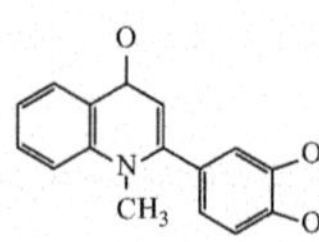

33

Fig. 3 (continued)

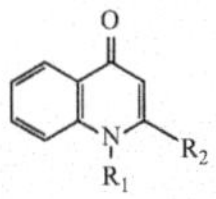

	R_1	R_2
33	Me	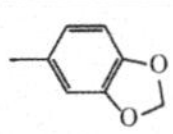
34	Me	*n*-nonyl
35	Me	*n*-decyl
36	H	*n*-undecyl
37	H	*n*-heptyl
38	H	*n*-octyl
39	H	*n*-nonyl
40	H	*n*-decyl
41	H	*n*-undecyl

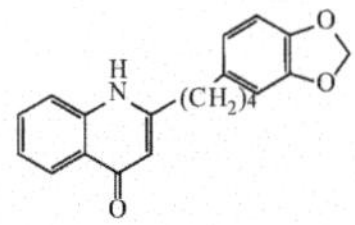

43

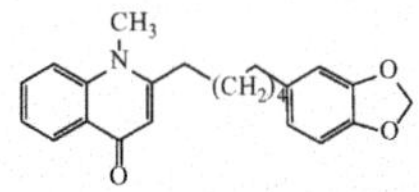

44

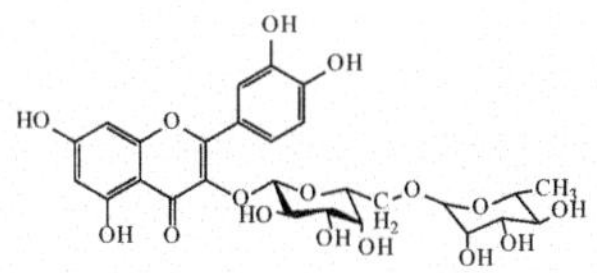

45

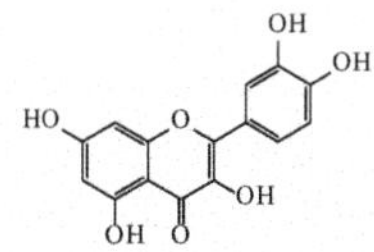

48

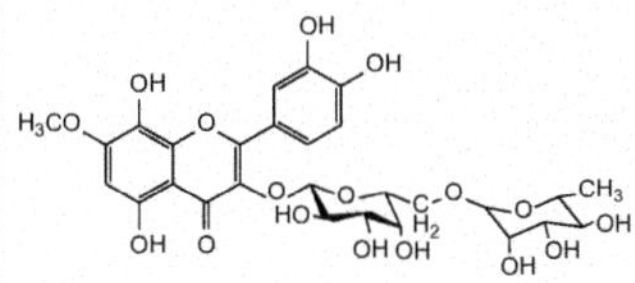

46

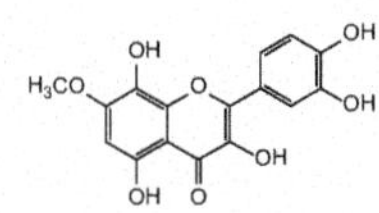

47

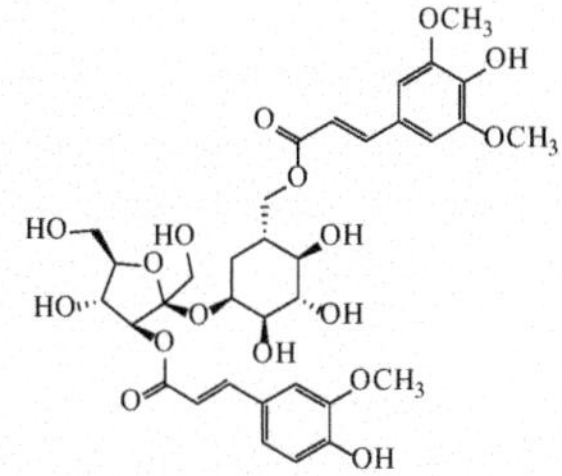

51

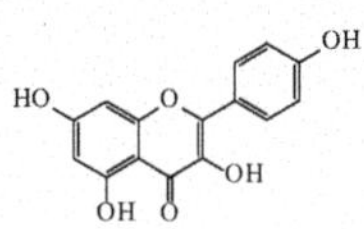

49

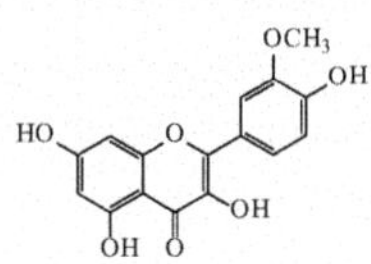

50

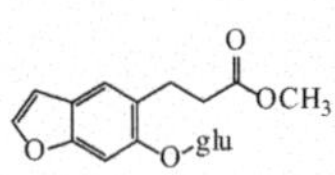

52

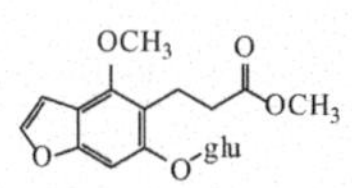

53

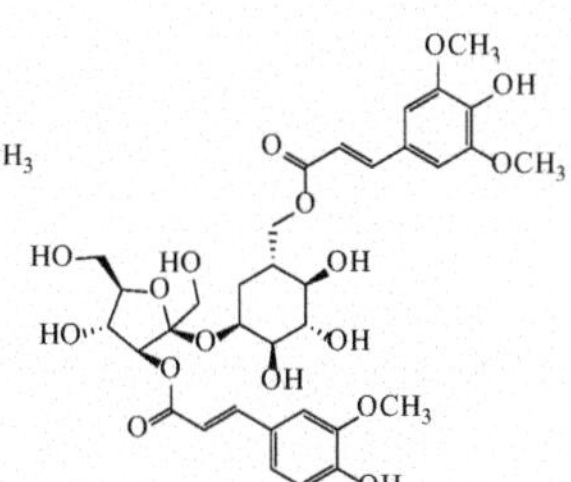

54

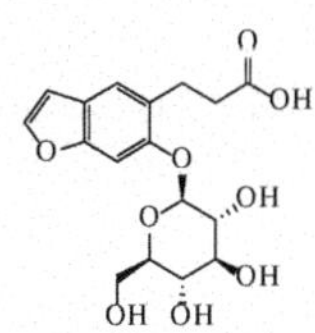

55

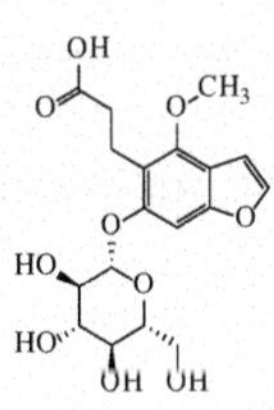

56

Fig. 3 Secondary compounds present in *R. graveolens*

common secondary compounds present in *R. graveolens*; the details of these compounds are presented in the following sections.

4.1.1 Coumarins

The main types of coumarins identified in *R. graveolens* are: simple coumarin, furanocoumarins, dihydrofuranocoumarins, and isocoumarins among others [36, 39, 40].

The coumarins isolated from the roots of the *R. graveolens* for the first time were gravelliferone methyl ether (**1**) chalepensin (**2**), daphnoretin (**3**) [41], 3-(1′,1′-dimethylallyl)-herniarin (**4**), and gravelliferon (**5**) [42, 43]. The furanocoumarins (FCs) with linear structure are produced in different parts of the species; studies of field-grown plants showed that FCs' concentration was proportionally related to both the plant's phenological stages and the organs in which the substance is stored [44]. Fruits accumulate higher concentrations of FCs, followed by leaves, roots, and stems. On the leaf surface, concentrations and proportions of FCs as psoralens (psoralen **(6)**, xanthotoxin **(7)**, and bergapten **(8)** vary during aging [45, 46]. Xanthotoxin prevails in absolute amounts regardless of the species' age [47].

Other furanocoumarins as rutamarin (**9**), rutarentin (**10**) [48], chalepin (**11**) [49], isopimpinellin (**12**), and a biscoumarin, O-methyl-daphnoetin (**13**) [50] has been found in many parts of the plant species in different amounts. Abyshev et al. [51] isolated from the epigeal part of *R. graveolens*, the furonocoumarins, bergapten (**8**), chalepin, and chalepensin (**2**). The novel and, besides, recently a new coumarin derivative, 8-methoxy chromen-2-one (**14**) was isolated by Debasis et al. [52].

4.1.2 Alkaloids

Rue is a valuable natural source of alkaloids. The alkaloids found in this species can be classified as furoquinolines, acridones, quinolines, and alkylquinolinones [37, 53, 54].

The furoquinoline alkaloids such as dictamine (**15**), kokusagenin (**16**), peleteine (**17**), skimmianine (**18**), and fagarine (**19**) occur, in different amounts, on the leaves, shoots, roots, and flowers [37, 55]. In the roots and leaves, one of the main alkaloids present was skimmianine **(18)** [56]. According to Mancuso et al. [50] kokusagenin (**16**) and skimmianine (**18**) are almost equally distributed in the plant's aerial parts.

Furacridone (**20**) and rutacridone (**21**) were the first representatives from acridone alkaloids, to be extracted from the roots of *R. graveolens* [57], from that a series of rutacridone derivatives have been obtained such as rutacridone epoxide (**22**), gravacridonol (**23**) [58–61], gravacridondiol (**24**), gravacridondiol acetate (**25**) [62], gravacridontriol (**26**), glycosides gravacridondiol-*O*-18-β-D-glucoside (**27**), and gravacridontriol-*O*-18- β-D-glucoside (**28**), which accumulate in different parts of the roots [53, 63].

Distribution and compartmentalization studies revealed the accumulation of acridone alkaloids in intact roots of *R. graveolens* [60]. Rutacridone (**21**) was found in the differentiation zone where the root hairs are located, suggesting its accumulation in the roots trichomes. More hydrophilic acridones such as gravacridondiol (**24**) and 1-hydroxy-3-methoxy-*N*-methylacridone (**29**) [53] and other metabolites, in turn, experience a decrease in concentration in this specific root segment. Gravacridondiol glucoside (**27**) was shown to be the major compound in the root tips [53]. Arborinine (**30**) [55, 64] is another acridone alkaloid identified in *R. graveolens*.

Graveolinine (**31**) and (4S) 1,4-dihydro-4-methoxy-1,4-dimethyl-3-(3-methylbut-2-enyl) quinoline 2,7-diol (**32**) [65] are quinoline alkaloids isolated from the leaves of the species.

R. graveolens is also an important source of alkylquinolone alkaloids (AQs) of the series 4-quinolones. Graveoline (**33**), an alkaloid of the 2-aryl-4-(1H)-quinolone series, has been identified by Grundon and Okeley [56], Oliva et al. [66], and Ghosh [67]. At a subsequent time, analogues of 2-n-alkyl **34**–**42** [37], such as 2-[4′-(3′,4′-methylenedioxyphenyl) butyl]-4-quinolone (**43**), [68] and 1-methyl-2-[6′-(3″,4″-methylenedioxyphenyl] hexyl]-4-quinolone (**44**) [69] were also identified.

According to Eun-Tae et al. [54], the AQs without methyl groups (HAQs) and 1-methyl-AQSs (MeAQs) with 2-nonyl substituents were the most abundant in leaves while in root, the one with 2-undecyl group were the dominant metabolites.

4.1.3 Volatile Oils

The volatile oils obtained from *R. graveolens* fruits, leaves, roots, flowers, or stems have a yellowish color, as well as an intense and penetrating odor. They are composed mainly of oxygenated compounds (ketones, alcohols, acetates), sesquiterpenes and monoterpenes hydrocarbons, aromatics hydrocarbons, and coumarins.

Within the series of methyl nonyl ketones, 2-undecanone and 2-nonanone are the predominant constituents in essential oils extracted from the aerial parts of *R. graveolens,* even if it is grown in different parts of the world such as Malaysia [70], Italy [71], Venezuela [72, 73], Egypt [74], Iran [75], Algeria [76], Ukraine [77], and Brazil [78].

Monoterpenes and sesquiterpenes have been identified in this plant species include, α-pinene, limonene, 1,8-cineol, α-thujene, camphene, terpinolene, camphor, *trans-p*-menth-2-en-1-ol, β-phellandrene, germacrene-B, 3-carene, D-cadinene, β-caryophyllen, β-humulene, elemol, geijerene, and geyrene [71, 76]. Pregeijerene and geijerene, sesquiterpenes compounds, are the major constituents of the essential oil from *R. graveolens* roots [79, 80]. Xanthotoxin was also found in the oil extracted from aerial parts and roots of the plant [71]. Aromatic hydrocarbons such as trans-anethole are also found in essential oil [75].

The composition of the essential oil is affected by climatic, seasonal, and geographic conditions; harvest period; chemotypes; and/or extraction procedure [81].

4.1.4 Flavonoid Glycosides and Flavonoids

R. graveolens is a rich source of flavonoid glycosides, for example, rutin (**45**) [82, 83].

A yellow pigment containing gossypetin 7-methyl ether 3-rutinoside (**46**) and gossypetin 7-methyl ether (**47**) has been isolated from the flowers of *R. graveolens* [84].

The leaves and flowers of this plant species have been reported to possess flavonoid, quercetin (**48**), kaempferol (**49**), and isorhamnetin (**50**) [83, 84].

4.1.5 Miscellaneous Compounds

Besides the constituents listed above, other metabolites have also been reported in *R. graveolens*, such as glycosides and phenolic compounds. Chien-Chih et al. [85] have reported six glycosides, 3′-sinapoyl-6-feruloylsucrose (**51**), methylcnidioside A (**52**), methylpicraquassioside A (**53**), 3′,6-disinapoylsucrose (**54**), cnidioside A (**55**), picraquassioside A (**56**), isolated from aerial parts of *R. graveolens* plants. Phenolic acids gentisic acid, caffeic acid, ferulic acid, and p-coumaric acid were isolated from the leaves of this plant species [86].

5 Pharmacological Studies

R. graveolens is found abundantly around the different parts of the world with wide and diversified use for medicinal purposes, which have stimulated the development of several studies evaluating the biological activity to support its therapeutic use in both allopathic and homeopathic approaches.

In proof of view of therapeutic efficacy of homeopathic preparations, Rosi Cruvinel et al. [87] found that administration of *R. graveolens* 6 CH interferes favorably in the feeding, water intake, and weight gain in experiments carried out with chickens.

In vivo test with patients (18–60 years old, candidates for surgery of anterior cruciate ligament) has shown that homeopathic complex containing *Arnica montana* 5 CH, *Bryonia alba* 5 CH, *Hypericum perforatum* 5 CH, and *R. graveolens* 3 DH showed similar results to placebo group in reducing the morphine consumption 24 h after the surgery to fix the knee ligament [88].

Study by Khan [89] indicated the potential for topical homeopathic preparations with *Symphytum officinale, Thuya occidentalis, R. graveolens, Rosemary officinalis, Bellis perennis, Hypericum perforatum, Calendula officinalis, and Tagetes* sp. to treat lesions of foot.

Arora et al. [90] evaluated the in vitro action of the mother tincture (MT), *R. graveolens* 30C, 200C, 1 M, and 10 M against cell lines deriving from tumors of human colorectal carcinoma (COLO-205), showing that homeopathic preparations had highly significant effects in the respective cancer cell lines, cytotoxicity and decrease in cell proliferation, thus indicating potential of these preparations as anticarcinogenic.

In search of anticancer effect of homeopathic preparations based on *R. graveolens* on COLO-205 cell line, another study showed that the mother tincture (MT) and *R. graveolens* 30C were able to cause a decrease in cell viability with reduced clonogenecity and migration capabilities, causing also with morphological and biochemical changes indicative of cell death by apoptosis, demonstrating the potential of these preparations in the treatment of colon carcinoma [91].

Patients with locally advanced solid tumors or metastases and previously treated with conventional anticancer drugs showed a transitory improvement in quality of life when given homeopathic preparation of *R. graveolens* 9C by oral administration; however, there was no influence on the tumor progression [92].

The ultradiluted potencies and mother tincture (MT) of *R. graveolens* showed anticancer activity when avoided the proliferation and cytotoxic effects on normal kidney epithelial cell model [93].

To study the antineoplastic properties of homeopathic medicines, homeopathic preparations of *R. graveolens* were administered orally in mice and evaluated its effect on immune system. It was found that higher potencies caused significant enhancement of hematological parameters including the total white blood cell count, bone marrow cellularity, and the number of α-esterase positive cells. Other parameters of immune response such as antibody titer circulating and the number of plaque forming cells, enhanced proliferation of B and T lymphoid cells were also observed, which suggest the immune modulatory activity of homeopathic preparations in high dilutions [94].

Besides the homeopathic preparations of *R. graveolens* to check out pharmacological properties, this plant species has been also analyzed to find out the new alternatives or therapeutic supplements for employment in allopathy using isolated compounds as well as using the whole components (phytocomplex) to produce herbal medicines. A number of studies have been carried out to show the activities such as contraceptive, anti-inflammatory, antimicrobial, and analgesic, and these will be discussed in the following sections.

In view of development of new contraceptive drugs, Guerra and Andrade [95] developed an in vivo study using pregnant albino primiparous female rats subjected to intramuscular or oral administration of *R. graveolens* extract and noted marked contraceptive effect with loss of implantation of eggs. Similarly, Gandhi et al. [96] demonstrated the efficacy of oral administration of powder of aerial parts from *R. graveolens* and its extracts (petroleum ether and methanolic) in the model female rats.

Evaluation study on the pregnancy of the ethanol extract of aerial parts of *R. graveolens* developed in model in vivo (CF1 mice) showed that the extract did not cause preimplantation embryonic loss or reabsorptions, but it led to fetal death [97].

Sailani and Moeini [98] evaluated the effect of ethanol extracts of *R. graveolens* and *Cannabis sativa* on spermatogenesis in the adult male Wistar rats and found that the extracts cause decrease in spermatogenesis. In the search for male contraceptive drugs, Harat et al. [99] developed an in vitro study with human sperm using aqueous extract of *R. graveolens* and proved the potential effect of extract. Oral administration (5 g/kg) of an aqueous extract of *R. graveolens* in male rats resulted in reduced sperm motility only after one hour of administration without any change in other sperm characteristics, indicating the potential of this plant extract in male contraception [100]. Intraperitoneal administration of aqueous extract of *R. graveolens* in rat resulted in reduced number of spermatogonia, thus suggesting that this species can be used in birth control [101].

Nasirinezhad et al. [102] investigated the effect of aqueous extract of *R. graveolens* on the reproductive system of immature female mice, BALB/C line. They reported that the extract of *R. graveolens* can interfere with reproductive system function in immature female mice by alterations in sex hormonal level and ovarian morphology and thus it might be useful as a source of antifertility substance.

In research of the anti-inflammatory effect of *R. graveolens* extract and rutin (flavonoid present in the species), Raghav et al. [103] showed that the extract has better anti-inflammatory activity than the rutin using different models such as murine macrophage cells (J-774) challenged with lipopolysaccharide (LPS), induction of inflammatory response by nitric oxide, and other mediators.

Li et al. [104] performed a study with extracts of this plant species to analyze their NGF-potentiating activities on the NGF-mediated neurite outgrowth from PC12D and proved that the methanol extract of the leaves of *R. graveolens* markedly increased the proportion of neurite-bearing cells.

The isolated compound from methanolic extract of *R. graveolens*, identified as 3-(1′-allyl-1′-dimethyl)-6-hydroxy-7-methoxy-coumarin has been measured in iNOS, COX-2 genes, and some cytokines pro-inflammatory. This compound revealed ability to inhibit protein and mRNA expression of iNOS and IL-1β in LPS challenged macrophages, showing also anti-oxidant activity [105].

The ability of methanol extract of *R. graveolens* has been demonstrated to inhibit inflammation and oxidative stress in adjuvant induced arthritis in rats [106]. Polyphenols and alkaloid fractions obtained from *R. graveolens* extract showed anti-inflammatory activity in a model of acute and chronic inflammation in rats [107].

The in vivo studies of methanol extract of *R. graveolens* in hypercholesteremic rats showed reducing oxidative damage, inflammation, and aortic pathology and indicated that the species has potential for therapeutic use in clinical conditions associated with atherosclerosis [108].

The evaluation of anti-inflammatory effect of skimmianine quinoline alkaloid isolated from *R. graveolens* extract proved decrease in mRNA levels of TNF- α and IL-6 as well as the levels of NO and PGE.sub 2, COX-2 and 5-LOXm activities, thus proving anti-inflammatory action of the substance by several mechanisms involved in the response cascade of events [109]. Methanol extract of leaves of *R. graveolens* showed antinociceptive, anti-inflammatory, and antipyretic activities in mice [110]. Kataki et al. [111] have reported the antioxidant and anti-inflammatory activities of methanol extract of leaves of *R. graveolens* in vitro and in vivo models and showed potent inhibitory effects on the arachidonic acid pathways. The antioxidant effect of *R. graveolens* extract has also been also evaluated in two models including free radical scavenging using DPPH and inhibition of lipid peroxidation by the ferric thiocyanate method [112].

Aqueous extract of *R. graveolens* leaves has been reported to possess antimicrobial effect against *Fusarium solani, Pyrenochaeta lycopersici, Trichoderma viride, Penicillium sp., Thielaviopsis basicola, and Verticillium dahliae, Bacillus cereus, Staphylococcus aureus, and Listeria monocytogenes* [66, 113]. In another report by Ivanova et al. [114], the methanol, petroleum ether, ethyl acetate, and water–methanol extracts of aerial parts of *R. graveolens* were found to possess cytotoxic as well as antimicrobial activity against *Streptococcus pyogenes, Listeria monocytogenes,* and *Bacillus subtilis*.

Extract and essential oils obtained from *R. graveolens* were evaluated against *Pseudomonas aeruginosa* strains, *Staphylococcus aureus, Candida albicans,* and *Candida krusei* isolated from patients suffering with acute external otitis. All the tested strains were found to be resistant to the extract, while essential oil (4%) of *R. graveolens* inhibited four *Staphylococcus* and all *Candida* strains (inhibition halos between 10 and 13 mm diameter) [115].

Oils obtained from *R. graveolens* inhibited the growth of Gram positive (*Staphylococcus aureus* and *Enterococcus faecalis*) and Gram negative (*Escherichia coli* and *Klebsiella pneumoniae*) bacteria [116]. However, the aqueous leaf extract and essential oils isolated from this plant species was found to be ineffective against *Trichophyton mentagrophytes* and *Pseudomonas aeruginosa* [117, 118]. Hydro-alcoholic and aqueous extracts of *R. graveolens* have shown to be ineffective against *Enterococcus faecalis* [119, 120].

In search of new therapies to fight against *Helicobacter pylori, R. graveolens* extract was found to exhibit strong inhibitory activity against IL-8 secretion [121]. Its essential oils possess antifungal activity, more significantly against *Aspergillus fumigatus* and *Cladosporium herbarum* [122]. Essential oil obtained from fresh leaves by hydrodistillation in a Clevenger-type apparatus and characterized by GC-FID and GC-MS showed antibacterial activity against Gram-positive and Gram-negative bacteria, especially *Bacillus cereus* and *Staphylococcus aureus* [78]. Aerial parts essential oil showed activity against different strains of *Legionella pneumophila* [123].

Figueroa-Valverde et al. [124] reported that methanol extract of the leaves (0.5 g/kg) is able to induce hypoglycemic effect, which is attributed to the presence of flavonoids in this species. In vivo experiments in hyperglycemic rats demonstrated that oral administration of the methanol extract resulted in decreased concentration of blood glucose [125].

Van Huyssteen et al. [126] conducted a study on popular practice in Africa for the treatment of *Diabetes mellitus* and reported that hydroalcoholic extract of aerial parts of *R. graveolens* produced the highest increase in glucose utilization in C2C12 muscle cells.

Effect on lipid and glucose levels as well as hematological parameters were studied upon administration of *R. graveolens* extract in rats with diabetes induced by injecting streptozotocin. This study showed that *R. graveolens* extract caused a significant decrease in cholesterol and LDL-C but no alterations were in the levels of glucose, triglycerides, VLDL-C, and HDL [127].

In vivo assay using acetic-acid-induced writhing and hot-plate-induced thermal stimulation in mice has demonstrated the antinociceptive activity of *R. graveolens* extract [128].

Park et al. [129] performed in vivo studies using various experimental models of pain and found that antinociceptive effect of *R. graveolens* extract was mediated by opioidergic and α 2-adrenergic receptors but not by serotonergic receptors.

Comparative studies between *R. graveolens* and *Matricaria chamomilla* (70% ethanol as solvent and fractions of petroleum ether, ethyl acetate, and *n-butanol*) showed that both of these plant species can be effectively used as analgesics, which is attributed to the presence of flavonoids and alkaloids [130].

Gilbert et al. [131] carried out in vitro assays to find out the antiparasitic activity *of R. graveolens* and proved that ether extract of leaves was active against *Strongyloides stercoralis*, a nematode that causes strongyloidiasis.

Mendes et al. [132] studied the molluscicidal activity of hexane and ethanol extracts from leaves and stem of *R. graveolens* and found to be effective against the intermediate host of *Schistosoma mansoni.*

In vitro study performed with hydroalcoholic extract of aerial parts of *R. graveolens* showed the reductions in viability, potential for invasion, and multiplication rate of the parasites *Leishmania amazonensis and Trypanosoma cruzi* and indicated that these herbal extracts may be potential candidates for developing drugs to treat leishmaniasis and Chagas disease [133]. Queiroz et al. [134] reported that aqueous extract of aerial parts of *R. graveolens* possessed leishmanicide activity against promastigotes and amastigotes of *Leishmania amazonensis.*

In view of new drugs for the treatment of neurodegenerative diseases such as Parkinson's and Alzheimer's, in vivo model for assessing the inhibition of the oxidative deamination of tyramine by monoamine oxidase (MAO) isolated from rat liver have been studied and they found that ethyl acetate extracts and oil extracted from leaves of *R. graveolens* have a significant ability to inhibit this enzyme [135]. In another study [136], hexane extract of this plant species caused potent inhibition of acetylcholinesterase (AChE) and butyrylcholinesterase (BChE) demonstrating potential to be used to treat Alzheimer's disease. Russo et al. [137] also indicate that *R. graveolens* has potential for therapeutic use in this disease, which is attributed to the presence of coumarin in the hexane extract with anti-AChE activity.

In order to find new anticonvulsant drugs, Keihanian et al. [138] investigated the effect of *R. graveolens* extract in vivo model of seizures induced by pentilentetrazole (PTZ) in mice and demonstrated that its ability to reduce seizures. In vivo study developed by Amabeoku and Ahmad [139] showed anticonvulsant activity of the methanol extract of the leaves of *R. graveolens* probably involving potentiation of gamma aminobutyric acid (GABA). Study by Bohuslavizki [140] indicated the action of infused *R. graveolens* in blocking potassium channels, thus signaling the potential of this plant species in the treatment of Encephalomyelitis disseminata.

Aqueous and ethanol extract of *R. graveolens* leaves have been shown to exhibit binding affinity to GABA-benzodiazepine receptor in the flumazenil-binding assay, which was attributed to the presence of furanocoumarines [141]. Adsersen et al. [142] reported that the extract of *R. graveolens* exhibited moderate inhibition of the acetylcholinesterase.

In vitro assays performed with human cell lines HeLa, MCF7, and A43 have shown the anticancer potential of furanoacridones and acridone alkaloids (arborinine and evoxanthine) isolated from *R. graveolens* [64]. Fadlalla et al. [143] investigated the effect of the methanol extract of *R. graveolens* on colon, breast, and prostate cancer cells and found that it inhibited the proliferation and survival of cancer cells via multiple targets. In vitro studies by Ghosh et al. [144] have demonstrated that graveoline, a compound isolated from *R. graveolens*, showed cytotoxicity against A375 skin melanoma cells, causing cell death by apoptosis and autophage. In vitro (skin melanoma cells) and in vivo (7,12-dimethylbenze (a) anthracene induced skin cancer in Swiss albino mice) assessment of ethanol extract of *R. graveolens* demonstrated the potential of this plant species to treat skin cancer without causing any acute or chronic toxicity [145]. The anticancer potential of aqueous extract of this species has been also evaluated in model of different glioblastoma cell lines (U87MG, C6, and U138 [146].

Al-Nimer and Ali [147] studied the effect of aqueous leaves extract of *R. graveolens* on nitric oxide (NO)–peroxynitrite (ONOO-) cycle biochemistry and found that the improved bioavailability of nitric oxide (NO), indicating that its use in coronary artery disease with nitrate tolerance.

Essential oils obtained by hydrodistillation of aerial parts of *R. graveolens* and isolated constituents showed repellent and larvicidal activity against *Aedes* L. and could be useful in mosquito control [148].

Methanolic extracts from aerial parts of *R. graveolens* have demonstrated significant ability to inhibit aldehyde oxidase, partially purified from liver homogenates of mature male guinea pigs by heat treatment and ammonium sulfate precipitation [149].

Extract from the aerial parts of *R. graveolens* obtained by different extractive procedures (extraction in Soxhlet percolation and ultrasound) were investigated for cytotoxicity (G-929 cells) and tyrosine inhibition capacity (spectrometry at 481 nm). It has been observed that extraction procedure interferes with the activity of the compounds. Extraction obtained by percolation method showed less cytotoxicity and higher percentage of inhibition of tyrosine [150]. Hypotensive effect of aqueous extract of *R. graveolens* under in vivo system in normotensive rats has been shown by Chiu et al. [151]. Studies by Khori et al. [152] indicated that the methanol extract of *R. graveolens* and its alkaloidal fraction has potential to treat supraventricular tachyarrhythmia.

Ueng et al. [153] evaluated the effect of aqueous extract of aerial parts of *R. graveolens* and isolated substances (rutin and furanocoumarins) on activities of enzymes; cytochrome P450 (P450/ CYP), uridine diphosphate (UDP)-glucuronosyltransferase, and reduced nicotinamide adenine dinucleotide (phosphate) (NAD (P) H): quinone oxidoreductase. They observed that oral administration of the extract in mice/rats caused an increase in the levels of CYP1A and CYP2B in a dose dependent manner and by inducing increased hepatic UDP-glucuronosyltransferase activity.

In vivo studies have indicated that hydroalcoholic extract of *R. graveolens* caused relaxation in the rings of the trachea of rat and the effect is mediated by noncompetitive antagonistic mechanism [154].

A recent review on genus *Ruta* by Hammiche and Azzouz [155] reported the diversity of biological activities of *R. graveolens*, especially the anticancer activity, action on pigmentation of the skin to stimulate melanin synthesis signaling for use in vitiligo or psoriasis, effect on the central nervous system by inhibiting the activity of MAO-B, anti-inflammatory, antimicrobial, cytotoxic, hypotensive, antiviral, and antiplasmodial affects.

6 Toxicological Studies

To study the safety assessment parameters in the use of *R. graveolens*, some studies have been carried out to assess the toxicity. There are a few clinical case reports of poisoning in humans.

Simon et al. [156] present a patient developed severe dermatitis as a result of contact with *R. graveolens*, which is attributed to the toxic bioactive compounds furanocoumarin constituents: 5-methoxypsoralen (bergapten), 8-methoxypsoralen (xanthotoxine), and furanoquinoline dictamnine. Sharma et al. [157] reported the case of 26-year-old pregnant woman led to abortion after consuming *R. graveolens* tea in combination with chocolate and cinnamon. In another case report, use of decoction of *R. graveolens* by 78-year-old woman led to poisoning by bradycardia, renal failure with acute hyperkalemia, and coagulopathy [158]. Adams et al. [159] reported the case of acute phytophotodermatitis in 2-year-old child after contact with this plant species. A 12-year-old patient who developed lesion with linear distribution in the lower limbs has been diagnosed with phototoxic reaction, attributed to the use of alcoholic extract of *R. graveolens* [160].

Experimental study in goats has indicated that oral administration of *R. graveolens* leaves (1 and 5 g/kg) led to toxicity in animals with pathological changes in various organs. The changes included alterations in serum aspartate and copper, iron, zinc, manganese, calcium, and phosphorus with animal death record [161]. Lyophilized hydroalcoholic extract of aerial parts of *R. graveolens* resulted in fetal death when administered orally (1000 mg/kg/day) during the pregnancy [97].

In the research of potential antiparasitic property against *Vampirolepis nana* (gastrointestinal parasite), hydroalcoholic extract of the leaves and seeds of *R. graveolens* was administered by gavage to Swiss albino mice and it was proven mild to moderate hepatotoxic with weak antiparasitic effect [162].

Intraperitoneal administration of *R. graveolens* extract (30 and 100 mg) in Wistar rats for 3 days resulted in morphological changes in the liver and led to hepatotoxicity [163].

Aqueous extract (20%) of *R. graveolens* leaves when administered orally in rats caused a significant decrease in the number of normal embryos and increase in cases of persistent late embryonic development thus demonstrating toxic effect of the extract [164]. Shama et al. [165] reported that ethanolic and aqueous extracts of *R. graveolens* seeds resulted in change in body weight, biochemical, and hematology parameters and thus caused toxicity when administered orally to male Wistar rats at 200 mg/kg/day (for 4 weeks). The ethnobotanical study carried out in health care service of Provincia del Chaco at Argentina has reported the potential dangers and poisoning effects (which may lead to death) associated with the use of medicinal plants (highlighted *R. graveolens*) [166].

7 Biotechnological Studies in *R. graveolens*

7.1 In Vitro Regeneration and Mass Propagation

From the above literature, it has been evidenced that *R. graveolens* is a multipurpose herb. Biotechnology using in vitro technique provides a viable tool for mass multiplication and germplasm conservation of aromatic and medicinal plants of interest [167, 168]. Faisal et al. [169] developed a protocol for rapid clonal multiplication of *R. graveolens* through high frequency shoot induction from nodal explants and reported the successful outdoor establishment of regenerated plants. Various concentrations and combinations of plant growth regulators, (PGRs) viz., BA, Kn, IAA, and NAA have been tried. The highest shoot regeneration frequency (98.5%) was reported on MS medium containing BA (10 μM) and NAA (2.5 μM). The regenerated shoots were found to be rooted best on MS medium with 0.5 μM IBA [169]. There are other reports on tissue culture studies of *R. graveolens* to monitor alkaloids and coumarins content in callus, shoots, and regenerated plants [170–172].

7.2 Use of Elicitors

Commercial production of secondary compounds is generally hampered by their low yield. Elicitation is a strategy to enhance the secondary metabolite production and is considered an integral part of any large-scale process for secondary metabolite production. Elicitor can be biotic and abiotic depending on its origin. The mechanism of both the elicitors is different and quite complex. Chitin and chitosan are the elicitors inducing phytoalexin accumulation in plant tissue. Orlita et al. [173] studied the effect of chitin and chitosan on alkaloids and coumarins in in vitro shoot cultures of *R. graveolens* and found that elicitation induced a significant increase in quantity of all the metabolites. The application of abiotic elicitors saccharin and benzothiadiazole has been reported to increase the production of simple coumarins, linear furanocoumarins, dihydrofuranocoumarins, and furoquinolone alkaloids by several times in in vitro shoots of *R. graveolens*. [174]

Effect of polyamines, spermine and putrescine, have been studied on growth and furanocoumarins in *R. graveolens* cultures. Spermine was found to increase the multiple shoots formation and furanocoumarins production by 2.5 and 1.47-fold, respectively [175].

7.3 Scale Up Studies

Bioreactors of different types are generally used for the large scale production of shoots and bioactive compounds for commercialization. In *R. graveolens*, bioreactor system has been reported to be an effective technique for large scale production of shoots [176]. A one-step protocol with improved regeneration efficiency for multiple shoots induction employing liquid culture system has been reported by Diwan and Malpathak [175]. They scaled up selected shoot line, RS2 from 250 mL to 5 L culture vessels, with 1.53-fold increase in biomass without affecting the productivity of cultures. Gontier et al. [177] developed bioreactor for economic and efficient production of furocoumarins from *R. graveolens* shoots. The traditional systems of in vitro propagation have been compared with simple aerated bioreactor for large scale shoot biomass in same species [176]. Shoots cultured in the simple aerated bioreactor systems showed 4.1-fold increase in 2 l bioreactor vessel and 5.2-fold increase in 5 l bioreactor as compared to 500 mL conical flask culture.

7.4 Genetic Transformation and Plant Hairy Root Culture

Transgenic or hairy roots obtained after transformation with *Agrobacterium rhizogenes* present the fast growing system as compared to nondifferentiated plant cells and therefore have become a promising source of bioactive compounds of interest in several plant species. Chemical synthesis of furocoumarins in *R. graveolens* is very expensive and does not always lead to biologically active compounds [178]. To study the distribution and compartmentalization of alkaloids in intact roots, Kuzovkina [179] established hairy root cultures. There are several factors determining the efficient transformation starting from explant selection to *Agrobacterium rhizogenes* strain and media used. Certain tissue or plant organs are more appropriate for transformation in its particular developmental stage. Usually young tissues are more responsive/ susceptible to transformation as compared to mature ones. The commonly used explants for transformation with *A. rhizogenes* are young seedlings or their parts, roots, and shoots [180]. Hypocotyls, callus, and shoots were inoculated with *A. rhizogenes* strains (LBA 9402 and A4) in *R. graveolens* [181]. Hypocotyls were found to be more responsive with *A. rhizogenes* strain LBA 9402 due to low level of coumarins and furanocoumarins as compared to other explants. Hairy roots obtained as a result of transformation indicated high level of coumarins, furanocoumarins, and alkaloids. The content of pinnarin, rutacultin, bergapten, isopimpinelin, and xanthotoxin in hairy cultures has been reported to be twofold higher than in shoot cultures. The two novel coumarins, osthole and osthenol have been also found, which are known to possess many biological activities including anti-inflammatory, anti-oxidative, and anti-tumorigenic [181]. Genetic transformation system has been developed for *R. graveolens* by co-cultivation of hypocotyls with *A. tumefaciens* strain C58C1Rif^{R} containing a plasmid harboring neomycin phosphotransferase and β-glucuronidase encoding genes [182]. The stable transgene integration was confirmed by growth on selection medium for *npt*II, by PCR and southern blot analysis [182].

8 Conclusions

R. graveolens is considered as an important plant species with various pharmacological activities. Although it is a plant used with medicinal purpose in many parts of the world, safety assessment parameters indicate that there are a few clinical case reports of poisoning in humans with the use of *R. graveolens*. It may damage the important organs of the body, when taken in high dosage. Therefore, it is important to consume the correct concentration and dose of this plant species within safer limits. The pharmacological and toxicological studies indicate the correct way to use this plant (or compounds derived from it) to produce safe medicines that can be used around the world. Regarding the improvement in production of secondary compounds from *Ruta graveolens*, there is still gap in the knowledge about the biosynthetic pathway of various bioactive compounds, and future studies could be focused to solve this problem in order to rise the production of bioactive compounds from *R. graveolens* by using strategies like immobilization, two phase culture system, as well as metabolic engineering approach to improve the production.

Acknowledgments Authors would like to acknowledge the Foundation for the Support of Research, Scientific, and Technological Development of the state of Maranhão – FAPEMA for financial support.

References

1. Malik S, Bhushan S, Sharma M, Ahuja PS (2014) Biotechnological approaches to the production of shikonins: a critical review with recent updates. Crit Rev Biotechnol 16:1–14
2. Malik S, Biba O, Gruz J, Arroo RRJ, Strnad M (2014) Biotechnological approaches for producing aryltetralin lignans from *Linum* species. Phytochem Rev 13:893–913
3. Malik S, Cusidó RM, Mirjalili MH, Moyano E, Palazón J, Bonfill M (2011) Production of the anticancer drug taxol in *Taxus baccata* suspension cultures: a review. Process Biochem 46(1):23–34
4. Malik S, Bhushan S, Sharma M, Ahuja PS (2011) Physico-chemical factors influencing the shikonin derivatives production in cell suspension cultures of *Arnebia euchroma* (Royle) Johnston, a medicinally important plant species. Cell Biol Int 35(2):152–157
5. Malik S, Bhushan S, Verma SC, Sinha AK, Sharma M, Ahuja PS (2008) Effect of pH on cell growth and shikonin derivative formation in *Arnebia euchroma* suspension cultures. Med Aromat Plant Sci Biotechnol 2(1):43–49
6. Stevens PF (2001 onwards) Angiosperm Phylogeny Website. Version 12, July 2012. Accessed 10 Mar 2016
7. Groppo M, Pirani JR, Salatino MLF, Blanco SR, Kallunki JA (2008) Phylogeny of Rutaceae based on two noncoding regions from cpDNA. Am J Bot 95(8):985–1005
8. Engler A (1896) Rutaceae. Nat. Pflanzenfam. Engelmann, Leipzig
9. Engler A (1931) Rutaceae. In: Engler A, Prantl K (eds) Die natürlichen Pflanzenfamilien, 2nd edn. Engelmann, Leipzig
10. Takhtajan A (1987) Systema magnoliophytorum. Officina Editoria NAUKA, Leningrad
11. Takhtajan A (1997) Diversity and classification of flowering plants. Columbia University Press, New York

12. Thorne RF (1992) An updated phylogenetic classification of the flowering plants. Aliso 13:365–389
13. Cronquist A (1993) An integrated system of classification of flowering plants. Columbia University Press, New York
14. Wei L, Wang YZ, Li ZY (2012) Floral ontogeny Rutae Rutaceae and its systematic implications. Plant Biol 14:190–197
15. Barrera CAC, Suárez LEC, Londoño CQ (2013) Pharmacognosy and pharmacobotany of species of genera Esenbeckia and Raputia (Rutaceae). Rev Cuba Plantas Med 18:638–653
16. Orlando JFF (2011) Estudo da composição química e atividade biológica do óleo essencial de *Ruta graveolens* Linneau (Rutaceae). Dissertation, Federal University of Paraíba
17. Kalil NA, Sperb D, Lichtenfels E (1994) Efeito da pilocarpina na regeneração hepática pós hepatectomia parcial em ratos. Acta Cir Bras 13:222–226
18. The Plant List (2013) Version 1.1. http://www.theplantlist.org/ (Accessed 1 Jan). Accessed 10 Mar 2016
19. Oliva A, Meepagala KM, Wedge DE, Harries D, Hale AL, Aliotta G, Duke SO (2003) Natural fungicides from *Ruta graveolens* L. leaves, including a new quinolone alkaloid. J Agric Food Chem 12:890–896
20. Orlanda JFF (2011) Estudo da composição química e atividade biológica do óleo essencial de *Ruta graveolens* Linneau (Rutaceae). Dissertation, Universidade Federal da Paraíba
21. Mejri J, Abderrabba M, Mejri M (2010) Chemical composition of the essential oil of *Ruta chalepensis* L: influence of drying, hydro-distillation duration and plant parts. Ind Crop Prod 32:671–673
22. Kannan R, Babu UV (2012) Identity and pharmacognosy of *Ruta graveolens* Linn. Anc Sci Life 32:16–19
23. Pollio A, de Natale A, Appetitil E, Aliotta G, Touwaide A (2008) Continuity and change in the Mediterranean medical tradition: *Ruta* sp. (Rutaceae) in Hippocratic medicine and present practices. J Ethnopharmacol 116:469–482
24. Prabhu S, Vijayakumar S, Morvin Yabesh JE, Ravichandran K, Sakthivel B (2014) Documentation and quantitative analysis of the local knowledge on medicinal plants in Kalrayan hills of Villupuram district, Tamil Nadu, India. J Ethnopharmacol 157:7–20
25. Jaric S, Mitrovic M, Djurdjevic L, Kostic O, Gajic G, Pavlovic D, Pavlovic P (2011) Phytotherapy in medieval Serbian medicine according to the pharmacological manuscripts of the Chilandar Medical Codex (15–16th centuries). J Ethnopharmacol 137:601–619
26. Heinrich M, Rimpler H, Barrera A (1992) Indigenous phytotherapy of gastrointestinal disorders in a lowland Mixe community (Oaxaca, Mexico). J Ethnopharmacol 36:63–80
27. WHO (World Health Organization) (2002) Traditional medicine: growing needs and potential. WHO Policy Perspect Med 2:1–6
28. Pandikumar P, Chellappandian M, Mutheeswaran S, Ignacimuthu S (2011) Consensus of local knowledge on medicinal plants among traditional healers in Mayiladumparai block of Theni District, Tamil Nadu, India. J Ethnopharmacol 134:354–362
29. Tribess AB, Pintarelli GM, Bini LA, Camargo A, Funez LA, de Gasper AD, Zeni ALB (2015) Ethnobotanical study of plants used for therapeutic purposes in the Atlantic Forest region, Southern Brazil. J Ethnopharmacol 164:136–146
30. Cavender AP, Albán M (2009) Review open access the use of magical plants by curanderos in the Ecuador highlands. J Ethnobiol Ethnomed 5:1–20
31. Quave CL, Pieroni A (2005) Ritual Healing in Arbëreshë Albanian and Italian Communities of Lucania, Southern, Italy. J Folk Res 42:57–95
32. Coelho-Ferreira M (2009) Medicinal knowledge and plant utilization in an Amazonian coastal community of Marudá, Pará State (Brazil). J Ethnopharmacol 103:261–275
33. Hale AL, Meepagala KM, Oliva A, Aliotta G, Duke SO (2004) Phytotoxins from the leaves of *Ruta graveolens*. J Agric Food Chem 52:3345–3349

34. Thring TSA, Weitz FM (2006) Medicinal plant use in the Bredasdorp/Elim region of the Southern Overberg in the Western Cape Province of South Africa. J Ethnopharmacol 103:261–275
35. Lovell CR (1995) Beware of the plant: cutaneous reactions to plants in children. Curr Padiactr 5:262–266
36. Petit-Paly G, Rideau M, Chenieux JC (1982) Etude de quelques Rutacees a alcaloides II. *Ruta graveolens*: revue botanique, chimique et pharmacologique (Etude particuliere des alcaloides quaternaires quinoleiques). Plantes med phytother 16:55–72
37. Kostova I, Ivanova A, Mikhova B, Klaiber I (1999) Alkaloids and coumarins from *Ruta graveolens*. Monatsh Chem 130:703–707
38. Hammiche V, Merad R, Azzouz M (2013) In Plantes toxiques à usage médicinal du pourtour méditerranéen. Springer, Paris, pp 197–222
39. Stashenko EE, Acosta R, Martınez JR (2000) High-resolution gas-chromatographic analysis of the secondary metabolites obtained by subcritical-fluid extraction from Colombian rue (*Ruta graveolens* L.). J Biochem Biophys Methods 43:379–390
40. Hussain SA, Numan IT, Hamad MN (2012) Comparative study of the analgesic activity of two Iraqi medicinal plants, *Ruta graveolens* and *Matricaria chamomilla* extracts. J Intercult Ethnopharmacol 1:79–83. doi:10.5455/jice.20120529080606
41. Reisch J, Szendrei K, Minker E, Novak I (1968) Chalepensin, gravelliferon-ethylether und 3-(1′, 1′-dimethylallyl)-herniarin aus den wurzeln von *Ruta graveolens*. Tetrahedron Lett 41:4395–4396
42. Reisch J, Szendrei K, Minker E, Novak I (1970) Neue C3-substituierte cumarin-derivate aus *Ruta graveolens* L.: 3-(l′,1′ -dimethylallyl)-daphnetin-dimethylether,8-methoxy-gravelliferon,3-(1′.l′-dimethylallyl)-8-(3″,3″-dimethylallyl)-xanthyletin. Tetrahedron Lett 49:4305–4308
43. Reisch J, Szendrei K, Minker E, Novak J (1968) 3-(1′,1′-Dimethylallyl)-6-(3′,3′-dimethylallyl)-umbelliferon (gravelliferon) aus den Wurzeln von *Ruta graveolens*. Experientia 24:992–994
44. Milesi S, Massot B, Gontier E, Bourgaud F, Guckert A (2001) *Ruta graveolens* L.: a promising species for the production of furanocoumarins. Plant Sci 61:189–199
45. Zobel AM, Brown SA (1989) Histological localization of furanocoumarins in *Ruta graveolens*. Can J Bot 67:915–921
46. Zobel AM, Brown SA (1990) Dermatitis-inducing furanocoumarins on the leaf surfaces of eight species of rutaceous and umbelliferous plants. J Chem Ecol 16:693–700
47. Zobel AM, Alicja M, Brown S (1991) Psoralens in senescing in leaves of *Ruta graveolens*. J Chem Ecol 17:1801–1810
48. Gray AI, Waterman PG (1978) Coumarins in the Rutaceae. Phytochemisry 17:845–864
49. Sampaio OM, das Silva MFGF (2012) Avaliação de furanocumarinas como inibidores da fotossíntese através de ensaios de fluorescência da clorofila. Quim Nova 35:2115–2118
50. Mancuso G, Borgonovo G, Scaglioni L, Bassoli A (2015) Phytochemicals from *Ruta graveolens* activate TAS2R bitter taste receptors and TRP channels involved in gustation and nociception. Molecules 20:18907–18922
51. Abyshev AZ, Gindin VA, Kerimov Yu B, Ismailov ES, Agaev EM, Isaev NY (1992) Furocoumarins of *Ruta graveolens*. Khim Prir Soedin 34:438–439
52. Debasis S, Raghav SK, Gautam H, Das HR (2015) A novel coumarin derivative, 8-methoxy chromen-2-one alleviates collagen induced arthritis by down regulating nitric oxide, NFkB and proinflammatory cytokines. Int Immunopharmacol 29:891–900
53. Kuzovkina I, Al'termana I, Schneiderb B (2004) Specific accumulation and revised structures of acridone alkaloid glucosides in the tips of transformed roots of *Ruta graveolens*. Phytochemistry 65:1095–1100
54. Eun-Tae O, Ji-Ho L, Chang-Soo K, Young-Soo K (2014) Alkylquinolone alkaloid profiles in *Ruta graveolens*. Biochem Syst Ecol 57:384–387

55. Mester I (1983) Structure diversity and distribution of alkaloids in the Rutales. In: Waterman PG, Grundon MF (eds) Chemistry and chemical taxonomy of the Rutales. Academic, London, pp 47–51
56. Grundon MF, Okely HM (1979) A new quinoline alkaloid from *Ruta graveolens*. Phytochemistry 18:1768–1769
57. Reisch J, Rozsa Z, Szendrei K, Novak I, Minker E (1976) Acridon-alkaloid-glukoside aus *Ruta graveolens*. Phytochemistry 15:240–241
58. Rozsa Z, Kusovkina IN, Reisch J, Novak I, Szendrei K, Minker E (1976) Gravacridondiol and its glucoside from the tissue culture of *Ruta graveolens* L. Fitoterapia 47:147–149
59. Nahrstedt A, Eilert U, Wolters B (1981) Rutacridone-epoxide, a new acridone alkaloid from *Ruta graveolens*. Z Naturforsch 36C:200–203
60. Rozsa Z, Reisch J, Szendrei K, Minker E (1981) Rutacridone-epoxide and gravacridonol from the roots of *Ruta graveolens*. Fitoterapia 42:93–95
61. Meepagala KM, Schrader KK, Wedge DE, Duke SO (2005) Algicidal and antifungal compounds from the roots of *Ruta graveolens* and synthesis of their analogs. Phytochemistry 66:2689–2695
62. Paulini H, Waibel R, Kiefer J, Schimmer O (1991) Gravacridondiol acetate, a new dihydrofuroacridone alkaloid from *Ruta graveolens*. Planta Med 57:82–83
63. Eilert U, Wolters B, Constabel F (1986) Ultrastructure of acridone alkaloid idioblasts in roots and cell cultures of *Ruta graveolens*. Can J Bot 64:1089–1096
64. Réthy B, Zupkó I, Minorics R, Hohmann J, Ocsovszki I, Falkay G (2007) Investigations of cytotoxic activity on human cancer cell lines of arborinine and furanoacridones isolated from *Ruta graveolens*. Planta Med 73:41–48
65. Salib JY, El-Toumy SA, Hassan EM, Shafik NH, Abdel-Latif SM, Brouard I (2014) New quinoline alkaloid from *Ruta graveolens* aerial parts and evaluation of the antifertility activity. Nat Prod Lett: Formerly Nat Prod Lett 28:1335–1345
66. Oliva A, Lahoz E, Contillo R, Aliotta G (1999) Fungistatic activity of *Ruta graveolens* extract and its allelochemicals. J Chem Ecol 25:519–526
67. Ghosh S, Bishayee K, Khuda-Bukhsh AR (2014) Graveoline isolated from ethanolic extract of *Ruta graveolens* triggers apoptosis and autophagy in skin melanoma cells: a novel apoptosis-independent autophagic signaling pathway. Phytother Res 28:1153–1162. doi:10.1002/ptr.5107
68. Reisch J, Novak I, Szendrei K, Minker E (1967) Chemistry of natural substances. XV. 2-[4-(3,4-methylenedioxyphenyl)butyl]-4-quinolone, an alkaloid from *Ruta graveolens*. Naturwissenschaften 54:517
69. Oliva A, Meepagala KM, Wedge DE, Harries D, Hale AL, Aliotta G, Duke SO (2003) Natural fungicides from *Ruta graveolens* L. leaves, including a new quinolone alkaloid. J Agric Food Chem 51:890–896
70. Yaacob-Karim B, Mazenah AC, Joulain D (1989) Essential oil of *Ruta graveolens* L. Kebangsaan Malaysia, Selangor, Malay. J Essent Oil Res 1:203–207
71. De Feo V, De Simone F, Senatore F (2002) Potential allelochemicals from the essential oil of *Ruta graveolens*. Phytochemistry 61:573–578
72. Gina M, Rojas LB, Usubillaga A (2008) Estudio del aceite esencial de *Ruta graveolens* L.que crece en el Estado Mérida, Venezuela. Rev Fac Farm 50:7–9
73. Rojas J, Mender T, Rojas Gullien E, Buitrago A, Lucena M, Cardenas N (2011) Estudio comparativo de la composición química y actividad antibacteriana del aceite esencial de *Ruta graveolens* L. recolectada en los estados Mérida y Miranda, Venezuela. Av Quim 6:89–93
74. El-Sherbeny SE, Hussein MS, Khalil MY (2007) Improving the production of *Ruta graveolens* L. plants cultivated under different compost levels and various sowing distance. Am-Eurasian J Agric Environ Sci 2:271–281
75. Soleimani M, Aberoomand-Azar P, Saber-Tehrani M, Rustaiyan A (2009) Volatile composition of *Ruta graveolens* L. of North of Iran. World Appl Sci J 7:124–126

76. Haddouchi F, Chaouche TM, Zaouali Y, Ksouri R, Attou A, Benmansour A (2013) Chemical composition and antimicrobial activity of the essential oils from four *Ruta* species growing in Algeria. Food Chem 141:253–258
77. Melnyk MV, Grytsyk AR, Stalyus LV (2015) Research of component composition of *Ruta graveolens*L. herb essential oil. Pharma Innov J 4:41–43
78. Orlanda JFF, Nascimento AR (2015) Chemical composition and antibacterial activity of *Ruta graveolens* L. (Rutaceae) volatile oils from São Luís, Maranhão, Brazil. S Afr J Bot 99:103–106
79. Kubeczka KH (1974) Pregeijeren, hauptk omponente des therischen wurzelijls von *Ruta graveolens* L. Phytochemistry 13:2017–2018. doi:10.1016/0031-9422(74)85155-1
80. Tattje DH, Bos R (1972) Volatile oil from *Ruta graveolens* L. II Geijerene, principle component of the root oil. Pharm Week 1107:261–265
81. Ferhat M, Kabouche A, Kabouche Z (2014) Comparative compositions of essential oils of three *Ruta* species growing in different soils. J Mater Environ Sci 5:735–738
82. Kong YC, Lau CP, Wat KH, Ng KH, But PPH, Cheng KF, Waterman PG (1989) Antifertility principle of *Ruta graveolens*. Planta Med 55:176–178
83. Saieed P, Reza RM, Abbas D, Seyyedvali R, Aliasghar H (2006) Inhibitory effects of *Ruta graveolens* L. Extract on guinea pig liver aldehyde oxidase. Chem Pharm Bull 54:9–13
84. Harborne JB, Boardley M (1983) Identification of two gossypetin monomethyl ethers as yellow flower pigments in the Rutaceae. Z Naturforsch C 38:148–150
85. Chien-Chih C, Yu-Lin H, Fei-In H, Chun-Wen W, Jun-Chih O (2001) Water-soluble glycosides from *Ruta graveolens*. J Nat Prod 64:990–992
86. Proestos C, Boziaris IS, Nychas GJE, Komaitis M (2006) Analysis of flavonoids and phenolic acids in Greek aromatic plants: investigation of their antioxidant capacity and antimicrobial activity. Food Chem 95:664–671. doi:10.1016/j.foodchem.2005.01.049
87. Rosi Cruvinel HM, Chaves AH, Goloubeff B (2006) Effects of homeophatic medication *Arnica montana* and *Ruta graveolens* on weight gain, food and water consumption and food conversion in broilers. Int J High Dilution Res. http://www.feg.unesp.br/~ojs/index.php/ijhdr/article/view/218/239. Accessed 18 Dec 2015
88. Paris A, Gonnet N, Chaussard C, Belon P, Rocourt F, Saragaglia D, Cracowski JL (2008) Effect of homeopathy on analgesic intake following knee ligament reconstruction: a phase III monocentre randomized placebo controlled study. Br J Clin Pharmacol 65:180–187
89. Khan T (2008) The evidenced-based approach to treating podiatric conditions with Marigold therapy and homeopathy. Eur J Integr Med 1:21–22
90. Arora S, Aggarwal A, Singla P, Jyoti S, Tandon S (2013) Anti-proliferative effects of homeopathic medicines on human kidney, colon and breast cancer cells. Homeopathy 102:274–282
91. Arora S, Tandon S (2015) DNA fragmentation and cell cycle arrest: a hallmark of apoptosis induced by *Ruta graveolens* in human colon cancer cells. Homeopathy 104:36–47
92. Freyer G, You B, Villet S, Tartas S, Fournel-Federico C, Trillet-Lenoir V, Hamizi S, Colomban O, Chavernoz N, Falandry C (2014) Open-label uncontrolled pilot study to evaluate complementary therapy with *Ruta graveolens* 9C in patients with advanced cancer. Homeopathy 103:232–238
93. Arora S, Tandon C, Tandon S (2014) Evaluation of the cytotoxic effects of CAM therapies: an *in vitro* study in normal kidney cell lines. Sci World J. doi:10.1155/2014/452892. Accessed 18 Dec 2015
94. Remya V, Kuttan G (2015) Homeopathic remedies with antineoplastic properties have immunomodulatory effects in experimental animals. Homeopathy 104:211–219
95. Guerra MO, Andrade ATL (1978) Contraceptive effects of native plants in rats. Contraception 18:191–199
96. Gandhi M, Lal R, Sankaranarayanan A, Lal Sharma P (1991) Post-coital antifertility activity of *Ruta graveolens* in female rats and hamsters. J Ethnopharmacol 34:49–59

97. De Freitas TG, Augusto PM, Montanari T (2005) Effect of *Ruta graveolens* L. on pregnant mice. Contraception 71:74–77
98. Sailani M, Moeini H (2007) Effect of *Ruta graveolens* and *Cannabis sativa* alcoholic extract on spermatogenesis in the adult wistar male rats. Indian J Urol 23:257–260
99. Harat ZN, Sadeghi MR, Sadeghipour HR, Kamalinejad M, Eshraghian MR (2008) Immobilization effect of *Ruta graveolens* L. on human sperm: a new hope for male contraception. J Ethnopharmacol 115:36–41
100. Halvaei I, Roodsari HRS, Harat ZN (2012) Acute Effects of *Ruta graveolens* L. on sperm parameters and DNA integrity in rats. J Reprod Infertil 13:33–38
101. Bazrafkan M, Panahi M, Saki G, Ahangarpour A, Zaeimzadeh N (2010) Effect of aqueous extract of *Ruta graveolens* on spermatogenesis of adult rats. Int J Pharm 6:926–929
102. Nasirinezhad F, Khoshnevis FM, Parivar K, Amin G (2009) Antifertility effect of aqueous extract of airal part of *Ruta graveolens* on immature female Balb/C mice. Physiol Pharmacol 13:279–287
103. Raghav SK, Gupta B, Agrawal C, Goswami K, Das HR (2006) Anti-inflammatory effect of *Ruta graveolens* L. in murine macrophage cells. J Ethnopharmacol 104:234–236
104. Li P, Matsunaga K, Ohizumi Y (1999) Enhancement of the nerve growth factor-mediated neurite outgrowth from PC12D cells by Chinese and Paraguayan medicinal plants. Biol Pharm Bull 22:752–755
105. Raghav SK, Gupta B, Shrivastava A, Das H (2007) Inhibition of lipopolysaccharide-inducible nitric oxide synthase and IL-1β through suppression of NF-kB activation by 3-(1′-1′-dimethyl-allyl)-6-hydroxy-7-methoxy-coumarin isolated from Ruta graveolens L. Eur J Pharmacol 560:69–80
106. Ratheesh M, Shyni G, Helen A (2009) Methanolic extract of *Ruta graveolens* L. inhibits inflammation and oxidative stress in adjuvant induced model of arthritis in rats. Inflammopharmacology 17:100–105
107. Ratheesh M, Shyni G, Sindhu G, Helen A (2010) Protective effects of isolated polyphenolic and alkaloid fractions of *Ruta graveolens* L. on acute and chronic models of inflammation. Inflammation 33:18–24
108. Ratheesh M, Shyni GL, Sindhu G, Helen A (2011) Inhibitory effect of *Ruta graveolens* L. on oxidative damage, inflammation and aortic pathology in hypercholesteromic rats. Exp Toxicol Pathol 63:285–290
109. Ratheesh M, Sindhu G, Helen A (2013) Anti-inflammatory effect of quinoline alkaloid skimmianine isolated from *Ruta graveolens* L. Inflamm Res 62:367–376
110. Loonat F, Amabeoku GJ (2014) Antinociceptive, anti-inflammatory and antipyretic activities of the leaf methanol extract of *Ruta graveolens* L. (Rutaceae) in mice and rats. Afr J Tradit Complement Altern Med 11:173–181
111. Kataki MS, Kakoti BB, Bhuyan B, Rajkumari A, Rajak P (2014) Garden rue inhibits the arachidonic acid pathway, scavenges free radicals, and elevates FRAP: role in inflammation. Chin J Nat Med 12:172–179
112. Mohammadi-Motamed S, Shahidi-Motlagh S, Bagherzadeh H, Forouz SA, Tafazoli H (2014) Evaluation of antioxidant activity of *Ruta graveolens* L. extract on inhibition of lipid peroxidation and DPPH radicals and the effects of some external factors on plant extract's potency. Res J Pharmacogn 1:45–50
113. Alzoreky NS, Nakahara K (2003) Antibacterial activity of extracts from some edible plants commonly consumed in Asia. Int J Food Microbiol 80:223–230
114. Ivanova A, Mikhova B, Najdenski H, Tsvetkova I, Kostova I (2005) Antimicrobial and cytotoxic activity of *Ruta graveolens*. Fitoterapia 76:344–347
115. Nogueira JCR, Diniz MFMD, Lima EO (2008) Atividade antimicrobiana *in vitro* de produtos vegetais em otite externa aguda *in vitro* antimicrobial activity of plants in acute otitis externa. Braz J Otorhinolaryngol 74:18–124
116. Rojas J, Mender T, Rojas L, Gullien E, Buitrago A, Lucena M, Cardenas N (2011) Estudio comparativo de la composición química y actividad antibacteriana del aceite esencial de

Ruta graveolens L. recolectada en los estados Mérida y Miranda, Venezuela. Av Quím 6:89–93
117. Frias DF, Kozusny–Andreani DI (2009) Verificacion de la actividad antifungica de extractos de plantas y aceite de eucalipto para el control de *Aspergillus niger*. Ces Med Vet Zootec 12:1–8
118. Owlia P, Saderi H, Rasooli I, Sefidkon F (2009) Antimicrobial characteristics of some herbal oils on *Pseudomonas aeruginosa* with special reference to their chemical compositions. Iran J Pharm Res 8:107–114
119. Honarmand H, Saeidinia A, Delavar SF (2012) Evaluation of antimicrobial effect of hydro-alcoholic extract of *Ruta graveolens* on *Enterococcus fecalis*. Planta Med 78:11
120. Fallah Delavar S, Saeidinia A, Honarmand D (2013) Effect of methanolic extract of *Ruta graveolens* on *Enterococcus fecalis* growth. Planta Med 79:13
121. Zaidi SF, Muhammad JS, Shahryar S, Usmanghani K, Gilani A, Jafri W, Sugiyama T (2012) Anti-inflammatory and cytoprotective effects of selected Pakistani medicinal plants in *Helicobacter pylori*-infected gastric epithelial cells. J Ethnopharmacol 141:403–410
122. Haddouchi F, Chaouche TM, Zaouali Y, Ksouri R, Attou A, Benmansour A (2013) Chemical composition and antimicrobial activity of the essential oils from four *Ruta* species growing in Algeria. Food Chem 14:253–258
123. Chaftar N, Girardot M, Quellard N, Labanowski J, Ghrairi T, Hani K, Frère J, Imbert C (2015) Activity of six essential oils extracted from Tunisian plants against *Legionella pneumophila*. Chem Biodivers 12:1565–1574
124. Figueroa-Valverde L, Díaz-Cedillo F, Camacho-Luis A, Ramos ML (2009) Induced effects by *Ruta graveolens* L., Rutaceae, *Cnidoscolus chayamansa* McVaugh, Euphorbiaceae, and *Citrus aurantium* L., Rutaceae, on glucose, cholesterol and triacylglycerides levels in a diabetic rat model. Rev Bras Farmacogn 19:898–907
125. Meléndez M, Rico A (2011) Antihyperglycaemic effect of the Rue (*Ruta graveolens*) methanol extract in an experimental model of hyperglycemic rats. Rev Fac Cienc Vet 52:119–126
126. Van Huyssteen M, Milne PJ, Campbell EE, Van De Venter M (2011) Antidiabetic and cytotoxicity screening of five medicinal plants used by traditional African health practitioners in the Nelson Mandela Metropole, South Africa. Afr J Tradit Complement Altern Med 8:150–158
127. Toserkani A, Razi Jalali M, Najafzaheh H (2012) Changes of lipid profiles, glucose, and hemogram after administration of *Ruta graveolens* extract in diabetic rats. Comp Clin Pathol 21:1587–1592
128. Atta AH, Alkofahi A (1998) Anti-nociceptive and anti-inflammatory effects of some Jordanian medicinal plant extracts. J Ethnopharmacol 60:117–124
129. Park SH, Sim YB, Kim SM, Lee JK, Lim SS, Kim JK, Suh HW (2010) Antinociception effect and mechanism of *Ruta graveolens* L. in mice. J Korean Soc Appl Biol Chem 53:593–597
130. Hussain SA, Numan IT, Hamad MN (2012) Comparative study of the analgesic activity of two Iraqi medicinal plants, *Ruta graveolens* and *Matricaria chamomilla* extracts. J Intercult Ethnopharmacol 1:79–83
131. Gilbert B, Mors WB, Baker PM, Tomassini TCB, Coulart EG, De Holanda JC, Ribeiro Da Costa J, Lopes JNG, Dos Santos FD (1972) Antihelmintic activity of essential oils and their chemical components. An Acad Bras Cienc 44:423–428
132. Mendes NM, Pereira JP, Souza C, Oliveira MLL (1984) Preliminary laboratory tests of the molluscicide activity of some species of Brazilian flora. Rev Saude Publica 18:348–354
133. Teixeira TL, Teixeira SC, Silva CV, Souza MA (2014) Potential therapeutic use of herbal extracts in trypanosomiasis. Pathog Glob Health 108:30–36
134. Queiroz AC, Freire Dias TLM, Matta CBB, Silva LHAC, Araujo-Junior JX, Araujo GB, Moura FBP, Alexandre-Moreira MS (2014) Antileishmanial activity of medicinal plants used in endemic areas in northeastern Brazil. J Evid Based Complement Altern Med 2014:1–9
135. Stafford GI, Pedersen PD, Jäger AK, Van Staden J (2007) Monoamine oxidase inhibition by southern African traditional medicinal plants. S Afr J Bot 73:384–390

136. Wszelaki N, Kuciun A, Kiss AK (2010) Screening of traditional European herbal medicines for acetylcholinesterase and butyrylcholinesterase inhibitory activity. Acta Pharm 60:119–128
137. Russo P, Frustaci A, Del Bufalo A, Fini M, Cesario A (2013) From traditional European medicine to discovery of new drug candidates for the treatment of dementia and Alzheimer's disease: acetylcholinesterase inhibitors. Curr Med Chem 20:976–983
138. Keihanian F, Vajari MR, Saeidynia A, Elmieh A (2012) Effect of *Ruta graveolens* hydro-alcoholic extract on pentylenetetrazole-induced seizure in male mice. Planta Med 78:1–11
139. Amabeoku GJ, Ahmad AH (2013) Involvement of gamma aminobutyric acid in the anticonvulsant effect of the leaf methanol extract of *Ruta graveolens* L. (Rutaceae) in mice. Int J Pharm 9:134–142
140. Bohuslavizki KH, Koppenhöfer E, Hänsel W, Möller WD (1988) A new approach for the treatment of demyelinating diseases? J Neuroimmunol 20:251–252
141. Jäger AK, Gauguin B, Adsersen A, Gudiksen L (2006) Screening of plants used in Danish folk medicine to treat epilepsy and convulsions. J Ethnopharmacol 105:294–300
142. Adsersen A, Gauguin B, Gudiksen L, Jäger AK (2006) Screening of plants used in Danish folk medicine to treat memory dysfunction for acetylcholinesterase inhibitory activity. J Ethnopharmacol 104:418–422
143. Fadlalla K, Watson A, Yehualaeshet T, Turner T, Samuel T (2011) *Ruta graveolens* extract induces DNA damage pathways and blocks Akt activation to inhibit cancer cell proliferation and survival. Anticancer Res 31:233–241
144. Ghosh S, Bishayee K, Khuda-Bukhsh AR (2014) Graveoline isolated from ethanolic extract of *Ruta graveolens* triggers apoptosis and autophagy in skin melanoma cells: a novel apoptosis-independent autophagic signaling pathway. Phytother Res 28:1153–1162
145. Ghosh S, Sikdar S, Mukherjee A, Khuda-Bukhsh AR (2015) Evaluation of chemopreventive potentials of ethanolic extract of *Ruta graveolens* against A375 skin melanoma cells in vitro and induced skin cancer in mice *in vivo*. J Integr Med 13:34–44
146. Gentile MT, Ciniglia C, Reccia MG, Volpicelli F, Gatti M, Thellung S, Florio T, Melone MAB, Colucci-D'amato L, Condorelli G (2015) *Ruta graveolens* L. induces death of glioblastoma cells and neural progenitors, but not of neurons, via ERK 1/2 and AKT activation. PLoS One. http://www.ncbi.nlm.nih.gov/pmc/articles/PMC4364962/pdf/pone.0118864.pdf. Accessed 15 Dec 2015
147. Al-Nimer M, Ali E (2010) Aqueous extract of *Salvia officinalis* and *Ruta graveolens*: potential source of reactive nitrogen species. Int J Green Pharm 4:189–192
148. Tabanca N, Demirci B, Kiyan H, Ali A, Bernier U, Wedge DE, Khan I, Başer K (2012) Repellent and larvicidal activity of *Ruta graveolens* essential oil and its major individual constituents against *Aedes aegypti*. Planta Med 78:90
149. Pirouzpanah S, Saieed P, Rashidi MR, Reza RM, Delazar A, Abbas D, Razavieh SV, Seyyedvali R, Hamidi A, Aliasghar H (2006) Inhibitory effects of *Ruta graveolens* L. extract on guinea pig liver aldehyde oxidase. Chem Pharm Bull 54:9–13
150. Muñoz DK, Londoño LJA, Arango AGJ, Sierra RJA, Bravo MKE (2007) Effect of extraction technique of *Ruta graveolens* on antityrosinase activity and correlation amongst inhibitory activity, phenolic compounds content and cytotoxicity. Vitae 14:71–77
151. Chiu KW, Fung AYL (1997) The cardiovascular effects of green beans (*Phaseolus aureus*), common rue (*Ruta graveolens*), and kelp (*Laminaria japonica*) in rats. Gen Pharmacol 29:859–862
152. Khori V, Nayebpour M, Semnani S, Golalipour MJ, Marjani A (2008) Prolongation of AV nodal refractoriness by *Ruta graveolens* in isolated rat hearts. Potential role as an anti-arrhythmic agent. Saudi Med J 29:357–363
153. Ueng YF, Chen CC, Huang YL, Lee IJ, Yun CH, Chen YH, Huang CC (2015) Effects of aqueous extract of *Ruta graveolens* and its ingredients on cytochrome P450, uridine diphosphate (UDP)-glucuronosyltransferase, and reduced nicotinamide adenine dinucleotide (phosphate) (NAD(P)H)-quinone oxidoreductase in mice. J Food Drug Anal 23:516–528

154. Águila L, Ruedlinger J, Mansilla K, Ordenes J, Salvatici R, De Campos RR, Romero F (2015) Relaxant effects of a hydroalcoholic extract of *Ruta graveolens* on isolated rat tracheal rings. Biol Res 48:28
155. Hammiche V, Azzouz M (2013) Les rues: ethnobotanique, phytopharmacologie et toxicité. Phytothérapie 11:22–30
156. Schempp CM, Schöpf E, Simon JC (1999) Dermatitis bullosa striata pratensis durch *Ruta graveolens* L. (Gartenraute). Bullous phototoxic contact dermatitis caused by *Ruta graveolens* L. (garden rue), Rutaceae. Case report and review of literature. Hautarzt 50:432–434
157. Sharma AN, Nelson LS, Hoffman RS (2001) *Ruta graveolens* as an ethnic abortifacient. J Toxicol Clin Toxicol 39:312
158. Seak CJ, Lin CC (2007) *Ruta graveolens* intoxication. Clin Toxicol (Phila) 45:173–175
159. Furniss D, Adams T (2007) Herb of grace: an unusual cause of phytophotodermatitis mimicking burn injury. J Burn Care Res 28:767–769
160. Unzueta Roch JLD, Pizarro Pizarro I, Astiz Blanco MI, Hernández Delgado MJ (2014) Fitofotodermatitis por *Ruta graveolens*. Rev Pediatr Aten Primaria 16:327–329
161. El Agraa SEI, El Badwi SMA, Adam SE (2002) Preliminary observations on experimental *Ruta graveolens* toxicosis in nubian goats. Trop Anim Health Prod 34:271–281
162. Freire R, Borba H, Coelho C (2010) *Ruta graveolens* L. toxicity in *Vampirolepis nana* infected mice. Indian J Pharmacol 42:345–350
163. Serrano-Gallardo LB, Soto-Domínguez A, Ruiz-Flores P, Nava-Hernández MP, Morán-Martínez J, García-Garza R, Martínez-Pérez EF (2013) Effect of aqueous extract of *Ruta graveolens* from the north of Mexico on the liver of wistar rat. Int J Morphol 31:1041–1048
164. Benavides V, Trujillo G, D'Arrigo G, Paredes U, Pino J (2014) Preliminar toxicological assesement of *Ruta graveolens*, *Origanum vulgare* and *Persea americana* on the preimplantational mouse embryos. Rev Peru Biol 7:87–89
165. Shama IYA, Ahmed NA, Eltayeb AM, Saad H, Taha KA (2014) Toxicity of *Ruta graveolens* seeds' extracts on male wistar. Int J Anim Vet Adv 6:92–96
166. Marinoff MA, Martínez JL, Urbina MA (2009) Precauciones en el empleo de plantas medicinales. B Latinoam Caribe Pl 8:184–187
167. Malik S, Kumar R, Vats SK, Bhushan S, Sharma M, Ahuja PS (2009) Regeneration in *Rheum emodi* Wall: a step towards conservation of an endangered medicinal plant species. Eng Life Sci 2:130–134
168. Malik S, Mirjalili MH, Fett-Neto AG, Mazzafera P, Bonfill M (2013) Living between two worlds: two-phase culture systems for producing plant secondary metabolites. Crit Rev Biotechnol 33(1):1–22
169. Faizal M, Ahmad N, Anis M (2005) *In Vitro* regeneration and mass propagation of *Ruta graveolens*: a multipurpose shrub. Hort Sci 40:1478–1480
170. Baumert A, Maier W, Schumann B, Groger D (1991) Increased accumulation of acridone alkaloids by cell suspension cultures of *Ruta graveolens* in response to elictors. J Plant Physiol 139:224–228
171. Eilert U, Wolters B (1989) Elicitor induction of S-adenosyl-Lmethionine: anthranilic acid *N*-methyltransferase activity in cell suspension and organ cultures of *Ruta graveolens* L. Plant Cell Tiss Organ Cult 18:1–18
172. Massot B, Milesi S, Gontier E, Bourgaud F, Guckert A (2000) Optimized culture conditions for the production of furanocoumarins by micropropagated shoots of *Ruta graveolens*. Plant Cell Tiss Org Cult 62:11–19
173. Orlita A, Sidwa-Gorycka M, Kumirska J, Malinski E, Siedlecka EM, Gajdus J, Lojkowska E, Stepnowski P (2008) Identification of *Ruta graveolens* L. metabolites accumulated in the presence of abiotic elicitors. Biotechnol Prog 24:128–133
174. Orlita A, Sidwa-Gorycka M, Malinski E, Czerwicka M, Kumirska J, Golebiowski M, Lojkowska E, Stepnowski P (2008) Effective biotic elicitation of *Ruta graveolens* L. shoot cultures by lysates from *Pectobacterium atrosepticum* and *Bacillus* sp. Biotechnol Lett 30:541–545

175. Diwan R, Malpathak N (2008) Novel technique for scaling up of micropropagated *Ruta graveolens* shoots using liquid culture systems: a step towards commercialization. New Biotechnol 25(1):85–91
176. Kengar AA, Paratkar GT (2015) Large scale production of *Ruta graveolens* L. shoots using aerated bioreactor. Eur J Biotechnol Biosci 3(7):35–38
177. Gontier E et al (2005) Development and validation of an efficient low cost bioreactor for furanocoumarin production with *Ruta graveolens* shoot cultures. In: Anne Kathrine HE, Walter P (eds) Liquid culture systems for *in vitro* plant propagation, vol. 38. Springer, AA Dordrecht, The Netherlands. pp 509–524
178. Bohlmann J, Eilert U (1994) Elicitor induced secondary metabofism in *Ruta graveolens L.* Role of chorismate utilizing enzymes. Plant Cell. Tiss Org Cult 38:189–198
179. Sidwa-Gorycka M, Krolicka A, Orlita A, Malinski E, Golebiowski M, Kumirska J, Chromik A, Biskup E, Stepnowski P, Lojkowska E (2009) Genetic transformation of *Ruta graveolens* L. by *Agrobacterium rhizogenes*: hairy root cultures a promising approach for production of coumarins and furanocoumarins. Plant Cell Tiss Org Cult 97:59–69
180. Kuzovkina IN (1992) Cultivation of genetically transformed plant roots: possibilities and prospects of use in plant physiology. Sov Plant Physiol 39:1208–1214
181. Tepfer D (1984) Transformation of several species of higher plants by *Agrobacterium rhizogenes*: sexual transmission of the transformed genotype and phenotype. Cell 37:959–967
182. Lievre K, Hehn A, Tran TLM, Gravot A, Thomasset B, Bourgaud F, Gontier E (2005) Genetic transformation of the medicinal plant *Ruta graveolens* L. by an *Agrobacterium tumefaciens*-mediated method. Plant Sci 168:883–888

Secondary Metabolite Profile of Transgenic Centaury (*Centaurium erythraea* Rafn.) Plants, Potential Producers of Anticancer Compounds

8

Milana Trifunović-Momčilov, Dijana Krstić-Milošević, Snežana Trifunović, Ana Podolski-Renić, Milica Pešić, and Angelina Subotić

Abstract

The genus *Centaurium* includes about 50 plant species found throughout the northern hemisphere. Plant species *Centaurium erythraea* Rafn., commonly known as common centaury, has been used for centuries for medical purposes. Centaury is used to treat anemia, jaundice, and gout and to cure febrile conditions and regulate blood sugar. So far, centaury species were genetically transformed mostly using *Agrobacterium rhizogenes*. Only one report has described *A. tumefaciens*-mediated gene delivery for the production of transgenic centaury plants. Genetic transformation of centaury using *AtCKX* genes did not influence the quality but influenced the quantity of xanthones in shoots and roots. The majority of *AtCKX* transformed centaury lines grown in vitro produced increased eustomin and/or demethyleustomin content than untransformed control plants. This work clearly demonstrates, for the first time, the effect of centaury secoiridoids and xanthones on colorectal cancer cell line (DLD1) and its resistant counterpart (DLD1-TxR). The xanthone eustomin showed the most significant cell growth inhibition effects. Since xanthones are increasingly being used for their pharmacological properties, *AtCKX* transgenic centaury plants could be used as a useful source of plant material for the production of novel drugs.

M. Trifunović-Momčilov (✉) • D. Krstić-Milošević (✉) • A. Podolski-Renić (✉) • M. Pešić (✉) • A. Subotić (✉)
Institute for Biological Research "Siniša Stanković", University of Belgrade, Belgrade, Serbia
e-mail: milanag@ibiss.bg.ac.rs; dijana@ibiss.bg.ac.rs; ana.podolski@ibiss.bg.ac.rs; camala@ibiss.bg.ac.rs; heroina@ibiss.bg.ac.rs

S. Trifunović (✉)
Faculty of Chemistry, University of Belgrade, Belgrade, Serbia
e-mail: snezanat@chem.bg.ac.rs

S. Jha (ed.), *Transgenesis and Secondary Metabolism*, Reference Series in Phytochemistry, DOI 10.1007/978-3-319-28669-3_5

Keywords
Centaurium erythraea Rafn. • *AtCKX* genes • Secondary metabolites • Xanthones • Cytotoxicity

Abbreviations

CKX	Cytokinin oxidase/dehydrogenase
HPLC	High-pressure liquid chromatography
MS	Mass spectrometry
MSM	Murashige and Skoog medium
NMRS	Nuclear magnetic resonance spectroscopy
SRB	Sulforhodamine B
UVS	Ultraviolet–visible spectroscopy

Contents

1 Introduction

Plants are very important source of many chemical compounds for medicinal use and pharmaceutical industry. It is estimated that a quarter of drugs prescribed nowadays as a therapy in modern medicine are produced by numerous plant species. Due to over-exploitation of plants in nature and the destruction of their natural habitats, many medicinal plant species are endangered and disappeared. Very important biotechnological tools such as in vitro techniques represent one of the alternative ways for commercial production of numerous plants especially medicinal plant species [1]. The use of tissue culture enables regeneration and multiplication of plants on various types of explants in short period of time. Three basic methods, micropropagation, organogenesis, and somatic embryogenesis, are commonly used for plant regeneration in vitro [2]. Plant tissue culture also enabled the increased production of secondary metabolites, very important natural compounds. It was also noticed that the accumulation of secondary metabolites depends on nutrition medium composition and in vitro culture conditions [3]. Another biotechnological tool such as genetic engineering also enables production of plants with the improved attributes such as increased level of secondary metabolites [4–8]. Thus, plant tissue culture is able to provide the production of secondary metabolites and sustainable development of natural product drugs. The application and development of these techniques enable the preservation and improvement of some properties of many plant species endangered in natural habitat, including *Centaurium erythraea*.

1.1 General Characteristics of Centaury

Centaurium is the largest genus in the Gentianaceae family, which comprises about 50 species divided into several sections according to morphological, ecological, and taxonomic characteristics. All plant species of the genus *Centaurium* belong to the Gentianaceae family, order Gentianales, and class Magnoliopsida [9]. The genus name comes from the Greek word *kentayrion*, because the discovery of medicinal properties was attributed to the centaur Chiron from Greek mythology. Another explanation of the origin of the genus name comes from the literal translation of its name (*centum* = hundred and *aureum* = gold), which means a plant worth hundreds of gold coins. *Centaurium* species are growing in the northern hemisphere including Great Britain, Scandinavia, the Mediterranean, Southwest Asia, and North Africa [10]. It is considered that genus *Centaurium* originates from the Mediterranean region and Central America. Many species can also be found in North America, California, and Mexico [11]. Plant species of genus *Centaurium* usually inhabit different types of habitats and soil (limestone, alumina, sand, and marshland).

Centaurium erythraea Rafn. (syn. *C. umbellatum* Gillib and *C. minus* Moench), known as a common centaury, is an annual or biennial medicinal plant with 10–50 cm in height with a quadrangular tree branched in the upper part (Fig. 1a). The leaves on the shoot are opposite, oblong-oval to lanceolate-elliptical, and mainly arranged in rosette. The pink–red-colored flowers formed compacted inflorescence (Fig. 1b). In natural habitat, centaury blooms from July to September. *C. erythraea* is cosmopolitan plant species that inhabits dry grasslands and scrubs, saline soils, and mountain slopes [12].

Plant species of the genus *Centaurium* are widely used in the traditional medicine since ancient times. Centaury is used to cure febrile conditions and to regulate blood sugar. It is also used to treat anemia, jaundice, and gout [13]. Methanol extract of centaury has been used in the treatment of gastrointestinal tract diseases, increases appetite, and stimulates digestion [14]. In medicinal purposes, the entire herb (*Centaurii herba*) is used. Centaury plants are usually harvested in the flowering time from natural habitat. Due to uncontrolled collection from natural habitats, *C. erythraea* is nowadays rapidly disappearing and is listed as an endangered species.

Fig. 1 *Centaurium erythraea* Rafn. (**a**) Plant in natural habitat and (**b**) detail of *pink–red*-colored flowers

Generally, the common centaury is the most investigated species from *Centaurium* genus. This plant species can be relatively easily propagated in vitro and pass the whole natural life cycle including vegetative growth, flowering, fruiting, and production of viable seeds in the culture. The previous investigation showed that centaury revealed vigorous regenerative potential in vitro [15–19]. Centaury also served as a model system for studies in developmental biology [20–23]. Thus, all of these reasons make *C. erythraea* a very suitable model system for investigation of numerous physiological processes in vitro.

1.2 Genetic Transformation of Plant Species in the Gentianaceae Family

Genetic manipulation of plant genomes introducing desired transgenes has become a very suitable tool in plant biology. To date, numerous widely used medicinal and aromatic plant species were genetically modified. Plant transformation using *Agrobacterium* represents the most extensively used method. The most of Gentianaceae species, including *C. erythraea*, were genetically transformed using *A. rhizogenes* [15, 24]. Although genetic transformation using *A. tumefacien*s enables a precise and stable incorporation of transgenes, this technique of transformation is not applied to many types of Gentianaceae family. An unsuccessful genetic transformation of *Gentiana triflora* x *G. scabra* using

Table 1 Previous investigations considering genetic transformation of plant species in Gentianaceae family

Plant species	Genetic transformation technique	References
Gentiana scabra Bunge var. *buergeri* Maxim	*A. rhizogenes* (strain MAFF03-01724)	[30]
Gentiana acaulis	*A. rhizogenes* (strain A4M70GUS)	[31]
Gentiana cruciata	*A. rhizogenes* (strain A4M70GUS)	[31]
	A. rhizogenes (strains A4, 15834, 8196, R1000)	[32]
Gentiana lutea	*A. rhizogenes* (strain ATCC15834)	[31]
Gentiana purpurea		
Gentiana triflora x *G. scabra*	*A. rhizogenes* (strain ATCC43057)	[25]
	A. tumefaciens – **unsuccessful** (strains LBA4404, EHA101)	
Gentiana punctata	*A. rhizogenes* (strain A4M70GUS)	[33]
	A. tumefaciens (strain C58C1)	[27]
Gentiana triflora x *G. scabra*	**Biolistics**	[34]
	A. rhizogenes (strain ATCC43057)	[35]
Gentiana macrophylla	*A. rhizogenes* (strains A4GUS, R1000, LBA9402, ATCC11325)	[36]
	A. rhizogenes (strain R1000)	[37]
Gentiana dahurica Fish.	*A. tumefaciens* (strain GV3130)	[28]
Gentiana dinarica Beck.	*A. rhizogenes* (strains A4M70GUS, 15834/PI)	[38]
Eustoma grandiflorum Grise.	*A. rhizogenes* (strain MAFF02-10266)	[39]
	A. tumefaciens (strain A722)	[26]
	A. rhizogenes (strain MAFF0301724)	[40]
	Biolistics	[41]
	A. rhizogenes (strain NCPPB1855)	[42]
	A. tumefaciens (strain A281, EHA105)	[43]
	A. tumefaciens (soj A722)	[44]
	A. tumefaciens (strain A722)	[45]

(*continued*)

Table 1 (continued)

Plant species	Genetic transformation technique	References
	Biolistics	[46]
Swertia japonica	*A. rhizogenes* (strain ATCC15834)	[47]
Centaurium erythraea Rafn.	*A. rhizogenes* (strain A4M70GUS)	[15]
	A. rhizogenes (strain LBA9402)	[24]
	A. tumefaciens (strain GV3101)	[29]

A. tumefaciens (LBA4404 and EHA101 strains) was described about 20 years ago [25]. Genetic transformation with *A. tumefaciens* has been successfully confirmed in only three plant species belonging to the Gentianaceae, *Eustoma grandiflorum* Grise. [26], *Gentiana punctata* [27], and *G. dahurica* Fish [28]. Considering genus *Centaurium*, only one report has described *A. tumefaciens*-mediated gene delivery for the production of centaury transgenic plants [29]. All previously mentioned investigations have mostly described the efficiency of plant regeneration and selection of transformed plant tissues. A detailed review of the literature considering genetic transformation of plant species belonging to the Gentianaceae family is presented in Table 1.

1.3 Secondary Metabolite Characteristic for the *Centaurium* Species

Primary metabolites represent the compounds (carbohydrates, fats, proteins, vitamins, and mineral nutrients) resulted in various biochemical reactions in the cell. These compounds serve as the basic compounds for further biosynthesis. Primary metabolism is universal, conservative, and essential for a normal functioning of plant organism. In contrast, secondary metabolism is unique, flexible, and also unnecessary for the usual growth and development but very important for the survival of the whole organism. Plant secondary metabolites include various chemical compounds, such as antibiotics, steroids, and alkaloids, and a number of other compounds involved in the interaction with plants, animals, and microorganisms. Most of these natural products are produced in later stages of plant development. It is known that plant secondary metabolites serve as excellent protectants against various bacteria, fungi, and insects and also serve as attractants for pollinators and seed-dispersing animals [48]. Plants produce a huge number of secondary metabolites. Up to now, more than 100,000 secondary metabolites have been detected, but the discovery of new metabolites is rapidly growing [49]. Products of secondary metabolism, exclusively produced by plants, are widely used in medical, pharmaceutical, and cosmetic industry, agriculture, food industry, etc. Although most of the secondary metabolites have been synthesized during the last 60 years, the plants are still the most important source of biologically active compounds. Plant species belonging to the Gentianaceae family are characterized by the presence of secondary metabolites with medicinal properties. *Centaurium* species produced two types of bioactive compounds, iridoids and xanthones.

1.3.1 Iridoids

Iridoids are natural compounds that belong to the class of monoterpenes. Iridoids are widespread in the plant world and can be found in about 60 plant families, including the Gentianaceae family. To date, more than 1200 different iridoids and secoiridoids are identified [50]. The basic structure of iridoids makes cyclopentane-(c)-pyran ring (Fig. 2). Based on the chemical structure, iridoids belong to the monocyclic monoterpenes. Iridoids are derived from geranial and can be further metabolized to loganin [51]. The most important biosynthetic pathway, especially for plant species in the Gentianaceae, goes in the direction of the transformation of loganin to secologanin which can be further transformed to sweroside (Fig. 3a), swertiamarin (Fig. 3b), and gentiopicrin (Fig. 3c), secoiridoids commonly found among *Centaurium* species [52, 53].

Iridoid compounds belong to bitter-tasting secoiridoid glucosides [54, 55]. These bioactive compounds have been reported to possess hepatoprotective, choleretic, pancreatic, antimicrobial, antimutagenic, and antioxidative activities [56–62]. They are often used as stimulators of the secretion of gastric and intestinal juices and therefore increase appetite and improve digestion [63]. It was also shown that numerous iridoid glycosides can serve as diuretics [64].

1.3.2 Xanthones

Xanthones represent the natural compounds belonging to the class of plant polyphenols [65]. According to the chemical structure, xanthones and flavonoids are very similar. Both groups of these natural compounds share benzo-γ-pyrone structure (Fig. 4). In contrast to the widely distributed flavonoids, xanthones determined only in few families. The highest number of xanthones are identified in Gentianaceae (more than 110 xanthones) and Clusiaceae (syn. Guttiferae) families

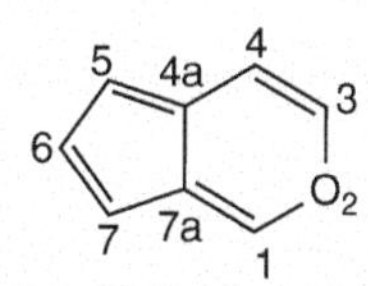

Fig. 2 The basic structure of iridoids (cyclopentane-(c)-pyran ring)

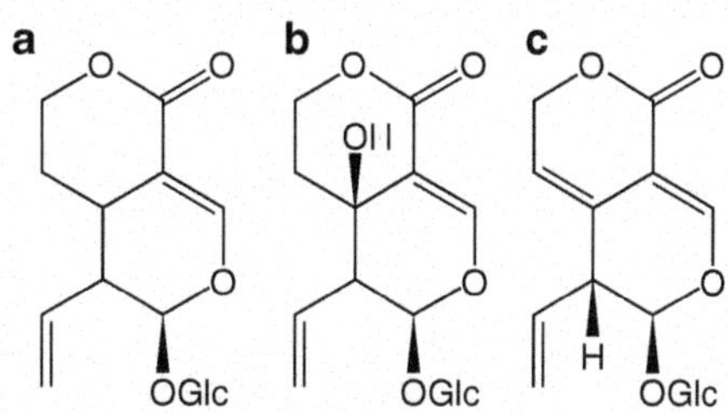

Fig. 3 Secoiridoid characteristic for *Centaurium* species. (**a**) Sweroside, (**b**) swertiamarin, and (**c**) gentiopicrin

[54]. *Centaurium* species include tetra-, penta-, and hexaoxygenated xanthones present in free form (aglycones) or in the form of *O*-glucosides [54]. All substituted xanthones originated from 1,3,5 or 1,3,7-trihydroxyxanthones.

The biosynthesis of xanthones originates mainly from shikimate-acetate pathway. The main steps in the xanthone biosynthesis involve benzophenone intermediate followed by intramolecular coupling to form the xanthone ring. The initial molecules in the biosynthesis of xanthones are benzoic acid and 3-hydroxybenzoic acid, originated directly from the shikimate pathway [66]. In centaury plants 3-hydroxybenzoic acid is used as the initial compound for tetrahydroxyxanthone biosynthesis [67]. The xanthone biosynthesis is obtained via conversion of 3-hydroxybenzoic acid to 3-hydroxybenzoyl-CoA and tetrahydroxybenzophenone, further converted to trihydroxyxanthone. Two methoxylated xanthones, 1-hydroxy-3,5,6,7,8-pentamethoxyxanthone known as eustomin (Fig. 5a) and 1,8-dihydroxy-3,5,6,7-tetramethoxyxanthone known as demethyleustomin (Fig. 5b), are the characteristic xanthone compounds for *C. erythraea* and *C. pulchellum* species [68–72].

To date, more than 100 different xanthone compounds were isolated from 110 plant species [54]. In recent years xanthones have become more interesting since they exhibit different pharmacological effects such as cardioprotective [73], antimalarial [74], antimicrobial [75], antioxidant, antimutagenic, and anticancer [76]. It is known that xanthones act as antidepressants [77, 78]. Earlier investigations also showed HIV-inhibitory potential of xanthones [79, 80]. Nowadays, the effects of natural and synthetic xanthones on the central nervous system and the growth of tumor cells are investigated. In addition to their pharmacological values, xanthones are very useful systematic markers in hemotaxonomy [54, 81].

Literature data about biological effects of xanthones isolated from centaury is very poor. Conclusions about numerous biological activities of centaury have mainly

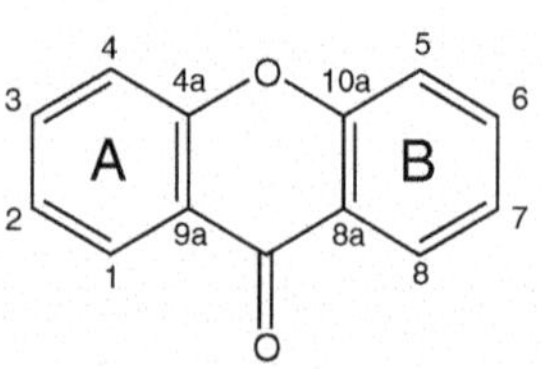

Fig. 4 The basic structure of xanthones

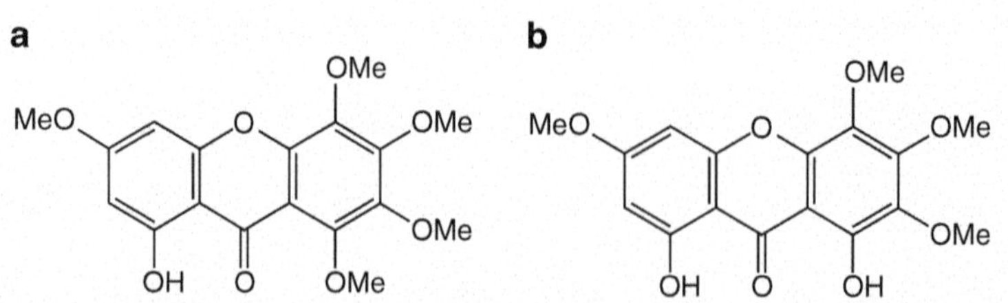

Fig. 5 Xanthones characteristic for *Centaurium* species. (**a**) Eustomin and (**b**) demethyleustomin

been derived from studies that considered the effects of the whole plant extracts. Only available data are related to antimutagenic activity of eustomin and demethyleustomin in *Salmonella typhimurium* [82]. Due to these biological characteristics, xanthones today represent a significant source of plant material used in the production of new drugs. It is very important to notice that xanthones have a wide range of biological effect and almost nontoxic and no side effects. Therefore, numerous researches in recent years are directed in determining the mechanisms of xanthone's pharmacological effects on a biochemical level.

1.4 Secondary Metabolites of Centaury In Vitro

Secondary metabolites can be isolated from plants grown in the natural habitat or from cultivated plant species. Breeding of plants in order to increase the production of active substances is often not successful and economically profitable. For this reason it has been necessary to find alternative methods for the production of useful compounds produced by plants. One of these alternative methods certainly represents plant tissue culture. This method provides a homogeneous and continuous source of plant material, independent from climatic conditions and many pathogens. In the middle of the twentieth century, it was shown, for the first time, that plants grown in vitro produced small amount of secondary metabolites. Thereafter, the composition of the culture media was intensively investigated as well as the selection of highly productive lines, and a very high level of secondary metabolite production was achieved. Twenty years later increased content of at least 30 components was discovered in plants grown in vitro compared to plants grown in natural habitat [83].

Centaury represents the plant species that produce secondary metabolites in vitro. The main secondary metabolite products of centaury are bitter glycosides (sweroside, swertiamarin, and gentiopicrin) and xanthones (eustomin and demethyleustomin). These secondary metabolites were detected and identified in centaury shoots and roots grown in vitro as well as from natural habitat. In centaury shoots and roots collected from natural habitat, swertiamarin was the dominant component, whereas gentiopicrin represents the dominant secoiridoid in both plant organs grown in vitro. On the other hand, secoiridoid swertiamarin was detected only in traces [68, 69, 84, 85]. Both plants grown in vitro and plants from natural habitat produced an increased amount of xanthones in centaury roots compared to the shoots. However, a significantly increased amount of xanthones was detected in centaury plants grown in vitro compared to the plants from natural habitat [70].

It is already known that the accumulation of secondary metabolites in plants cultured in vitro depends on the composition of culture media, type, and quantity of exogenously applied plant growth regulators, mineral elements, as well as carbohydrates [86]. The influence of plant growth regulators on the production of secondary metabolites in centaury plants grown in vitro was previously investigated [70]. It was shown that the amount of the xanthone eustomin was concentration dependent of the cytokinin benzylaminopurine (N^6-benzyladenine (BAP)) in the nutrient medium. The shoots grown on nutrient medium supplemented even with the lowest concentration of BA (3 μmol/l) produced an increased amount of xanthones compared to shoots from natural habitat. It was shown that sucrose concentration in nutrition media also significantly affected the production of centaury secondary metabolites. An increased amount of secoiridoids and xanthones was detected in centaury plants grown in nutrient media supplemented with different concentrations of sucrose [70].

Literature data considering secondary metabolites isolated from genetically transformed centaury plants almost do not exist. It was noted that centaury roots transformed using *A. rhizogenes* (strain A4M70GUS) produced only xanthones [87]. Hairy roots of centaury plants retained the capacity for the production of xanthone compounds while did not produce secoiridoids. Also hairy roots produced greater amounts of xanthones than control roots grown in nature. Hairy roots regenerated transgenic centaury plants contained both secoiridoids and xanthones. In transgenic plants xanthone compounds detected a few times greater than in wild-growing plants. On the other hand, genetic transformation of centaury using *A. rhizogenes* (strain LBA9402) showed that hairy roots were also able to regenerate adventitious shoots, further developed into plants [24]. It was determined that transgenic centaury shoots produced a double amount of total secoiridoids compared to control untransformed shoots. The highest accumulation of secoiridoids was detected in shoots at the vegetative stage of 10-week-old transformed regenerants. It was also indicated that transformed centaury plants could serve as valuable source of secoiridoids.

1.5 Methods of Transformation

Until today there are two reports considering genetic transformation of centaury using *A. rhizogenes* and only one report of transformation using *A. tumefaciens*.

In vitro-grown centaury seedlings were inoculated with *A. rhizogenes* strain A4M70GUS, harboring with GUS construct integrated into the TL region of pRiA4 plasmid [15]. Hairy roots appeared 14 days after inoculation. Root tips from hairy root culture were transferred to MS medium [89] and subcultured every month. Adventitious shoots were initiated under these conditions. When regenerated shoots were 2–3 mm long, they were excited and transferred to hormone-free MS medium for the shoot elongation and root initiation.

An agropine-type strain of *A. rhizogenes* LBA 9402 with pRi 1855 plasmid was used for the generation of centaury hairy roots [24]. Centaury explants were wounded with sterile needle dipped into a bacterial culture. The infected explants were incubated on hormone-free agar-solidified MS medium. Eight weeks after transformation, root tips were transferred on a liquid medium. Adventitious centaury buds, emerging from the hairy roots, were excised and further cultured on liquid or solid MS medium supplemented with 0.1 mg l^{-1} IAA and 1.0 mg l^{-1} BAP.

Genes coding for two *Arabidopsis* CKX isoforms, *AtCKX1* and *AtCKX2*, were introduced separately into a binary cloning vector, immobilized into *A. tumefaciens* strain GV3101 [88], and introduced into root explants of *C. erythraea* Rafn. as previously described [29]. Four *AtCKX1* and *AtCKX2* transgenic lines were selected for further analyses. All selected *AtCKX* transgenic lines satisfied the two criteria. The first one was unquestionable expression of *AtCKX1* or *AtCKX2* transgenes, in shoots as well as in centaury roots, previously confirmed by qRT-PCR reactions. The second criterion was significantly increased or at the control level CKX activity at least, in both investigated plant organs [29]. All transgenic lines including control plants grown in vitro were cultured for 4 weeks on a half-strength MS medium (½MS, [90]) solidified with 0.7% agar and supplemented with 3% sucrose and 100 mg l^{-1} *myo*-inositol, without plant growth regulators. The shoots spontaneously rooted after additional 3–4 weeks on the same medium. All in vitro-cultured plants were grown at 25 ± 2°C and at 16 h/8 h photoperiod ("Tesla" white fluorescent lamps, 65 W, 4500 K; light flux of 47 μmol$s^{-1}m^{-2}$).

1.6 Methods for Extraction and Quantification of Secondary Metabolites

The air-dried and powered *AtCKX* transgenic and control *C. erythraea* 4-week-old shoots and roots were extracted with methanol for 48 h at room temperature in the dark. The ratio between the plant material and solvent was 1:20, w/v. Analyses were carried out on an Agilent series 1100 HPLC instrument with a DAD detector and a reverse phase Zorbax SB-C18 analytical column (150 × 4.6 mm, 5 μm). Mobile phase consisted of 1% v/v solution of orthophosphoric acid in water (solvent A) and acetonitrile (solvent B). The flow rate was 1 ml/min^{-1}. Injection volume of sample was 5 μl, and the gradient elution was as follows: 98–90% A, 0–5 min; 90–85% A, 5–10 min; 85% A, 10–13 min; 85–70% A, 13–15 min; 70–10% A, 15–20 min; 10% A, 20–22 min; and 10–0% A, 22–25 min. Detection wavelengths were set at 260 and 320 nm.

For quantification of swertiamarin and gentiopicrin, commercial standard manufacturers Cfm Oscar Tropitzsch (Germany) were used. The amounts of the secoiridoids were calculated using external standard and calibration curves. All analyses of secoiridoids in shoots as well as in centaury roots were repeated at least two times. The results are presented as mg/g of dry sample weight.

The xanthones, eustomin and demethyleustomin, were isolated from aerial parts of *C. erythraea* plants grown in nature. Their structures were confirmed by spectroscopic techniques: UV, 1D and 2D NMR spectroscopy, and MS spectrometry. Quantification was performed using HPLC, and the amounts of these compounds were calculated using calibration curves. The results are presented as mg/g of dry weight (DW).

1.7 Methods for Culturing Colorectal Cancer Cell Lines

DLD1 cell line was purchased from the American Type Culture Collection (Rockville, MD, USA). DLD1-TxR cells are multidrug resistant variant selected from DLD1 cells after continuous exposure to stepwise increasing concentrations of PTX

(60–600 nM) for a period of 10 months [90]. DLD1 and DLD1-TxR cells were maintained in RPMI 1640 containing 10% FBS, 2 mm L-glutamine, 10,000 U/ml penicillin, 10 mg/ml streptomycin, and 25 μg/ml amphotericin B solution. The cells were maintained at 37 C in a humidified 5% CO_2 atmosphere and subcultured at 72-h intervals using 0.25% trypsin/EDTA.

1.8 Methods for Detection of Chemosensitivity by SRB Assay

The effects of methanol extracts of control and transgenic *AtCKX1*-29 centaury line grown in vitro as well as pure secoiridoids and xanthones were assessed by sulforhodamine B test.

Cells grown in 25-cm^2 tissue flasks were trypsinized, seeded into flat-bottomed 96-well tissue culture plates (1000 cell/well), and incubated overnight. Then, the cells were treated 72 h with various concentrations of methanol extracts and pure compounds. Sulforhodamine B (SRB) binds to the cell proteins enabling detection of differences in cell viability between untreated and treated cells. Briefly, the cells in 96-well plates were fixed in 50% trichloroacetic acid (50 μl/well) for 1 h at 4 C, rinsed in tap water, and stained with 0.4% (w/v) SRB in 1% acetic acid (50 μl/well) for 30 min at room temperature. The cells were then rinsed three times in 1% acetic acid to remove the unbound stain. The protein-bound stain was extracted with 200 μl 10 mm trisbase (pH 10.5) per well. The optical density was read at 540 nm with correction at 670 nm in an LKB 5060-006 μ plate reader (Vienna, Austria). IC50 values were defined as the concentration of the drug that inhibited cell growth by 50% and calculated by linear regression analysis using Excel software.

1.9 Statistical Analysis

All analyses of centaury secondary metabolites were repeated at least two times. Statistical analyses were performed using StatGraphics software version 4.2 (STSC Inc. and Statistical Graphics Corporation, 1985–1989, USA). The data were subjected to analysis of variance (ANOVA), and comparisons between the mean values were made using the least significant difference (LSD) test calculated at a confidence level of $p \leq 0.05$. GraphPad Prism 6 software was employed for statistical analysis of results obtained by SRB assay. Three independent experiments were carried out. Kruskal–Wallis test for multiple comparisons was applied in order to assess significant differences between untreated control and different concentrations of extracts and compounds.

2 Secondary Metabolites of *AtCKX* Transgenic Centaury Plants and Anticancer Activity

2.1 Identification and Quantification of Secondary Metabolite Using HPLC Method

The chromatograms of methanol extracts of control and *AtCKX* transgenic centaury plants showed a similar peak profile (Fig. 6). The HPLC analyses confirmed the presence of bitter secoiridoids (swertiamarin and gentiopicrin) and xanthones (eustomin and demethyleustomin) as characteristic compounds for *C. erythraea* species.

No qualitative differences were observed in secondary metabolites between analyzed *AtCKX* transgenic centaury plants and control plants. It means that *AtCKX* transgenic centaury plants grown in vitro produce secondary metabolite characteristic for corresponding plant species grown in nature.

However, quantitative differences in secoiridoid content in centaury shoots and roots were obvious. Secoiridoid swertiamarin was the dominant component in centaury shoots from natural habitat, in shoots grown in vitro, and in all *AtCKX* transgenic centaury shoots (Figs. 7a and 8a). Control centaury shoots collected from nature and cultured in vitro showed no statistically significant differences in swertiamarin content. It was observed that a lower content of swertiamarin was detected in all analyzed *AtCKX* transgenic centaury lines, except line *AtCKX1*-5, compared to the control shoots in vitro and shoots from nature. Transgenic line *AtCKX1*-5 showed the highest production of swertiamarin (91.51 mg/g DW), compared to the control shoots in vitro (64.77 mg/g DW) and control shoots from nature (68.41 mg/g DW). In shoots of transgenic line *AtCKX1*-29, the production of swertiamarin significantly decreased, and the content of this bitter glycoside reduced to 1.34 mg/g DW. The content of secoiridoid gentiopicrin was twofold higher in control shoots collected from the natural habitat compared to the control shoots grown in vitro. It was also noticed that in shoots of two *AtCKX1* lines, 31 and 39, and two *AtCKX2* lines, 17 and 31, the production of gentiopicrin was significantly higher compared to the control shoots from nature. However, in the shoots of some transgenic lines, such as *AtCKX1*-5 and *AtCKX1*-29, very low gentiopicrin content was determined, while in the shoots of transgenic line *AtCKX2*-29, gentiopicrin was detected only in traces.

It was interesting to note that in roots of control and all *AtCKX* transgenic centaury lines, swertiamarin was not the dominant secoiridoid (Figs. 7b and 8b). There was no significant difference in the amount of accumulated swertiamarin and gentiopicrin in the control roots from the natural habitat, while control roots grown in vitro produced two times more gentiopicrin compared to swertiamarin. Centaury roots collected from nature accumulated more swertiamarin (9.47 mg/g DW) than roots cultured in vitro (6.36 mg/g DW). On the other hand, all analyzed *AtCKX*

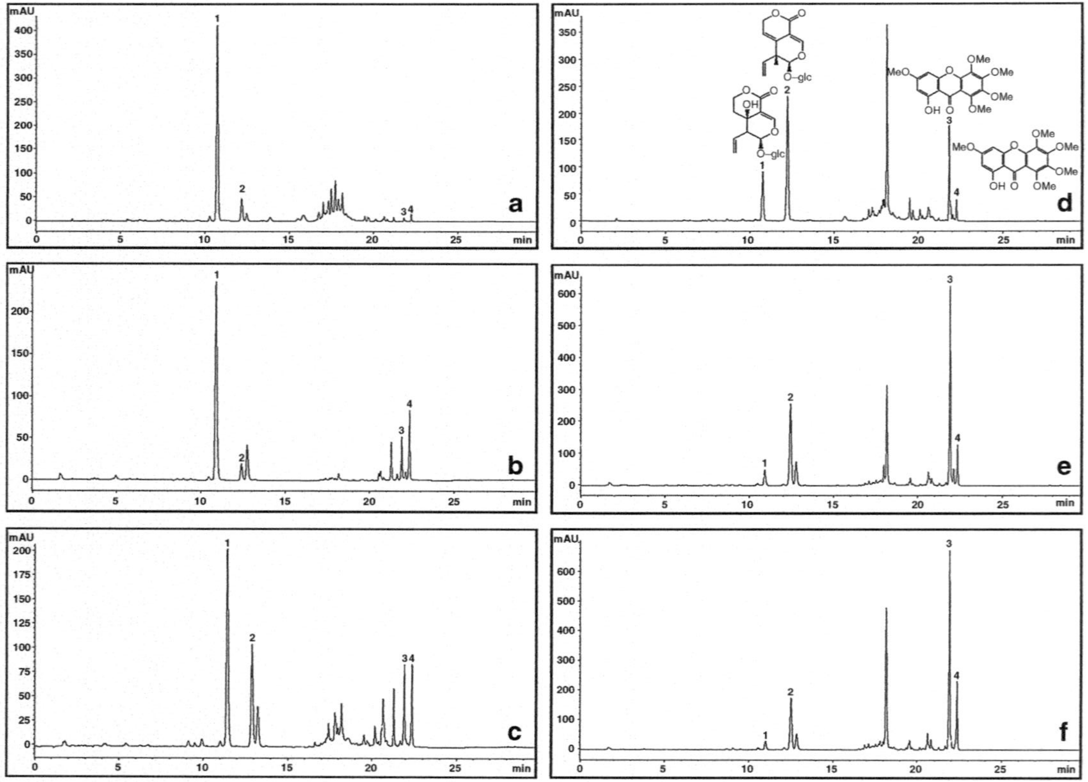

Fig. 6 HPLC profiles of methanol extracts of centaury (λ = 260 nm). (**a**) Aerial parts grown in nature, (**b**) control shoots cultured in vitro, (**c**) transgenic shoots – line *AtCKX1*-29, (**d**) roots from plants grown in nature, (**e**) control roots cultured in vitro, (**f**) transgenic roots – line *AtCKX1*-29. Peaks: 1 – swertiamarin, 2 – gentiopicrin, 3 – eustomin, 4 – demethyleustomin

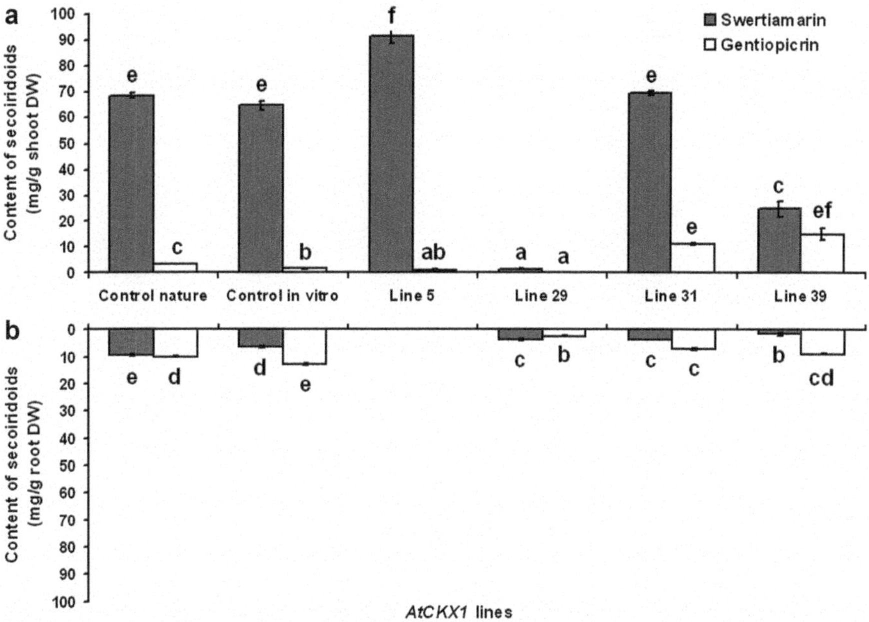

Fig. 7 Content of secoiridoids (swertiamarin and gentiopicrin) in shoots (**a**) and roots (**b**) of 4-week-old *AtCKX1* transgenic *Centaurium erythraea* lines. Data represent the mean ± standard error. Means marked with the same letter are not significantly different from the control according to LSD test ($p \leq 0.05$)

transgenic centaury lines produced a decreased content of swertiamarin compared to control roots grown in vitro. In the roots of only two transgenic centaury lines, *AtCKX1*-5 and *AtCKX2*-31, swertiamarin was detected in traces. Gentiopicrin represent the dominant secoiridoid in the control roots grown in vitro (12.75 mg/g DW) compared to the roots collected from the natural habitat (6.36 mg/g DW). However, the roots of all *AtCKX* transgenic centaury lines produced a decreased content of gentiopicrin compared to the roots from nature. In several centaury lines *AtCKX1*-5, *AtCKX2*-17, and *AtCKX2*-31, gentiopicrin was detected only in traces.

It can be summarized that the highest amount of swertiamarin (higher than in both control samples) was determined in shoots of only one transgenic centaury line *AtCKX1*-5. An increased production of gentiopicrin was also detected in shoots. The shoots of four transgenic centaury lines *AtCKX1*-31, *AtCKX1*-39, *AtCKX2*-17, and *AtCKX2*-31 produced more gentiopicrin than both control samples. On the other hand, it was noted that the amount of swertiamarin and gentiopicrin decreased in roots of all analyzed *AtCKX* transgenic centaury lines compared to the roots collected from the nature as well as the roots grown in vitro.

Quantitative differences in xanthone content in both shoots and roots were also determined. The most significant difference was detected between xanthones from control shoots and those grown in vitro (Figs. 9a and 10a). Eustomin content was two times higher in control shoots grown in vitro compared to plants collected from natural

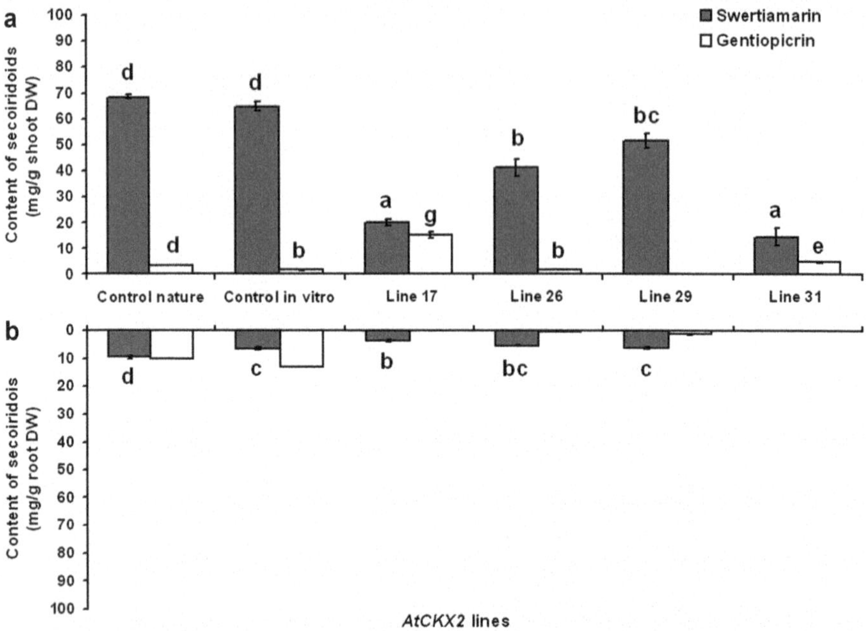

Fig. 8 Content of secoiridoids (swertiamarin and gentiopicrin) in shoots (**a**) and roots (**b**) of 4-week-old *AtCKX2* transgenic *Centaurium erythraea* lines. Data represent the mean ± standard error. Means marked with the same letter are not significantly different from the control according to LSD test ($p \leq 0.05$)

habitat. Importantly, the majority of shoots of the analyzed *AtCKX* transgenic lines produced more eustomin than shoots from nature (0.6 mg/g DW) but less than the control shoots in vitro (1.18 mg/g DW). Only one centaury transgenic line *AtCKX1*-29 contained more eustomin than control shoots grown in vitro (1.48 mg/g DW). Considering demethyleustomin, the content of this xanthone in transgenic centaury lines was higher compared to the control shoots from nature (0.24 mg/g DW). On the other hand, shoots of *AtCKX* transgenic lines produced less demethyleustomin in comparison with control shoots in vitro (1.3 mg/g DW). The shoots of only one transgenic *AtCKX1*-29 centaury line produced significantly an increased amount of demethyleustomin (1.47 mg/g DW) compared to both controls (Fig. 9a).

The production of xanthones was stimulated in control roots in vitro since the eustomin content was about ten times higher (3.31 mg/g DW), and demethyleustomin content was about three times higher (0.9 mg/g DW) (Figs. 9b and 10b). The roots of most of the analyzed *AtCKX* transgenic centaury lines produced an increased amount of both xanthones compared to control roots from nature but a decreased amount to control roots in vitro. Significantly, increased content of both xanthones compared to controls was determined only in roots of

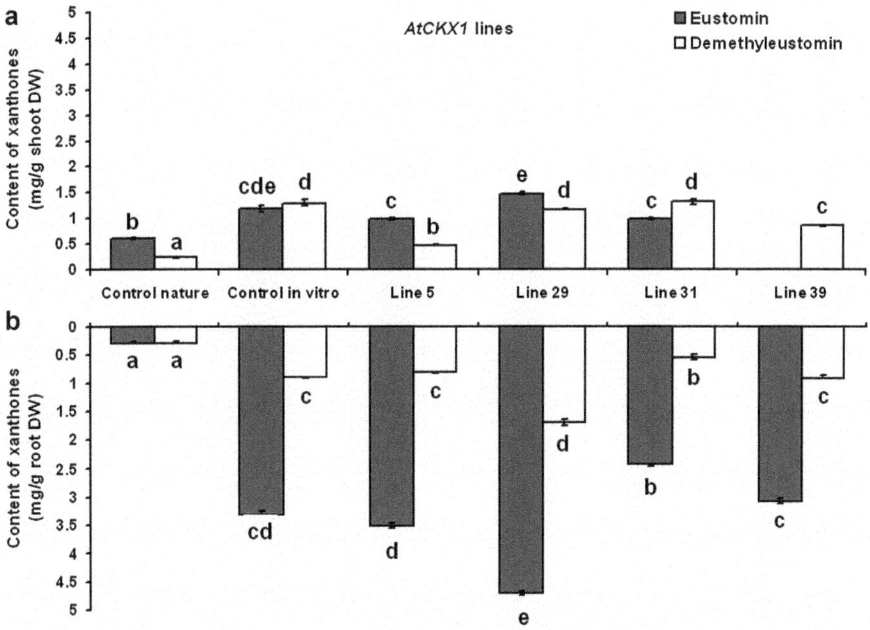

Fig. 9 Content of xanthones (eustomin and demethyleustomin) in shoots (**a**) and roots (**b**) of 4-week-old *AtCKX1* transgenic *Centaurium erythraea* lines. Data represent the mean ± standard error. Means marked with the same letter are not significantly different from the control according to LSD test ($p \leq 0.05$)

AtCKX1-29 transgenic centaury line (eustomin content was 4.71 mg/g DW and demethyleustomin content was 1.7 mg/g DW).

AtCKX transgenic centaury lines in both plant organs showed different xanthone contents compared to non-transformed plants from natural habitat and in vitro conditions. The eustomin and demethyleustomin content of certain lines varied compared to the xanthone content obtained in most other transgenic lines. The most increased xanthone content compared to both controls was detected only in shoots and roots of *AtCKX1*-29 transgenic line.

2.2 Anticancer Activity

The shoot and root methanol extracts of control and selected transgenic centaury line *AtCKX1*-29 were further tested for their capacity to inhibit the in vitro growth of two human colorectal carcinoma cell lines (DLD1 and DLD1-TxR) by SRB assay (Table 2). The results are expressed as concentrations of compounds that inhibited cell growth by 50% (IC50). The greatest efficacy was obtained by shoot and root methanol extracts of transgenic centaury line *AtCKX1*-29 in DLD1 cells.

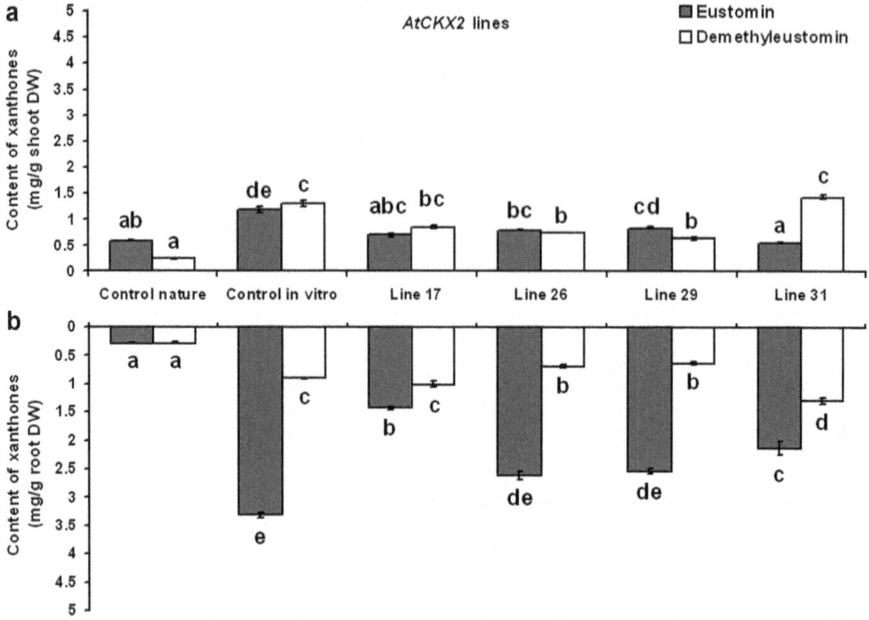

Fig. 10 Content of xanthones (eustomin and demethyleustomin) in shoots (**a**) and roots (**b**) of 4-week-old *AtCKX2* transgenic *Centaurium erythraea* lines. Data represent the mean ± standard error. Means marked with the same letter are not significantly different from the control according to LSD test ($p \leq 0.05$)

Importantly, the shoot and root methanol extracts of in vitro transgenic line *AtCKX1*-29 were significantly more potent than methanol extracts of in vitro control centaury shoots and roots. Nevertheless, the inhibitory potential of applied centaury methanol extracts was decreased in multidrug-resistant DLD1-TxR cell line in comparison with its sensitive counterpart DLD1.

Next, the effect of pure secoiridoids (gentiopicrin and swertiamarin) and pure xanthones (eustomin and demethyleustomin) in sensitive DLD1 and multidrug-resistant DLD1-TxR cancer cell lines has been examined (Fig. 11). The growth of DLD1 was inhibited in the presence of gentiopicrin, swertiamarin, and demethyleustomin. Demethyleustomin significantly inhibited the growth of DLD1 cells with 50 μM. However, 50 μM either of secoiridoids or demethyleustomin was not enough to inhibit the cell growth by 50%. Furthermore, the growth of multidrug-resistant cancer cells DLD1-TxR was not suppressed by these compounds. Eustomin showed the best inhibitory potential in both cell lines with concentration-dependent effect. IC_{50} values for eustomin were 8 μM and 23 μM in DLD1 and DLD1-TxR cells, respectively. Eustomin exerted significant efficacy on cell growth inhibition with 25 μM and 50 μM.

Table 2 Anticancer effects of methanol extracts isolated from control and transgenic centaury line *AtCKX1*-29 expressed by IC_{50} value (μg/ml)

Centaury methanol extracts	Control centaury shoots	Shoots of transgenic centaury line *AtCKX1*-29	Control centaury roots	Roots of transgenic centaury line *AtCKX1*-29
Sensitive DLD1 cancer cell lines	97.7 ± 4.7	43.0 ± 2.1^{a}	90.3 ± 7.3	41.7 ± 5.9^{a}
Multidrug resistant DLD1-TxR cancer cell lines	192.0 ± 0.6^{b}	$175.3 \pm 1.4^{a,\ b}$	162.3 ± 3.4^{b}	$104.3 \pm 11.3^{a,\ b}$

[a]Statistically significant difference in efficacy between control and transgenic centaury line AtCKX1–29 ($p \leq 0.01$)
[b]Statistically significant difference in sensitivity between DLD1 and DLD1-TxR ($p \leq 0.01$)

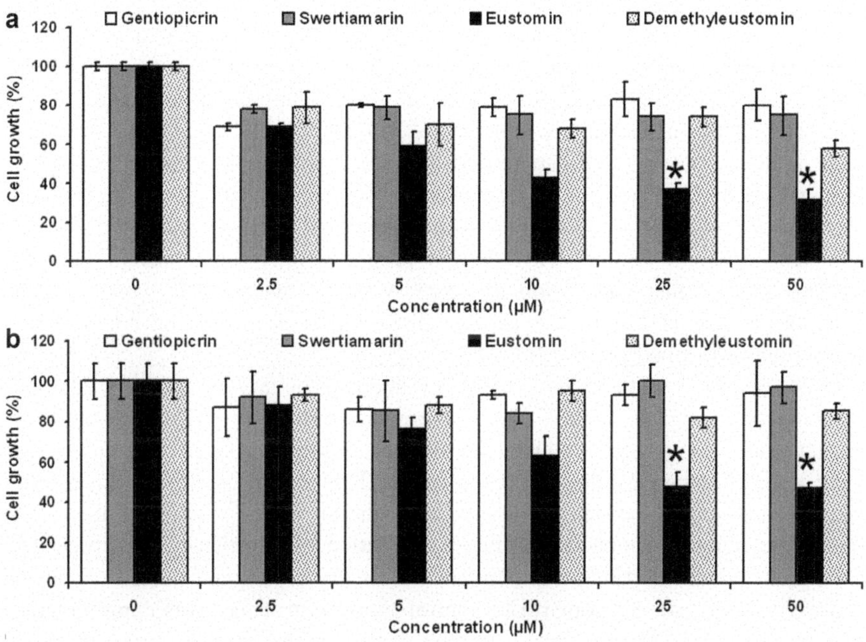

Fig. 11 Cell growth inhibition effects of two centaury secoiridoids (gentiopicrin and swertiamarin) and two xanthones (eustomin and demethyleustomin) in sensitive DLD1 (**a**) and multidrug-resistant DLD1-TxR cancer cell lines (**b**). Data represent the mean ± standard deviation. Means marked with an asterisk differ significantly from untreated cells (0 μM) ($p \leq 0.05$)

3 Secondary Metabolite Production in Genetically Transformed Centaury Plants

So far, there are no literature data considering genetic transformation of *C. erythraea* using *A. tumefaciens* and consequently no data of transgenic centaury secondary metabolites. Overexpression of *AtCKX1* and *AtCKX2* transgenes was successfully confirmed in *C. erythraea* plants grown in vitro [29]. In addition, changes in cytokinin content and altered cytokinin homeostasis in *AtCKX1*- and *AtCKX2*-overexpressing centaury plants grown in vitro were investigated [91]. These results provided very important and novel information about the influence of altered endogenous cytokinin levels on the secondary metabolite production of *AtCKX* transgenic centaury plants.

According to the presented results, genetic transformation of centaury using *AtCKX* genes may influence secondary plant metabolism. Based on the HPLC analyses of the centaury methanol extracts, the presence of bitter secoiridoids (gentiopicrin and sweroside) and xanthones (eustomin and demethyleustomin) was confirmed. Unlike roots, in all analyzed centaury shoots, both control and transgenic *AtCKX* shoots, swertiamarin was the dominant secoiridoid. These results correspond with previous reports only partially. Previous researches showed that in centaury shoots and roots collected from nature secoiridoid swertiamarin dominated, but in control centaury shoots and roots grown in vitro, gentiopicrin dominated, while swertiamarin was detected only in traces [68, 69, 84, 85].

It is already known that genetic transformation could affect the production of centaury secondary metabolites. Regarding secondary metabolites of genetically transformed centaury plants, it has been shown that centaury roots transformed with *A. rhizogenes* (strain A4M70GUS) produced only xanthones [87]. Other authors showed that inoculation of centaury shoots with *A. rhizogenes* (strain LBA9402) induced two times higher total secoiridoids production compared to control untransformed shoots [24].

Since this work represents the first report about secondary metabolite production of centaury transgenic *AtCKX* plants, these results are very valuable. Although centaury shoots and roots produced the same secoiridoids, changes in their content were noted. Centaury shoots from the natural habitat and shoots grown in vitro showed no significant difference in the swertiamarin content. Almost in all *AtCKX* transgenic centaury shoots, the reduced production of swertiamarin was detected compared to the control shoots in vitro and shoots collected from nature. The content of swertiamarin and gentiopicrin in control roots from nature was approximately equal, while in control in vitro roots, the content of gentiopicrin was two times higher than the content of swertiamarin. Also it was noticed that all analyzed *AtCKX* transgenic roots produced a decreased swertiamarin content compared to control roots from nature and roots grown in vitro. In certain transgenic lines, swertiamarin was detected only in traces. The same ratio was obtained in the gentiopicrin content. All *AtCKX* transgenic centaury roots produced less gentiopicrin compared to the roots grown in vitro and roots originated from nature. In some transgenic lines, gentiopicrin was also detected only in traces. Based on the content of centaury swertiamarin and gentiopicrin, it was concluded that in vitro culture conditions influenced and reduced the production of gentiopicrin [85]. However, the results presented here indicated that in vitro culture conditions stimulated the production of gentiopicrin but only in centaury roots. Contrarily, increased gentiopicrin content was detected only in centaury shoots grown in vitro compared to the shoots from the natural habitat [84].

It was also observed that most of transgenic *AtCKX* shoots and roots contained higher amounts of xanthones than control shoots from the natural habitat. However, transgenic *AtCKX* shoots and roots produced less amount of xanthones compared to controls grown in vitro. Only *AtCKX1*-29 transgenic line produced an increased content of eustomin and demethyleustomin compared to both controls. It is interesting to note that roots of this transgenic line showed an increased activity of cytokinin oxidase/dehydrogenase, the only known catabolic enzyme that specifically catalyzes cytokinin degradation in plant tissues. Also the shoots and roots of transgenic line *AtCKX1*-29 showed the reduced levels of bioactive forms of cytokinins [91].

Centaury control shoots grown in vitro produced twofold higher eustomin and fivefold higher demethyleustomin amount in comparison with the control shoots originated from the natural habitat. The content of both eustomin and demethyleustomin in control roots grown in vitro was increased threefold when compared to control roots from nature. The obtained increase corresponds with a previously reported enhancement of xanthone production in centaury in vitro compared to the plants from natural habitat [70].

4 Conclusions

Based on the presented results, it can be said that the genetic transformation of centaury with *AtCKX* genes influenced the secondary metabolism of this valuable medicinal plant species. Considering secoiridoid content in transgenic *AtCKX* centaury plants in vitro, it is very difficult to make a comprehensive conclusion. It was shown that swertiamarin content in shoots of all centaury transgenic lines was decreased compared to the shoots grown in vitro, which produced approximately the equal swertiamarin content with centaury shoots from the natural habitat. On the other hand, it is interesting to note that only in shoots of the majority of transgenic centaury lines gentiopicrin production was stimulated. Obtained results also indicated that in vitro culture conditions stimulated the production of both xanthones in centaury shoots and roots. It is important to note that all centaury transgenic and control plants in vitro were grown under the same temperature and light intensity conditions. It is known that in vitro cultivation, although under controlled temperature and light conditions, stimulates increased secondary metabolite accumulation in response to abiotic stress. It can be also assumed that stress caused by genetic transformation itself additionally affected metabolic changes, resulting in an increased content of gentiopicrin and xanthones in *AtCKX* transgenic centaury plants. In the metabolic pathway of iridoids, gentiopicrin originated from swertiamarin, so an increased accumulation of gentiopicrin in *AtCKX* transgenic centaury shoots corresponds with the reduced swertiamarin content. Most likely, it can be assumed that the possible increased activity of the enzyme converts swertiamarin into gentiopicrin and affected on elevated gentiopicrin content. Importantly, some of *AtCKX* centaury transgenic lines produced increased levels of eustomin and/or demethyleustomin content. The anticancer activity of transgenic centaury line *AtCKX1*-29 methanol extracts was more potent than that observed with methanol extracts of control plants. The increased inhibitory effect corresponded to the higher content of eustomin (xanthone with considerable anticancer activity) in transgenic centaury line *AtCKX1*-29 methanol extracts.

So far there are no literature data considering cytotoxicity of centaury methanol extract and/or single secoiridoids or xanthones isolated from centaury. Notwithstanding numerous pharmaceutical properties of centaury extract, the effect of centaury xanthones on sensitive and resistant cancer cell lines has never been examined. The results presented here are very valuable and certainly open new possibilities for further investigations considering the secondary metabolite production in centaury. Thus, the present investigation showed that some *AtCKX* transgenic centaury lines can be used as an alternative source for isolation of valuable pharmacologically active compounds without exploitation of the plants from their natural habitat.

Acknowledgments This work was supported by the Ministry of Education, Science, and Technological Development of the Republic of Serbia (grant No. ON173015, ON172053, and III41031).

References

1. Sridhar TM, Aswath CR (2014) Review on medicinal plants propagation: a comprehensive study on role of natural organic extracts in tissue culture medium. Am J Plant Sci 5:3073–3088
2. Kumar N, Reddy MP (2011) In vitro plant propagation: a review. J For Sci 27:61–72
3. Karruppusamy S (2009) A review on trends in production of secondary metabolites from higher plants by in vitro tissue, organ and cell cultures. J Med Plants Res 3:1222–1239
4. Naina NS, Gupta PK, Mascarenhas AF (1989) Genetic transformation and regeneration of transgenic neem (*Azadirachta indica*) plants using *Agrobacterium tumefaciens*. Curr Sci 58:184–187
5. Yun DJ, Hashimoto T, Yamada Y (1992) Metabolic engineering of medicinal plants: transgenic *Atropa belladonna* with an improved alkaloid composition. Proc Natl Acad Sci U S A 89:11799–11803
6. Cucu N, Gabriela N, Gavrila L (2002) Genetically modified medicinal plants. II. Transfer and expression of a marker kanamycin resistance gene in *Atropa belladonna* plants. Rom Biotechnol Lett 7:869–874
7. Koroch AR, Kapteyn J, Juliani HR, Simon JE (2003) In vitro regeneration and *Agrobacterium* transformation of *Echinacea purpurea* leaf explants. Trends New Crop News 39:415–418
8. Siahsar B, Rahimi M, Tavassoli A, Raissi A (2011) Application of biotechnology in production of medicinal plants. Am Eurasian J Agric Environ Sci 11:439–444
9. Takhtajan AL (1980) Outline of the classification of flowering plants (*Magnoliophyta*). Bot Rev 46:225–359
10. Chevallier A (2000) Encyclopedia of herbal medicine. Dorling Kindersley, London
11. Jägger E (1978) *Centaurium* hill. In: Meusel H (ed) Vergleichende Chrologie der Zentraleuropäischen, Flora, vol 2. Gustav Ficher, Jena, pp 352–353

12. Melderis A (1972) *Centaurium*. In: Flora Europea, vol 3. Cambridge University Press, Cambridge, UK, pp 56–59
13. Tucakov J (1990) Healing with plants – phytotherapy. Rad, Belgrade
14. Skrzypczak L, Wesolovska M, Skrzypczak E (1993) *Gentiana* species: in vitro culture, regeneration and production of secoiridoid glucosides. In: Bajaj YPS (ed) Biotechnology in agriculture and forestry, vol 21, Medicinal and aromatic plants IV. Springer, Berlin/Hiedelberg, pp 172–186
15. Subotić A, Budimir S, Grubišić D, Momčilović I (2003/2004) Direct regeneration of shoots from hairy root cultures of *Centaurium erythraea* inoculated with *Agrobacterium rhizogenes*. Biol Plant 47:617–619
16. Subotić A, Janković T, Jevremović S, Grubišić D (2006) Plant tissue culture and secondary metabolites productions of *Centaurium erythraea* Rafn., a medical plant. In: Teixeira da Silva JA (ed) Floriculture, ornamental and plant biotechnology: advances and topical issues, vol II, 1st edn. Global Science Books, London, pp 564–570
17. Subotić A, Jevremović S, Grubišić D, Janković T (2009) Spontaneous plant regeneration and production of secondary metabolites from hairy root cultures of *Centaurium erythraea* Rafn. In: Jain SM, Saxena PK (eds) Protocols for in vitro cultures and secondary metabolite analysis of aromatic and medicinal plants, vol 547, Methods in molecular biology. Springer, Berlin, pp 205–217
18. Subotić A, Jevremović S, Grubišić D (2009) Influence of cytokinins on in vitro morphogenesis in root cultures of *Centaurium erythraea* – valuable medicinal plant. Sci Hortic 120:386–390
19. Subotić A, Jevremović S, Trifunović M, Petrić M, Milošević S, Grubišić D (2009) The influence of gibberellic acid and paclobutrazol on induction of somatic embryogenesis in wild type and hairy root cultures of *Centaurium erythraea* Gillib. Afr J Biotechnol 8:3223–3228
20. Trifunović M, Tadić V, Petrić M, Jontulović D, Jevremović S, Subotić A (2014) Quantification of arabinogalactan proteins during in vitro morphogenesis induced by β-D-glucosyl Yariv reagent in *Centaurium erythraea* root culture. Acta Physiol Plant 36:1187–1195
21. Filipović BK, Simonović AD, Trifunović MM, Dmitrović SS, Savić JM, Jevremović SB, Subotić AR (2015) Plant regeneration in leaf culture of *Centaurium erythraea* Rafn. Part 1: the role of antioxidant enzymes. Plant Cell Tiss Org Cult 121:703–719
22. Simonović AD, Filipović BK, Trifunović MM, Malkov SN, Milinković VP, Jevremović SB, Subotić AR (2015) Plant regeneration in leaf culture of *Centaurium erythraea* Rafn. Part 2: the role of arabinogalactan proteins. Plant Cell Tiss Org Cult 21:721–739
23. Trifunović-Momčilov M, Motyka V, Dragićević IČ, Petrić M, Jevremović S, Malbeck J, Holík J, Dobrev PI, Subotić A (2015) Endogenous phytohormones in spontaneously regenerated *Centaurium erythraea* Rafn. plants grown in vitro. J Plant Growth Regul. doi:10.1007/s00344-015-9558-x
24. Piatczak E, Krolicka A, Wysokinska H (2006) Genetic transformation of *Centaurium erythraea* Rafn by *Agrobacterium rhizogenes* and the production of secoiridoids. Plant Cell Rep 25:1308–1315
25. Hosokawa K, Matsuki R, Oikawa Y, Yamamura S (1997) Genetic transformation of gentian wild-type *Agrobacterium rhizogenes*. Plant Cell Tiss Org Cult 51:137–140
26. Deroles SC, Ledger SE, Miller RM, Davies KM, Given NK (1993) Transformation of *Eustoma grandiflorum* (Lisianthus). In: Bajaj YPS (ed) Biotechnology in agriculture and forestry. Plant protoplasts and genetic engineering III, vol. 22. Springer-Verlag Berlin Heidelberg, pp 202–212
27. Vinterhalter B, Momčilović I, Vinterhalter D (2000) High biomass producing root cultures of *Gentiana punctata* L. Transformed with *Agrobacterium tumefaciens* C58C1 (pArA4b). Arch Biol Sci 52:85–90
28. Sun SB, Meng LS (2010) Genetic transformation of *Gentiana dahurica* Fisch. by *Agrobacterium tumefaciens* using zygotic embryo-derived callus. Acta Physiol Plant 32:629–634
29. Trifunović M, Cingel A, Simonović A, Jevremović S, Petrić M, Dragićević IČ, Motyka V, Dobrev PI, Zahajská L, Subotić A (2013) Overexpression of *Arabidopsis* cytokinin oxidase/

dehydrogenase genes *AtCKX1* and *AtCKX2* in transgenic *Centaurium erythraea* Rafn. Plant Cell Tiss Org Cult 115:139–150

30. Suginuma C, Akihana T (1995) Transformation of gentian with *Agrobacterium rhizogenes*. Acta Hortic 392:153–160
31. Momčilović I, Grubišić D, Kojić M, Nešković M (1997) *Agrobacterium rhizogenes*-mediated transformation and plant regeneration of four *Gentiana* species. Plant Cell Tiss Org Cult 50:1–6
32. Hayta S, Gurel A, Akgun IH, Altan F, Ganzera M, Tanyolac B, Bedir E (2011) Induction of *Gentiana cruciata* hairy roots and their secondary metabolites. Biologia 66:618–625
33. Vinterhalter B, Orbović V, Vinterhalter D (1999) Transgenic root cultures of *Gentiana punctata* L. Acta Soc Bot Pol 68:275–280
34. Hosokawa K, Matsuki R, Oikawa Y, Yamamura S (2000) Production of transgenic gentian plants by particle bombardment of suspension-culture cells. Plant Cell Rep 19:454–458
35. Mishiba K, Nishihara M, Abe Y, Nakatsuka T, Kawamura K, Komada K, Takesawa T, Abe J, Yamamura S (2006) Production of dwarf potted gentian using wild-type *Agrobacterium rhizogenes*. Plant Biotechnol 23:33–38
36. Tiwari RK, Trivedi M, Gunag ZC, Guo GQ, Zheng GC (2007) Genetic transformation of *Gentiana macrophylla* with *Agrobacterium rhizogenes*: growth and production of secoiridoid glucoside gentiopicroside in transformed hairy root culture. Plant Cell Rep 26:199–210
37. Zhang HL, Xue SH, Pu F, Tiwari RK, Wang XY (2010) Establishment of hairy root lines and analysis of gentiopicroside in the medicinal plant *Gentiana macrophylla*. Russ J Plant Physiol 57:110–117
38. Vinterhalter B, Krstić-Milošević D, Janković T, Pljevljakušić D, Ninković S, Smigocki A, Vinterhalter D (2015) *Gentiana dinarica* Beck. hairy root cultures and evaluation of factors affecting growth and xanthone production. Plant Cell Tiss Org Cult 121:667–679
39. Handa T (1992) Regeneration and characterization of prairie gentian (*Eustoma grandiflorum*) plants transformed by *Agrobacterium rhizogenes*. Plant Tissue Cult Lett 9:10–14
40. Handa T, Sujimura T, Kato E, Kamada H, Takayanagi K (1995) Genetic transformation of *Eustoma grandiflorum* with *rol* genes. Acta Hortic 392:209–218
41. Semeria L, Vaira AM, Accotto GP, Allavena A (1995) Genetic transformation of *Eustoma grandiflorum* Griseb. By microprojectile bombardment. Euphytica 85:125–130
42. Giovannini A, Pecchioni N, Allavena A (1996) Genetic transformation of lisianthus (*Eustoma grandiflorum* Griseb) by *Agrobacterium rhizogenes*. J Genet Breed 50:35–39
43. Semeria L, Ruffoni B, Rabaglio M, Ganga A, Vaira AM, Accotto GP (1996) Genetic transformation of *Eustoma grandiflorum* by *Agrobacterium tumefaciens*. Plant Cell Tiss Org Cult 47:67–72
44. Ledger SE, Deroles SC, Manson DG, Bradley JM, Gicen NK (1997) Transformation of lisianthus. Plant Cell Rep 16:853–858
45. Schwinn KE, Davies KM, Deroles SC, Markham KR, Miller RM, Brdaley JM, Manson DG, Given NK (1997) Expression of an *Antirrhinum majus* UDP-glucose:flavonoid-3-o-glucosyltransferase transgene alters flavonoid glycosylation and acylation in lisianthus (*Eustoma grandiflorum* Grise.). Plant Sci 25:53–61
46. Takahashi M, Nishihara M, Yamamura S, Nishizava S, Irifune K, Morikawa H (1998) Stable transformation of *Eustoma grandiflorum* by particle bombardment. Plant Cell Rep 17:504–507
47. Ishimaru K, Sudo H, Satake M, Matsunaga Y, Takemoto S, Shimomura K (1990) Amarogentin, amaroswerin and four xanthones from root culture of *Swertia japonica*. Phytochemistry 29:1563–1565
48. Freeman BC, Beattie GA (2008) An overview of plant defenses against pathogens and herbivores. Plant Health Instr. doi:10.1094/PHI-I-2008-0226-01
49. Oksman-Caldentey KM, Inzé D (2004) Plant cell factories in the post-genomic era: new ways to produce designer secondary metabolites. Trends Plant Sci 9:433–440
50. Rodriguez S, Marston A, Wolfender JL, Hostettmann K (1998) Iridoids and secoiridoids in the *Gentianaceae*. Curr Org Chem 2:627–648

51. Coscia CJ, Guarnaccia R (1968) Natural occurrence and biosynthesis of a cyclopentanoid monoterpene carboxylic acid. Chem Commun 3:138–140
52. Inouye H, Uesato S (1986) Biosynthesis of iridoids and secoiridoids. Prog Chem Org Nat Prod 50:169–236
53. Jensen SR (1991) Plant iridoids, their biosynthesis and distribution in angiosperms. In: Harborne JB, Tomas-Barbaran FA (eds) Ecological chemistry and biochemistry of plant terpenoids. Clarendon, Oxford, UK, pp 133–158
54. Jensen SR, Schripsema J (2002) Chemotaxonomy and pharmacology of *Gentianaceae*. In: Struve L, Albert V (eds) *Gentianaceae* – systematics and natural history. Cambridge University Press, Cambridge, UK, pp 573–631
55. Van der Sluis WG, Van der Nat JM, Spek ÁL, Ikeshiro Y, Labadie RP (1983) Secoiridoids and xanthones in the genus *Centaurium*. Part VI: gentiogenal, a conversion product of gentiopicrin (gentiopicroside). Planta Med 49:211–215
56. Bhattacharya SK, Reddy PKSP, Ghosal S, Singh AK, Sharma P (1976) Chemical constituents of *Gentianaceae*. Indian J Pharm Sci 65:1547–1549
57. Liang J, Han D, Li H, Yuan X (1982) Isolation and identification of swertiamarin, the active principle in *Swertia patens* Burk. Yaoxue Tongbao 17:242–243
58. Yamamura J, Kobayashi M, Matsuda H, Aoki S (1991) Anticholinergic action of *Swertia japonica* and an active constituent. J Ethnopharmacol 33:31–35
59. Zhou J (1991) Bioactive glycosides from Chinese medicines. Mem Inst Oswaldo Cruz Rio J 86:231–234
60. Kondo Y, Takano F, Hojo H (1994) Suppression of chemically and immunologically induced hepatic injuries by gentiopicroside in mice. Planta Med 60:414–416
61. Kumarasamy Y, Nahar L, Cox PJ, Jaspars M, Sarker SD (2003) Bioactivity of secoiridoid glycosides from *Centaurium erythraea*. Phytomedicine 10:344–347
62. Kumarasamy Y, Nahar L, Sarker SD (2003) Bioactivity of gentiopicroside from the aerial parts of *Centaurium erythraea*. Fitoterapia 74:151–154
63. Martindale (1982) Bitters. In: Reynolda JEF (ed) The extra pharmacopoeia, 28th edn. The Pharmaceutical Press, London
64. Haloui M, Louedec L, Michel JB, Lyoussi B (2000) Experimental diuretic effect of *Rosmarinus officinalis* and *Centaurium erythraea*. J Ethnopharmacol 71:465–472
65. Mandal S, Das PC, Joshi PC (1992) Naturally occurring xanthones from terrestrial flora. J Indian Chem Soc 69:611–636
66. Abd el Mawla AMA, Schmidt W, Beerhues L (2001) Cinnamic acid is a precursor of benzoic acids in cell cultures of *Hypericum androsaemum* L. but not in cell cultures of *Centaurium erythraea* Rafn. Planta 212:288–293
67. Wang C-Z, Maier UH, Keil M, Zenk MH, Bacher A, Rondich F, Eisenreich W (2003) Phenylalanine-independent biosynthesis of 1,3,5,8-tetrahydroxyxanthone. Eur J Biochem 270:2950–2958
68. Van der Sluis WG (1985) Secoiridoids and xanthones in the genus *Centaurium* Hill *(Gentianaceae)* – a pharmacognostical study. Drukkerij Elinkwijk bv, Utrecht
69. Van der Sluis WG (1985) Chemotaxonomical investigations of the Genera *Blackstonia* and *Centaurium* (*Gentianaceae*). Plant Syst Evol 149:253–286
70. Janković T, Krstić D, Šavikin-Fodulović K, Menković N, Grubišić D (2000) Xanthone compounds of *Centaurium erythraea* grown in nature and cultured in vitro. Pharm Pharmacol Lett 10:23–25
71. Valentão P, Andrade PB, Silva E, Vincente A, Santos H, Bastos ML, Seabra R (2002) Methoxylated xanthones in the quality control of small centaury (*Centaurium erythraea*) flowering tops. J Agric Food Chem 50:460–463
72. Krstić D, Janković T, Šavikin-Fodulović K, Menković N, Grubišić D (2003) Secoiridoids and xanthones in the shoots and roots of *Centaurium pulchellum* cultured in vitro. In Vitro Cell Dev Biol Plant 39:203–207

73. Jiang DJ, Dai Z, Li YJ (2004) Pharmacological effects of xanthones as cardiovascular protective agents. Cardiovasc Drug Rev 22:91–102
74. Saxena S, Pant S, Jain DC, Bhakuni RS (2003) Antimalarial agents from plant sources. Curr Sci 85:1314–1329
75. Rukachaisirikul V, Kamkaew M, Sukavisit D, Phongpaichit S, Sawangchote P, Taylor WC (2003) Antibacterial Xanthones from the Leaves of *Garcinia nigrolineata*. J Nat Prod 66:1531–1535
76. Pinto MMM, Sousa ME, Nascimento MSJ (2005) Xanthone derivatives: new insights in biological activities. Curr Med Chem 12:2517–2538
77. Schaufelberger D, Hostettmann K (1988) Chemistry and Pharmacology of *Gentiana lutea*. Planta Med 54:219–221
78. Moureau F, Wouters J, Vercauteren DP, Colin S, Edrvard G, Durant F, Ducrey F, Koenig J, Jarreau FX (1994) A reversible monoamine oxidase inhibitor, toloxatone: spectrophotometric and molecular orbital studies of the interaction with flavin adenine dinucleotide (FAD). Eur J Med Chem 29:269–277
79. Wang JN, Hou CY, Liu YL (1994) Swertia francheside, an HIV-reverse transcriptase inhibitor and the first flavone-xanthone dimer, from *Swertia franchetiana*. J Nat Prod 57:211–217
80. Groweiss A, Cardellina JH, Boyd MR (2000) HIV-Inhibitory Prenylated Xanthones and Flavones from *Maclura tinctoria*. J Nat Prod 63:1537–1539
81. Hostettmann K, Hostettmann M (1989) Xanthones. In: Dey PM, Harbone JB (eds) Methods in plant biochemistry, vol 1 – plant phenolics. Academic Press, London, San Diego, New York, Berkeley, Boston, Sydney, Tokyo, Toronto, pp 493–508
82. Schimmer O, Mauthner H (1996) Polymethoxylated xanthones from the herb of *Centaurium erythraea* with strong antimutagenic properties in *Salmonella typhimurium*. Planta Med 62:561–564
83. Zafar R, Aeri V, Datta A (1992) Application of plant tissue and cell culture for production of secondary metabolites. Fitoterapia 63:33–43
84. Janković T, Krstić D, Šavikin-Fodulović K, Menković N, Grubišić D (1997) Comparative investigation of secoiridoid compounds of *Centaurium erythraea* grown in nature and cultured in vitro. Pharm Pharmacol Lett 7:30–32
85. Piatczak E, Wielanek M, Wysokinska H (2005) Liquid culture system for shoot multiplication and secoiridoid production in micropropagated plants of *Centaurium erythraea* Rafn. Plant Sci 168:431–437
86. Mulabagal V, Tsay HS (2004) Plant cell cultures – an alternative and efficient source for the production of biologically important secondary metabolites. Int J Appl Sci Eng 2:29–48
87. Janković T, Krstić D, Šavikin-Fodulović K, Menković N, Grubišić D (2002) Xanthones and secoiridoids from hairy root cultures of *Centaurium erythraea* and *C. pulchellum*. Planta Med 68:944–946
88. Werner T, Motyka V, Strnad M, Schmülling T (2001) Regulation of plant growth by cytokinin. Proc Natl Acad Sci U S A 98:10487–10492
89. Murashige T, Skoog F (1962) A revised medium for rapid growth and bioassays with tobacco tissue cultures. Physiol Plant 15:473–479
90. Podolski-Renić A, Anđelković T, Banković J, Tanić N, Ruždijić S, Pešić M (2011) The role of paclitaxel in the development and treatment of multidrug resistant cancer cell lines. Biomed Pharmacother 65:345–353
91. Trifunović M, Motyka V, Cingel A, Subotić A, Jevremović S, Petrić M, Holík J, Malbeck J, Dobrev PI, Dragićević IČ (2015) Changes in cytokinin content and altered cytokinin homeostasis in *AtCKX1* and *AtCKX2*-overexpressing centaury (*Centaurium erythraea* Rafn.) plants grown in vitro. Plant Cell Tiss Org Cult 120:767–777

Syzygium cumini (L.) Skeels: Cardiometabolic Properties and Potential Tissue Culture-Based Improvement of Secondary Metabolites Production

9

Sonia Malik, Eduardo Bezerra Almeida Jr., and Antonio Marcus de Andrade Paes

Abstract

Syzygium cumini (L.) Skeels (family Myrtaceae), commonly known as jambolão, jambolan, or jamun, has been suggested as a potential source of bioactive molecules against diabetes and associated cardiometabolic diseases. A wide variety of secondary compounds, mainly, terpenes, and phenolic compounds, such as phenolic acids, flavonoids, and tannins, are present in different parts of this plant species. This chapter describes about the various pharmacological properties of *S. cumini*, including antihyperglycemic, antihyperlipidemic, anti-inflammatory, and antioxidant activities, which make it a very interesting species for multitarget therapeutic purposes. Geographical distribution, botanical description, as well as potential of this plant species for in vitro culture have been discussed. Review of literature shows that despite the recalcitrant nature of this plant species, attempts have been made to standardize the protocol for its

S. Malik (✉)
Graduate Program in Health Sciences, Biological and Health Sciences Center, Federal University of Maranhão, São Luís, MA, Brazil
e-mail: 777soniamalik@gmail.com; sonia.unicamp@gmail.com

E.B. Almeida Jr.
Graduate Program in Health Sciences, Biological and Health Sciences Center, Federal University of Maranhão, São Luís, MA, Brazil

Laboratory of Botanical Studies, Department of Biology, Biological and Health Sciences Center, Federal University of Maranhão, São Luís, MA, Brazil
e-mail: ebaj25@gmail.com

A.M. de Andrade Paes
Graduate Program in Health Sciences, Biological and Health Sciences Center, Federal University of Maranhão, São Luís, MA, Brazil

Laboratory of Experimental Physiology, Department of Physiological Sciences, Biological and Health Sciences Center, Federal University of Maranhão, São Luís, MA, Brazil
e-mail: antonio.marcus@ufma.br

S. Jha (ed.), *Transgenesis and Secondary Metabolism*, Reference Series in Phytochemistry, DOI 10.1007/978-3-319-28669-3_9

micropropagation. Although cells and tissues of *S. cumini* have capacity for in vitro production of bioactive compounds, but basic studies for their mechanisms of production as well as full understanding of biosynthesis pathways are required to be known to exploit the potential of this plant species. Further research towards employing novel elicitors, two-phase culture system, and metabolic engineering may also help in improving the performance of *S. cumini* under in vitro culture system.

Keywords

Antihyperglycemic • Anti-inflammatory • Antioxidant • Bioactive compounds • Jamun • In vitro culture • Medicinal plant • Metabolic syndrome • Micropropagation • Myrtaceae • Pharmacology

Contents

1 Introduction

Myrtaceae is a monophyletic family from Myrtales order which comprises about 132 genera and over 5,600 species mainly composed of trees and shrubs [1]. In Brazil, it comprises of about 24 genera and 927 identified species [2]. *Syzygium cumini* (L.) Skeels, formerly classified as *Eugenia jambolana* Lam., is one of the important plant species possessing highest pharmacological and economic potential. This plant species has been proposed as a potential source of bioactive molecules against diabetes and its associated cardiometabolic diseases [3–6]. *S. cumini* is native from Indian subcontinent but widely distributed in tropical and subtropical areas (Fig. 1), where it may be either categorized as an exotic [7] or naturalized [2] species. It is popularly known as jamun or jambul in India, Indian black berry in England, faux pistachier in France, black plum in the United States, jambolan in Spanish-spoken countries, and jambolão in Brazil [8].

Seeds of *S. cumini* have been used for centuries by Indian Ayurvedic medicine practitioners for diverse therapeutic purposes [9]. However, it was only during the

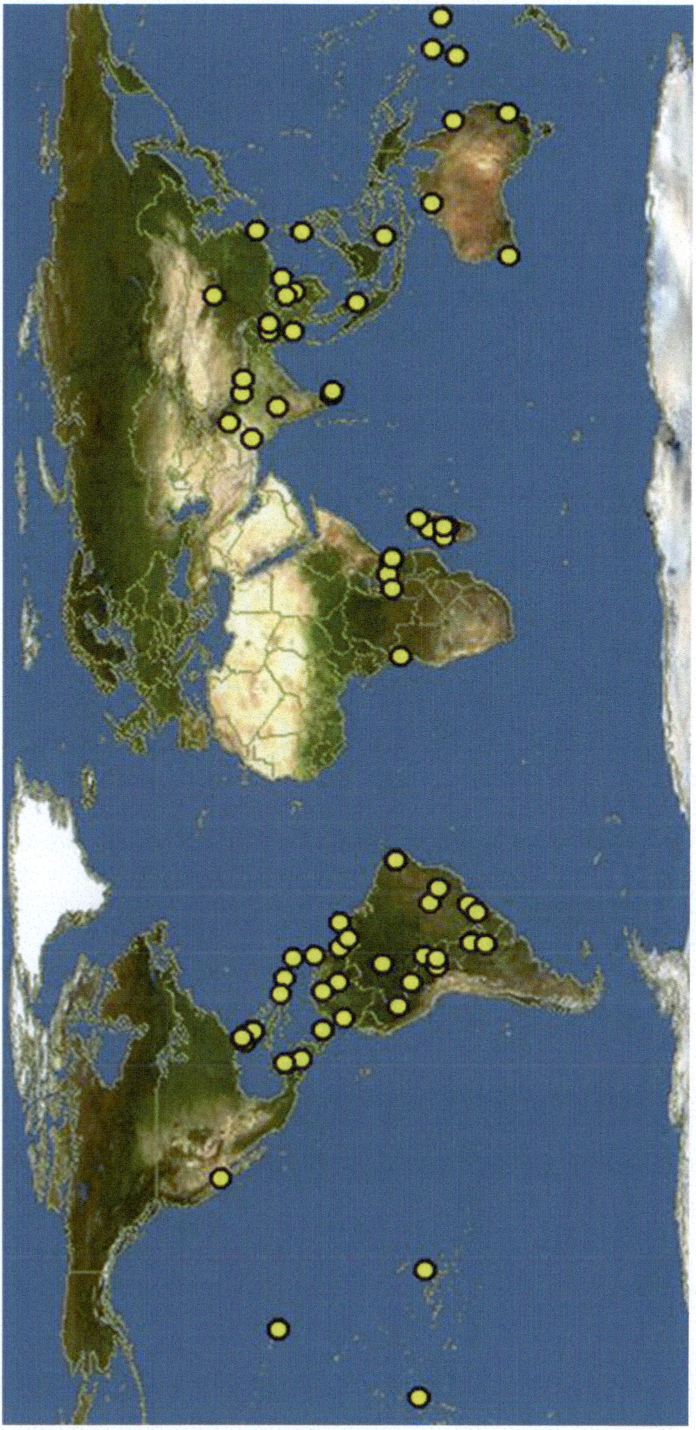

Fig. 1 Phytogeographic domains of *Syzygium cumini*. Illustration showing tropical and subtropical distribution of *S. cumini* (Source: US National Plant Germplasm System: GRIN Taxonomy. Accessed via http://www.gbif.org/species/101390909 on 2016-03-24)

nineteenth century that British traders imported it to Europe because of its unique property to decrease urine sugar content in diabetic patients. The first morphological, phytochemical, pharmacological, and clinical reports on the species and its antidiabetic effect were published in the 1880s and readily attracted the attention of scientific community. As a result the use of jambolan seeds had already been included in the therapeutic and phytotherapeutic standard literature by the year 1990 [6]. Ever since, *S. cumini* became a worldwide cultivated medicinal plant. Currently, most of the knowledge on the pharmacological properties of *S. cumini* has come from experimental studies, with very limited data coming from clinical trials. On the other hand, clinical trials demand large amounts of extracts or its isolated and identified secondary metabolites, whose obtention would be considered extremely laborious or ecologically unsustainable.

Biotechnological methods by employing plant cell, tissue, and organ culture techniques offer an important alternative for micropropagation as well as in vitro production of bioactive molecules of interest without harvesting the plants from nature [10–14]. In *S. cumini*, micropropagation technique has been employed for mass multiplication. The other advantage of using in vitro method is to obtain plants throughout the year without any risk of disease. However, there are several problems associated with standardization of protocol for in vitro establishment and culture of tree species like *S. cumini.* This chapter will address the difficulties in raising in vitro cultures in *S. cumini* as well as future challenges. The chapter commences with the short description of plant species, followed by pharmacological properties and tissue culture potential.

2 Botany

S. cumini presents as trees ranging from 10 to 15 m high; evergreen, often twisted trunk, branches (Fig. 2a). Leaves simple, opposite, shortly petiolate, as petiole 16–23 mm length; leaf blade from 5.5 to 13 cm length × 3.5–5.5 cm width, elliptic or elliptic-lanceolate, apex acuminate, base attenuate, margin entire, prominent midrib on the abaxial and furrowed face on the adaxial face, secondary veins 35–45, texture cartácea type, discolor (each side of the leaf exhibits a different shade), venation pattern brochidodromous, numerous oleíferos channels present in the form of translucent glands (Fig. 2c). Inflorescence in panicles or panicles of cyme, 31–59 mm length, terminal, and extra-axillary lateral; bracteoles 0.5–0.6 mm length × 0.3–0.4 mm width, elliptic, caducous before or after anthesis. Sessile flowers; bisexual, actinomorphic, dichlamydeous, sepals with indistinct lobes, persistent calyx tube; petals 4–5, measuring 2.5–3.5 mm length; stamens numerous, free each other, whitish; ovary inferior, two-locular, plain, stiletto 5–5.5 mm length (Fig. 2b). Drupaceous fruit, ellipsoid, 13–17 mm length × 7–8 mm width, plain, varying between red and black (Fig. 2d). Seed ovoid, about 1.5 cm length (Fig. 2e). Description is based on the following examined material: São Luís, Maranhão, Brazil, Campus da UFMA, fl., E.B. Almeida Jr. & I.F. Amorim 1094, 11/IV/2014 (MAR 4574); Campus da UFMA, fl. fr., A. Kely & E.B. Almeida Jr. 01, 10/III/2014 (MAR 4694); Campus da UFMA, próximo ao prédio de anatomia, fl., E.B. Almeida

Fig. 2 Botanical aspects of *Syzygium cumini* (L.) Skeels: (**a**) tree picture, (**b**) detail of flower and stamens, (**c**) leaves, (**d**) fruit (color varying between *red* and *black*), (**e**) open fruit with detail for seed (Figures adapted: **a**, **b**, Mercadante, 2016; **c**, Upcavage, 2016; **d**, **e**, Santos 2015)

Fig. 3 Specimen of *Syzygium cumini* (L.) Skeels: Specimen recorded in Herbarium of Maranhão (*MAR*)

Jr. & A.N. Silva 1093, 11/IV/2014 (MAR 4573); and Campus Bacanga-CCBS, fl., D.N de. Melo 01, 09/V/2014 (MAR 7556) (Fig. 3).

3 Phytochemistry

S. cumini presents a wide variety of secondary metabolites thoroughly distributed on its different parts, from roots to fruits, mainly consisting of terpenes and phenolic compounds, such as phenolic acids, flavonoids, and tannins. Early assessment of its phytochemical profile, performed in late nineteenth century, led authors to propose jambulin, a yellow and unstable glucoside extracted from its seed, as responsible for the already well-known glucose-lowering effects of jambolan. Half a century later, Wastl and coworkers stated that no substance of glycosidic nature could be found in the plant even though jambolan's fame remained [6]. Table 1 summarizes the main compounds already identified in different parts of *S. cumini*. The leaf contains high levels of flavonoids, especially quercetin and myricetin derivatives, besides simple phenols like ellagic acid, ferulic acid, and gallic acid. The leaf presents a terpene-rich essential oil, which contains α-cadinol, pinocarvone, and pinocarveol. A number of

Table 1 Phytochemical compounds identified in different parts of *Syzygium cumini* tree

Part	Class	Compounds	References
Leaf	Flavonoids	Catechin, (*epi*)gallocatechin-(*epi*)gallocatechin-*O*-gallate, kaempferol, myricetin, myricetin deoxyhexoside, acetylated myricetin deoxyhexoside, acetylated methylmyricetin deoxyhexoside, myricetin 3-*O*-β-D-glucuronopyranoside, myricetin-4′-methyl ether 3-*O*-α-rhamnopyranoside, myricetrin 4″-*O*-acetate, myricetrin 4″-*O*-acetyl-2-*O*-gallate, myricitrin, quercetin-3-*O*-α-rhamno_ pyranoside	[17, 18]
	Phenolic acids	Caffeic acid, chlorogenic acid, ellagic acid, ferulic acid, gallic acid	[17, 19, 20]
	Tannins	Nilocitin, HHDP-glucose, pedunculagin I, casuarinin, trigalloylglucose, tetragalloylglucose, pentagalloylglucose	[17, 18]
	Terpenes	α-Pinene, α-cadinol, pinocarvone, pinocarveol, α-terpineol, myrtenol, eucarvone, muurolol, myrtenal, cineole, geranylacetone	[21, 22]
Seed	Flavonoids	Quercetin, rutin, 3,5,7,4′-tetrahydroxy flavanone	[23, 24]
	Phenolic acids	Caffeic acid, ellagic acid, ferulic acid, gallic acid	[23]
	Tannins	Corilagin, 3,6-HHDP-glucose, 4,6-HHDP-glucose, 1-galloylglucose, 3-galloyl glucose	[23]
	Terpenes	α-Terpineol, β-pinene, β-terpinene, betulinic acid, eugenol	[24, 25]
Fruit	Flavonoids	Myricetin, myricetin deoxyhexoside	[26]
	Phenolic acids	Ellagic acid, gallic acid	[26, 27]
	Tannins	HHDP-galloylglucose, trigalloylglucose	[26]
	Terpenes	Citronellol, geraniol, hotrienol, nerol, β-phenylethanol, phenylpropanal	[28]
	Anthocyanins	Cyanidin, delphinidin, petunidin	[29]
Flower	Flavonoids	Kaempferol, myricetin, dihydromyricetin, myricetin-3-L-arabinoside, isoquercetin, quercetin, quercetin-3-D-galactoside	[30]
	Phenolic acids	Ellagic acid	[31]
	Terpenes	Eugenol, oleanolic acid	[16]
Bark	Flavonoids	Myricetin, quercetin, kaempferol	[31]
	Phenolic acids	3,3′-di-*O*-methyl ellagic acid, 3,3′, 4-tri-*O*-methyl ellagic acid, gallic acid	[16, 23]
	Terpenes	β-Sitosterol, friedelin, betulinic acid	[16]

hydrolyzable tannins have also been identified in leaves, mainly consisting of galloyl monosaccharides, which resulted from esterification of glucose by three to five gallic acid units. Interestingly, it has been shown that the number of galloyl units determine antioxidant capacity of galloylglucose isomers [15]. The flower and seed show similar phytochemical profile, whereas there is some scarcity of flowers' studies. The seed particularly contains hydrolyzable tannins, terpenes, and phenolic acids.

The fruit is rich in anthocyanins such as cyanidin, delphinidin, and petunidin, which give it a typical bright purple color. The stem and barks essentially contain the same phenolic profile found in the leaves [3, 9, 16]. From all phytochemical classes mentioned above, (poly)phenolic compounds are considered as the most prominent compounds responsible for the pharmacological properties described in the next sections.

4 Pharmacological Properties

Traditional practitioners of Indian Ayurveda medicine have applied *S. cumini*-based remedies for countless illnesses over the centuries [9]. Its particular capacity to decrease urine sugar content in diabetic patients has consistently impelled scientific researchers to investigate its antidiabetic and metabolism-related properties since the late nineteenth century, even before the discovery of insulin [6]. Ever since, a number of in vitro, in vivo, and even clinical studies have validated multiple pharmacological properties for jambolan, as described in the following sections.

4.1 Antihyperglycemic Activity

Blood glucose-lowering activity of *S. cumini* has been assessed with extracts prepared from its different parts, although the spotlight has been mainly put on the seed. Administration of ethanolic extract of seed kernel (100 mg/kg/day) for 30 days to streptozotocin-induced diabetic rats decreased both glycemia and glycosuria [32]. Similarly, flavonoid-rich extract of seed further restored peripheral glucose tolerance in the same animal model, an effect ascribed to adipocytic activation of peroxisome proliferator-activated receptors (PPAR) alpha and gamma [33, 34], whose downstream pathways control fatty acid oxidation, adipocyte differentiation, and insulin sensitivity (Fig. 4). Quercetin, a flavonoid importantly found in jambolan seeds, has also been shown to improve adipocyte glucose uptake via PPAR gamma upregulation [35]. On the other hand, glucose-lowering effect of seed and bark extracts has also been attributed to direct stimulus of insulin release from pancreatic beta cells of rats, a property further extended to leaf extracts [3, 36] (Fig. 4). Among flavonoids found in *S. cumini*, quercetin and quercetin-3-rutinoside are able to improve insulin release through stimulation of calcium influx [37, 38]. More recently, Sanches and coworkers [18] demonstrated that a polyphenol-rich extract from jambolan leaves improved metabolic status of monosodium L-glutamate-induced obese rats by dual effect on peripheral insulin sensitivity and insulin release modulation (Fig. 4). These effects were mainly ascribed to myricetin and its derivatives, which have been shown to improve insulin signaling pathways in skeletal muscles [39] and adipocytes [40], as well as to protect pancreatic beta cells from cytokine-induced cell death [41].

Besides the aforementioned mechanisms, other insulin-dependent effects have been shown for jambolan extracts. Methanol extract of leaf was seen to elevate

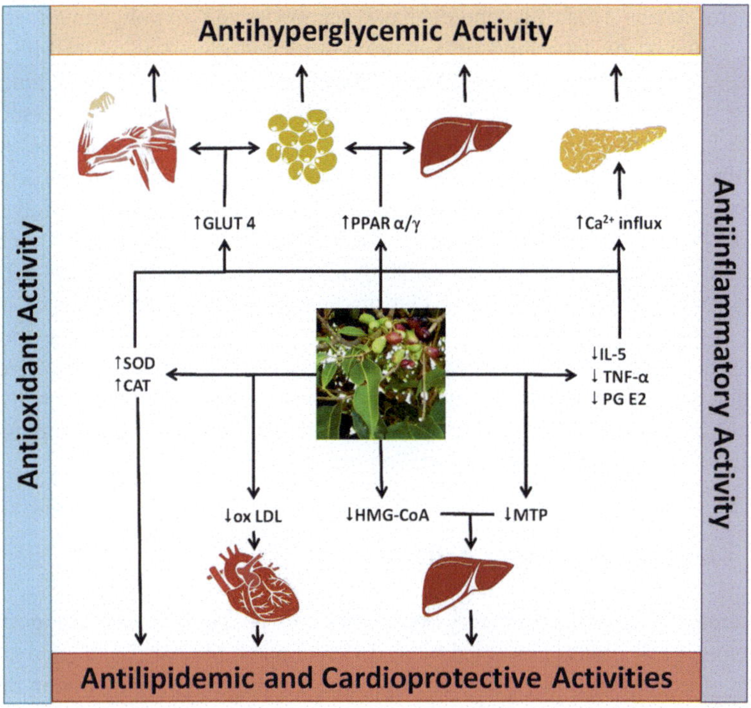

Fig. 4 Main pharmacological activities of *Syzygium cumini*. Extracts prepared from different parts of *S. cumini* have been shown to contain compounds with antihyperglycemic, antihyperlipidemic, anti-inflammatory, and antioxidant activities. Abbreviations: *SOD* superoxide dismutase, *CAT* catalase, *ox LDL* oxidized LDL, *HMG-CoA reductase* 3-hydroxy-3-methyl-glutaryl-CoA reductase, *MTP* microsomal triglyceride transfer protein, *GLUT4* glucose transporter 4, *IL-5* interleukin-5, *TNF-α* tumor necrosis factor-alpha, *PGE2* prostaglandin E2, *PPAR* peroxisome proliferator-activated receptors

mRNA expression of glucose transporter 4 and phosphatidylinositol-3 kinase, important mediators of insulin actions on adipocytes and skeletal muscle [42]. Moreover, there are studies demonstrating that *S. cumini* increases hepatic and muscular glycogen stores, as well as inhibits pancreatic alpha-amylase activity. As shown in Table 1, *S. cumini* leaf contains (*epi*)gallocatechins, important proanthocyanidins with inhibitory activity against alpha-amylase, a particularly interesting property for diabetes treatment [43]. Both properties directly impact carbohydrate metabolism and further support the use of this plant species for diabetes control.

4.2 Antihyperlipidemic Activity

Although less extensive than antidiabetic, studies on antilipidemic properties of *S. cumini* have demonstrated that flavonoid-rich extract of its seeds reduces blood levels of total cholesterol, LDL cholesterol, and triglycerides while increasing HDL cholesterol levels [33]. Flavonoids have acknowledged hypolipemiant activity mainly achieved by inhibition of 3-hydroxy-3-methyl-glutaryl-CoA (HMG-CoA) reductase, a key enzyme on cholesterol biosynthesis [44] (Fig. 4). Such ability to decrease serum lipoprotein levels also gives jambolan a cardioprotective-associated activity. In fact, α-hydroxy succinamic acid, a compound isolated from *S. cumini* fruit pulp aqueous extract, was able to improve endothelial dysfunction markers, like oxidized LDL, as a consequence of lowering serum lipoproteins and triglyceride levels associated to anti-inflammatory effects [45]. Moreover, pancreatic lipase inhibition has been considered the mechanism by which *S. cumini* extracts decrease gut absorption of triglycerides [46]. Our research group has been interested on the effects of polyphenol-rich extract of jambolan's leaves on triglyceride metabolism. Unpublished data shows that this extract restores normal blood triglyceride levels by inhibiting both expression and activity of the hepatic isoform of microsomal triglyceride transfer protein (MTP), which is responsible for very low density lipoprotein (VLDL) particle assembly and exportation into the bloodstream (Fig. 4).

4.3 Anti-inflammatory Activity

Assessment of anti-inflammatory properties of *S. cumini* has demonstrated antiedematogenic effects promoted by its different parts. Methanol and ethyl acetate extracts of leaf, as well as ethanolic extract of bark, reduced edema volume in both acute [47, 48] and chronic [49] inflammatory animal models. These effects have been attributed to inhibition of leukocyte migration consequent to lower production of secondary mediators like prostaglandin E2, serotonin, and histamine. Furthermore, *S. cumini* has also proven to be effective against allergies [50] and lethal sepsis [51] through modulation of inflammatory versus anti-inflammatory cytokine production, such as interleukin-5 and tumor necrosis factor-alpha (Fig. 4).

4.4 Antimicrobial Activity

Hydroethanolic extract of *S. cumini* leaf has been shown to be active against *Candida krusei* and multiresistant bacteria like *Pseudomonas aeruginosa*, *Klebsiella pneumoniae*, and *Staphylococcus aureus*. Besides leaves, aqueous and methanolic extracts of seed have also been shown to be effective against Gram-positive (*Bacillus subtilis* and *Staphylococcus aureus*) and Gram-negative (*Salmonella typhimurium*, *Pseudomonas aeruginosa*, *Klebsiella pneumoniae*, and *Escherichia coli*) bacteria, as well as fungi, which include *Candida albicans*, *Aspergillus flavus*, *Aspergillus fumigatus*, and *Aspergillus niger* [52]. Such antimicrobial activity has been ascribed to tannins and other phenolic compounds present in the extracts [53].

4.5 Antioxidant and Antigenotoxic Activities

Antioxidant capacity of *S. cumini* has been vastly assessed in both in vitro and in vivo studies, which have given it a role of prominent source of cardiometabolic bioactive compounds [3]. Considering in vitro studies, methanol extract of *S. cumini* leaf and branch has been shown to scavenge $OH^{\bullet}$ and $DPPH^{\bullet}$·radicals and decrease Fe^{3+} reduction in the FRAP assay in straight correlation with the high content of polyphenols and flavonoids present in the extract [54]. Similarly, antioxidant activity of ethyl acetate fraction prepared from methanolic extract of *S. cumini* leaf was also correlated with its polyphenolic composition, especially to ferulic acid and catechins present in the extract [19]. Finally, ethanolic extracts of fruit and seed, at different concentrations, showed strong antioxidant activity assessed by mean of distinct assays, such as $ABTS^{\bullet+}$, $DPPH^{\bullet}$, FRAP, and ORAC [55]. Noteworthy, $ABTS^{\bullet+}$ method has been shown to correlate with other antioxidant assays that simulate physiological radical reactions, reinforcing the potential activity of polyphenols and flavonoids contained in *S. cumini* [56].

In vivo studies have further reinforced the abovementioned findings. Pretreatment of jambolan extracts was found to increase the activity of antioxidant enzymes such as superoxide dismutase and catalase in animals challenged with oxidative stressors like urethane, 7,12-dimethylbenzanthracene [57], or cyclophosphamide [58] (Fig. 4). Additionally, these treatments also reduced oxidative chromosomal damage, which enable *S. cumini* extract for antigenotoxic purposes, an activity further supported by findings that seed extract was showed to be effective against implantation and proliferation of benzopyrene-induced gastric carcinomas [59]. Noteworthy, the plethoric polyphenolic composition of *S. cumini* is thought to underlie not only its antioxidant capacity but virtually all the pharmacological properties herein discussed, which makes it a very interesting species for multitarget therapeutic purposes.

5 Tissue Culture

The importance of tissue culture approach for the propagation and tree improvement in woody plant species has already been known and described well in literature [60, 61]. It is generally not easy to culture the explants derived from mature trees due to number of factors including recalcitrant nature, microbial contamination, phenolic exudation, limitation of episodic growth patterns, vitrification, and difficulty in root induction [62, 63]. There are not many reports of in vitro studies in *S. cumini* probably because of abovementioned problems in this woody plant species. However, attempts have been made by several researchers to propagate *S. cumini* under in vitro conditions for mass multiplication and to obtain plants at any time of the year without the risk of disease. In 1990, Yadav and his coworkers described the procedure for in vitro propagation from seedling-derived explants in *S. cumini* [64]. Due to difficulty in surface sterilization of explants from field-grown trees, in vitro raised seedlings were used for the micropropagation and regeneration studies [64–66]. Different explants, viz., epicotyl, nodal segments, shoot tips, zygotic embryos, leaves, and roots, have been reported to be used for in vitro studies of this plant species (Table 2) [65, 67]. In vitro propagation depends on the physiological state of explants as nodal explants were found to be more responsive than shoot tips [64]. Jain and Babber [65] reported the recurrent production (an average of 8.6 shoots/explant) of *S. cumini* from epicotyl segments bearing scaly leaves (nodes) cultured on Murashige and Skoog (MS) [68] medium supplemented with 6-benzylaminopurine (BAP; 4.4 μM). To see the development of strong root system, three different auxins, viz., α-naphthaleneacetic acid (NAA), indole-3-acetic acid (IAA), and indole-3-butyric acid (IBA), at 0.2–1 mg/l concentration were added to Knop's medium with 2% sucrose. The best rhizogenic response was seen on medium fortified with 1 mg/l IAA [67]. The rooted plantlets have been reported to transfer to soil and then to fields after an acclimatization of 7–8 months. The plants were found to thrive well for more than 3 years without any apparent phenotypic aberrations. The authors reported the development of an efficient protocol to raise plants of *S. cumini* at any time of the year.

Later in 2003, Jain and Babbar [67] tried to develop a protocol for in vitro regeneration of plants from explants derived from mature trees of *S. cumini* with the aim that for micropropagation, it is preferable to use explants from mature plants where superior characteristics are evident. Nodal explants from mature trees of *S. cumini* cultured on BAP (1 mg/l)-supplemented medium induced only the greening and opening of incipient shoot buds, but for further elongation and normal development of these buds, an additional supply of reduced nitrogen was required which was not required in case of explants taken from in vitro grown seedlings [67]. Therefore, MS medium supplemented with BAP (1 mg/l), casein hydrolysate (1.5 g/l), or glutamine (200 mg/l) was devised for shoot initiation and multiplication from ex vitro plants of *S. cumini*. For rooting, these in vitro shoots were transferred to Knop's medium containing NAA (2 mg/l).

Different cytokines, viz., 6-benzylaminopurine, kinetin, thidiazuron (TDZ), and 2-iP, have been tried either alone or in combination with an auxin NAA for optimum shoot multiplication of this plant species. BAP (7.5 shoots per node after 6 weeks of culture) was reported to be the most suitable cytokine followed by Kn and 2-iP for the shoot multiplication from axillary buds in this plant species [69]. The best rooting was reported on half-strength MS medium with NAA (0.5 μM) [69].

There are reports on indirect shoot regeneration and somatic embryos from callus in *S. cumini* [70, 71] (Iyer and Gopinath 2000; Yadav et al. 2014). Callus obtained from seedling explants produced shoot buds on MS medium with BAP (0.5–2.0 mg/l) singly or in combination with NAA (0.05–0.2 mg/l). About 85% of shoot buds were reported to form shoots and subsequently roots on medium containing BAP (1 mg/l) + NAA (0.05 mg/l) and IBA (1 mg/l) or NAA (1 mg/l) [70].

Table 2 Use of basal medium, plant growth regulators, additives, and response of different explants in *S. cumini* under in vitro conditions

Explant used	Culture medium[a] + sucrose (%)	Plant growth regulators[b]	Additives	Response	Reference (s)
Shoot tips, nodal segments from in vitro raised plants	MS +3	BAP (4.5 μM) or BAP (1.12 μM) + NAA (0.25 μM)	–	Shoot proliferation	[64]
In vitro shoots	MS +3	IBA (2.5, 5.0 μM) or NAA (2.7, 1.14 μM)	–	Root initiation	
In vitro shoots	MS +3	–	–	Root elongation	
Epicotyl segments from in vitro grown shoots	MS +4	BAP (1 mg/l)	–	Multiple shoot formation	[65]
In vitro grown shoots	Knop's +2	IAA (1 mg/l)	–	rooting	[65, 67]
Nodal or microcuttings from mature trees	MS +4	BAP (1 mg/l)	Casein hydrolysate (1.5 g/l), glutamine (200 mg/l)	Shoot bud elongation	
Seeds	Knop's + 1	–	–	Seed germination	[71]
Zygotic embryos	MS +3	–	Activated charcoal (0.25%)	Plantlet formation	
In vitro shoots	MS +7	IBA (4 mg/l)	–	Root initiation	
In vitro shoots	MS +10	IBA (4 mg/l)	–	Root elongation	
In vitro shoots	MS +3	2,4-D (5 mg/l) + NAA (1 mg/l)	Coconut milk (15%)	Callus formation	
Callus	MS +10	IBA (4 mg/l)	–	Somatic embryos	
Nodal segments from in vitro raised plants	1/2MS +3	BAP (4.4 μM)	–	Multiple shoots	[69]
In vitro shoots	1/2MS + 3	NAA (0.5 μM)	–	Root formation	

(*continued*)

Table 2 (continued)

Explant used	Culture medium[a] + sucrose (%)	Plant growth regulators[b]	Additives	Response	Reference (s)
In vitro seedlings	MS + 3	2,4-D (0.05–1.0 mg/l) + BAP (0.5–2.0 mg/l)	–	Callus formation	[70]
In vitro seedling-derived callus	MS + 3	BAP (1.0 mg/l) + NAA (0.05 mg/l)		Shoot regeneration	
In vitro seedling-derived callus	MS + 3	IBA (1.0 mg/l) or NAA (1.0 mg/l)		Root initiation	
In vitro seedling-derived callus	MS + 3	–		Root elongation	

[a]*MS* Murashige and Skoog [68], *Knop* Knop [73]
[b]*BAP* 6-benzylamino purine, *IAA* indole-3-acetic acid, *IBA* indole-3-butyric acid, *NAA* α-naphthaleneacetic acid, *2,4-D* 2,4-dichlorophenoxyacetic acid

In another report by Iyer and Gopinath [71], immature zygotic embryos were cultured on MS medium containing 3% sucrose and activated charcoal (0.25%) to obtain plantlets. Activated charcoal has been reported to use in in vitro studies of many plant species due to absorption of phenolic exudation and soil-like properties, which help in growth enhancement. Shoots when placed vertically on medium with enhanced concentration of sucrose (7%) and IBA (4 mg/l) were reported to form roots. On the other hand, callus was formed from shoots when placed horizontally on medium with 3% sucrose, 2,4-D (5 mgl), NAA (1 mg/l), and coconut milk (15%). Indirect somatic embryogenesis was obtained upon transfer of callus to medium with high concentration of sucrose (10%) and replacing the auxins 2,4-D and NAA with IBA (4 mg/l) [71].

The agar is the most frequently used gelling agent in tissue culture media due to its inertness, stability, and nontoxicity. However, the use of isabgol (the husk derived from the seeds of *Plantago ovata*) as a gelling agent has also been described for seed germination as well as shoot and root formation [72]. The efficiency and response of *S. cumini* cultures were compared with media gelled with agar or isabgol. The seeds cultured on agar or isabgol media showed the initiation of germination after 3 weeks of inoculation and growth, morphology, and the percentage of shoot as well as root formation were found to be almost similar without any significant difference in both the media [72].

6 Genetic Transformation Studies in *S. cumini* and Related Species of the Family Myrtaceae

Genetic transformation is an important advancement in biotechnology for introducing a gene of interest especially to tree and woody plant species without disturbing its genetic global organization. In family Myrtaceae, there are examples of genetic transformation studies in economically important plant species including *Eucalyptus tereticornis* Sm., *Eucalyptus camaldulensis* Dehnh, *Psidium guajava* L., and *Verticordia grandis* J.L. Drumm [74–77]. *Eucalyptus tereticornis* Sm. is one among the economically important plant species, mainly used for raising plantations due to its use as fuel wood and raw material for paper pulp. Precultured cotyledons and hypocotyls of this plant species were co-cultured with *A. tumefaciens* strain LBA 4404 harboring the binary vector pBI121 (containing *uid* A and *npt* II genes) to obtain transformed plants upon direct regeneration. Transgenic nature of plants was confirmed by polymerase chain reaction and southern hybridization [74]. In another report by Aggarwal et al. [75], a procedure for *A. tumefaciens* mediated T-DNA delivery into the tissues taken from selected *E. tereticornis* elite clones, and subsequent regeneration of transformed shoots has been developed, with the aim to introduce the useful properties in this commercially important tree species. Similarly agrobalistic method has been used for genetic transformation in *E. camaldulensis* Dehnh [78]. Stummer et al. [77] described a system for transformation and regeneration of transgenic *V. grandis* J.L. Drumm by using *A. rhizogenes*. In *S. cumini*, however, there are no reports on genetic transformation studies.

7 Conclusions

Literature studies show that *S. cumini* is an important multipurpose medicinal plant species possessing various pharmacological properties. These properties are attributed to the presence of various bioactive molecules. Despite therapeutic potential for a number of diseases, there is lack of research in understanding the mechanisms involved, clinical trials. There are scanty reports on establishment of micropropagation protocols due to recalcitrant nature of this tree species. However, in vitro studies depict that this plant species has potential for in vitro production of bioactive compounds of interest, but lack of fundamental research and poor understanding of biosynthesis pathways involved impede the genetic manipulation strategies. Further studies are required to be focused on proper understanding of pathways involved in bioactive compounds of interest. An increase in production of compounds of interest using cell culture technique would be helpful in obtaining compounds of interest in an eco-friendly way. The strategies like elicitation, two-phase culture system, immobilization, and metabolic engineering may help in bringing this medicinal plant species as one of the promising sources of herbal product to prevent several diseases.

Acknowledgments The authors are thankful to Foundation for the Support of Research, Scientific, and Technological Development of the State of Maranhão (FAPEMA), which has importantly funded the research on pharmacological properties of *S. cumini* through the grants #APP01128/10 and #APP00280/12.

References

1. Govaerts R, Sobral M, Ashton P, Barrie F (2008) World checklist of Myrtaceae. Royal Botanic Gardens, Richmond
2. Sobral MPC, Souza M, Mazine F, Lucas E (2016) Myrtaceae. In: List of species of the Brazilian Flora Available from: http://floradobrasil.jbrj.gov.br/jabot/floradobrasil/FB86017
3. Chagas VT, França LM, Malik S, Paes AMA (2015) *Syzygium cumini* (L.) skeels: a prominent source of bioactive molecules against cardiometabolic diseases. Front Pharmacol 6:1–8
4. Ayyanar MP, Subash-Babu IS (2013) *Syzygium cumini* (L.) Skeels., a novel therapeutic agent for diabetes: folk medicinal and pharmacological evidences. Complement Ther Med 21(3):232–243
5. Baliga MS, Fernandes S, Thilakchand KR, D'souza P, Rao S (2013) Scientific validation of the antidiabetic effects of *Syzygium jambolanum* DC (black plum), a traditional medicinal plant of India. J Altern Complement Med 19(3):191–197
6. Helmstadter A (2008) *Syzygium cumini* (L.) SKEELS (Myrtaceae) against diabetes – 125 years of research. Pharmazie 63(2):91–101
7. Lorenzi H (2003) Árvores exóticas no Brasil: madeireiras, ornamentais e aromáticas. Instituto Plantarum de Estudos da Flora, Nova Odessa
8. Corrêa MP, Penna LA, Florestal IBD (1984) Dicionário das plantas úteis do Brasil e das exóticas cultivadas. Ministério da Agricultura, Instituto Brasileiro de Desenvolvimento Florestal, Rio de Janeiro
9. Ayyanar M, Subash-Babu P (2012) *Syzygium cumini* (L.) Skeels: a review of its phytochemical constituents and traditional uses. Asian Pac J Trop Biomed 2(3):240–246

10. Malik S, Kumar R, Vats SK, Bhushan S, Sharma M, Ahuja PS (2009) Regeneration in *Rheum emodi* Wall.: a step towards conservation of an endangered medicinal plant species. Eng Life Sci 2:130–134
11. Malik S, Cusidó RM, Mirjalili MH, Moyano E, Palazón J, Bonfill M (2011) Production of the anticancer drug taxol in *Taxus baccata* suspension cultures: a review. Process Biochem 46(1):23–34
12. Malik S, Mirjalili MH, Fett-Neto AG, Mazzafera P, Bonfill M (2013) Living between two worlds: two-phase culture systems for producing plant secondary metabolites. Crit Rev Biotechnol 33(1):1–22
13. Malik S, Biba O, Gruz J, Arroo RRJ, Strnad M (2014) Biotechnological approaches for producing aryltetralin lignans from *Linum* species. Phytochem Rev 13:893–913
14. Malik S, Bhushan S, Sharma M, Ahuja PS (2014) Biotechnological approaches to the production of shikonins: a critical review with recent updates. Crit Rev Biotechnol 16:1–14
15. Ngoc TM, Hung TM, Thuong PT, Kim JC, Choi JS, Bae K, Hattori M, Choi CS, Lee JS, Min BS (2008) Antioxidative activities of galloyl glucopyranosides from the stem-bark of *Juglans mandshurica*. Biosci Biotechnol Biochem 72(8):2158–2163
16. Ramya SK, Neethirajan, Jayakumararaj R (2012) Profile of bioactive compounds in *Syzygium cumini*-a review. J Pharm Res 5(8):4548–4553.
17. Mahmoud II, Marzouk MS, Moharram FA, El-Gindi MR, Hassan AM (2001) Acylated flavonol glycosides from *Eugenia jambolana* leaves. Phytochemistry 58(8):1239–1244
18. Sanches JR, França LM, Chagas VT, Gaspar RS, Dos Santos KA, Gonçalves LM, Sloboda DM, Holloway AC, Dutra RP, Carneiro EM, Cappelli AP, Paes AM (2016) Polyphenol-rich extract of *Syzygium cumini* leaf dually improves peripheral insulin sensitivity and pancreatic islet function in monosodium L-glutamate-induced obese rats. Front Pharmacol 7:48
19. Ruan ZP, Zhang LL, Lin YM (2008) Evaluation of the antioxidant activity of *Syzygium cumini* leaves. Molecules 13(10):2545–2556
20. Timbola AK, Szpoganicz B, Branco A, Monache FD, Pizzolatti MG (2002) A new flavonol from leaves of *Eugenia jambolana*. Fitoterapia 73(2):174–176
21. Shafi PM, Rosamma MK, Jamil K, Reddy PS (2002) Antibacterial activity of *Syzygium cumini and Syzygium travancoricum* leaf essential oils. Fitoterapia 73(5):414–416
22. Mohamed AA, Ali SI, El-Baz FK (2013) Antioxidant and antibacterial activities of crude extracts and essential oils of *Syzygium cumini* leaves. PLoS one 8(4):e60269
23. Bhatia I, Bajaj K (1975) Chemical constituents of the seeds and bark of *Syzygium cumini*. Planta Med 28(08):346–352
24. Karthic K, Kirthiram KS, Sadasivam S, Thayumanavan B (2008) Identification of alpha amylase inhibitors from *Syzygium cumini* Linn seeds. Indian J Exp Biol 46(9):677–680
25. Williamson EM (2002) Major herbs of ayurveda. Churchill Livingstone, New York
26. Gordon A, Jungfer E, da Silva BA, Maia JG, Marx F (2011) Phenolic constituents and antioxidant capacity of four underutilized fruits from the Amazon region. J Agric Food Chem 59(14):7688–7699
27. Reynertson KA, Yang H, Jiang B, Basile MJ, Kennelly EJ (2008) Quantitative analysis of antiradical phenolic constituents from fourteen edible Myrtaceae fruits. Food Chem 109(4):883–890
28. Vernin G, Metzger J, Mondon JP, Pieribattesti JC (1991) Volatile constituents of the Jamrosa aroma *Syzygium jambos* L. Aston from Reunion Island. J Essent Oil Res 3(2):83–97
29. Veigas JM, Narayan MS, Laxman PM, Neelwarne B (2007) Chemical nature, stability and bioefficacies of anthocyanins from fruit peel of *Syzygium cumini* Skeels. Food Chem 105(2):619–627
30. Subramanian SS, Nair AGR (1972) Flavonoids of the flowers of *Eugenia jambolana*. Curr Sci 41(19):2
31. Baliga MS, Bhat HP, Baliga BRV, Wilson R, Palatty PL (2011) Phytochemistry, traditional uses and pharmacology of *Eugenia jambolana* Lam. (black plum): a review. Food Res Int 44(7):1776–1789

32. Ravi K, Rajasekaran S, Subramanian S (2003) Hypoglycemic effect of *Eugenia jambolana* seed kernels on streptozotocin-induced diabetes in rats. Pharm Biol 41(8):598–603
33. Sharma BC, Balomajumder RP (2008) Hypoglycemic and hypolipidemic effects of flavonoid rich extract from *Eugenia jambolana seeds on streptozotocin induced diabetic rats*. Food Chem Toxicol 46(7):2376–2383
34. Sharma AK, Bharti S, Kumar R, Krishnamurthy B, Bhatia J, Kumari S, Arya DS (2012) *Syzygium cumini* ameliorates insulin resistance and beta-cell dysfunction via modulation of PPAR, dyslipidemia, oxidative stress, and TNF-alpha in type 2 diabetic rats. J Pharmacol Sci 119(3):205–213
35. Fang XK, Gao J, Zhu DN (2008) Kaempferol and quercetin isolated from Euonymus alatus improve glucose uptake of 3T3-L1 cells without adipogenesis activity. Life Sci 82(11–12):615–622
36. Schossler DRC, Mazzanti CM, Luz SCA, Filappi A, Prestes D, Silveira AF, Cecim M (2004) *Syzygium cumini* and the regeneration of insulin positive cells from the pancreatic duct. Braz J Vet Res Anim Sci 41(4):236–239
37. Esmaeili MA, Zohari F, Sadeghi H (2009) Antioxidant and protective effects of major flavonoids from *Teucrium polium* on beta-cell destruction in a model of streptozotocin-induced diabetes. Planta Med 75(13):1418–1420
38. Bardy G, Virsolvy A, Quignard JF, Ravier MA, Bertrand G, Dalle S, Cros G, Magous R, Richard S, Oiry C (2013) Quercetin induces insulin secretion by direct activation of L-type calcium channels in pancreatic beta cells. Br J Pharmacol 169(5):1102–1113
39. Tzeng TF, Liou SS, Liu IM (2011) Myricetin ameliorates defective post-receptor insulin signaling via beta-endorphin signaling in the skeletal muscles of fructose-fed rats. Evid Based Complement Alternat Med: 150752
40. Wang Q, Wang ST, Yang X, You PP, Zhang W (2015) Myricetin suppresses differentiation of 3 T3-L1 preadipocytes and enhances lipolysis in adipocytes. Nutr Res 35(4):317–327
41. Ding Y, Zhang ZF, Dai XQ, Li Y (2012) Myricetin protects against cytokine-induced cell death in RIN-m5f beta cells. J Med Food 15(8):733–740
42. Anandharajan R, Jaiganesh S, Shankernarayanan NP, Viswakarma RA, Balakrishnan A (2006) In vitro glucose uptake activity of *Aegles marmelos* and *Syzygium cumini* by activation of Glut-4, PI3 kinase and PPAR in L6 myotubes. Phytomedicine 13(6):434–441
43. Wang H, Liu T, Song L, Huang D (2012) Profiles and alpha-amylase inhibition activity of proanthocyanidins in unripe *Manilkara zapota* (chiku). J Agric Food Chem 60(12):3098–3104
44. Havsteen BH (2002) The biochemistry and medical significance of the flavonoids. Pharmacol Ther 96(2–3):67–202
45. Tanwar RS, Sharma SB, Singh UR, Prabhu KM (2011) Antiatherosclerotic potential of active principle isolated from *Eugenia jambolana* in streptozotocin-induced diabetic rats. Evid Based Complement Alternat Med: 127641
46. Ravi KS, Rajasekaran SS (2005) Antihyperlipidemic effect of Eugenia jambolana seed kernel on streptozotocin-induced diabetes in rats. Food Chem Toxicol 43(9):1433–1439
47. Jain A, Sharma S, Goyal M, Dubey S, Jain S, Sahu J, Sharma A, Kaushik A (2011) Anti-inflammatory activity of *Syzygium cumini* leaves. Int J Phytomed 2(2):124
48. Muruganandan S, Srinivasan K, Chandra S, Tandan SK, Lal J, Raviprakash V (2002) Inhibitory role of *Syzygium cumini* on autacoid-induced inflammation in rats. Indian J Physiol Pharmacol 46(4):482–486
49. Kumar E, Mastan S, Reddy KR, Reddy GA, Raghunandan N, Chaitanya G (2008) Anti-arthritic property of the methanolic extract of *Syzygium cumini* seeds. Int J Integr Biol 4:55–61
50. Brito FA, Lima LA, Ramos MF, Nakamura MJ, Cavalher-Machado SC, Siani AC, Henriques MG, Sampaio AL (2007) Pharmacological study of anti-allergic activity of *Syzygium cumini* (L.) Skeels. Braz J Med Biol Res 40(1):105–115
51. Maciel MCG, Farias JC, Maluf MJ (2008) *Syzygium jambolanum* treatment improves survival in lethal sepsis induced in mice. BMC Complement Altern Med 8:57

52. Chandrasekaran M, Venkatesalu V (2004) Antibacterial and antifungal activity of *Syzygium jambolanum* seeds. J Ethnopharmacol 91(1):105–108
53. Oliveira GF, Furtado NAJC, Filho AAS, Martins CHG, Bastos JK, Cunha WR, Silva MLA (2007) Antimicrobial activity of *Syzygium cumini* (Myrtaceae) leaves extract. Braz J Microbiol 38(2):381–384
54. Eshwarappa RS, Iyer RS, Subbaramaiah SR, Richard SA, Dhananjaya BL (2014) Antioxidant activity of *Syzygium cumini* leaf gall extracts. Bioimpacts 4(2):101–107
55. Aqil F, Gupta A, Munagala R, Jeyabalan J, Kausar H, Sharma RJ, Singh IP, Gupta RC (2012) Antioxidant and antiproliferative activities of anthocyanin/ellagitannin-enriched extracts from *Syzygium cumini L.* (Jamun, the Indian Blackberry). Nutr Cancer 64(3):428–438
56. Floegel A, Kim DO, Chung SJ, Koo SI, Chun O (2011) Comparison of ABTS/DPPH assays to measure antioxidant capacity in popular antioxidant-rich US foods. J Food Compos Anal 24(7):1043–1048
57. Arun R, Prakash MV, Abraham SK, Premkumar K (2011) Role of *Syzygium cumini* seed extract in the chemoprevention of in vivo genomic damage and oxidative stress. J Ethnopharmacol 134(2):329–333
58. Tripathi P, Tripathi R, Patel RK, Pancholi SS (2013) Investigation of antimutagenic potential of *Foeniculum vulgare* essential oil on cyclophosphamide induced genotoxicity and oxidative stress in mice. Drug Chem Toxicol 36(1):35–41
59. Goyal PK, Verma P, Sharma P, Parmar J, Agarwal A (2010) Evaluation of anti-cancer and anti-oxidative potential of *Syzygium cumini* against benzo[a]pyrene (BaP) induced gastric carcinogenesis in mice. Asian Pac J Cancer Prev 11(3):753–758
60. Farnum P, Timmis R, Kulp JL (1983) Biotechnology of forest yield. Science 219:694–702
61. Bajaj YPS (1986) Biotechnology in agriculture and forestry, vol 1, Trees. Springer, New York
62. Bonga JM (1981) Vegetative propagation of mature trees by tissue culture. In: Rao AN (ed) Proceedings of COSTED symposium on tissue culture economically important plants, Singapore. pp 191–196
63. Zimmerman RH (1986) Regeneration in woody ornamentals and fruit trees. In: Vasil IK (ed) Cell culture and somatic cell genetics vol. 3: plant regeneration and genetic variability. Academic, New York, pp 243–258
64. Yadav U, Lal M, Jaiswal VS (1990) In vitro micropropagation of tropical fruit tree *Syzygium cuminii* L. Plant Cell Tissue Org Cult 21:87–92
65. Jain N, Babbar SB (2000) Recurrent production of plants of black plum, *Syzygium cuminii* (L.) Skeels, a myrtaceous fruit tree, from in vitro cultured seedling explants. Plant Cell Rep 19:519–524
66. Roy PK, Rahman MM, Roy SK (1996) In vitro propagation of *Syzygium cuminii* L. from selected elite trees. Acta Hortic 429:489–495
67. Jain N, Babbar SB (2003) Regeneration of 'juvenile' plants of black plum, *Syzygium cuminii* Skeels, from nodal exp of mature trees. Plant Cell Tiss Org Cult 73:257–263
68. Murashige T, Skoog F (1962) A revised medium for rapid growth and bio assay with tobacco tissue cultures. Physiol Plant 15:473–497
69. Randriamampionona D, Rafamantanana M, Rabemanantsoa C et al (2008) Ex situ conservation and clonal propagation of the Malagasy *Syzygium cuminii*, an antidiabetic plant. Belg J Bot 141(1):14–20
70. Yadav U, Bisht S, Siddique SS, Babu R (2014) Micropropagation of *Syzygium cuminii* L. Using in vitro technique. Int J Sci Eng Res 5(10):1517–1528
71. Iyer RI, Gpoinath PM (2000) Responses of zygotic embryos of *Syzygium cuminii* L. in culture and induction of somatic embryogenesis from pigmented callus. Indian J Plant Physiol 5(4):387–388
72. Babbar SB, Jain N (1998) 'Isubgol' as an alternative gelling agent in plant tissue culture media. Plant Cell Rep 17:318–322
73. Knop W (1865) Quantitative Untersuchung über die Ernährungsprozesse der Pflanzen. Landwirtsch Vers Stn 30:292–294

74. Prakash MG, Gurumurthi K (2009) Genetic transformation and regeneration of transgenic plants from precultured cotyledon and hypocotyl explants of *Eucalyptus tereticornis* Sm. using *Agrobacterium tumefaciens*. In Vitro Cell Dev Biol-Plant 45:429–434
75. Aggarwal D, Kumar A, Reddy MS (2011) *Agrobacterium tumefaciens* mediated genetic transformation of selected elite clone(s) of *Eucalyptus tereticornis*. Acta Physiol Plant 33:1603–1611
76. Mishra M, Jalil SU, Sharma N, Hudedamani U (2014) An *Agrobacterium* mediated transformation system of guava (*Psidium guajava* L.) with endochitinase gene. Crop Breed Appl Biotechnol 14:232–237
77. Stummer BE, Smith SE, Langridge P (1995) Genetic transformation of *Verticordia grandis* (Myrtaceae) using wild-type *Agrobacterium rhizogenes* and binary *Agrobacterium* vectors. Plant Sci 111:51–62
78. Mendonça EG, Stein VC, Balieiro FP, Lima CDF, Santos BR, Paiva LV (2013) Genetic transformation of *Eucalyptus camaldulensis* by agrobalistic method. Revista Árvore, Viçosa-MG 37(3):419–429

Part II

Hairy Roots and Secondary Metabolism

Secondary Products from Plant Cell Cultures: Early Experiences with *Agrobacterium rhizogenes*-Transformed Hairy Roots

10

Adrian J. Parr

Abstract

The enormous range of secondary metabolites produced within the plant kingdom includes many of scientific and commercial interest. There are frequently problems associated both with the study of secondary product biosynthesis in planta, and the reliability of agricultural systems for the production of commercially valuable products. In recent decades there has thus been an interest in developing in vitro cell culture systems that produce high levels of selected secondary metabolites. While callus and cell suspension cultures have been widely developed, their secondary metabolite productive capacities can be rather low and unpredictable. A major step forward came with the development of organ cultures, particularly the so-called hairy roots produced by transformation of plants with the bacterium *Agrobacterium rhizogenes*. The group at the Institute of Food Research, Norwich, was one of the first to exploit the technology, and this article describes their experiences with hairy roots and illustrates the range of approaches that can be taken to maximize their potential. In particular, because hairy root formation already involves a genetic transfer, they are especially good systems in which to study the effects of transgenesis. While some of the techniques described have now been extensively exploited, others have still not reached their full potential, and hopefully this article might serve to throw some light on possible future developments.

Keywords

Agrobacterium rhizogenes • Biotransformation • *Datura* • Hairy roots • Secondary products • Transgenesis • Tropane alkaloids

A.J. Parr (✉)
Institute of Food Research, Norwich Research Park, Colney, Norwich, UK
e-mail: adrian.parr@btinternet.com

S. Jha (ed.), *Transgenesis and Secondary Metabolism*, Reference Series in Phytochemistry, DOI 10.1007/978-3-319-28669-3_20

Abbreviations

ADC	Arginine decarboxylase
HCHL	*p*-hydroxycinnamoyl-CoA hydratase/lyase
ODC	Ornithine decarboxylase
T-DNA	Transferred DNA

Contents

1 Introduction

The production of large numbers of bioactive secondary metabolites by plants has long been of interest to man. Commercially, trade in herbs, spices, pigments, and fragrances which owe their utility to their secondary metabolite content has existed for millennia. More recently, there has also been growing interest in the science of secondary metabolite production. Understanding secondary product formation at both the biochemical and genetic levels thus gives insights into not only the metabolic pathways involved but also their control [1–3] and evolution [2, 4, 5].

Given the interest in secondary metabolites, there has been a growing movement over recent decades to develop in vitro production systems. Such systems mean that, from a scientific perspective, secondary metabolism can be more easily studied and manipulated, and from a commercial perspective, production can be freed from some of the geographical and political constrains that operate "in the field."

One of the first developments in the in vitro study of secondary metabolism was the establishment of cell culture systems, either callus material growing on a solid substrate in the presence of added phytohormones [6] or small cell aggregates (and more rarely free cells) growing in liquid suspension [7]. Such systems do, however, have a number of fundamental problems. Expression of the relevant biosynthetic pathway, if it occurs at all, is thus frequently rather variable and unpredictable [8, 9], and cell cultures are often phenotypically and genetically unstable [9–11]. Much work has gone on to exert some sort of empirical control over these problems – e.g. manipulation of nutrient and hormonal conditions to stimulate metabolite biosynthesis [12] and the application of elicitors to induce the production of metabolites that serve as phytoalexins [9, 12, 13]. Cell culture systems have thus now found a wide range of applications in modern-day science and biotechnology, but underlying problems are potentially never far away.

2 Hairy Root Systems

A significant development in the study and exploitation of secondary metabolism in vitro came with the development of organ cultures, most notably "hairy" root cultures produced by infection of plant tissues with the bacterium *Agrobacterium rhizogenes* [14–16]. In this process the bacterium, which carries a large root-inducing Ri-plasmid, interacts with host cells, and sections of plasmid DNA, the T-DNA, are then transferred to the host and integrated into its genome. By a complex process that is still incompletely understood, but which likely involves altered phytohormone production and changes in hormone perception [17], infected cells subsequently develop into roots (often with prominent root hairs, hence the term "hairy roots"). These can then be excised and grown indefinitely in vitro using a culture medium based simply on sucrose and mineral salts. Such hairy roots frequently possess many lateral branches and can thus grow and produce biomass at high rates. More importantly, they have been shown to stably express the general biosynthetic capacities characteristic of roots from the parent plants [15, 18]. This is seemingly true across a wide range of plant families and potentially in general, though in recent years some subtle differences have been reported [e.g., 19, 20], perhaps as a result of developmental effects or as biochemical consequences of the introduced T-DNA [20]. Hairy root systems thus offer a number of very considerable advantages for the study and exploitation of plant secondary metabolism. From a practical point of view, it should however be noted that the induction of hairy roots in certain genera is not always facile and indeed doesn't readily occur with monocots, though there are reports of success with *Zea mays* [21] and a few other species. Clearly hairy root systems are also only useful tools for the study of secondary products that are made in roots, but the number of such metabolites has been found to be rather greater than originally anticipated from early phytochemical work. Within intact plants, certain products may thus actually be synthesized in roots but be transported and stored elsewhere.

Many examples of detailed fundamental biochemical work on secondary metabolite biosynthesis in hairy roots now exist. Robins et al. [22], for instance, studied (i) the levels of enzyme activities present, (ii) the pool sizes of metabolic intermediates and end products, and (iii) the effects of feeding metabolic intermediates, in order to understand what limits flux into tropane alkaloids in root cultures of a *Datura candida* x *D. aurea* hybrid. They identified levels of ODC activity and the supply of activated acids for condensation with tropine as potentially rate-limiting steps [22]. A few years later, Robins et al. [23] fed 13C-labeled tropane alkaloid precursors to transformed roots of *Datura stramonium* and looked at incorporation of label into alkaloids using NMR. In this way they were able to deduce the likely mechanism by which littorine, a putative intermediate in the pathway, rearranges to hyoscyamine, one of the major end products in tropane alkaloid biosynthesis. In recent years, the use of hairy root systems for fundamental biochemical work has become even more widespread [e.g., 24], as has their use in genetic approaches to understand biosynthetic pathways.

Table 1 Examples of species used for the production of hairy roots during early work at IFR, Norwich, with typical secondary metabolites investigated either then or later

Species	Metabolite	Reference
Nicotiana rustica	Pyridine alkaloids, e.g., nicotine	[14]
N. umbratica		[11]
N. africana		[11]
N. hesperis		[28]
Datura stramonium	Tropane alkaloids, e.g., hyoscyamine	[29–31]
D. ferox	Tropane alkaloids, e.g., hyoscyamine	[29]
D. inoxia	Tropane alkaloids, e.g., hyoscyamine	[29]
D. wrightii	Tropane alkaloids, e.g., hyoscyamine	[29]
D. fastuosa	Tropane alkaloids, e.g., hyoscyamine	[29]
D. quercifolia	Tropane alkaloids, e.g., hyoscyamine	[29]
D. sanguinea	Tropane alkaloids, e.g., hyoscyamine	[29]
Hyoscyamus albus	Tropane alkaloids, e.g., hyoscyamine	[29]
H. desertorum	Tropane alkaloids, e.g., hyoscyamine	[29]
H. muticus	Tropane alkaloids, e.g., hyoscyamine	[29]
Scopolia stramonifolia	Tropane alkaloids, e.g., hyoscyamine	[29]
Nicandra physaloides	Hygrine	[32]
Beta vulgaris	Betacyanins, phenylpropanoids	[14, 33]
Catharanthus roseus	Indole alkaloids, e.g., ajmalicine	[34]
Cinchona ledgeriana		[11]
Phaseolus vulgaris		[11]

From a more biotechnological point of view (Table 1), the biosynthetic capacity of hairy roots has now also been exploited by a large number of workers for the bulk production of secondary products. This can be achieved by growing roots in appropriate culture vessels and either harvesting the roots after maximal biomass/metabolite content has been reached, or else continuously extracting products from the growth medium, should they be found to be excreted [25]. Because in planta the root plays a major role in conditioning the rhizosphere and controlling allelopathic or symbiotic interactions with other organisms, a considerable range of metabolites are indeed excreted by roots. Although on a laboratory scale, culture vessels are typically no more than Erlenmeyer flasks, scale up can be achieved and quasi-industrial scale vessels are possible. Quite early in the development of hairy root technology, the group in Norwich thus described the operation of a 500 l fermenter for trial studies with hairy roots of *Datura stramonium* [26], allowing the production of 40 kg (fresh weight) of tropane alkaloid-containing roots per run. More recent developments with fermenters, though rarely on such a large scale, are described by Ruffoni et al. [27].

In many ways the stability of hairy root cultures offers considerable practical advantages, but there is a potential downside in that levels of metabolite production are, to a large extent, predetermined. A low-producing plant will thus give low-producing hairy roots after transformation. Should levels of secondary metabolite production be insufficient for one's needs, some strategies do however exist for improving productivity. At their simplest, these involve searching for higher-producing parent plants. One example from early work at the Institute of Food Research involved the screening of 1000+ individual plants within the genera *Datura, Scopolia*, and *Hyoscyamus*, which allowed parent material possessing a range of potentially interesting traits in tropane alkaloid biosynthesis to be identified [29]. Many of the techniques of media manipulation that have been applied to cell cultures can, in addition, also be applied to hairy roots. The production of sesquiterpenoid phytoalexins could thus be induced in roots of *Datura stramonium* by the application of abiotic elicitors [30]. Another potential approach to improving the productivity of hairy roots is to combine the techniques of cell and organ culture more directly. Furze et al. [35] were able to produce suspension cultures from hairy roots of *Nicotiana rustica* by treatment with exogenous phytohormones. Protoplasts were then isolated and used as a source of single-cell-derived callus lines where a degree of genetic and epigenetic variability had ensued. At this point, hormones were removed and fully organized hairy roots allowed to reform as the calli grew on. Although individual regenerated roots appeared to have regained their stability, clear differences in the behavior of lines produced from different roots were now apparent [35], in contrast to the predictability of the parent root material. Some lines indeed showed improved alkaloid production characteristics (e.g., total levels of nicotine, and/or the proportion released into the culture medium) compared to the parent material [35].

3 Hairy Roots and Biotransformation

Early work on hairy roots concentrated initially on simply generating cultures producing useful levels of secondary metabolites, be they model compounds of scientific importance or specific metabolites of direct commercial relevance. Another approach to exploiting the roots' biosynthetic capacities is however possible – namely, their use in biotransformation. Selected metabolites are thus added to the culture medium, and the roots are used essentially as a biocatalyst to convert these compounds into others. The nature of the reactions that occur need to be determined analytically, which is in itself frequently of considerable scientific interest. Many biotransformations carried out by roots are typical xenobiotic detoxification processes (though of course they are frequently easier to carry out and to study than the equivalent reactions occurring in planta). Reactions such as hydroxylation and glycosylation are thus frequent [36, 37]. If the fed metabolites are, however, close in structure to metabolic intermediates found within the roots, then provided there is a degree of plasticity in the substrate specificity of the enzymes involved, they can be more extensively metabolized by the endogenous secondary metabolite biosynthetic pathways. At the Institute of Food Research, Boswell et al. thus fed *N*-ethyl putrescine to transformed root cultures of *Nicotiana rustica* and showed the extensive formation of N-ethyl nornicotine – the higher homologue of nicotine [38]. N-modified versions of hygrine and tropane alkaloids could similarly be produced by feeding appropriate substrate analogues to *Brugmansia* (=*Datura*) hairy roots [39]. The differing levels of incorporation of various analogues into the different alkaloids present also gave considerable insight into the substrate specificities of many of the enzymes involved in the biosynthetic pathways [39].

4 Hairy Roots and Transgenesis

Since the formation of hairy roots by necessity involves a genetic transfer, such cultures are also excellent systems for targeted genetic modification of secondary metabolite biosynthesis. This can be achieved either by inserting genes of interest into the T-DNA of the Ri-plasmid of the *A. rhizogenes* strain used for inducing hairy roots or, more conveniently, by introducing a second plasmid carrying the desired genes in mobilizable form into *A. rhizogenes* alongside the Ri-plasmid itself, using a binary strategy [40].

One approach to the use of genetic modification for manipulating secondary metabolite production is to modify the overall flux through preexisting pathways. This may involve either targeted upregulation or downregulation of specific routes. Fundamentally, the upregulation route is perhaps the easiest and can most simply be achieved by insertion and expression of genes related to rate-limiting steps within the relevant biosynthetic pathway. Prior biochemical work to identify such flux-controlling enzymatic steps is thus useful, though to some extent a trial-and-error approach is not without merit, as it may potentially generate information on flux control within the relevant biosynthetic pathways even if end-product overaccumulation is not achieved. The introduced genes can be either additional copies of host genes, orthologs from other species, or functionally equivalent genes from other pathways. "Foreign" genes need not necessarily come from a closely related species. Indeed the overexpression of genes of yeast or bacterial origin may sometimes offer technical advantages as well as potentially allowing the expression of enzymes with different control mechanisms (e.g., feedback control) to those normally operating in planta and thus perhaps producing stronger effects.

One of the first demonstrations of the potential for upregulation of pathways in hairy roots came from Hamill et al. [41]. Here, hairy roots of *Nicotiana rustica* were produced that also expressed a yeast (*Saccharomyces cerevisiae*) gene for ornithine decarboxylase (ODC), the key enzyme in the production of putrescine. Putrescine is the precursor of the characteristic tobacco pyridine alkaloids such as nicotine and nornicotine, and by potentially increasing substrate supply, Hamill et al. [41] hoped to increase alkaloid levels within the hairy roots. This they achieved quite readily, though concentrations of nicotine were generally only twice that of control roots produced by infection of the same parent plants with *A. rhizogenes* carrying a biosynthetically neutral gene. This was so despite a very strong promoter (the Cauliflower Mosaic Virus CaMV 35S promoter with an upstream duplicated enhancer sequence [42]) being used to drive expression of the introduced ODC gene. Analysis of enzyme activities in root lines showed only a relatively modest increase in peak ODC activity following introduction of the yeast gene, though activity was maintained at proportionally higher levels later in the growth cycle, when activity from the endogenous *Nicotiana* ornithine decarboxylase decreased sharply. Levels of OCD enzyme activity also did not always correlate closely with the increased nicotine levels produced. These facts illustrate two of the potential problems with genetic manipulation of biosynthetic pathways. Firstly, it is possible that either, or both, the expression of the introduced gene or its enzymatic activity may be subject to strong control within the plant, so that additional copies of a gene do not necessarily feed through into an equivalent increased enzyme activity. This issue is perhaps less important in the final stages of biosynthesis of secondary metabolites, which have little physiological impact on the plant. In the case discussed above, putrescine, in addition to being the precursor for nicotine, is however also the precursor for spermidine and spermine. These polyamines, along with putrescine itself, have important functions in plants [43], so it is perhaps unsurprising that the formation of putrescine seems to be strongly controlled in hairy roots. Secondly, as mentioned above, the rate-limiting steps in secondary metabolic pathways need to be taken into account. Primary substrate supply is often likely to have at least some input into product generation, but there may come a stage when increasing the activity of one key step simply shifts flux limitation elsewhere. Moreover the physical or enzymatic compartmentation of biosynthesis also needs to be born in mind. It is unlikely that the overexpression of a gene leading to, for instance, increased enzyme activity within the cytoplasm will have an effective action on biosynthesis that takes place within organelles. In the case of the biosynthesis of nicotine, it is also theoretically possible that the putrescine that acts as a precursor is derived from arginine, via arginine decarboxylase (ADC), and not from ornithine via ODC. Effects of overexpressing ODC might thus be due to a "bleed-through" into an ADC-dependent pathway perhaps involving multienzyme complexes, which could quite easily limit the effectiveness of any manipulation. Recent research by DeBoer et al. [44] in transformed roots of *N. tabacum* has provided evidence that this ADC route is, in fact, not very likely, but the concept still remains as a salutary warning.

Given that initial efforts to genetically manipulate biosynthetic pathways may be only partially successful, even if very strong promoters are used and lines derived from the most favorable transformation events are selected, various techniques can be used to increase the observed effects. On an empirical level, because hairy roots share several features with in vitro cell cultures, one way to try and maximize the effects of transgenesis would be simply to manipulate culture conditions, as, for example, has been done by Singh et al. [45] who looked at phenylpropanoid production in lines of *Beta vulgaris* carrying a gene which affected *p*-hydroxycinnamoyl-CoA metabolism. More fundamentally, it may also be worth investigating the manipulation of other steps in the biosynthetic pathway, which may show greater flux control. It may also be far more effective to try and influence the expression not so much of a single biosynthetic gene but rather of a regulatory gene such as one of the MYB family [46] which may then go on to influence the expression of the desired biosynthetic pathway as a whole. Much work has been done, in particular, on the genetic regulation of flavonoid biosynthesis, and this is just starting to be applied to hairy roots [47].

Another approach to genetic modification of secondary product production in hairy root cultures involves downregulation of biosynthetic routes. Even this "negative" regulation could be useful in increasing the production of certain compounds, by acting to reduce the flow of carbon away from products of interest. A number of genetic techniques for producing downregulation have been described over recent decades, with RNA-mediated interference, or RNAi, now typically being the most convenient option [48]. Over recent years there have been many examples of the use of downregulation in hairy roots to identify potential physiological functions of secondary metabolites [e.g., 49], and to examine flux control in pathways [e.g., 50], but as yet rather few more practical applications toward controlling the production of specific desired products have been reported.

Finally, it should be noted that hairy root cultures are, in practice, not totally divorced from the intact plant, and it is possible to regenerate plants from transformed roots of many species [28, 51] should the need arise to express the introduced genetic manipulation in soil-grown plants. Such regeneration is, however, not always facile [51]. Interestingly, in the genus *Nicotiana*, where regeneration can sometimes be so efficient as to even be spontaneous [28], there are signs that an ancient exposure to *Agrobacterium* may have occurred in the genus's evolutionary past [52]. Preexisting partial morphogenic compensation mechanisms may thus already exist in this and other(?) genera.

5 Hairy Roots and the Production of Unusual Metabolite Patterns or Novel Metabolites

Introduction of genes to modify the spectrum of products produced, or to create entirely novel products, is also a potential target for genetic manipulation in hairy roots. At its simplest, this may involve the introduction of, e.g., hydroxylases or methylases to "fine tune" the broad array of related secondary products that are typically produced by any one species. More fundamental changes in metabolism can however also be brought about, and one of the first demonstrations of the potential for such an approach came from Mitra et al. [31], who expressed a pseudomonad gene for *p*-Hydroxycinnamoyl-CoA hydratase/lyase (HCHL) in hairy root cultures of *Datura stramonium*. This enzyme catalyzes the side chain cleavage of a range of substituted 4-hydroxycinnamoyl-CoAs to produce the equivalent 4-hydroxybenzaldehyde derivatives, of which vanillin (3-methoxy, 4-hydroxybenzaldehyde) is a notable example of commercial importance.

4-Hydroxycinnamoyl-CoA derivatives are major metabolic intermediates in essentially all plant species, being involved in the biosynthesis of both simple phenylpropanoids and of lignin [53–55]; they also act as activated substrates for the conjugation of phenylpropanoids with other secondary metabolites or with cell wall carbohydrates [53, 56]. The substrates for *p*-hydroxycinnamoyl-CoA hydratase/lyase are thus likely to be readily available following introduction of the HCHL gene, which would then act to divert carbon flow into formation of chain-shortened products. Although a limited amount of phenylpropanoid chain-shortening occurs naturally in many plants, it is normally a very minor reaction in most species (except of course in the vanilla orchid *Vanilla planifolia* and related species). The introduction of HCHL activity thus has the potential to produce major shifts in metabolism. Although *D. stramonium* normally accumulates very little chain-shortened phenylpropanoids, hairy roots co-transformed with, and actively expressing, the HCHL gene were however found to contain up to 0.5% of the dry weight of 4-hydroxybenzoic acid and 4-hydroxybenzyl alcohol glucosides [31]. A major shift in secondary metabolite partitioning had thus been introduced. The accumulation not of free hydroxybenzaldehydes but rather of glucosides of the related acids and alcohols is an interesting if not entirely unexpected result. Oxidation/reduction and conjugate formation are very typical ways in which plants deal with exposure to unexpected chemicals [36, 37]. Given that vanillin is more commercially interesting than the 4-hydroxyacids/alcohols, the initial results were in some respects disappointing from a practical, if not from a scientific, perspective. In recent years further research has gone on to try and develop improved systems. Hairy roots of *Beta vulgaris*, a species that incorporates a lot of ferulic acid into cell walls [57] and thus is likely to have a ready supply of feruloyl-CoA (from which vanillin can be produced by HCHL action), were found to accumulate high levels of chain shortened products [33]. Levels reached an impressive 14% of the dry weight in one line [33], rather implying that continuous removal of phenylpropanoids had increased total flux through the pathway, possibly by affecting feedback inhibition systems. Even in this species, products were however still predominantly 4-hydroxy derivatives produced from coumaroyl-CoA rather than 3-methoxy, 4-hydroxy derivatives from feruloyl-CoA, perhaps as a consequence of selective substrate availability to the introduced HCHL enzyme. Accumulation of aldehydes, while detectable, was also still low [33].

In a slightly more speculative context, one of the potentially most exciting uses of GM technology and hairy roots would be to introduce several genes for sequentially related enzymes into an easily cultivated host species to try and create entire new biosynthetic routes for that species. This type of approach currently remains very much in its infancy. There are examples of where genes for "foreign" biosynthesis have been inserted into hairy roots, and in the absence of any metabolic connection, suitable precursors have then been fed exogenously [58]. Conceptually, this is however rather different to the creation of entire new biosynthetic pathways. Since the metabolic outcome of inserting foreign pathways into plants is to some extent unpredictable, not only do the gene isolation issues have to be resolved but it is perhaps not until the recent advances in the powerful unfocused analytical techniques of metabolomics [59] that such an approach has even become fully feasible from the analytical side. The application of metabolomics to understand the broader implications of the simple gene modifications that have so far been carried out with hairy roots indeed also has much to recommend it, and research in this area is now starting to appear [50].

6 Conclusions

The use of hairy roots for the study of secondary metabolite biosynthesis, and for the (attempted) production of commercially important target compounds, has now been in existence for nearly thirty years. The development of genetically engineered roots, containing not just the T-DNA involved in root formation but also specifically chosen metabolic or regulatory genes, has led to a great number of advances in recent years. At the same time, some of the concepts developed in the early years still remain very relevant, and a number of issues (e.g., optimal large-scale cultivation) remain only partially solved. It is hoped that the present article, focusing both on where the field started and where progress has taken it, might throw some light on potential future developments. In particular, continued research using transgenic root material, most notably systems with either suppressed or overexpressed regulatory gene function, would seem to offer many advantages.

References

1. Hahlbrock K, Grisebach H (1979) Enzymic controls in the biosynthesis of lignin and flavonoids. Annu Rev Plant Physiol Plant Mol Biol 30:105–130
2. Koes RE, Quattrocchio F, Mol JNM (1994) The flavonoid biosynthetic pathway in plants: function and evolution. Bioessays 16:123–132
3. Parr AJ (1992) Quantitation of pathway intermediates in tropane alkaloid biosynthesis: an approach to understanding how the pathway may be regulated. Phytochem (Life Sci Adv) 11:155–158
4. Gachon CMM, Langlois-Meurinne M, Saindrenan P (2005) Plant secondary metabolism glycosyltransferases: the emerging functional analysis. Trends Plant Sci 10:542–549
5. Qi X, Bakht S, Leggett M, Maxwell C, Melton R, Osbourn A (2004) A gene cluster for secondary metabolism in oat: implications for the evolution of metabolic diversity in plants. Proc Natl Acad Sci U S A 101:8233–8238
6. Schenk RU, Hildebrandt AC (1972) Medium and techniques for induction and growth of monocotyledonous and dicotyledonous plant cell cultures. Can J Bot 50:199–204
7. Gamborg OL, Miller RA, Ojima K (1968) Nutrient requirements of suspension cultures of soybean root cells. Exp Cell Res 50:151–158
8. Hall RD, Yeoman MM (1987) Intercellular and intercultural heterogeneity in secondary metabolite accumulation in cultures of *Catharanthus roseus* following cell line selection. J Exp Bot 38:1391–1398
9. Bourgaud F, Gravot A, Milesi S, Gontier E (2001) Production of plant secondary metabolites: a historical perspective. Plant Sci 161:839–851
10. Larkin PJ, Scowcroft WR (1981) Somaclonal variation – a novel source of variability from cell cultures for plant improvement. Theor Appl Genet 60:197–214
11. Aird ELH, Hamill JD, Rhodes MJC (1988) Cytogenetic analysis of hairy root cultures from a number of plant species transformed by *Agrobacterium rhizogenes*. Plant Cell Tiss Org 15:47–57

12. Yeoman MM, Yeoman CL (1996) Manipulating secondary metabolism in cultured plant cells. New Phytol 134:553–569
13. Zhao J, Zhu W-H, Hu Q (2001) Selection of fungal elicitors to increase indole alkaloid accumulation in *Catharanthus roseus* suspension cell culture. Enzyme Microb Technol 28:666–672
14. Hamill JD, Parr AJ, Robins RJ, Rhodes MJC (1986) Secondary product formation by cultures of *Beta vulgaris* and *Nicotiana rustica* transformed with *Agrobacterium rhizogenes*. Plant Cell Rep 5:111–114
15. Hamill JD, Parr AJ, Rhodes MJC, Robins RJ, Walton NJ (1987) New routes to plant secondary products. Nat Biotechnol 5:800–804
16. Flores HE, Hoy MW, Pickard JJ (1987) Secondary metabolites from root cultures. Trends Biotechnol 5:64–69
17. Britton MT, Escobar MA, Dandekar AM (2008) The oncogenes of *Agrobacterium tumefaciens* and *Agrobacterium rhizogenes*. In: Tzfira T, Citovsky V (eds) *Agrobacterium*: from biology to biotechnology. Springer, New York, pp 523–563
18. Parr AJ, Hamill JD (1987) Relationship between *Agrobacterium rhizogenes* transformed hairy roots and intact, uninfected *Nicotiana* plants. Phytochemistry 26:3241–3245
19. Abhyankara G, Reddy VD, Giri CC, Rao KV, Lakshmi VVS, Prabhakar S, Vairamani M, Thippeswamy BS, Bhattacharya PS (2005) Amplified fragment length polymorphism and metabolomic profiles of hairy roots of *Psoralea corylifolia* L. Phytochemistry 66:2441–2457
20. Bulgakov VP (2008) Functions of rol genes in plant secondary metabolism. Biotechnol Adv 26:318–324
21. Runo S, Macharia S, Alakonya A, Machuka J, Sinha N, Scholes J (2012) *Striga* parasitizes transgenic hairy roots of *Zea mays* and provides a tool for studying plant-plant interactions. Plant Methods 8:20
22. Robins RJ, Parr AJ, Payne J, Walton NJ, Rhodes MJC (1990) Factors regulating tropane-alkaloid production in a transformed root culture of a *Datura candida* × *D. aurea* hybrid. Planta 181:414–422
23. Robins RJ, Chesters NCJE, O'Hagan D, Parr AJ, Walton NJ, Woolley JG (1995) The biosynthesis of hyoscyamine: the process by which littorine rearranges to hyoscyamine. J Chem Soc Perkin Trans 1:481–485
24. Wu SJ, Liu YS, Wu JY (2008) The signaling role of extracellular ATP and its dependence on Ca2+ flux in elicitation of *Salvia miltiorrhiza* hairy root cultures. Plant Cell Physiol 49:617–624
25. Rhodes MJC, Hilton M, Parr AJ, Hamill JD, Robins RJ (1986) Nicotine production by "hairy root" cultures of *Nicotiana rustica*: fermentation and product recovery. Biotechnol Lett 8:415–420
26. Wilson PDG (1997) The pilot-scale cultivation of transformed roots. In: Doran PM (ed) Hairy roots: culture and applications. CRC Press, Amsterdam, pp 179–190
27. Ruffoni B, Pistelli L, Bertoli A, Pistelli L (2010) Plant cell cultures: bioreactors for industrial production. Adv Exp Med Biol 698:203–221
28. Hamill JD, Rhodes MJC (1988) A spontaneous, light independent and prolific plant regeneration response from hairy roots of *Nicotiana hesperis* transformed by *Agrobacterium rhizogenes*. J Plant Physiol 133:506–509
29. Parr AJ, Payne J, Eagles J, Chapman BT, Robins RJ, Rhodes MJC (1990) Variation in tropane alkaloid accumulation within the Solanaceae and strategies for its exploitation. Phytochemistry 29:2545–2550
30. Furze JM, Rhodes MJC, Parr AJ, Robins RJ, Whitehead IM, Threlfall DR (1991) Abiotic factors elicit sesquiterpenoid phytoalexin production but not alkaloid production in transformed root cultures of *Datura stramonium*. Plant Cell Rep 10:111–114
31. Mitra A, Mayer MJ, Mellon FA, Michael AJ, Narbad A, Parr AJ, Waldron KW, Walton NJ (2002) 4-Hydroxycinnamoyl-CoA hydratase/lyase, an enzyme of phenylpropanoid cleavage

from *Pseudomonas*, causes formation of C(6)-C(1) acid and alcohol glucose conjugates when expressed in hairy roots of *Datura stramonium* L. Planta 215:79–89

32. Parr AJ (1992) Alternative metabolic fates of hygrine in transformed root cultures of *Nicandra physaloides*. Plant Cell Rep 11:270–273
33. Ur Rahman L, Kouno H, Hashiguchi Y, Yamamoto H, Narbad A, Parr A, Walton N, Ikenaga T, Kitamura Y (2009) HCHL expression in hairy roots of *Beta vulgaris* yields a high accumulation of *p*-hydroxybenzoic acid (pHBA) glucose ester, and linkage of pHBA into cell walls. Bioresour Technol 100:4836–4842
34. Parr AJ, Peerless ACJ, Hamill JD, Walton NJ, Robins RJ, Rhodes MJC (1988) Alkaloid production by transformed root cultures of *Catharanthus roseus*. Plant Cell Rep 7:309–312
35. Furze JM, Hamill JD, Parr AJ, Robins RJ (1987) Variations in morphology and nicotine alkaloid accumulation in protoplast-derived hairy root cultures of *Nicotiana rustica*. J Plant Physiol 131:237–246
36. Sandermann H Jr (1992) Plant metabolism of xenobiotics. Trends Biochem Sci 17:82–84
37. Brazier-Hicks M, Offen WA, Gershater MC, Revett TJ, Lim E-K, Bowles DJ, Davies GJ, Edwards R (2007) Characterization and engineering of the bifunctional *N*- and *O*-glucosyltransferase involved in xenobiotic metabolism in plants. Proc Natl Acad Sci U S A 104:20238–20243
38. Boswell HD, Watson AB, Walton NJ, Robins DJ (1993) Formation of *N*-ethyl-S-nornicotine by transformed root cultures of *Nicotiana rustica*. Phytochemistry 34:153–155
39. Boswell HD, Dräger B, Eagles J, McClintock C, Parr A, Portsteffen A, Robins DJ, Robins RJ, Walton NJ, Wong C (1999) Metabolism of *N*-alkyldiamines and *N*-alkylnortropinones by transformed root cultures of *Nicotiana and Brugmansia*. Phytochemistry 52:855–869
40. Tempe J, Casse-Delbart F (1989) Plant gene vectors and genetic transformation: *Agrobacterium* Ri plasmids. In: Schell J, Vasil IK (eds) Cell culture and somatic cell genetics of plants, vol 6. Academic, New York, pp 25–49
41. Hamill JD, Robins RJ, Parr AJ, Evans DM, Furze JM, Rhodes MJC (1990) Over-expressing a yeast ornithine decarboxylase gene in transgenic roots of *Nicotiana rustica* can lead to enhanced nicotine accumulation. Plant Mol Biol 15:27–38
42. Kay R, Chan A, Daly M, McPherson J (1987) Duplication of CaMV 35S promoter sequences creates a strong enhancer for plant genes. Science 236:1299–1302
43. Martin-Tanguy J (2001) Metabolism and function of polyamines in plants: recent development (new approaches). Plant Growth Regul 34:135–148
44. DeBoer KD, Dalton HL, Edward FJ, Hamill JD (2011) RNAi-mediated down-regulation of ornithine decarboxylase (ODC) leads to reduced nicotine and increased anatabine levels in transgenic *Nicotiana tabacum* L. Phytochemistry 72:344–355
45. Singh P, Khan S, Pandey SS, Singh M, Banerjee S, Kitamura Y, Ur Rahman L (2015) Vanillin production in metabolically engineered *Beta vulgaris* hairy roots through heterologous expression of *Pseudomonas fluorescens* HCHL gene. Ind Crop Prod 74:839–848
46. Jin H, Martin C (1999) Multifunctionality and diversity within the plant MYB-gene family. Plant Mol Biol 41:577–585
47. Verdier J, Zhao J, Torres-Jerez I, Ge S, Liu C, He X, Mysore KS, Dixon RA, Udvardi MK (2012) MtPAR MYB transcription factor acts as an on switch for proanthocyanidin biosynthesis in *Medicago truncatula*. Proc Natl Acad Sci U S A 109:1766–1771
48. Gilchrist E, Haughn G (2010) Reverse genetics techniques: engineering loss and gain of gene function in plants. Brief Funct Genomics 9:103–110
49. Graham TL, Graham MY, Subramanian S, Yu O (2007) RNAi silencing of genes for elicitation or biosynthesis of 5-deoxyisoflavonoids suppresses race-specific resistance and hypersensitive cell death in *Phytophthora sojae* infected tissues. Plant Physiol 144:728–740
50. Asano T, Kobayashi K, Kashihara E, Sudo H, Sasaki R, Iijima Y, Aoki K, Shibata D, Saito K, Yamazaki M (2013) Suppression of camptothecin biosynthetic genes results in metabolic modification of secondary products in hairy roots of *Ophiorrhiza pumila*. Phytochemistry 91:128–139

51. Hu Z-B, Du M (2006) Hairy root and its application in plant genetic engineering. J Integr Plant Biol 48:121–127
52. Furner IJ, Huffman GA, Amasino RM, Garfinkel DJ, Gordon MP, Nester EW (1986) An *Agrobacterium transformation* in the evolution of the genus *Nicotiana*. Nature 319:422–427
53. Niggeweg R, Michael AJ, Martin C (2004) Engineering plants with increased levels of the antioxidant chlorogenic acid. Nat Biotechnol 22:746–754
54. Petersen M, Hans J, Matern U (2010) Biosynthesis of phenylpropanoids and related compounds. Annu Plant Rev 40:182–257
55. Whetten R, Sederoff R (1995) Lignin biosynthesis. Plant Cell 7:1001–1013
56. Yoshida-Shimokawa T, Yoshida S, Kakegawa K, Ishii T (2001) Enzymic feruloylation of arabinoxylan-trisaccharide by feruloyl-CoA: arabinoxylan-trisaccharide *O*-hydroxycinnamoyl transferase from *Oryza sativa*. Planta 212:470–474
57. Waldron KW, Ng A, Parker ML, Parr AJ (1997) Ferulic acid dehydrodimers in the cell walls of *Beta vulgaris* and their possible role in texture. J Sci Food Agric 74:221–228
58. Häkkinen ST, Moyano E, Cusidó RM, Palazón J, Piñol MT, Oksman-Caldentey K-M (2005) Enhanced secretion of tropane alkaloids in *Nicotiana tabacum* hairy roots expressing heterologous hyoscyamine-6b-hydroxylase. J Exp Bot 56:2611–2618
59. Sumner LW, Mendes P, Dixon RA (2003) Plant metabolomics: large-scale phytochemistry in the functional genomics era. Phytochemistry 62:817–836

Secondary Metabolite Production in Transgenic Hairy Root Cultures of Cucurbits

11

Kaliyaperumal Rekha and Muthu Thiruvengadam

Abstract

Cucurbits are important group of vegetables due to their nutritional significance and are also used for valuable traditional medicine. The infection of plants by *Agrobacterium rhizogenes* results in a hairy root (HR) phenotype characterized by rapid growth in hormone-free medium, an unusual ageotropism and extensive lateral branching. These genetically transformed root cultures (hairy roots) can produce levels of secondary metabolites comparable to that of intact plants. Hairy root cultures offer promise for high production and productivity of valuable secondary metabolites in many plants. High stability and productivity features allow the exploitation of HRs as valuable biotechnological tool for the production of plant secondary metabolites. While these chemical compounds are employed by plants for interactions with their environment, humans have long since explored and exploited plant secondary metabolites for medicinal and practical uses. The main constraint for commercial exploitation of hairy root cultivations is the development and scaling up of appropriate reactor vessels (bioreactors) that permit the growth of interconnected tissues normally unevenly distributed throughout the vessel. Emphasis has focused on designing appropriate bioreactors suitable to culture the delicate and sensitive plant hairy roots. To this end, hairy root culture presents an excellent platform for producing valuable secondary metabolites. For these reasons, this chapter describes the establishment of hairy roots and production of secondary metabolites from hairy roots of cucurbits and also phytochemicals uses for biological activity.

K. Rekha (✉) • M. Thiruvengadam (✉)
Department of Applied Bioscience, College of Life and Environmental Sciences, Konkuk University, Seoul, Republic of Korea
e-mail: krekhathiruvengadam@gmail.com; thiruv30@yahoo.com

S. Jha (ed.), *Transgenesis and Secondary Metabolism*, Reference Series in Phytochemistry, DOI 10.1007/978-3-319-28669-3_6

Keywords
Agrobacterium rhizogenes • Hairy roots • Secondary metabolites • Phenolic compounds • Triterpenoids • Charantin • Biological activity

Abbreviations

CaMV	Cauliflower mosaic virus
ELISA	Enzyme-linked immunosorbent assay
GFP	Green fluorescent protein
Gyp	Gypenosides
HIV	Human immunodeficiency virus
HRs	Hairy roots
HSV	Herpes simplex virus
PAL	Phenylalanine ammonia lyase
PCR	Polymerase chain reaction
RIPs	Ribosome-inactivating protein
RT PCR	Reverse transcription polymerase chain reaction
TLC	Thin layer chromatography
UHPLC	Ultra-high performance liquid chromatography
WHO	World Health Organization

Contents

1 Introduction

Cucurbits are the popular name to the plants of family Cucurbitaceae, which include over 130 genera and 800 species [1], among the economically most important plant families [2, 3]. It is a large group of plants which are medicinally valuable. Cucurbitaceae members are primarily established in the tropical regions of the world. Global production of cucumbers, including gherkins, was among the top ten vegetables produced globally (http://faostat.fao.org). Cucurbits are a prominent source of secondary metabolites, and many genera of this family received a great level of scientific interest because of the extensive range of pharmacological and nutraceutical properties [4]. It is reported that the bitter flavor of cucurbits is due to tetracyclic triterpenoids [5]. Various phytochemicals such as alkaloids and saponins are extracted from *Momordica*, *Citrullus*, *Cucurbita*, and *Lagenaria* [6]. The family proved itself as a strong source of food and medicine. Major species of importance include: *Citrullus lanatus* (watermelon), *Cucumis sativus* (cucumber), *Cucumis melo* (musk melon), *Cucumis anguria* (bur gherkin), *Cucurbita pepo* (pumpkin), *Momordica charantia* (bitter gourd), *Momordica dioica* (spine gourd), *Coccinia grandis* (ivy gourd), and *Praecitrullus fistulosus* (tinda). In recent years, consumption of cucurbits in the average diet has been highlighted for its contribution towards lowering the risks of several life-threatening diseases such as coronary heart disease, stroke, pulmonary disease, and different types of cancer.

Plant species are capable of producing different types of secondary products which can be harnessed by humans for their beneficial properties in a large domain of industrial or medicinal applications [7]. World Health Organization (WHO) estimates that up to 80% of people rely mainly on traditional herbs as remedies for their medicines [8]. Extracted from entire plants, secondary products are used by food and pharmaceutical industries, although most often numerous natural plant-derived molecules remain undiscovered or unexplored for their pharmacological properties [9]. Roots play most important roles in plants and they anchor plants to the ground, take up minerals and water from the soil, store nutrients for perennial plants, and produce a diverse array of chemicals for symbiotic interactions or defensive with other plants or microbes in the rhizosphere. These plant-produced chemicals have traditionally been referred to as secondary metabolites and more recently tagged as specialized metabolites. Bioactive compounds are extra nutritional constituents that naturally occur in small quantities in plant and food products [10]. Most common bioactive compounds include secondary metabolites such as antibiotics, mycotoxins, alkaloids, food grade pigments, plant growth factors, and phenolic compounds [10, 11]. Many secondary metabolites not only protect plants from pathogens, insects, and environmental stresses but also are valuable for human health. Many plant species, including crop plants, are capable of producing and releasing biologically active compounds (allelochemicals). Allelochemicals (e.g., phenolics, terpenoids, alkaloids, coumarins, tannins, steroids, and quinines) are released by the plant into the environment by root exudation, volatile emissions, leaching from the leaves and other aerial parts, and the decomposition of plant material [12, 13]. Plant roots release a range of compounds that are not directly involved in the growth and development of the plant but are very much important for plants during stress conditions (biotic/abiotic). These compounds include aliphatic acids, aromatic acids, fatty acids, sterols, phenolics, enzymes, and other secondary metabolites, including flavonoids [14, 15]. Many plant secondary metabolites of interest are accumulated in roots. However, plant cultivation is often time consuming and metabolite extraction from plant roots is destructive to plant growth.

Agrobacterium rhizogenes is a Gram negative soil-borne bacterium of the family Rhizobiaceae, which causes the hairy roots disease by infecting wounded higher plants. The transformed roots can be excised to establish axenic root cultures and indefinitely propagated in growth regulator free medium. The root exhibit fast, plagiotropic growth characterized by profuse lateral branching and rapid root tip elongation [16, 17]. Root loci (*rol*) genes harbored by the root-inducing (Ri) plasmid of this bacterium are incorporated into the host plant genome, causing hairy root. *Rol* genes are thought to affect growth and development of transformed roots and induce secondary metabolite synthesis by turning on the transcription defense genes [18, 19]. The *rolB* and *rolC* genes are absolutely essential for induction of hairy roots [20]. Fast growing and genetically stable hairy roots can be efficiently cultured in large scale bioreactors [21]. Besides, hairy root cultures are usually capable of producing the same compound(s) of identical chemistry found in wild-type roots of the naturally occurring parent plant without loss of structural integrity and/or quantity or concentration of the product, which is frequently observed in callus or cell suspension cultures [22]. *A. rhizogenes* to regulate the genes that were involved in the plant secondary metabolite production [23]. Hairy roots induced from different plant tissues generally grow fast, are genetically stable, and often, but not always, simulate the biochemical profiles of plant roots, which makes hairy roots an attractive system for producing valuable secondary metabolites. Plant roots can synthesize, store, and secrete a vast array of compounds, and transformed root cultures have a wide range of biosynthetic capacities [24]. Various advantages of hairy root culture over cell suspension culture include genotypic and biochemical stability, cytodifferentiation, and growth in hormone free medium. These factors play a vital role during secondary metabolite production. Fast growth, low doubling time, ease of maintenance of hairy roots, and their ability to synthesize a large range of chemical compounds offer an additional advantage as a continuous source for the production of valuable secondary metabolites [25]. A number of secondary metabolites have been reported to be produced from hairy root cultures [26]. Progress has been made on commercialization of hairy root products. ROOTec Bioactives Ltd., founded in 2005 in Switzerland, currently produces phytochemicals from hairy roots induced from 17 plant species in their proprietary mist bioreactors. In the future, more investigations could be directed toward determining the efficacy of crude hairy root extracts or hairy root-produced chemicals. There have been few reviews in the literature on a wide variety of hairy root applications in secondary metabolites of medicinal plants. Previously, very few studies of secondary metabolite production in hairy root cultures of cucurbits have been reported. First time, we focus this chapter on establishment of hairy roots and production of secondary metabolites in cucurbits.

2 Establishment of Hairy Root Cultures in Cucurbits

2.1 Hairy Root Induction of Cucurbits

Hairy roots were induced from various explants (leaf, cotyledon, hypocotyl, node, and root) after 3–4 weeks of culture. Control explants failed to induce hairy root formation. High induction of hairy roots was observed in leaves compared to other explants in *Gynostemma pentaphyllum*, *Cucumis anguria*, *Momordica charantia*, and *M. dioica* [27–30]. Cotyledon explants produced higher frequency of hairy root induction in *Cucumis melo* [31–33] and *C. sativus* [34, 35]. Transgenic frequency (80%) of the infected stems of *Luffa cylindrica* formed vigorous hairy roots within 4 weeks from the inoculation of the bacteria [36]. Table 1 shows the different strains of *Agrobacterium rhizogenes* (MAFF 03–01724, K599, R1000, C58C1, A4, 1855, MTCC 532, ATCC15834, R1601, KCTC 2703, and KCTC 2704) that were examined for their ability to induce hairy roots of various cucurbits such as melon, pumpkin, cucumber, sponge gourd, Chinese cucumber, southern ginseng, bitter melon, spine gourd, and bur gherkin [28–45]. Two strains of *A. rhizogenes* differed in their ability to induce hairy roots, with strain KCTC 2703 being more effective than KCTC 2704 [28–30]. Monocyclic phenolic compound, acetosyringone incorporated into the nutrient medium showed enhanced transformation frequency than the medium without it. Acetosyringone was used for co-cultivation in *C. sativus* [34] and *M. charantia* [41]. Acetosyringone is an amino acid derivative which served as a nutrient source for the invading *Agrobacterium* and enhanced the transformation rate. It was reported that acetosyringone would induce the *vir* gene of *Agrobacterium* cultures. The established hairy roots show typical morphological characteristics with rapid growth on phytohormone-free medium, lack of geotropism, and extensive lateral branching [27–30].

2.2 Molecular Confirmation of Hairy Roots in Cucurbits

The transformed root was confirmed by PCR to determine the presence of a T-DNA sequence in their genomes in *Gynostemma pentaphyllum* [27]. The PCR products from the hairy roots for *rolB* regions but not from untransformed roots of *G. pentaphyllum*. This finding indicated that the *rolB* genes from the Ri plasmid of *A. rhizogenes* were integrated into the genome of *G. pentaphyllum* hairy roots. The negative results of PCR amplification for the *virC* gene demonstrated that no bacterial DNA was involved in *rolB* amplification leading to false positives [27]. The transgenic nature of hairy roots was confirmed by PCR using *rolC* and *aux1* gene specific primers, and transgenicity was also confirmed by polymerase chain reaction (PCR), reverse-transcriptase PCR (RT-PCR), and sequencing in *C. anguria*, *M. charantia*, and *M. dioica* [28–30]. The integration of Ri T-DNA into the genome of plant cells caused the formation of hairy roots, in which *rol* and *aux* genes were harbored in *M. dioica* [30]. PCR analysis targeted the *A. rhizogenes* *rolC*, *aux1*, and *virD2* genes in *M. dioica*. The *rolC* and *aux1* genes, located on

Table 1 Secondary metabolites produced in hairy root cultures of cucurbits

Botanical name of plant	Common name	Agrobacterium strains and plasmids	Phytochemicals	Biological activities	References
Cucumis melo	Melon	MAFF 03–01724 (*pRi1724*)	Aroma essential oils (Z)-3-hexenol, (E)-2-hexenal, 1-nonanol, and (Z)-6-nonenol	Antimicrobial	[32]
Cucumis melo	Oriental melon	K599 (*t-PA*)			[33]
Cucumis melo	Oriental melon	K599 or DCAR2 (*pKGWFS7.0-35SP*)			[37]
Cucurbita pepo	Pumpkin	R1000 (*pRiA4b* and MSU440 *pRiA4*)			[44]
Cucurbita pepo	Pumpkin	C58C1 *(pArA4abc)* and C58C1 *(pArA4b)*			[45]
Cucumis sativus	Cucumber	A4 (*pCAMBIA 1301*)			[34]
Cucumis sativus	Cucumber	A4 (*pARC8*)			[42]
Cucumis sativus	Cucumber	A4T C58 with *pTiC58 pRiA4*			[35]
Cucumis sativus	Cucumber	A4*rolABC*			[43]
Luffa cylindrica	Sponge gourd	1855	Ribosome-inactivating protein (RIPs)	Antiviral	[36]
Luffa cylindrica	Sponge gourd	1855	Ribosome-inactivating protein	Cytotoxic	[79]
Trichosanthes kirilowii var. japonica	Chinese cucumber	ATCC15834	Karasurins, ribosome-inactivating proteins (RIPs)	Antiviral	[39]
Trichosanthes kirilowii var. japonica	Chinese cucumber	R1601	Triterpenoids (bryonolic acid, chondrillasterol)	Cytotoxic, growth inhibition of B-16 melanoma cells	[40]
Trichosanthes kirilowii var japonicum	Chinese cucumber	ATCC 15834	Defense related proteins		[38]

Gynostemma pentaphyllum	Poor man's ginseng or Southern ginseng	ATCC 15834	Triterpene saponins (gypenosides)	Antitumor, cholesterol lowering, immunopotentiating, antioxidant, hypoglycemic, antidiabetic	[27]
Gynostemma pentaphyllum	Poor man's ginseng or Southern ginseng	ATCC 15834	Triterpene saponins (gypenosides)	Antitumor, cholesterol lowering, immunopotentiating, antioxidant, hypoglycemic, antidiabetic	[66]
Momordica charantia	Bitter melon	KCTC 2703 and KCTC 2704	Hydroxybenzoic acids, hydroxycinnamic acids, flavonols	Antioxidant, antibacterial, antifungal	[29]
Momordica charantia	Bitter melon	ATCC 15834	Charantin	Antidiabetic	[41]
Momordica dioica	Spine gourd	KCTC 2703	Hydroxybenzoic acids, hydroxycinnamic acids, flavonols	antioxidant, antibacterial, antifungal, and antiviral	[30]
Cucumis anguria	Bur gherkin	KCTC 2703	Hydroxybenzoic acids, hydroxycinnamic acids, flavonols	antioxidant, antibacterial, antifungal	[28]

independent T-DNAs (TL-DNA and TR-DNA, respectively) of the Ri plasmid of *A. rhizogenes* strain, are diagnostic for T-DNA integration into the host genome. The *virD2* gene, located outside the T-DNA, is diagnostic for the presence of any remaining *Agrobacteria* in the root tissue [30]. The *rol* and *aux* genes are essential for the induction of hairy roots, and they act as a potential activator of secondary metabolites in cucurbits [28–30]. The *virD2* gene was used to verify the complete absence of *A. rhizogenes* in the hairy roots lines of *C. anguria* and *M. dioica*. This result indicates that pRi T-DNA fragments of *A. rhizogenes* were successfully integrated into the genome of *C. anguria* and *M. dioica* without bacterial residues [28, 30]. The obtained full length coding sequence of *rolC* gene of *M. charantia* and *M. dioica* [29, 30]. The use of PCR combined with DNA sequencing instead of Southern blotting for the characterization of transgenic plants has the advantage that the newly inserted genes can be detected at an earlier stage with less DNA and less plant material [29, 30]. The presence of pRi T-DNA in pumpkin long-term hairy root cultures was determined by Southern hybridization [31]. Integration of the T-DNA region of Ri-plasmids into the plant genome was confirmed by both opine assay on paper electrophoresis and PCR-based detection of *rol* genes in *Trichosanthes kirilowii* var. *japonica* [39].

Successful integration of the T-DNA into chromosomal DNA of the KMH-009 was first examined by PCR amplifying the *rolC* gene located on the integrated T-DNA in *Cucumis melo* [32]. An immunoblot analysis of the Oriental melon transgenic hairy root extract revealed 97 kDa single bands coincident with the molecular weight of the GFP GUS fusion proteins. ELISA demonstrated that the highest level of GFP-GUS fusion protein expression was 0.47% of the total soluble protein in a transgenic hairy root of Oriental melon [37]. The integration of T-DNA containing a *gus* reporter gene in hairy root lines was confirmed at low copy numbers ranging from 1 to 4 copies using quantitative real-time PCR, and histochemical staining of cucumber hairy roots showed overexpression of the *gus* gene when driven with the CaMV 35S promoter in *C. sativus* [34]. The presence of GUS activity and its localization were observed in all of the tissues of the root, especially in transgenic cucumber hairy root lines with the CaMV 35S and CaMV 35ST/AMV promoters. The transgenic cucumber hairy roots lines with the CaMV 35S promoter or the CaMV 35ST promoter showed localized GUS activity only in the vascular bundles in *C. sativus* [34]. Quantification of the copy number of the *gus* gene using absolute quantification in real-time PCR revealed a low copy number of the GUS gene per genome [34]. The transgenic plants looked normal and were positive for the neomycin phosphotransferase II. Southern blot analysis of the transgenic plants revealed that all plants contained vector DNA, but only some of them contained DNA from the Ri plasmid [42]. Enzyme-linked immunosorbent assay (ELISA) revealed the highest levels of the recombinant t-PA accumulation in transgenic hairy roots carrying the t-PA transgene under the control of single and dual *rolD* promoters as compared to triple and quadruple *rolD* promoters [33].

Previously, it was reported that changes in secondary metabolite production in hairy roots and Ri plants correlate with changes in the phenotype induced by the insertion of *rol* genes and with the quantity of the polypeptide encoded by the *rolC* gene [29, 30]. Interestingly, both the capacity to grow and produce nicotine in hairy roots and Ri plants of *Nicotiana tabacum* cv. Xanthi were higher after integration of the three *rol* genes (*A*, *B*, *C*) together than with *rolC* alone. In addition, the level of nicotine accumulation was positively correlated with the levels of the polypeptide encoded by the *rolC* gene, as detected by immunoassays [28–30]. The *rolA* gene appears to be an activator of growth and secondary metabolism. Although the r*olB* gene has emerged as the most powerful stimulator, its use is presently disputed owing to its growth-suppressing effect. More positively, the self-activation of *rolC* gene seems to be promising [28–30].

2.3 Growth Kinetics of Hairy Root Culture in Cucurbits

The time profile of the growth of hairy roots in liquid culture was reported in *C. anguria*, *M. charantia*, and *M. dioica* [28–30]. The sixth day was the lag period of hairy root growth; then it began to increase gradually during the eleventh day. The exponential growth stage during the 21 days was followed by the stationary phase during the 15–25 days. The higher fresh mass (FM) and dry mass (DM) was observed at 20, 21, and 22 days of culture of *C. anguria*, *M. charantia,* and *M. dioica* [28–30]. The culture duration of 49 days of hairy roots increased about 120-fold compared to inoculum and the gypenoside content in *G. pentaphyllum* [27]. Hairy root cultures showed a sigmoidal growth curve, and crude extracts showed a progressively increasing translational inhibitory activity that reached the maximum value during the early stationary phase of *L. cylindrica* [36].

2.4 Effects of Sucrose Concentration and Different Media on Biomass Accumulation of Hairy Root Cultures of Cucurbits

Sucrose is the most significant carbon source for plant tissue cultures and helps as the chief energy source and an important constituent in secondary metabolite biosynthesis in cucurbits [29]. The amount of sucrose usually affects the accumulation of secondary metabolites in cultures. About 3% of sucrose produced the higher amount of biomass accumulation and metabolite production in *C. anguria*, *M. charantia*, and *M. dioica* [28–30, 41]. About 2% of sucrose induced hairy root induction in *Trichosanthes kirilowii* var. *japonica* [39]. Many previous reports focus on the composition of medium nutrients to achieve maximum accumulation of metabolites in cultured cells [46]. The different media, full and half strength MS, B5, NN, and LS were employed in hairy root culture and the results shown that MS medium was superior for biomass accumulation in *G. pentaphyllum*, *C. anguria*, *M. charantia*, *M. dioica*, *Cucumis melo*, *C. sativus*, and *T. kirilowii* [27–34, 38, 41, 42]. However, other media like B5 was also used to induce hairy roots in *L. cylindrica* [36]. Hairy root induction of cucurbits using carbohydrate source is by sucrose and nutrients media is by MS or B5.

3 Production of Valuable Secondary Metabolites from Hairy Root Cultures of Cucurbits

Increased secondary metabolite production in hairy roots cultured in vitro, over their wild-type counterparts, may be seen as one of the most exciting spin-offs of biotechnology. Due to their great richness in secondary products, such as triterpenoids and phenolic compounds, plants represent an immense source of therapeutic and/or industrial compounds. For example, plant-derived biomolecules, such as saponins (*G. pentaphyllum*), triterpenoids of bryonolic acid, and chondrillasterol (*Trichosanthes kirilowii* var. *japonica*), ribosome-inactivating protein (*Luffa cylindrica*), charantin (*Momordica charantia*), hydroxybenzoic acids, hydroxycinnamic acids, and flavonols (*C. anguria*, *M. charantia* and *M. dioica*), are efficient in the treatment of different pathology types relating to cancer, cardiovascular and metabolic disorders, and/or other infectious diseases (Table 1 and Fig. 1). Many plant metabolites are commercially available as drugs, flavors, food additives, cosmetics, fragrances, and insecticides. Here, several important phytochemicals from hairy roots of cucurbits are discussed (Fig. 1).

3.1 Phenolic Compounds of Hairy Root Cultures in Cucurbits

Phenolic compounds are secondary metabolites, ubiquitous in plants and plant derived foods. They show a large diversity of structures, including rather simple molecules (e.g., vanillin, gallic acid, caffeic acid) and polyphenols such as stilbenes, flavonoids, and polymers derived from these various groups [47]. Phenolic compounds are classified into three major groups based on the number and binding position of exchangeable hydroxyl groups on aromatic compounds: simple phenol and phenolic acid group, hydroxycinnamic acid derivative group, and flavonoid group. A majority of the plant phenolic metabolites are derived from the aromatic amino acids that are synthesized from the shikimate pathway. Phenolics are collectively valued for their wide variety of health-promoting activities. Flavonoids are phenylpropanoid metabolites, most of which are synthesized from *p*-coumaroyl-CoA and malonyl-CoA, and share their precursors with the biosynthetic pathway for lignin biosynthesis [48]. Flavonoids are low-molecular-weight compounds having approximately 15 atoms of carbon, which are organized in a C6−C3−C6 configuration [49]. More than 9000 flavonoids have thus far been identified in plants [50]. Phenolic acids are considered as simple phenolics, and they are categorized into two groups, i.e., the hydroxybenzoic and hydroxycinnamic acids. Hydroxybenzoic acids have C6−C1 arrangement, such as protocatechuic, vanillic, gallic, syringic, and *p*-hydroxybenzoic acids, while hydroxycinnamic acids have C6−C3 arrangement, such as *p*-coumaric, sinapic, caffeic, and ferulic acids. Flavonoids and phenolic acids are considered as potentially health-promoting substances and have a number of physiological properties, such as antithrombotic,

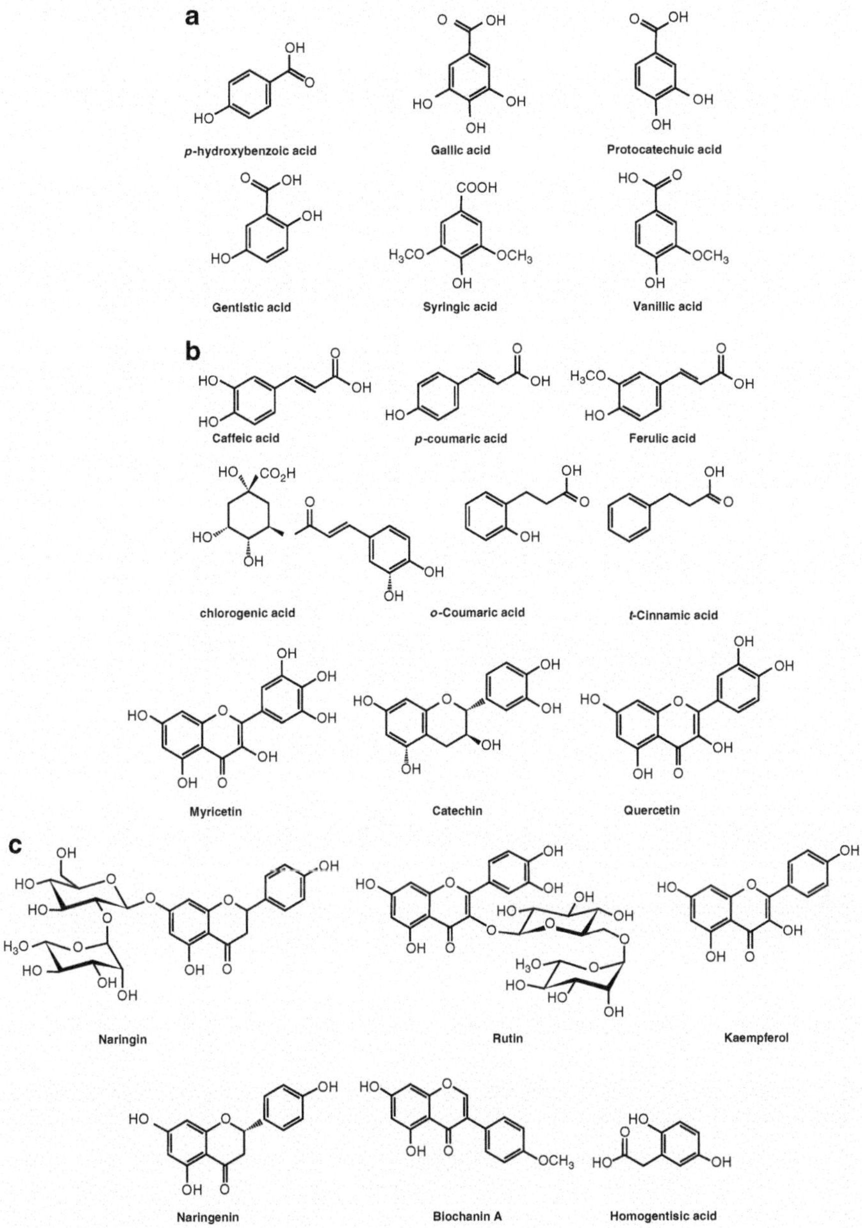

Fig. 1 (continued)

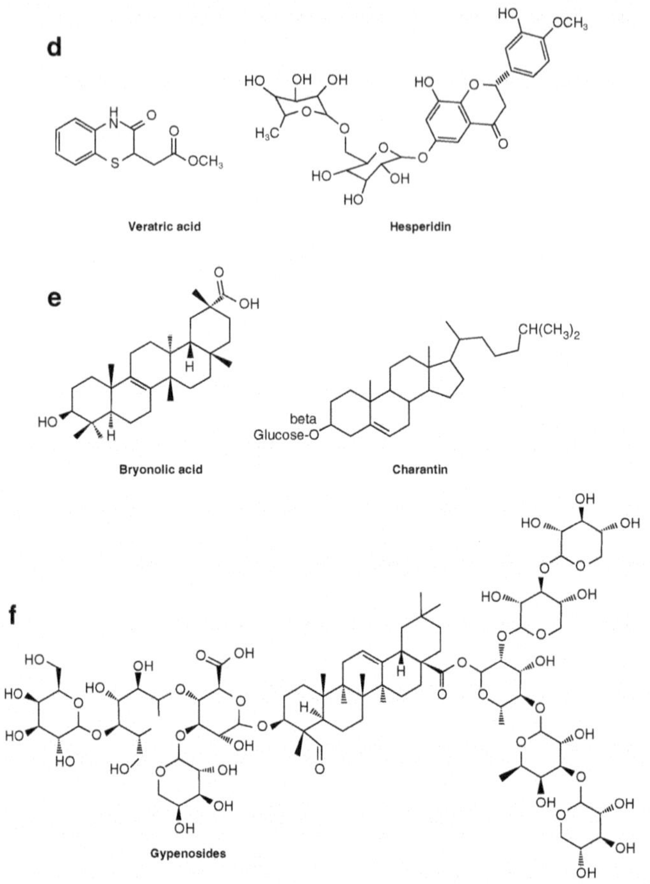

Fig. 1 Chemical structures of selected plant secondary metabolites produced in hairy root cultures of cucurbits

antiallergenic, antimicrobial, cardioprotective, anti-inflammatory, antioxidant, artherogenic, and vasodilatory effects [51–53]. Phenolic compounds are synthesized via the phenylpropanoid pathway that begins with conversion of phenylalanine to cinnamic acid by phenylalanine ammonia lyase (PAL). In the last few years, great attention has been paid to the bioactive compounds due to their ability to promote benefits for human health, such as the reduction in the incidence of some degenerative diseases like cancer and diabetes [54, 55], reduction in risk factors of cardiovascular diseases [10, 56], antioxidant, antimutagenic, antiallergenic, anti-inflammatory, and antimicrobial effects [49, 57], among others. Due to these countless beneficial characteristics for human health, researches have been intensified aiming to find fruits, vegetables, plants, agricultural, and agro-industrial residues as sources of bioactive phenolic compounds.

3.1.1 Individual Phenolic Compounds of Hairy Root Cultures in Cucurbits

The qualitative and quantitative analysis of phenolic compounds from hairy roots and untransformed (roots from in vitro seedling) root extracts of *C. anguria*, *M. charantia*, and *M. dioica* [28–30] were studied using ultra-HPLC. The phenolic compounds in the *C. anguria*, *M. charantia*, and *M. dioica* extracts were identified by comparisons of the retention time and UV spectra of authentic standards and the quantitative data were calculated from calibration curves [28–30]. Both transgenic and nontransgenic roots contained flavonols, hydroxycinnamic, and hydroxybenzoic acids. Hairy roots contained higher amounts of flavonols compared to nontransgenic roots of *C. anguria*, *M. charantia*, and *M. dioica* [28–30]. Myricetin, quercetin, catechin, kaempferol, and rutin levels were higher in hairy roots compared to nontransgenic roots of *C. anguria* [28]. The contents of naringenin and biochanin A were lower in concentrations in hairy roots than nontransgenic roots of *C. anguria* [28]. Myricetin, quercetin, catechin, kaempferol, rutin, biochanin A, and naringenin levels were higher in hairy roots compared to nontransgenic roots of *M. charantia* [29]. Naringin was presented in nontransgenic roots, but it was absent in hairy roots of *M. charantia* [29]. Quercetin, kaempferol, catechin, and rutin levels were higher in hairy roots compared to nontransgenic roots of *M. dioica* [30]. Myricetin, naringenin, and biochanin A contents were lower in concentrations in hairy roots than nontransgenic roots of *M. dioica* [30]. Kaempferol, myricetin, naringin, quercetin, and rutin have antimicrobial activity against human pathogenic microorganisms with some mechanisms of action such as inhibition of nucleic acid synthesis, cytoplasmic membrane function, and energy metabolisms [58].

Caffeic acid and chlorogenic acid were major hydroxycinnamic acid derivatives in hairy roots and nontransformed roots compared to *p*-coumaric acid, ferulic acid, *o*-coumaric acid, and *t*-cinnamic acid in *C. anguria* [30]. Caffeic acid, ferulic acid, *o*-coumaric acid, and *t*-cinnamic acid levels decreased in hairy roots compared to nontransformed roots of *C. anguria* [30]. Chlorogenic acid and *p*-coumaric acid contents were higher in hairy roots than nontransformed roots of *C. anguria* [30]. Chlorogenic acid containing plant materials have been shown to have antiviral, antifungal, and strong antibacterial activities [59]. Chlorogenic acid, *p*-coumaric acid, and ferulic acid levels were higher in hairy roots compared to nontransformed roots of *M. dioica* [30]. Caffeic acid, *o*-coumaric acid, and *t*-cinnamic acid contents were lower in hairy roots than nontransformed roots of *M. dioica* [30]. Caffeic acid, *p*-coumaric acid, *o*-coumaric acid, chlorogenic acid, and *m*-coumaric acid levels were higher and ferulic acid content was lower in hairy roots than nontransgenic roots of *M. charantia* [29]. Protocatechuic acid, *β*-resorcylic acid, syringic acid, gentisic acid, and salicylic acid levels were higher and gallic acid, *p*-hydroxybenzoic acid, and vanillic acid were lower in hairy roots than nontransgenic roots of *C. anguria* [28]. Gallic acid, *p*-hydroxybenzoic acid, gentisic acid, and salicylic acid levels were higher and protocatechuic acid, *β* -resorcylic acid, and vanillic acid were lower in hairy roots compared to nontransformed roots of *M. dioica* [30]. Gentisic acid has an effective role in the anticarcinogenetic activity [60]. Gallic acid, protocatechuic acid, β-resorcylic acid, vanillic acid, syringic acid, gentisic acid, and salicylic acid levels were higher and *p*-hydroxybenzoic acid was lower in hairy roots compared to nontransformed roots of *M. charantia* [29]. Veratric acid was higher and vanillin, hesperidin, and homogentisic acid were lower in hairy roots compared to nontransformed roots of *C. anguria* [28]. Vanillin was higher and veratric acid, hesperidin, and homogentisic acid levels were lower in *M. charantia* and *M. dioica* [29, 30].

3.1.2 Total Phenolic Compounds of Hairy Root Cultures in Cucurbits

Previous studies have revealed that polyphenolic compounds are commonly found in both edible and nonedible plants and that they have multiple biological effects, including antioxidant activity [61]. Flavonoids and other phenolic substances may play a preventive role in the development of cancer and heart disease [62]. Biological activities related to antibacterial and antioxidant activities may be correlated with total polyphenol and flavonoid contents [63]. The total phenolic and flavonoid contents were higher in hairy roots compared to untransformed roots of *C. anguria*, *M. charantia*, and *M. dioica* [28–30].

3.2 Triterpenoids of Hairy Root Cultures in Cucurbits

Gypenosides (Gyp) are the major components of *Gynostemma pentaphyllum* Makino, a Chinese medicinal plant. Phytochemical studies of *G. pentaphyllum* have identified approximately 90 dammarane-type saponin glycosides, known as gypenosides, which are responsible for its pharmacological activities [64]. Saponins are a class of chemical compounds found in particular abundance in various plant species. More specifically, they are amphipathic glycosides grouped phenomenologically by the soap-like foaming they produce when shaken in aqueous solutions and structurally by having one or more hydrophilic glycoside moieties combined with a lipophilic triterpene derivative. Triterpenoid saponins are triterpenes which belong to the group of saponin compounds. Triterpenes are a type of terpene containing 30 carbon atoms. Triterpenes are assembled from a five-carbon isoprene unit through the cytosolic mevalonate pathway to make a thirty-carbon compound. Cucurbitacins are triterpenoids that confer a bitter taste in cucurbits such as cucumber, melon, watermelon, squash, and pumpkin. These compounds discourage most pests on the plant and have also been shown to have antitumor properties [65]. Gypenoside content was higher compared to roots of control parent plant of *Gynostemma pentaphyllum* hairy root cultures [27] which was significantly higher than that previously reported for hairy root cultures [66]. Transformed roots can synthesize and store significant quantities of secondary metabolites. Although the hairy roots under these conditions produced approximately 30% to 40% less gypenosides than commercial sources of *G. pentaphyllum*, the growing time was much shorter when compared to field-grown plants. With hairy root cultures, product quality and quantity are easy to control because natural variances in seasonal climates and geographical environments are excluded and culture conditions and process variables are easily optimized [67]. The hairy root cultures have been considered as a potential alternative for production of gypenosides. Several strategies for the enhancement of biomass and gypenosides have been adopted like the effects of medium compositions, culture conditions, and elicitations [27].

3.3 Ribosome-Inactivating Protein of Hairy Root Cultures in Cucurbits

Ribosome-inactivating proteins (RIPs) are widely distributed plant enzymes that inhibit protein synthesis by virtue of their N-glycosidic activity, selectively cleaving an adenine residue from a highly conserved and surface-exposed stem loop structure in the 28S rRNA [68]. This cleavage prevents the binding of the EF-2/GTP complex, with the subsequent arrest of protein synthesis leading to autonomous cell death [69]. RIPs are either enzymatically active single polypeptides (type I) or heterodimers (type II). A type II RIP consists of an A chain, functionally equivalent to a type I RIP, which is attached to a sugar-binding B chain [70]. Besides RNA N-glycosidase activity, some RIPs have ribonuclease, DNase, DNA glycosylase, and apurinic/apyrimidic lyase activities [71, 72]. In addition, RIPs from *Trichosanthes kirilowii* cell cultures have been demonstrated to possess chitinase activity [73]. Certain type I RIPs display a variety of antimicrobial activities, including antifungal, antibacterial [74], and broad-spectrum antiviral effects against different plant and animal viruses [75], including human immunodeficiency virus [76]. RIPs have been studied as potential tumor cytotoxic agents, both in their native form and after conjugation with monoclonal antibodies. RIP activity of *L. cylindrica* plantlets, grown in vitro on MS medium, was evaluated in crude extracts from different parts and organs and compared to the inhibitory activity shown by extracts from seeds and from transformed roots [36]. The inhibitory activity, as far as normal, nontransformed tissues are concerned, is in agreement with what was already known from previous reports of *L. cylindrica* [77, 78]. RIP-producing hairy roots promise to be much more stable than conventional in vitro grown calluses and cell suspensions [36]. This study tested the sc-RIP extracts from the seeds and hairy root tissue cultures of *Luffa cylindrica* (established by transformation with *Agrobacterium rhizogenes* strain 1855) for inhibitory effects on the growth of in vitro melanotic and amelanotic human melanoma cell lines [79]. The results reported that RIPs can be produced and purified from hairy root cultures, in good agreement with what has been recently reported [38] for hairy root lines of *Trichosanthes kirilowii*. Ribosome-inactivating proteins (RIPs) from plants catalytically damage eukaryotic ribosomes, making them unable to perform the elongation step of protein synthesis. Type 1 RIPs are single-chain proteins, whereas type 2 RIPs consist of two polypeptide chains and possess a galactose-specific binding domain to cell surfaces. Type 1 RIPs are more common and have been identified and purified from more than 30 plants. Interest in type 1 RIPs has been growing due to their widespread physiological activities as abortifacient agents and immunotoxins [39]. The antiviral activity of RIPs has also focused attention on their potential use as anti-HIV agents [39]. There have been few reports on the production of RIPs by plant tissue or cell cultures. A low level of trichosanthin was reported to accumulate in transformed hairy root cultures of *Trichosanthes kirilowii* var. *japonica* [38]. Trichosanthin was also identified in cell extracts of the transformed callus tissues resulting from infection by *Agrobacterium rhizogenes* but not in the untransformed callus of *T. kirilowii* [39]. The major protein in the basic protein fraction was tentatively identified as a class III chitinase based on the N-terminal amino acid sequence. This is consistent with the report [38], who identified two major extracellular basic proteins and one intracellular basic protein produced by *T. kirilowii* var. *japonica* hairy roots as class III chitinases. However, the N-terminal sequence of HR-PB 1 was very similar to but not identical with the sequence of any of these proteins.

3.4 Charantin of Hairy Root Cultures in Cucurbits

A molecule of charantin consists of aglycone or a steroidal portion, which is highly soluble in relatively nonpolar solvent such as chloroform and dichloromethane. However, the glucosides attached to its molecules make it slightly soluble in polar organic solvents such as ethanol or methanol. Conventionally, isolation of this compound involves extraction with mixtures of these solvents using Soxhlet apparatus. Chloroform is highly toxic and carcinogenic, and its use has now been replaced with its much less toxic relative, dichloromethane, which still carry some health risks. Chronic exposure to dichloromethane has been linked to cancer of lungs, liver, and pancreas in laboratory animals. It is a mutagen and may cause birth defect if women were exposed to it during pregnancy [80]. This compound could be used to treat diabetes and can potentially replace treatment by injection of insulin which has not been successful in stimulating the pancreas of the diabetic patients to lower blood sugar to the desired level [81]. In some cases, the injected patient shows signs of side effects. Plant derived compounds that show antidiabetic property such as charantin and others are now being widely accepted as an alternative medicine for diabetes mellitus, and they are free from side effects [82]. Charantin, a naturally occurring steroidal glycoside, is widely distributed throughout the plant of *Momordica charantia*. The presence of charantin was confirmed by performing thin layer chromatography (TLC) in hairy roots as well as in fruit and leaf [41]. The charantin content was lower in hairy roots compared to leaf and fruit of *M. charantia* [41].

3.5 Aroma Essential Oils of Hairy Root Cultures in Cucurbits

The typical cucumber flavor results from the enzymatic action of LOX on linolenic and linoleic acids, which introduces molecular oxygen at C13 or C9, forming 13-hydroperoxylinolenic acid (13-HPOT) or 9-hydroperoxylinolenic acid (9-HPOT). HPL cleaves 13-hydroperoxide (13-HPO) and 9-HPO to produce the C6 and C9 aldehydes that are responsible for the cucumber flavor [83]. These aldehydes can then be reduced to the corresponding C6 alcohols by alcohol dehydrogenase (ADH). Studies have reported that only the oxylipin metabolic pathway contributes to aldehyde and alcohol content and hence flavor [84]. To date, 78 volatile compounds have been identified in cucumber fruits, including aldehydes, alcohols, esters, alkanes, furfurans, and others [85], and (E,Z)-2,6-nonadienal and (E)-2-nonenal are the main aroma compounds [86]. The hairy roots could newly synthesize some essential oils such as (Z)-3-hexenol, (E)-2-hexenal, 1-nonanol, and (Z)-6- nonenol, which were reported to be important aroma volatiles in melon [87]. The stable production of the fruity aroma volatiles by the KMH-009 was assessed by comparing the yields of the compounds from the hairy roots repeatedly subcultured for more than 3 years. The data revealed that the essential oils for aroma scent were constantly synthesized with no relation to the increased number of times for subculture, and the constant production by this clone was successfully maintained in the hairy roots repeatedly subcultured of *Cucumis melo* [32]. The volatile compounds were extracted and identified by GLC-mass spectrometry. Some essential oils such as (Z)-3-hexenol, (E)-2-hexenal, 1-nonanol, and (Z)-6-nonenol were stably synthesized by these hairy roots despite the increased number of subcultures. The productivity of these compounds by the best hairy root line was shown to be considerably higher than naturally ripened melon fruits [32].

4 Cucurbit Hairy Root Cultures of Secondary Metabolites Using for Biological Activities

Phenolic compounds have multiple additional roles in plants, including attracting insects for seed dispersion and pollination. They are also part of the natural defense system against insects, fungi, viruses, and bacteria, and they can act as plant hormone controllers. Moreover, in recent years, phenolic compounds have been intensively investigated because of their potential health-promoting effects [88, 89]. They have been reported to possess many useful properties for human health, including anti-inflammatory, enzyme inhibition, antimicrobial, antiallergic, vascular, and cytotoxic antitumor activity, but the most important action of phenolics is their antioxidant activity [58, 89, 90]. It has been demonstrated recently that quercetin and kaempferol synergistically suppress cell proliferation in human gut cancer lines [91]. The translational inhibitory activity found in extracts from our hairy root cultures is the highest that has been found in various tissues of *L. cylindrica*, including seeds [36].

4.1 Antioxidant Activity

The antioxidant potential of hairy roots and nontransformed roots were determined using free radicals scavenging, reducing potential, phosphomolybdenum assays, and chelating effects on ferrous ions. The highest antioxidant activity was exhibited in hairy roots compared to nontransformed roots in *C. anguria*, *M. charantia*, and *M. dioica* [28–30]. Reducing capacity of extracts suggests that hairy roots were more potential when compared to untransformed roots in *C. anguria*, *M. charantia*, and *M. dioica* [28–30]. The antioxidant capacity shown by phosphomolybdenum method was higher in the hairy root extract than nontransformed root extract of *C. anguria*, *M. charantia*, and *M. dioica* [28–30]. The percentage of metal scavenging capacity of transgenic hairy roots was higher than nontransgenic roots of *M. charantia*, *C. anguria*, and *M. dioica* [28–30]. Hairy roots exhibited higher antioxidant activity in *M. charantia* [29].

4.2 Antimicrobial Activity

The hairy roots and nontransformed roots of *C. anguria*, *M. charantia*, and *M. dioica* revealed varying antibacterial activity, as exposed by the growth inhibition zones [28–30]. The results from the disc diffusion method indicated that both hairy roots and nontransformed root extracts had comparable antibacterial effects against Gram positive and Gram-negative bacteria. Hairy roots exhibited highest activity with both Gram-positive and Gram-negative bacteria compared to nontransformed roots of *C. anguria*, *M. charantia*, and *M. dioica* [28–30]. Gram-positive (*S. aureus*) bacteria exhibited greater inhibition compared to Gram-negative (*P. aeruginosa* and *E. coli*) bacteria in *M. charantia*, *C. anguria*, and *M. dioica* [28–30]. By using the disc diffusion method against the fungal strains, it can be seen that extracts of *M. charantia*, *C. anguria*, and *M. dioica* hairy roots and nontransformed roots exhibited good antifungal activity [28–30]. Hairy roots exhibited greater inhibition of fungus (*F. oxysporum* and *A. niger*) in hairy roots than nontransgenic roots of *C. anguria*, *M. charantia*, and *M. dioica* [28–30]. Hairy roots exhibited higher antibacterial and antifungal activity compared to nontransformed roots [92, 93]. Flavonoid derivatives have also been reported to possess antiviral activity against a wide range of viruses such as HSV, HIV, Coxsackie B virus, corona virus, cytomegalovirus, poliomyelitis virus, rhinovirus, rotavirus, poliovirus, sindbis virus, and rabies virus [94]. Cytotoxicity activity and quantitative assay of virus yields using plaque assay were carried out for hairy roots and nontransgenic roots of *M. dioica* [30]. Hairy roots exhibited higher antiviral activity compared to nontransgenic root extracts of *M. dioica* [30]. *M. charantia* was reported to possess several antiviral activities including hepatitis B virus, dengue virus, human immunodeficiency virus (HIV), and influenza A subtypes including H1N1, H3N2, and H5N1 [95]. Hairy roots have potential antiviral activity compared with nontransgenic roots of various plants like *Phylanthus amarus* [96] and *Daucus carota* [97].

4.3 Anticancer Activity

Gypenosides (Gyp) are compounds found in the crude extracts from *G. pentaphyllum* and they have been shown to exert various biological effects such as anti-inflammatory and antioxidative [98], antihyperlipidemic, anticardiovascular [99], and anticancer [100–102]. Our previous studies have shown that gypenosides induced apoptosis in human colon cancer colo 205 cells [103] and human tongue cancer SCC-4 cells through endoplasmic reticulum stress and mitochondria-dependent pathways [104]. Although gypenosides have been shown to induce cell cycle arrest and apoptosis in several human cancer cell lines, there is no available information to address whether gypenosides induce DNA damage or affects DNA repair genes in SAS human oral cancer cells.

4.4 Antidiabetic Activity

Diabetes mellitus is an endocrine metabolic disorder in which the body does not produce sufficient insulin or lack of responsiveness to insulin, resulting in hyperglycemia (high blood glucose level). The classical symptoms include polyuria, polydipsia, weight loss, lethargy, polyphagia, visual blurring, frequent or recurring infections, cuts and bruises that are slow to heal, tingling and/or numbness in hands and/or feet, drowsiness, nausea, and decreased endurance during exercise [105]. A number of potential medicinal components from bitter gourd, such as α and β momorcharin, momordin, and cucurbitacin B, have been isolated. A number of reported clinical studies have shown that bitter gourd extract from fruits, seeds, and leaves contain several bioactive compounds that have hypoglycemic activity in both diabetic animals and humans [106]. Fruits, seeds, and leaves extract of *Momordica charantia* possess hypoglycemic activity in antihyperglycemic activity in alloxan [107] or streptozotocin [108]. The major compounds that have been isolated and identified as hypoglycemic agents include charantin, polypeptide-p, and vicine. Charantin is a steroidal glycoside shown to possess powerful hypoglycemic properties when administered orally and intravenously in diabetic rabbits [109]. Hairy roots produced higher amount of charantin which used for antidiabetics [41].

5 Large Scale Productions of Secondary Metabolites in Bioreactors Using Hairy Root Cultures

The vast potential of hairy root cultures as a stable source of biologically active chemicals has focused the attention of the scientific community for its exploitation. Scaling up of hairy roots in novel bioreactors can provide the best conditions for optimum growth and secondary metabolite production, comparable to or higher than that in native roots. Though the need for developing bioreactors suitable for the hairy root cultivation has long been recognized, root cultures present unique challenges [110]. The complex fibrous structure of the roots makes the growth analysis and development of a large-scale culture system difficult. Hairy root growth is not homogeneous, which affects the reactor performance. Furthermore, the hairy root morphology is quite plastic as the roots respond to the changes in the local environment. Changes in morphology, including changes in the density and length of the root hairs, directly affect the secondary metabolite production from hairy roots

[111]. Thus, bioreactor design for root cultures is a balancing act between the biological needs of the tissues, without inducing an additional, undesirable biological response [112]. Reviews on hairy roots briefly discuss the importance of the use of bioreactors for hairy root cultures [26]. Mechanical agitation causes wounding of hairy roots and leads to callus formation. Due to branching, the roots form an interlocked matrix that exhibits resistance to nutrient flow. Hairy roots are hetrotrophic, respiratory organisms that rely on oxygen for energy generation and other metabolic functions. Substantial progress has been made in understanding the mechanisms of oxygen limitation, one of the principle challenges for large-scale growth of hairy root cultures [113]. Because of the solid phase nature of the roots and the development of oxygen gradients within root tissues, relatively small reductions in the dissolved oxygen concentration in the medium can lead to a significant decrease in growth rate and may also affect the synthesis of certain secondary metabolites. In fact, hairy roots can be oxygen limited even in shake flask cultures [114]. Restriction of nutrient oxygen delivery to the central mass of tissue gives rise to a pocket of senescent tissues. Mass transfer resistances near the liquid and solid boundary affect the oxygen delivery to the growing hairy roots. Thus, exploitation of hairy root culture as a source of bioactive chemicals depends on the development of suitable bioreactor system where several physical and chemical parameters (nutrient availability, nutrient uptake, oxygen, and hydrogen depletion in the medium, mixing, and shear sensitivity) must be taken into account. The design of bioreactors for hairy root cultures should also take into consideration factors such as the requirement for a support matrix and the possibility of flow restriction by the root mass in certain parts of the bioreactor. Several bioreactor designs have been reported for hairy root culture taking into consideration the above factors that permit the growth of interconnected tissue unevenly distributed throughout the culture vessel. Reactors used to culture hairy roots can roughly be divided into three types: liquid-phase, gas-phase, or hybrid reactors that are a combination of both [22]. Previously, there are no reports on the large scale production of hairy roots using bioreactor system in cucurbits and the production of phenolic compounds.

6 Elicitors Increase the Secondary Metabolites Production in Hairy Root Cultures

Various biotic and abiotic elicitors applied to hairy root cultures and their stimulating effects on the accumulation of secondary metabolites. According to their origin, elicitors can be divided into different types: (a) biotic and (b) abiotic. Abiotic elicitors can be considered as substances of nonbiological origin, being predominantly inorganic compounds such as salts or physical factors [115, 116]. Inorganic chemicals like salts or metal ions have been used to increase the production of bioactive compounds by their modification of plant secondary metabolism. Among the many elicitors applied to hairy root cultures, the most common and effective elicitors are fungal cell extracts, polysaccharides from fungal and plant cells, and heavy metal salts. With the crude fungal cell extracts, it is essential to observe the preparation conditions carefully for achieving reproducible effects. In addition to the chemical agents, UV-radiation, hyperosmotic stress, and temperature shift have been shown effective for some plant species/metabolites. Elicitor type, dose, and treatment schedule are major factors determining the effects on the secondary metabolite production. In addition to the accumulation of products in roots, elicitor treatments often stimulate the release of intracellular products. Although elicitation is mainly effective to increase specific product yield on per unit mass of roots, the incorporation of nutrient feeding strategies can be applied to enhance the volumetric product yield. The integration of in situ product recovery from the roots/liquid medium is another synergistic strategy with the elicitor treatment to improve the process. So far, there are no reports on the elicitation of hairy roots and production of phenolic compounds from hairy root cultures of cucurbits. Further, researchers can use the elicitation to improve the contents of secondary metabolites in cucurbits.

7 Conclusions

Hairy root technology has been significantly improved in various fields for past few years. Overall, the major groups of secondary metabolites have already been produced from hairy roots of cucurbits. Compared to plant cell suspension cultures, hairy root cultures appear to be potential systems for continuous production of valuable secondary metabolites because of their fast growth rates, ease of maintenance, genetic and biosynthetic stability, and ability to synthesize a vast array of compounds. Environmental factors, such as light, oxygen, and temperature, as well as abiotic and biotic stress factors, such as phytohormones, heavy metals, and fungal elicitors, have all been applied to hairy roots for increased yield of phytochemicals. In addition to these external stimuli, secondary metabolic pathways have also been modified for enhanced metabolite production, such as overexpression of biosynthetic genes and transcription factors, and suppression of catabolic or competing pathway genes. A better understanding of the biosynthetic pathway and regulation architecture of valuable secondary metabolites is crucial for genetic engineering and fully realizing the biosynthetic potential of hairy roots. The discovery of new genes that participate in the metabolic pathways from hairy root studies increases the tremendous potential of such cultures. It is also predicted that this model of pharmaceutical production is relatively safe. Driven by the demand for productive, robust, and stable hairy root cultures for the production of active agents for the food, cosmetics, and pharmaceutical industry, the development of a direct available measuring method for the biomass concentration of hairy root cultures in liquid medium still does not exist. Transgenic hairy roots grew rapidly than nontransgenic roots in standardized liquid culture conditions and produced greater amount of biomass and phenolic compounds. The higher amount of secondary metabolites possibly contributes to greater biological activity of hairy roots in cucurbits. The genetic and biochemical stability of the hairy roots as well as its high productivity offers an effective platform for further studies on the biosynthetic pathways of phytochemicals. This prediction is strengthened by the observation that emerging private companies have converted this technology to allow production at a commercial scale. Plant biologists can work closely with engineers to tackle the challenges with scaling up hairy root cultures, such as optimal biomass growth and adaptation of the extraction methods to industrial-scale metabolite production. Looking forward, establishment of hairy roots guided by bioassays, augmented by elicitations and genetic manipulations, and coupled with efficient metabolite extractions will streamline the process and allow full exploitation of hairy roots as a production platform of valuable secondary metabolites.

Acknowledgements This paper was supported by the KU Research Professor Program of Konkuk University, Seoul, Republic of Korea.

References

1. Jeffrey C (2005) A new system of Cucurbitaceae. Bot Zhurn 90:332–335
2. Ali MA, Pandey AK (2006) Cucurbitaceae of Bihar diversity and conservation. In: Trivedi PC (ed) Global biodiversity status and conservation. Pointer Publisher, Jaipur, pp 257–263
3. Pandey AK, Varma SK, Ali MA (2006) The genus Luffa in India: diversity and conservation. In: Trivedi PC (ed) Global biodiversity status and conservation. Pointer Publisher, Jaipur, pp 260–270
4. Dhiman K, Gupta A, Sharma DK, Gill NS, Goyal A (2012) A review on the medicinally important plants of the family Cucurbitaceae. Asian J Clin Nutr 4:16–26
5. Whitaker TW, Davis GN (1962) Cucurbits. Interscience Publishers, New York
6. Kunwar RM, Bussmann RW (2008) Ethnobotany in the Nepal Himalaya. J Ethnobiol Ethnomed 4:24
7. Zwenger S, Basu C (2008) Plant terpenoids: applications and future potentials. Biotechnol Mol Biol Rev 3(1):1–7
8. Khan MY, Saleh A, Kumar V, Rajkumar S (2009) Recent advances in medicinal plant biotechnology. Indian J Biotechnol 8:9–22
9. Poulev A, O'Neal JM, Logendra S, Pouleva RB, Timeva V, Garvey AS, Gleba D, Jenkins IS, Halpern BT, Kneer R, Cragg GM, Raskin I (2003) Elicitation a new window into plant chemodiversity and phytochemical drug discovery. J Med Chem 46:2542–2547
10. Kris-Etherton PM, Hecker KD, Bonanome A, Coval SM, Binkoski AE, Hilpert KF (2002) Bioactive compounds in foods: their role in the prevention of cardiovascular disease and cancer. Am J Med 113:71S–88S
11. Nigam PS (2009) Production of bioactive secondary metabolites. In: Nigam PS, Pandey A (eds) Biotechnology for agro-industrial residues utilization. Springer, Netherlands, pp 129–145
12. Einhelling FA, Leather GR (1988) Potentials for exploiting allelopathy to enhance crop production. J Chem Ecol 14:1829–1844
13. Weir TL, Park SW, Vivanco JM (2004) Biochemical and physiological mechanisms mediated by allelochemicals. Curr Opin Plant Biol 7:472–479
14. Jochum GM, Mudge KW, Thomas RB (2006) Elevated temperatures increase leaf senescence and root secondary metabolite concentrations in the understory herb *Panax quinquefolius* (Araliaceae). Am J Bot 94:819–826
15. Pang J, Cuin T, Shabala L, Zhou M, Mendham N, Shabala S (2007) Effect of secondary metabolites associated with anaerobic soil conditions on ion fluxes and electrophysiology in barley roots. Plant Physiol 145:266–276

16. Tepfer D, Tempé J (1981) Production of d'agropine par des racines transformes sous I'action d'Agrobacterium rhizogenes souche A4. CR Acad Sci 292:153–156
17. Chilton MD, Tepfer DA, Petit A, David C, Casse-Delbart F, Tempé J (1982) *Agrobacterium rhizogenes* inserts T-DNA into the genomes of the host-plant root cells. Nature 295:432–434
18. Shkryl YN, Vermeichik GN, Bulgakov VP, Tchernoded GK, Mischenko NP, Fedoreyev SA, Zhuraulev YN (2008) Individual and combined effect of *rolA*, *B* and *C* genes on anthraquinone production in *Rubia cordifolia* on transformed Calli. Biotechnol Bioeng 100:118–125
19. Bulgakov VP (2008) Function of *rol* gene in plant secondary metabolism. Biotechnol Adv 26:318–324
20. Georgiev M, Pavlov A, Bley T (2007) Hairy root type plant in vitro systems as sources of bioactive substances. Appl Microbiol Biotechnol 74:1175–1185
21. Kuz′ma Ł, Skrzypek Z, ska Wysokin′ H (2006) Diterpenoids and triterpenoids in hairy roots of *Salvia sclarea*. Plant Cell Tiss Org 84:171–179
22. Kim Y, Wyslouzil BE, Weathers PJ (2002) Secondary metabolism of hairy root cultures in bioreactors. In Vitro Cell Dev Biol Plant 38:1–10
23. Sudha CG, Sherina TV, Anu Anand VP, Reji JV, Padmesh P, Soniya EV (2013) *Agrobacterium rhizogenes* mediated transformation of the medicinal plant *Decalepis arayalpathra* and production of 2-hydroxy-4-methoxy benzaldehyde. Plant Cell Tiss Org Cult 112:217–226
24. Flores HE, Curtis WR (1992) Approaches to understanding and manipulating the biosynthetic potential of plant roots. Ann N Y Acad Sci 665:188–209
25. Bourgaud F, Gravot A, Milesi S, Gontier E (2001) Production of plant secondary metabolites: a historical perspective. Plant Sci 161:839–851
26. Giri A, Narasu ML (2000) Transgenic hairy roots: recent trends and applications. Biotechnol Adv 18:1–22
27. Chang CK, Chang KS, Lin YC, Liu SY, Chen CY (2005) Hairy root cultures of *Gynostemma pentaphyllum* (Thunb.) Makino: a promising approach for the production of gypenosides as an alternative of ginseng saponins. Biotechnol Lett 27(16):1165–1169
28. Yoon JY, Chung IM, Thiruvengadam M (2015) Evaluation of phenolic compounds, antioxidant and antimicrobial activities from transgenic hairy root cultures of gherkin (*Cucumis anguria* L.). S Afr J Bot 100:80–86
29. Thiruvengadam M, Praveen N, Kim SH, Chung IM (2014) Establishment of *Momordica charantia* hairy root cultures for the production of phenolic compounds and the determination of their biological activities. Plant Cell Tissue Org Cult 118:545–557
30. Thiruvengadam M, Rekha K, Chung IM (2016) Induction of hairy roots by *Agrobacterium rhizogenes*-mediated transformation of spine gourd (*Momordica dioica* Roxb. ex. willd) for the assessment of phenolic compounds and biological activities. Sci Hortic 198:132–141
31. Katavić V, Jelaska S (1991) The influence of plant growth regulators on callus induction in pumpkin (*Cucurbita pepo* L.) hairy roots. Int J Dev Biol 35(3):265–268
32. Matsuda Y, Toyoda H, Sawabe A, Maeda K, Shimizu N, Fujita N, Fujita T, Nonomura T, Ouchi S (2000) A hairy root culture of melon produces aroma compounds. J Agric Food Chem 48(4):1417–1420
33. Kang S, Ajjappala H, Seo HH, Sim JS, Yoon SH, Koo BS, Kim YH, Lee S, Hahn BS (2011) Expression of the human tissue-plasminogen activator in hairy roots of oriental melon (*Cucumis melo*). Plant Mol Biol Rep 29:919–926
34. Anuar MR, Ismail I, Zainal Z (2011) Expression analysis of the 35S CaMV promoter and its derivatives in transgenic hairy root cultures of cucumber (*Cucumis sativus*) generated by *Agrobacterium rhizogenes* infection. Afr J Biotechnol 10(42):8236–8244
35. Mclnnes E, Morgan AJ, Mulligan BJ, Davey MR (1991) Roots induced on cucumber cotyledons by the agropine Ri plasmid TR-DNA exhibit the transformed phenotype. Plant Cell Rep 9:647–650

36. di Toppi LS, Gorini P, Properzi G, Barbieri L, Spanò L (1996) Production of ribosome-inactivating protein from hairy root cultures of *Luffa cylindrica* (L.) Roem. Plant Cell Rep 15:910–913
37. Pak HK, Sim JS, Rhee Y, Ko HR, Ha SH, Yoon MS, Kang CH, Lee S, Kim YH, Hahn BS (2009) Hairy roots induction in Oriental melon (*Cucumis melo*) by *A. rhizogenes* and production of the root-knot nematode. Plant Cell Tiss Org Cult 98:219–228
38. Savary BJ, Flores HE (1994) Biosynthesis of defense-related proteins in transformed root cultures of *Trichosanthes kirilowii* Maxim. var japonicum (Kitam.)'. Plant Physiol 106:1195–1204
39. Mizukami H, Kondo T, Ogihara Y (1997) Basic proteins produced by hairy root cultures of *Trichosanthes kirilowii* var. japonica. Plant Biotechnol 14(1):35–39
40. Takeda T, Kondo T, Mizukami H, Ogihara Y (1994) Bryonolic acid production in hairy roots of *Trichosanthes kirilowii* Max. var Japonica Kitam. Transformed with *Agrobacterium rhizogenes* and its cytotoxic activity. Chem Pharm Bull (Tokyo) 42(3): 730–732
41. Swarna J, Ravindhran R (2012) *Agrobacterium rhizogenes*–mediated hairy root induction of *Momordica charantia* Linn. and the detection of charantin, a potent hypoglycaemic agent in hairy roots. Research J biotechnol 7(4):227–231
42. Trulson AJ, Simpson RB, Shahin EA (1986) Transformation of cucumber (*Cucumis sativus* L.) plants with *Agrobacterium rhizogenes*. Theor Appl Genet 73(1):11–15
43. Amselem J, Tepfer M (1992) Molecular basis for novel root phenotypes induced by *Agrobacterium rhizogenes* A4 on cucumber. Plant Mol Biol 19(3):421–432
44. Ilina EL, Logachov AA, Laplaze L, Demchenko NP, Pawlowski K, Demchenko KN (2012) Composite *Cucurbita pepo* plants with transgenic roots as a tool to study root development. Ann Bot 110(2):479–489
45. Katavić V, Jelaska S, Bakran-Petricioli T, David C (1991) Host-tissue differences in transformation of pumpkin (*Cucurbita pepo* L.) by *Agrobacterium rhizogenes*. Plant Cell Tiss Org Cult 24:35–42
46. Ramachandra Rao S, Ravishankar GA (2000) Biotransformation of protocatechuic aldehyde and caffeic acid to vanillin and capsaicin in freely suspended and immobilized cell cultures of *Capsicum frutescens*. J Biotechnol 76:137–146
47. Cheynier V (2012) Phenolic compounds: from plants to foods. Phytochem Rev 11(2):153–177
48. Stafford HA (1990) Flavone and flavonone pathway. In: Flavonoids Metabolism. CRC Press, Boca Raton
49. Balasundram N, Sundram K, Samman S (2006) Phenolic compounds in plants and agri industrial by-products: antioxidant activity, occurrence, and potential uses. Food Chem 99:191–203
50. Ferrer JL, Austin MB, Stewart CJ, Noel JP (2008) Structure and function of enzymes involved in the biosynthesis of phenylpropanoids. Plant Physiol Biochem 46:356–370
51. Samman S, Naghii MR, Lyons WPM, Verus AP (1998) The nutritional and metabolic effects of boron in humans and animals. Biol Trace Elem Res 66:227–235
52. Middleton EJ, Kandaswami C, Theoharides TC (2000) The effects of plant flavonoids on mammalian cells: implications for inflammation, heart disease, and cancer. Pharmacol Rev 52:673–751
53. Puupponen-Pimia R, Nohynek L, Meier C, Kahkonen M, Heinonen M, Hopia A, Oksman-Caldentey KM (2001) Antimicrobial properties of phenolic compounds from berries. J Appl Microbiol 90:494–507
54. Conforti F, Menichini F, Formisano C, Rigano D, Senatore F, Arnold NA, Piozzi F (2009) Comparative chemical composition, free radical-scavenging and cytotoxic properties of essential oils of six Stachys species from different regions of the Mediterranean Area. Food Chem 116:898–905
55. Kim GN, Shin JG, Jang HD (2009) Antioxidant and antidiabetic activity of Dangyuja (*Citrus grandis* Osbeck) extract treated with *Aspergillus saitoi*. Food Chem 117:35–41

56. Jiménez JP, Serrano J, Tabernero M, Arranz S, Díaz-Rubio ME, García-Diz L, Goñi I, Saura-Calixto F (2008) Effects of grape antioxidant dietary fiber in cardiovascular disease risk factors. Nutrition 24:646–653
57. Parvathy KS, Negi PS, Srinivas P (2009) Antioxidant, antimutagenic and antibacterial activities of curcumin-β-diglusoside. Food Chem 115:265–271
58. Cushnie TPT, Lamb AJ (2005) Antimicrobial activity of flavonoids. Int J Antimicrob Agents 26:343–356
59. Karunanidhi A, Thomas R, Belkum A, Neela V (2013) In vitro antibacterial and antibiofilm activities of chlorogenic acid against clinical isolates of *Stenotrophomonas maltophilia* including the trimethoprim/sulfamethoxazole resistant strain. BioMed Res Int 2013:392058
60. Sharma S, Khan N, Sultana S (2004) Study on prevention of two-stage skin carcinogenesis by Hibiscus rosa sinensis extract and the role of its chemical constituent, gentisic acid, in the inhibition of tumour promotion response and oxidative stress in mice. Eur J Cancer Prev 13:53–63
61. Wojdyło A, Oszmian J, Czemerys R (2007) Antioxidant activity and phenolic compounds in 32 selected herbs. Food Chemi 105(3):940–949
62. Kähkönen MP, Hopia AI, Vuorela HJ, Rauha JP, Pihlaja K, Kujala TS, Heinonen M (1999) Antioxidant activity of plant extracts containing phenolic compounds. J Agric Food Chem 47:3954–3962
63. Daglia M (2011) Polyphenols as antimicrobial agents. Curr Opin Biotechnol 23:174–181
64. Yin F, Hu LH, Pan RX (2004) Novel dammarane-type glycosides from *Gynostemma pentaphyllum*. Chem Pharm Bull 52:1440–1444
65. Shang Y, Ma Y, Zhou Y, Zhang H, Duan L, Chen H, Zeng J, Zhou Q, Wang S, Gu W, Liu M, Ren J, Gu X, Zhang S, Wang Y, Yasukawa K, Bouwmeester HJ, Qi X, Zhang Z, Lucas WJ, Huang S (2014) Plant science. Biosynthesis, regulation, and domestication of bitterness in cucumber. Science 346(6213):1084–1088
66. Fei HM, Mei KF, Shen X, Ye YM, Lin ZP, Peng LH (1993) Transformation of *Gynostemma pentaphyllum* by *Agrobacterium rhizogenes* and saponin production in hairy root cultures. Acta Bot Sin 35:626–631
67. Payne GF, Bringi V, Prince C, Shuler ML (1992) Plant cell and tissue culture in liquid systems. Hanser Publishing, New York
68. Endo Y, Tsurugi K (1988) The RNA N-glycosidase activity of ricin A-chain. The characteristics of the enzymatic activity of ricin A-chain with ribosomes and with rRNA. J Biol Chem 263(18):8735–8739
69. Osborn RW, Hartley MR (1990) Dual effects of the ricin A chain on protein synthesis in rabbit reticulocyte lysate. Inhibition of initiation and translocation. Eur J Biochem 193(2):401–407
70. Nielsen K, Boston RS (2001) Ribosome-inactivating proteins: a plant perspective. Annu Rev Plant Physiol Plant Mol Biol 52:785–816
71. Li MX, Yeung HW, Pan LP, Chan SI (1991) Trichosanthin, a potent HIV-1 inhibitor, can cleave supercoiled DNA in vitro. Nucleic Acids Res 19:6309–6312
72. Roncuzzi L, Gasperi-Campani A (1996) DNA-nuclease activity of the single-chain ribosome-inactivating proteins dianthin 30, saporin 6 and gelonin. FEBS Lett 392:16–20
73. Remi Shih NR, McDonald KA, Jackman AP, Girbes T, Lglesias R (1997) Bifunctional plant defense enzymes with chitinase and ribosome inactivating activities from *Trichosanthes kirilowii* cell cultures. Plant Sci 130:145–150
74. Vivanco JM, Flores HE (2000) Control of root formation by plant growth regulators. In: Basra AS (ed) Plant growth regulators in agriculture and horticulture: their role and commercial uses. Food Products Press, New York, pp 1–25: an Imprint of The Haworth Press
75. Chen ZC, White RF, Antoniw JF, Lin Q (1991) Effect of pokeweed antiviral protein (PAP) on the infection of viruses. Plant Pathol 40:416–620
76. Zarling JM, Moran RA, Haffar O, Sias J, Richmann DD, Spina CA, Myers DE, Kuelbeck V, Ledbetter JA, Uckun FM (1990) Inhibition of HIV replication by pokeweed antiviral protein targeted to CD4+ cells by monoclonal antibodies. Nature 347:92–95

77. Kamenosono M, Nishida H, Funatsu G (1988) Isolation and characterization of two luffins, protein-biosynthesis inhibitory proteins from the seeds of *Luffa cylindrica*. Agric Biol Chem 52:1223–1227
78. Watanabe K, Suemasu Y, Funatsu G (1989) Identification of lysine residue at or near active site of luffin-a, a ribosome inactivating protein from seeds of *Luffa cylindrica*. J Biochem 106:977–981
79. Poma A, Galeota K, Miranda M, Spanò L (1997) A Ribosome-inactivating protein principle from hairy roots and seeds of *Luffa cylindrica* (L) Roem and its cytotoxicity on melanotic and amelanotic melanoma cell lines. Int J Pharm 35(3):212–214
80. Jamwal KS, Anand KK (1962) Preliminary screening of some reputed abortifacient indigeneous plants. Ind J Pharm 24:218–220
81. El-Said SM, Al-Barak AS (2011) Extraction of insulin like compounds from bitter melon plants. Amer J Drug Disc Devel 1:1–7
82. Sharma SR, Dwivedi SK, Vershney VP (1996) Antihyperglycaemic and insulin release effects of *Aegle mamelos* leaves in sterptozotocin-diabetic rats. Phytother Res 10:426–428
83. Matsui K, Ujita C, Fujimoto S, Wilkinson J, Hiatt B, Knauf V, Kajiwara T, Feussner I (2000) Fatty acid 9- and 13-hydroperoxide lyases from cucumber. FEBS Lett 481(2):183–188
84. Finkelstein JA, Schiffman SS (1999) Workshop on taste and smell in the elderly: an overview. Physiol Behav 66(2):173–176
85. Hao LN, Chen SX, Wang CY, Chen Q, Wan XH, Shen XQ (2013) Aroma components and their contents in cucumbers from different genotypes. J Northwest A & F Univ (Nat Sci Ed) 6:25–30
86. Forss D, Dunstone E, Ramshaw E, Stark W (1962) The flavor of cucumbers. J Food Sci 27:90–93
87. Buttery RG, Seifert RM, Ling LC, Soderstrom EL, Ogawa JM, Turnbaugh JG (1982) Additional aroma components of honeydew melon. J Agric Food Chem 30:1208–1211
88. Crozier A, Jaganath IB, Clifford MN (2009) Dietary phenolics: chemistry, bioavailability and effects on health. Nat Prod Rep 26:1001–1043
89. de Pascual-Teresa S, Moreno DA, Garcia-Viguera C (2010) Flavanols and anthocyanins in cardiovascular health: a review of current evidence. Int J Mol Sci 11:1679–1703
90. Chu YH, Chang CL, Hsu HF (2000) Flavonoid content of several vegetables and their antioxidant activity. J Sci Food Agric 80:561–566
91. Ackland ML, Van de Waarsenburg S, Jones R (2005) Synergistic antiproliferative action of the flavonols quercetin and kaempferol in cultured human cancer cell lines. In Vivo 19:69–76
92. Jain N, Light ME, Van Staden J (2008) Antibacterial activity of hairy-root cultures of *Maytenus senegalensis*. South Afr J Bot 74:163–166
93. Thiruvengadam M, Praveen N, Kim EH, Kim SH, Chung IM (2014) Production of anthraquinones, phenolic compounds and biological activities from hairy root cultures of *Polygonum multiflorum* Thunb. Protoplasma 251:555–566
94. Özcelik B, Kartal M, Orhan I (2011) Cytotoxicity, antiviral and antimicrobialactivities of alkaloids, flavonoids, and phenolic acids. Pharm Biol 49:396–402
95. Pongthanapisith V, Ikuta K, Puthavathana P, Leelamanit W (2013) Anti-viral protein of *Momordica charantia* L. inhibits different subtypes of influenza A. Evid Based Complement Alter Nat Med 2013:729081
96. Bhattacharyya R, Bhattacharya S (2004) Development of a potent in vitro source of *Phyllanthus amarus* roots with pronounced activity against surface antigen of the hepatitis B virus. In Vitro Cell Dev Biol Plant 40:504–508
97. Luchakivskaya YS, Olevinskaya ZM, Kishchenko EM, Spivak NY, Kuchuk NV (2012) Obtaining of hairy root: callus and suspension cell cultures of carrot (*Daucus carota* L.) able to accumulate human interferon alpha-2b1. Cytol Genet 46:15–20
98. Shang L, Liu J, Zhu Q, Zhao L, Feng Y, Wang X, Cao W, Xin H (2006) Gypenosides protect primary cultures of rat cortical cells against oxidative neurotoxicity. Brain Res 1102(1): 163–174

99. Circosta C, De Pasquale R, Occhiuto F (2005) Cardiovascular effects of the aqueous extract of *Gynostemma pentaphyllum* Makino. Phytomedicine 12(9):638–643
100. Wang QF, Chen JC, Hsieh SJ, Cheng CC, Hsu SL (2002) Regulation of Bcl-2 family molecules and activation of caspase cascade involved in gypenosides-induced apoptosis in human hepatoma cells. Cancer Lett 183(2):169–178
101. Zhou Z, Wang Y, Zhou Y (1996) The effect of *Gynostemma pentaphyllum* mak (GP) on carcinogenesis of the golden hamster cheek pouch induced by DMBA. Chin J Stomatol 31(5):267–270
102. Wang C, Wang X, Li Y, Deng S, Jiang Y, Yue L (1995) A preliminary observation of preventive and blocking effect of *Gynostemma pentaphyllum* (Thunb) Makino on esophageal cancer in rats. J West China Univ Medi Sci 26(4):430–432
103. Chen JC, Lu KW, Lee JH, Yeh CC, Chung JG (2006) Gypenosides induced apoptosis in human colon cancer cells through the mitochondria-dependent pathways and activation of caspase-3. Anticancer Res 26(6B):4313–4326
104. Chen JC, Lu KW, Tsai ML, Hsu SC, Kuo CL, Yang JS, Hsia TC, Yu CS, Chou ST, Kao MC, Chung JG, Wood WG (2009) Gypenosides induced G0/G1 arrest via CHk2 and apoptosis through endoplasmic reticulum stress and mitochondria-dependent pathways in human tongue cancer SCC-4 cells. Oral Oncol 45(3):273–283
105. Jimam N, Wannang N, Omale S, Gotom B (2010) Evaluation of the hypoglycemic activity of *Cucumis metuliferus* (Cucurbitaceae) fruit pulp extract in normoglycemic and alloxan-induced hyperglycemic rats. J Young Pharm 2(4):384–387
106. Yibchok-Anun S, Adisakwattana S, Yao CY, Sangvanich P, Roengsumran S, Haw HW (2006) Slow acting protein extract from fruit pulp of *Momordica charantia* with insulin secretagogue and insulinomimetic activities. Biol Pharm Bull 29:1126–1131
107. Kar A, Choudhary BK, Bandyopadhyay NG (2003) Comparative evaluation of ypoglycaemic activity of some Indian medicinal plants in alloxan diabetic rats. J Ethnopharmacol 84:105–108
108. Grover JK, Rathi SS, Vats V (2002) Amelioration of experimental diabetic neuropathy and gastropathy in rats following oral administration of plant (*Eugenia jambolana*, *Mucuna pruriens* and *Tinospora cordifolia*) extracts. Indian J Exp Biol 40:273–276
109. Raman A, Lau C (1996) Anti-diabetic properties and phytochemistry *Momordica charantia* L. (Curcurbitaceae). Phytomed 2:349–362
110. Wilson PDG, Hilton MG, Robins RJ, Rhodes MJC (1987) Fermentation studies of transformed root cultures. In: Moody GW, Baker PB (eds) International conference on bio-reactors and biotransformation. Elsevier, London, pp 38–51
111. Carvalho EB, Curtis WR (1998) Characterization of fluid-flow resistance in root cultures with a convective flow tubular bioreactor. Biotechnol Bioeng 60:375–384
112. Wyslouzil BE, Waterbury RG, Weathers PJ (2000) The growth of single roots of *Artemisia annua* in nutrient mist reactors. Biotechnol Bioeng 70(2):143–150
113. Curtis WR, Carvalho EB, Tescione LD (2001) Advances and challenges in bioreactor design for the production of chemicals from plant tissue culture. In: Sorvari S, Karhu S, Kanervo E, Pihakaski S (eds) Proceedings of the fourth international symposium on in vitro culture and horticultural breeding. Acta Horticulturae, vol 560. ISHS-International Society for Horticultural Science, Tempere, pp 247–253
114. Kanokwaree K, Doran PM (1997) The extent to which external oxygen transfer limits growth in shake flask culture of hairy roots. Biotechnol Bioeng 55:520–526
115. Gorelick J, Bernstein N (2014) Elicitation: an underutilized tool in the development of medicinal plants as a source of therapeutic secondary metabolites, Chapter 5. In: Sparks DL (ed) Advances in agronomy, vol 124. Elsevier, Amsterdam, pp 201–230
116. Namdeo AG (2007) Plant cell elicitation for production of secondary metabolites: a review. Pharmacogn Rev 1:69–79

"Hairy Root" Technology: An Emerging Arena for Heterologous Expression of Biosynthetic Pathway Genes in Medicinal Plants

12

Suchitra Banerjee, Sailendra Singh, and Pallavi Pandey

Abstract

Plant-based secondary metabolites play an important role towards the drug development process, but their lower yield in the source plant and uncertainty in supply owing to miscellaneous intervening factors have necessitated biotechnological intervention for devising suitable and economical alternative production systems. The progression of innovative biotechnological tools in tandem with the understanding of the plant metabolic pathways at both biochemical and cellular levels through combining the accumulating knowledge of next-generation sequencing has opened up new avenues for metabolic engineering of biosynthetic pathways. In this context, hairy root (HR) cultures have emerged as a promising platform for tailoring the metabolic flux of a given plant system towards the overproduction of desired metabolites by heterologously or homologously expressing the rate-limiting genes. A rational utilization of such cultures of diverse medicinal plants for heterologous expression of targeted pathway genes has started gaining attention over the years in order to overcome the co-suppression related to normally encountered disadvantages of homologous overexpression. The potential and appropriateness of this approach have gathered the maximum momentum during recent years even though such studies have come into existence for more than two decades ago. The present review summarizes the overall reported advances made in the area of hairy root-mediated heterologous expression of rate-limiting key genes of diverse biosynthetic pathways which remained mainly concentrated on tropane alkaloid, terpene indole alkaloid, and mevalonate and phenylpropanoid pathways. Successful implementation of the entire procedure is also found to be reigned by several other underlying factors, amongst which characteristic/exclusivity of plant families,

S. Banerjee (✉) • S. Singh • P. Pandey
Department of Plant Biotechnology, Central Institute of Medicinal and Aromatic Plants (CSIR-CIMAP), Lucknow, Uttar Pradesh, India
e-mail: suchitrabanerjee07@yahoo.com; sk_rmsu@yahoo.com; pallavicimap@gmail.com

S. Jha (ed.), *Transgenesis and Secondary Metabolism*, Reference Series in Phytochemistry, DOI 10.1007/978-3-319-28669-3_7

A. rhizogenes strains' specificities, explant types, promoters' specifications and media constituents which are some of the prominent deciding factors that differed amongst the reported observations and have been outlined in this review.

Keywords
Agrobacterium rhizogenes • Biosynthetic pathway • Heterologous plant gene expression • Hairy root culture • Medicinal plants

Contents

1 Introduction

Agrobacterium rhizogenes-mediated "hairy root" (HR) technology has traversed a long way since its judicious utilization as an efficient maneuvering technique for plant transformation by Ackermann in 1970 to reach the central stage of plant-based bioactive molecule production process, meant for sustenance of the whole underlying legacy. Going back to the history of plant-derived drug discovery trend, one can undeniably recount the strength of its progression through a rational drug discovery process since the beginning of the nineteenth century, which gradually thrived by the use of pure compounds instead of primarily used extracts and partly purified natural products [1]. The domineering impact of plant-derived natural compounds in modern drug development succession can very well be gauged by the statistical facts illustrating their indispensability in more than half of the 1073 new chemical entities getting approval between 1981 and 2010 [2]. At the same time, it also cannot be ignored that diverse prevailing complexities have emerged lately which recurrently hindered the access to authentic raw plant materials (with unvarying yield/quality parameters) and consequently discouraged pharmaceutical companies from phytomolecule-based drug development process [1]. In that context, it has recently been eloquently illustrated by Atanasov et al. [3] that most of the existing encumbrances have arose owing to many unavoidable circumstances (such as demographical, anthropological, environmental, and ecological impediments, coupled with other yield and quality-linked fluctuation hassles), which unquestionably calls for intense attention towards replenishing renewed interest in phytomolecule-based drug development process by overcoming the aforesaid challenges.

Devising an undeviating process of generating alternative production source with tailor-made production profile of plant-based metabolites has progressively turned into a reality all over the world in this decade through the rational advent of the "hairy root" technology [4]. The multifarious beneficial traits of such cultures not only bestow genetic and biochemical stability with rapid growth potential and hormone autotrophy but also offer equivalent or higher biosynthetic potential mimicking that of the parent plant [5]. Hairy root cultures have therefore undeniably fulfilled the prerequisite qualification of cellular differentiation related to sanguinity in terms of higher amount of product accumulation [6] and accordingly have immeasurably intervened in the frontline ventures of medicinal plant research to meet the growing demand of pharmaceutical industries. In this context, the advancements in the modern post-genomic technological insights in elucidating the metabolic pathways and associated regulatory/rate-limiting genes have progressively opened up new avenues for tailoring the metabolic flux of a given plant system towards the production of high-demand, low-yield molecules in greater quantities by pathway engineering in hairy root cultures [4]. In any given scenario, overproduction of a desired metabolite through such organ culture can be achieved by following either of the two realistic ventures, i.e., overexpression of rate-limiting gene for maximizing the metabolic flux towards the desired end product or by diverting the metabolic channel through the incorporation of appropriate participating genes from heterologous system [4, 7]. Although heterologous expression of plant genes in microbes has gathered some attention over the years, several practical impediments in terms of unavailability of the plant-based precursor substrate, complexity in posttranslational modifications, toxicity resulting from the accumulation of plant products, etc. appeared as stumbling block to be solved in the future [8]. On the contrary, the existence of the entire metabolic pathway with physiologically active genetic/enzymatic makeup in hairy root clones makes it extremely lucrative alternative to achieve the maximum reward through heterologous expression of pathway genes towards improved metabolite accumulation [9]. Indisputably, the incorporation of pathway genes from different source plant with the same function has significantly facilitated in rectifying the co-suppression-mediated disadvantage of overexpressing the pathway genes of the same plant system [4]. For the last three decades, the HR cultures of several medicinal and aromatic plants have been widely explored in the arena of heterologous gene expression based on medicinal plant research as depicted in Fig. 1. Literature survey revealed that the maximum number of published reports could be documented in the year 2011–2015 (ten) followed by 2006–2010 (nine), 2001–2005 (eight), 1996–2000 (three), and 1990–1995 (two). More than 85% of the published reports came into existence in the last 15 years showing the significance and potential of such HR-based technology towards the manipulation of any biosynthetic pathway genes for the production of

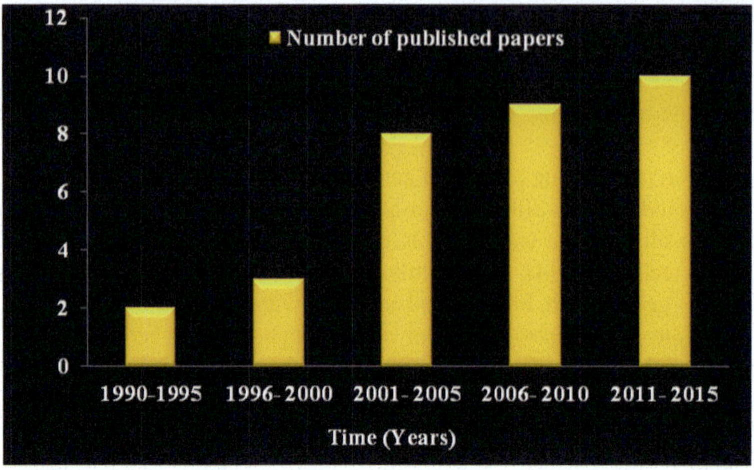

Fig. 1 Representative rate of publications relating to hairy root-mediated heterologous gene expression carried out for tailoring biosynthetic pathways

therapeutically active metabolites through heterologous expression of pathway genes from a separate plant system (Fig. 1).

Accumulating knowledge in this direction needs a comprehensive evaluation for generating future directions in medicinal plant research for tailoring the yield potentials towards a desired biomolecule production in hairy root cultures, to which the present review is committed. The types of metabolites, contributions of the underlying pathways, influence of the rate-liming genes, gross role of plant families and even the hairy root inducing *A. rhizogenes* strains, and culture media's attributes will be figured out in this review as each of these attributes imparts a decisive role towards the final outcome of such endeavor. This review is an attempt to scrutinize each and every aspect in great detail to delineate and mend any misapprehension under the present topic by offering a comprehensive view of global collaborative wisdom.

2 Biosynthetic Pathways and Genes Involved

The impact of post-genomic technologies on gene discovery and metabolic pathway elucidation can very well be perceived in recent years on the basis of accumulating knowledge in this arena. The crucial steps involved in a successful modulation of biosynthetic pathways through heterologous expression of pathway genes have been impressively elucidated under five headings in a recent review [4]. This information has clearly indicated that a detailed knowledge of the biosynthetic pathways and the role of decisive genes have undoubtedly facilitated the advancement of metabolic engineering as a potential approach to increase the yield of specific metabolites by heterologously or homologously expressing the rate-limiting genes [10]. In this context, the heterologous expression of targeted genes has started gaining attention over the years in order to overcome the disadvantage of overexpression of specific

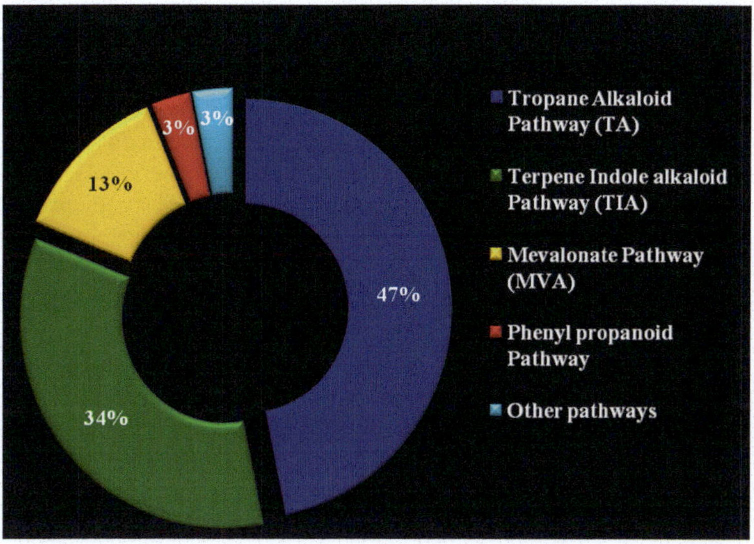

Fig. 2 Prevailing biosynthetic pathways engineered through HR-based heterologous gene expressions – tropane alkaloid (*TA*) pathway, terpene indole alkaloid (*TIA*) pathway, mevalonate (*MVA*) pathway, phenylpropanoid pathway, and other pathways (galacturonate pathway)

genes from the same plant source owing to the phenomenon of co-suppression of genes [11]. Although the prospect of utilizing hairy roots as expression system for modulating the biosynthetic pathways through heterologous gene incorporation was investigated as early as 1993, the expediency of this approach has gathered the maximum momentum during recent years.

It is evident from literature that plant alkaloids (terpene and tropane) constitute the largest group of natural products, offering many pharmacologically active secondary metabolites [10]. Correspondingly the major bulk of information related to successful pathway modulation through heterologous gene expression through hairy roots remained concentrated on these two groups of metabolites (Fig. 2). Out of 32 reported results in this area involving hairy root as expression system, 15 belong to the tropane alkaloids (47%), 11 to the terpene indole alkaloids (34%), and 4 to the mevalonate pathway (13%), while only two were related to other pathways (6%) in which one had covered the phenylpropanoid pathway (Fig. 2).

2.1 Tropane Alkaloid (TA) Pathway (*Putrescine N-methyltransferase* (*PMT*) *and Hyoscyamine- 6β-hydroxylase* (*H6H*) *Genes*)

The tropane alkaloids (TAs), found mostly in the members of the Solanaceae family, constitute a major group of pharmaceutically significant metabolites, amongst which hyoscyamine or atropine and scopolamine are widely reputed for their various pharmacological functions [12]. The supply of these molecules to the

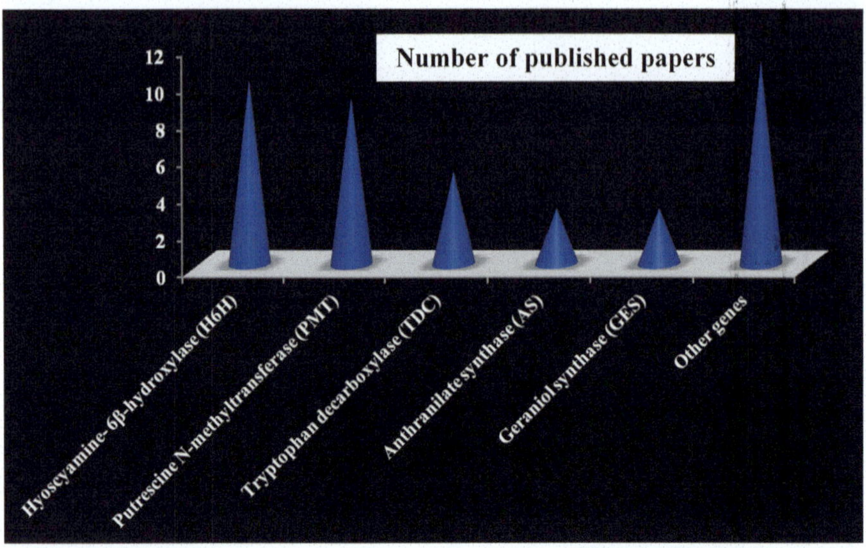

Fig. 3 An assortment of the major biosynthetic pathway genes utilized for the hairy root-mediated heterologous gene expression. [Other pathway genes include deoxy xylulose synthase (*DXS*), deoxy xylulose reductase (*DXR*), 3-hydroxy-3-methylglutaryl-coenzyme A reductase (*HMGR*), farnesyl diphosphate synthase (*FPS*), squalene synthase (*SS*), transcription factor (*AtPAP1*), cytochrome P450, and D-galacturonic acid reductase (GalUR)]

pharmaceutical industries is mainly dependent on the plant resources as their chemical synthesis is tedious and not economically feasible [13]. Therefore, to develop an alternative approach for their production, TA biosynthetic pathway has been manipulated through HR-mediated expression of exogenous genes from different medicinal plants. Throughout the TA pathway, there are a number of key enzymes that regulate the flux through the pathway, but the major contributing enzymes were noted to be hyoscyamine-6β-hydroxylase (*H6H*) and putrescine *N*-methyltransferase (*PMT*) genes (Fig. 3). To the best of our knowledge, till date the reported studies relating to these aforementioned genes (*PMT* and *H6H*) contributed about 47% of the total published papers (15 papers). These genes were heterologously expressed (either individually or collectively) in different medicinal plant systems via HR culture for the enhancement in the production of the targeted tropane alkaloids – hyoscyamine and scopolamine (Fig. 3).

The *PMT* gene regulates the N-methylation of putrescine with its further rerouting towards the biosynthesis of alkaloids, while the bifunctional enzyme *H6H* hydroxylates hyoscyamine to 6β-hydroxy hyoscyamine, which subsequently undergoes epoxidation to form scopolamine [12]. Interestingly, the *PMT* and *H6H* genes of the tropane pathway have been the most explored biosynthetic enzymes of the Solanaceae family for the enhancement in the alkaloid productivities through heterologous expression (Table 1). To the best of our knowledge, about 49% of the total published papers reporting heterologous expression of pathway genes are covered by

Table 1 Hairy root-mediated heterologous expression of pathway genes involving different medicinal plants during the last 25 years (arranged in chronological sequence starting from recent reports)

Sl. No.	Engineered plant species	Pathway involved	Gene (s)	Gene source	Promoter	Overall effect	References
1	*Nicotiana tabacum*	Terpene indole alkaloid (TIA)	Geraniol synthase (*GES*)	*Valeriana officinalis*	*CaMV 35S*	Transgenic HR revealed a maximum free geraniol content: (31.3 μg/g DW)	[26]
2	*N. tabacum* *N. benthamiana*	Terpene indole alkaloid (TIA)	Geraniol synthase (*GES*)	*V. officinalis*	*CaMV 35S*	Geraniol production reported in diverse forms of in vitro cultures	[27]
3	*N. tabacum*	Terpene indole alkaloid (TIA)	Geraniol synthase (*GES*) + Geranyl pyrophosphate synthase (*GPPS*)	*V. officinalis* *Arabidopsis thaliana*	*CaMV 35S*	*GES* HR line accumulated significant amounts of geraniol/ geraniol glycosides (151 ± 24 ng/mg DW), while *GES* + *GPPS* expression clones showed lower levels of accumulation than that in the single gene transgenic	[28]
4	*Salvia sclarea*	Mevalonate (MVA)	Deoxy xylulose synthase (*DXS*) + Deoxy xylulose reductase (*DXR*)	*A. thaliana*	*CaMV 35S*	Enhancement in the content of bioactive abietanic diterpenes (aethiopinone, 1-oxoaethiopinone, savipisone, ferruginol, carnosic acid, 1-oxo-ferruginol)	[39]
5	*Platycodon grandiflorum*	Phenylpropanoid (PP)	Transcription factor (*AtPAP1*)	*A. thaliana*	*CaMV 35S*	The best transgenic HR line showed 9.89 times (approx) higher content of chlorogenic acid (CGA) [421.31 μg/100 mg DW]	[43]

(*continued*)

Table 1 (continued)

Sl. No.	Engineered plant species	Pathway involved	Gene (s)	Gene source	Promoter	Overall effect	References
6	*Rauvolfia serpentina*	Terpene indole alkaloid (TIA)	Tryptophan decarboxylase (*TDC*)	*Catharanthus roseus*	*CaMV 35S*	The best transgenic HR clone revealed better yield potentials with respect to **(i) Reserpine:** (0.1202 ± 0.002% DW) **(ii) Ajmalicine:** (0.0064 ± 0.003% DW)	[30]
7	*P. grandiflorum*	Mevalonate (MVA)	3-Hydroxy-3-methylglutaryl-coenzyme A reductase (*HMGR*)	*Panax ginseng*	*CaMV 35S*	**Phytosterol levels increased by** 1.1–1.6-fold in the transgenic HR lines	[42]
8	*Scopolia parviflora*	Tropane alkaloid (TA)	Putrescine *N*-methyltransferase (*PMT1 and PMT2*) + Hyoscyamine-6β-hydroxylase (*H6H*)	*N. sylvestris Hyoscyamus niger*	*CaMV 35S*	Enhancement in the contents of scopolamine and hyoscyamine in the transgenic HR clones	[12]
9	*Atropa belladonna*	Tropane alkaloid (TA)	Putrescine *N*-methyltransferase (*PMT*) + Hyoscyamine-6β-hydroxylase (*H6H*)	*N. tabacum H. niger*	*CaMV 35S*	**Significant enhancement in the contents of both alkaloids in transgenic HR clones** Compared to non-transgenic and HR normal (intact) root culture **(i) Scopolamine:** five and four times higher, respectively **(ii) Hyoscyamine:** eleven and twenty-four times higher, respectively	[25]

10	*C. roseus*	Terpene indole alkaloid (TIA)	Deoxy xylulose synthase (*DXS*) + geraniol-10-hydroxylase (*G10H*) + anthranilate synthase (*AS*)	*A. thaliana*	*GAL4-UAS*	**DXS expression:** (I) increased: ajmalicine (67%), serpentine (26%), and lochnericine (49%) (II) decreased: tabersonine (66%) and horhammericine (54%) **Co-overexpression of DXS and G10H:** Increased: ajmalicine (16%), lochnericine (31%), and tabersonine (13%) **Co-expression of DXS with AS:** increased: horhammericine (30%), lochnericine (27%), and tabersonine by (34%)	[29]
11	*Withania coagulans*	Mevalonate (MVA)	Squalene synthase (SS)	*A. thaliana*	*CaMV 35S*	The transgenic HR roots accumulated higher content of the total withanolide (ranged from 0.68 to 4.63 μg/gm DW) as compared to the control (0.11 μg/gm DW) after 4 weeks of cultivation	[40]
12	*Centella asiatica*	Terpene indole alkaloid (TIA)	Farnesyl diphosphate synthase (*FPS*)	*P. ginseng*	*CaMV 35S*	Compared to non-transgenic root culture, transgenic HR revealed enhanced **(i)** Squalene content (1.1–1.5-fold) **(ii)** Total sterol contents (threefold)	[36]

(*continued*)

Table 1 (continued)

Sl. No.	Engineered plant species	Pathway involved	Gene (s)	Gene source	Promoter	Overall effect	References
13	*Solanum lycopersicum* cv. Money Maker	Galacturonate pathway	D-galacturonic acid reductase (GalUR)	*Fragaria ananassa*	*CaMV 35S*	The total L-ascorbic acid (AsA) content increased by 2.5-fold in a transgenic HR clone treated with D-galacturonic acid (D-GalUA)	[44]
14	*Brassica rapa*	Terpene indole alkaloid (TIA)	Cytochrome P450 (CYP79B2, CYP79B3, CYP83B1)	*A. thaliana*	*CaMV 35S*	Compared to the control HR lines with rol ABC, the incorporation of CYP83B1 in conjunction with CYP79B2 or CYP79B3 enhanced the accumulation of glucobrassicin or 4-methoxy glucobrassicin levels. However, no influence in their accumulation could be noted through individual overexpression of either of these genes	[37]
15	*Nicotiana* cell culture derived from the transgenic HR	Tropane alkaloid (TA)	Hyoscyamine-6β-hydroxylase (*H6H*)	*H. muticus*	*CaMV 35S*	**Scopolamine content increased in the transgenic HR:** **Shake flask:** 21.6 ± 1.1 mg/L **Bioreactor:** 35.5 ± 0.8 mg/L	[23]
16	*H. niger*	Tropane alkaloid (TA)	Putrescine *N*-methyltransferase (*PMT*)	*N. tabacum*	*CaMV 35S*	Compared to the control, a fivefold higher PMT activity in the transgenic HR with no increase in alkaloid yield	[14]
17	*Duboisia leichhardtii*	Tropane alkaloid (TA)	Hyoscyamine-6β-hydroxylase (*H6H*)	*H. niger*	*ParAt*	Enhanced content of scopolamine in the transgenic HR clones	[21]

18	*Atropa baetica*	Tropane alkaloid (TA)	Hyoscyamine-6β-hydroxylase (*H6H*)	*H. niger*	*CaMV 35S*	Superior alkaloid content in transgenic HR **Scopolamine:** 5.63 mg/g DW **Hyoscyamine:** 0.2 mg/g DW **6β-Hydroxyhyoscyamine:** 1.01 mg/g DW	[24]
19	*C. roseus*	Terpene indole alkaloid (TIA)	Tryptophan decarboxylase (*TDC*) + Anthranilate synthase (*AS*)	*C. roseus* *A. thaliana*	*CaMV 35S*	Enhanced content of tryptamine and tryptophan in transgenic HR	[33]
20	*N. tabacum* *H. muticus*	Tropane alkaloid (TA)	Hyoscyamine-6β-hydroxylase (*H6H*)	*H. niger*	*CaMV 35S*	Transgenic *N. tabacum* HR showed more conversion of hyoscyamine to scopolamine over that in *H. muticus*	[22]
21	*S. parviflora*	Tropane alkaloid (TA)	Putrescine *N*-methyltransferase (*PMT1 and 2*)	*N. sylvestris*	*CaMV 35S*	Enhanced content in transgenic HR clones: (i) Hyoscyamine: 3.5-fold (4.2 mg/g DW) (ii) Scopolamine: fivefold (6.4 mg/g DW)	[18]
22	*C. roseus*	Terpene indole alkaloid (TIA)	Tryptophan decarboxylase (*TDC*) + Anthranilate synthase (*AS*)	*C. roseus* *A. thaliana*	*CaMV 35S*	Enhanced content of tryptophan and serpentine	[32]
23	*H. niger*	Tropane alkaloid (TA)	Hyoscyamine-6β-hydroxylase (*H6H*) + Putrescine *N*-methyltransferase (*PMT*)	*H. niger* *N. tabacum*	*CaMV 35S*	Compared to the control, the transgenic HR with both genes showed increase in scopolamine ninefold and twofold higher than only *H6H* expressing HR lines	[10]
24	*Duboisia hybrid* (*D. myoporoides* x *D. leichhardtii*)	Tropane alkaloid (TA)	Hyoscyamine-6β-hydroxylase (*H6H*)	*H. niger*	*CaMV 35S*	Scopolamine content in transgenic HR: three times (74.50 mg/l) higher than the control	[20]

(*continued*)

Table 1 (continued)

Sl. No.	Engineered plant species	Pathway involved	Gene (s)	Gene source	Promoter	Overall effect	References
25	*Datura metel* *H. muticus*	Tropane alkaloid (TA)	Putrescine *N*-methyltransferase (*PMT*)	*N. tabacum*	*CaMV 35S*	Only hyoscyamine content increased in *H. muticus* transgenic HR clone while both hyoscyamine and scopolamine contents increased in *D. metel*	[16]
26	*Duboisia hybrid* (*D. myoporoides* x *D. leichhardtii*)	Tropane alkaloid (TA)	Putrescine *N*-methyltransferase (*PMT*)	*N. tabacum*	*CaMV 35S*	Enhanced N-methylputrescine level: two- to fourfold with no increase in the content of other metabolites	[15]
27	*A. belladonna*	Tropane alkaloid (TA)	Putrescine *N*-methyltransferase (*PMT*)	*N. tabacum*	*CaMV 35S*	Compared to a wild-type HR, the transgenic HR revealed fivefold higher expression of PMT level with no change in the total alkaloid content	[17]
28	*Solanum aviculare*	Mevalonate (MVA)	3-Hydroxy-3-methylglutaryl-coenzyme A reductase (*HMGR*)	*Artemisia annua*	*CaMV 35S*	Solasodine content enhanced 4.2 times over that in the control HR.	[41]

29	*Cinchona officinalis "Ladgeriana"*	Terpene indole alkaloid (TIA)	Tryptophan decarboxylase (*TDC*) + Strictosidine synthase (*STR*)	*C. roseus*	*CaMV 35S*	Enhancement in the contents of tryptamine, strictosidine, quinine, and quinidine	[34]
30	*H. muticus*	Tropane alkaloid (TA)	Hyoscyamine-6β-hydroxylase (*H6H*)	*H. niger*	*CaMV 35S*	Compared to control HR, the best transgenic clone produced 100 times higher content of scopolamine	[13]
31	*A. belladonna*	Tropane alkaloid (TA)	Hyoscyamine-6β-hydroxylase (*H6H*)	*H. niger*	*CaMV 35S*	Compared to control HR, the transgenic clone revealed fivefold enhancement in the scopolamine content	[19]
32	*Peganum harmala*	Terpene indole alkaloid (TIA)	Tryptophan decarboxylase (*TDC*)	*C. roseus*	*CaMV 35S*	The tryptamine level remained unchanged but the indole alkaloid (serotonin) content enhanced by tenfold in the transgenic HR clone. Supplementation with tryptophan further enhanced the serotonin content	[31]

individual or collective expression of the aforementioned genes (*PMT* and *H6H*) in different medicinal plant via HR cultures contributing a major share in the production of the targeted tropane alkaloids – hyoscyamine and scopolamine (Fig. 3).

The exclusive role of *PMT* gene has been studied earlier by different research groups, one of which is through its isolation from *N. tabacum* and successful expression in *H. niger* HR cultures, leading to the fivefold higher PMT activity in the transgenic HR than that in the control, while no increase could be noted in the overall content of the alkaloids [14; Table 1]. Similarly, the same gene (*PMT*) from *N. tabacum* has also been expressed earlier in *Duboisia hybrid* (*D. myoporoides* x *D. leichhardtii*) HR cultures enhancing the *N*-methylputrescine level up to twofold to fourfold, but no enhancement could be observed in the content of other metabolites [15]. In contrast, the production of both hyoscyamine and scopolamine was found to be increased by expressing the *PMT* gene from *N. tabacum* in *Datura metel* HR cultures [16], while the *Hyoscyamus muticus* HR clone with *PMT* gene of the same origin failed to produce the latter, and only the content of the former was augmented [16; Table 1]. Ahead of these results, Sato et al. [17] have fruitfully expressed the *PMT* gene from *N. tabacum* in *Atropa belladonna* HR culture which resulted in a fivefold increase of *PMT* level in transgenic HR as compared to wild-type HR clone but the total alkaloid content remained unchanged (Table 1).

Likewise, the effect of introducing putrescine *N*-methyltransferase (*PMT1* and *PMT2*) under the control of a CaMV 35S promoter from *N. sylvestris* into *Scopolia parviflora* HR cultures has also been investigated by Lee et al. [18]. The transgenic HR line produced over 3.5 and 5 times more hyoscyamine and scopolamine, respectively, than the wild-type HR cultures, substantiating the fact that improvement in the production of such metabolites as the end products could assuredly be achieved through such heterologous expression of the upstream genes of tropane alkaloid pathway [18].

So far as the *H6H* gene is concerned, it was observed that out of the 32 published information under the present topic, ten reports have predominantly demonstrated the efficacy of this specific gene in enhancing the production of scopolamine in different members of Solanaceae family (Fig. 3). The *H6H* gene from *Hyoscyamus niger* has been cloned and introduced into *A. belladonna* HR culture where a fivefold higher scopolamine yield could be documented as compared to the wild-type HR culture, and the content of 6β-hydroxyhyoscyamine was also increased in the transformed roots [19; Table 1]. The *H6H* gene of the same source has also been cloned and introduced into *H. muticus*, where the best transgenic HR clone produced 100 times higher content of scopolamine than that of the control clone. This observation convincingly illustrated the prospect of utilizing hairy root as a promising and reliable tool for the heterologous gene expression system for enhancing the scopolamine productivity [13; Table 1].

In another study, the *H6H* gene from *H. niger* was introduced into the genome of a scopolamine-rich *Duboisia hybrid* (*D. myoporoides* x *D. leichhardtii*) under the regulation of the CaMV35S promoter [20]. The scopolamine levels (74.50 mg/l) in the engineered HR lines improved up to three times as compared to wild-type hairy roots, but no significant changes could be documented in the HR-regenerated plants [20; Table 1). Another report by Rahman et al. [21] has demonstrated that the incorporation of the *H6H* gene from *H. muticus* under the control of "ParAt" promoter in *Duboisia leichhardtii* HR culture has enhanced the content of scopolamine in the *H6H*-positive HR clones [21; Table 1].

Hakkinen and co-workers [22] have reengineered the *N. tabacum* and *H. muticus* HR cultures by the incorporation of the *H6H* gene from *H. niger* for the enhanced secretion of tropane alkaloids. Both the transgenic hairy roots were examined for their potentials towards the production of scopolamine and other tropane alkaloids after the exogenous supply of the precursor (hyoscyamine) in the culture medium. Amongst them, *N. tabacum* HR clone showed more proficiency towards the uptake of hyoscyamine from the culture medium and a higher bioconversion rate towards the scopolamine production compared to that of the *H. muticus* HR clone (Table 1). Apart from the secretion of maximum scopolamine (85%) into the culture medium, the addition of hyoscyamine also influenced the accumulation of nicotine in the resultant transgenic *N. tabacum* HR line [22].

In another study, the competence of dedifferentiated cell cultures derived from the *N. tabacum* hairy roots carrying the *H. muticus H6H* gene has also been investigated for the bioconversion of hyoscyamine to scopolamine [23]. They have shown that *N. tabacum* cell suspensions, which normally do not produce scopolamine, are able to produce this alkaloid after overexpressing the gene *H6H* in tandem with the precursor feeding of hyoscyamine (Table 1). The transgenic cell suspension cultures showed a higher potential for the conversion of exogenously added hyoscyamine at two different concentrations (100 mg/l and 200 mg/l) to the culture medium [23]. The total scopolamine production (21.6 ± 1.1 mg/l) in the shake flask could be obtained at the higher concentration. The overall scopolamine accumulation could further be enhanced significantly by upscaling the transgenic cell culture from the shake flask ($21.6 + 1.1$ mg/l) to 5 l turbine-stirred tank bioreactor (35.5 ± 0.8 mg/l), indicating the commercial prospect of the entire effort [23].

The effect of the *H6H* gene of *H. niger* on the production potentials of *Atropa baetica* HR clone for scopolamine, hyoscyamine, and 6β-hydroxyhyoscyamine has also been investigated by Zarate et al. [24]. The transgenic roots overexpressing *H6H* revealed higher content of scopolamine (5.63 mg/g DW), hyoscyamine (0.2 mg/g DW), and 6β-hydroxyhyoscyamine (1.01 mg/g DW). The scopolamine level increased up to ninefold as compared to the plants [24; Table 1]. The heterologous co-expression of *PMT* and *H6H* genes from *N. tabacum* and *H. niger*, respectively, in *H. niger* HR culture revealed nine times higher content of scopolamine in the transgenic HR clone than that of the wild type (Table 1) and only two times higher than that in the HR lines expressing *H6H* gene exclusively [10]. Subsequently, the cloning and co-expression of the same *PMT* and *H6H* genes from the abovementioned sources in *Atropa belladonna* HR culture showed a distinct effect by producing five and four times higher scopolamine content and 11 and 24 times higher hyoscyamine content than that in the non-transgenic HR and normal (intact) root culture, respectively [25; Table 1].

In another contemporary investigation, the genes for the two key enzymes of tropane alkaloid pathway (i.e., *PMT1* and *PMT2*) from *Nicotiana sylvestris* and *H6H* from *Hyoscyamus niger* were expressed in the metabolically engineered hairy root cultures of *Scopolia parviflora* [12]. The morphology of the *S. parviflora* transformed roots with *PMT1*, *PMT2*, and *H6H* genes was noted to be distinct, and the overall effects of these genes led to the enhancement in the content of scopolamine and hyoscyamine in the transformed roots [12; Table 1]. Supplementation with different phytohormones further enhanced the accumulation of scopolamine and hyoscyamine in these transgenic HR lines [12].

2.2 Terpenoid Indole Alkaloid (TIA) Pathway

TIAs are one of the major classes of phytomolecules renowned for their various applications in the pharmaceutical, fragrance, and cosmetic industries [26]. There are several TIAs found in nature governed by numerous pathway genes, but very few have been explored in terms of HR-mediated heterologous gene expression for improving the productivity of targeted metabolites involving distinct medicinal plants (Fig. 2). The success stories are discussed below.

2.2.1 Geraniol Synthase (*GES*) Gene

Geraniol is synthesized from the monoterpenoid precursor geranyl diphosphate by the geraniol synthase (*GES*) gene, which is an intermediate metabolite in the monoterpenoid pathway, interconnecting with the indole pathway to form pharmaceutical products of TIAs [27]. Geraniol possesses several pharmacological importances by suppressing different types of cancer such as colon, pancreatic, hepatic, and prostate tumors and also reduces serum cholesterol levels [27]. The low yield from the natural plant sources and intricate chemical structure of geraniol has attracted considerable recent interest towards developing biotechnology-based production alternatives to meet the escalating demand of this valuable phytomolecule. Literature survey revealed that out of the total 32 reported observations on hairy root-mediated heterologous expression of different pathway genes, three reports are available concerning the expression of geraniol synthase (*GES*) gene till date (Fig. 3).

The heterologous expression of *GES* gene under the control of cauliflower mosaic virus (CaMV 35S) promoter from *Valeriana officinalis* has been achieved in *Nicotiana tabacum* and *N. benthamiana* HR cultures which resulted in the production of geraniol in different in vitro cultures [27; Table 1]. The maximum geraniol content could be noted in the stable transgenic plants grown in vitro (48 μg/g FW), and the least could be noted in the hairy root culture (9 μg/g FW), while the other expression systems showed intermediary results in reducing the order as follows: transient expression system (27 μg/g FW), transgenic plants under hydroponic conditions in the greenhouse, and cell suspension cultures (16 μg/g FW) [27].

Similarly, Ritala et al. [26] have successfully expressed the *GES* gene from *V. officinalis* in the HR clone of another species of *Nicotiana*, i.e., *N. tabacum* which resulted in the production of twenty transgenic HR clones having a maximum free geraniol content of 31.3 μg/g DW [26; Table 1]. In another study by Masakapalli et al., [28], enhancement in the biosynthesis of geraniol in the HR cultures of *N. tabacum* through metabolic engineering has also been accomplished by expressing (i) either *GES* alone from *V. officinalis* or (ii) by combining *GES* with geranyl pyrophosphate synthase (*GPPS*) gene from *Arabidopsis thaliana* (*GES* + *GPPS*). The transgenic HR line expressing *GES* accumulated significant amounts of geraniol and geraniol glycosides (151 ± 24 ng/mg DW), while HR clones with *GES* + *GPPS* accumulated lower levels of geraniol/geraniol glycosides compared to that in the former [28; Table 1). With regard to the growth and biomass accumulation potentials of both the transgenic HR lines, no significant differences were observed corroborating the possibility of increasing the accumulation of a useful secondary metabolite through such heterologous gene expression approach. The main conclusion of this study was that the simultaneous manipulation of the precursors of the TIA pathway through metabolic engineering can lead to a superior result in terms of enhancement in the production of the geraniol and geraniol glycosides [28].

2.2.2 Tryptophan Decarboxylase (*TDC*) and Other Related Genes

The terpene indole alkaloids (TIAs) are formed by the condensation of the indole moiety – tryptamine and the monoterpenoid – secologanin by strictosidine synthase (*STR*) to form strictosidine, the precursor to a wide variety of TIAs. Tryptamine is synthesized by tryptophan through tryptophan decarboxylase (*TDC*) activity, and anthranilate synthase (*AS*) catalyzes the first committed step in the synthesis of tryptophan [29]. Literature survey revealed that out of the total 32 reported observations on hairy root-mediated heterologous expression of different pathway genes, so far six reports are available concerning the expression of *TDC* gene (Fig. 3).

In order to manipulate the reserpine and ajmalicine flux, an attempt has been carried out to heterologously express the single *TDC* gene from *Catharanthus roseus* in *Rauvolfia serpentina* HR cultures for enhancing the production of such important alkaloids [30]. The best transgenic HR clone (RT4) bearing the exogenous *TDC* gene, cultivated in Gamborg's B_5 medium, accumulated higher reserpine (0.1202 ± 0.002% DW) and ajmalicine (0.0064 ± 0.003% DW) content as compared to the control non-transgenic HR clone after 10 weeks of cultivation [30; Table 1].

The same *TDC* gene from *C. roseus* has also been expressed in the hairy root culture of *Peganum harmala* by Berlin and co-workers [31]. The tryptamine and β-carboline alkaloid levels (other tryptamine-derived alkaloids) in the transgenic lines did not change, while the content of indole alkaloid serotonin was enhanced up to tenfolds as compared to that in the non-transgenic root. Supplementation of tryptophan to the cultures further enhanced the serotonin content in the transgenic HR clones [31].

In another study, the *TDC* from *C. roseus* and *AS* from *A. thaliana* were concurrently co-expressed in *C. roseus* HR cultures, resulting in the enhancement in the content of tryptophan and serpentine [32]. Similarly, Hong et al. [33] also successfully expressed both these genes of the same abovementioned sources in

C. roseus HR culture for the production of tryptamine and tryptophan. Geerlings and co-workers [34] isolated the two key enzymes *TDC* and strictosidine synthase (*STR*) from *C. roseus*, cloned under the control of constitutive CaMV35S promoter and successfully co-expressed both genes in the HR cultures of a woody plant – *Cinchona officinalis "Ladgeriana"* (Table 1). This resulted in improving the yields of tryptamine and strictosidine in the transgenic HR lines (1200 and 1950 μg/g DW, respectively). Concurrently, the levels of quinine and quinidine were also enhanced (500 and 1000 μg/g DW, respectively) in the same transgenic HR clone [34]. It is however disappointing to note that after one year of culture establishment, the transgenic *C. officinalis* HR clones completely lost their capacities of accumulating these alkaloids even though the presence of the *TDC* and *STR* genes was observed with the total loss of activity of the *TDC* gene [34]. On the contrary, another study has demonstrated the long-term stability of a transgenic *C. roseus* HR clone containing the inducible expression of a feedback-insensitive *Arabidopsis* anthranilate synthase (*AS*) gene after several years of culture establishment [35]. Stabilized production of an array of metabolites in these transgenic hairy root cultures was obtained after 5 years, which indicated that an initial transient stage might be mandatory to reap the maximum benefit out of such cultures before the metabolite profile reaches a stabilized state [35].

The regulatory mechanism behind the overexpression of several key genes in the upstream of the TIA pathway has also been investigated through heterologous approach in order to increase the flux towards the secondary metabolite production within hairy root cultures of *C. roseus* [29]. The inducible overexpression of *DXS* alone or together with "Anthranilate-synthase A" (*ASA*) subunit or *DXS* with geraniol-10-hydroxylase (*G10H*) under the control of *GAL4-UAS* promoter in the *C. roseus* HR cultures demonstrated variable outcomes. The transgenic HR clone with *DXS* gene alone revealed increment in the content of ajmalicine (67%), serpentine (26%), and lochnericine (49%), while the tabersonine and horhammericine contents were decreased by 66% and 54%, respectively [29]. On the other hand, co-expression of *DXS* with *G10H* resulted in the enhancement of the overall productivity of ajmalicine (16%), lochnericine (31%), and tabersonine (13%) [29; Table 1], while co-expression of *DXS* and *AS* demonstrated improved contents of horhammericine (30%), lochnericine (27%), and tabersonine (34%) (Table 1). These results clearly exemplified that the modulation of the complete range of metabolites could be achieved through the functional expression of single gene or through multiple genes within the pathway depending upon the targeted flux towards the final products [29].

Kim et al. [36] have demonstrated that the cloning and expression of farnesyl diphosphate synthase (*FPS*) from *P. ginseng* in *Centella asiatica* have played a significant role towards enhancing the production of total sterols in its HR cultures (Table 1). The content of squalene and total sterol in the best transgenic HR lines of *C. asiatica* was increased to 1.1–1.5-fold and three times, respectively, as compared to that in the control HR line [36].

Indole glucosinolates (IGs) are found widely in the Brassicales, which play an important role as a defense compounds in plants. Zang and co-workers [37] attempted to produce IG by heterologously expressing three cytochrome P450 (CYP79B2, CYP79B3, and CYP83B1) genes from *A. thaliana* under the control of CaMV35S promoter in the hairy root clone of *Brassica rapa.* The incorporation of CYP83B1 along with CYP79B2 or CYP79B3 enhanced the accumulation of glucobrassicin or 4-methoxy glucobrassicin than the control HR line with *rol* ABC. However, overexpression of either of these genes alone did not influence the accumulation level [37; Table 1].

2.3 Mevalonate (MVA) Pathway

The biosynthesis of isoprenoid in the plant systems initiates from the common precursor prenyl diphosphate (prenyl-PP) and is synthesized through two different pathways, i.e., 2-*C*-methyl-D-erythritol 4-phosphate (MEP) pathway and the mevalonate (MVA) pathway, occurring in plastids and cytoplasm, respectively [38]. Diterpenoids are the important phytomolecules which not only are required for the survival of the plant but also bear desirable pharmacological significance. In *Salvia sclarea*, the abietanic diterpenes are normally synthesized in the roots, but at a very lower level [39]. Recently, the two upstream genes of the MVA pathway – deoxy xylulose synthase (*DXS*) and deoxy xylulose reductase (*DXR*) – have been isolated from a Brassicaceae family member (*A. thaliana*) and have been successfully co-expressed in the HR culture of a Lamiaceae plant – *Salvia sclarea* [39; Table 1]. This has led to the increment in the content of all the targeted bioactive abietanic diterpenes (i.e., aethiopinone, 1-oxoaethiopinone, savipisone, ferruginol, carnosic acid, and 1-oxo-ferruginol), which has clearly exhibited that the heterologous expression of MVA pathway genes can play a decisive role towards enhancing the production of pharmaceutically important diterpenes [39].

The squalene synthase (SS) from *A. thaliana* has been cloned and successfully expressed in the HR of *Withania coagulans* resulted in the enhancement of the total withanolides (ranged from 0.68 to 4.63 μg/g DW) as compared to the control HR (0.11 μg/g DW) after 4 weeks [40].

In another study, the gene for the 3-hydroxy-3-methylglutaryl-coenzyme A reductase (*HMGR*), which catalyzes the rate-limiting step in the MVA pathway, has been isolated from *Artemisia annua* and cloned under the control of CaMV 35S promoter in *Solanum aviculare* HR lines [41]. The expression of this construct in the resultant *S. aviculare* HR lines resulted in the increment of the solasodine content by 4.2 times higher as compared to the control HR line [41]. On the other hand, heterologous expression of the same *HMGR* gene from *Panax ginseng* in *Platycodon grandiflorum* HR clones has demonstrated only 1.1–1.6-fold higher phytosterol levels in transgenic HR lines [42].

2.4 Phenylpropanoid Pathway

The phenylpropanoid pathway serves as a starting point for the production of many important compounds, such as the flavonoids, coumarins, and lignans, and is required for the biosynthesis of lignin. However, only a single published report is available till date under the current topic (Fig. 2). Tuan and co-workers (2014) have made a sole attempt for the production of chlorogenic acid (CGA) in the *Platycodon grandiflorum* HR culture by expressing *A. thaliana* transcription factor (*AtPAP1*) cloned under the control of CaMV 35S promoter (Table 1). The best HR line produced ten times higher CGA (421.31 μg/100 mg DW) as compared to that of the control HR line, which emphasized that HR culture of *P. grandiflorum* can be a promising alternative route for the production of CGA through heterologous expression of the *AtPAP1* gene [43].

2.5 Other Pathways

A sole report was documented with regard to the galacturonate pathway, where D-galacturonic acid reductase (GalUR) gene from *Fragaria ananassa* under the control of *CaMV 35S* has been heterologously expressed in the HR culture of *Solanum lycopersicum* cv. Money Maker [44; Table 1). As compared to the wild-type HR clones, the content of the total L-ascorbic acid (AsA) improved up to 2.5-fold in the transgenic HR clone upon treatment with D-galacturonic acid (D-GalUA) [44].

3 Other Contributing Factors

The successful implementation of the entire procedure is reigned by several underlying factors, which play a crucial role in implementing the positive gene transfer events and also substantially influence the overall expression mechanism of heterologously transferred plant genes. The characteristic and exclusivity of plants families (depending upon their degrees of susceptibilities), the types of *A. rhizogenes* strains, explant types, promoters' specifications, and media constituents are some of the prominent deciding factors that differed amongst the reported observations and have been outlined below.

4 Involvement of Plant Families

It is evident from literature survey that certain plant families played crucial role as the front-runner with respect to improving the production of targeted metabolites through hairy root (HR)-based heterologous gene expression of pathway genes. In this context, the major contribution could be noted on account of the Solanaceae family (60%) followed by Apocynaceae and Brassicaceae (Fig. 4).

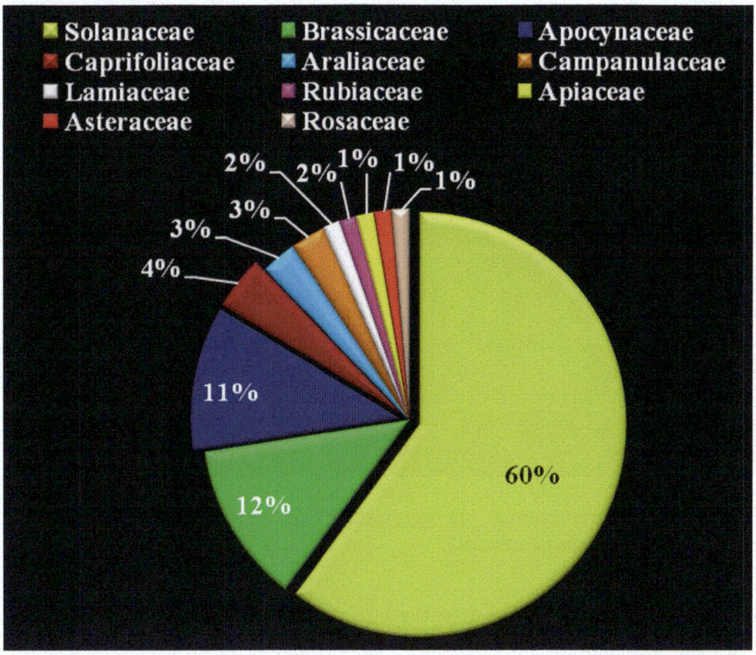

Fig. 4 Preferential involvement of plant families in HR-mediated heterologous gene expression studies

Different species of *Nicotiana* (i.e., *N. tabacum*, *N. benthamiana*, *N. sylvestris*), *Atropa* (i.e., *A. belladonna*, *A. baetica*), *Duboisia* (i.e., *D. leichhardtii*, *D. hybrid*, *D. myoporoides* x *D. leichhardtii*, *Hyoscyamus* (i.e., *H. niger*, *H. muticus*), and *Solanum* (i.e., *S. aviculare*, *S. lycopersicum*) along with *Scopolia parviflora*, *Withania coagulans*, and *Datura metel* constituted the major plant species of Solanaceae family which dominated the entire HR-based heterologous gene expression progression (60%) relating to tropane alkaloid biosynthesis (Fig. 4). The second most explored family is Brassicaceae, in which the major attention remained mostly focused on *Arabidopsis thaliana* and *Brassica rapa* HR cultures, which contributed about 12% of the total reported literatures for heterologous expression (Fig. 4).

On the other hand, Apocynaceae family occupied the third position in this list (11%) in which *Rauvolfia serpentina* and *Catharanthus roseus* HR cultures have been explored as the expression system for the pathway modulation (Fig. 4). The HR cultures of several other plant families too contributed towards the HR-mediated heterologous gene expression (17%) in which Caprifoliaceae, *Valeriana officinalis* (4%); Araliaceae, *Panax ginseng* (3%); Campanulaceae, *Platycodon grandiflorum* (3%); Lamiaceae, *Salvia sclarea* (2%); Rubiaceae, *Cinchona officinalis* (2%); Apiaceae, *Centella asiatica* (1%); Asteraceae, *Artemisia annua* (1%); and Rosaceae, *Fragaria ananassa* (1%) were investigated towards improving the yield potentials relating to targeted metabolites (Fig. 4).

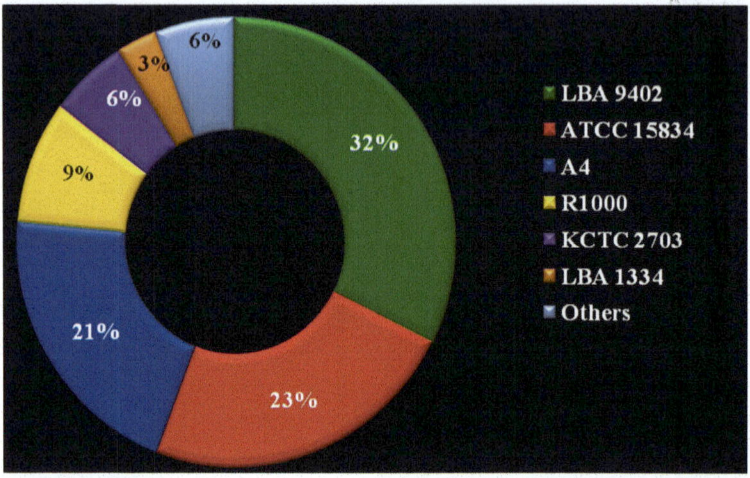

Fig. 5 Influence of *Agrobacterium rhizogenes* strains on the frequency of affirmative results relating to heterologous gene expression

5 Preference of *A. rhizogenes* Strain

An in-depth analysis concerning the use of different strains of *A. rhizogenes* has clearly indicated the preferential utilization of six different strains with varying degrees of success through their gainful exploitation in heterologous expression of pathway genes in HR cultures (Fig. 5). In this context, the LBA 9402 strain attained maximum deliberations (32%) followed by the ATCC 15834 (23%), A4 (21%), R1000 (9%), KCTC 2703 (6%), and LBA 1334 (3%) strains (Fig. 5). Exclusive susceptibilities of the targeted medicinal plant systems towards the specific bacterial strain might be the underlying reason for such diversity.

6 Popularity of Explant Types

A review of the published literature has clearly indicated that preference of explant types has played an implicit role in all the successful reports of the HR-mediated heterologous pathway gene expression studies. Noticeably, leaf has most prevalently been used as the universally preferred explants with 71% reported frequency of uses (Fig. 6). The use of seedlings attained the second preference with 11% reported cases, while the utilization of either stem or both stem and leaf attained equal preference (total 7% reported usages). It is however worth mentioning that Rahman and co-workers have observed better performance of leaf explants over that of stem explants in their heterologous gene expression study where both have been tested. The use of cotyledon has gained minimum application in similar studies with 4% of use amongst the reported cases (Fig. 6).

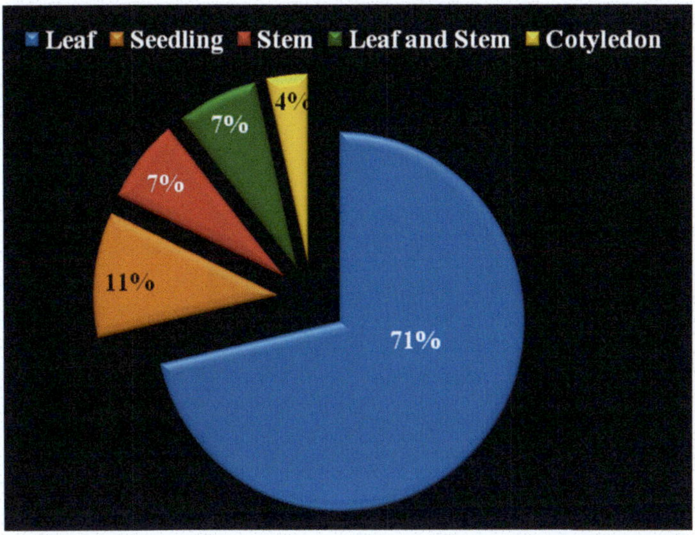

Fig. 6 Preferential utilization of different explants for HR-mediated heterologous gene expression

7 Choices of Promoters' Specifications

Literature survey indicates that although promoters play a crucial role in such heterologous gene expression studies, a single constitutive promoter (i.e., cauliflower mosaic virus 35S promoter – CaMV 35S) has fulfilled the underlying requirement of most the presently reviewed studies and showed 94% of prevalence amongst the reported HR-mediated heterologous plant gene expression studies (Fig. 7). Two individual reports are available that deviate from such usually accepted practice and have used ParAt and GAL4-UAS as the preferred promoters [21, 29].

8 Media Choices

Careful analysis of the available published data relating to the preferential selection of media composition for successful execution of the entire heterologous gene expression procedure evidently revolved around the two most reputed media formulations with distinct divergence in their utilization (Fig. 8). The Gamborg's B5 medium has been used in majority (69%) of the studies, while the Murashige and Skoog's medium has been utilized in 31% of the reported investigations which indicates that the vitamin and other additives of the former medium might be having some influence on the overall success of the entire procedure.

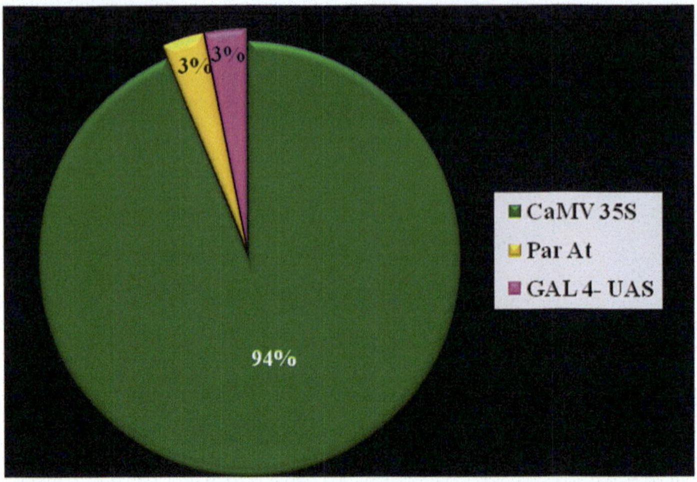

Fig. 7 Options of different promoters for HR-mediated heterologous gene expression

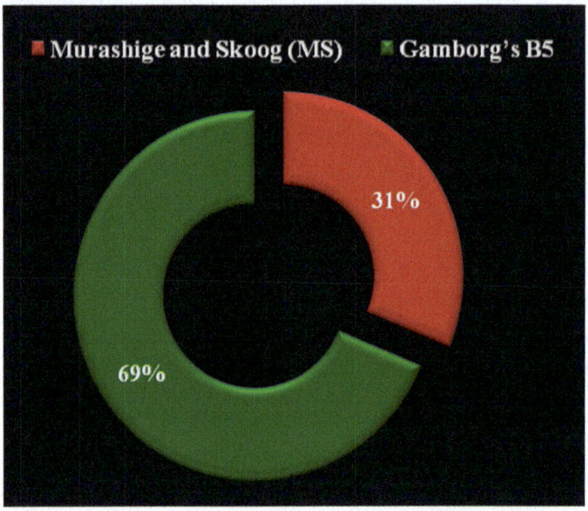

Fig. 8 Documented media preferences for HR-mediated heterologous gene expression

9 Conclusions

The advancement of next-generation sequencing and innovative biotechnological tools along with the understanding of the plant metabolic pathways at both the biochemical and cellular levels has opened up a new avenue for metabolic engineering of any biosynthetic pathways, which is quite evident through the increasing number of reported observations in the most recent years. The higher commercial demand with lower supply of plant-based secondary metabolites has necessitated the search for appropriate and inexpensive expression systems other than the host plant, where hairy root cultures have emerged as a promising alternate approach for the production of metabolites. This review summarizes the HR-mediated heterologous expression of several rate-limiting key genes of TIA, TA, and MVA biosynthetic pathway. The maximum utilization of HR cultures of Solanaceae family has been explored for enhancing the productivity of tropane alkaloids by expressing *H6H* and *PMT* genes. The detailed literature survey indicated that several factors including the targeted genes and their sources as well as recipient plant systems, promoter types, explant types, bacterial strain specificity, and media choices may ultimately decide the fate of such efforts which undoubtedly can bridge the gap between the demand and supply of several plant-based vital bioactive molecules. Further refinement of this HR technology for heterologous expression of biosynthetic pathway genes in medicinal plants can become a powerful approach in the coming years through focused efforts towards translating this technology into industrial applications for the large-scale production of therapeutically important phytomolecules, which lacks any other alternative production source other than the resource mother plant.

Acknowledgments The authors wish to express their sincere thanks to the Director, CSIR-CIMAP, for providing the facilities to carry out similar types of research at the Institute. SS and PP are thankful to the Council of Scientific and Industrial Research (CSIR, New Delhi, India) and the Department of Science and Technology (DST, New Delhi, India), respectively, for financial supports in the form of fellowships. Further acknowledgment also goes to the Academy of Scientific and Innovative Research (AcSIR-CIMAP).

References

1. David B, Wolfender JL, Dias DA (2015) The pharmaceutical industry and natural products: historical status and new trends. Phytochem Rev 14:299–315
2. Newman DJ, Cragg GM (2012) Natural products as sources of new drugs over the 30 years from 1981 to 2010. J Nat Prod 75:311–335
3. Atanasov AG, Waltenberger B, Wenzig EMP, Linder T, Wawrosch C, Uhrin P, Temml V, Wang L, Schwaiger S, Heiss EH, Rollinger JM, Schuster D, Breuss JM, Bochkov V, Mihovilovic MD, Kopp B, Bauer R, Dirscha VM, Stuppner H (2015) Discovery and resupply of pharmacologically active plant-derived natural products: a review. Biotechnol Adv 33:1582–1614
4. Muller JL, Jahn L, Lippert A, Puschel J, Walter A (2014) Improvement of hairy root cultures and plants by changing biosynthetic pathways leading to pharmaceutical metabolites: strategies and applications. Biotechnol Adv 32:1168–1179
5. Banerjee S, Singh S, Rahman LU (2012) Biotransformation studies using hairy root cultures – a review. Biotechnol Adv 30:461–468
6. Verpoorte R, Contin A, Memelink J (2002) Biotechnology for the production of plant secondary metabolites. Phytochem Rev 1:13–25
7. Zhou ML, Zhu XM, Shao JR, Tang YX, Wu YM (2011) Production and metabolic engineering of bioactive substances in plant hairy root culture. Appl Microbiol Biotechnol 90:1229–1239
8. Miralpeix B, Rischer H, Hakkinen ST, Ritala A, Tuulikki SL, Oksman-Caldentey KM, Capell T, Christou P (2013) Metabolic engineering of plant secondary products: which way forward? Curr Pharmaceu Design 19:5622–5639

9. Ikram NKBK, Zhan X, Pan XW, King BC, Simonsen HT (2015) Stable heterologous expression of biologically active terpenoids in green plant cells. Front Plant Sci 6:129–139
10. Zhang L, Ding R, Chai Y, Bonfill M, Moyano E, Oksman-Caldentey KM, Xu T, Pi Y, Wang Z, Zhang H, Kai G, Liao Z, Sun X, Tang K (2004) Engineering tropane biosynthetic pathway in *Hyoscyamus niger* hairy root cultures. Proc Natl Acad Sci U S A 101:6786–6791
11. Matzke MA, Matzke AJM (1995) How and why do plants inactivate homologous (Trans) genes? Plant Physiol 107:679–685
12. Kang YM, Park DJ, Min JY, Song HJ, Jeong MJ, Kim YD, Kang SM, Karigar CS, Choi MS (2011) Enhanced production of tropane alkaloids in transgenic *Scopolia parviflora* hairy root cultures over-expressing putrescine *N*-methyl transferase (PMT) and hyoscyamine-6-β-hydroxylase (H6H). In Vitro Cell Dev Biol Plant 47:516–524
13. Jouhikainen K, Lindgren L, Jokelainen T, Hiltunen R, Teeri TH, Oksman-Caldentey KM (1999) Enhancement of scopolamine production in *Hyoscyamus muticus* L. hairy root cultures by genetic engineering. Planta 208:545–551
14. Zhang L, Yang B, Lu B, Kai G, Wang Z, Xia Y, Ding R, Zhang H, Sun X, Chen W, Tang K (2007) Tropane alkaloids production in transgenic *Hyoscyamus niger* hairy root cultures over-expressing *Putrescine N-methyltransferase* is methyl jasmonate-dependent. Planta 225:887–896
15. Moyano E, Fornale S, Palazon J, Cusido RM, Bagni N, Pinol MT (2002) Alkaloid production in *Duboisia* hybrid hairy root cultures overexpressing the pmt gene. Phytochemistry 59:697–702
16. Moyano E, Jouhikainen K, Tammela P, Palazon J, Cusido RM, Pinol MT, Teeri TH, Oksman-Caldentey KM (2003) Effect of pmt gene overexpression on tropane alkaloid production in transformed root cultures of *Datura metel* and *Hyoscyamus muticus*. J Exp Bot 54:203–211
17. Sato F, Hashimoto T, Hachiya A, Tamura K, Choi KB, Morishige T, Fujimoto H, Yamada Y (2001) Metabolic engineering of plant alkaloid biosynthesis. Proc Natl Acad Sci U S A 98:367–372
18. Lee OS, Kang YM, Jung HY, Min JY, Kang SM, Karigar SC, Prasad DT, Bahk JD, Choi MS (2005) Enhanced production of tropane alkaloids in *Scopolia parviflora* by introducing the PMT (putrescine *N*-methyltransferase) gene. In Vitro Cell Dev Biol Plant 41:167–172
19. Hashimoto T, Yun DJ, Yamada Y (1993) Products of tropane alkaloids in genetically engineered root cultures. Phytochemistry 32:713–718
20. Palazon J, Moyano E, Cusido RM, Bonfilla M, Bagni N, Oksman-Caldentey KM, Pinol MT (2003) Alkaloid production in *Duboisia* hybrid hairy roots and plants overexpressing the h6h gene. Plant Sci 165:1289–1295
21. Rahman LU, Kitamura Y, Yamaguchi J, Mukaia M, Akiyamaa K, Yamamotoa H, Muranaka T, Ikenaga T (2006) Exogenous plant *H6H* but not bacterial *HCHL* gene is expressed in *Duboisia leichhardtii* hairy roots and affects tropane alkaloid production. Enz Microbial Technol 39:1183–1189
22. Hakkinen ST, Moyano E, Cusido RM, Palazon J, Pinol MT, Oksman-Caldentey KM (2005) Enhanced secretion of tropane alkaloids in *Nicotiana tabacum* hairy roots expressing heterologous hyoscyamine-6b-hydroxylase. J Exp Bot 56:2611–2618
23. Moyano E, Palazon J, Bonfill M, Osuna L, Cusido RM, Oksman-Caldentey KM, Pinol MT (2007) Biotransformation of hyoscyamine into scopolamine in transgenic *tobacco* cell cultures. J Plant Physiol 164:521–524
24. Zarate R, Jaber-Vazdekis NE, Medina B, Ravelo AG (2006) Tailoring tropane alkaloid accumulation in transgenic hairy roots of *Atropa baetica* by over-expressing the gene encoding hyoscyamine 6β-hydroxylase. Biotechnol Lett 28:1271–1277
25. Yang C, Chen M, Zeng L, Zhang L, Liu X, Lan X, Tang K, Liao Z (2011) Improvement of tropane alkaloids production in hairy root cultures of *Atropa belladonna* by overexpressing pmt and h6h genes. Plant Omics J 4:29–33
26. Ritala A, Dong L, Imseng N, Seppanen-Laakso T, Vasilev N, van der Krol S, Rischer H, Maaheimo H, Virkki A, Brandli J, Schillberg S, Eibl R, Bouwmeester H, Oksman-Caldentey KM (2014) Evaluation of tobacco (*Nicotiana tabacum* L. cv. Petit Havana SR1) hairy roots for

the production of geraniol, the first committed step in terpenoid indole alkaloid pathway. J Biotechnol 176:20–28

27. Vasilev N, Schmitz C, Dong L, Ritala A, Imseng N, Hakkinen ST, van der Krol S, Eibl R, Oksman-Caldentey KM, Bouwmeester H, Fischer R, Schillberg S (2014) Comparison of plant-based expression platforms for the heterologous production of geraniol. Plant Cell Tiss Org Cult 117:373–380
28. Masakapalli SK, Ritala A, Dong L, van der Krol AR, Oksman-Caldentey KM, Ratcliffe RG, Sweetlove LJ (2014) Metabolic flux phenotype of tobacco hairy roots engineered for increased geraniol production. Phytochemistry 99:73–85
29. Peebles CAM, Sander GW, Hughes EH, Peacock R, Shanks JV, San KY (2011) The expression of 1-deoxy-D-xylulose synthase and geraniol-10-hydroxylase or anthranilate synthase increases terpenoid indole alkaloid accumulation in *Catharanthus roseus* hairy roots. Met Eng 13:234–240
30. Mehrotra S, Srivastava V, Rahman LU, Kukreja AK (2013) Overexpression of a *Catharanthus* tryptophan decarboxylase (tdc) gene leads to enhanced terpenoid indole alkaloid (TIA) production in transgenic hairy root lines of *Rauwolfia serpentina*. Plant Cell Tiss Org Cult 115:377–384
31. Berlin J, Rugenhagen C, Dietze P, Fecker LF, Goddijn OJM, Hoge JHC (1993) Increased production of serotonin by suspension and root cultures of *Peganum harmala* transformed with a tryptophan decarboxylase cDNA clone from *Catharanthus roseus*. Transg Res 2:336–344
32. Hughes EH, Hong SB, Gibson SI, Shanks JV, San KY (2004) Metabolic engineering of the indole pathway in *Catharanthus roseus* hairy roots and increased accumulation of tryptamine and serpentine. Met Eng 6:268–276
33. Hong SB, Peebles CA, Shanks JV, San KY, Gibson SI (2006) Expression of the *Arabidopsis* feedback-insensitive anthranilate synthase holoenzyme and tryptophan decarboxylase genes in *Catharanthus roseus* hairy roots. J Biotechnol 122:28–38
34. Geerlings A, Hallard D, Caballero AM, Cardoso IL, van der Heijden R, Verpoorte R (1999) Alkaloid production by a *Cinchona officinalis* 'Ledergriana' hairy root culture containing constitutive expression constructs of tryptophan decarboxylase and strictosidine synthase cDNAs from *Catharanthus roseus*. Plant Cell Rep 19:191–196
35. Peebles CAM, Sander GW, Li M, Shanks JV, San KY (2009) Five year maintenance of the inducible expression of anthranilate synthase in *Catharanthus roseus* hairy roots. Biotechnol Bioeng 102:1521–1525
36. Kim OT, Kim SH, Ohyama K, Muranaka T, Choi YE, Lee HY, Kim MY, Hwang B (2010) Upregulation of phytosterol and triterpene biosynthesis in *Centella asiatica* hairy roots overexpressed *ginseng* farnesyl diphosphate synthase. Plant Cell Rep 29:403–411
37. Zang YX, Kim DH, Park BS, Hong SB (2009) Metabolic engineering of indole glucosinolates in Chinese cabbage hairy roots expressing *Arabidopsis* CYP79B2, CYP79B3 and CYP83B1. Biotechnol Bioprocess Eng 14:467–473
38. Vranová E, Coman D, Gruissem W (2013) Network analysis of the MVA and MEP pathways for isoprenoid synthesis. Annu Rev Plant Biol 64:665–700
39. Vaccaro M, Malafronte N, Alfieri M, Tommasi ND, Leone A (2014) Enhanced biosynthesis of bioactive abietane diterpenes by overexpressing *AtDXS* or *AtDXR* genes in *Salvia sclarea* hairy roots. Plant Cell Tiss Org Cult 119:65–77
40. Mirjalili MH, Moyano E, Bonfill M, Cusido RM, Palazón J (2011) Overexpression of the *Arabidopsis thaliana* squalene synthase gene in *Withania coagulans* hairy root cultures. Biol Plant 5:357–360
41. Argolo ACC, Charlwood BV, Pletsch M (2000) The regulation of solasodine production by *Agrobacterium rhizogenes*-transformed roots of *Solanum aviculare*. Planta Med 66:448–451
42. Kim YK, Kim JK, Kim YB, Lee S, Kim SU, Park SU (2013) Enhanced accumulation of phytosterol and triterpene in hairy root cultures of *Platycodon grandiflorum* by overexpression of *Panax ginseng* 3-Hydroxy-3-methylglutaryl-coenzyme A Reductase. J Agric Food Chem 61:1928–1934

43. Tuan PA, Kwon DY, Lee S, Arasu MV, Al-Dhabi NA, Park Nam I, Park SU (2014) Enhancement of chlorogenic acid production in hairy roots of *Platycodon grandiflorum* by over-expression of an *Arabidopsis thaliana* transcription factor *AtPAP1*. Int J Mol Sci 15:14743–14752
44. Oller ALW, Agostini E, Milrad SR, Medina MI (2009) In situ and de novo biosynthesis of vitamin C in wild type and transgenic tomato hairy roots: a precursor feeding study. Plant Sci 177:28–34

Agrobacterium rhizogenes-Mediated Transformation in Medicinal Plants: Genetic Stability in Long-Term Culture

13

Dipasree Roychowdhury, Mihir Halder, and Sumita Jha

Abstract

Variations at morphological, cytogenetical, cytochemical, biochemical, and molecular levels have been reported in cell, callus cultures, clonally propagated plants, and in regenerated plants in some plant species. Metabolic instability in the genetically manipulated transgenic cell lines with respect to secondary metabolite production has been reported by different authors in long-term in vitro culture, although in a few cases the transgenic nature of the cell lines was retained. Transgenic hairy root cultures are another promising way of production of commercially valuable secondary metabolites in vitro, which open up a new dimension of the role of plant tissue culture in secondary metabolite production. Hairy root cultures and the plants regenerated from transformed roots are well known for their cytogenetical, morphological, and biochemical stability when compared to cell suspension cultures and callus cultures. But there are very few studies on the stability of hairy roots under long-term cultural condition. Variability in Ri-transformed root cultures and regenerated Ri-transformed plants has also been reported in a few species. In the present review, the stability/instability of hairy root cultures and Ri-plants maintained in vitro for long term is discussed in detail.

Keywords

Agrobacterium rhizogenes • Genetic stability • Cytological stability • Clonal fidelity • In vitro culture • Ri-transformed roots • Ri-transformed plants

D. Roychowdhury (✉) • M. Halder (✉)
Department of Botany, University of Calcutta, Kolkata, India
e-mail: dipasree.roychowdhury@yahoo.com; mhalder16@gmail.com

S. Jha (✉)
Department of Botany, Centre of Advanced Study, University of Calcutta, Kolkata, India
e-mail: sumitajha.cu@gmail.com; sjbot@caluniv.ac.in

S. Jha (ed.), *Transgenesis and Secondary Metabolism*, Reference Series in Phytochemistry, DOI 10.1007/978-3-319-28669-3_8

Contents

1 Introduction

Genetic instability is common in the cell cultures of many species [1–5]. Variations at the morphological level, cytological level (chromosome number and structure), cytochemical level (genome size), biochemical level (secondary metabolites and isozymes), and molecular levels in cells, callus cultures, and in regenerated plants have been reported [6–12]. Variation in clonally propagated plants as well as in plants regenerated directly from explants is also demonstrating such variation [8, 13, 14]. In few species, stability of callus cultures and plants regenerated from callus has been reported in long-term cultures [15–18]. Cytological abnormalities include changes in chromosome number, chromosome rearrangements, deletions, duplications, sequence change play major roles in the development of such variations [4, 6, 7, 19]. The role of gene activation and silencing, epigenetic changes in the development of variants in long-term tissue culture has been reported [13, 20].

Plant cell and tissue cultures are promising renewable alternative source of commercially valuable secondary metabolites, generally for those complex molecules for which chemical synthesis is not viable economically. Interest regarding these biotechnological techniques increased when it was found that undifferentiated cell suspension cultures or differentiated organ cultures synthesize and accumulate specific secondary metabolites at a similar or higher amount as compared to parent plants [21, 22]. However, for production of secondary metabolites in commercially useful amounts, optimization of culture conditions, selection of fast growing high-yielding cell lines, precursor feeding, elicitation, immobilization, and genetic transformation techniques have been utilized in different species [23, 24].

Plant cell suspension cultures of medicinally important species have been used as a promising alternative "chemical factory" for the production of industrially important pharmaceuticals. The major drawback of cell suspension culture is the spontaneous changes in the physiological, morphological, biochemical, and the cytological behavior of cells that leads to instability in secondary metabolite production during long-term culture [25, 26]. Cytological heterogeneity may be represented by mixture of euploid and aneuploid cells with normal cells in cell suspension culture and callus cultures [5, 27–29]. Variations may arise at the structural level of chromosomes without affecting chromosome number. Detection of such kind of cytological instability is comparatively difficult as the culture cells show normal chromosome number. The high degree of such change has been documented in long-term tissue culture. Karyotypic changes due to chromosomal translocations have been observed [4, 30]. Culture induced phenotypic, biochemical, and genetic variations were observed in regenerated plants and progenies of the regenerated plants. Various analyses have revealed that these variants show genetic behavior similar to the naturally occurring mutant [4, 31]. Regeneration of plants may involve de-differentiation and redifferentiated during which different variables can develop.

Transgenic cell cultures of *Vitis amurensis* and *Catharanthus* have been developed by genetic manipulation for production of target secondary metabolites [32, 33]. Instability in transgenic cell cultures with respect to secondary metabolite production is reported by different authors [32, 33]. Metabolic instability of such several transgenic cell lines of *Catharanthus* overexpressing key enzymes of TIA pathway was observed in the long-term in vitro culture when studied for 30 months, although the transgenic nature of the cell lines was retained [32]. Drastic decrease in metabolite content in the transformed cell lines is also reported in grapes by Dubrovina and Kiselev [33]. *rolB* transformed cell lines were established in *V. amurensis* accumulating high levels of resveratrol showed a decline in the resveratrol content in long-term culture with regular subculturing. Stable expression of *rolB* gene in the transgenic cell cultures has been reported after 5 years of transformation by the same group [34]. In in vitro micropropagated clones of transgenic birch, decrease in expression of foreign genes with increase in number of subcultures has been reported [35]. Instability of the foreign gene is not restricted to transgenic cell cultures but also reported in some hairy root cultures and transgenic plants [36]. Integration of foreign gene/genes into the host plant genome might cause alteration in its structure and thus affecting the host plant and/or expression of the transgene by gene silencing or negative influence of the flanking plant DNA and location in the chromosome [37, 38].

Transgenic hairy root cultures have opened up a new dimension to application of organ cultures for synthesis, accumulation, and regulation of secondary metabolites production in vitro due to their rapid growth in simple media without phytohormone and easier maintenance in long-term culture. *Agrobacterium rhizogenes*, a Gram negative soil bacterium is responsible for causing hairy root disease in higher plants by transfer of T-DNA present in Ri-plasmid of the bacteria to the plant host genome. Root loci A-D (*rol A-D*) of the T-DNA are responsible for the development of hairy root phenotype [22, 39]. Such hairy roots are also capable of spontaneously regenerating plants (Ri-transformed plants) in a number of species [22, 40]. The hairy roots and Ri-transformed plants derived from them are known to synthesize important plant secondary metabolites in enhanced or similar levels to the nontransformed plants [22]. Bulgakov et al. [41] suggested that *rol* genes (*rolB* and *rolC*) of Ri plasmid are potential activators of secondary metabolism, activate phytoalexin production and suppress intracellular ROS level, the combination of defense responses and effect of ROS suppression affects sensitivity towards auxin, growth and metabolism of the transformed tissue [41].

Hairy root cultures are well known for their cytogenetical, morphological, and biochemical stability compared to cell suspension cultures and callus cultures. But there are very few studies on the stability of hairy roots under long-term cultural condition. In this chapter, we have discussed regarding the stability and variability of hairy root cultures and regenerated plants of different species on the basis of morphological, biochemical, and cytological analysis reported as far as possible. Genetic stability on the basis of integration and expression of T-DNA genes in transgenic cultures and clonal fidelity of hairy root cultures and Ri-transformed plants are also discussed in this review.

2 Characterization of Ri-transformed Roots

Hairy root cultures are characterized by profuse lateral branching and rapid root tip elongation with plagiotropic growth in hormone free medium [42]. Apart from the wild type transformed hairy roots, foreign genes can be also inserted into the hairy roots in many plants [43]. The altered morphological characteristics of the hairy root cultures enable marker free selection of the transformed root lines which are advantageous over use of *A. tumefaciens*-mediated transformation [40, 42].

A. rhizogenes-mediated transformation and transformed root cultures have been established in numerous plant species, including many medicinally important plants [22, 44] as an alternative source for the production of important plant secondary metabolites [22]. Transformed root lines are reported to vary in morphology, biomass accumulation, integration of T-DNA genes, and biosynthetic ability in a species [45–52]. Such variations in between the hairy root lines of the same plant species are suggested to be due to variation in the integration of T-DNA genes of Ri-plasmid into the plant genome [50, 53, 54].

Several reports are available on the effects of the TR and TL T-DNAs on the growth, morphology, and secondary metabolite production in transformed roots [46–50, 55, 56]. For example, in *Catharanthus roseus*, *rolAB*$^{+}$/*ags*$^{+}$ and *rolAB*$^{+}$/*ags*$^{-}$ root clones belonged to four different morphological types, whereas *rolAB*$^{-}$/*ags*$^{+}$ root clones were of slow growing and callusing morphology [46]. In coffee hairy roots, *rolB* and *rolC* genes were reported to be systematically integrated; however, the presence of *rolA* and *rolD* genes could not be related to the morphological variability. The correlation could not be found between the presence of T-DNA genes and hairy root lines showing altered morphology in *Coffea arabica* [49]. Similarly, in *Tylophora indica* no direct correlation was found between the presence of T-DNA genes and transformed root morphology [52]. Differential loss of T-DNA genes in hairy root lines of *C. roseus* and the effect of such loss on the morphology and the biosynthetic ability of the root lines are reported by Taneja et al. [50]. Ten Ri-transformed root lines studied for the presence of 23 ORFs of the

T-DNA showed loss of few ORFs that drastically affected the growth, morphology, and alkaloid biosynthesis in transformed root lines [50]. The induction and development of the hairy root phenotypes are affected by synergistic activity of the *rol* genes expressed simultaneously [57–59].

The study of stability of such variant Ri-transformed root lines in the long-term in vitro culture is reported only in a few plant species. The details of such reports on morphological, biosynthetic, cytological, and molecular stability of *A. rhizogenes*-transformed root cultures in various species are discussed in detail in the following section of this review.

2.1 Morphological Stability of Ri-transformed Roots

There are contradictory reports regarding the extent of stability in transformed roots maintained in vitro during prolonged culture. Stability in growth and morphological characters over prolonged culture is reported in different species like – *Lycopersicon esculentum* [60], *Cinchona officinalis* "Ledgeriana" [61], *Aesculus hippocastanum* L. [62], *Pimpinella anisum* [63], Coffee [49], *T. indica* [52], *Arachis hypogaea* [51], *Plumbago zeylanica* L. [64], *Rauvolfia serpentina* [65, 66], etc. Morphology of different Ri-transformed root lines of different plant species is known to vary to a large extent (Fig. 1).

Long-term stability of Ri-transformed root cultures in *L. esculentum* is reported by Lipp Jao and Brown [60]. Transformed root clones established with *A. rhizogenes* strain R1601 were maintained in solid and liquid media. The hairy roots were found to retain their growth and characteristic phenotype for 50 passages over 25 months in liquid culture and for 12 passages over 12 months in solid culture. The growth rate of these root clones was not affected by the presence or absence of selective agent in the solid media, thus not necessary for maintaining the transformed state in long-term culture as suggested by the authors [60].

Stability in growth and morphology of hairy root cultures of *C. officinalis* "Ledgeriana" after 1 year of in vitro culture is reported [61]. Similar unchanged growth rate of hairy root cultures of *A. hippocastanum* L. after 4 years of in vitro culture is reported by Zdravković-Korać et al. [62].

Santos et al. [63] studied the morphological stability of hairy root lines of *P. anisum* in four different media. Root lines growing on SH medium showed stable morphological characters over the period of study compared to the other three media where marked changes were observed [63].

Alpizar et al. [49] reported that exogenous auxin supplementation was essential for proliferation of hairy root lines of coffee. Sixty two root clones varied significantly in two growth parameters (total root length and frequency of fine roots), and the characteristic phenotypes were stable over the subcultures for over 3 years [49].

In *T. indica,* four morphologically distinct phenotypes observed in *A. rhizogenes* strain A4 transformed root lines were stable when maintained for more than 4 years on hormone free MS basal medium [52].

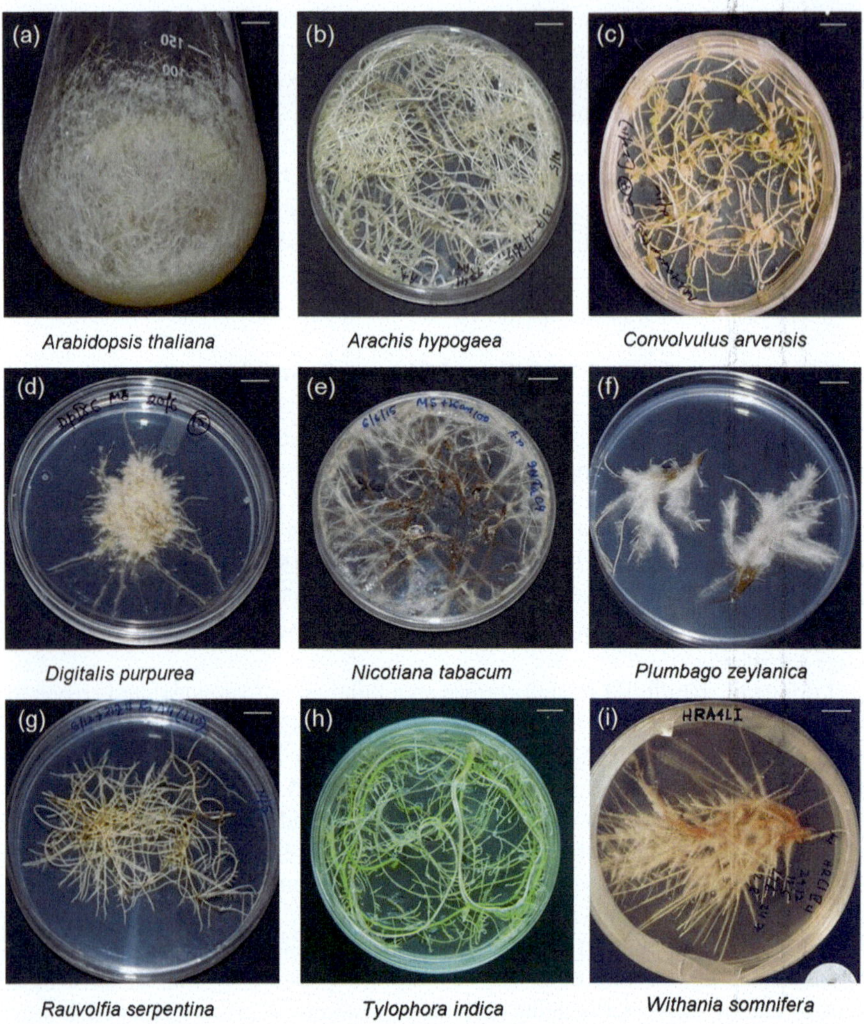

Fig. 1 Ri-transformed root lines of different plant species transformed with wild type *Agrobacterium rhizogenes* strains, maintained for long term in vitro culture for more than 5–8 years (Bar = 1 cm) – (**a**, **c**–**g**) LBA 9402 transformed root cultures of *Arabidopsis thaliana* (Photo P. Paul), *Convolvulus arvensis* (Photo A. Majumder), *Digitalis purpurea* (Photo A. Basu), *Nicotiana tabacum* (Photo M. Halder), *Plumbago zyelanica* (Photo A. Basu), and *Rauvolfia serpentina* (Photo Smita Ray), respectively. (**b**, **h**, **i**) A4 transformed root culture of *Arachis hypogaea* var. JL-24 (Photo M. Halder), *Tylophora indica* (Photo D. Roychowdhury), and *Withania somnifera* (Photo Swagata Ray), respectively

In peanut (*A. hypogaea*) cv. JL-24, Halder and Jha [51] reported strain independent variability among 30 Ri-transformed root lines on the basis of morphology, biomass accumulation, and *trans*-resveratrol content. Root lines and clones maintained in vitro for over 3 years showed stable morphological characters as well as growth index value [51].

Growth and morphology of LBA9402 transformed root lines of *P. zeylanica* cultured on solid modified MS medium were observed to be stable for 2 years in in vitro culture [64].

Ray et al. [66] reported retention of stable phenotype of Ri-transformed root lines of *R. serpentina* over 3 years of in vitro culture. The phenotype of transformed root lines included creamish roots with a high degree of branching, plagiotropic growth, and devoid of extensive root hairs [66]. Long-term stability of hairy root cultures of *R. serpentina* is also reported by Pandey et al. [65] for more than 6 years of in vitro culture. The growth kinetic analysis of *R. serpentina* hairy roots exhibited higher growth potential following long-term cultivation with either of the two carbohydrates, viz., sucrose and table sugar [65].

However, instability in morphological phenotype and growth kinetics of hairy roots in prolonged in vitro culture is reported in certain species like carrot [36], *Brugmansia candida* [67], *Duboisia myoporoides* [68], *Hyoscyamus muticus* [69], etc.

Instability in phenotype and expression of transgenes were observed in hairy root lines of *Daucus carota* in long-term culture [36]. Different phenotypes and growth patterns were observed between hairy root clones of carrot and between the subcultures of single root clones [36]. Significant decrease in growth of hairy root cultures is reported in *B. candida* on prolonged subculture for 5 years [67]. Yukimune et al. [68] performed repeated selection in hairy root cultures of *D. myoporoides* and observed that the morphology of the transformed roots with improved scopolamine content differed after the repeated selection, viz., fine root lines with extensive lateral branching. In *H. muticus*, variation was observed between the different hairy root lines and hairy root clones derived from protoplast culture [69].

Thus, it is evident from the above reports that the majority of the plant species studied to observe the long-term morphological characters and growth of hairy roots in culture conditions showed stability. And only certain species are found to behave differently in the long-term in vitro cultures with a decrease in growth kinetics and change in morphology. Type of growth media can be a deciding factor for the phenotypic stability of *A. rhizogenes*-transformed root cultures as observed in *P. anisum* [63].

2.2 Biochemical Stability of Ri-transformed Roots

The utility of hairy root culture for synthesizing important secondary metabolites greatly depends on the biochemical stability of the high yielding roots in the long-term in vitro culture. The stable production of important plant metabolites by the Ri-transformed hairy roots after prolonged culture period is reported in different plants like *Beta vulgaris*, *Nicotiana rustica* [70], *Datura stramonium* [71, 72], *D. innoxia* [73], *H. muticus* [69, 74], *P. anisum* [63], *T. indica* [52], *R. serpentina* [65], etc.

Biosynthetic stability of hairy roots of *B. vulgaris* and *N. rustica* after long-term cryopreservation is reported by Benson and Hamill [70]. The total betalain (total betaxanthin and betacyanin pigments) in *B. vulgaris* and total alkaloid production in *N. rustica* was found to be stable after the recovery of the cryopreserved hairy roots suggesting biosynthetic stability of these Ri-transformed root lines [70].

Biosynthetic stability of hairy root lines in the long-term in vitro culture is also reported in *D. stramonium* [71]. Very high stability in total alkaloid production for over 5 years, i.e., over 75 subcultures, was observed in the high alkaloid producing root lines [71]. This finding was advantageous over the high level of instability observed in normal root cultures of *D. stramonium* upon subculturing [75]. Baíza et al. [72] also reported biochemical stability of the hairy root lines of *D. stramonium* over a period of 6 years. On contrary, only one root line of *D. stramonium* showed great instability in alkaloid production through time, with regard to the production of both hyoscyamine and scopolamine [72, 75]. Dechaux and Boitel-Conti [73] reported a decrease in scopolamine levels similar to control levels after 1 year of subculture in hairy roots of *D. innoxia* overexpressing *H6H* gene from *H. niger.*

Stable secondary metabolite production after 5 years of in vitro culture is reported in hairy roots of *H. muticus*. All the studied Ri-transformed root lines of *H. muticus* showed stable alkaloid production during the long-term culture [69]. Stability of secondary metabolite production in hairy roots of *H. muticus* during two and a half years of culture is also reported by Jouhikainen et al. [74].

Hairy root cultures of *P. anisum,* maintained in the SH medium, both under darkness and photoperiod conditions showed stable essential oil production [63].

Hairy root cultures of *T. indica* established by using *A. rhizogenes* strain A4 showed biochemical stability over a period of 4 years. The transformed root lines of different phenotypes accumulated tylophorine at higher levels as compared to the to nontransformed roots, and the potential of enhanced growth and tylophorine content was stably maintained in long-term in vitro culture [52].

R. serpentina hairy roots (Ri-A4 transformed roots) produce considerably higher reserpine content than the normal roots after 10 weeks of growth [76]. Ray et al. [66] reported higher reserpine (approx 3 mg g^{-1}DW) content in the LBA 9402 root lines. Pandey et al.[65] reported the production of all three major terpene alkaloid in *R. serpentina,* viz., reserpine, ajmaline, yohimbine since the establishment of cultures with yohimbine concentration being highest followed by ajmaline and reserpine. After 6 years of maintenance on both sucrose and table sugar containing media, the selected root culture of *R. serpentina* maintained similar trend of production, yohimbine appeared to be the major alkaloid. Root cultures showed differential yields of the alkaloids in the two carbon sources used. The hairy roots of *R. serpentina* showed 193.8% more reserpine production in the table sugar supplemented media compared to media supplemented with sucrose. However, the yield of yohimbine and ajmaline was 38.32% and 61.98% higher in the sucrose supplemented media than the table sugar supplemented one. These hairy root cultures of *R. serpentina* showed escalation of secondary metabolite production in long-term culture over the 6 years of study [65].

Increase in secondary metabolite production over a long period of subculture of Ri-transformed roots is reported in plant species like *B. candida* [67] and *D. myoporoides* [68]. Marconi et al. [67] reported increasing secondary metabolite production after 5 years of in vitro culture of *B. candida* hairy roots, possibly due to the prolonged in vitro culture mediated stress. In the long-term in vitro culture, a pronounced increase in the production of scopolamine was observed in such root lines [67]. In *D. myoporoides,* repeated selection in transformed root cultures was done and it was observed that scopolamine content of the hairy root lines obtained at each selection increased with the number of selections [68].

On the contrary, complete loss of capacity to accumulate alkaloids after 1 year of in vitro culture of hairy roots is reported in *C. officinalis* "Ledgeriana" [61]. These transgenic hairy roots containing tryptophan decarboxylase (Tdc) and strictosidine synthase (Str) from *C. roseus*, two key enzymes in terpenoid indole and quinoline alkaloid biosynthesis, showed increased levels of quinine and quinidine initially after transformation [61].

2.3 Cytological Stability of Ri-transformed Roots

Several reports suggested that long-term in vitro tissue culture causes somaclonal variations and chromosomal abnormality, including both numerical and structural alterations [1]. Unorganized cultures are more prone to chromosomal instability – aneuploidy and polyploidy [4, 10, 25] – than organized cultures. There are very few reports regarding the chromosomal status (genetic stability) of long-termed hairy root culture.

Ambros et al. [77] reported chromosomal localization by in situ hybridization of Ri-T-DNA in five different *Crepis capillaries* ($2n = 6$) transformed root lines, and all the root lines showed normal diploid chromosome number. The first report of chromosomal stability in *A. rhizogenes*-transformed root cultures was by Aird et al. [78]. The hairy root lines were studied after 6–18 months of in vitro culture and were found to show stable, normal chromosome number in all the seven plant species, viz., *C. roseus* ($2n = 16$), *D. stramonium* ($2n = 24$), *N. rustica* ($2n = 48$), *N. umbratica* ($2n = 46$), *N. africana* ($2n = 46$), *Phaseolus vulgaris* ($2n = 22$), and *B. vulgaris* ($2n = 18$) [78]. Transformed root cultures of *D. stramonium* showed typical karyotype and chromosome number same as in the control plants [72]. Baíza et al. [72] studied the karyotypic stability of three lines of hairy roots with stable secondary metabolite production of *D. stramonium* compared to instability of nontransformed root cultures. The transformed root cultures growing in the absence of any phytohormones exclusively contained diploid cells with $2n = 24$, whereas the nontransformed root cultures whose growth needs exogenous hormone supplementation showed presence of mixoploidy and aneusomaty. This cytological stability of the hairy root lines of *D. stramonium* was maintained irrespective of the age of the transformed cultures [72]. In *Swainsona galegifolia,* untransformed root cultures were found to contain 90% diploid ($2n = 32$) cells, while *A. rhizogenes* transformed roots showed stability of diploid chromosome number [79].

Chromosomal stability in transformed hairy root cultures of *Artemisia annua* L. with normal diploid chromosome number $2n = 18$ in all the LBA9402 transformed clones studied was reported by Mukherjee et al. [80]. Addition of growth regulators induced disorganization and dedifferentiation accompanied by loss of chromosomal stability in such cultures. Redifferentiation and rhizogenesis could be induced in such cultures in phytohormone free media and 90% of the regenerated roots showed diploid chromosome number ($2n = 18$) [80]. In *R. serpentina*, chromosome and karyotype analysis in transformed root lines maintained over 3 years in vitro was reported by Ray et al. [66] concluding that the karyotype of transformed root lines was similar to the roots of parent *R. serpentina* plants [66].

Cytological instability in hairy root cultures has been observed in some plant species such as *Trifolium pretense, T. repens, Lotus corniculatus* [81], *Artemisia cina* [82], *Vicia faba* [83], *Lycopersicon peruvianum* [84], potato [85], and *Onobrychis viciaefolia* [86].

Webb et al. [81] reported instability in anatomy, morphology, and cytology of Ri-transformed root lines of three legume species, namely *T. pretense, T. repens*, and *L. corniculatus* established using wild type strain C58C1 with pRi15834 of *A. rhizogenes*.

In *A. cina,* hairy root cultures established with three different *A. rhizogenes* strains showed instability in chromosome numbers along with the nontransformed ones after 6–12 months in culture. Nontransformed roots had a diploid chromosome number $2n = 32$ in 53.7% of the cells, while the rest 46.3% cells showed chromosome number ranging from $2n = 22$ to 64. The chromosome numbers of transformed roots were affected by different strains of *A. rhizogenes* used. Hairy roots established with strain 07–20001 showed the highest normal chromosome number (62.4%) followed by strain ATCC15834 (61.9%) and strain A4 (43%). The chromosome number range in transformed root lines was $2n = 11$ to $2n = 66$ [82].

In *V. faba,* 65 transformed root clones were studied, out of which 50% were **polyploids** and 6% were aneuploids or showed structural rearrangements. Polyploid root clones included octaploids, i.e., $2n = 8\times = 48$ [83]. Detailed karyotypic analysis of *L. peruvianum* hairy roots showed a diploid chromosome number with structural rearrangements [84]. Vries-Uijtewaal et al. [85] showed that hairy root clones developed from mono-haploid or di-haploid genotype were either diploid or tetraploid in potato. Ri-transformed roots of O. *viciaefolia* showed normal chromosome number $2n = 4\times = 28$ and spontaneously formed shoot buds developing into plants. However, the percentage of hairy root cells with a normal chromosome number reduced drastically with time, i.e., from 85% to 23% after 4 months and only 4% after 8 months. After 12 months of maintenance, roots with a normal chromosome number $2n = 28$ was scarcely found with an increase in cells with 14 chromosomes [86]. The elimination of the normal chromosome number in hairy roots of *O. viciaefolia* was accompanied with loss of regeneration potential.

2.4 Integration and Expression of T-DNA Genes in Ri-transformed Roots

The T-DNA genes that are integrated and expressed during prolonged culture is a measure for assessment of transgenic nature of the hairy roots of different plants after repeated subcultures. Ri-transformed roots of different plant species have been reported to show the genetic stability of the Ri-T-DNA genes in long-term in vitro culture like in *B. vulgaris*, *N. rustica* [70], *L. esculentum* [60], *C. officinalis* "Ledgeriana"[61], *A. hippocastanum* L. [62], *D. innoxia* [73], coffee [49], *T. indica* [52], *A. hypogaea* [51], *P. zeylanica* L. [64], *R. serpentina* [65], etc.

Post-freeze molecular stability of T-DNA genes has been reported in Ri-transformed roots of *B. vulgaris* and *N. rustica* after short term and long-term recovery from cryopreservation [70]. The TL-DNA fragment was stably retained in hairy roots of both the species recovered from cryopreservation [70].

Molecular stability of hairy root lines of tomato is reported by Lipp Jao and Brown [60]. Transformed root lines of *L. esculentum* maintained for 50 passages in liquid culture and 12 passages in solid culture were reported to show the presence of the *nptII* gene by PCR and dot blot hybridization analyses. These root lines were also found to show NPTII enzyme activity in the long-term cultures [60].

The hairy root lines of *C. officinalis* "Ledgeriana" showing stable growth and morphology were found to completely lose the capacity of alkaloid production after 1 year. These root lines, however, showed the transgenic nature on PCR analysis, *gus* assay, and southern hybridization of the transgenes (*Tdc* and *Str* probes from *C. roseus*) after 1 year of maintenance. In Northern blot analysis, it was found that *Tdc* gene was not being expressed; only signal was detected for the GUS probe [61]. Therefore, hairy roots of *C. officinalis* showed stable integration of transgenes after 1 year, but the expression of one transgene was lost in long-term culture [61].

Transformed root lines of *A. hippocastanum* showed stable integration of *rolA, rolB, rolC*, and *rolD* genes by PCR analysis after 4 years of in vitro culture [62]. In *D. innoxia,* stable transgene expression is reported after 1 year of the subculture of the hairy roots by Dechaux and Boite-Conti [73].

Stable integration of *rol* genes has been reported in Ri-transformed root lines of coffee [49]. Fifty five Ri-transformed root lines maintained for over 3 years in vitro and characterized at the molecular level to study intra- and inter-clonal variability of the root lines showed to be TL^+/TR^- [49].

In *T. indica* transformed root lines were of two types; majority (87%) were TL^+/TR^-, while rest were TL^+/TR^+. Stable integration and expression of T-DNA genes of each clone were maintained in long-term in vitro culture [52].

Stable integration and expression of different *rol* genes were observed in 30 Ri-transformed root lines of *A. hypogaea* induced following infection with different strains of *A. rhizogenes*, after 2 years of maintenance in basal medium establishing genetic stability root lines and clones [51].

In Ri-transformed root lines of *P. zeylanica,* the *rolA*, *rolB*, *rolC*, and *rolD* genes were stably integrated, retained, and expressed at the transcription level as revealed by PCR and RT-PCR analysis of 12–18-month-old cultures [64].

In *R. serpentina*, stable integration and expression of Ri-T-DNA genes have been reported in long-term in vitro culture by Pandey et al. [65]. The hairy root lines showed positive results for *rolB* and *rolC* after 6 years of maintenance on either of the carbon source used in media, viz., sucrose or table sugar [65].

Contradictorily, in hairy roots of certain plant species, instability of integrated transgene is reported. For example, instability in expression of transgene in in vitro culture of carrot hairy roots was reported by Guivarc'h et al. [36]. Reversible inactivation of the transgene expression was noted in one of the hairy root clones studied with a high copy number of the transgene [36]. In potato, frequent spontaneous deletion of Ri-T-DNA in hairy roots and regenerated plants was observed [87].

The study on the clonal fidelity of hairy root cultures in long-term in vitro culture can be an important criterion to assess the genetic stability of the transformed cultures. DNA fingerprinting profiles of 15 hairy root lines of *T. indica* along with nontransformed roots generated with 11 OPA primers after every 1 year for 4 years showed genetic stability and clonal fidelity of the clones of root lines in the long-term in vitro culture [52]. The study showed similarity in between the transformed roots and nontransformed ones at the genetic level in terms of the primers used, and the fingerprinting profiles for each primer did not vary with age [52].

3 Characterization of Ri-transformed Plants

Regeneration of Ri-transformed plants from transformed roots has been reported in many plant species, including medicinally important plants as reviewed earlier [22, 40]. These Ri-transformed plants exhibit unique characteristic features distinctly different from the nontransformed plants known as the "hairy root syndrome" [88]. Ri-transformed plants are morphologically characterized by stunted growth with reduced shoot and internodes length, accompanied with the increase in the number of nodes, internodes and leaves (Fig. 2). However, the morphology of leaves is different from the nontransformed plants, i.e., the leaves are smaller in size and wrinkled in appearance. The decrease in shoot length is compensated by an increase in number of axillary branches. The root system is pronounced with decrease in the length of main root and an increase in the number of lateral branches giving rise to highly branched and extensive root system in the Ri-transformed plants. The roots are plageotropic and often seem to grow above the surface of the media. Presence of adventitious roots in the Ri-transformed plants is another feature noted. Apart from these characters, in some species, altered floral morphology, early flowering, and reduced pollen and seed production have also been reported in the transgenic plants regenerated from hairy roots. Additionally, conversion of biennial species to annuals was observed in Ri-transformed plants [42, 89, 90]. Hairy root syndromes of Ri-transformed plants are due to the insertion and expression of *rol* genes of the TL-DNA of Ri-plasmid. The alterations in morphological characters are caused by *rolA*, *rolB*, and *rolC* genes [42, 88].

Biochemically, the Ri-transformed plants are capable of synthesizing all the important secondary metabolites at a level comparable to or higher than the parent

Fig. 2 Spontaneous regeneration of plants from root cultures of *Nicotiana tabacum* on phytohormone-free MS medium (Photos M. Halder) (**a**) nontransformed root culture showing spontaneous regeneration of shoot (Bar = 1 cm; *arrow* indicating regenerating shoot), (**b**) regenerated nontransformed plant growing on MS basal medium maintained for more than 3 years by regular subculture (Bar = 1 cm), (**c**) Ri-transformed root culture showing spontaneous regeneration of shoot (Bar = 1 cm, *arrow* indicating regenerating shoot), (**d**) regenerated Ri-transformed plant growing on MS basal medium maintained for more than 3 years by regular subculture (Bar = 1 cm)

plants. Enhanced biomass of the transformed plants itself makes the use of Ri-transformed plants advantageous over nontransformed plants for the production of biologically active plant secondary metabolites. Many species are reported to produce medicinally important metabolites at higher concentration compared to their nontransformed counterparts, in addition to higher production of target metabolite due to increased biomass. Such plants are often reported retaining these desired characters, even after transfer to the field. However, the genetic stability of transformed plants has been reported in a very few species as discussed in the following segments of this review.

3.1 Morphological Stability of Ri-transformed Plants

Studies on stability of phenotype of transformed plants regenerated from transformed roots are very few in number, although, we have seen quite a number of reports are available for the hairy root cultures as discussed above. The altered morphology of the Ri-transformed plants are reported to be stable under long-term in vitro *culture* in very few species like Kiwi [91], *T. indica* [92, 93], *Bacopa monnieri* [94], and *C. roseus* [95].

Morphological stability of *rolABC* transformed plants and *rolB* transformed plants of Kiwi for more than 6 years has been reported by Rugini et al. [91]. The *rolB* transformed plants of Kiwi were morphologically similar to the control plants, whereas the *rolABC* plants showed a typical hairy root phenotype. These agronomic traits were maintained for over 6 years and in 50% of the plants of T1 generation [91].

Ri-transformed plants of *T. indica,* established by spontaneous regeneration from A4-transformed roots, have been reported to show typical Ri-transformed phenotype [96]. The plants showed a stable transformed phenotype in long-term culture (6 years) [92]. However, a detailed study on the fate of integrated T-DNA *rol* genes during regeneration via somatic embryogenesis in *T. indica* showed 19 out of 23 Ri-transformed plants with typical Ri-transformed phenotype and rest four plants with morphology similar to the nontransformed plants. These 23 plants with typical or variant morphology retained their characteristic phenotype for more than 3 years of in vitro maintenance [93].

In *B. monnieri* (Linn.), most of the clones of Ri-transformed plants were morphologically stable and did not show any alteration in the morphological characters including flowering [94].

Transgenic *C. roseus* plants are characterized with broad dark green leaves, shorter internodal length, rooting from its lower nodes, and flower with filamentous corolla compared to nontransformed plants [95, 97]. During 5 years of maintenance of these transgenic plants, drastic changes were observed: like reduction in plant height up to 1/4th of the initial height, highly proliferating root system, and absence of flowering and narrower yellowish leaves [95].

In several other plant species, the alteration in phenotype in transformed plants have been retained after transfer to the greenhouse, such as in *Datura arborea* [98], *Limonium* hybrid [99], and *Kalanchoe blossfeldiana* [100]. After 6 months of growing in the greenhouse, height of some clones of *D. arborea* was similar to untransformed plants [98]. In Ri-transformed plants of *T. indica* [92], *Pelargonium graveolens* [101], and *Plumbago rosea* [102], leaf wrinkling was not observed after transfer to the field.

In field evaluation of transgenic plants in *Brassica*, 40% plants showed severe Ri-phenotype, 40% showed moderate phenotype, however 20% of the transgenics showed normal phenotype [103]. Similarly, Ri-transformed plants regenerated spontaneously from the hairy roots of *R. serpentina* showed varied morphological characters. Out of 30 successfully acclimatized transgenic plants, 90% plants showed phenotypic similarity with nontransformed controls, and only 10% plants showed stunted growth [76]. The phenotypic variation may be due to the presence of multiple copies of the transgenes as revealed by Southern hybridization in *R. serpentina* [76]. Transformed plants exhibiting normal phenotypes regenerated from hairy roots are reported in many other species like tomato [104], *Stylosanthes humilis* [105], tobacco [106], *Brassica napus* [107], *L. corniculatus* [81], Cauliflower [108, 109], and Broccoli [109]. The absence of the transformed phenotype may not be correlated with the absence of specific *rol* gene(s), but it can also occur due to independent segregation or co-segregation of T-DNAs. In addition, the expression of transgene can be influenced synergistically by the number of integration, site of integration (repetitive DNA or transcriptionally active region of the host genome), and orientation of multiple copy insertion of foreign gene [76, 110–112].

Segregation of phenotypic characteristics has been reported in *S. tuberosum* among Ri-transformed plants regenerated through callus formation, producing phenotypes similar to the controls [87, 113]. Deletion of a single copy of TL-DNA in two shoot lines showed disappearance of Ri-characters [87]. Similar phenotypes were observed in the transformed sister shoot lines of potato [113].

In *H. muticus* L., somaclonal variation among plants regenerated from a single hairy root clone via protoplast culture has been reported [114, 115]. Half of regenerated plants showed typical Ri-transformed phenotype, while the rest of the plant lines were morphologically similar to nontransformed controls [114, 115].

3.2 Biochemical Stability of Ri-transformed Plants

Reports on analysis of secondary metabolites in plants are comparatively very few compared to the extensive reports available for hairy roots [22, 94]. Stable secondary metabolite production in Ri-transformed plants maintained in vitro have been reported in some plant species like *B. monnieri* [94],*C. roseus* [95], *P. graveolens* [101], *R. serpentina* [76], and *T. indica* [92, 93].

In *P. graveolens*, the Ri-transformed plants are reported to synthesize essential oils after 5 months of transferring to the soil [101].

In *T. indica,* Ri-transformed plants showed augmented production of the major alkaloid tylophorine compared to nontransformed plants [96], retaining their ability to synthesize improved level of tylophorine after 6 years of maintenance in vitro and there after 1 year of field transfer [92, 93].

Hairy root lines of *R. serpentina* and plants regenerated from such roots contain high reserpine [76]. Reserpine content of Ri-transformed plant was considerably higher compared to the control nontransformed plants grown in fields and was stably maintained over more than 2 years of cultivation [76].

In *C. roseus* transgenic plants, the tryptophan and alkaloid profile are reported to be almost constant during the 5 years of maintenance [95]. These transgenic plants were found to show better tryptophan content (357.5 ± 25.6 μg g^{-1}DW in Ri-transformed plants compared to 114.5 ± 7.5 μg g^{-1}DW in control) and alkaloid profile compared to nontransformed plants [95].

Biochemical stability of Ri-transformed plants of *B. monnieri* long-term in vitro culture is reported by Paul et al. [94] in Ri-crypt co-transformed plants. The plants showed stable, high content of bacosides compared to nontransformed plants for 4 years of in vitro culture [94].

3.3 Cytological Stability of Ri-transformed Plants

There is a lacuna in information on chromosomal analysis of plants regenerated from transformed roots as compared to hairy roots in long-term culture. Among the few reports available, *T. indica* transformed plants showed chromosomal stability after 6 years of maintenance in vitro with a normal chromosome number $2n = 22$ as parent plant [92, 116]. On the other hand, Webb et al. [81] reported changes in cytology along with morphology and physiology in plants regenerated from hairy roots of *L. corniculatus*.

3.4 Integration and Expression of T-DNA Genes in Ri-transformed Plants

Among the handful of reports available on long-term genetic stability of Ri-transformed plants, genetic stability have been reported in kiwi [91], *T. indica* [92, 93], *B. monnieri* [94], and *C. roseus* [95]. Genetic stability in transgenic kiwi plants transformed with *rolABC* and *rolB* has been reported by Rugini et al. [91]. In *T. indica*, most of the plant lines are reported to show genetic stability over 3–6 years of in vitro culture and after 1 year of field transfer, however, some variants are also reported [92, 93, 96]. Transgenic plants of *B. monnieri* showed stable integration as well as stable expression of *rolA*, *rolB*, *rolC*, and *rolD* genes after 4 years of maintenance in vitro [94].

Stable retention of three important *rol* genes – *rolA, rolB*, and *rolC* genes – have been reported in Ri-transformed plants of *C. roseus* even in the fifth year, indicating the stable nature of the Ri-transgenes [95].

Instability in the integration or expression of transgene in transformed plant lines regenerated from hairy roots is reported in *S. tuberosum* L. cv. *Bintje* [87], *H. muticus* [114, 115], and *T. indica* [93].

In *S. tuberosum* L. cv. *Bintje*, it was suggested that spontaneous deletions of TL-DNA and TR-DNA can occur during long-term root culture and regeneration of Ri-transformed plants [87].

In *H. muticus*, 50% of the Ri-transformed plants regenerated from the root line via protoplast culture showed the presence of *rolA, rolB*, and *rolC* similar to the root line from which they regenerated. Whereas, the latter 50% plants regenerated from the same root line were found *rolA−/rolB−/rolC−*. The transgenic morphology of the regenerated plants of *H. muticus* could be directly correlated with the presence of *rol* genes [114, 115].

Four variant plant lines of *T. indica* showing morphology similar to nontransformed plants were observed to show the presence and expression of only *rolA* gene. Such plant lines were regenerated through somatic embryogenesis from callus lines spontaneously developed from root lines with stable integration and expression of all the four *rol* genes. The variant plant lines showed the integration and expression of only the *rolA* gene throughout the period of study [93]. Loss of the *rol* genes in *H. muticus* and *T. indica* could be due to the deletion of the *rol* genes during regeneration [87, 117, 118].

Studies on the clonal fidelity of Ri-transformed plants are very few. Marker studies performed with 15 ISSR primers and cluster analysis in transgenic *C. roseus* plants revealed the relationship between the Ri-plants, Ti-plants, and their respective controls. The Dhawal Ti-plants were more close to Nirmal control plants followed by Nirmal Ri-plants [95].

4 Conclusions

Stability of Ri-transformed roots and regenerated plants is an important requirement if such cultures are to be used commercially. From the different reports it is evident that a detailed study on the extent of stability of transformed phenotype and genotype is far from adequate. The majority of the Ri-transformed cultures reported so far in different species show morphological, biosynthetic and genetic stability (in terms of T-DNA integration, expression, and cytological studies) when maintained in vitro by regular subculture. In some species, this stability is not obtained in Ri-transformed root and Ri-transformed plants. Apart from the variation between different clones of hairy roots and Ri-plants of a single species, instability in morphology, growth kinetics, biosynthetic potential, and loss of integrated T-DNA gene or its expression are reported to take place during long-term in vitro culture. Thus, transgenic root lines and regenerated plants should be analyzed for stability of the desired characters in long-term in vitro culture before being considered for scale up studies.

References

1. Bayliss MW (1980) Chromosomal variation in plant tissues in culture. Int Rev Cytol Suppl 11A:113–144
2. Scowcroft WR (1985) Somaclonal variation: the myth of clonal uniformity. In: Hohn B, Dennis ES (eds) Genetic flux in plants. Springer Verlag, Berlin
3. Scowcroft WR, Brettell RIS, Ryan SA, Davies PA, Pallotta MA (1987) Somaclonal variation and genomic flux. In: Green CE, Somers DA, Hackett WP, Biesboer DD (eds) Plant tissue and cell culture. Alan R Liss, New York
4. Winfield M, Davey MR, Karp A (1993) A comparison of chromosome instability in cell suspensions of diploid, tetraploid and hexaploid wheats. Heredity 70:187–194
5. Wilson SA, Roberts SC (2012) Recent advances towards development and commercialization of plant cell culture processes for the synthesis of biomolecules. Plant Biotechnol J 10:249–268

6. Jha S, Sen S (1987) Karyotype variability in regenerated plants of *Urginea indica* Kunth. Cytologia 52:615–626
7. Jha S (1989) Cytological analysis of embryogenic callus and regenerated plants of *Urginea indica* Kunth. Indian squill. Caryologia 42:165–173
8. Rani V, Raina SN (2000) Genetic fidelity of organized meristem-derived micropropagated plants: a critical reappraisal. In Vitro Cell Dev Biol Plant 36:319–330
9. Hao YJ, Deng XX (2002) Occurrence of chromosomal variations and plant regeneration from long-term-cultured *Citrus* callus. In Vitro Cell Dev Biol Plant 38:472–476
10. Kumar PS, Mathur VL (2004) Chromosomal instability in callus culture of *Pisum sativum*. Plant Cell Tiss Org Cult 78:267–271
11. Orbovic V, Calovic M, Viloria Z, Nielsen B, Gmitter FG Jr, Castle WS, Grosser JW (2008) Analysis of genetic variability in various tissue culture-derived lemon plant populations using RAPD and flow cytometry. Euphytica 161:329–335
12. Jain SM, Ahloowalia BS, Brar DS (2013) Somaclonal variation and induced mutations in crop improvement. Springer Science & Business Media, Dordrecht
13. Kaeppler SM, Kaeppler HF, Rhee Y (2000) Epigenetic aspects of somaclonal variation in plants. Plant Mol Biol 43:179–188
14. Devi SP, Kumaria S, Rao SR, Tandon P (2015) Genetic fidelity assessment in micropropagated plants using cytogenetical analysis and heterochromatin distribution: a case study with *Nepenthes khasiana* Hook f. Protoplasma 252:1305–1312
15. Sengupta J, Jha S, Sen S (1988) Karyotype stability in long-term callus derived plants of *Crepis tectorum* L. Biol Plant 30:247–251
16. Franklin CI, Mott RL, Vuke TM (1989) Stable ploidy levels in long-term callus cultures of loblolly pine. Plant Cell Rep 8:101–104
17. Jha S, Sen J, Sen S (1989) Stable regenerants from long-term callus cultures of *Ruscus hypophyllum* L. Cytologia 54:687–691
18. Nayak S, Sen S (1995) Rapid and stable propagation of *Ornithogalum umbellatum* L. in long term culture. Plant Cell Rep 15:150–153
19. Jha TB, Jha S (1989) *In vitro* regeneration and cytological study of *Allium hookeri* Thw. Indian J Exp Biol 27:363–365
20. Landey RB, Cenci A, Guyot R, Bertrand B, Georget F, Dechamp E, Herrera JC, Aribi J, Lashermes P, Etienne H (2015) Assessment of genetic and epigenetic changes during cell culture ageing and relations with somaclonal variation in *Coffea arabica*. Plant Cell Tiss Org Cult 122:517–531
21. Hamill JD, Lidgett AJ (1997) Hairy root cultures- opportunities and key protocols for studies in metabolic engineering. In: Doran PM (ed) Hairy roots. Gordon and Breach/Harwood Academic, London
22. Roychowdhury D, Majumder A, Jha S (2013) *Agrobacterium rhizogenes*-mediated transformation in medicinal plants: prospects and challenges. In: Chandra S et al (eds) Biotechnology for medicinal plants. Springer, Berlin/Heidelberg
23. DiCosmo F, Misawa M (1995) Plant cell and tissue culture: alternatives for metabolite production. Biotechnol Adv 13:425–435
24. Rao SR, Ravishankar GA (2002) Plant cell cultures: chemical factories of secondary metabolites. Biotechnol Adv 20:101–153
25. Deus-Neumann B, Zenk MH (1984) Instability of indole alkaloid production in *Catharanthus roseus* cell suspension cultures. Planta Med 50:427–431
26. Holden MA, Holden PR, Yeoman MM (1988) Elicitation of secondary product formation in *Capsicum frutescens* cultures. In: Robins RJ, Rhodes MJC (eds) Manipulating secondary metabolism in culture. Cambridge University Press, Cambridge
27. Evans DA, Gamborg OL (1982) Chromosome stability of cell suspension cultures of *Nicotiana* spp. Plant Cell Rep 1:104–107

28. Karp A, Wu QS, Maddock SE, Jones MGK (1990) Chromosome instability in bread wheat (*Triticum aestivum*) cell suspensions and their dividing protoplasts. In: Bajaj YPS (ed) Wheat, vol 13. Springer, Berlin/Heidelberg
29. Anrini M, Jha S (2009) Characterization of podophyllotoxin yielding cell lines of *Podophyllum hexandrum*. Caryologia 62:220–235
30. Lapitan NLV, Sears RG, Gill BS (1984) Translocations and other karyotypic structural changes in wheat x rye hybrids regenerated from tissue culture. Theor Appl Genet 68:547–554
31. Phillips RL, Kaeppler SM, Olhoft P (1994) Genetic instability of plant tissue cultures: breakdown of normal controls. Proc Natl Acad Sci 91:5222–5226
32. Whitmer S, Canel C, Heijden RVD, Verpoorte R (2003) Long-term instability of alkaloid production by stably transformed cell lines of *Catharanthus roseus*. Plant Cell Tiss Org Cult 74:73–80
33. Dubrovina AS, Kiselev KV (2012) Effect of long-term cultivation on resveratrol accumulation in a high-producing cell culture of *Vitis amurensis*. Acta Physiol Plant 34:1101–1106
34. Kiselev KV, Dubrovina AS, Bulgakov VP (2009) Phenylalanine ammonia-lyase and stilbene synthase gene expression in rolB transgenic cell cultures of *Vitis amurensis*. Appl Microbiol Biotechnol 82:647–655
35. Zeng F, Qian J, Luo W, Zhan Y, Xin Y, Yang C (2010) Stabilityof transgenes in long-termmicropropagation of plants of transgenicbirch (*Betula platyphylla*). Biotechnol Lett 32:151–156
36. Guivarc'h A, Boccara M, Prouteau M, Chriqui D (1999) Instability of phenotype and gene expression in long-term culture of carrot hairy root clones. Plant Cell Rep 19:43–50
37. Kooter JM, Matzke MA, Meyer P (1999) Listening to the silent genes: transgene silencing, gene regulation and pathogencontrol. Trends Plant Sci 4:340–347
38. Mette MF, Van der Winden J, Matzke MA, Matzke AJM (1999) Production of aberrant promoter transcripts contributes to methylation and silencing of unlinked homologous promoters in trans. EMBO J 18:241–248
39. Slightom JL, Durand-Tardif M, Jouanin L, Tepfer D (1986) Nucleotide sequence analysis of TL-DNA of *Agrobacterium rhizogenes* agropine type plasmid. J Biol Chem 261:108–121
40. Christey MC (2001) Use of Ri-mediated transformation for production of transgenic plants. In Vitro Cell Dev Biol Plant 37:687–700
41. Bulgakov VP, Shkryl YN, Veremeichik GN, Gorpenchenko TY, Inyushkina YV (2011) Application of *Agrobacterium rol* genes in plant biotechnology: a natural phenomenon of secondary metabolism regulation. In: Alvarez M (ed) Genetic transformation. INTECH Open Access Publisher
42. Tepfer D (1984) Genetic transformation of several species of higher plants by *Agrobacterium rhizogenes*: phenotypic consequences and sexual transmission of the transformed genotype and phenotype. Cell 37:959–967
43. Hamill JD, Evans D, Robins RJ, Rhodes MJC, Prescott A, Martin C (1988) Foreign gene insertion into hairy roots with binary vectors and *Agrobacterium rhizogenes*: potential for genetic manipulation of plant secondary metabolism in culture. In: Robins RJ, Rhodes MJC (eds) Manipulating secondary metabolism in culture. Cambridge University Press, Cambridge
44. Sevón N, Oksman-Caldentey KM (2002) *Agrobacterium rhizogenes*-mediated transformation: root cultures as a source of alkaloids. Planta Med 68:859–868
45. Aoki T, Matsumoto H, Asako Y, Matsunaga Y, Shimomura K (1997) Variation of alkaloid productivity among several clones of hairy roots and regenerated plants of *Atropa belladonna* transformed with *Agrobacterium rhizogenes* 15834. Plant Cell Rep 16:282–286
46. Batra J, Dutta A, Singh D, Kumar S, Sen J (2004) Growth and terpenoid indole alkaloid production in *Catharanthus roseus* hairy root clones in relation to left- and right-termini-linked Ri T-DNA gene integration. Plant Cell Rep 23:148–154

47. Chaudhuri KN, Ghosh B, Tepfer D, Jha S (2005) Genetic transformation of *Tylophora indica* with *Agrobacterium rhizogenes* A4: growth and tylophorine productivity in different transformed root clones. Plant Cell Rep 24:25–35
48. Bandyopadhyay M, Jha S, Tepfer D (2007) Changes in morphological phenotypes and withanolide composition of Ri-transformed roots of *Withania somnifera*. Plant Cell Rep 26:599–609
49. Alpizar E, Dechamp E, Lapeyre-Montes F, Guilhaumon C, Bertrand B, Jourdan C, Lashermes P, Etienne H (2008) *Agrobacterium rhizogenes*-transformed roots of Coffee (*Coffea arabica*): conditions for long-term proliferation and morphological and molecular characterization. Ann Bot 101:929–940
50. Taneja J, Jaggi M, Wankhede DP, Sinha AK (2010) Effect of loss of T-DNA genes on MIA biosynthetic pathway gene regulation and alkaloid accumulation in *Catharanthus roseus* hairy roots. Plant Cell Rep 29:1119–1129
51. Halder M, Jha S (2016) Enhanced *trans*-resveratrol production in genetically transformed root cultures of Peanut (*Arachis hypogaea* L.). Plant Cell Tiss Org Cult 124:555–572
52. Roychowdhury D, Basu A, Jha S (2015) Morphological and molecular variation in Ri-transformed root lines are stable in long term cultures of *Tylophora indica*. Plant Growth Regul 75:443–453
53. Mano Y, Ohkawa H, Yamada Y (1989) Production of tropane alkaloids by hairy root cultures of *Duboisia leichhardtii* transformed by *Agrobacterium rhizogenes*. Plant Sci 59:191–201
54. Jouanin L, Guerche D, Pamboukdjian N, Tourneur C, Casse-Delbart F, Tourneur J (1987) Structure of T-DNA in plants regenerated from roots transformed by *Agrobacterium rhizogenes* strain A4. Mol Gen Genet 206:387–392
55. Piñol MT, Palazòn J, Serrano M (1996) Effects of Ri T-DNA from *Agrobacterium rhizogenes* on growth and hyoscyamine production in *Datura stramonium* root cultures. Bot Acta 109:133–138
56. Bulgakov VP, Khodakovskaya MV, Labetskaya NV, Chernoded GK, Zhuravlev YN (1998) The impact of plant *rolC* oncogene on ginsenoside production by ginseng hairy root cultures. Phytochemistry 49:1929–1934
57. Spena A, Schmülling T, Koncz C, Schell J (1987) Independent and synergistic activity of *Rol A, B* and *C* loci in stimulating abnormal growth in plants. EMBO J 6:3891–3899
58. Schmülling T, Schell J, Spena A (1988) Single genes from *Agrobacterium rhizogenes* influence plant development. EMBO J 7:2621–2629
59. Spanó L, Mariotti D, Cardarelli M, Branca C, Constantino P (1988) Morphogenesis and auxin sensitivity of transgenic tobacco with different complements of Ri T-DNA. Plant Physiol 87:479–483
60. Lipp Joao KH, Brown TA (1994) Long-term stability of root cultures of tomato transformed with *Agrobacterium rhizogenes* R1601. J Exp Bot 45:641–647
61. Geerlings A, Hallard D, Martinez Caballero A, Lopes Cardoso I, van der Heijden R, Verpoorte R (1999) Alkaloid production by a *Cinchona officinalis* 'ledgeriana' hairy root culture containing constitutive expression constructs of tryptophan decarboxylase and strictosidine synthase cDNA from *Catharanthus roseus*. Plant Cell Rep 19:191–196
62. Zdravković-Korać S, Ćalić D, Druart PH, Radojević L (2003) The horse chestnut lines harboring the rol genes. Biol Plant 47:487–491
63. Santos PM, Figueiredo AC, Oliveira MM, Barroso JG, Pedro LG, Deans SG, Younus AKM, Scheffer JJC (1999) Morphological stability of *Pimpinella anisum* hairy root cultures and time-course study of their essential oils. Biotechnol Lett 21:859–864
64. Basu A, Joshi RK, Jha S (2015) Genetic transformation of *Plumbago zeylanica* with *Agrobacterium rhizogenes* strain LBA 9402 and characterization of transformed root lines. Plant Tiss Cult Biotechnol 25:21–35
65. Pandey P, Kaur R, Singh S, Chattopadhyay SK, Srivastava SK, Banerjee S (2014) Long-term stability in biomass and production of terpene indole alkaloids by hairy root culture of

Rauvolfia serpentina and cost approximation to endorse commercial realism. Biotechnol Lett 36:1523–1528

66. Ray S, Samanta T, Majumder A, Bandyopadhyay M, Jha S (2014) Cytogenetic characterization of *Agrobacterium rhizogenes* transformed root lines of *Rauvolfia serpentina*. Nucleus 57:105–112
67. Marconi PL, Setten LM, Cálcena EN, Alvarez MA et al (2008) Changes in growth and tropane alkaloid production in long-term culture of hairy roots of *Brugmansia candida*. J Integr Biosci 3:38–44
68. Yukimune Y, Hara Y, Yamada Y (1994) Tropane alkaloid production in root cultures of *Duboisia myoporoides* obtained by repeated selection. Biosci Biotechnol Biochem 58:1443–1446
69. Sevón N, Hiltunen R, Oksman-Caldentey KM (1998) Somaclonal variation in transformed roots and protoplast-derived hairy root clones of *Hyoscyamus muticus*. Planta Med 64:37–41
70. Benson EE, Hamill JD (1991) Cryopreservation and post freeze molecular and biosynthetic stability in transformed roots of *Beta vulgaris* and *Nicotiana rustica*. Plant Cell Tiss Org Cult 24:163–172
71. Maldonado-Mendoza IE, Ayora-Talavera T, Loyola-Vargas VM (1993) Establishment of hairy root cultures of *Datura stramonium* characterization and stability of tropane alkaloid production during long periods of subculturing. Plant Cell Tiss Org Cult 33:321–329
72. Baíza AM, Quiroz-Moreno A, Ruíz JA, Loyola-Vargas VM (1999) Genetic stability of hairy root cultures of *Datura stramonium*. Plant Cell Tiss Org Cult 59:9–17
73. Dechaux C, Boite-Conti M (2005) A strategy for overaccumulation of scopolamine in *Datura innoxia* hairy root cultures. Acta Biol Cracov Bot 47:101–107
74. Jouhikainen K, Lindgren L, Jokelainen T, Hiltunen R, Teeri TH, Oksman-Caldentey KM (1999) Enhancement of scopolamine production in *Hyoscyamus muticus* L. hairy root culture by genetic engineering. Planta 208:545–551
75. Maldonado-Mendoza IE, Ayora-Talavera TRD, Loyola-Vargas VM (1992) Tropane alkaloid production in *Datura stramonium* root cultures. In Vitro Cell Dev Biol Plant 28:67–72
76. Mehrotra S, Goel MK, Rahman LU, Kukreja AK (2013) Molecular and chemical characterization of plants regenerated from Ri-mediated hairy root cultures of *Rauwolfia serpentina*. Plant Cell Tiss Org Cult 114:31–38
77. Ambros PF, Matzke AJM, Matzke MA (1986) Localization of *Agrobacterium rhizogenes* T-DNA in plant chromosomes by *in situ* hybridization. EMBO J 5:2073–2077
78. Aird ELH, Hamill JD, Rhodes MJC (1988) Cytogenetic analysis of hairy root cultures from a number of plant species transformed by *Agrobacterium rhizogenes*. Plant Cell Tiss Org Cult 15:47–57
79. Ermayanti TM, McComb JA, O'Brien PA (1992) Cytological analysis of seedling roots, transformed root cultures and roots regenerated from callus of *Swainsona galegifolia* (Andr.) R. Br. J Exp Bot 44:375–380
80. Mukherjee S, Das S, Jha S (1994) Chromosomal stability in transformed hairy root cultures of *Artemissia annua* L. Cell Chromos Res 17:71–76
81. Webb KJ, Jones S, Robbins MP, Minchin FR (1990) Characterization of transgenic root cultures of *Trifolium repens*, *Trifolium pratense* and *Lotus corniculatus* and transgenic plants of *Lotus corniculatus*. Plant Sci 70:243–254
82. Ermayanti TM, Octavia Y, Hafizh EA (2004) Cytological analysis of root cultures of *Artemissia cina*. Ann Bogoriensesns 9:50–58
83. Ramsay G, Kumar A (1990) Transformation of *Vicia faba* cotyledon and stem tissues by *Agrobacterium rhizogenes*: infectivity and cytological studies. J Exp Bot 41:841–847
84. Banerjee-Chattopadhyay S, Schwemmin AM, Schwemmin DJ (1985) A study of karyotypes and their alterations in cultured and *Agrobacterium* transformed roots of *Lycopersicon peruvianum* Mill. Theor Appl Genet 71:258–262

85. Vries-Uijtewaal ED, Gilissen LJW, Flipse E, Sree Ramulu K, De Groot B (1988) Characterization of root clones obtained after transformation of monohaploid and diploid potato genotypes with hairy root inducing strains of *Agrobacterium*. Plant Sci 58:193–202
86. Xu Z-Q, Jia J-F (1996) The reduction of chromosome number and the loss of generation ability during subculture of hairy root cultures of *Onobrychis viciaefolia* transformed by *Agrobacterium rhizogenes*. Plant Sci 120:107–112
87. Hänisch ten Cate CH, Loonen AE, Ottaviani MP, Ennik L, van Eldik G, Stiekema WJ (1990) Frequent spontaneous deletions of Ri T-DNA in *Agrobacterium rhizogenes* transformed potato roots and regenerated plants. Plant Mol Biol 14:735–741
88. Nilsson O, Olsson O (1997) Getting to the root: the role of the *Agrobacterium rhizogenes rol* genes in the formation of hairy roots. Physiol Plant 100:463–473
89. Kamada H, Saitou T, Harada H (1992) No requirement of vernalization for flower formation in Ri-transformed *Cichorium* plants. Plant Tiss Cult Lett 9:206–208
90. Sun LY, Touraud G, Charbonnier C, Tepfer D (1991) Modification of phenotype in Belgian endive (*Cichorium intybus*) through genetic transformation by *Agrobacterium rhizogenes*: conversion from biennial to annual flowering. Transgenic Res 1:14–22
91. Rugini E, Caricato G, Muganu M, Taratufolo C, Camilli M, Cammilli C (1997) Genetic stability and agronomic evaluation of six year-old transgenic kiwi plants for *rolABC* and *rolB* gene. Acta Hortic 447:609–610
92. Roychowdhury D, Ghosh B, Chaubey B, Jha S (2013) Genetic and morphological stability of six-year-old transgenic *Tylophora indica* plants. Nucleus 56:81–89
93. Roychowdhury D, Chaubey B, Jha S (2015) The fate of integrated Ri T-DNA *rol* genes during regeneration via somatic embryogenesis in *Tylophora indica*. J Bot 2015:16
94. Paul P, Sarkar S, Jha S (2015) Effects associated with insertion of *cryptogein* gene utilizing Ri and Ti plasmids on morphology and secondary metabolites are stable in *Bacopa monnieri* transformed plants grown *in vitro* and *ex vitro*. Plant Biotechnol Rep 9:231–245
95. Verma P, Sharma A, Khan SA, Mathur AK, Shanker K (2015) Morphogenetic and chemical stability of long-term maintained *Agrobacterium*-mediated transgenic *Catharanthus roseus* plants. Nat Prod Res 29:315–320
96. Chaudhuri KN, Ghosh B, Tepfer D, Jha S (2006) Spontaneous plant regeneration in transformed roots and calli from *Tylophora indica*: changes in morphological phenotype and tylophorine accumulation associated with transformation by *Agrobacterium rhizogenes*. Plant Cell Rep 25:1059–1066
97. Choi PS, Kim YD, Choi KM, Chung HJ, Choi DW Jr (2004) Plant regeneration from hairy-root cultures transformed by infection with *Agrobacterium rhizogenes* in *Catharanthus roseus*. Plant Cell Rep 22:828–831
98. Giovannini A, Pecchioni N, Rabaglio M, Allavena A (1997) Characterization of ornamental *Datura* plants transformed by *Agrobacterium rhizogenes*. In Vitro Cell Dev Biol Plant 33:101–106
99. Mercuri A, Bruna S, Benedetti LD, Burchi G, Schiva T (2001) Modification of plant architecture in *Limonium* spp. induced by rol genes. Plant Cell Tiss Org Cult 65:247–253
100. Christensen B, Sriskandarajah S, Müller R (2009) Biomass distribution in *Kalanchoe blossfeldiana* transformed with *rol*-genes of *Agrobacterium rhizogenes*. Hortic Sci 44:1233–1237
101. Saxena G, Banerjee S, ur Rahman L, Verma PC, Mallavarapu GR, Kumar S (2007) Rose-scented geranium (*Pelargonium* sp.) generated by *Agrobacterium rhizogenes* mediated Ri-insertion for improved essential oil quality. Plant Cell Tiss Org Cult 90:215–223
102. Satheeshkumar K, Jose B, Soniya EV, Seeni S (2009) Isolation of morphovariants through plant regeneration in *Agrobacterium rhizogenes* induced hairy root cultures of *Plumbago rosea* L. Indian J Biotechnol 8:435–441
103. Christey MC, Braun RH, Reader JK (1999) Field performance of transgenic vegetable Brassicas (*Brassica oleracea* and *B. rapa*) transformed with *Agrobacterium rhizogenes*. SABRAO J Breed Genet 31:93–108

104. Shahin EA, Sukhapinda K, Simpson RB, Spivey R (1986) Transformation of cultivated tomato by a binary vector in *Agrobacterium rhizogenes*: transgenic plants with normal phenotypes harbour binary vector T-DNA, but no Ri-plasmid T-DNA. Theor Appl Genet 72:770–777
105. Manners JM, Way H (1989) Efficient transformation with regeneration of the tropical pasture legume *Stylosanthes humilis* using *Agrobacterium rhizogenes* and a Ti plasmid binary vector system. Plant Cell Rep 8:341–345
106. Hatamoto H, Boulter ME, Shirsat AH, Croy EJ, Ellis JR (1990) Recovery of morphologically normal transgenic tobacco from hairy roots co-transformed with *Agrobacterium rhizogenes* and a binary vector plasmid. Plant Cell Rep 9:88–92
107. Boulter ME, Croy E, Simpson P, Shields R, Croy RRD, Shirsat AH (1990) Transformation of Brassica napus L. (oilseed rape) using *Agrobacterium tumefaciens* and *Agrobacterium rhizogenes*- a comparison. Plant Sci 70:91–99
108. Braun RH, Reader JK, Christey MC (2000) Evaluation of cauliflower transgenic for resistance to *Xanthomonas campestris* pv. *campestris*. Acta Hortic 539:137–143
109. Puddephat IJ, Robinson HT, Fenning TM, Barbara DJ, Morton A, Pink DAC (2001) Recovery of phenotypically normal transgenic plants of *Brassica oleracea* upon *Agrobacterium rhizogenes* mediated cotransformation and selection of transformed hairy roots by GUS assay. Mol Breed 7:229–242
110. Stam M, Joseph NMM, Kooter JM (1997) The silence of genes in transgenic plants. Ann Bot 79:3–12
111. Li XG, Chen SB, Lu ZX, Chang TJ, Zeng QC, Zhu Z (2002) Impact of copy number on transgene expression in tobacco. Acta Bot Sin 44:120–123
112. Tang W, Ronald JN, Douglas AW (2007) Genetic transformation and gene silencing mediated by multiple copies of a transgene in eastern white pine. J Exp Bot 58:545–554
113. Hänisch ten Cate CH, Ennik E, Roest S, Sree Ramulu K, Dijkhuis P, De Groot B (1988) Regeneration and characterization of plants from potato root lines transformed by *Agrobacterium rhizogenes*. Theor Appl Genet 75:452–459
114. Sevón N, Oksman-Caldentey KM, Hiltunen R (1995) Efficient plant regeneration from hairy root-derived protoplasts of *Hyoscyamus muticus*. Plant Cell Rep 14:738–742
115. Sevón N, Dräger B, Hiltunen R, Oksman-Caldentey KM (1997) Characterization of transgenic plants derived from hairy roots of *Hyoscyamus muticus*. Plant Cell Rep 16:605–611
116. Samaddar T, Nath S, Halder M, Sil B, Roychowdhury D, Sen S et al (2012) Karyotype analysis of three important traditional Indian medicinal plants, *Bacopa monnieri, Tylophora indica* and *Withania somnifera*. Nucleus 55:17–20
117. Trulson AJ, Simpson RB, Shahin EA (1986) Transformation of cucumber (*Cucumis sativus* L.) plants with *Agrobacterium rhizogenes*. Theor Appl Genet 73:11–15
118. Atkinson RG, Gardner R (1991) *Agrobacterium*-mediated transformation of pepino and regeneration of transgenic plants. Plant Cell Rep 10:208–212

Biotransformation Through Hairy Roots: Perspectives, Outcomes, and Major Challenges

14

Vikas Srivastava, Shakti Mehrotra, and Sonal Mishra

Abstract

The *Agrobacterium rhizogenes* mediated hairy root cultures have shown great potential in bringing out structural and/or conformational alteration of a chemical moiety through its enzymatic paraphernalia. The inherent enzymes involved in bioconversions perform various kinds of reactions such as hydroxylation, glycosylation, oxidoreduction, and hydrolysis. Hairy root cultures of variety of plant systems have been explored for the bioconversion of a wide range of substrates into the molecules of improved pharmaceutical properties. Some specific inherent properties of hairy roots like genetic/biochemical stability, hormonal independence, efficient enzyme paraphernalia, and low-cost cultural requirements contribute to the superiority of HRCs over other in vitro production systems. Rational use of hairy root cultures as proficient biotransformation system may result in the synthesis of molecules that have desired physico-chemical properties, adequate solubility, and reduced toxicity, thus more desired in pharmaceutical industries. The present chapter elaborates a comprehensive discussion on hairy root mediated biotransformation, types of reactions, and products formed concomitantly with their commercial importance.

V. Srivastava (✉)
National Institute of Plant Genome Research, New Delhi, India

Department of Plant Sciences, Central University of Jammu, Samba, Jammu (J & K), India
e-mail: vikassrivastava25@gmail.com

S. Mehrotra (✉)
Plant Biotechnology Division, Central Institute of Medicinal and Aromatic Plants, Lucknow
Uttar Pradesh, India
e-mail: shaktimehrotra22@gmail.com

S. Mishra (✉)
National Institute of Plant Genome Research, New Delhi, India
e-mail: mishrasonal.nipgr@gmail.com

S. Jha (ed.), *Transgenesis and Secondary Metabolism*, Reference Series in Phytochemistry, DOI 10.1007/978-3-319-28669-3_10

Keywords
Biosynthetic pathway • Biotransformation • Hairy root cultures • Secondary metabolism • Precursor feeding • Co-culture system

Abbreviations

4-OHBA	4-hydroxybenzalacetone
BG	Betuligenol
DA	Deoxy artemisinin
DHA	Dehydroabietic acid
HBA	*p*-Hydroxy benzyl alcohol
HQ	Hydroquinone or 1–4-benzendiol
HRCs	Hairy root cultures
PCS	Plant cell suspension
PF	Precursor feed or supplementation
RK	Raspberry ketone
SA	Salicylic acid
SM	Secondary metabolism
TBA	3,4,5-trimethoxy benzaldehyde
THPB	Tetrahydroprotoberberines
YE	Yeast extract

Contents

1 Introduction

Biotransformation is defined as structural and/or conformational alteration of a chemical moiety by enzymatic militia of biological system. When referred to the molecules of pharmaceutical and other commercial importance, biotransformation is considered as a flawless methodology aimed to create specific structural changes in rather complex structured bioactive molecules. The transformed products show properties like reduced toxicity, adequate solubility, and improved pharmacokinetics which is more desired in pharmaceutical industries. Thus, this class of molecules with more desired physiochemical and therapeutic properties gain increased commercial demand [1]. During biotransformation of chemical compounds, the biological system functions as suitable biological matrices by offering their inherent enzymatic paraphernalia to catalyze various types of chemical reactions. Majority of these reactions include hydroxylation, glycosylation, oxidoreduction, hydrolysis, etc. Nowadays, biotransformation is increasingly utilized to develop repositories of analogs of variety of naturally occurring phytomolecules with the help of enzymes derived from microbes, plant/animal cells, tissues, and their cultured counterparts [1, 2]. Thus, distinguished from therapeutic protein production, biotransformation not only enables to develop advance version of existing molecules but also to enhance their desired inherent potential. Another phenomenon that sometime comes parallel to this process is fermentation. However, distinguished from fermentation where usually several catalytic steps involve between substrate and product, biotransformation is restricted to one or two. Additionally, the basic backbone structures of the substrate and the product resemble one another in biotransformation but not necessarily in fermentation [3].

The applicability of biotransformation has been envisaged through different biological systems and enzymes thereof, and many studies irrespective of rank and cellular organization have been extensively explored ranging from bacteria (microbe) to higher plants and animals [1]. In animals, this approach is generally confined to adjudge only toxicity or metabolic fate of administered drug (pharmacokinetic studies). On the other hand, the microbial and plant based biotransformation systems are useful for the production of significant molecules which can be of high therapeutic, flavor, or nutritional values. Nevertheless, these two potent candidate systems have their own limitations and advantages. The microbes, though superior in production and processing, have inherent drawback of easy cross-contamination, genetic instability, and comparatively low biotransformation prospects. Plants, on other side, due to complex biological behavior offer broader enzymatic potential for biotransformation and also possess genetic firmness. A comparison of different biotransformation systems has been summarized in Table 1.

The biotransformation procedures carried out by cultured plant cells and organs are independent of seasonal and pathological constraints and offer sustainable use of resources under controlled and defined culture conditions. Plant in vitro systems like plant cell suspensions (PCSs) and hairy root cultures (HRCs) are being utilized for biotransformation of variety of compounds because of their quick response to any variability in culture environment and simple to follow extraction procedures. The

Table 1 Comparative account on plant cell suspension, hairy root, and microbial biotransformation culture

Plant-based biotransformation systems		Microbial biotransformation system
HRCs	Cell suspension	
High growth rate	High growth rate	Very high growth rate
Easy maintenance and relatively low cost	Comparatively high cost	Low cost
Genetically stable	Prone to instability	High mutability (short life-span)
Broader enzyme biosynthetic potential	Broader enzyme biosynthetic potential	Narrow enzyme potential (low organizational status)
Less responsive to the variations in culture conditions	Highly responsive to variation in condition (callus)	Sensitive to culture condition
Capable of virtually infinite growth under defined conditions	Capable of virtually infinite growth under defined conditions	Capable of virtually infinite growth under defined conditions
Not easy downstream processing	Easy downstream processing	Easy downstream processing
Comparatively difficult to upscale	Easy to follow established upscaling methods	Easy to upscale
Susceptible to microbial contamination	Highly susceptible to microbial contamination	High risk of contamination
Free of hazardous chemicals	Free of hazardous chemicals	Free of hazardous chemicals
High prospects of genetic manipulations and its maintenance	High prospects of genetic manipulations but maintenance is not easy	Easy to genetically manipulate
Low yields	Low yields	High yielding

upcoming text elaborates a comprehensive discussion on hairy root mediated biotransformation, types of reactions, and products formed concomitantly with their commercial importance. Additionally, a pictorial presentation of biotransformation procedure is also being given in Fig. 1.

2 Hairy Root Cultures (HRCs): Suitable Matrix for Biotransformation

Plant in vitro systems in which plant cells, tissues, and organs are cultured under defined sterile conditions offer attractive alternatives for the production of desired metabolites of commercial importance [4–7]. Hairy root cultures (HRCs) have been extensively exploited not only for secondary metabolite production but also to understand the physiology and molecular biology existing behind [8–11]. Under natural conditions, hairy roots are disease manifestations developed by plants that are wounded and subsequently infected by *Agrobacterium rhizogenes* [12]. During

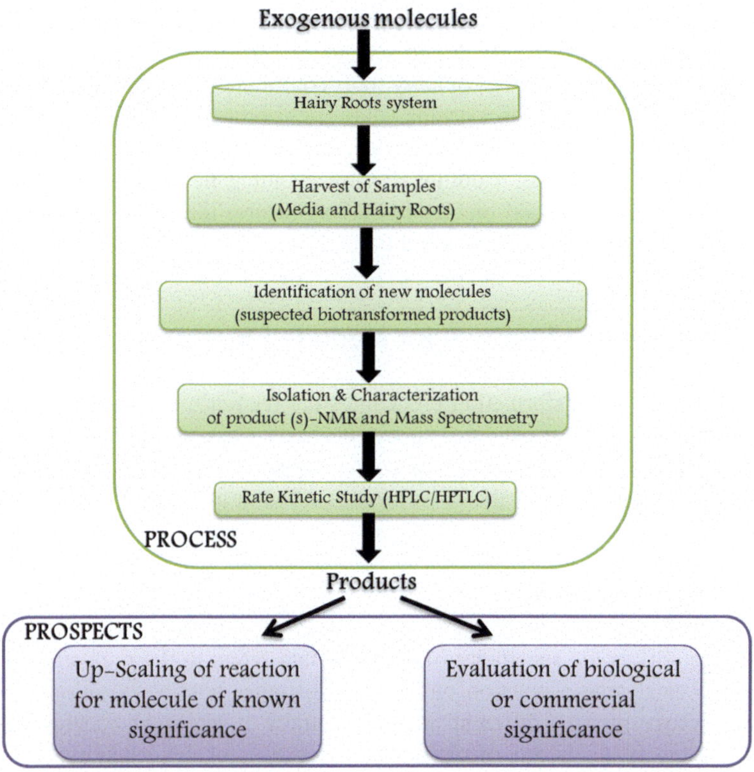

Fig. 1 Process and prospects of hairy roots mediated biotransformation of exogenous molecules

the infection process, root loci (*rol*) genes present in the root inducing (Ri) plasmid of *A. rhizogenes* are incorporated into the nuclear genome of the host plant, eventually causing neoplastic roots and root hair proliferation [12, 13]. The laboratory exploitation of this natural genetic transformation phenomenon and development of HRCs of diverse plant systems offer relatively easy "establish and explore" approach for various purposes including secondary metabolite production [11]. Owing to the characteristic property of fast growth and ability to produce existing and/or novel secondary metabolites in detectable range, the hairy roots have gained attention of global scientific fraternity as a cost effective in vitro production system for desired bioactive compounds [14]. Incessant research on HRCs of various plant species from past three decades have proved their worth as an excellent system as a whole to investigate various aspects of plant secondary metabolism (SM) [8, 11, 14]. Additionally, with time, this system has also emerged as a suitable matrix to explore a broad range of biotechnological applications [11, 14–16].

Hairy root based biotransformation is one such application where structural and/or conformational modification of chemical compounds is achieved by utilizing the enzymatic machinery of root tissue [2, 13]. Because of the defined in vitro conditions, such structural diversifications in parent compounds and subsequent development of their analogs are independent of seasonal and pathological constraints. In addition to fast growth and secondary metabolite production potential, some specific inherent properties of hairy roots contribute to the superiority of HRCs over other in vitro production systems [11]. These properties include genetic/biochemical stability, hormonal independence, efficient enzyme paraphernalia, and low-cost culture necessities. Additionally, by avoiding hazardous chemicals and reagents, hairy root mediated biotransformation ensures the sustainable utilization of resources under definite culture conditions [2, 11, 17].

3 Properties of Hairy Root Mediated Biotransformation

HRCs mediated biotransformation procedures are recognized with many features, because of which these procedures supersedes over classical method of synthetic chemistry. Moreover, its suitability for context specific metabolite production also occurs due to some novel attributes well exemplified in literature.

3.1 Regio-Selectivity

Regio-selectivity is the specificity of enzymes that refers to the preference of one location of catalyzing a reaction out of over all other possible locations. Regio-specific glycosylation of different position isomers of hydroxyl benzoic acid by *Panax ginseng* HRCs is one of the relevant examples of this category [18]. This study showed glycosyl esterification of *para-* and *meta-*hydroxy benzoic acid but not of *ortho-*hydroxy benzoic acid. Likewise, HRCs of *Polygonum multiflorum* depicted efficient regio-selective glucosylation of phenolic hydroxyl groups of compounds into its derivatives [19]. In the same way, *Atropa belladonna* hairy roots also show regio-specific glucosylation during biotransformation of betuligenol (BG) into 4-(*p*-hydroxyphenyl) but-2-yl-β-D-glucopyranoside or betuloside [20] (Scheme 1).

Further, in a recent study, HRCs of *P. ginseng* exhibited high degree of regio-selective glycosylation for tetrahydroprotoberberine (THPBs) substrates [21]. These HRCs showed greater catalytic efficiency on THPBs with phenolic hydroxyl at C-9 as compared to the positions other than C-9. In this way, HRCs provide an easy way by avoiding the tedious methods of chemistry involving protection of functional equivalent/groups to attain desired derivatization to get specific products with desired properties.

3.2 Reaction Flexibility

Biotransformation reactions are flexible in nature in which one group is prone to be transformed into any of its possible derivatives. With this feature, exogenous

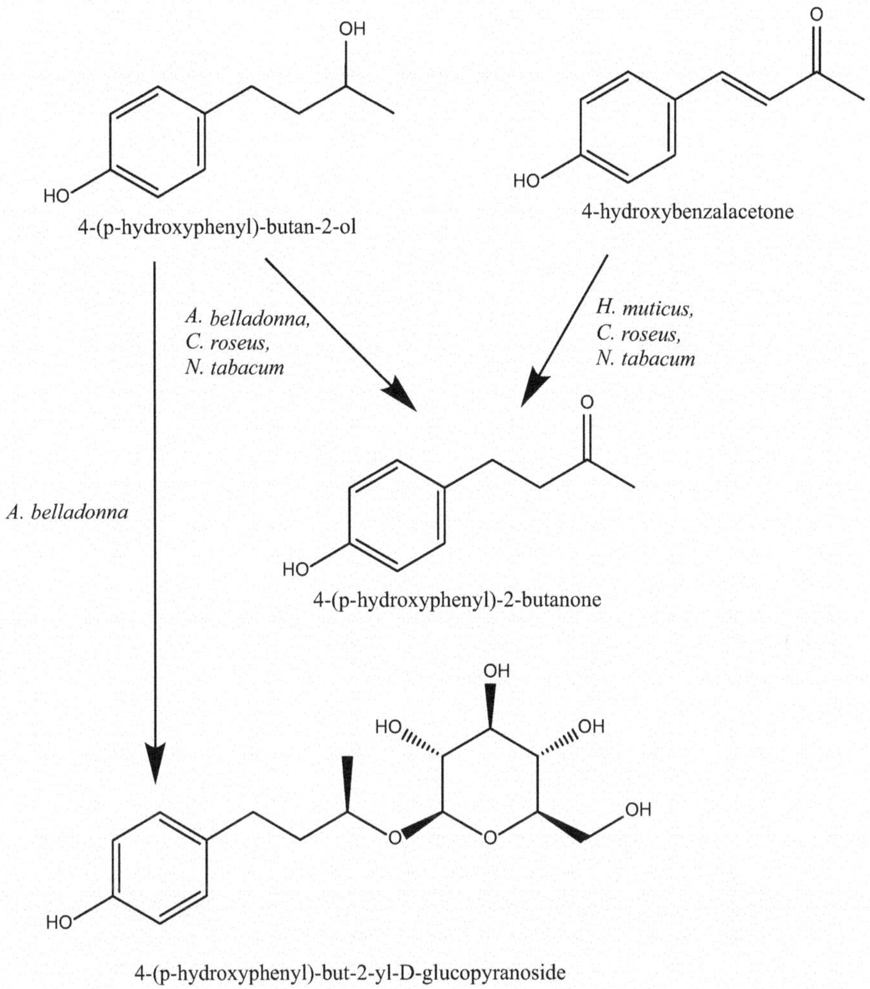

Scheme 1 Biotransformation of 4-(*p*-hydroxyphenyl)-butan-2-ol (BG) and 4-hydroxy benzalacetone(4-OHBA) utilizing HRCs (Biotransformation of 4-OHBA has not been investigated in *A. belladonna* HRCs) [20, 50]

molecule can undergo any of the possible transformation events. In an example of reaction flexibility, both oxidation and reduction of formyl group of 3, 4, 5-trimethoxy benzaldehyde (TBA) was observed by *A. belladonna* hairy roots [22]. Same hairy roots also performed oxidation and glucosylation at secondary alcohol during biotransformation [20]. Besides, this kind of reaction flexibility have also been well presented in studies with *Pharbitis nil* and *P. multiflorum* HRCs mediated biotransformation reactions [19, 23, 24].

3.3 Stereoselectivity

The stereoselectivity in a reaction refers to the preferential stereoisomer formation over another, as a result of inherent reaction specificity, substrate, enzymes, etc. More specifically, according to the type of reactions and products the terms enantioselectivity and diastereoselectivity are referred. In context of HRCs mediated biotransformation, steric nature of substrate and products remains tightly maintained and only one of the isomer in racemic mixtures of substrate is attacked by respective enzymes. Moreover, in enantioselective reactions, where numbers of enantiomers are possible, production of only one enantiomer prevails. *Daucas carota* HRCs based stereoselective reduction of aromatic ketones, keto esters, and a simple aliphatic ketone is one such example where all these substrate reduced to corresponding derivatives with excellent stereoselectivity [25]. In another example, HRCs of *P. multiflorum* depicted efficient regio and stereoselective glucosylation of phenolic hydroxyl groups of compounds into corresponding compounds [26]. Similarly, enantioselective biotransformation of prochiral heteroaryl and prochiral alkylaryl ketones was achieved through HRCs of *Brassica napus* and *Raphanus sativus*, respectively, to their corresponding chiral alcohols [27, 28]. This advocates the property of stereoselectivity in HRCs based bioconversion of different substrates. However, on contrary, in some examples such as bioconversion of tetrahydroprotoberberins (THPBs) through *P. ginseng* HRCs, the conversion reactions exhibit nonstereoselectivity towards their substrates [21].

3.4 Activation of Molecules

There are molecules which are chemically inactive because of lack of functional groups in its backbone. The biotransformation processes have the ability to decorate the back bone of any molecule through the addition of chemically reactive groups (like hydroxylation etc.). This kind of property can be observed in cell suspension cultures of *Catharanthus roseus*, where Shimoda et al. [29] had demonstrated hydroxylation of 2-hydroxybenzoic acid into 2, 5-dihydroxybenzoic acid and subsequent glucosylation of this newly formed phenolic hydroxyl group into 5-*O*-(β-D-glucopyranosyl)-2,5-dihydroxybenzoic acid. Though also possible in HRCs based biotransformation of compounds yet, such modifications are not much explored in hairy root systems. However, the example of glycosylation of primary alcohol derived after aldehyde reduction in *Pharbitis nil* HRCs based biotransformation of benzaldehyde- and acetophenone-type derivatives would be relevant to state [30].

3.5 Nonextreme Conditions and No Hazardous Chemicals Required

The in vitro culture procedures, including HRCs, necessitate the maintenance of controlled, aseptic, and defined conditions. All kinds of HRCs based bioconversion reactions takes place under very nominal in vitro conditions including optimum pH, temperature, etc. Thus, a thermolabile compound can be converted into product of interest using low energy / effort consumption, without undesired decomposition or isomerisation of substrate. Additionally, the biotransformation procedures do not require any hazardous solvents/reagents except the organic solvent(s), which are generally used to solubilize the substrate or making fraction/extract, which are volatile in nature and can easily separate out from reaction products. Therefore, with reduced usage of toxic chemicals the molecules thus produced are safer to use and does not require much toxicity analysis [1, 2]. In this way, biotransformation procedures carried out utilizing inherent enzymatic potential of plant cell and/or organ cultures are gaining recognition as environment friendly, sustainable, and safe methods to produce valuable compounds.

4 HRCs Mediated Various Biotransformation Strategies

4.1 Biotransformation Through Pathway Precursor Feeding

Hairy roots, due to their easy maintenance, offer a suitable tissue system for the elucidation of various metabolic pathways and investigation of various intermediates and precursor compounds in a pathway [11, 15]. Till date various SM biosynthetic pathways in different related or unrelated plant species have been partially or completely explicated. Tropane alkaloid (TA), terpenoid indole alkaloid (TIA), and benzyl isoquinoline alkaloid (BIA) pathways are such examples which are well understood not only in terms of their reaction steps but also for precursor, intermediates, and enzyme/genes required for various steps [8, 15]. Further, this has also paved the way for enhancement in product flux via precursor feeding (PF) [31, 32]. Sometimes, in spite of presence of adequate enzymatic activity and chemical ambience, the production of metabolite of interest is not satisfactory due to scarcity of its near or distant precursors. To address the problem, advantage of feeding precursor to the immediate environment of the system has been explored in many plants to get the metabolite of interest through biotransformation. The earlier studies, where addition of hyoscyamine to the growth medium containing HRCs of TA bearing plants individually, resulted improved production of scopolamine are the classical examples of precursor feeding (Scheme 2) [33]. In these studies, hyoscyamine was biotransformed into scopolamine utilizing enzymatic epoxidation (through H6H of TA pathway) reaction.

In a relatively recent study, silymarin accumulation was enhanced due to feeding of precursor compound in *Silybum marianum* HRCs. After feeding of *L*-phenylalanine as precursor in these HRCs, the accumulation of silymarin and naringenin was found to be significantly affected [34]. At present, PF is also being utilized as an integrative approach along with other enhancement methods like elicitation, metabolic engineering, and upscaling for desired metabolite production [35–37].

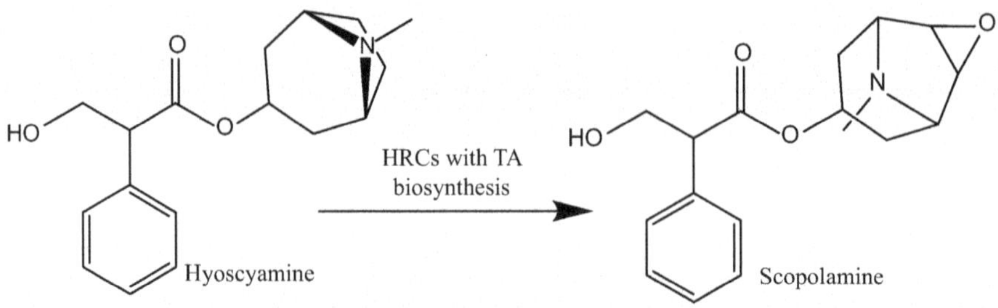

Scheme 2 Biotransformation of hyoscyamine to scopolamine in hairy root cultures of TA pathway bearing plants [33]

4.2 Biotransformation Using Co-culture Approach

Fundamentally, a co-culture system is a defined cultivation setup where two or more different populations of cells/tissues are grown together with distinct degree of contact. In recent years, in plant tissue culture with particular reference to secondary metabolite production through in vitro systems, the concept of co-culture systems have gained noteworthy interest [38]. Utilizing in vitro systems, the production of desired bioactive molecule is accomplished by acquiring the simplest form of any co-culture scheme by growing two complementary systems in one matrix (culture medium). The two systems complement each other in a way that one produces the substrate for the other to metabolize [39]. Here, the medium in which two systems grow behaves as a "mean of translocation." Apart from being beneficial for studying natural interaction between two or more related on unrelated cells/tissues/ populations, this co-culture system has shown its suitability in secondary metabolite production through hairy roots co-cultured with in vitro growing shoots and/or cell suspensions of intra and/or inter species [39, 40]. For example, in vitro shoot and hairy root cultures of *Genista tinctoria* were co-cultured to produce detectable amounts of two pharmaceutically important secondary metabolites, viz. daidzin and daidzein. In this co-culture system, hairy roots produced isoliquiritigenin, a daidzein precursor absent in intact plant. The external supplementation of ABA in the medium resulted in release of isoliquiritigenin from G.tinctoria HRCs in the medium from which it was used by the shoots to convert it into significant amounts of daidzin and daidzein [40]. Further, successful production of podophyllotoxin from co-culture of *Linum flavum* hairy roots and *Podophyllum hexandrum* cell suspensions is also a relevant example [41]. In an earlier study *P. hexandrum* cell suspension cultures were found able to convert only coniferin out of seven precursors from the phenylpropanoid routing into podophyllotoxin, an important therapeutic compound widely used in drugs meant for the treatment of cancers [42]. Keeping this in mind, the co-culture system of *L. flavum* hairy roots and *P. hexandrum* cell suspension was developed [41]. Hairy roots of *L. flavum* naturally produce coniferin which is released in growth medium and utilized by the cells of *P. hexandrum* to produce podophyllotoxin. Similarly, the co-culture of *Ammi majus* hairy roots along with *Ruta graveolens* shoots for the synthesis of furanno coumarin has been reported [43]. Sometimes hairy roots are also co-cultured in association with microbes and plant nematodes. For example, enhanced tanshinone production from *Salvia militiorrhiza* hairy roots was observed in a root-bacteria (*Bacillus cereus*) co-culture [44]. The results suggested the stimulation of tanshinone production by elicitors of bacterial origin in the root cultures. In a co-culture of *Solanum tuberosum* and root knot nematode *Meloidogyne chitwoodi* enhanced production of volatiles is reported [45]. Nevertheless, the latter two examples show perspectives of co-culture system involving HRCs in association with other organisms. These systems need thoughtful attention for their utilization in HRCs based biotransformation. Further, though such examples present a coordinated way of production and/or transformation of substrate molecule through two partners, yet, there are certain points that are needed to be scientifically attended. For example co-culturing requires extensive optimization and stabilization of growth of culture partners for incessant production. Additionally, deep insight of physiological behavior of partners and degree of their contact/separation also require attention for culture stability. Further, extending towards upscaling, the optimization of culture densities, volume of medium, the interaction of partners at different time scale, etc. are the points of consideration.

4.3 Biotransformation Using Nonspecific/Exogenous Molecules

It has been seen that many compounds are present in immensity in source plant system, but those are not biologically very active and valuable from commerce point of view. On the other hand, small structural diversification in these molecules can add or enhance their commercial significance [2]. Hairy roots have the ability to biochemically transform exogenously added foreign substrates into their corresponding derivatives (Table 2) [1, 2].

This system has been widely reported to utilize the desired regio-, stereo-, and enantioselectivity of the reaction and also substrate specificity. The kind of reaction explored depends upon the functional groups in the substrate and the type of enzyme present in host system. HRCs of various plant species have been utilized for the bioconversion of a variety of exogenous compounds supplemented in their growth medium through distinguished reaction types [2, 17]. A comprehensive account on HRCs of different plant species, transformed products, and reaction types has been earlier reviewed by Banerjee et al. [2]. Bioconversion of exogenous molecules is most widely applied in HRCs based biotransformation processes and subject of choice for generation/derivatization of potential molecules. The substrate for this purpose ranges from synthetic to natural origin. In a recent example of HRCs based biotransformation of nonspecific exogenous molecule, bioconversion of artemisinin into its derivative compounds is reported [46]. These derivative compounds reveal significant antimalarial activity and tumor necrosis factor (TNF) lowering potential.

The HRCs of representative genera and species from various plant families like Asteraceae, Solanaceae, Campanulaceae, Brassicaceae, Apiaceae, Lamiaceae, etc. have been exploited for biotransformation of exogenous molecules with almost no

Table 2 Biotransformation of exogenous molecules utilizing hairy root cultures

Plant	Reaction explored	Substrate	Product	Reference
Atropa belladonna	Oxidation and reduction	3,4,5-trimethoxy benzaldehyde	3,4,5-trimethoxy benzoic acid and 3,4,5-trimethoxy benzyl alcohol	[22]
	Reduction	3,4,5-trimethoxy acetophenone	1-(3,4,5-trimethoxy phenyl)-ethanol	[22]
Atropa belladonna	Oxidation and glycosylation	Betuligenol	Raspberry ketone and betuloside	[20]
Atropa belladonna; Hyoscyamus muticus; Ocimum basilicum	–	Artemisinin	3-α-hydroxy-1-deoxyartemisinin; 4-hydroxy-9,10-dimethyloctahydrofuro-(3,2-i) - isochromen-11(4H)-one	[46]
Nicotiana tabacum (expressing 35S-*Ri*ZS1- 4-OHBA reductase)	Reduction	4-hydroxybenzalacetone (4-OHBA)	Raspberry ketone	[50]
Catharanthus roseus	Reduction and oxidation	4-hydroxybenzalacetone and betuligenol	Raspberry ketone	[50]
N. tabacum SR1	Reduction and oxidation	4-hydroxybenzalacetone and betuligenol	Raspberry ketone	[50]
Hyoscyamus muticus	Reduction	4-hydroxybenzalacetone	Raspberry ketone	[50]
Rhodiola kirilowii	Glycosylation	Cinnamyl alcohol	Rosarin and rosavins	[66]
Polygonum multiflorum	Glycosylation	Esculetin and eight synthetic hydroxyl coumarins	Corresponding glycosides	[62]
Datura tatula L.	Glucosylation	p-hydroxybenzyl alcohol	Gastrodin	[57]
Polygonum multiflorum	Glycosylation	7-hydroxy-4-phenylcoumarin and 5,7-dihydroxy-4-phenylcoumarin	4-phenylcoumarin-7-*O*-*β*-D-glucopyranoside and 7-hydroxy-4-phenylcoumarin-5-*O*-*β*-D-glucopyranoside	[61]

Panax ginseng	Glycosylation and hydrolysis	Tetrahydroprotoberberines (THPBs)	Corresponding glycosides	[21]
Polygonum multiflorum	Glycosylation	7-Hydroxy-2,3-dihydrocycl openta [c]chromen-4-one, 7,9-dihydroxy-2,3-dihydrocyclopenta[c]chromen-4-one and 6,7-dihydroxy-2,3-dihydrocyclopenta[c]chromen-4-one	7-*O*-β-D-glucopyranosyl-2,3-dihydrocyclopenta[c]chromen-4-one, 9-*O*-β-D-glucopyranosyl-7-hydroxy-2,3-dihydrocyclop enta[c]chromen-4-one and 6-*O*-β-D-glucopyranosyl-7-hydroxy-2,3-dihydrocyclop enta[c]chromen-4-one	[60]
Nicotiana tabacum	Glycosylation	Dehydroabietic acid (DHA)	DHA-18-*O*-glucoside	[50]
Catharanthus roseus	Hydroxylation and glycosylation	DHA	17-hydroxy-DHA, DHA-17-*O*-α- glucoside and DHA-17-*O*-β-glucoside	[50]

dependency on additional biosynthetic pathway (secondary metabolic pathway) [2]. Among the plant systems explored, the maximum utilization of HRCs of *P. multiflorum* [24, 26], *P. ginseng* [18, 21, 47], *P. nil* [23, 30], and *A. belladonna* [17, 20, 22, 46] for the bioconversion of variety of substrate has been reported. These reports advocate the wide potential of hairy roots to perform variety of reactions on structurally unrelated exogenous molecules to produce corresponding compounds with more desired chemical properties (Table 2).

5 Reactions Carried Out by Hairy Roots

5.1 Reduction

A variety of carbonyl substrates undergo hairy root based enzymatic reduction and produce primary or secondary alcohols. The HRCs of *A. belladonna* [22], *P. multiflorum* [26], and *P. nil* [30] clearly exhibit aldehyde reduction. Besides, HRCs of *A. belladonna* [17, 21] also shares the property of acetophenone/ketone reduction with HRCs of *Daucas carota* [25], *Brassica napus* [27], and *Raphanus sativa* [27, 28]. Additionally, HRCs of few plant species have demonstrated biotransformation of artemisinin to its reduced derivative, i.e., deoxy artemisinin (e.g., *Cyanotis arachnoidea* and *Rheum palmatum*) [48, 49]. Recently, the biotransformation of 4-OHBA (p-hydroxybenzalacetone) into RK (raspberry ketone) utilizing cell cultures of plants of different families have also demonstrated reduction reactions [50].

5.2 Oxidation

The oxidation reactions in biotransformation procedures are confined mainly to aldehyde and alcohol substrate. The oxidation of aldehyde, a reaction mostly favored by microbes, is normally less observed in hairy roots based conversions. Despite, *A. belladonna* HRCs are known to perform this reaction against 3,4,5-trimethoxy benzaldehyde (TBA) leading to production of 3,4,5-trimethoxy benzoic acid [22]. Additionally, the same HRCs have also exhibited oxidation of betuligenol (BG; secondary alcohol) to raspberry ketone (RK) [20]. Furthermore, HRCs of *C. roseus* and *N. tabacum* also demonstrated oxidation of BG into RK (Scheme 1) [50]. The other significant example of oxidation of substrate was depicted by HRCs of *Levisticum officinale* [51]. This report describes that exogenous supplementation of geraniol in growth medium as a substrate induced production of six new volatiles out of which some were oxidized products. Likewise, hairy roots of *Anisodus tanguiticus* exhibited conversion of dehydroepiandrosterone (DHEA), as a substrate for oxidation into its corresponding compounds [52].

5.3 Esterification

The esterification reactions involve formation of ester bond in substrate. HRCs of *P. ginseng* have demonstrated esterification of digitoxigenin into different esters such as digitoxigenin stearate, digitoxigenin palmitate, digitoxigenin laurate, and digitoxigenin myristate [53].

5.4 Acetylation

Acetylation of exogenously supplemented monoterpenes (menthol and geraniol) was observed in HRCs of *Anethum graveolens* [54]. In this case, menthyl acetate was formed as biotransformation product of menthol, whereas geraniol was transformed into ten new products which were alcohols (linalool, alpha-terpineol, and citronellol), aldehydes (neral and geranial), esters (citronellyl, neryl) and geranyl acetates etc. Substrate specificity in biotransformation was also observed with *L. officinale* HRCs [51]. In comparison to hairy roots of *A. graveolens* [54] which acetylate both monoterpenes (menthol and geraniol) in *L. officinale* acetylation was restricted to geraniol only [51].

5.5 Hydroxylation

The bioconversions utilizing hydroxylation reactions are well explored in plant cell suspension cultures [1, 55]. However, HRCs has also been documented to acquire biotransformation utilizing hydroxylation reactions. Derivatization of trans-cinnamic acid into *p*-coumaric acid has been achieved by hydroxylation utilizing HRCs of *Lobelia sessilifolia, L. cardinalis, Campanula medium,* and *Fragaria xananassa* [56].

5.6 Glycosyl Conjugation

The biochemical reactions in which a carbohydrate moiety (saccharide unit) conjugates with any acceptor molecule thereby making it hydrophilic (water soluble) comes under glycosyl conjugation. The acceptor molecule can be protein, lipid, or any other organic molecule. It is later most category of acceptor which is normally targeted in biotransformation. Two kinds of glycosyl conjugation have been observed in plant hairy root systems. First is glycosyl esterification, which occurs as a consequence of esterification of carboxylic acids and sugar moieties, and second is glycosylation, which occurs due to ether formation of alcohols and sugar moieties. The glycosylation and glycosyl esterification of *p*-hydroxy benzoic acid has been noticed in *P. ginseng* HRCs (Scheme 3) [18].

Besides, glycosylation has also been observed in HRCs of various plants utilizing diverse substrate backbone [2]. It is the most explored biochemical reaction through

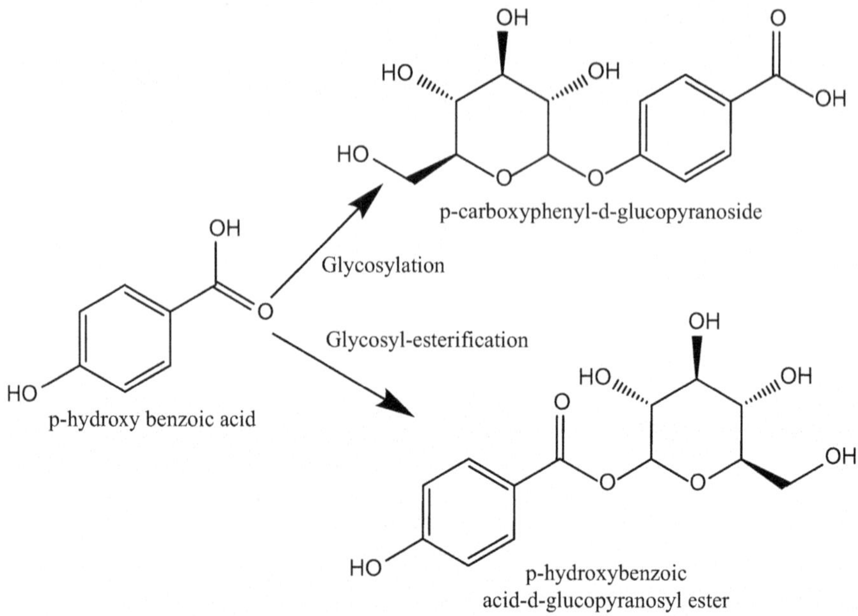

Scheme 3 Biotransformation of *p*-hydroxy benzoic acid utilizing *Panax ginseng* HRCs showing glycosylation and glycosyl esterification [18]

HRCs mediated biotransformation [2, 17]. The production of gastrodin as a consequence of glycosylation of 4-hydroxybenzyl alcohol was obtained by HRCs of *Datura tatula* [57]. Glucosylation of thymol into glucoside-5-methyl-2-(1-methylethyl) phenyl-β-D-glucopyranoside has been observed in HRCs of *P. multiflorum* [58].

6 Transgenic Hairy Roots and Biotransformation

Approaches like clonal selection, physical elicitation, permeabilization, alteration of chemical conditions, etc. are normally used to optimize the production potential of an in vitro system. Transfer and overexpression of candidate genes of required reactions in a pathway is one such approach that has gained much interest in hairy root mediated research. Being unique in their genetic and long lasting biosynthetic stability, HRCs have been sought as a competent means of biomass production concomitant with their use for transforming various substrates into their analogs. Besides, its recognition as a proficient system to study transfer and expression of gene segment relevant to metabolic pathways has also paved the way for transgene functional characterization. Additionally, in the context of biotransformation, use of well established HRCs appears more expedient as most of the studies do not require transgenic plant development. To be precise, with reference to the biotransformation of exogenous/inherent compound utilizing enzymatic machinery of HRCs, identification and characterization of enzymes and related genes involved in biotransformation will lead to their isolation and further utilization through engineering approaches. Enhancement of biotransformation potential of selected HRCs can be done either by modulating expression of gene already existing in the system or through heterologous overexpression. Vanillin production in transgenic HRCs of *Beta vulagaris* through heterologous expression of *Pseudomonas fluorescens* HCHL gene is a relevant example to discuss [59]. The study demonstrated bioconversion of inherently available ferulic acid into vanillin through bacterial HCHL overexpression. Besides, this study also showed enhancement of vanillin production after ferulic acid PF. Similarly, in a recent study, raspberry ketone, a natural flavor compound of high commercial value, was produced through transgenic HRCs of *Nicotiana tabacum* [50]. The study revealed that overexpression of *Rubus idaeus* ketone/zingerone synthase1 (*Ri*ZS1), a NADPH-dependent reductase, which is found to be responsible for the production of raspberry ketone from *p*-hydroxybenzalacetone (4-OHBA), resulted in accumulation of ketone. On the whole, these results persuade the use of infinite opportunities to develop and explore transgenic HRCs for biotransformation of a wide range of chemical compounds through activation or suppression of functional genes [11, 14].

7 Recent Reports on HRCs Based Biotransformation and Production of High Value Compounds

The research area of HRCs mediated biotransformation is still in its infancy as hairy roots of only few plant systems have been explored for this purpose [2]. The progress in this field can be measured in terms of establishment of new candidate HRCs systems solely for bioconversions as well as through the widening of substrate range. Out of many successful biotransformation reports, the significance of some of the biotransformed products has also been observed as listed in Table 3.

A detailed account on different hairy root systems, biotransformation reactions carried out there with, and transformed products has been reviewed by Banerjee et al. 2012 [2]. The upcoming text, therefore, discusses recent reports illustrating different hairy root systems used for biotransformation of various substrates and thereby production of commercially more valuable corresponding compounds (Tables 2 and 3).

HRCs of Atropa belladonna: The biotransformation capability of *A. belladonna* HRCs has been investigated utilizing three carbonyl substrates [22]. Out of three substrate utilized, TBA and 3,4,5-trimethoxy acetophenone got biotransformed; however, no biotransformation product was detected for 3,4,5-trimethoxy benzoic acid. The TBA was transformed into 3,4,5-trimethoxy benzoic acid and 3,4,5-trimethoxy benzyl alcohol after oxidation and reduction of substrate, respectively. The reduction of 3,4,5-trimethoxy acetophenone was also achieved resulting into 1-(3,4,5-trimethoxyphenyl)-ethanol. Utilizing this HRCs biotransformation of other substrates of different backbone was also investigated [20, 46]. The

Table 3 Production of value added molecules through HRCs mediated biotransformation

Transformed products	Therapeutic significance	Parent molecule	Hairy roots system	Reference
Gastrodin	Anti-inflammatory, anticonvulsion, analgesic, and antianoxemia	p-hydroxybenzyl alcohol (HBA)	*Datura tatula*	[57]
		4-hydroxybenzaldehyde and 4-hydroxybenzyl alcohol	*Polygonum multiflorum*	[26]
Raspberry ketone	Flavor molecule, antiobese, antibacterial, anticancer, and depigmenting activities	Betuligenol	*Atropa belladonna*	[20]
		Betuligenol and 4-OHBA	*Nicotiana tabacum; Catharanthus roseus*	[50]
		4-OHBA	*Hyoscyamus muticus*	[50]
Betuloside	Anti-inflammatory and hepatoprotective	Betuligenol	*Atropa belladonna*	[20]
Arbutin	Skin-lightening agent	Hydroquinone	*Brugmansia candida,*	[67]
			Physalis ixocarpa	[68]
		1–4-benzendiol	*Polygonum multiflorum*	[26]
3-α-hydroxy-1-deoxyartemsinin	Antiplasmodial and TNF lowering potential	Artemisinin	*Atropa belladonna; Hyoscyamus muticus; Ocimum basilicum*	[46]
Rosavin	Antidepressant and anxiolytic	Cinnamyl alcohol	*Rhodiola kirilowii*	[66]

biotransformation of 4-(*p*-hydroxyphenyl)-butan-2-ol or BG exhibited oxidation and glucosylation of substrate into raspberry ketone and betuloside, respectively (Scheme 1). Both of the derivatives were found to have significant pharmaceutical interest (Table 3).

HRCs of A. belladonna, *Hyoscyamus muticus*, and *Ocimum basilicum:* Pandey et al. (2015) have reported the biotransformation of antimalarial molecule artemisinin utilizing HRCs of *A. belladonna, H. muticus*, and *O. basilicum* [46]. The biotransformation reaction resulted into 3-α-hydroxy-1-deoxyartemisinin and 4-hydroxy-9,10-dimethyloctahydrofuro-(3,2-i)-isochromen-11(4H)-one. The HRCs of *H. muticus* and *A. belladonna* were found superior, and 3-α-hydroxy-1-deoxyartemisinin exhibit antiplasmodial activity with notable TNF lowering potential.

HRCs of Polygonum multiflorum: The glycosylation of 3,4-cyclocondensed coumarins (7-Hydroxy-2,3-dihydrocyclopenta[c]chromen-4-one,7,9-dihydroxy-2,3-dihydrocyclopenta[c]chromen-4-one, 6,7-dihydroxy-2,3-dihydrocyclopenta[c]chromen-4-one) have been evaluated [60]. This led to the formation of its corresponding glycosides. Moreover, this study also highlights the use of elicitation with salicylic acid (SA), yeast extract (YE), and $HgCl_2$ for enhancement of biotransformation reactions. Out of tested elicitors, the biotransformation reaction was enhanced by SA (11%) and YE (12%).

The glycosylation of two 4-phenylcoumarins (7-hydroxy-4-phenylcoumarin and 5,7-dihydroxy-4-phenylcoumarin) was also achieved utilizing *P. multiflorum* HRCs [61]. Additionally, other substrates were also analyzed for biotransformation utilizing *P. multiflorum* HRCs [62–65].

HRCs of P. ginseng: The biotransformation of tetrahydroprotoberberines (THPBs) was investigated utilizing *P. ginseng* HRCs. This substrate though alkaloid structurally also bear phenolic hydroxyl groups. The study demonstrates efficient glycosylation of THPBs having single phenolic hydroxyl and has presented no stereoselectivity. Furthermore, the culture also demonstrated hydrolysis of other substrate followed by glycosylation [21].

HRCs of Rhodiola kirilowii: The biotransformation of cinnamyl alcohol into corresponding glycosides (rosarin and rosavin) was achieved utilizing HRCs of *R. kirilowii* [66].

8 Conclusions

The hairy root research till now has made its significant contribution in plant biotechnology, and HRCs are being globally explored to investigate diverse aspects of hairy root biotechnology right from root initiation to translational prospects [8, 9, 11, 14]. This is the reason behind continuous inflow of reports involving hairy root establishment in diverse plant species of which number even crossed 0.5 K as per recent reviews [11, 14]. With such immense interest, the curiosity related to the application of HRCs also transited towards metabolism, elicitation, phytoremediation, metabolic engineering, and biotransformation from hairy root based traditional SM production. Despite, the HRCs are also proving worthwhile to study elemental research related to root biology [11, 14].

The biotransformation though an old practice since the beginning of civilization simultaneously recognized as an efficient step towards green chemistry. Biotransformation exists in diverse life forms under natural conditions, and exploitation of this ability of living entities into the production process of wider utility compounds has been judiciously focussed. In context of biotransformation mediated by different plant in vitro systems, the indubitable superiority of HRCs in terms of cost, maintenance, stability, and enzymatic potential has been well established [2, 17]. The HRCs based biotransformation procedures involve different strategies which are precursor feeding, co-culture system, and exogenous molecule supplementation. Whichever the strategy involved, the rational is either to enhance inherent potential of machinery or to get rid of substrate (precursor) limitation. The reaction potential here provides the additional advantage of regiospecificity, enantioselectivity, and almost nominal culture conditions. The literature suggested that diverse substrates which include phenolics, carbonyl compounds, coumarins, terpenes, etc. have been explored for biotransformation utilizing HRCs. Besides, the production of significant molecules has also been achieved which validates the concept of "value addition" through biotransformation (Table 3).

Altogether, many successful hairy root systems have been established that can be considered for having potential prospects for commercial scale setups. However, the incorporation of integrative approaches like elicitation and gene overexpression is also needed to explore broader utility of HRCs based transformation. Further, optimization of various other factors like selection of suitable hairy root system with desired enzymatic pool, time scale management for different reactions to deal with a reaction flexible system, optimization of growth medium composition, and consumption require special attention to develop a successful HRCs based biotransformation system.

References

1. Giri A, Dhingra V, Giri CC, Singh A, Ward OP, Narasu ML (2001) Biotransformations using plant cells, organ cultures and enzyme systems: current trends and future prospects. Biotechnol Adv 19(3):175–199
2. Banerjee S, Singh S, Rahman LU (2012) Biotransformation studies using hairy root cultures – a review. Biotechnol Adv 30(3):461–468. doi:10.1016/j.biotechadv.2011.08.010
3. Vasic-Racki D (2006) History of industrial biotransformations – dreams and realities. Industrial biotransformations. Wiley, New York. doi:10.1002/9783527614165.ch2
4. Hussain MS, Fareed S, Ansari S, Rahman MA, Ahmad IZ, Saeed M (2012) Current approaches toward production of secondary plant metabolites. J Pharm Bioallied Sci 4(1):10–20. doi:10.4103/0975-7406.92725
5. Rao SR, Ravishankar GA (2002) Plant cell cultures: chemical factories of secondary metabolites. Biotechnol Adv 20(2):101–153
6. Karuppusamy S (2009) A review on trends in production of secondary metabolites from higher plants by *in vitro* tissue, organ and cell cultures. J Med Plant Res 3(13):1222–1239
7. Wink M, Alfermann AW, Franke R, Wetterauer B, Distl M, Windhövel J, Krohn O, Fuss E, Garden H, Mohagheghzadeh A, Wildi E (2005) Sustainable bioproduction of phytochemicals by plant in vitro cultures: anticancer agents. Plant Gen Resour 3(2):90–100. doi:10.1079/PGR200575
8. Georgiev MI, Agostini E, Ludwig-Müller J, Xu J (2012) Genetically transformed roots: from plant disease to biotechnological resource. Trends Biotechnol 30(10):528–537. doi:10.1016/j.tibtech.2012.07.001
9. Guillon S, Trémouillaux-Guiller J, Pati PK, Rideau M, Gantet P (2006) Harnessing the potential of hairy roots: dawn of a new era. Trends Biotechnol 24(9):403–409. doi:10.1016/j.tibtech.2006.07.002
10. Guillon S, Trémouillaux-Guiller J, Pati PK, Rideau M, Gantet P (2006) Hairy root research: recent scenario and exciting prospects. Curr Opin Plant Biol 9(3):341–346. doi:10.1016/j.pbi.2006.03.008

11. Mehrotra S, Srivastava V, Rahman LU, Kukreja AK (2015) Hairy root biotechnology – indicative timeline to understand missing links and future outlook. Protoplasma 252(5):1189–1201. doi:10.1007/s00709-015-0761-1
12. Chandra S (2012) Natural plant genetic engineer *Agrobacterium rhizogenes*: role of T-DNA in plant secondary metabolism. Biotechnol Lett 34(3):407–415. doi:10.1007/s10529-011-0785-3
13. Chandra S, Chandra R (2011) Engineering secondary metabolite production in hairy roots. Phytochem Rev 10(3):371–395. doi:10.1007/s11101-011-9210-8
14. Ono NN, Tian L (2011) The multiplicity of hairy root cultures: prolific possibilities. Plant Sci 180(3):439–446. doi:10.1016/j.plantsci.2010.11.012
15. Mehrotra S, Rahman LU, Kukreja AK (2010) An extensive case study of hairy-root cultures for enhanced secondary-metabolite production through metabolic-pathway engineering. Biotechnol Appl Biochem 56(4):161–172. doi:10.1042/BA20100171
16. Mehrotra S, Goel MK, Rahman LU, Kukreja AK (2013) Molecular and chemical characterization of plants regenerated from Ri-mediated hairy root cultures of *Rauwolfia serpentina*. Plant Cell Tissue Organ Cult 114(1):31–38. doi:10.1007/s11240-013-0302-6
17. Srivastava V (2015) Studies on hairy roots as bioreactors for normal and elicited biotransformation of exogenous molecules. Ph.D thesis, University of Lucknow, India
18. Chen X, Zhang J, Liu JH, Yu BY (2008) Biotransformation of p-, m-, and o-hydroxybenzoic acids by *Panax ginseng* hairy root cultures. J Mol Catal B Enzym 54(3):72–75. doi:10.1016/j.molcatb.2007.12.015
19. Yan CY, Ma WL, Yan WW, Yu RM (2008) Biotransformation of furannoligularenone by hairy root cultures of *Polygonum multiflorum*. J Chin Med Mater 31(5):633–635
20. Srivastava V, Kaur R, Chattopadhyay SK, Banerjee S (2013) Production of industrially important cosmaceutical and pharmaceutical derivatives of betuligenol by *Atropa belladonna* hairy root mediated biotransformation. Ind Crop Prod 44:171–175
21. Ge HX, Zhang J, Lu LL, Yu BY (2014) Biotransformation of tetrahydroprotoberberines by *Panax ginseng* hairy root culture. J Mol Catal B Enzym 110:133–139. doi:10.1016/j.molcatb.2014.10.001
22. Srivastava V, Negi AS, Ajayakumar PV, Khan SA, Banerjee S (2012) *Atropa belladonna* hairy roots: orchestration of concurrent oxidation and reduction reactions for biotransformation of carbonyl compounds. Appl Biochem Biotechnol 166(6):1401–1408. doi:10.1007/s12010-011-9533-3
23. Kanho H, Yaoya S, Itani T, Nakane T, Kawahara N, Takase Y, Masuda K, Kuroyanagi M (2004) Glucosylation of phenolic compounds by *Pharbitis nil* hairy roots: I. Glucosylation of coumarin and flavone derivatives. Biosci Biotechnol Biochem 68(10):2032–2039. doi:10.1271/bbb.68.2032
24. Yan C, Ma W, Liang J, Lu H, Zhang D, Yu R (2008) Biotransformation of eight active constituents by hairy roots of *Polygonum multiflorum* Thunb. J Chin Biotechnol 28(4):78–81
25. Caron D, Coughlan AP, Simard M, Bernier J, Piché Y, Chênevert R (2005) Stereoselective reduction of ketones by *Daucus carota* hairy root cultures. Biotechnol Lett 27(10):713–716. doi:10.1007/s10529-005-5187-y
26. Yan CY, Yu RM, Zhang Z, Kong LY (2007) Biotransformation of 4-hydroxybenzen derivatives by hairy root cultures of *Polygonum multiflorum* Thunb. J Integr Plant Biol 49(2):207–212. doi:10.1111/j.1744-7909.2007.00400.x
27. Orden AA, Bisogno FR, Cifuente DA, Giordano OS, Sanz MK (2006) Asymmetric bioreduction of natural xenobiotic diketones by *Brassica napus* hairy roots. J Mol Catal B Enzym 42(3):71–77. doi:10.1016/j.molcatb.2006.06.010
28. Orden AA, Magallanes-Noguera C, Agostini E, Kurina-Sanz M (2009) Anti-Prelog reduction of ketones by hairy root cultures. J Mol Catal B Enzym 61(3):216–220. doi:10.1016/j.molcatb.2009.07.007

29. Shimoda K, Kubota N, Sano T, Hirakawa H, Hirata T (2004) A novel hydroxylase from *Catharanthus roseus* participating in the hydroxylation of 2-hydroxybenzoic acid. J Biosci Bioeng 98(2):67–70. doi:10.1016/S1389-1723(04)70244-3
30. Kanho H, Yaoya S, Kawahara N, Nakane T, Takase Y, Masuda K, Kuroyanagi M (2005) Biotransformation of benzaldehyde-type and acetophenone-type derivatives by *Pharbitis nil* hairy roots. Chem Pharm Bull 53(4):361–365. doi:10.1248/cpb.53.361
31. Zhou ML, Zhu XM, Shao JR, Tang YX, Wu YM (2011) Production and metabolic engineering of bioactive substances in plant hairy root culture. Appl Microbiol Biotechnol 90(4):1229–1239. doi:10.1007/s00253-011-3228-0
32. Morgan JA, Shanks JV (2000) Determination of metabolic rate-limitations by precursor feeding in *Catharanthus roseus* hairy root cultures. J Biotechnol 79:137–145
33. Hashimoto T, Yamada Y (1983) Scopolamine production in suspension cultures and redifferentiated roots of *Hyoscyamus niger*. Planta Med 47(4):195–199
34. Shinde AN, Nutan M, Devanand PF (2009) Enhanced production of phytoestrogenic isoflavones from hairy root cultures of *Psoralea corylifolia* L. using elicitation and precursor feeding. Biotechnol Bioprocess Eng 14:288–294. doi:10.1007/s12257-008-0238-6
35. Łuczkiewicz M, Adam K (2005) *Genista tinctoria* hairy root cultures for selective production of isoliquiritigenin. Z Naturforsch 60c:867–875
36. Kim Y, Barbara EW, Weathers PJ (2002) Secondary metabolism of hairy root cultures in bioreactors. In Vitro Cell Dev Biol Plant 38:1–10. doi:10.1079/IVP2001243
37. Sung LS, Huang SY (2000) Headspace ethylene accumulation on *Stizolobium hassjoo* hairy root culture producing L-3,4-dihydroxyphenylalanine. Biotechnol Lett 22:875–878. doi:10.1023/A:1005679800393
38. Goers L, Freemont P, Polizzi KM (2014) Co-culture systems and technologies: taking synthetic biology to the next level. J Roy Soc Inter 11(96). doi:10.1098/rsif.2014.0065
39. Subroto MA, Kwok KH, Hamill JD, Doran PM (1996) Co-culture of genetically transformed roots and shoots for synthesis, translocation, and biotransformation of secondary metabolites. Biotechnol Bioeng 49(5):481–494. doi:10.1002/(SICI)1097-0290
40. Łuczkiewicz M, Kokotkiewicz A (2005) Co-cultures of shoots and hairy roots of *Genista tinctoria* L. for synthesis and biotransformation of large amounts of phytoestrogens. Plant Sci 169(5):862–871. doi:10.1016/j.plantsci.2005.06.005
41. Lin HW, Kwok KH, Doran PM (2003) Production of podophyllotoxin using cross-species c-culture of *Linum flavum* hairy roots and *Podophyllum hexandrum* cell suspensions. Biotechnol Prog 19(5):1417–1426. doi:10.1021/bp034036h
42. Uden-Van W, Pras N, Malingré TM (1990) On the improvement of the podophyllotoxin production by phenylpropanoid precursor feeding to cell cultures of *Podophyllum hexandrum* Royle. Plant Cell Tissue Organ Cult 23(3):217–224. doi:10.1007/BF00034435
43. Sidwa-Gorycka M, Królicka A, Kozyra M, Głowniak K, Bourgaud F, Łojkowska E (2003) Establishment of a co-culture of *Ammi majus* L. and *Ruta graveolens* L. for the synthesis of furanocoumarins. Plant Sci 165(6):1315–1319. doi:10.1016/S0168-9452(03)00343-1
44. Wu JY, Ng J, Shi M, Wu SJ (2007) Enhanced secondary metabolite (tanshinone) production of *Salvia miltiorrhiza* hairy roots in a novel root–bacteria co-culture process. Appl Microbiol Biotechnol 77(3):543–550. doi:10.1007/s00253-007-1192-5
45. Faria JM, Sena I, Maleita CM, da Silva IV, Ascensão L, Abrantes I, Bennett RN, Mota M, Figueiredo AC (2014) In vitro co-culture of *Solanum tuberosum* hairy roots with *Meloidogyne chitwoodi*: structure, growth and production of volatiles. Plant Cell Tissue Organ Cult 118(3):519–530. doi:10.1007/s11240-0504-6
46. Pandey P, Singh S, Tewari N, Srinivas KVNS, Shukla A, Gupta N, Vasudev PG, Khan F, Pal A, Bhakuni RS, Tandon S (2015) Hairy root mediated functional derivatization of artemisinin and their bioactivity analysis. J Mol Catal B Enzym 113:95–103. doi:10.1016/j.molcatb.2015.01.007
47. Cai J, Ding J, Hua Y, Nan L (2005) Establishment of biotransformation system of the gastrodin biosynthesis by hairy root of *Panax ginseng*. J Plant Resour Environ 14(2):29–31

48. Zhou L, Ruan D, He Z, Zhu H, Yang C, Wang J (1997) Biotransformation of artemisinin by hairy roots of *Cyanotis arachnoidea*. Acta Bot Yunnanica 20(2):229–232, ISSN: 0253–2700 CN: 53-1039/Q
49. Lixin L, Yanfang S, Xiaofeng L, Dean G (2002) Biotransformation of artemisinin by hairy root cultures of *Rheum palmatum* L. J Chin Pharm Sci 11(4):122–124
50. Häkkinen ST, Seppänen-Laakso T, Oksman-Caldentey KM, Rischer H (2015) Bioconversion to raspberry ketone is achieved by several non-related plant cell cultures. Front Plant Sci 6:1035. doi:10.3389/fpls.2015.01035
51. Nunes IS, Faria JM, Figueiredo AC, Pedro LG, Trindade H, Barroso JG (2009) Menthol and geraniol biotransformation and glycosylation capacity of *Levisticum officinale* hairy roots. Planta Med 75(4):387–391. doi:10.1055/s-0028-1112217
52. Liu Y, Cheng KD, Zhu P, Feng WH, Meng C, Zhu HX, He HX, Ma XJ (2004) Biotransformation of dehydroepiandrosterone by hairy root cultures of *Anisodus tanguticus*. Acta Pharm Sin 39(6):445–448
53. Kawaguchi K, Hirotani M, Yoshikawa T, Furuya T (1990) Biotransformation of digitoxigenin by ginseng hairy root cultures. Phytochem 29(3):837–843. doi:10.1016/0031-9422(90)80029-G
54. Faria JM, Nunes IS, Figueiredo AC, Pedro LG, Trindade H, Barroso JG (2009) Biotransformation of menthol and geraniol by hairy root cultures of *Anethum graveolens*: effect on growth and volatile components. Biotechnol Lett 31(6):897–903. doi:10.1007/s10529-009-9934-3
55. Suga T, Hirata T (1990) Biotransformation of exogenous substrates by plant cell cultures. Phytochem 29(8):2393–2406. doi:10.1016/0031-9422(90)85155-9
56. Ishimaru K, Yamanaka M, Terahara N, Shimomura K, Okamoto D, Yoshihira K (1996) Biotransformation of phenolics by hairy root cultures of five herbal plants. Jpn J Food Chem 3:38–42
57. Peng CX, Gong JS, Zhang XF, Zhang M, Zheng SQ (2010) Production of gastrodin through biotransformation of p-hydroxybenzyl alcohol using hairy root cultures of *Datura tatula* L. Afr J Biotechnol 7(3):211–216
58. Dong QF, Jia JZ, Zhu JH, Yu RM (2009) Biotransformation of thymol by hairy roots of transgenic *Polygonum multiflorum*. J Chin Med Mater 32(10):1495–1499
59. Singh P, Khan S, Pandey SS, Singh M, Banerjee S, Kitamura Y, Rahman LU (2015) Vanillin production in metabolically engineered *Beta vulgaris* hairy roots through heterologous expression of *Pseudomonas fluorescens* HCHL gene. Ind Crop Prod 74:839–848. doi:10.1016/j.indcrop.2015.05.037
60. Liu L, Zhang M, Zhou L, Zhu J, Guan S, Yu R (2012) Biotransformation of 3, 4-cyclocondensed coumarins by transgenic hairy roots of *Polygonum multiflorum*. Afr J Pharm Pharmacol 6(44):3047–3054. doi:10.5897/AJPP11.858
61. Zhou L, Liang Y, Wang W, Tan H, Xiao M, Fan C, Yu R (2011) Biotransformation of 4-phenylcoumarin by transgenic hairy roots of *Polygonum multiflorum*. J Med Plant Res 5:4274–4278. doi:10.1007/s00709-013-0554-3
62. Zhou L, Tian T, Xue B, Song L, Liu L, Yu R (2012) Biosynthesis of coumarin glycosides by transgenic hairy roots of *Polygonum multiflorum*. BioSci Biotechnol Biochem 76(5):1008–1010. doi:10.1271/bbb.110347
63. Liang Y, Deng WJ, Zhou LB, Yan CY, Yu RM (2011) Biotransformation of 4-methyl-7-allyloxy coumarin by transgenic hairy roots of *Polygonum multiflorum*. J Chin Med Mater 34(6):901–904
64. Deng W, Zhou L, Yu R (2011) Biotransformation of daphnetin by suspension transgenic hairy roots of *Polygonum multiflorum*. Chin J Chin Mat Med 36(3):351–355
65. Dong QF, Jia JZ, Zhu JH, Yu RM (2009) Biotransformation of thymol by hairy roots of transgenic *Polygonum multiflorum*. J Chin Med Mat 32(10):1495–1499
66. Grech-Baran M, Sykłowska-Baranek K, Krajewska-Patan A, Wyrwał A, Pietrosiuk A (2014) Biotransformation of cinnamyl alcohol to rosavins by non-transformed wild type and hairy root cultures of *Rhodiola kirilowii*. Biotechnol Lett 36(3):649–656. doi:10.1007/s10529-013-1401-5

67. Casas DA, Pitta-Alvarez SI, Giulietti AM (1998) Biotransformation of hydroquinone by hairy roots of *Brugmansia candida* and effect of sugars and free-radical scavengers. Appl Biochem Biotechnol 69:127–136. doi:10.1007/BF02919394
68. Bergier K, Polaszczyk B, Gajewska E, Wielanek M, Krolicka A, Sklodowska M (2008) Glucosylation of hydroquinone to arbutin by hairy roots of *Physalis ixocarpa*. Zesz Probl Postep Nauk Rol 524:339–347

Part III

Metabolic Engineering

Metabolic Engineering of Lignan Biosynthesis Pathways for the Production of Transgenic Plant-Based Foods

15

Honoo Satake, Tomotsugu Koyama, Erika Matsumoto, Kinuyo Morimoto, Eiichiro Ono, and Jun Murata

Abstract

Lignans are major phytochemicals biosynthesized in several plants including *Sesamum*, *Linum*, *Forsythia*, and *Podophyllum* genus, and a great variety of lignans have received wide attentions as leading compounds of novel drugs for tumor treatment and healthy diets to reduce of the risks of lifestyle-related diseases. Recent genome and transcriptome studies have characterized multiple novel lignan-biosynthetic enzymes, and thus have opened new avenues to transgenic metabolic engineering of various nonmodel dietary or medicinal plants. *Forsythia* and *Linum* are the most useful and prevalent natural and agricultural sources for the development of both transgenic foods and medicinal compounds. Over the past few years, transiently gene-transfected or transgenic *Forsythia* and *Linum* plants or cell cultures have been shown to be promising platforms for the sustainable and efficient production of beneficial lignans. In this chapter, we present the essential knowledge and recent advances regarding metabolic engineering of lignans based on their biosynthetic pathways and biological activities and the perspectives in lignan production via metabolic engineering.

Keywords

Lignan • Biosynthesis • Metabolic engineering • Transgenic plant, *Forsythia*, *Linum*

H. Satake (✉) • T. Koyama (✉) • E. Matsumoto (✉) • K. Morimoto (✉) • J. Murata (✉)
Bioorganic Research Institute, Suntory Foundation for Life Sciences, Seika, Kyoto, Japan
e-mail: satake@sunbor.or.jp; koyama@sunbor.or.jp; matsumoto@sunbor.or.jp; sial_glc@yahoo.co.jp; murata-j@sunbor.or.jp

E. Ono (✉)
Research Institute, Suntory Global Innovation Center (SIC) Ltd., Seika, Kyoto, Japan
e-mail: Eiichiro_Ono@suntory.co.jp

S. Jha (ed.), *Transgenesis and Secondary Metabolism*, Reference Series in Phytochemistry, DOI 10.1007/978-3-319-28669-3_11

Abbreviations

CAD	Cinnamylalcohol dehydrogenase
CCR	Cinnamoyl-CoA reductase
DIR	Dirigent protein
ER	Estrogen receptor
MAPK	Mitogen-activated protein kinase
MeJA	Methyl jasmonate
MOMT	Matairesinol *O*-methyltransferase
PAL	Phenylalanine ammonialyase
PIP	Pinoresinol-lariciresinol/isoflavone/phenylcoumaran benzylic ether reductase
PLR	Pinoresinol-lariciresinol reductase
PTOX	Podophyllotoxin
RNAi	RNA interference
SA	Salicylic acid
SDG	Secoisolariciresinol diglucoside
SIRD	Secoisolariciresinol dehydrogenase

Contents

1 Introduction

Functional foods, dietary supplements, and drug compounds are largely derived from specialized metabolites, previously called secondary metabolites of plants, including alkaloids, flavonoids, isoflavonoids, and lignans. Recently, the rapid increase in the number of elderly individuals has required various medical costs, which may eventually cause a serious disruption in essential medical care systems and national financial burdens. To address these issues, the consistent and appropriate intake of dietary supplements and the efficient development of clinical drugs are the most promising and effective ways to increase the healthy life expectancy and prevent lifestyle-related diseases. In this context, intensive efforts should be made on the development of functional foods and supplements as well as of clinical agents.

Lignans are naturally occurring phenylpropanoid dimers (C6-C3 unit; e.g., coniferyl alcohol), in which the phenylpropane units are linked by the central carbons of the side chains (Fig. 1). These specialized metabolites are classified

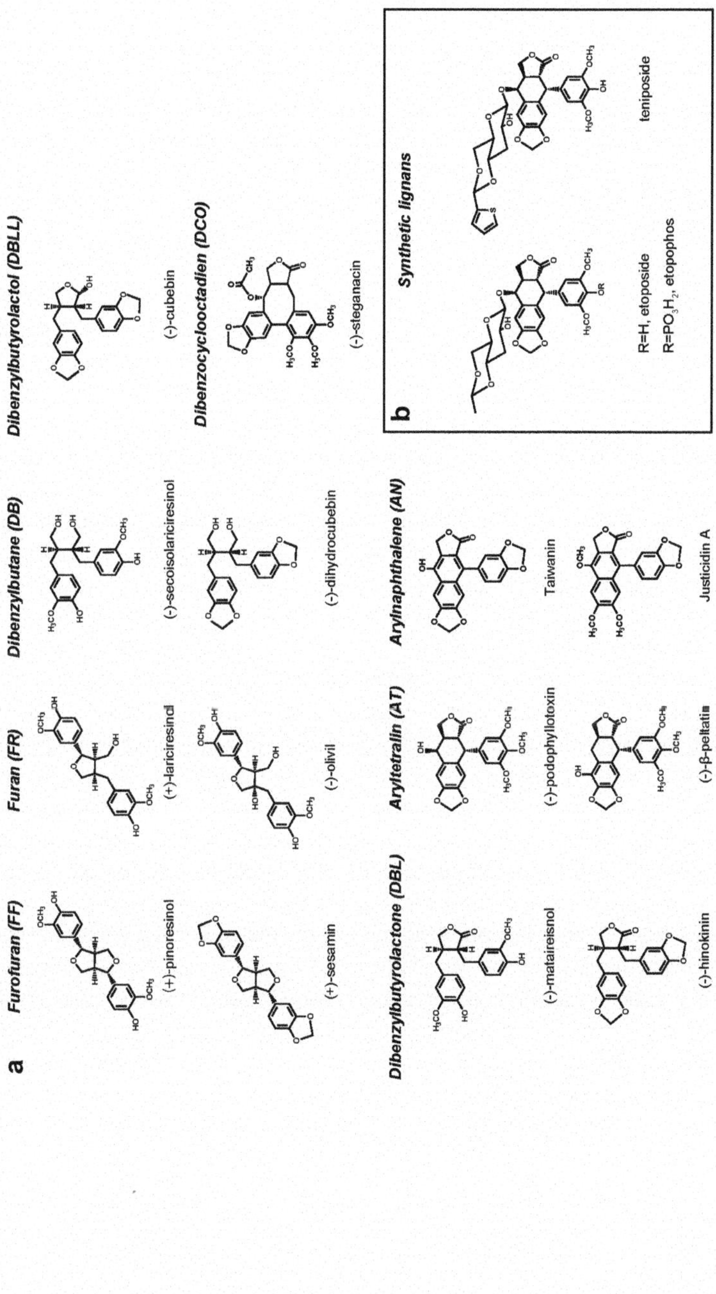

Fig. 1 Chemical structures of typical lignans in dietary and medicinal sources (**a**) and synthetic podophyllotoxin derivatives (**b**)

into eight groups based on their structural patterns, including their carbon skeletons, the way in which oxygen is incorporated into the skeletons, and the cyclization pattern: furofuran, furan, dibenzylbutane, dibenzylbutyrolactone, aryltetralin, arylnaphthalene, dibenzocyclooctadiene, and dibenzylbutyrolactol [1, 2]. Lignans have been shown to exhibit not only various pharmaceutical activities but also preventive or reductive effects on extensive life-related diseases (see the following section) [3–15], indicating the prominent potentials as functional foods and supplements. Indeed, a sesame lignan, sesamin, is already commercially available as functional supplements for antihypertension and protection of the liver based on reduction of lipid oxidation. Unfortunately, plant sources of lignans are frequently limited because of the high cost of plant hunting and collection, poor cultivation systems, long growth phase, and the low lignan content. For instance, sesamin is extracted from sesame seed oil, the most abundant source of this compound. Nevertheless, sesamin at most constitutes 0.4–0.6% (w/w) of sesame seed oil. Moreover, sesame seeds are cultivated only once per year, limiting the ability to obtain large amounts of this compound. Likewise, podophyllotoxin (PTOX), a precursor of semisynthetic antitumor drugs (Fig. 1), is isolated from the roots and rhizomes of *Podophyllum hexandrum*, which is distributed in very limited regions, and is now endangered due to overharvesting and environmental disruption [16]. Moreover, lignans and their precursors are not or faintly biosynthesized in prevalent model plants such as *Arabidopsis thaliana* and *Nicotiana tabacum*, given that lignan biosynthetic pathways for plant specialized metabolites involve multiple enzymatic steps that are absent in most plant species including these model plants. Thus, transformation of a whole set of the biosynthetic genes would be required for the generation of transgenic model plants that could produce phytochemicals, indicating the critical shortcomings of these plants as biological lignan producers [12, 13, 17]. Similarly, transgenic microbial or animal cells are also not suitable for any biological lignan producers. In addition, the complicated chemical structures of lignans and the related compounds (Fig. 1a) make stereoselective organic synthesis impractical and costly for lignan producing large supplies of these compounds. These drawbacks indicate the requirement for efficient, stable, and sustainable systems for lignan production using lignan-rich plants or its precursor compound-rich plants.

There has been a growing body of reports on the molecular characterization of the enzymes involved in the biosynthesis of lignans, lignan production using lignan-rich plants or cultured plant cells [18–23]. These recent findings have allowed us to attempt the metabolic engineering of lignan biosynthesis using transgenic lignin-rich plants such as *Linum* and *Forsythia*. In this chapter, we provide current knowledge of lignan production via metabolic engineering and perspectives in the development of metabolic engineering-based lignan production.

2 Lignan Biological Activity on Mammals

Lignans exhibit a wide variety of bioactivities on plants, insects, and mammals [12, 13, 24–29], but special attentions have been paid to their unique antitumor-associated activities and reduction of the risks of lifestyle-related diseases. The modes of actions of lignans on mammals are classified in two ways: the pharmacological actions of specific metabolites of lignans by intestinal microflora and those of intact lignans. Many of lignans and their glycosides, including pinoresiniol, sesamin, lariciresinol, secoisolariciresinol, and matairesinol, are metabolized by intestinal microflora to yield enterodiol and enterolactone, which are defined as enterolignans or mammalian lignans [30–32]. These lignan metabolites are believed to elicit the modest estrogen (mammalian female steroid hormone)-like activity in mammals. For example, enterolignans bind to the mammalian estrogen receptors (ER), ERα or ERβ, which are key regulatory nuclear receptors in the sexual maturation of genital organs [33, 34]. Consequently, enterolignans, combined with other intestinal flora generating metabolites of isoflavones and coumestans, are also called phytoestrogens.

Intact lignans have also been detected in the sera of mammals fed with lignan-rich diets, suggesting that nonmetabolized lignans are also taken up by the mammalian digestive system, and exhibit ER-independent activities in vivo and in vitro, including tumor growth suppression, angiogenesis inhibition, and reduction of diabetes [6, 35–40]. Furthermore, lignans have been shown to manifest positive effects on other lifestyle-related diseases. Administration of flaxseed lignan complexes improved hyperglycemia and reduced markers of type II diabetes in elderly patients and various animal models [41, 42]. In particular, secoisolariciresinol diglucosides (SDG), secoisolariciresinol, enterodiol, and enterolactone inhibited pancreatic α-amidase activity in a noncompetitive manner [43]. Sesamin and its metabolites also exhibited antihypertensive activities [44–46]. Moreover, the antioxidative activity of sesamin is believed to be involved in protecting the liver from oxidation by alcohols, lipids, and oxygen radicals [44, 47–49]. In human intestinal Caco 2 cells, pinoresinol suppressed expression of Cox-2, an inducible prostaglandin synthase that is responsible for the synthesis of prostaglandin H, a precursor of any other prostaglandins, leading to the decrease in the production of inflammatory factors, such as interleukin-6 and prostaglandin E2 [35]. In contrast, matairesinol increased levels of prostaglandin E2 [35]. These findings proved that pinoresinol and matairesinol have opposite activities in these cells [35].

Of the most prominent epidemiological significance is that intake of lignan-rich foods, such as flaxseeds and sesame seeds, has been shown to reduce breast cancer risk and to improve the breast cancer-specific survival of postmenopausal women [39, 50–55]. Moreover, serum enterolactone levels were positively correlated with the improvement of prognosis in postmenopausal women with breast cancer [56]. These epidemiological studies suggest the unique and beneficial suppressive activity of lignans against breast cancer risks in elderly women.

Dietary lariciresinol was found to suppress tumor growth and angiogenesis in nude mice implanted with human MCF-7 breast cancer via the induction of apoptosis and the upregulation of ERβ expression [40]. SDG elicited potent inhibition of cell proliferation and induction of the apoptosis of breast cancer cells via the downregulation of ER- and growth factor-mediated gene expression in athymic mice [57]. Sesamin was found to reduce signaling downstream of mitogen-activated protein kinase (MAPK) [58]. Additionally, the inhibitory effect of sesamin on breast tumor growth is likely to be more potent than SDG [58]. These pharmacological effects, combined with the abundance of lignans in flax or sesame seeds and oils, confirm that the seeds and oils are promising functional diets for the prevention of breast cancer.

PTOX and its structurally related natural derivatives exhibit the suppressive activity on mitotic spindle assembly by binding to tubulin, resulting in cell cycle arrest at metaphase [22]. The PTOX semi-synthetic derivatives, etoposide, teniposide, and etopophos (Fig. 1b), are clinically utilized to treat certain types of cancers, including testicular/small-cell lung cancer, acute leukemia, Hodgkin's and non-Hodgkin's lymphoma [58, 59]. These PTOX-derived antitumor drugs induce apoptosis of tumor cells by binding to topoisimease II, a key enzyme for cell division [58, 59]. In addition, other new synthetic PTOX derivatives, including GP-11, NK-611, TOP-53, GL-331, and NPF, are undergoing phase I or II clinical trials as novel cancer drugs [22, 59]. Consistent with the difficulty in efficient chemical synthesis of PTOX due to its complicated structure, these findings reinforce the importance of PTOX as a natural seed material for the production of various anticancer drugs.

Altogether, these epidemiological and physiological studies demonstrate that lignans exert beneficial effects as dietary compounds or medicinal agents for the prevention of lifestyle-related diseases, such as cancer, hypertension, and diabetes. Of particular interest is that respective lignans exhibit specific bioactivities in mammals, strongly suggesting the requirements for the efficient, stable, and sustainable production of these compounds of interest. In other words, these findings not only endorse the high usefulness of lignin-rich sesame and flax seeds as unique functional foods but also shed light on the importance of the development of novel lignan production systems using transgenic lignan-rich plants.

3 Lignan Biosynthesis Pathways

Two major lignan biosynthesis pathways have thus far been identified. Both of the pathways originate from the coupling of achiral *E*-coniferyl alcohol, leading to the generation of pinoresinol, a basal lignan (Fig. 2). A pinoresinol synthase has yet to be identified. However, a dirigent protein (DIR) was shown to participate in the stereo-specific dimerization of *E*-coniferyl alcohol [60]. In several plants including *Sesamum*, pinoresinol is metabolized into piperitol, followed by further conversion

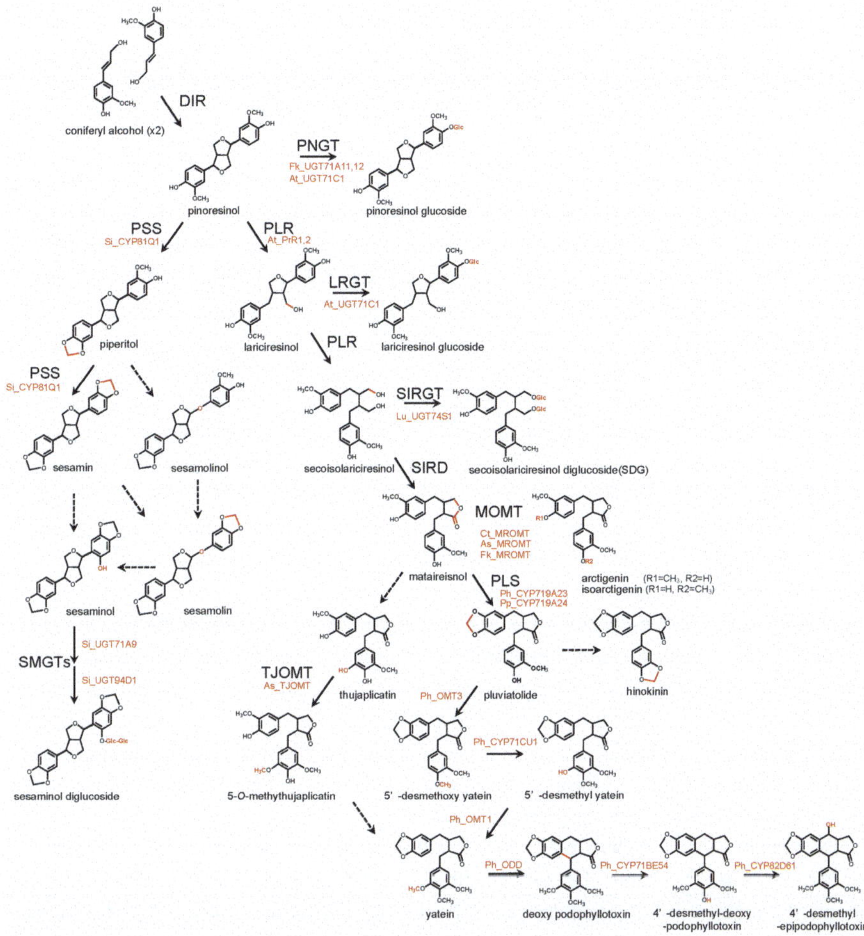

Fig. 2 Biosynthesis pathways of major lignans. Chemical conversions at each step are indicated in *red. Solid and broken lines* represent identified and unidentified enzyme-catalyzed reactions, respectively

into (+)-sesamin by a cytochrome P450 family enzyme, CYP81Q1, which is responsible for the formation of two methylenedioxy bridges [61]. The CYP81Q1 gene is expressed almost exclusively in sesame seeds, which is compatible with sesamin production at the highest level in sesame seeds [61]. Sesamin is anticipated to be further metabolized into sesaminol and sesamolin (Fig. 2), although the relevant enzymes remain to be characterized. Sesaminol is glucosylated at its 2-hydroxyl group by the homologous enzymes UGT71A8 (*S. radiatum*), 9 (*S. indicium*), and 10 (*S. alatum*) [62]. Moreover, the resultant sesaminol 2-*O*-monoglucoside is further glucosylated by UGT94D1, which is specific to the glucosylation of sesaminol 2-*O*-monoglucoside at 6-position of the conjugated glucose conjugated by UGT71A18 [62].

No genes homologous to CYP81Q1 have been detected in diverse lignan-rich plant species including *Forsythia*, *Linum*, or *Podophyllum* [63–69]. This is in good agreement to the findings that these plants fail to biosynthesize sesamin and its derivatives. Instead, pinoresinol is stepwisely reduced to lariciresinol and then secoisolariciresinol by pinoresinol-lariciresinol reductase (PLR), a member of the pinoresinol-lariciresinol/isoflavone/phenylcoumaran benzylic ether reductase (PIP) family in extensive plant species including *Forsythia*, *Linum*, and *Podophyllum* [70–75]. PLR converts pinoresinol to secoisolariciresinol via lariciresinol (Fig. 2). Pinoresinol also undergoes glucosylation by UGT71A18, a UDP-glucose-dependent glucosyltranferase [76]. Such glycosylation is highly likely to suppress the chemical reactivity of a phenolic hydroxyl group of pinoresinol and to potentiate high water solubility of pinoresinol aglycone, resulting in large and stable amounts of pinoresinol [1, 2, 11, 12]. Indeed, approximately 90% of pinoresinol is accumulated in its glucosylated form in *Forsythia* spp. [77, 78]. Thus, PLR-catalyzed metabolism and UGT71A18-directed glucosylation are reciprocally competitive pathways (Fig. 2), given that both of them share pinoresinol as a substrate. *PLR* shows opposite seasonal alteration in gene expression against *UGT71A18*; in *Forsythia* leaves in Japan, *PLR* gene is intensely expressed from April to August but poorly from September to November, whereas gene expression of *UGT71A18* is observed at high level from September to November but at faint or no level from April to August in Japan [78]. These findings indicate that PLR and UGT71A18 participate in the competitive regulation of lignan biosynthesis via pinoresinol metabolism. In *A. thaliana*, AtPrR1 and 2 are only responsible for the reduction of pinoresinol to lariciresinol [74], and lariciresinol and pinoresinol are glucosylated by another novel UDP-glucose-dependent glucosyltranferase, UGT71C1 [79], revealing the diversity of lignan metabolism among plant species.

Secoisolariciresinol, like pinoresinol and lariciresinol, undergoes two metabolic pathways (Fig. 2). First, Secoisolariciresinol is converted into matairesinol by secoisolariciresinol dehydrogenase (SIRD) [80]. Second, a novel UDP-glucose-dependent glucosyltranferase in *Linum*, UGT74S1, generates secoisolariciresinol monoglucoside and SDG [81]. Matairesinol is metabolized to arctigenin (Fig. 2) by matairesinol *O*-methytransferase (MOMT) via methylation of a phenolic hydroxyl group in various plants including *F. koreana*, *Carthamus tinctorius*, and *Anthriscus sylvestris* [82, 83]. Additionally, 70–90% of matairesinol is glucosylated throughout the year in the *Forsythia* leaves [78], although characterization of matairesinol-glucosylating enzymes awaits further study. As shown in Fig. 2, the biosynthetic pathways downstream of matairesinol are complexed and relatively species-specific. In *Linum*, *Anthriscus*, and *Podophyllum* plants, matairesinol is also converted into hinokinin, yatein, or PTOX via multiple biosynthetic pathways [1, 2, 12, 13, 60]. In *A. sylvestris*, AsTJOMT exclusively methylates the 5-hydroxyl group of thujaplicatin, an intermediate of the PTOX biosynthesis pathway [84].

The homologous enzymes, CYP719A23 (from *P. hexandrum*) and CYP719A24 (from *P. peltatum*) participate in the conversion of matairesinol into pluviatolide, a more downstream intermediate of PTOX (Fig. 2), via methylenedioxy bridge formation [63, 64]. Quite recently, six novel genes, which were also detected by

NGS-based transcriptome, have been characterized from *P. hexandrum* and shown to be responsible for the PTOX biosynthesis [23]. CYP71CU1 was found to hydroxydise (−)-5′-desmethoxy-yatein into (−)-5′-desmethyl-yatein followed by *O*–methylation by OMT1to (−)-yatein (Fig. 2). (−)- yatein is converted into (−)-deoxy-podophyllotoxin, which is demethylated to (−)-4′-desmethyl-deoxy-podophyllotoxin by CYP71BE54 (Fig. 2). CYP82D61 was shown to participate in the production of (−)-4′-desmethyl-epipodophyllotoxin via hydroxylation of (−)-4′-desmethyl-deoxy-podophyllotoxin (Fig. 2). Notably, (−)-4′-desmethyl-epipodophyllotoxin, which is an aglycone of an antitumor drug, etoposide, was detected in transgenic tobacco transected with these six genes [23]. Taken into account that (−)-4′-desmethyl-epipodophyllotoxin is synthesized from PTOX in the industrial production of etoposide, this study leads to the development of the novel procedure for the production etoposide using transgenic tobacco as well as explored total biosynthesis pathways of PTOX and its related lignans [23].

Over the past few years, the genomes or transcriptomes of lignan-rich plants including *Linum* [68, 85, 86], *Sesamum* [65–67], and *Podophyllum* [23, 63, 64, 69] have been documented, followed by *in silico* detection of functional genes. Particularly, next-generation sequencers (NGS)-based *de novo* transcriptome has been shown to be a powerful procedure for molecular characterization of lignan-biosynthetic genes at the first step, as described above. These findings are highly likely to remarkably enhance the molecular and functional characterization of unknown lignan biosynthetic enzymes. In addition, it is suggested that a *Podophyllum* endophyte may produce PTOX [87]. NGS analyses of the genome, metagenome, and transcriptome of *Podophyllum* and its endophytes are expected to provide crucial clues to understanding the PTOX biosynthesis pathways.

4 Metabolic Engineering of Lignan Biosynthesis

A growing body of studies has revealed that lignan biosynthesis is altered by genetic modification, light, and elicitors. This section presents an overview and discussion of recent progress in major lignan metabolic engineering using plants, plant cells, and organ cultures.

4.1 Metabolic Engineering of Transgenic Plants and Cells

Stable transfection or gene silencing, namely authentic transgenic metabolic engineering of a lignan biosynthetic enzyme gene is expected to directly alter the lignan production cascades in host plants, organs, and cells. To date, *Forsythia* and *Linum* cell, organ cultures, and plants are attempted to generate the transgenic plants among lignan-rich plants. *Agrobacterium*-based gene introduction was employed for transformation of both *Forsythia* and *Linum*, which is also essentially common to generation of transgenic model plants [17, 77, 88–94]. Figure 3 demonstrates the typical procedure for *Agrobacterium*-based transformation of *Forsythia*. In the

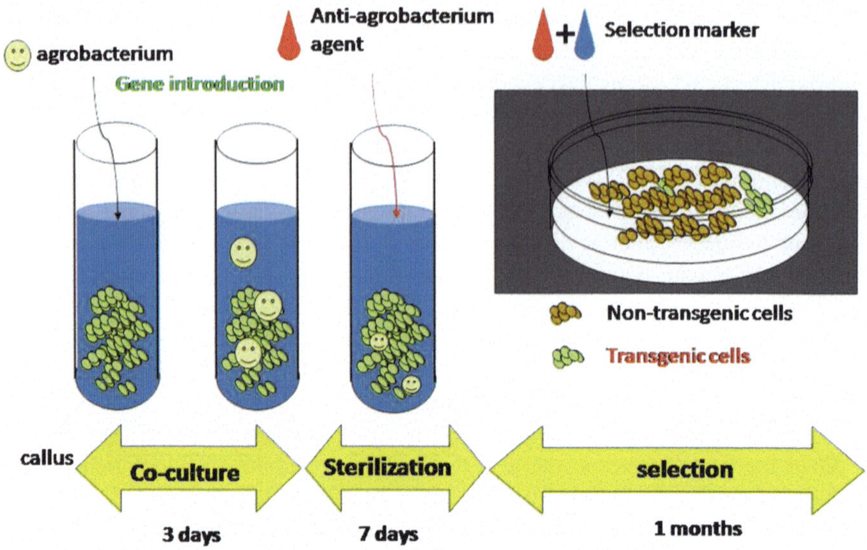

Fig. 3 Scheme for generation of transgenic *Forsythia* mediated by *Agrobacterium*. This process is common for the generation of transgenic *Forsythia* plants and suspension cultures

subsection, we present the recent progress in transformation of *Forsythia* and *Linum* and metabolic engineering of lignan biosynthetic pathways using these plants.

4.1.1 Transgenic *Forsythia* Cells

Forsythia is a perennial plant commonly known as the golden bell flower and is used for a variety of Chinese medicines and health diets [1, 2, 5, 7, 12, 13]. As shown in Fig. 2, *Forsythia* biosynthesizes pinoresinol, lariciresinol, secoisolariciresinol, matairesinol, and arctigenin, with >90% of pinoresinol, >80% of matairesinol, and 40–80% of arctigenin accumulated in glucosylated forms [12, 13, 17, 77, 78]. Seasonal changes in amounts of major *Forsythia* lignans and the relevant gene expression were also reported. All of the lignans in the leaf continuously increased from April to June, reached the maximal level in June, and then decreased [78]. *PLR* was stably expressed from April to August, whereas the *PLR* expression was not detected from September to November [78]. In contrast, the UGT71A18 expression was detected from August to November but not from April to July. The *SIRD* expression was prominent from April to May, not detected in June to July, and then increased again from September to November [78]. These expression profiles of the lignan-synthetic enzymes are essentially correlated with the alteration in lignan amounts.

Several transgenic *Forsythia* plants and cells have been documented for the past 5 years [17, 77, 89]. It is noteworthy that the regeneration efficiency of callus (shoot formation and rooting) and optimal condition for them differ among the variety of *Forsythia* species (*F. koreana*, *F. intermedia*, and *F. suspense*) (Fig. 4). For instance, *F. koreana*, *F. intermedia*, and *F. suspense* explants regenerated more than 100, 36, and 4 shoots per leaf, respectively [89]. Likewise, *F. intermedia* calli, unlike

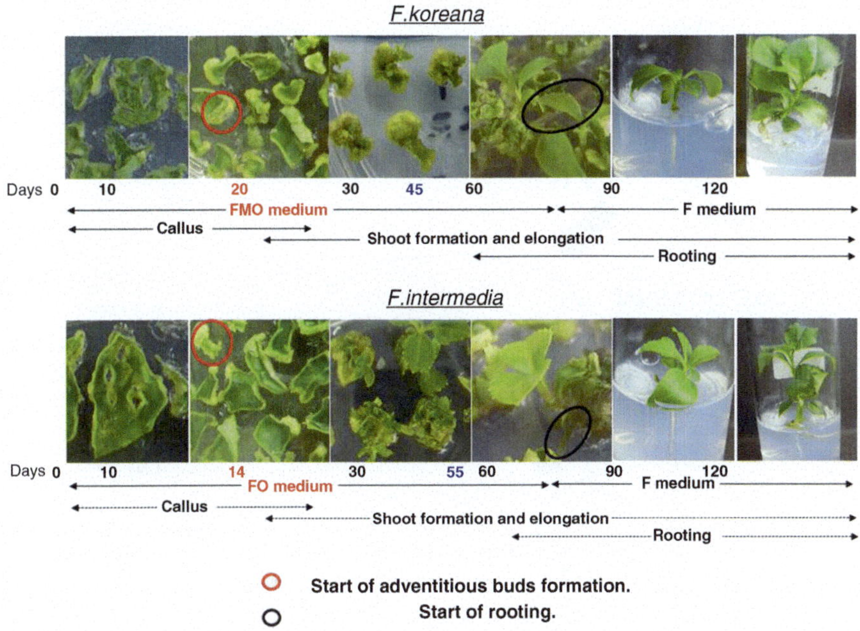

Fig. 4 Different conditions in regeneration between *Forsythia* varieties (*F. koreana* and *F. intermedia*). Note that different media is used for regeneration of the respective *Forsythia* species. Culturing periods also vary between these species

F. koreana, calli, were much more effectively regenerated on the F0 medium than on the FM0 medium [89]. Two transgenic *F. intermedia* and one transgenic *F. koreana* have acquired hygromycin resistance, but none of them have exhibited metabolic alteration in lignan biosynthesis [89]. Moreover, the greatest drawback in generation of transgenic *Forsythia* lies in extremely low transformation efficiencies. These findings strongly suggest the potential requirement for innovation of transgenic *Forsythia* plant generation. In other words, a high-efficient transgenic method for *Forsythia* is expected to remarkably enhance transgenic metabolic engineering-based lignan production using transgenic *Forsythia*.

The transgenic metabolic engineering of *Forsythia* culture cells was originally reported in 2009. *F. koreana* suspension cells stably transfected with a PLR-RNA interference (RNAi) sequence (*PLR*-RNAi) showed complete loss of matairesinol and an approximately 20-fold increase in total pinoresinol (pinoresinol aglycone and glucoside), compared with the wild type cells [77]. Furthermore, *Forsythia* transgenic cells CPi-Fk, which are stably double transfected with *PLR*-RNAi and the sesamin-producing enzyme *CYP81Q1* (Fig. 2), produced sesamin (0.01 mg/g dry weight of the cell [DW]) (Fig. 5), although sesamin is not biosynthesized in native *Forsythia* [77]. This is the first success in the metabolic engineering leading to an exogenous lignan using transgenic plant cells, demonstrating that the *Forsythia* cell culture system is an efficient and promising platform for producing both endogenous and exogenous lignans by transgenic metabolic engineering. A striking feature is

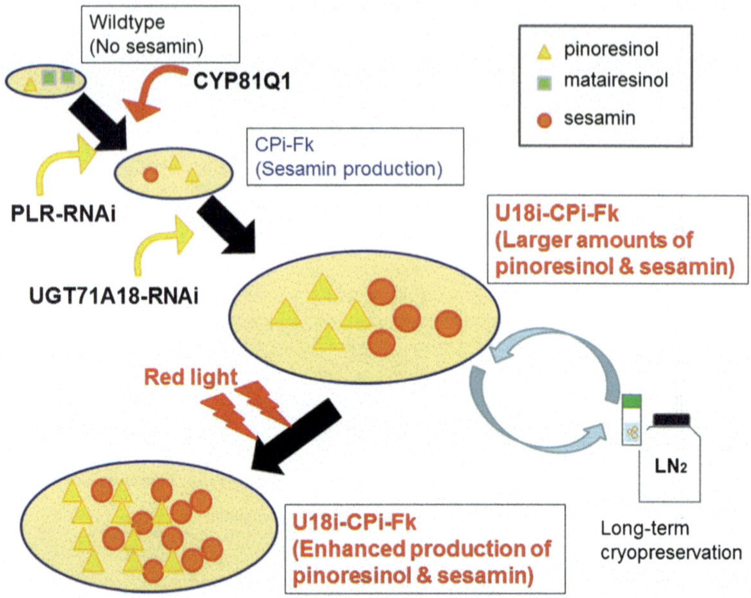

Fig. 5 Metabolic engineering of *Forsythia* suspension cell cultures. The double-transgenic *Forsythia* suspension cells, CPi-Fk, acquired the ability to produce sesamin by stable transfection of *PLR*-RNAi and an exogenous (*Sesamum*) *CYP81Q1* gene. The triple-transgenic cells, U18i-CPi-Fk, were generated by the introduction of UGT71A18-RNAi into CPi-Fk and exhibit higher productivity of pinoresinol and sesamin than CPI-Fk. The lignan productivity is approximately three- to fivefold upregulated under red LED. U18i-CPi-Fk can also be stocked in liquid nitrogen for a long period, and re-thawed U18i-CPi-Fk exhibit as high productivity of sesamin as noncryopreserved U18i-CPi-Fk

that light irradiation has been shown to improve the production of both endogenous and exogenous lignans by CPi-Fk cells. Irradiation of CPi-Fk cells for 2 weeks with white fluorescent, blue LED, and red LED light increased sesamin production 2.3-fold, 2.7-fold, and 1.6-fold, respectively, compared with cells cultured in the dark [94]. Likewise, irradiation of CPi-Fk cells increased pinoresinol (aglycone and glucoside) production 1.5- to 3.0-fold [94]. Intriguingly, expression of the pinoresinol-glucosylating enzyme UGT71A18 was also downregulated in CPi-Fk cells under blue LED or red LED light [94]. This reduction of the expression of UGT71A18 is also likely to contribute to the increase of sesamin production [94], given that pinoresinol glucoside is not metabolized into sesamin by CYP81Q1 [12, 75], and 90% of pinoresinol is glucosylated in *Forsythia* wildtype cells [12, 13, 17, 77, 78]. In other words, these findings suggested that suppression of UGT71A18 by RNAi might contribute to an increase in productivity of pinoresinol and sesamin in CPi-Fk. This presumption was substantiated in our subsequent study. Quite recently, we have created more efficient, stable, and sustainable sesamin production system using triple-transgenic *Forsythia koreana* cell suspension cultures, U18i-CPi-Fk, compared to CPi-Fk [17]. These transgenic cells were generated by stable transfection of CPi-Fk with an RNAi sequence against the pinoresinol-glucosylating enzyme UGT71A18. *UGT71A18* expression was not detected in the triple-transgenic *Forsythia* cells [17]. Moreover, U18i-CPi-Fk accumulated approximately fivefold higher amounts of pinoresinol aglycone than CPi-Fk, and the ratio of pinoresinol aglycone to total pinoresinol in U18i-CPi-Fk is 81.81 ± 6.43%, which is approximately 6.5-fold greater than that in CPi-Fk (13.19 ± 2.35%). These results proved that UGT71A18-RNAi contributed a great deal to the increase in the ratio of pinoresinol aglycone to total pinoresinol. Notably, U18i-CPi-Fk has also been shown to display 1.4-fold higher production of sesamin than CPi-Fk [17], confirming that the suppression of UGT71A18 gene expression is effective for improvement of the sesamin production. Furthermore, pinoresinol aglycone was 3.4-fold and 2.8-fold greater produced under white fluorescent and red LED, respectively, than under the dark condition. Consistently, sesamin production in U18i-CPi-Fk was approximately threefold (31.02 ± 3.45 μg/g DW) upregulated specifically under red LED, whereas white fluorescent or blue light failed to affect sesamin production [17]. It should be noteworthy that the light types effective for the sesamin production differed between CPi-Fk and U18i-CPi-Fk; the sesamin production was potentiated exclusively by blue LED light in CPi-Fk [94], whereas red LED light upregulated the sesamin production in U18i-CPi-Fk [17]. The molecular mechanism underlying such different sensitivity of these transgenic *Forsythia* cells remains unclear, but the suppression of UGT71A18 gene is likely to alter other biosynthetic pathways than pinoresinol glucosylation, which ultimately may affect light sensitivity of the sesamin production. In addition, upregulation of lignan production by light was also observed in other natural plants or cells. In *Linum* species, suspension of *L. album* cells produced twofold more PTOX under red light than those in the dark [95]. Irradiation of *S. indicum* leaves 3–5 weeks after sowing with blue LED light increased sesamin content twofold, compared with white fluorescent light, whereas irradiation with red LED light reduced sesamin content twofold [9, 96]. In combination, these findings highlight the different specificity of light types to lignan production among lignan compounds and host plant species.

U18i-CPi-Fk has also been found to possess another prominent advantage over CPi-Fk, that is, long-term and reproducible storage [17]. Universal procedures for long-term stock of plant cell cultures, unlike those of seeds, bacteria, or animal cells, have not been well established, and cryopreservation procedures for a particular plant species are not always applicable to other plant cells [97–99]. Indeed, we failed to establish any procedure for long-term stock of CPi-Fk, and thus observed a decrease in the growth rate of CPi-Fk cells after 2 years of culture and, eventually, proliferation loss. Nevertheless, we have developed a procedure for sodium alginate-based long-term storage of U18i-CPi-Fk in liquid nitrogen [17]. Moreover, production of sesamin in U18i-CPi-Fk re-thawed after 6-month cryopreservation was equivalent to that of noncryopreserved U18i-CPi-Fk, proving the reproducible functionalities of U18i-CPi-Fk [17]. Altogether, the high lignan (pinoresinol and sesamin) productivity and establishment of the freeze stocks of U18i-CPi-Fk endorses the marked usefulness of U18i-CPi-Fk as a stable and sustainable platform of lignan production (Fig. 5).

4.1.2 Transgenic *Linum*

Linum spp. (flax, Linaceae) are annual flowering plants comprised of approximately 200 species. This genus has received pharmaceutical and medicinal attention due to the presence of various lignans, including PTOX and its related compounds, which are practically applied for the semisynthesis of antitumor drugs for breast and testicular cancers described above. Since *Linum* is also known to biosynthesize PTOX and its derivatives, and the procedures for tissue and cell culture are well established, optimal conditions and stimulating factors for production of various lignans, including (−)-podophyllotoxin, by *Linum* calli, suspension cell cultures, and roots have been extensively investigated as described in the followings. Recently, flax seeds and oils have also received attentions as functional foods due to the contents of lignans beneficial for human health [14, 15]. Accordingly, metabolic engineering of *Linum* is also highly likely to contribute a great deal to the development of novel lignan production and transgenic foods.

In various *Linum* species, the effects of RNAi on production of endogenous lignans via metabolic engineering were investigated. *PLR*-RNAi-transgenic plants of *L. usitatissimum* showed the high accumulation of pinoresinol diglucoside and loss of SDG in the seed coat [90]. Intriguingly, these *PLR*-RNAi-transgenic *L. usitatissimum* plants produced the 8–5′ linked neolignans, dehydrodiconifnyl alcohol and dihydro-dehydrodiconifnyl alcohol, while these neolignans were not biosynthesized in the wildtype plants [90]. These findings indicate that RNAi occasionally affects some biosynthetic pathways in an indirect fashion.

4.1.3 Transient Transformation of *Linum*

Hairy roots of *L. perenne* transiently transfected with *PLR*-RNAi reduced the production of the major endogenous lignan, justicidin B, to 25%, compared with the untreated hairy roots [72]. Likewise, transient transfection of *L. corymbulosum* hairy roots with PLR-RNAi resulted in a marked reduction of hinokinin [73]. Combined with the justicidin B and hinokinin biosynthetic pathways, in which PLR converts pinoresinol into secoisolariciresinol (Fig. 2), these findings indicate that PLR-directed conversion of pinoresinol into secoisolariciresinol is a rate-limiting step of justicidin B and hinokinin biosynthesis, at least in the hairy roots of *L. perenn and L. corymbulosum,* respectively. Therefore, identification and genetic manipulation of justicidin B and hinokinin synthase will contribute a great deal to the establishment of procedures for the direct metabolic engineering of these lignans. Taken together, these findings reinforce the potential of *Forsythia* and *Linum* transgenic or transiently gene-transfected cells and plants as the metabolic engineering-based platforms for on-demand production of both endogenous and exogenous lignans. The draft genome and transcriptome of *L. usitatissimum* [68, 85, 86] will also accelerate the identification of the enzymes involved in the biosynthesis of *Linum* lignans, leading to the efficient lignan production using gene-modified plant sources.

4.1.4 What Should Be Considered for Lignan Production via Metabolic Engineering Using Gene-Modified Plants?

To establish gene-modified plant platforms for lignan production, we should consider two crucial factors: the type of host and the use of transgenic or transiently transfected hosts. Host types can be classified into plants, organs, and cell cultures. For example, although the amount of sesamin produced by U18i-CPi-Fk cells is ~100-fold lower than that by native sesame seeds, U18i-CPi-Fk-based lignan metabolic engineering has several advantages. Furthermore, the *Forsythia* transgenic cells are propagated tenfold in 2 weeks in standard culture medium [17] and can be cultivated at all times and locations which is also endorsed by the fact that U18i-CPi-Fk can be stocked for a long time [17]. In contrast, sesame seeds are cultivated in limited regions only once a year. Moreover, the conditions used in the culturing of U18i-CPi-Fk cells, including temperature, light wavelength and intensity, and medium components, can be altered to optimize sesamin production [17, 77]. *Forsythia* plants have much greater biomass, with higher amount of lignans, than suspension cell cultures, and these plants can grow and propagate from small explants without flowering or seed formation [89]. However, efficient generation of transgenic *Forsythia* plants still requires further basic research due to the markedly low transformation efficiency by any known gene transfection methods [89]. In addition, as mentioned above, the optimal culturing and regeneration conditions were found to vary among *Forsythia* species [89]. In other words, the development of a procedure for efficient generation of transgenic *Forsythia* would surely enhance novel lignan production system.

The generation of both stable (namely transgenic) and transient transfectants of *Linum* species are well established [72, 73, 90, 93], and thus the amounts of precursors or intermediates of targeted lignans are major determinants for the employment of cell cultures, organ cultures, or plants as host platforms. Additionally, gene-modified host plants may fail to normally grow or to produce lignans of interest due to cytotoxicity of lignans, although the underlying molecular mechanisms have not fully been elucidated. Therefore, generation of lignan-producing plants using multiple plant species is occasionally required.

The second factor involves construction of either transgenic or transiently transfected hosts. Transgenic plants and cell cultures, once generated, are sustainably used for lignan production and readily upscaled, whereas generation of transgenic plants, in particular nonmodel plants, may be time- and cost-consuming. Moreover, cultivation of transgenic plants in general requires a closed facility for transgenic plants. Transiently transfected plants require repeated transfections, and transient transfection of multiple genes may dramatically decrease the transfection efficiency. Furthermore, massive transient transfection methods for industrial use remains to be fully developed [100]. Further research on lignan metabolic engineering, using transgenic or transiently gene-transfected plants, organ cultures, and cell cultures, is expected to lead to the establishment of both universal and molecular species-specific strategies for gene-regulated metabolic engineering of lignan biosynthesis pathways.

4.2 Metabolic Engineering by Elicitation

Plant defense systems are triggered upon injury or infection via signaling by the phytohormones, methyl jasmonate (MeJA) and salicylic acid (SA), and treatment with elicitors, including fungi, their extracts, and the glycan components, MeJA and SA, also mimic such activation. Moreover, lignans, at least in part, are likely to be implicated with host defense systems [12, 13, 22, 101]. In combination, elicitors are expected to enhance lignan biosynthesis [13, 22, 102]. As summarized in Table 1, the effects of various elicitors on lignan production have been examined in a wide variety of cell cultures and hairy roots of *Forsythia, Juniperus,* and *Podophyllum* (Table 1).

MeJA and SA were found to increase the production of PTOX and the structurally related lignan production or the gene expression of lignan biosynthetic enzymes responsible for biosynthesis of conifenyl alcohol, phenylalanine ammonialyase (PAL), cinnamoyl-CoA reductase (CCR), and cinnamylalcohol dehydrogenase (CAD) in cell suspension cultures of *L. album* [103, 104] and *L. nodiflorum* [103], *Podophyllum hexandrum* [105], and callus of *L. austriacum* callus culture [106]. These phytohormones also increased the PTOX production or the relevant gene expression in hairy roots of *L. tauricum* [107]. Additionally, an increase in production of pinoresinol and matairesinol by MeJA was observed in *Forsythia intermedia* cell suspension culture [108].

Chitosan, chitin oligomers, and other glycans also enhanced PTOX production or gene expression of lignan biosynthetic enzymes in *Juniperus chinensis* callus culture [109], *L. austriacum* callus culture [106], and *L. album* cell suspension culture and hairy roots [110–112]. In particular, comparisons of chitin tetramer, pentamer, and hexamer and chitosan tetramer and pentamer showed that treatment of *L. album* hairy roots with chitosan hexamer for 5 days most potently enhanced PTOX and lariciresinol production, as well as upregulating the expression of *PAL, CCR, CAD,* and *PLR* genes [112]. In summary, treatment with these elicitors resulted in two- to sevenfold increase in PTOX synthesis and expression of genes encoding enzymes involved in the early steps of lignan biosynthesis in various plant cells and hairy roots.

Fungal co-culturing, extracts, and filtrate exhibited unique effects on the metabolic engineering of lignan production (Table 1). *Botrytis cinerea, Phoma exigua,* and *Fusarium oxysporum* extracts triggered the accumulation of monolignols and enhanced PAL activity and gene expression of *PAL, CCR,* and *CAD* in *L. usitatissimum* cell suspension cultures [113]. Treatment of *L. album* cell cultures with *Fusarium graminearum* extract for 5 days increased PTOX sevenfold and PAL, CCR, and CAD mRNAs > tenfold compared with untreated cells. These results confirmed that this extract is a more potent elicitor of PTOX production and PAL, CCR, and CAD expression than treatment with chitosan, chitin, or MeJA treatment for 3 days [110, 111, 114].

Rhizopus stolonifer and *Rhizoctonia solani* extract stimulated 8.8-fold and 6.7-fold greater accumulation of lariciresinol, instead of PTOX, in *L. album* cell cultures after 5-days treatment as compared with untreated cells, and the highest (6.5-fold)

Table 1 List of major elicitors and their effects on lignan biosynthesis

Elicitor	Target	Effect	Refs
Chito-oligosaccharides (1 mg)	*Juniperus chinensis* callus culture	Increased PTOX production	[104]
Methyl jasmonate (MeJA) (100 μM)	*Forsythia intermedia* cell suspension culture	Increased pinoresinol and matairesinol production	[103]
Mannan (0.1 mg mL^{-1}) β-1,3-glucan (0.1 mg mL^{-1}) Ancymidol (10^{-7} M)	*L. austriacum* callus culture	Enhanced activity of tyrosine ammonia-lyase (TAL), coumarate 3-hydroxylase (C3H), polyphenoloxidase (PPO) and PAL Increased PTOX, 6-MPTOX, dPTOX, α- and β-peltatins production Increased PTOX and α-peltatins production Increased PTOX, 6-MPTOX, dPTOX and α- peltatins production	[101]
Indanoyl-isoleucine (5–100 μM) Coronalon, (10–50 μM) MeJA(100 μM)	*L. nodiflorum* cell suspension culture	Increased deoxypodophyllotoxin production Enhanced activity of 6-hydroxylase and β-peltatin 6-*O*-methyltransferase, Increased 6-MPTOX and 5′-d-6-MPTOX production	[97]
MeJA (100 μM)	*L. album* cell suspension culture	Increased PTOX production	[98]
Botrytis cinerea extract (3% v/v) *Phoma exigua* extract (3% v/v) *Fusarium oxysporum* extract (3% v/v)	*L. usitatissimum* cell suspension culture	Rapid stimulation of the monolignol pathway, enhanced PAL activity, and expression of genes encoding PAL, CCR, and CAD	[108]
MeJA(50–200 μM)	*L. tauricum* hairy root culture	Increased 6MPTOX and 4′-DM6MPTOX production	[102]
Salicylic acid (SA) (10 μM)	*L. album* cell suspension culture	Enhanced *PAL*, *CCR*, and *CAD* gene expression and PTOX production	[99]
Chitin (100 mg l^{-1}) Chitosan (100–200 mg l^{-1}) MeJA (100–200 μM)	*L. album* cell suspension culture	Increased lariciresinol and/or PTOX production	[105]
Fusarium graminearum extract(1%v/v) *Sclerotinia sclerotiorum* extract (1%v/v) *Rhizopus stolonifer*	*L. album* cell suspension culture	Enhanced *PAL*, *CCR*, *CAD*, and *PLR* gene expression Increased PTOX and lariciresinol production	[105, 109]

(*continued*)

Table 1 (continued)

Elicitor	Target	Effect	Refs
extract(1% v/v) *Rhizoctonia solani* extract(1% v/v)			
MeJA(10–100 μM)	*Podophyllum hexandrum* cell suspension culture	Changes in cell proteome Increased PTOX production	[109]
Fusarium graminearum extract (1%v/v) *Sclerotinia sclerotiorum* extract (1%v/v) *Trichoderma viride* extract (1%v/v) Chitosan (100 mg l^{-1})	*L. album* hairy root culture	Enhanced *PAL*, *CCR*, *CAD*, and *PLR* gene expression Increased PTOX, 6MPTOX, and lariciresinol production	[106]
Chitosan and chitin oligomers (100 mg l^{-1})	*L. album* cell suspension culture	Enhanced *PAL*, *CCR*, *CAD*, and *PLR* gene expression Increased PTOX, 6MPTOX, and lariciresinol production	[107]
Fusarium graminearum culture filtrate (1% v/v)	*L. album* cell suspension culture	Increased phenolic compound, PTOX, and lariciresinol production Enhanced PAL activity	[110]

PLR gene induction was observed in *L. album* cell cultures treated with *Rhizopus stolonifer* extract for 2 days [114]. Similar data were obtained in *L. album* hairy roots with the same fungal extracts [111] or *L. album* cell suspension culture with *Fusarium graminearum* culture filtrate [115], but the latter manifested less lignan production. These studies revealed that fungal extract exhibited the species-specific effects on the lignan biosynthesis pathways, although investigation of the molecular basis awaits further study.

As described above, the regulation of gene expression has thus far been restricted to enzymes responsible for the upstream of lignan biosynthesis pathways. Therefore, the effects of these elicitors on lignans and the relevant biosynthetic genes downstream of PLR, such as SIRD or 719A23 (Fig. 2), would provide a clue to understanding the molecular mechanisms underlying upregulation of PTOX production and to identifying more effective elicitors for lignan production.

5 Conclusions

In this chapter, we have provided diverse recent advances in metabolic engineering for lignan production by plants, including: (i) the molecular characterization of novel genes encoding enzymes for biosynthesis pathways of dietary and medicinal lignans;

(ii) the production of both endogenous and exogenous lignans by transient or stable transfection of lignan biosynthetic genes into cultured cells, tissues, and plants; (iii) the long-term stock and following reproduction of the cell functionality of a transgenic *Forsythia* lignan producing cells, U18i-CPI-Fk; and (iv) the upregulation of productivity of lignans in cells and plants by exogenous stimuli such as light and elicitors in a plant species- and lignan-specific fashion. Taken together, combination of transgenesis, light, and elicitors will be a promising strategy for further improvement of the lignan productivity. For example, elicitation of U18i-CPi-Fk under red LED light is expected to increase the amounts of sesamin and/or pinoresinol. Moreover, bioinformatic integration of the aforementioned experimental data is likely to enable the systematic prediction of optimal lignan production strategy: hosts (cells, organ cultures, plants), light conditions, elicitor types, and transfection types. For example, three *Forsythia* varieties, *F. koreana*, *F. intermedia,* and *F. suspensa*, displayed differential growth and regeneration in a medium component-dependent fashion or selection marker antibiotics-dependent fashions [89], and *Linum* spp. showed genus-specific sensitivities to different elicitors (Table 1).

Public acceptance of transgenic dietary products is not yet sufficient all over the world. Nevertheless, it should be noted that lignans produced by transgenic hosts are chemically identical to natural ones and free from any recombinant genes or proteins. Thus, public acceptance for lignans produced by transgenic plants should also be more easily garnered than that for transgenic foods themselves. In this context, we will pay more attention to the establishment of scaling-up and following industrialization of the lignan production systems as well as the development of efficient generation of transgenic plants in the near future [116–118]. Large-scale lignan production by transgenic plants must be carried out in a closed cultivation system to prevent contamination of the environment by transgenic plants. Recently, various closed plant factories have been emerging, which completely shut off a gene flow into the outer environment and enables the transgenic plants-based molecular breeding of genes or compounds of interest under optimal and sterile conditions [116–118]. Such advances in the metabolic engineering of lignan biosynthesis will surely pave the way for the conversion of conventional agricultural lignan production to innovative industrial production of various beneficial lignans and, ultimately, contribute a great deal to the improvement of quality of life and national financial burdens for medical care via extension of our healthy life expectancy owing to the preventive effects of lignans on diverse diseases.

Acknowledgments This work was, in part, supported by the Plant Factory Project of the Ministry of Economy, Technology, and Industry of Japan.

References

1. Umezawa T (2003) Diversity in lignan biosynthesis. Phytochem Rev 2:371–390
2. Suzuki S, Umezawa T (2007) Biosynthesis of lignans and norlignans. J Wood Sci 53:273–284

3. Macías FA, López A, Varela RM, Torres A, Molinillo JMG (2004) Bioactive lignans from a cultivar of *Helianthus annuus*. J Agric Food Chem 52:6443–6447
4. Lee J, Choe E (2006) Extraction of lignan compounds from roasted sesame oil and their effects on the autoxidation of methyl linoleate. J Food Sci 71:C430–C436
5. Guo H, Liu A-H, Ye M, Yang M, Guo D-A (2007) Characterization of phenolic compounds in the fruits of *Forsythia suspense* by high-performance liquid chromatography coupled with electrospray ionization tandem mass spectrometry. Rapid Commun Mass Spectrom 21:715–729
6. Peñalvo JL, Adlercreutz H, Uehara M, Ristimaki A, Watanabe S (2008) Lignan content of selected foods from Japan. J Agric Food Chem 56:401–409
7. Piao X-L, Jang M-H, Cui J, Piao X (2008) Lignans from the fruits of *Forsythia suspensa*. Bioorg Med Chem Lett 18:1980–1984
8. Hata N, Hayashi Y, Okazawa A, Ono E, Satake H, Kobayashi A (2010) Comparison of sesamin contents and CYP81Q1 gene expressions in aboveground vegetative organs between two Japanese sesame (*Sesamum indicum* L.) varieties differing in seed sesamin contents. Plant Sci 178:510–516
9. Hata N, Hayashi Y, Ono E, Satake H, Kobayashi A, Muranaka T, Okazawa A (2013) Differences in plant growth and leaf sesamin content of the lignan-rich sesame variety "Gomazou" under continuous light of different wavelengths. Plant Biotechnol 30:1–8
10. Okazawa A, Hori K, Okumura R, Izumi Y, Hata N, Bamba T, Fukusaki E, Ono E, Satake H, Kobayashi A (2011) Simultaneous quantification of lignans in *Arabidopsis thaliana* by highly sensitive capillary liquid chromatography-electrospray ionization-ion trap mass spectrometry. Plant Biotechnol 28:287–293
11. Schmidt TJ, Klaes M, Sendker J (2012) Lignans in seeds of *Linum* species. Phytochemistry 82:89–99
12. Satake H, Ono E, Murata J (2013) Recent advances in metabolic engineering of lignan biosynthesis pathways for the production of transgenic plant-based foods and supplements. J Agric Food Chem 61:11721–11729
13. Satake H, Koyama T, Bahabadi SE, Matsumoto E, Ono E, Murata J (2015) Essences inmetabolic engineering of lignan biosynthesis. Metabolites 5:270–290
14. Kajla P, Sharma A, Sood DR (2015) Flaxseed-a potential functional food source. J Food Sci Technol 52:1857–1871
15. Goyal A, Sharma V, Upadhyay N, Gill S, Sihag M (2014) Flax and flaxseed oil: an ancient medicine & modern functional food. J Food Sci Technol 51:1633–1653
16. Chaurasia OP, Ballabh B, Tayade A, Kumar R, Kumar GP, Singh SB (2012) *Podophyllum* L.: an endangered and anticancerous medicinal plant–an overview. Indian J Tradit Knowl 11:234–241
17. Murata J, Matsumoto E, Morimoto K, Koyama T, Satake H (2015) Generation of triple-transgenic *Forsythia* cell cultures as a platform for the efficient, stable, and sustainable production of lignans. PLoS One 10:e0144519
18. Ionkova I (2007) Biotechnological approaches for the production of lignans. Phcog Rev 1:57–68
19. Ionkova I, Antonova I, Momekov G, Fuss E (2010) Production of podophyllotoxin in *Linum linearifolium* in vitro cultures. Pharmacogn Mag 6:180–185
20. Ionkova I (2011) Anticancer lignans – from discovery to biotechnology. Mini Rev Med Chem 11:843–856
21. Lata H, Mizuno CS, Moraes RM (2009) The role of biotechnology in the production of the anticancer compound podophyllotoxin. Methods Mol Biol 547:387–402
22. Malik S, Biba O, Grúz J, Arroo RRJ, Strnad M (2014) Biotechnological approaches for producing aryltetralin lignans from *Linum* species. Phytochem Rev 13:893–913
23. Lau W, Sattely ES (2015) Six enzymes from mayapple that complete the biosynthetic pathway to the etoposide aglycone. Science 349:1224–1228

24. Oliva A, Moraes RA, Watson B, Duke SO, Dayan FE (2002) Aryltertralin lignans inhibit plant growth by affecting formation of mitotic microtubular organizing centers. Pestic Biochem Phys 72:45–54
25. Harmatha J, Dinan L (2003) Biological activities of lignans and stilbenoids associated with plant-insect chemical interaction. Phytochem Rev 2:321–330
26. Schroeder FC, del Campo ML, Grant JB, Weibel DB, Smedley SR, Bolton KL, Meinwald J, Eisner T (2006) Pinoresinol: a lignol of plant origin serving for defense in a caterpillar. Proc Natl Acad Sci U S A 103:15497–15501
27. Cutillo F, D'Abrosca B, DellaGreca M, Fiorentino A, Zarrelli A (2003) Lignans and neolignans from *Brassica fruticulosa*: effects on seed germination and plant growth. J Agric Food Chem 51:6165–6172
28. Nishiwaki H, Kumamoto M, Shuto Y, Yamauchi S (2011) Stereoselective syntheses of allstereoisomers of lariciresinol and their plant growth inhibitory activities. J Agric Food Chem 59:13089–13095
29. Carillo P, Cozzolino C, D'Abrosca B, Nacca F, DellaGreca M, Fiorentio A, Fuggi A (2011) Effects of the allelochemicals dihydrodiconiferylalcohol and lariciresinol on metabolism of *Lactuca sativa*. Open Bioact Compd J 3:18–24
30. Heinonen S, Nurmi T, Liukkonen K, Poutanen K, Wähälä K, Deyama T, Nishibe S, Adlercreutz H (2001) In vitro metabolism of plant lignans: new precursors of mammalian lignans enterolactone and enterodiol. J Agric Food Chem 49:3178–3186
31. Lampe JW, Atkinson C, Hullar MA (2006) Assessing exposure to lignans and their metabolites in humans. J AOAC Int 89:1174–1181
32. Liu Z, Saarinen NM, Thompson LU (2006) Sesamin is one of the major precursors of mammalian lignans in sesame seed (*Sesamum indicum*) as observed in vitro and in rats. J Nutr 136:906–912
33. Mueller SO, Simon S, Chae K, Metzler M, Korach KS (2004) Phytoestrogens and their human metabolites show distinct agonistic and antagonistic properties on estrogenreceptor alpha (ERα) and ERβ in human cells. Toxicol Sci 80:14–25
34. Penttinen P, Jaehrling J, Damdimopoulos AE, Inzunza J, Lemmen JG, van der Saag P, Pettersson K, Gauglitz G, Mäkelä S, Pongratz I (2007) Diet-derived polyphenol metabolite enterolactone is a tissue-specific estrogen receptor activator. Endocrinology 148:4875–4886
35. During A, Debouche C, Raas T, Larondelle Y (2012) Among plant lignans, pinoresinolhas the strongest antiinflammatory properties in human intestinal Caco-2 cells. J Nutr 142:1798–1805
36. Adlercreutz H (2007) Lignans and human health. CRC Crit Rev Clin Lab Sci 44:483–525
37. Bergman Jungeström M, Thompson LU, Dabrosin C (2007) Flaxseed and its lignansinhibit estradiol-induced growth, angiogenesis, and secretion of vascular endothelial growth factor in human breast cancer xenografts in vivo. Clin Cancer Res 13:1061–1067
38. Power KA, Saarinen NM, Chen JM, Thompson LU (2006) Mammalian lignans enterolactone and enterodiol, alone and in combination with the isoflavone genistein, do not promote the growth of MCF-7 xenografts in ovariectomized athymic nude mice. Int J Cancer 118:1316–1320
39. Mense SM, Hei TK, Ganju RK, Bhat HK (2008) Phytoestrogens and breast cancer prevention: possible mechanisms of action. Environ Health Perspect 116:426–433
40. Saarinen NM, Wärri A, Dings RPM, Airio M, Smeds AI, Mäkelä S (2008) Dietary lariciresinol attenuates mammary tumor growth and reduces blood vessel density inhuman MCF-7 breast cancer xenografts and carcinogen-induced mammary tumors in rats. Int J Cancer 123:1196–1204
41. Adolphe JL, Whiting SJ, Juurlink BHJ, Thorpe LU, Alcorn J (2010) Health effects with consumption of the flax lignan secoisolariciresinol diglucoside. Br J Nutr 103:929–938
42. Barre DE, Mizier-Barre KA, Stelmach E, Hobson J, Griscti O, Rudiuk A, Muthuthevar D (2012) Flaxseed lignan complex administration in older human type 2diabetics manages central obesity and prothrombosis-an invitation to further investigation into poly pharmacy reduction. J Nutr Metab 585170

43. Hano C, Renouard S, Molinié R, Corbin C, Barakzoy E, Doussot J, Lamblin F, Lainé E (2013) Flaxseed (*Linum usitatissimum* L.) extract as well as(+)-secoisolariciresinol diglucoside and its mammalian derivatives are potentinhibitors of α-amylase activity. Bioorg Med Chem Lett 23:3007–3012
44. Sirato-Yasumoto S, Katsuta M, Okuyama Y, Takahashi Y, Ide T (2001) Effect of sesame seeds rich in sesamin and sesamolin on fatty acid oxidation in rat liver. J Agric Food Chem 49:2647–2651
45. Nakano D, Itoh C, Takaoka M, Kiso Y, Tanaka T, Matsumura Y (2002) Antihypertensive effect of sesamin. IV. Inhibition of vascular superoxide production by sesamin. Biol Pharm Bull 25:1247–1249
46. Nakai M, Harada M, Nakahara K, Akimoto K, Shibata H, Miki W, Kiso Y (2003) Novel antioxidative metabolites in rat liver with ingested sesamin. J Agric Food Chem 51:1666–1670
47. Akimoto K, Kitagawa Y, Akamatsu T, Hirose N, Sugano M, Shimizu S, Yamada H (1993) Protective effects of sesamin against liver damage caused by alcohol or carbon tetrachloride in rodents. Ann Nutr Metab 37:218–224
48. Tada M, Ono Y, Nakai M, Harada M, Shibata H, Kiso Y, Ogata T (2013) Evaluation of antioxidative effects of sesamin on the in vivo hepatic reducing abilities by a radiofrequency ESR method. Anal Sci 29:89–94
49. Liu C-M, Zheng G-H, Ming Q-L, Cheng C, Sun J-M (2013) Sesamin protects mouse liver against nickel-induced oxidative DNA damage and apoptosis by the PI3K/Akt pathway. J Agric Food Chem 61:1146–1154
50. Saarinen NM, Wärri A, Airio M, Smeds A, Mäkelä S (2007) Role of dietary lignans inthe reduction of breast cancer risk. Mol Nutr Food Res 51:857–866
51. Velentzis LS, Cantwell MM, Cardwell C, Keshtgar MR, Leathem AJ, Woodside JV (2009) Lignans and breast cancer risk in pre- and post-menopausal women: meta-analysesof observational studies. Br J Cancer 100:1492–1498
52. Velentzis LS, Keshtgar MR, Woodside JV, Leathem AJ, Titcomb A, Perkins KA, Mazurowska M, Anderson V, Wardell K, Cantwell MM (2011) Significant changes in dietary intake and supplement use after breast cancer diagnosis in a UK multicentre study. Breast Cancer Res Treat 128:473–482
53. Buck K, Zaineddin AK, Vrieling A, Linseisen J, Chang-Claude J (2010) Meta-analysesof lignans and enterolignans in relation to breast cancer risk. Am J Clin Nutr 92:141–153
54. Buck K, Zaineddin AK, Vrieling A, Heinz J, Linseisen J, Flesch-Janys D, Chang-Claude J (2011) Estimated enterolignans, lignan-rich foods, and fibre in relationto survival after postmenopausal breast cancer. Br J Cancer 105:1151–1157
55. Zaineddin AK, Buck K, Vrieling A, Heinz J, Flesch-Janys D, Linseisen J, Chang-Claude J (2012) The association between dietary lignans, phytoestrogen-rich foods, and fiber intake and postmenopausal breast cancer risk: a German case–control study. Nutr Cancer 64:652–665
56. Buck K, Vrieling A, Zaineddin AK, Becker S, Hüsing A, Kaaks R, Linseisen J, Flesch-Janys D, Chang-Claude J (2011) Serum enterolactone and prognosis of postmenopausal breast cancer. J Clin Oncol 29:3730–3738
57. Chen JM, Saggar JK, Corey P, Thompson LU (2009) Flaxseed and pure secoisolariciresinol diglucoside, but not flaxseed hull, reduce human breast tumor growth (MCF-7) in athymic mice. J Nutr 139:2061–2066
58. Truan JS, Chen JM, Thompson LU (2012) Comparative effects of sesame seed lignan and flaxseed lignan in reducing the growth of human breast tumors (MCF-7) at high levels of circulating estrogen in athymic mice. Nutr Cancer 64:65–71
59. Yousefzadi M, Sharifi M, Behmanesh M, Moyano E, Bonfill M, Cusido RM, Palazon J (2010) Podophyllotoxin: current approaches to its biotechnological production and future challenges. Eng Life Sci 10:281–292
60. Davin LB, Lewis NG (2003) A historical perspective on lignan biosynthesis: monolignol, allylphenol and hydroxycinnamic acid coupling and downstream metabolism. Phytochem Rev 2:257–288

61. Ono E, Nakai M, Fukui Y, Tomimori N, Fukuchi-Mizutani M, Saito M, Satake H, Tanaka T, Katsuta M, Umezawa T, Tanaka Y (2006) Formation of two methylenedioxy bridges by a *Sesamum* CYP81Q protein yielding a furofuran lignan, (+)-sesamin. Proc Natl Acad Sci U S A 103:10116–10121
62. Noguchi A, Fukui Y, Iuchi-Okada A, Kakutani S, Satake H, Iwashita T, Nakao M, Umezawa T, Ono E (2008) Sequential glucosylation of a furofuran lignan, (+)-sesaminol, by *Sesamum indicum* UGT71A9 and UGT94D1. Plant J 54:415–427
63. Marques JV, Kim K-W, Lee C, Costa MA, May GD, Crow JA, Davin LB, Lewis NG (2013) Next generation sequencing in predicting gene function in podophyllotoxin biosynthesis. J Biol Chem 288:466–479
64. Marques JV, Dalisay DS, Yang H, Lee C, Davin LB, Lewis NG (2014) A multi-omics strategy resolves the elusive nature of alkaloids in *Podophyllum* species. Mol BioSyst 10:2838–2849
65. Wang L, Yu S, Tong C, Zhao Y, Liu Y, Song C, Zhang Y, Zhang X, Wang Y, Hua W, Li D, Li D, Li F, Yu J, Xu C, Han X, Huang S, Tai S, Wang J, Xu X, Li Y, Liu S, Varshney RK, Wang J, Zhang X (2014) Genome sequencing of the high oil crop sesame provides insight into oil biosynthesis. Genome Biol 15:R39
66. Wang L, Han X, Zhang Y, Li D, Wei X, Ding X, Zhang X (2014) Deep resequencing reveals allelic variation in Sesamum indicum. BMC Plant Biol 14:225
67. Wu K, Yang M, Liu H, Tao Y, Mei J, Zhao Y (2014) Genetic analysis and molecular characterization of Chinese sesame (*Sesamum indicum* L.) cultivars using insertion-deletion (InDel) and simple sequence repeat (SSR) markers. BMC Genet 15:35
68. Babu PR, Rao KV, Reddy VD (2013) Structural organization and classification of cytochrome P450 genes in flax (*Linum usitatissimum* L.). Gene 513:156–162
69. Bhattacharyya D, Sinha R, Hazra S, Datta R, Chattopadhyay S (2013) De novotranscriptome analysis using 454 pyrosequencing of the Himalayan Mayapple, *Podophyllum hexandrum*. BMC Genomics 14:748
70. Dinkova-Kostova AT, Gang DR, Davin LB, Bedgar DL, Chu A, Lewis NG (1996) (+)-pinoresinol/(+)-lariciresinol reductase from *Forsythia intermedia*. J Biol Chem 271:29473–29482
71. Gang DR, Kasahara H, Xia Z-Q, Vander-Mijnsbrugge K, Bauw G, Boerjan W, Van Montagu M, Davin LB, Lewis NG (1999) Evolution of plant defense mechanisms. Relationships of phenylcoumaran benzylic ether reductases to pinoresinol-lariciresinol and isoflavone reductases. J Biol Chem 274:7516–7527
72. Hemmati S, Schmidt TJ, Fuss E (2007) (+)-pinoresinol/(−)-lariciresinol reductase from *Linum perenne Himmelszelt* involved in the biosynthesis of justicidin B. FEBS Lett 581:603–610
73. Bayindir Ü, Alfermann AW, Fuss E (2008) Hinokinin Biosynthesis in *Linum corymbulosum* Reichenb. Plant J 55:810–820
74. Nakatsubo T, Mizutani M, Suzuki S, Hattori T, Umezawa T (2008) Characterization of *Arabidopsis thaliana* pinoresinol reductase, a new type of enzyme involved in lignan biosynthesis. J Biol Chem 283:15550–15557
75. Wankhede DP, Biswas DK, Rajkumar S, Sinha AK (2013) Expressed sequence tags and molecular cloning and characterization of gene encoding pinoresinol/lariciresinolreductase from *Podophyllum hexandrum*. Protoplasma 250:1239–1249
76. Ono E, Kim H-J, Murata J, Morimoto K, Okazawa A, Kobayashi A, Umezawa T, Satake H (2010) Molecular and functional characterization of novel furofuran-class lignan glucosyltransferases from *Forsythia*. Plant Biotechnol 27:317–324
77. Kim H-J, Ono E, Morimoto K, Yamagaki T, Okazawa A, Kobayashi A, Satake H (2009) Metabolic engineering of lignan biosynthesis in *Forsythia* cell culture. Plant Cell Physiol 50:2200–2209
78. Morimoto K, Satake H (2013) Seasonal alteration in amounts of lignans and their glucosides and gene expression of the relevant biosynthetic enzymes in the *Forsythia suspense* leaf. Biol Pharm Bull 36:1519–1523

79. Okazawa A, Kusunose T, Ono E, Kim H-J, Satake H, Shimizu B, Mizutani M, Seki H, Muranaka T (2014) Glucosyltransferase activity of *Arabidopsis* UGT71C1 towards pinoresinol and lariciresinol. Plant Biotechnol 31:561–566. doi:10.5511/plantbiotechnology.14.0910a
80. Xia Z-Q, Costa MA, Pélissier HC, Davin LB, Lewis NG (2001) Secoisolariciresinol dehydrogenase purification, cloning, and functional expression. J Biol Chem 276:12614–12623
81. Ghose K, Selvaraj K, McCallum J, Kirby CW, Sweeney-Nixon M, Cloutier SJ, Deyholos M, Datla R, Fofana B (2014) Identification and functional characterization of a flax UDP-glycosyltransferase glucosylating secoisolariciresinol (SECO) into secoisolariciresinol monoglucoside (SMG) and diglucoside (SDG). BMC Plant Biol 14:82
82. Umezawa T, Ragamustari SK, Nakatsubo T, Wada S, Li L, Yamamura M, Sakakibara N, Hattori T, Suzuki S, Chiang VL (2013) A lignan *O*-methyltransferase catalyzing the regioselective methylation of matairesinol in *Carthamus tinctorius*. Plant Biotechnol 30:97–109
83. Ragamustari SK, Yamamura M, Ono E, Hattori T, Suzuki S, Suzuki H, Shibata D, Umezawa T (2014) Substrate-enantiomer selectivity of matairesinol *O*-methyltransferases. Plant Biotechnol 31:257–267
84. Ragamustari SK, Nakatsubo T, Hattori T, Ono E, Kitamura Y, Suzuki S, Yamamura M, Umezawa T (2013) A novel *O*-methyltransferase involved in the first methylation step of yatein biosynthesis from matairesinol in *Anthriscus sylvestris*. Plant Biotechnol 30:375–384
85. Wang Z, Hobson N, Galindo L, Zhu S, Shi D, McDill J, Yang L, Hawkins S, Neutelings G, Datla R, Lambert G, Galbraith DW, Grassa CJ, Geraldes A, Cronk QC, Cullis C, Dash PK, Kumar PA, Cloutier S, Sharpe AG, Wong GKS, Wang J, Deyholos MK (2012) The genome of flax (*Linum usitatissimum*) assembled de novo from short shotgun sequence reads. Plant J 72:461–473
86. Barvkar VT, Pardeshi VC, Kale SM, Kadoo NY, Gupta VS (2012) Phylogenomic analysis of UDP glycosyltransferase1 multigene family in *Linum usitatissimum* identified genes with varied expression patterns. BMC Genomics 2012:13,175
87. Eyberger AL, Dondapati R, Porter JR (2006) Endophyte fungal isolates from *Podophyllum peltatum* produce podophyllotoxin. J Nat Prod 69:1121–1124
88. Schmitt J, Petersen M (2002) Pinoresinol and matairesinol accumulation in a *Forsythia* x *intermedia* cell suspension culture. Plant Cell Tissue Organ Cult 68:91–98
89. Morimoto K, Ono E, Kim H-J, Okazawa A, Kobayashi A, Satake H (2011) The construction of transgenic *Forsythia* plants: comparative study of three *Forsythia* species. Plant Biotechnol 28:273–280
90. Renouard S, Tribalatc M-A, Lamblin F, Mongelard G, Fliniaux O, Corbin C, Marosevic D, Pilard S, Demailly H, Gutierrez L, Hano C, Mesnard F, Lainé E (2014) RNAi-mediated pinoresinol lariciresinol reductase gene silencing in flax (*Linum usitatissimum* L.) seed coat: consequences on lignans and neolignans accumulation. J Plant Physiol 171:1372–1377
91. Rosati C, Cadic A, Renou J-P, Duron M (1996) Regeneration and agrobacterium-mediated transformation of *Forsythia* x *intermedia* "Spring Glowly". Plant Cell Rep 16:114–117
92. Rosati C, Simoneau P, Treutter D, Poupard P, Cadot Y, Cadic A, Duron M (2003) Engineering of flower color in forsythia by expression of two independently-transformed dihydroflavonol 4-reductase and anthocyanidin synthase genes of flavonoid pathway. Mol Breed 12:197–208
93. Murase K, Sugai Y, Hayashi S, Suzuki Y, Tsuji K, Takayama S (2015) Generation of transgenic *Linum perenne* by *Agrobacterium* mediated transformation. Plant Biotechnol 32:349–352
94. Morimoto K, Kim H-J, Ono E, Kobayashi A, Okazawa A, Satake H (2011) Effects of light on production of endogenous and exogenous lignans by *Forsythia koreana* wildtype and transgenic cells. Plant Biotechnol 28:331–337
95. Yousefzadi M, Sharifi M, Behmanesh M, Ghasempour A, Moyano E, Palazon J (2012) The effect of light on gene expression and podophyllotoxin biosynthesis in *Linum album* cell culture. Plant Physiol Biochem 56:41–56

96. Hata N, Hayashi Y, Okazawa A, Ono E, Satake H, Kobayashi A (2012) Effect of photoperiod on growth of the plants, and sesamin content and CYP81Q1 gene expression in the leaves of sesame (*Sesamum indicum* L.). Environ Exp Bot 75:212–219
97. Kobayashi T, Niino T, Kobayashi M (2005) Simple cryopreservation protocol with an encapsulation technique for tobacco BY-2 suspension cell cultures. Plant Biotechnol 22:105–112. doi:10.5511/plantbiotechnology.22.105
98. Ogawa Y, Sakurai N, Oikawa A, Kai K, Morishita Y, Mori K et al (2011) High-throughput cryopreservation of plant cell cultures for functional genomics. Plant Cell Physiol 53:943–952. doi:10.1093/pcp/pcs038, PMID: 22437846
99. Sarasan V, Cripps R, Ramsay MM, Atherton C, Mcmichen M, Prendergast G (2006) Conservation *in vitro* of threatened plants – progress in the past decade. In Vitro Cell Dev Biol Plant 42:206–214. doi:10.1079/IVP2006769
100. Chen Q, Lai H, Hurtado J, Stahnke J, Leuzinger K, Dent M (2013) Agroinfiltration as an effective and scalable strategy of gene delivery for production of pharmaceutical proteins. Adv Tech Biol Med 1:103
101. Zhao J, Davis LC, Verpoorte R (2005) Elicitor signal transduction leading to production of plant secondary metabolites. Biotechnol Adv 23:283–333
102. Berim A, Spring O, Conrad J, Maitrejean M, Boland W, Petersen M (2005) Enhancement of lignan biosynthesis in suspension cultures of *Linum nodiflorum* by coronalon, indanoyl-isoleucine and methyl jasmonate. Planta 222:769–776
103. Van Fürden B, Humburg A, Fuss E (2005) Influence of methyl jasmonate on podophyllotoxin and 6-methoxypodophyllotoxin accumulation in *Linum album* cell suspension cultures. Plant Cell Rep 24:312–317
104. Yousefzadi M, Sharifi M, Behmanesh M, Ghasempour A, Moyano E, Palazon J (2010) Salicylic acid improves podophyllotoxin production in cell cultures of *Linum album* by increasing the expression of genes related with its biosynthesis. Biotechnol Lett 32:1739–1743
105. Bhattacharyya D, Sinha R, Ghanta S, Chakraborty A, Hazra S, Chattopadhyay S (2012) Proteins differentially expressed in elicited cell suspension culture of *Podophyllum hexandrum* with enhanced podophyllotoxin content. Proteome Sci 10:34
106. Vardapetyan HR, Kirakosyan AB, Oganesyan AA, Penesyan AR, Alfermann WA (2003) Effect of various elicitors onlignan biosynthesis in callus cultures of *Linum austriacum*. Russ J Plant Physl 50:297–300
107. Ionkova I (2009) Effect of methyl jasmonate on production of ariltetralin lignans in hairy root cultures of *Linum tauricum*. Pharmacogn Res 1:102–105
108. Schmitt J, Petersen M (2002) Influence of methyl jasmonate and coniferyl alcohol on pinoresinol and matairesinol accumulation in a *Forsythia* × *intermedia* suspension culture. Plant Cell Rep 20:885–890
109. Muranaka T, Miyata M, Ito K, Tachibana S (1998) Production of podophyllotoxin in *Juniperus chinensis* callus cultures treated with oligosaccharides and a biogenetic precursor. Phytochemistry 49:491–496
110. Bahabadi SE, Sharifi M, Safaie N, Murata J, Yamagaki T, Satake H (2011) Increased lignan biosynthesis in the suspension cultures of *Linum album* by fungal extracts. Plant Biotechnol Rep 5:367–373
111. Bahabadi SE, Sharifi M, Chashmi NA, Murata J, Satake H (2014) Significant enhancement of lignans accumulation in hairy root cultures of *Linum album* using biotic elicitors. Acta Physiol Plant 36:3325–3331
112. Bahabadi SE, Sharifi M, Murata J, Satake H (2014) The effect of chitosan and chitin oligomers on gene expression and lignan production in *Linum album* cell cultures. J Med Plant 13:46–53
113. Hano C, Addi M, Bensaddek L, Crônier D, Baltora-Rosset S, Doussot J, Maury S, Mesnard B, Chabbert B, Hawkins S, Lainé E, Lamblin F (2006) Differential accumulation of monolignol-drived compounds in elicited flax (*Linum usitatissimum*) cell suspension cultures. Planta 223:975–989

114. Bahabadi SE, Sharifi M, Behmanesh M, Safaie N, Murata J, Araki R, Yamagaki T, Satake H (2012) Time-course changes in fungal elicitor-induced lignan synthesis and expression of the relevant genes in cell cultures of *Linum album*. J Plant Physiol 169:487–491
115. Tahsili J, Sharifi M, Safaie N, Bahabadi SE, Behmanesh M (2014) Induction of lignans and phenolic compounds in cell culture of Linum album by culture filtrate of *Fusarium graminearum*. J Plant Interact 9:412–417
116. Chun C, Kozai T (2001) A closed transplant production system, a hybrid of scaled-up micropropagation system and plant factory. J Plant Biotechnol 3:59–66
117. Hirai T, Fukukawa G, Kakuta H, Fukuda N, Ezura H (2010) Production of recombinant miraculin using transgenic tomatoes in a closed cultivation system. J Agric Food Chem 58:6096–6101
118. Kato K, Yoshida R, Kikuzaki A, Hirai T, Kuroda H, Hiwasa-Tanase K, Takane K, Ezura H, Mizoguchi T (2010) Molecular breeding of tomato lines for mass production of miraculin in a plant factory. J Agric Food Chem 58:9505–9510

The Elucidation and Metabolic Engineering of Terpenoid Indole Alkaloid Pathway in *Catharanthus roseus* Hairy Roots

16

Jiayi Sun and Christie A. M. Peebles

Abstract

Catharanthus roseus (Madagascar periwinkle) produces many pharmaceutically important chemicals such as vinblastine, vincristine, serpentine, and ajmalicine. They are synthesized through the highly branched and complex terpenoid indole alkaloid (TIA) pathway in *C. roseus*. Among the compounds produced in this pathway, vinblastine and vincristine are efficient anticancer drugs widely used in the clinic and are only found in *C. roseus*. Due to low accumulation of these TIAs within the plant and the infeasibility of production using chemical synthesis at industrial scale, the market price of these drugs remains high. In addition, inconsistent production along with increased demand causes these drugs to often show up on the FDA drug shortage list. With advanced knowledge of molecular biology, metabolic engineering, and bioinformatics, the construction of a robust and efficient alternative production platform in *C. roseus* hairy roots by manipulating the TIA pathway has become a promising strategy in recent years. In addition, recent advances in high-throughput sequencing technology have greatly sped up the elucidation of the TIA pathway. This review highlights present efforts to discover TIA pathway genes, their compartmentalization, and regulation in *C. roseus* and summarizes TIA pathway engineering efforts in *C. roseus* hairy roots.

Keywords

Plant secondary metabolism • Pathway regulation • Medicinal plants • High-throughput sequencing • Transcription factors • Functional genomics

J. Sun (✉) • C.A.M. Peebles (✉)
Chemical and Biological Engineering Department, Colorado State University, Fort Collins, CO, USA
e-mail: jiayisun@rams.colostate.edu; christie.peebles@colostate.edu; cpeebles@engr.colostate.edu

S. Jha (ed.), *Transgenesis and Secondary Metabolism*, Reference Series in Phytochemistry, DOI 10.1007/978-3-319-28669-3_14

Contents

1 Introduction

Plants produce a staggering variety of metabolites with diverse structures and therefore are an abundant source of compounds for use in the pharmaceutical, food, flavorings, fragrance, and supplement industry. Many plant secondary metabolites are not only believed to help plants defend against herbivores and pathogens but are also of significant importance in providing medicine and structure inspiration for human health. *Catharanthus roseus* is an important medicinal plant from Apocynaceae family. As a rich source of alkaloids, *C. roseus* can produce more than 130 identified terpenoid indole alkaloids (TIAs) [1]. Many of these TIAs exhibit strong pharmacological activities.

The most striking biological activity found in compounds from this family is the antineoplastic effect of dimeric alkaloids such as anhydrovinblastine, vinblastine, and vincristine from *C. roseus* [1]. These dimeric alkaloids together with a number of related semisynthetic compounds (vinorelbine, vindesine, vinflunine) are named *Vinca* alkaloids and have been used in the clinic for the treatment of various cancers. The *Vinca* alkaloids are microtubule-depolymerizing agents [112]. They arrest tumor cells during mitosis by binding at the surface between two tubulin heterodimers next to the exchangeable guanosine 5-triphosphate-biding site, which depolymerizes the microtubules leading eventually to apoptosis. This unique mode of action makes *Vinca* alkaloids powerful and effective chemotherapy drugs. Besides antitumor activity, *Vinca* alkaloids also exhibit antidiabetic and antioxidant properties [2]. In addition, ajmalicine and serpentine have been widely used for the treatment of hypertension because they act as an α-adrenergic receptor antagonist. Serpentine also shows anti-arrhythmic and anti-neuroinflammatory activity [3].

Despite the great medicinal importance of TIAs, relatively little is known about the biosynthesis, regulation, and transport of these compounds in *C. roseus*. The TIA biosynthesis pathway is highly complex and branched. The pathway involves at least 35 intermediates and 30 enzymes from indole pathway, methylerythritol 4-phosphate pathway, terpenoid pathway, and alkaloid pathway. Currently the number of unknown enzymes in TIA pathway makes it infeasible to heterologously construct this pathway in microorganisms for the production of valuable TIAs. In addition, these valuable TIAs accumulate at very low levels in *C. roseus* due to the high degree of compartmentalization of the pathway intermediates which involves in at least seven intra- and intercellular compartments and the tight transcriptional regulation within plant. This review highlights the current effort in deciphering the mystery in TIA biosynthesis, compartmentalization, and regulation in *C. roseus*.

In addition, TIAs accumulate to very low levels in *C. roseus*. A significant effort in metabolic engineering has been placed on increasing TIA production in *C. roseus* plants, dedifferentiated cell cultures, and hairy root cultures. Hairy roots have been considered as an ideal biological system for TIA production due to its superior genetic and chemical stability compared to cell cultures and scalability to large scale without restriction by environmental factors as the whole plan does. This review also summarizes TIA pathway engineering in *C. roseus* hairy roots.

2 TIA Biosynthetic Pathway in *C. roseus*

2.1 Monoterpenoid Pathway

One important precursor for TIA biosynthesis is secologanin, which is a monoterpenoid (also called iridoid or secoiridoid) synthesized through the terpenoid pathway (Scheme 1). This pathway broadly exists in many other plants and leads to the accumulation of iridoids (including secologanin) as end products. Many terpenoids possess anticancer, antimicrobial, and anti-inflammatory activities [4, 5]. In *C. roseus*, the terpenoid pathway consists of ten enzymatic steps and has recently been fully elucidated [6]. It starts from the coupling of two MEP (methylerythritol 4-phosphate) pathway products isopentenyl pyrophosphate (IPP) and dimethylallyl pyrophosphate (DMAPP) [7]. This coupling forms the product geranyl diphosphate, a monoterpene, which is then converted to geraniol by geraniol synthase (GES) [8]. Geraniol is successively oxidized by geraniol 10-hydroxylase (G10H)/8-oxidase (G8O) [9] and 8-hydroxygeraniol oxidoreductase (8-HGO) [6] to form 8-oxogeranial. Iridoid synthase (IS) is responsible for the reductive cyclization of 8-oxogeranial to two iridoid ring scaffolds [10]. These two scaffolds are further oxidized consecutively by the same enzyme iridoid oxidase (IO) to generate

7-deoxyloganetic acid [11]. The recently discovered 7-deoxyloganetic acid glucosyltransferase (7-DLGT) can use UDP-glucose as the sugar donor to catalyze 7-deoxyloganetic acid to 7-deoxyloganic acid [6] followed by the conversion to loganic acid by 7-deoxyloganic acid hydroxylase (7-DLH) [12]. The two last steps (loganic acid O-methyltransferase, LAMT, and secologanin synthase, SLS) lead to the final production of secologanin [13, 14]. In the terpenoid pathway, four enzymes (G10H, IO, 7-DLH, and SLS) belong to the cytochrome P450 family. These enzymes require the transfer of electrons from NADPH by cytochrome P450 reductase (CPR) in the endoplasmic reticulum.

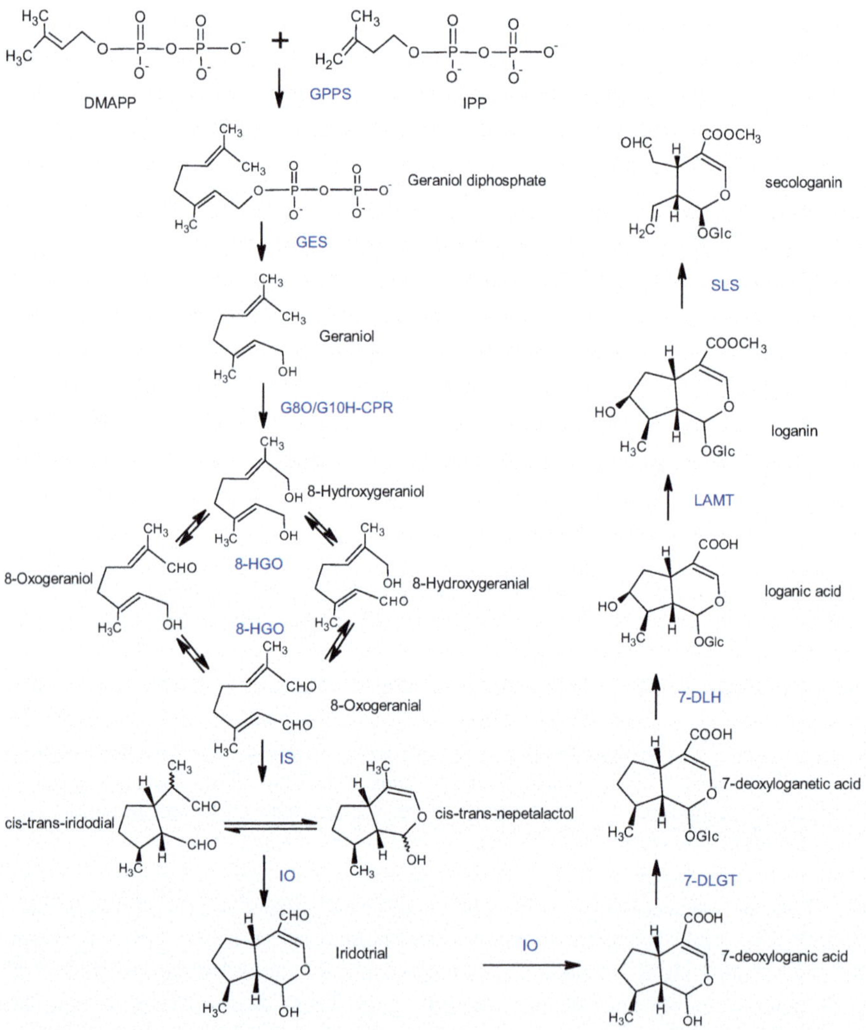

Scheme 1 Terpenoid pathway in *C. roseus*

2.2 Indole Pathway

Tryptamine is generated by the indole pathway and serves as the indole moiety for TIAs. The indole pathway begins with chorismate derived from the shikimate pathway. Chorismate is the common precursor for all three aromatic amino acids. The first enzyme in the indole pathway, anthranilate synthase (AS), directs chorismate toward tryptophan biosynthesis (Scheme 2). Anthranilate is converted by several uncharacterized steps in *C. roseus* to generate tryptophan. AS, which is strongly feedback inhibited by tryptophan and tryptamine, exists as heterotetramers composed of two alpha and two beta subunits. The binding site of tryptophan in the alpha subunit accounts for the feedback inhibition [15]. The last step in indole pathway is the conversion of tryptophan to tryptamine by tryptophan decarboxylase (TDC) [16]. TDC is also feedback inhibited by tryptamine [17].

2.3 Alkaloid Pathway

From the condensation of tryptamine and secologanin, the first alkaloid strictosidine is synthesized by strictosidine synthase (STR) (Scheme 3). This step switches the carbon flux from primary metabolism to secondary metabolism [18]. Strictosidine is then hydrolyzed by strictosidine glucosidase (SGD) to an unstable intermediate strictosidine aglycone [19]. The highly reactive ring-opened dialdehyde intermediate can be converted to TIAs with dramatically diverse structures in different TIA-producing plants. In *C. roseus*, three main types of TIAs are produced including *Corynanthe*-type alkaloids (ajmalicine and serpentine), *Aspidosperma*-type alkaloids (tabersonine and its derivatives), and iboga-type alkaloids (catharanthine) [20]. Catharanthine, tabersonine, and their derivatives uniquely accumulate in *C. roseus*. However, the mechanisms and the enzymatic steps by which strictosidine, a glycoside, rearranges into tabersonine, catharanthine, and ajmalicine remain largely obscure. Based on in vivo feeding studies of isotopically labeled substrates, on isolation of discrete intermediates from plants, and on biomimetic model reactions, *Corynanthe* alkaloids are believed to be synthesized from the major intermediate cathenamine, while *Aspidosperma* alkaloids are mainly derived from the intermediate stemmadenine [21].

Notably, tabersonine is converted to different metabolites in different organs of *C. roseus* [20] (Scheme 4). In leaves, tabersonine is catalyzed to vindoline by seven steps with all seven enzymes being fully elucidated. The first two enzymes,

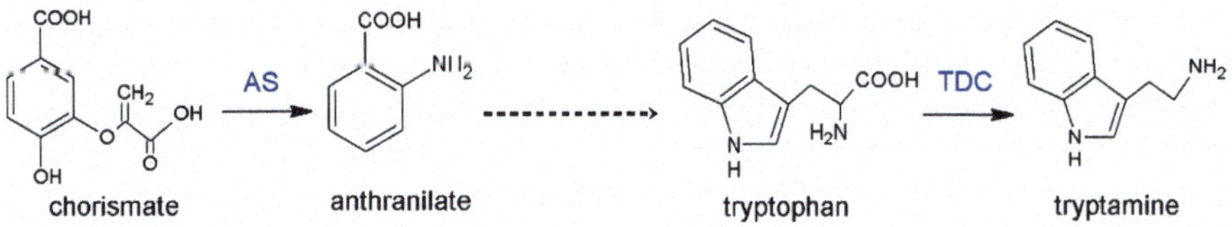

Scheme 2 Indole pathway in *C. roseus*

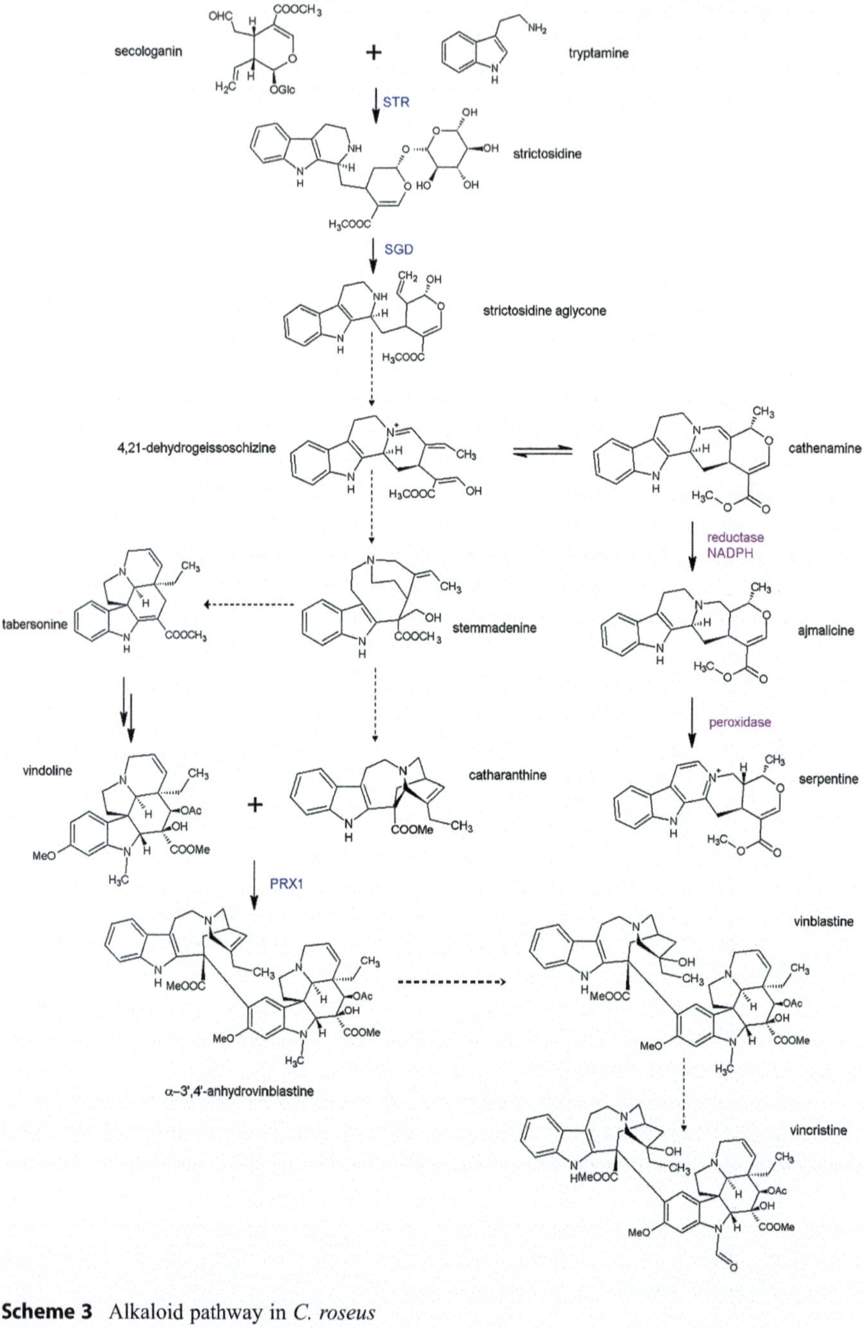

Scheme 3 Alkaloid pathway in *C. roseus*

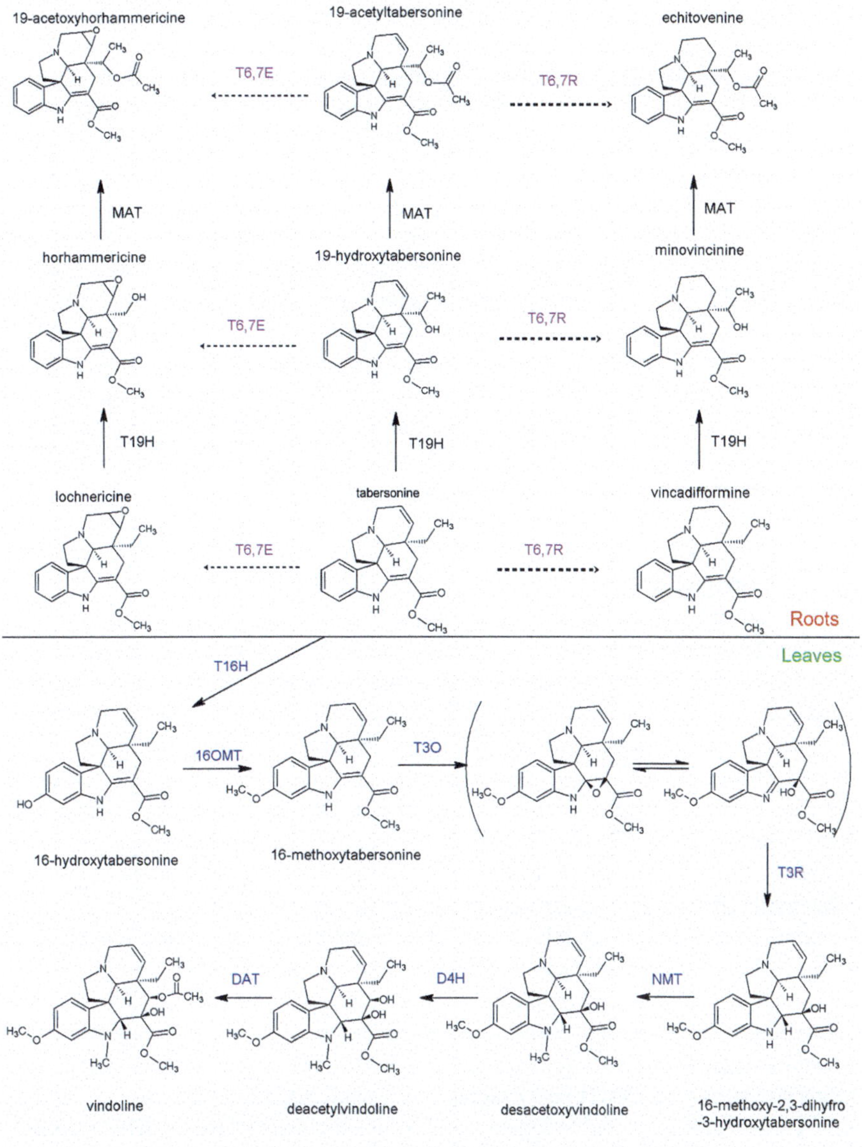

Scheme 4 Tabersonine metabolic pathway in leaves or hairy roots of *C. roseus*

tabersonine 16-hydroxylase (T16H) [22] and 16-*O*-methyltransferase (16OMT) [23], successively hydrolyze and *O*-methylate tabersonine to give rise to 16-methoxytabersonine. The recently elucidated tabersonine 3-oxygenase (T3O) and tabersonine 3-reductase (T3R) filled the missing steps between

16-methoxytabersonine and 16-methoxy-2,3-dihydro-3-hydroxytabersonine [24]. The last three enzymes N-methyltransferase (NMT) [25], desacetoxyvindoline 4-hydroxylase (D4H) [26], and deacetylvindoline-4-O-acetyltransferase (DAT) [27] lead to the final biosynthesis of vindoline. In contrast, the conversion of tabersonine in roots is realized by an alternative metabolic pathway which leads to the accumulation of epoxide derivatives of tabersonine such as lochnericine, hörhammericine, and 19-*O*-acetylhörhammericine (Scheme. 4). Tabersonine 6,7-epoxidase (T6,7E) activity was detected and characterized in hairy roots. This suggests that T6,7E is the dominate enzyme in roots responsible for diverting tabersonine to root-specific metabolites [28]. Two identified enzymes tabersonine-19-hydroxylase (T19H) [29] and minovincine 19-hydroxy-*O*-acetyl transferase (MAT) [30] are involved in this sub-pathway as well (Scheme 4). As a result, no vindoline, vinblastine, and vincristine could be detected in *C. roseus* hairy roots. The coupling of vindoline and catharanthine in leaves leads to the formation of α-3′, 4′- anhydrovinblastine (AVLB) in presence of peroxidase 1 (PRX1). The pathway between AVLB to vinblastine and vincristine has not been elucidated.

3 Compartmentalization of the TIA Pathway

The TIA biosynthesis pathway is spatially segregated at inter- and intracellular levels. The localization of enzymes has been studied using in situ RNA hybridization coupled with epi-fluorescence imaging, immunolocalization, X-ray crystallography, or transcriptome profiling methods [31]. The shikimate and indole pathway genes were identified in chloroplasts and root plastids using sucrose density gradient centrifugation [32] (Fig. 1). The last enzyme in the indole pathway, TDC, was found in the cytosol [32] (Fig. 1). Over the past 10 years, transcript localization studied by RNA in situ hybridization revealed that leaf internal phloem-associated parenchyma (IPAP) cells host the entire MEP pathway and the first eight enzymes in the monoterpenoid pathway which lead to the biosynthesis of loganic acid (Fig. 1) [6, 10, 33, 34]. Immunocytolabeling studies demonstrated that MEP pathway proteins could also be detected in other tissues such as epidermis, mesophyll of young leaves, and roots [33]. The subcellular localization of MEP pathway and early terpenoid pathway enzymes converting IPP and DMAPP to geraniol is in the plastid stroma and stromules [34] (Fig. 1). Geraniol is transferred from the plastid stromules to the cytosol and catalyzed to loganic acid by six enzymes (G10H, 10HGO, IS, IO, 7-DLGT, and 7-DLH) in the cytosol. Among these six enzymes, G10H, IO, and 7-DLH belong to ER-anchored cytochrome P450 proteins which require CPR to transfer electrons from NADPH to the P450 enzymes. The subsequent reactions to secologanin catalyzed by LAMT and SLS occur in the leaf epidermis (Fig. 1). Additional studies are needed to determine whether this translocation takes place passively through the symplasm or is controlled by plasmodesmata-localized proteins or plasma membrane transporters. Even though the compartmentalization of the terpenoid pathway has been extensively studied in *C. roseus* leaves, mRNA transcripts from this pathway also showed relatively high levels in hairy roots. The

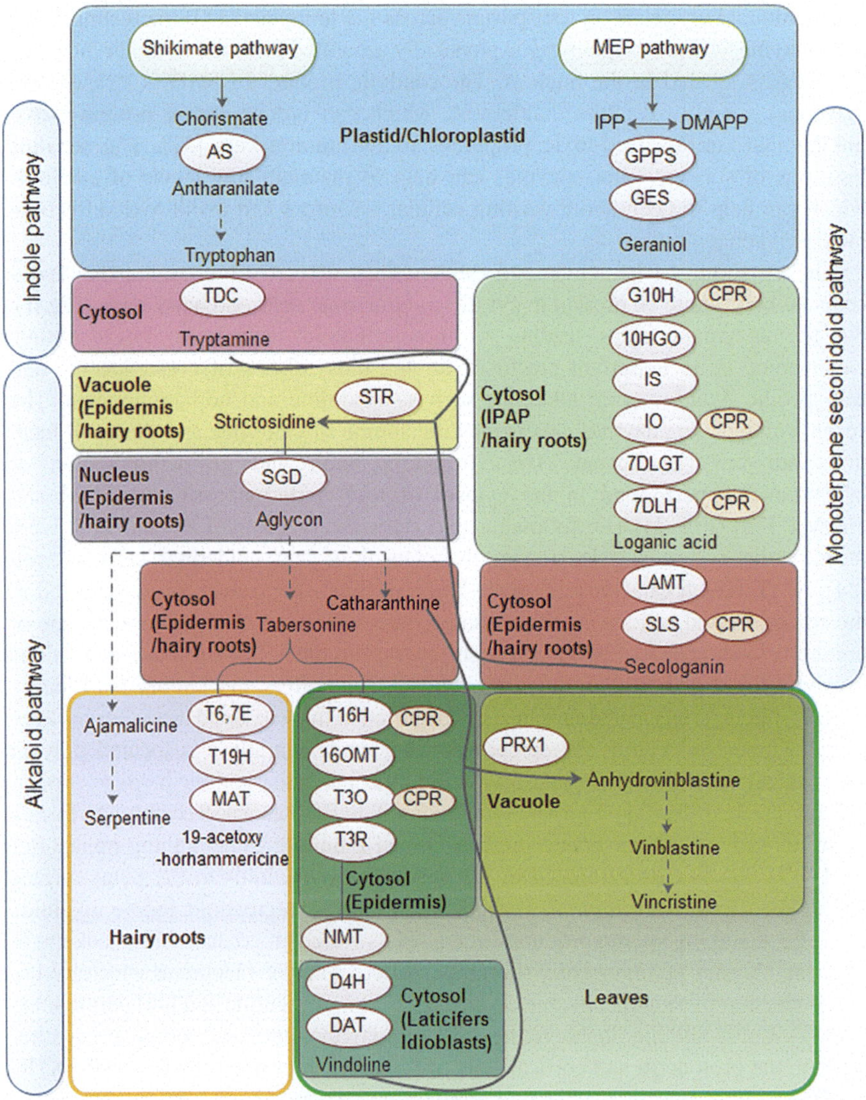

Fig. 1 TIA pathway compartmentalization

end product of this pathway secologanin has also been found to accumulate in *C. roseus* hairy roots [35, 36].

TIA biosynthesis is more complex at subcellular level compared to the indole and terpenoid pathway. The first committed enzyme STR in alkaloid pathway contains a signal peptide directing it to vacuole (Fig. 1). The coupling of secologanin and tryptamine by STR occurs in the vacuolar rather than the cytosol where secologanin and tryptamine are produced [18] (Fig. 1). Accordingly, the vacuolar assembly of strictosidine likely relies on transporters across the tonoplast [18]. Interestingly, the next enzyme SGD in the pathway is physically separated from the vacuolar localized STR and is located in the nucleus. The catalytic product of SGD is strictosidine aglycone, a highly reactive dialdehyde, which can induce strong protein cross-linking that confers feed-toxic properties to its bio-attackers [37]. The separate locations of substrate and enzymes can help to maintain low levels of alkaloids which can help the cell avoid wasting cellular resources and avoid toxic effects of increased concentrations.

The metabolites tabersonine and catharanthine derived from strictosidine aglycone are believed to be present in cytosol of epidermal cells and hairy roots (Fig. 1), but the enzymatic steps leading to these alkaloids remain to be identified. Tabersonine is an important precursor for the biosynthesis of organ-specific and cell-specific *Aspidosperma* alkaloids such as vindoline and hörhammericine. The seven leaf-specific enzymes catalyzing tabersonine to vindoline and their cell locations are fully elucidated. The 16-hydroxylation and 16-methoxylation of tabersonine is performed in the cytosol of leaf epidermal cells by T16H and 16OMT [38] (Fig. 1). The following two steps catalyzed by T3O and T3R were recently discovered and located in the cytosol of leaf epidermal cells as well [24]. NMT, which catalyzes the methylation of the T3R product, is not detected in the epidermis and resides in the remaining parts of the leaf. No further cell type or organelle localization of NMT has been reported to date. The subsequent reactions involving D4H and DAT occur in the cytosol of laticifers and idioblasts of leaves (Fig. 1). Interestingly, catharanthine is transported to the wax exudates of leaves after being synthesized within the epidermal cells, while vindoline is located only in internal leaf cells. Spatial separation of vindoline and catharanthine helps to explain the low levels of bisindole terpenoid alkaloids found in leaves. Recently De Luca's group cloned and characterized the function of a unique catharanthine transporter (CrTPT2) that is closely related to the secretion of alkaloids to the plant surface [39]. When leaves are injured or digested, these monoterpenoid indole alkaloids could be mixed together to produce dimeric TIAs which are toxic to bio-attackers. In contrast, *C. roseus* roots or hairy roots have different active TIA genes which lead to the production of root-specific metabolites including serpentine and hörhammericine (Fig. 1). Catharanthine, lochnericine, hörhammericine, and tabersonine are synthesized in the protoderm and cortical cells of *C. roseus* root tips [30, 40]. TDC, STR, and MAT transcripts are localized in protoderm and cortical cells around root apical meristem [41]. The complexity of compartmentalization of the pathway and limited knowledge about the transportation of pathway intermediates represent significant challenges to metabolic engineers trying to increase TIA biosynthesis in *C. roseus*.

4 TIA Pathway Regulation

TIA biosynthesis is controlled by complex regulation at the transcriptional, post-translational, and transportation levels in *C. roseus*. The most direct regulation occurs at the transcriptional level by transcription factors in a coordinate manner in response to various developmental signals (auxin, cytokinins, abscisic acid, ethylene, salicylic acid, and jasmonic acid) and environmental stimuli (UV light, fugal extracts, and nitric oxide). These factors affect TIA production synergistically or antagonistically through different signal transduction mechanisms which are still poorly understood.

4.1 Plant Growth Regulators that Influence TIA Production

Plant growth hormones play a crucial role on diverse developmental processes ranging from seed germination, plant growth, morphogenesis, and even secondary metabolism. The synthetic auxin, 2,4-dichlorophenoxyacetic acid (2,4-D), can negatively regulate terpene biosynthesis. Feeding 2,4-D to *C. roseus* cells significantly decreased alkaloid accumulation and downregulated *G10H* [42] in terpenoid pathway and *1-deoxy-D-xylulose 5 phosphate synthase (DXS)*, *1-deoxy-D-xylulose 5-phosphate reductoisomerase (DXR)*, *2-C-methyl-D-erythritol 2,4-cyclodiphosphate synthase (MECS)*, and *4-hydroxy-3-methylbut-2-enyl diphosphate synthase (HDS)* genes in MEP pathway [33, 43, 44]. In contrast, *SLS* in the late terpenoid pathway and *STR* in the alkaloid pathway are not influenced by 2,4-D [44]. Little is known about the 2,4-D-dependent signal transduction mechanisms involved in the regulation of TIA biosynthesis. Cytoplasmic calcium concentration, a ubiquitous secondary messenger in plants, was shown to play a role in the transduction pathway leading to the inhibitory effect of 2,4-D on TIA biosynthesis through multiple Ca^{2+}-release pathways [45]. In contrast, a cytokinin, trans-zeatin, can stimulate TIA biosynthesis and upregulate *G10H* transcription [42]. Furthermore, the feeding of trans-zeatin and ethylene synergistically increased MEP pathway gene expression which suggested these two hormones stimulate the TIA biosynthesis by two distinct mechanisms [42].

Other plant phytohormones including salicylic acid and ethylene also stimulate accumulation of catharanthine, vindoline, and vinblastine in *C. roseus*, while abscisic acid (ABA) and gibberellic acid showed decreased concentrations of these same TIAs [46]. The signal transduction pathway that connects these growth hormones with TIA biosynthesis is unknown.

4.2 Plant Stress Hormones that Influence TIA Production

Jasmonic acid (JA) is a phytohormone that plays an important role in plant defense against biological attack and surrounding environmental stress by eliciting the production of secondary metabolites. Feeding JA or its derivative methyl jasmonate (MeJA) can increase TIA production and increase transcripts of all known TIA pathway genes in *C. roseus* [47]. JA can be synthesized through the octadecanoid (ODA) pathway in plants [48]. Plants that have undergone stress or are fed with elicitors will have an altered flux of Ca^{2+} into the cell via unknown mechanisms. This influx of Ca^{2+} will trigger a series of signal transduction reactions to induce endogenous JA biosynthesis. CrMPK3, a *C. roseus* kinase in mitogen-activated protein kinase (MAPK) cascade, was found to involve in signal transduction upstream of ODA pathway in response to abiotic stress [49]. Exogenously applied JA or endogenous JA triggered by various stresses can serve as secondary messenger to induce JA perception and recognition of so-called transcriptional regulons [6]. In *C. roseus*, multiple transcriptional regulators have been discovered to be involved in JA-mediated TIA biosynthesis.

Among these TIA transcriptional regulators, ORCAs (octadecanoid-responsive *Catharanthus* AP2/ERF-domain proteins) are extensively studied and shown to positively regulate TIA biosynthesis. ORCAs are rapidly upregulated by JA elicitation in *C. roseus*. They directly bind to the JA- and elicitor-responsive element (JERE) found in the promoter of several TIA pathway genes such as TDC and STR. ORCA1, ORCA2, and ORCA3 from this family have been identified in *C. roseus*. They control overlapping but distinct set of TIA pathway genes due to their binding specificity. Engineering the overexpression of *ORCA3* can trigger *ASα*, *DXS*, *STR*, and *SLS* upregulation in *C. roseus* hairy roots [35] and can regulate *ASα*, *TDC*, *STR*, *CPR*, *D4H*, and *DXS* transcription positively in *C. roseus* cell cultures [115]. ORCA2 can boost *ASα*, *TDC*, *LAMT*, and *STR* transcriptional levels in *C. roseus* hairy roots [36]. ORCA1 only has marginal effect on JA- and fungal elicitor-induced STR gene expression [50]. Notably, some TIA genes are not controlled by any of the ORCAs. Analysis of the *G10H* promoter showed that it contains unique binding sites for several transcriptional factors but does not include the JERE binding site for ORCAs, indicating that *G10H* is regulated by a different transcriptional cascade [51].

The basic helix-loop-helix (bHLH) transcription factor (TF) CrMYC2 was identified as the major activator of ORCA genes in *C. roseus*. The CrMYC2 gene is an immediate-early JA-responsive gene that contains a JAZ-interacting domain [52]. JA-triggered JAZ complex degradation can release the attached TFs such as MYC2 allowing for the activation of these TFs [53]. The TF CrMYC1 helps regulate *STR* transcription and is responsive to JA [54]. Another recently discovered bHLH transcription factor, iridoid synthesis 1 (BIS1), was reported to be induced by JA elicitation. BIS1 transactivates the expression of all genes encoding the enzymes catalyzing the successive conversion of the ubiquitous terpenoid precursor geranyl diphosphate to the iridoid loganic acid [55]. Interestingly, BIS1 controls a different set of TIA genes from ORCA3. This suggests that BIS1 and ORCA3 are distinct elements of the TIA regulatory circuit. The overall control of the iridoid branch of the TIA pathway by BIS1 made overexpressing BIS1 in *C. roseus* cell suspension culture sufficient to boost production of iridoids and TIAs [55].

ZCT proteins (ZCT1, ZCT2, and ZCT3) encoded by zinc finger gene family bind in a zinc-dependent manner to the promoter region of some TIA pathway genes (STR and TDC) and act as repressors in the JA signaling pathway [56]. There are several mechanisms by which the ZCTs exert their repressive function. The ZCT proteins may prevent the association of a transcriptional activator with the promoters or suppress the function of a DNA-bound transcriptional activator protein.

Alternatively, ZCT proteins could have negative effects on the basal transcription machinery or could induce the formation of an inactive chromatin structure at the promoter region. Although both ORCA activators and ZCT repressors can bind to the RV element of the STR promoter, the ZCT proteins can repress the activating activity of the ORCAs without competing for the same binding sites [57].

Interestingly, extensive studies have shown that the treatment with stress hormones or the overexpression of TIA pathway gene usually induces the upregulation of a combination of activators and repressors. This complicated regulation may serve to fine-tune the level and timing of gene expression in order to balance between growth and defense. Both *ORCA* activators and *ZCT* repressors were upregulated after feeding with jasmonate or elicitor [35].

WRKYs, another TF family, potentially have a significant role in JA signaling to enhance TIA levels. Transcriptome analyses showed that the *C. roseus* WRKY TF family is comprised of at least 48 members, and 12 of them were jasmonate responsive [58]. Among these WRKYs, only WRKY1 and WRKY2 have been elucidated and engineered in *C. roseus*. Overexpression of *WRKY1* upregulated several key TIA pathway genes especially *TDC* and pathway repressors *ZCTs* and downregulated pathway activators *ORCA2*, *ORCA3*, and MYC2 in *C. roseus* hairy roots. Overexpression of *WRKY2* also triggered the expression of multiple pathway genes and both pathway activators (*ORCA2* and *ORCA3*) and repressors (*ZCT1* and *ZCT3*). The contrasting effect of engineering WRKY1 and WRKY2 on expression of other TF genes revealed the complex interactions between different TFs.

A MYB-like protein BPF1 was reported to bind to the BA region of *STR* promoter rather than RV region where ORCAs bind through yeast one-hybrid analyses. Importantly, BPF1 upregulates STR transcription through a JA-independent signal transduction pathway induced by other elicitors. Engineering of BPF1 in *C. roseus* hairy roots results in the upregulation of multiple genes in indole pathway, terpenoid pathway, and downstream alkaloid pathway [59]. Similarly, the upregulation of both transcriptional activators and repressors was detected in response to BPF1 overexpression, which again indicates the tight regulation of TIA pathway. GBFs, the G-box binding factor family (GBF1, GBF2, and GBF3) which binds to the G-box sequence located in the NR region immediately upstream of the RV region, act as repressors of *STR* gene expression. Whether GBFs are activated via JA transduction pathway still remains to be discovered [60].

4.3 The Influence of UV Light Stress on TIA Production

Environment stress also stimulates the production of secondary metabolites including TIAs that serve as putative protective function. Near-ultraviolet light in the UV-B spectrum boost the production of dimeric alkaloids in *C. roseus* seedlings [61] and monomeric alkaloids in cell suspension cultures [62]. UV-B exposure of hairy roots increases lochnericine, serpentine, and ajmalicine production, but decreases hörhammericine production [63].

5 Gene Discovery within the TIA Pathway

Complete elucidation of the TIA biosynthesis pathway is of great importance in establishing efficient platforms for enhanced TIA production by genetic manipulation of *C. roseus* or heterologous expression of the pathway in microbial organisms. However, there are still many unknown steps in the *C. roseus* TIA pathway as described above. This section will focus on summarizing the strategies used in elucidating the TIA pathway genes in *C. roseus* hairy roots (Table 1).

5.1 Traditional Approaches in TIA Pathway Discovery

Radioisotope labeling of pathway intermediates is usually combined with elicitor or metabolic inhibitor treatment to identify the types of chemical conversions a pathway may undergo. By tracing the radioisotope label, the synthesis or degradation of the metabolites and chemical transformations undergone during the process can be

Table 1 Strategies used to discover TIA pathway genes

Gene discovery using traditional strategies		
Approach used for gene identification	Gene names	Reference
Screen for candidate genes from cDNA library using primers generated from protein purification combined with peptide sequencing	STR	[73]
	D4H	[26]
	DAT	[74]
	G10H	[9]
	MAT	[30]
	16OMT	[22]
	PRX1	[68]
Screen for candidate genes from cDNA library using degenerate primers based on known plant genes	T16H	[70]
	SGD	[69]
	SLS	[13]
Gene discovery using high-throughput sequencing method		
Co-regulated gene clustering	T19H	[29]
	NMT	[25]
	IS	[6, 10]
	8HGO	[6]
	7-DLGT	[6, 75]
Orthology	7DLS/IO	[6, 11]
	DL7H/7-DLH	[6, 12]
Epidermome	T3O	[24]
	T3R	[24]
	LAMT	[14]

monitored [64]. Feeding the tissue or cell culture with a carefully chosen precursor helps to locate biosynthetic origins, intermediates, and important rate-limiting steps [65]. For example, feeding ^{13}C-glucose to *C. roseus* cell suspension cultures strongly suggests the terpene moiety in TIA pathway is ultimately derived from the triose phosphate/pyruvate or MEP pathway [7]. Feeding experiments of various ^{3}H-, ^{2}H-, and ^{13}C-labeled terpene intermediates to *C. roseus* suggested secologanin is synthesized via cyclization of 10-oxogeranial/10-oxoneral to iridodial [66]. Elicitor treatment can cause differential gene expression in treated cultures and untreated cultures. The functionally related genes will show similar expression patterns which provide the basis for TIA pathway elucidation [67]. Inhibitors that directly block specific enzymes can be used to identify enzymes, precursors, intermediates, or products of a particular pathway. For example, ^{14}C-tabersonine labeling and P-450-dependent enzyme inhibitor feeding indicated that tabersonine is converted to lochnericine by selective epoxidation at positions 6 and 7 by a cytochrome P-450-dependent monooxygenase in hairy roots [28].

From the above methods, the potential enzyme activity that may be involved in the biosynthetic pathway can be predicted. Traditionally, two major strategies are used to find and identify TIA pathway genes. The first is to fractionate the proteins present to isolate the protein(s) having the activity of interest, sequence the isolated protein peptide, design primers for screening of cDNA library based on the sequenced peptide information, and confirm functional expression in an appropriate host. This strategy was used to identify the major class III peroxidase CrPRX1 in *C. roseus* as the enzyme catalyzing the dimerization reaction of vindoline and catharanthine leading to α-3',4'-anhydrovinblastine (Table 1) [68]. The second strategy is a homology-based cloning approach which uses degenerate primers that anneal to a conserved sequence of a homologous gene from other organisms to PCR amplify candidate gene sequence from a cDNA library of *C. roseus*. The obtained candidate genes can be screened by cloning them into a heterologous expression system and investigating their function by feeding appropriate substrates and examining the products. The pathway genes SGD [69], SLS [13], and T16H [70] were elucidated using this method. However, these two approaches are slow and labor intensive. Because of this, they are limited to elucidating the function of only one or a few genes involved in a pathway at a time. Approximately 20 genes were discovered during the last three decades using these traditional strategies.

5.2 Sequencing Approaches in TIA Pathway Discovery

With high-throughput sequencing becoming more available and affordable, comparative transcriptome approaches have greatly increased the rate of TIA pathway elucidation in the non-model system *C. roseus*. Transcriptome sequencing usually results in thousands of transcripts. Narrowing them down to a reasonable number of candidates is crucial for further functional elucidation. Candidate genes could be selected based on enriched expression profile within appropriate cell type, orthology comparisons between close species, or on differential expression conditions.

Given the fact that most of TIA biosynthesis genes are cell type and organ specific, and some are minimally expressed genes, the expression profile within specific cells or tissues is necessary to identify the TIA pathway genes in these particular transcriptomes. The technique called carborundum abrasion (CA) was developed by De Luca's group to isolate RNA from epidermal cells of leaves where most TIA pathway reactions occur [14]. The first *C. roseus* cDNA library from these specific cells was created and called the "epidermome." Exploiting the *C. roseus* epidermome is a powerful method for candidate gene discovery as revealed by numerous occurrences of ESTs from known TIA biosynthetic genes and regulators of this pathway. For instance, the full-length candidate gene for LAMT was obtained by reconstituting more than 50 ESTs encoding O-methyltransferase in epidermome and was further functionally characterized [14]. More recently, the last two unknown enzymes in the vindoline pathway, T3O and T3R, were elucidated by selecting the candidate genes that encode "hydratase" and "alcohol dehydrogenase" activity from epidermome EST database [24].

Orthologous comparison of high-throughput transcriptomics is another strategy used in identifying gene candidates in the TIA pathway. This method relies on the fact that some TIA pathway intermediates can be synthesized in several close-related plant species, while some metabolites are unique to *C. roseus*. The candidate genes can be selected by searching for genes that are only expressed in plant species producing the desired metabolite, but not in other species that do not produce it. For example, enzymes in secologanin biosynthesis, 7DLS/IO [11] and DL7H [12], have been successfully identified by bioinformatics searching for homologous cytochrome P450 (CYP) candidate genes from annotated EST databases of seven secologanin-producing plant species from the Apocynaceae family, one secologanin/quinolone alkaloid-producing species from the Rubiaceae family, and one secologanin-producing species from the Caprifoliaceae family (www.phytometasyn.ca).

In addition, hierarchical clustering of differential gene expression generated by high-throughput sequencing of the transcriptome under a variety of conditions including precursor feeding, elicitor treatment, overexpression, or silencing of gene expression is also an important method of selecting candidate TIA pathway genes [71, 72]. This strategy is based on the postulate that genes from the same biosynthetic pathway show similar expression patterns among different treatments, and these co-regulated genes could be clustered by bioinformatic tools. For example, the O'Connor group identified iridoid synthase (IS) by selecting the NAD(P)H-dependent enzymes from the MPGR *C. roseus* transcriptome database (http://medicinalplantgenomics.msu.edu/). These enzymes were narrowed down based on their co-regulation pattern with geraniol 10-hydrolase (G10H) [10]. Similarly, TIA pathway genes tabersonine 19-hydroxylase (T19H) [29] and 2,3-dihydro-3-hydroxytabersonine-*N*-methyltransferase (NMT) [25] located downstream of tabersonine were elucidated by screening for cytochrome P450 family proteins which were co-expressed with known alkaloid pathway genes in *C. roseus*. Recently,

Karel Miettinen and his collaborators completed the missing genes in seco-iridoid pathway in *C. roseus* using multiple *C. roseus* databases including CathaCyc (http://omictools.com/cathacyc-s3424.html), RNA-seq data from cell suspensions overexpressing either pathway regulator ORCA2 or ORCA3, and proteomics database on epidermal and mesophyll protoplasts of *C. roseus* leaves [6].

6 Metabolic Engineering of *C. roseus* Hairy Roots for TIA Production

TIAs accumulated in *C. roseus* are of significant pharmaceutical importance. These include the important anticancer drugs vinblastine and vincristine which are exclusively synthesized by *C. roseus*. They have been listed on the FDA drug shortage list since 2012. Due to the complex chemical structure and chirality of these compounds, they are too expensive to produce through chemical synthesis. The industrial production relies on the extraction of these TIAs or their precursors from leaves of mature plants. However, vinblastine and vincristine accumulate in the plant at such a low level that only approximately 0.01% and 0.003%, respectively, can be extracted from dry leaf material [14]. Therefore, alternative production platforms must be developed to produce these drugs inexpensively, consistently, and at large scale. Since biological systems have potential to be scalable, selective, and engineered, rational design of metabolic engineering strategies could make it possible to create a stable production platform leading to the enhanced biosynthesis of these medicinally valuable TIAs [76].

6.1 Why Hairy Roots

Three *C. roseus* TIA production systems have been extensively investigated for TIA production since the early 1960s including whole plant, cell suspension, and hairy root culture. Among these, hairy roots have the greatest potential for industrial TIA production due to many advantages over the other two systems [77]. First, hairy roots have high growth rate with unlimited branching. Second, the biochemical and genetic stability of transgenic hairy roots can be maintained for years [78], while the instability of cell culture limits its application at industrial scale for secondary metabolite production [79]. Third, hairy roots as highly differentiated tissue can produce similar concentrations of secondary metabolites as the whole plant, while many alkaloids have significantly lower concentrations in undifferentiated cell cultures [80]. Fourth, hairy root culturing is not affected by land limitations, pests, or environmental factors such as drought. Lastly, hairy roots can be grown and scaled in bioreactors such as airlift bioreactors, basket bubble reactors, and airlift mesh-draft reactor with wire helixes [81].

6.2 Stable *C. roseus* Hairy Root Engineering

Wild-type hairy roots are naturally induced by the infection of pathogenic bacterium *Agrobacterium rhizogenes* which carries the Ri (root-inducing) plasmid. Transformed hairy roots can be generated by infecting seedlings or explants with a disarmed strain of *A. rhizogenes* carrying both an Ri plasmid and a plasmid containing a T-DNA region with the gene of interest and a selection marker. T-DNA can be randomly incorporated into the nuclear chromosome of the plant after infection [82]. This results in significant clonal variation between independent transformation events. Evidence of this is well documented in literature and is seen by differences in morphology and basal TIA concentrations between hairy root lines. Due to the inherent clonal variation, an inducible promoter is preferred to control transgene expression which allows for comparison of the effects of increased expression to that of the control within the same background. In addition, the use of an inducible promoter system can also help to optimize TIA production by adding the inducer during late exponential growth to separate growth and TIA accumulation. To date, three inducible promoter systems have been successively used in hairy root engineering. A glucocorticoid-inducible promoter is the most commonly used in *C. roseus* hairy roots [83]. An ethanol inducible promoter has also been shown to exhibit low background expression and high inducible expression in *C. roseus* hairy roots [84]. A recent report of *CrBPF1* engineering in *C. roseus* hairy roots used an estradiol-inducible expression system which provides another promoter option for inducing transgene expression [59].

6.3 TIA Pathway Gene Engineering

In order to channel metabolic flux toward TIA production, significant effort has been placed on overexpressing crucial genes in the terpenoid pathway, the indole pathway, or the downstream TIA pathway in *C. roseus* hairy roots (Table 2). AS, the rate-limiting step of indole biosynthesis pathway, is tightly controlled by feedback regulation of both tryptophan and tryptamine. AS holoenzymes are composed of two alpha and two beta subunits. The binding site of tryptophan for feedback inhibition is present in the alpha subunit [85]. Constitutive overexpression of *ASβ* subunit coupled with inducible overexpression of a feedback-resistant *ASα* subunit from *Arabidopsis* in *C. roseus* hairy roots resulted in the increased concentration of tryptophan, tryptamine, and ajmalicine, while the concentration of lochnericine, hörhammericine, and tabersonine decreased [86]. Feeding a terpenoid precursor loganin to an *ASαβ* overexpressing hairy root line helped enhance the downstream accumulation of catharanthine, ajmalicine, lochnericine, and tabersonine compared to unfed line. These increases are limited compared to the large increase in tryptophan [87]. Overexpressing *TDC* from the indole pathway did not change tryptamine accumulation but did result in increased serpentine production [88]. Overexpressing both *TDC* and *ASα* only enhanced tryptamine accumulation but did not change

Table 2 Metabolic engineering of *C. roseus* hairy root studies and the major results

Engineered gene/ genes	Enhanced TIAs	Reduced TIAs	Upregulated genes	Downregulated genes	Ref.
AtASα + CrASβ	Tryptophan, tryptamine, and ajmalicine	Lochnericine, hörhammericine, and tabersonine	TDC, DXS, G10H, SLS, LAMT, ORCA2, and ZCT2	T19H and MAT	[86]
CrTDC	Serpentine	NC	NA	NA	[88]
AtASα + CrTDC	Tryptamine	NC	NA	NA	[88]
CrDXS	Ajmalicine, lochnericine, and serpentine	Tabersonine and hörhammericine	NA	NA	[89]
CrDXS + CrG10H	Ajmalicine, lochnericine, and tabersonine	NC	NA	NA	[89]
CrDXS + AtASα	Tabersonine, lochnericine, and hörhammericine	NC	NA	NA	[89]
CrPrx	Ajmalicine and serpentine		TDC, G10H, SGD		[90]
CrDAT		Hörhammericine		MAT	[40]
CrORCA2	Serpentine, 16-hydroxytabersonine, and 19-hydroxytabersonine	Secologanin, strictosidine, tabersonine, and hörhammericine	*ASα*, *TDC*, *LAMT*, *STR*, *ORCA3*, and *ZCTs*	SGD and DAT	[36]
CrORCA3	Serpentine and ajmalicine	Tabersonine, lochnericine, hörhammericine	*AS*, *DXS*, *SLS*, *STR*, and *ZCTs*	SGD	[35]

(*continued*)

Table 2 (continued)

Engineered gene/ genes	Enhanced TIAs	Reduced TIAs	Upregulated genes	Downregulated genes	Ref.
CrWRKY1	Serpentine	Catharanthine	TDC and ZCTs	ORCA2, ORCA3, and MYC2	[91]
CrBPF1		Serpentine	G10H	T19H	[59]
CrORCA3 + CrG10H	Catharanthine	NA	NA	NA	[92]
CrORCA3 + CrSGD	Serpentine, ajmalicine, catharanthine, tabersonine, lochnericine, and hörhammericine	ND	AS, TDC, G10H, SLS, STR, T19H, ORCA2, and ZCTs	ND	[93]

NA not analyzed. *NC* no change. *ND* not detected

downstream alkaloid concentrations. Thus, single *AS* overexpression is sufficient to boost production of the indole precursor tryptamine [88].

Overexpressing the terpenoid pathway gene *DXS* in *C. roseus* hairy roots alone led to a significant increase in ajmalicine, serpentine, and lochnericine and a significant decrease in tabersonine and hörhammericine [89]. Notably, co-overexpression of two pathway genes (*DXS* and *G10H* or *DXS* and *AS*) revealed a promising result in increasing downstream TIAs. Simultaneously, overexpressing *DXS* and *G10H* caused a significant increase in ajmalicine, lochnericine, and tabersonine. This suggests that modifying the expression of these two genes may overcome the regulation around the DMAPP and IPP branch point through a push-and-pull mechanism to direct the flux toward secologanin and downstream alkaloid biosynthesis [89]. Moreover, overexpression of *DXS* in combination with the limiting step *AS* in the indole pathway resulted in an increase in several downstream metabolites by the quick turnover of both terpenoid and indole precursors. This alleviated the decrease of some TIA metabolites from overexpressing *DXS* or *AS* [85, 89].

Overexpressing some TIA pathway-specific genes has also been investigated in recent years (Table 2). The last enzyme gene in the vindoline pathway *DAT* was overexpressed in *C. roseus* hairy roots. Unsurprisingly, it did not lead to the accumulation of vindoline due to the absence of the expression of early vindoline pathway genes in hairy roots. Finally, hörhammericine concentrations increased due to inhibition of MAT by DAT [40], and overexpression of a peroxidase gene *CrPrx* in *C. roseus* hairy root lines elevated levels of ajmalicine and serpentine and upregulated pathway genes *TDC*, *G10H*, and *SGD* [90]. The above studies to overexpress pathway genes had limited success in increasing overall alkaloid production in hairy root cultures of *C. roseus*. Increases in some alkaloids were often accompanied by decreases in other alkaloids. Although overexpressing multiple pathway genes gave promising results, the increases in TIA production are limited in magnitude.

6.4 TIA Pathway Regulator Engineering

Since transcription factors control the expression of multiple pathway genes, overexpressing positive regulators is one way to coordinately upregulate multiple pathway genes which could lead to the increase in TIA metabolites. So far, increases and decreases in TIA metabolites were observed in hairy root lines overexpressing transcription factors (Tables 2). *ORCA3* overexpression in *C. roseus* hairy roots leads to a slight increase of serpentine and ajmalicine. In contrast, downstream metabolites tabersonine, lochnericine, and hörhammericine were decreased. The RT-qPCR analysis of hairy roots overexpressing *ORCA3* showed *AS*, *DXS*, *SLS*, *STR*, and *ZCT* expression were increased; *G10H*, *TDC*, and *CPR* mRNA levels remain unchanged; and *SGD* showed decrease [35]. The *ORCA2* engineered hairy roots produced more serpentine, 16-hydroxytabersonine, and 19-hydroxytabersonine but less secologanin, strictosidine, tabersonine, and hörhammericine than the control. ORCA2 can boost *ASα*, *TDC*, *LAMT*, *STR*, *ORCA3*, and *ZCT* transcriptional levels in *C. roseus* hairy roots [36]. Another jasmonate-induced transcription factor gene *WRKY1* was overexpressed in *C. roseus* hairy roots and resulted in an increase in serpentine but a repression in catharanthine. *TDC* and *ZCTs* were upregulated, while *ORCA2*, *ORCA3*, and *MYC2* were downregulated after overexpression *WRKY1* [91]. The effects of overexpressing BPF1 in *C. roseus* hairy roots were assessed recently. Overexpression of *BPF1* had only modest effects on the levels of measured metabolite [59].

Overexpression of a single positive transcription factor in hairy roots also led to mixed results in alkaloid concentrations and had limited success in increasing alkaloid production. This may be caused by two reasons. First, different transcription factors regulate shared but distinct set of pathway genes. The pathway genes that are not under the control of the engineered transcription factor may become the rate-limiting step of the pathway. In *ORCA3* transgenic hairy roots, *G10H* mRNA remained unchanged, and *SGD* was downregulated after inducing *ORCA3* overexpression. Previous analysis of the *G10H* promoter showed that it contains unique binding sites for several transcriptional factors but did not include a binding site for ORCA3, indicating the *G10H* may be regulated by a different transcriptional cascade [51]. Based on this observation, *G10H* and *ORCA3* were co-overexpressed in *C. roseus* hairy roots. There was a threefold increase in catharanthine compared to the control [92]. In addition, combining *ORCA3* and *SGD* overexpression resulted in a significant increase in all measured alkaloids [93]. Thus, combining overexpression of a positive regulator and a pathway gene which is not controlled by this regulator provided a way to enhance alkaloid production. A second cause of single TF engineering failure is the upregulation of negative transcription factors (ZCTs) in most reported TF engineering studies. Future studies could focus on eliminating the negative regulation of the TIA pathway after gene modification in *C. roses* engineering.

7 Future Directions

7.1 Further Elucidation of the TIA Pathway

C. roseus transcriptome sequencing offers a great resource for the selection of candidate genes with hypothesized functions. Such transcriptome databases from RNA-seq analysis were released by three main initiatives, the Medicinal Plant Genomics Resource (MPGR) [94], CathaCyc and Online Resource for Community Annotation of Eukaryotes (OrcAE) [95] and PhytoMetaSyn (PMS) [96], as well as other independent studies [97, 98]. Although over 10 TIA pathway genes have been discovered using these resources over the past 5 years, many unknown genes in this pathway need to be further deciphered. The combination of these datasets might greatly improve such gene discovery by providing high-quality, full-length mRNA for further functional characterization and by allowing the comparison of more experimental or tissue conditions for co-expression analysis. However, it is challenging to merge these databases. The apparent allelic diversity between individual plants used for generating these transcriptomes can cause failure when utilizing bioinformatic tools to combine them through de novo assembly or bring complexity to downstream bioinformatic analysis. Since single nucleotide polymorphisms (SNP) are inevitable, a better de novo assembly algorithm should be developed to solve this problem, or the length of indels should be carefully chosen when running de novo assembly.

In addition, transcriptome-based gene discovery of the TIA pathway in *C. roseus* is also limited by the erroneous assembly of close transcripts and isoforms. Recent studies described two isoforms of *T16H* [23] (*T16H1* and *T16H2* that share 86.5% nucleic acid similarity and 82% amino acid sequence identity) encoded by two distinct genes displaying different tissue-specific expression patterns and two *SLS* isoforms (*SLS1* and *SLS2* sharing 97% nucleotide sequence identity) showing different organ preference. Unfortunately, none of the three major *C. roseus* transcriptome resources is able to correctly identify these isoforms, which may result from improper de novo assembly or insufficient sequencing depth of samples. Courdavault's group, together with collaborators, constructed a more exhaustive and consensus transcriptome resource for *C. roseus* by re-processing previously available transcriptome data with adjusted bioinformatic analysis strategies [99]. They first generated assemblies using trinity for every available sample to take advantage of the diversity of experimental and tissue conditions. Next they combined the assemblies and tested different thresholds within the algorithm to cluster homologous contigs. Special care was taken to reduce the redundancy without affecting transcript quality. Optimization of this consensus assembly was performed by monitoring reconstruction quality of all known TIA genes, with a particular emphasis on the *T16H* isoforms and *SLS* isoforms. This new transcriptome dataset allowed a precise estimation of *T16H* and *SLS* isoform abundance, but further improvements and complementary resources should be advanced by developing advanced bioinformatic tools and exploring more experimental conditions.

With the availability of multiple *C. roseus* transcriptome resources containing a tremendous amount of information, gene discovery can go beyond uncovering TIA biosynthetic enzymes. Candidate genes for TIA pathway transcription factors and metabolite transporters can also be identified from these databases. For instance, Goossens' group identified a jasmonate-regulated basic helix-loop-helix (bHLH) transcription factor bHLH iridoid synthesis 1 (BIS1), which mainly regulates the terpenoid pathway and may act in a complementary manner to the ORCAs. They identified the *bHLH* candidates by screening for transcripts from multiple available RNA-seq database with several criteria: (1) TFs could be induced by JA in both cell suspensions and seedlings but not induced by ORCA overexpression in cell suspensions, (2) TFs showed expression patterns similar to terpenoid pathway genes including *G8O*, and (3) TFs belong to clade IVa of the bHLH family [55]. In general, WRKY TFs [58], MYB-related TFs, and jasmonate ZIM domain (JAZ) [100] proteins are all potential targets for *C. roseus* TIA pathway regulators to look for from the substantial transcriptome databases. Besides transcription factors, a unique catharanthine transporter (CrTPT2), which mediates transporting catharanthine to the surface of leaves, was identified from the inspection of a leaf epidermis-enriched transcript database that was shown to preferentially express TIA pathway genes. In all, the transcriptome-based gene discovery approaches exhibit great power and potential in elucidation of the TIA pathway from catalytic enzymes and transcription factors to metabolite transporters. The improvement of the *C. roseus* unigene set and the down-selection strategies for candidate gene selection should be further exploited.

Meanwhile, work to develop the genomic sequence of *C. roseus* is ongoing. The O'connor and the Buell groups generated a genome assembly for *C. roseus* that provides a near-comprehensive representation of the genetic space. This assembly revealed the genomic context of key points within the TIA biosynthetic pathway including cluster genes, tandem gene duplication, expression sub-functionalization, and putative neo-functionalization [101]. The availability of this genome assembly for *C. roseus* can greatly facilitate high-resolution co-expression analysis and accelerate the discovery of the remaining unknown TIA genes.

7.2 Further Development of Metabolic Engineering

7.2.1 Multiple Gene Engineering

The TIA biosynthesis pathway consists of more than 30 enzymes and is regulated by at least ten regulators. Genetic modifications of one or two genes or regulators of the pathway may not be sufficient to alleviate all the metabolic bottlenecks. However, multiple gene transformation in plants is difficult due to the technical limitations of existing methods. Traditional *Agrobacterium*-mediated transformation has advanced greatly over the last decade in some model plants to generate stable transgenic lines [102]. Unfortunately, little progress has been reported for the multiple genetic modifications in the recalcitrant plant *C. roseus*. Currently, the maximum number of manipulated genes in *C. roseus* hairy roots or cell suspension cultures is two. The multigene overexpression could be achieved either by constructing a vector containing multiple genes or transferring multiple single gene vectors in one host cell [103]. Thus, various plasmids suitable for multigene cloning developed in other plants could be tested in *C. roseus* hairy roots. A binary vector *pHUGE* for *Agrobacterium*-mediated transformation capable of transferring up to nine genes at once was reported in *Rhizobium symbiotic* [104].

7.2.2 Metabolic Reprogramming of Biosynthesis for Production of Unnatural Alkaloids in Plant Culture

TIAs from *C. roseus* have served as an inspiration for novel pharmaceuticals as well. Analogs of natural TIAs (also called unnatural product) can reveal improved or new medicinal properties. Metabolic engineering strategies to produce unnatural plant-derived products are largely unprecedented. In the following example, however, protein engineering has been used to alter the substrate specificity of a TIA pathway enzyme to produce new compounds. Strictosidine synthase (STR) is considered to be the key bottleneck in the TIA pathway accepting tryptamine and secologanin as substrates. An STR mutant was developed based on the reported crystal structure of STR allowing for broadened substrate specificities which in turn resulted in a greater variety of strictosidine analogs in hairy root cell cultures [105]. This strategy can also be combined with the suppression of plant native pathways to yield additional unnatural alkaloids. For instance, silencing the tryptamine biosynthesis and feeding an unnatural tryptamine analog in *C. roseus* root cultures led to the production of a variety of novel products derived from this unnatural starting substrate [106]. Despite the under-explored use of this metabolic engineering strategy, it holds the potential to effectively tailor the production of desired unnatural alkaloids practically and efficiently as more unknown genes are identified.

7.2.3 Heterologous Expression of TIA Pathway

Although the whole TIA pathway in *C. roseus* has not been fully uncovered, reconstitution of the pathway in microbial host is a promising approach for rapid and inexpensive production of these complex compounds and their precursors. The heterologous expression of the plant pathway in yeast, *E. coli*, or other organisms can bypass the tight regulation of the secondary pathway within the plant allowing for the high and constant yield. Additionally, it can avoid the complex mixture of metabolites in plants which makes separation of products challenging, laborious, and expensive. This strategy has been evidenced by successful production of other plant-derived terpenes, flavonoids, and benzylisoquinoline alkaloids in microorganisms [107–109]. In addition, expression of part of the TIA pathway in yeast has been achieved. Soon after the complete elucidation of terpenoid biosynthesis of the TIA pathway [6], O'Connor's group reported de novo strictosidine production in *Saccharomyces cerevisiae* host by introduction of 14 known monoterpene pathway genes along with an additional seven genes and three genes deletions that enhance secondary metabolism [110]. Moreover, followed by the completion of the tabersonine to vindoline pathway, De Luca group assembled the seven-step vindoline pathway in yeast to assess vindoline production by feeding tabersonine precursor [24]. These studies provide an important foundation for developing the production of more complex *C. roseus* TIAs in microorganisms in the future.

7.2.4 Combination of Gene Overexpression and Gene Silencing

Many crucial TIA pathway genes and positive transcription factors have been engineered in *C. roseus* hairy roots for the overproduction of alkaloids, yet they have not been very successful in pushing metabolic flux toward TIA metabolite accumulation [35, 36, 86]. Co-overexpression of two TIA pathway genes or paring overexpression of the positive transcription factor with a pathway gene that is not controlled by this regulator resulted in a significant but uneven increase in overall TIA levels [89, 93, 111]. Multiple studies showed that the overexpression of TIA genes in *C. roseus* hairy roots can upregulate both positive transcription factors and negative transcription factors. This may be the primary reason only small increases in TIA metabolites have been observed. These tight regulatory mechanisms in plants can fine-tune TIA biosynthesis to help the plant modulate their energy and resources to balance between growth and defense. Determining how to bypass this tight regulation in *C. roseus* hairy roots has become a big challenge for further enhancing TIA metabolite production.

8 Conclusions

The advances in high-throughput sequencing technology, in situ RNA hybridization, and reverse transgenic techniques have been used to fill many important gaps in our understanding of TIA pathway in *C. roseus* in recent years. Unfortunately there are still large gaps in our knowledge of the TIA pathway. In addition, metabolic engineering of *C. roseus* hairy roots has made some progresses in enhancing TIA productions, but these increases are still restricted by the tight regulation. Therefore, further elucidation of the regulatory network of the TIA pathway and more creative strategies to bypass the tight regulation are required for building a robust and efficient production platform in *C. roseus* hairy roots.

References

1. van der Heijden R, Jacobs DI, Snoeijer W, Hallared D, Verpoorte R (2004) The *Catharanthus* alkaloids: pharmacognosy and biotechnology. Curr Med Chem 11:607–628
2. Tiong SH, Looi CY, Hazni H, Arya A, Paydar M, Wong WF et al (2013) Antidiabetic and antioxidant properties of alkaloids from *Catharanthus roseus* (l.) g. Don. Molecules 18:9770–9784
3. Wink M, Schmeller T, Latz-Bruning B (1998) Modes of action of allelochemical alkaloids: interaction with neuroreceptors, DNA, and other molecular targets. J Chem Ecol 24:1881–1937
4. Tundis R, Loizzo MR, Menichini F, Statti GA, Menichini F (2008) Biological and pharmacological activities of iridoids: recent developments. Mini Rev Med Chem 8:399–420
5. Viljoen A, Mncwangi N, Vermaak I (2012) Anti-inflammatory iridoids of botanical origin. Curr Med Chem 19:2104–2127
6. Miettinen K, Dong L, Navrot N, Schneider T, Burlat V, Pollier J et al (2014) The seco-iridoid pathway from *Catharanthus roseus*. Nat Commun 5:3606
7. Contin A, van der Heijden R, Lefeber AWM, Verpoorte R (1998) The iridoid glucoside secologanin is derived from the novel triose phosphate/pyruvate pathway in a *Catharanthus roseus* cell culture. FEBS Lett 434:413–416
8. Simkin AJ, Miettinen K, Claudel P, Burlat V, Guirimand G, Courdavault V et al (2013) Characterization of the plastidial geraniol synthase from Madagascar periwinkle which initiates the monoterpenoid branch of the alkaloid pathway in internal phloem associated parenchyma. Phytochemistry 85:36–43
9. Collu G, Unver N, Peltenburg-Looman AM, van der Heijden R, Verpoorte R, Memelink J (2001) Geraniol 10-hydroxylase, a cytochrome P450 enzyme involved in terpenoid indole alkaloid biosynthesis. FEBS Lett 508:215–220
10. Geu-Flores F, Sherden NH, Courdavault V, Burlat V, Glenn WS, Wu C et al (2012) An alternative route to cyclic terpenes by reductive cyclization in iridoid biosynthesis. Nature 492:138–142

11. Salim V, Wiens B, Masada-Atsumi S, Yu F, De Luca V (2014) 7-deoxyloganetic acid synthase catalyzes a key 3 step oxidation to form 7-deoxyloganetic acid in *Catharanthus roseus* iridoid biosynthesis. Phytochemistry 101:23–31
12. Salim V, Yu F, Altarejos J, De Luca V (2013) Virus-induced gene silencing identifies *Catharanthus roseus* 7-deoxyloganic acid-7-hydroxylase, a step in iridoid and monoterpene indole alkaloid biosynthesis. Plant J 76:754–765
13. Irmler S, Schroder G, St-Pierre B, Crouch NP, Hotze M, Schmidt J et al (2000) Indole alkaloid biosynthesis in *Catharanthus roseus*: new enzyme activities and identification of cytochrome P450 CYP72A1 as secologanin synthase. Plant J 24:797–804
14. Murata J, Roepke J, Gordon H, De Luca V (2008) The leaf epidermome of *Catharanthus roseus* reveals its biochemical specialization. Plant Cell 20:524–542
15. Poulsen C, Bongaerts RJ, Verpoorte R (1993) Purification and characterization of anthranilate synthase from *Catharanthus roseus*. Eur J Biochem 212:431–440
16. De Luca V, Marineau C, Brisson N (1989) Molecular cloning and analysis of cdna encoding a plant tryptophan decarboxylase: comparison with animal dopa decarboxylases. Proc Natl Acad Sci U S A 86:2582–2586
17. Noe W, Mollenschott C, Berlin J (1984) Tryptophan decarboxylase from *Catharanthus roseus* cell suspension cultures: purification, molecular and kinetic data of the homogenous protein. Plant Mol Biol 3:281–288
18. Mcknight TD, Roessner CA, Devagupta R, Scott AI, Nessler CL (1990) Nucleotide-sequence of a cdna-encoding the vacuolar protein strictosidine synthase from *Catharanthus roseus*. Nucleic Acids Res 18:4939
19. Luijendijk TJC, Stevens LH, Verpoorte R (1998) Purification and characterisation of strictosidine beta-d-glucosidase from *Catharanthus roseus* cell suspension cultures. Plant Physiol Biochem 36:419–425
20. O'Connor SE, Maresh JJ (2006) Chemistry and biology of monoterpene indole alkaloid biosynthesis. Nat Prod Rep 23:532–547
21. El-Sayed M, Choi YH, Frederich M, Roytrakul S, Verpoorte R (2004) Alkaloid accumulation in *Catharanthus roseus* cell suspension cultures fed with stemmadenine. Biotechnol Lett 26:793–798
22. Levac D, Murata J, Kim WS, De Luca V (2008) Application of carborundum abrasion for investigating the leaf epidermis: molecular cloning of *Catharanthus roseus* 16-hydroxytabersonine-16-*o*-methyltransferase. Plant J 53:225–236
23. Besseau S, Kellner F, Lanoue A, Thamm AM, Salim V, Schneider B et al (2013) A pair of tabersonine 16-hydroxylases initiates the synthesis of vindoline in an organ-dependent manner in *Catharanthus roseus*. Plant Physiol 163:1792–1803
24. Qu Y, Easson ML, Froese J, Simionescu R, Hudlicky T, De Luca V (2015) Completion of the seven-step pathway from tabersonine to the anticancer drug precursor vindoline and its assembly in yeast. Proc Natl Acad Sci U S A 112:6224–6229
25. Liscombe DK, Usera AR, O'Connor SE (2010) Homolog of tocopherol c methyltransferases catalyzes n methylation in anticancer alkaloid biosynthesis. Proc Natl Acad Sci U S A 107:18793–18798
26. Vazquez-Flota F, De Carolis E, Alarco AM, De Luca V (1997) Molecular cloning and characterization of desacetoxyvindoline-4-hydroxylase, a 2-oxoglutarate dependent-dioxygenase involved in the biosynthesis of vindoline in *Catharanthus roseus* (l.) g. Don. Plant Mol Biol 34:935–948
27. Power R, Kurz WG, De Luca V (1990) Purification and characterization of acetylcoenzyme a: deacetylvindoline 4-*o*-acetyltransferase from *Catharanthus roseus*. Arch Biochem Biophys 279:370–376
28. Rodriguez S, Compagnon V, Crouch NP, St-Pierre B, De Luca V (2003) Jasmonate-induced epoxidation of tabersonine by a cytochrome O450 in hairy root cultures of *Catharanthus roseus*. Phytochemistry 64:401–409

29. Giddings LA, Liscombe DK, Hamilton JP, Childs KL, DellaPenna D, Buell CR et al (2011) A stereoselective hydroxylation step of alkaloid biosynthesis by a unique cytochrome P450 in *Catharanthus roseus*. J Biol Chem 286:16751–16757
30. Laflamme P, St-Pierre B, De Luca V (2001) Molecular and biochemical analysis of a Madagascar periwinkle root-specific minovincinine-19-hydroxy-*o*-acetyltransferase. Plant Physiol 125:189–198
31. Verma P, Mathur AK, Srivastava A, Mathur A (2012) Emerging trends in research on spatial and temporal organization of terpenoid indole alkaloid pathway in *Catharanthus roseus*: a literature update. Protoplasma 249:255–268
32. De Luca V, Cutler AJ (1987) Subcellular localization of enzymes involved in indole alkaloid biosynthesis in *Catharanthus roseus*. Plant Physiol 85:1099–1102
33. Oudin A, Mahroug S, Courdavault V, Hervouet N, Zelwer C, Rodriguez-Concepcion M et al (2007) Spatial distribution and hormonal regulation of gene products from methyl erythritol phosphate and monoterpene-secoiridoid pathways in *Catharanthus roseus*. Plant Mol Biol 65:13–30
34. Courdavault V, Papon N, Clastre M, Giglioli-Guivarc'h N, St-Pierre B, Burlat V (2014) A look inside an alkaloid multisite plant: the *Catharanthus* logistics. Curr Opin Plant Biol 19:43–50
35. Peebles CA, Hughes EH, Shanks JV, San KY (2009) Transcriptional response of the terpenoid indole alkaloid pathway to the overexpression of orca3 along with jasmonic acid elicitation of *Catharanthus roseus* hairy roots over time. Metab Eng 11:76–86
36. Li CY, Leopold AL, Sander GW, Shanks JV, Zhao L, Gibson SI (2013) The ORCA2 transcription factor plays a key role in regulation of the terpenoid indole alkaloid pathway. BMC Plant Biol 13:155
37. Guirimand G, Courdavault V, Lanoue A, Mahroug S, Guihur A, Blanc N et al (2010) Strictosidine activation in apocynaceae: towards a "nuclear time bomb"? BMC Plant Biol 10:182
38. Guirimand G, Guihur A, Poutrain P, Hericourt F, Mahroug S, St-Pierre B et al (2011) Spatial organization of the vindoline biosynthetic pathway in *Catharanthus roseus*. J Plant Physiol 168:549–557
39. Yu F, De Luca V (2013) Atp-binding cassette transporter controls leaf surface secretion of anticancer drug components in *Catharanthus roseus*. Proc Natl Acad Sci U S A 110:15830–15835
40. Magnotta M, Murata J, Chen J, De Luca V (2007) Expression of deacetylvindoline-4-o-acetyltransferase in *Catharanthus roseus* hairy roots. Phytochemistry 68:1922–1931
41. Moreno-Valenzuela OA, Minero-Garcia Y, Brito-Argaez L, Carbajal-Mora E, Echeverria O, Vazquez-Nin G et al (2003) Immunocytolocalization of tryptophan decarboxylase in *Catharanthus roseus* hairy roots. Mol Biotechnol 23:11–18
42. Papon N, Bremer J, Vansiri A, Andreu F, Rideau M, Creche J (2005) Cytokinin and ethylene control indole alkaloid production at the level of the mep/terpenoid pathway in *Catharanthus roseus* suspension cells. Planta Med 71:572–574
43. Chahed K (2000) 1-deoxy-d-xylulose 5-phosphate synthase from periwinkle: cdna identification and induced gene expression in terpenoid indole alkaloid-producing cells. Plant Physiol Biochem 38:559
44. Courdavault V (2005) Characterisation of caax-prenyltransferases in *Catharanthus roseus*: relationships with the expression of genes involved in the early stages of monoterpenoid biosynthetic pathway. Plant Sci 168:1097
45. Poutrain P, Mazars C, Thiersault M, Rideau M, Pichon O (2009) Two distinct intracellular ca^{2+}-release components act in opposite ways in the regulation of the auxin-dependent mia biosynthesis in *Catharanthus roseus* cells. J Exp Bot 60:1387–1398
46. Pan QF, Chen Y, Wang Q, Yuan F, Xing SH, Tian YS et al (2010) Effect of plant growth regulators on the biosynthesis of vinblastine, vindoline and catharanthine in *Catharanthus roseus*. Plant Growth Regul 60:133–141

47. Zhou ML, Hou HL, Zhu XM, Shao JR, Wu YM, Tang YX (2011) Molecular regulation of terpenoid indole alkaloids pathway in the medicinal plant, *Catharanthus roseus*. J Med Plants Res 5:663–676
48. Peebles CA, Shanks JV, San KY (2009) The role of the octadecanoid pathway in the production of terpenoid indole alkaloids in *Catharanthus roseus* hairy roots under normal and UV-B stress conditions. Biotechnol Bioeng 103:1248–1254
49. Raina SK, Wankhede DP, Jaggi M, Singh P, Jalmi SK, Raghuram B et al (2012) CrMPK3, a mitogen activated protein kinase from *Catharanthus roseus* and its possible role in stress induced biosynthesis of monoterpenoid indole alkaloids. BMC Plant Biol 12:134
50. Menke FL, Champion A, Kijne JW, Memelink J (1999) A novel jasmonate- and elicitor-responsive element in the periwinkle secondary metabolite biosynthetic gene str interacts with a jasmonate- and elicitor-inducible ap2-domain transcription factor, ORCA2. EMBO J 18:4455–4463
51. Suttipanta N, Pattanaik S, Gunjan S, Xie CH, Littleton J, Yuan L (2007) Promoter analysis of the *Catharanthus roseus* geraniol 10-hydroxylase gene involved in terpenoid indole alkaloid biosynthesis. Biochim Biophys Acta 1769:139–148
52. Zhang H, Hedhili S, Montiel G, Zhang Y, Chatel G, Pre M et al (2011) The basic helix-loop-helix transcription factor crMYC2 controls the jasmonate-responsive expression of the ORCA genes that regulate alkaloid biosynthesis in *Catharanthus roseus*. Plant J 67:61–71
53. De Geyter N, Gholami A, Goormachtig S, Goossens A (2012) Transcriptional machineries in jasmonate-elicited plant secondary metabolism. Trends Plant Sci 17:349–359
54. Chatel G, Montiel G, Pre M, Memelink J, Thiersault M, Saint-Pierre B et al (2003) CrMYC1, a *Catharanthus roseus* elicitor- and jasmonate-responsive bHLH transcription factor that binds the g-box element of the strictosidine synthase gene promoter. J Exp Bot 54:2587–2588
55. Van Moerkercke A, Steensma P, Schweizer F, Pollier J, Gariboldi I, Payne R et al (2015) The bHLH transcription factor bis1 controls the iridoid branch of the monoterpenoid indole alkaloid pathway in *Catharanthus roseus*. Proc Natl Acad Sci U S A 112:8130–8135
56. Pauw B, Hilliou FAO, Martin VS, Chatel G, de Wolf CJF, Champion A et al (2004) Zinc finger proteins act as transcriptional repressors of alkaloid biosynthesis genes in *Catharanthus roseus*. J Biol Chem 279:52940–52948
57. Roberts SGE (2000) Mechanisms of action of transcription activation and repression domains. Cell Mol Life Sci 57:1149–1160
58. Schluttenhofer C, Pattanaik S, Patra B, Yuan L (2014) Analyses of *Catharanthus roseus* and *Arabidopsis thaliana* wrky transcription factors reveal involvement in jasmonate signaling. BMC Genomics 15:502
59. Li CY, Leopold AL, Sander GW, Shanks JV, Zhao L, Gibson SI (2015) CrBPF1 overexpression alters transcript levels of terpenoid indole alkaloid biosynthetic and regulatory genes. Front Plant Sci 6:818
60. Siberil Y, Benhamron S, Memelink J, Giglioli-Guivarc'h N, Thiersault M, Boisson B et al (2001) *Catharanthus roseus* G-box binding factors 1 and 2 act as repressors of strictosidine synthase gene expression in cell cultures. Plant Mol Biol 45:477–488
61. Hirata K, Asada M, Yatani E, Miyamoto K, Miura Y (1993) Effects of near-ultraviolet light on alkaloid production in *Catharanthus roseus* plants. Planta Med 59:46–50
62. Ramani S, Chelliah J (2007) UV-B-induced signaling events leading to enhanced-production of catharanthine in *Catharanthus roseus* cell suspension cultures. BMC Plant Biol 7:61
63. Binder BYK, Peebles CAM, Shanks JV, San KY (2009) The effects of uv-b stress on the production of terpenoid indole alkaloids in *Catharanthus roseus* hairy roots. Biotechnol Prog 25:861–865
64. Sriram G, Fulton DB, Shanks JV (2007) Flux quantification in central carbon metabolism of *Catharanthus roseus* hairy roots by 13c labeling and comprehensive bondomer balancing. Phytochemistry 68:2243–2257
65. Morgan JA, Shanks JV (2000) Determination of metabolic rate-limitations by precursor feeding in *Catharanthus roseus* hairy root cultures. J Biotechnol 79:137–145

66. Uesato S (1986) Mechanism for iridane skeleton formation in the biosynthesis of secologanin and indole alkaloids in *lonicera tatarica*, *Catharanthus roseus* and suspension cultures of rauwolfia serpentina. Phytochemistry 25:839
67. Zhao J, Davis LC, Verpoorte R (2005) Elicitor signal transduction leading to production of plant secondary metabolites. Biotechnol Adv 23:283–333
68. Costa MM, Hilliou F, Duarte P, Pereira LG, Almeida I, Leech M et al (2008) Molecular cloning and characterization of a vacuolar class iii peroxidase involved in the metabolism of anticancer alkaloids in *Catharanthus roseus*. Plant Physiol 146:403–417
69. Geerlings A, Ibanez MM, Memelink J, van Der Heijden R, Verpoorte R (2000) Molecular cloning and analysis of strictosidine beta-d-glucosidase, an enzyme in terpenoid indole alkaloid biosynthesis in *Catharanthus roseus*. J Biol Chem 275:3051–3056
70. Schroder G, Unterbusch E, Kaltenbach M, Schmidt J, Strack D, De Luca V et al (1999) Light-induced cytochrome p450-dependent enzyme in indole alkaloid biosynthesis: tabersonine 16-hydroxylase. FEBS Lett 458:97–102
71. Zhou ML (2010) Transcriptional response of the catharanthine biosynthesis pathway to methyl jasmonate/nitric oxide elicitation in *Catharanthus roseus* hairy root culture. Appl Microbiol Biotechnol 88:737
72. Goklany S (2009) Assessing the limitations to terpenoid indole alkaloid biosynthesis in *Catharanthus roseus* hairy root cultures through gene expression profiling and precursor feeding. Biotechnol Prog 25:1289
73. Kutchan TM, Hampp N, Lottspeich F, Beyreuther K, Zenk MH (1988) The cDNA clone for strictosidine synthase from rauvolfia serpentina. DNA sequence determination and expression in *Escherichia coli*. FEBS Lett 237:40–44
74. St-Pierre B, Laflamme P, Alarco AM, De Luca V (1998) The terminal *o*-acetyltransferase involved in vindoline biosynthesis defines a new class of proteins responsible for coenzyme a-dependent acyl transfer. Plant J 14:703–713
75. Asada K, Salim V, Masada-Atsumi S, Edmunds E, Nagatoshi M, Terasaka K et al (2013) A 7-deoxyloganetic acid glucosyltransferase contributes a key step in secologanin biosynthesis in *Madagascar periwinkle*. Plant Cell 25:4123–4134
76. O'Connor SE (2012) Strategies for engineering plant natural products: the iridoid-derived monoterpene indole alkaloids of *Catharanthus roseus*. Method Enzymol 515:189–206
77. Gelvin SB (2003) Agrobacterium-mediated plant transformation: the biology behind the "gene-jockeying" tool. Microbiol Mol Biology Rev 67:16–37
78. Peebles CA, Sander GW, Li M, Shanks JV, San KY (2009) Five year maintenance of the inducible expression of anthranilate synthase in *Catharanthus roseus* hairy roots. Biotechnol Bioeng 102:1521–1525
79. Zhou ML, Shao JR, Tang YX (2009) Production and metabolic engineering of terpenoid indole alkaloids in cell cultures of the medicinal plant *Catharanthus roseus (Madagascar periwinkle)*. Biotechnol Appl Biochem 52:313–323
80. Pasquali G, Porto DD, Fett-Neto AG (2006) Metabolic engineering of cell cultures versus whole plant complexity in production of bioactive monoterpene indole alkaloids: recent progress related to old dilemma. J Biosci Bioeng 101:287–296
81. Guillon S, Tremouillaux-Guiller J, Pati PK, Rideau M, Gantet P (2006) Harnessing the potential of hairy roots: dawn of a new era. Trends Biotechnol 24:403–409
82. Taneja J, Jaggi M, Wankhede DP, Sinha AK (2010) Effect of loss of t-DNA genes on mia biosynthetic pathway gene regulation and alkaloid accumulation in *Catharanthus roseus* hairy roots. Plant Cell Rep 29:1119–1129
83. Hughes EH, Hong SB, Shanks JV, San KY, Gibson SI (2002) Characterization of an inducible promoter system in *Catharanthus roseus* hairy roots. Biotechnol Prog 18:1183–1186
84. Peebles CA, Gibson SI, Shanks JV, San KY (2007) Characterization of an ethanol-inducible promoter system in *Catharanthus roseus* hairy roots. Biotechnol Prog 23: 1258–1260

85. Hong SB, Peebles CA, Shanks JV, San KY, Gibson SI (2006) Expression of the *Arabidopsis* feedback-insensitive anthranilate synthase holoenzyme and tryptophan decarboxylase genes in *Catharanthus roseus* hairy roots. J Biotechnol 122:28–38
86. Peebles CA, Hong SB, Gibson SI, Shanks JV, San KY (2005) Transient effects of overexpressing anthranilate synthase alpha and beta subunits in *Catharanthus roseus* hairy roots. Biotechnol Prog 21:1572–1576
87. Peebles CA, Hong SB, Gibson SI, Shanks JV, San KY (2006) Effects of terpenoid precursor feeding on *Catharanthus roseus* hairy roots over-expressing the alpha or the alpha and beta subunits of anthranilate synthase. Biotechnol Bioeng 93:534–540
88. Hughes EH, Shanks JV (2002) Metabolic engineering of plants for alkaloid production. Metab Eng 4:41–48
89. Peebles CA, Sander GW, Hughes EH, Peacock R, Shanks JV, San KY (2011) The expression of 1-deoxy-d-xylulose synthase and geraniol-10-hydroxylase or anthranilate synthase increases terpenoid indole alkaloid accumulation in *Catharanthus roseus* hairy roots. Metab Eng 13:234–240
90. Jaggi M, Kumar S, Sinha AK (2011) Overexpression of an apoplastic peroxidase gene crprx in transgenic hairy root lines of *Catharanthus roseus*. Appl Microbiol Biotechnol 90:1005–1016
91. Suttipanta N, Pattanaik S, Kulshrestha M, Patra B, Singh SK, Yuan L (2011) The transcription factor crWRKY1 positively regulates the terpenoid indole alkaloid biosynthesis in *Catharanthus roseus*. Plant Physiol 157:2081–2093
92. Wang CT, Liu H, Gao XS, Zhang HX (2010) Overexpression of g10h and orca3 in the hairy roots of *Catharanthus roseus* improves catharanthine production. Plant Cell Rep 29:887–894
93. Sun J, Peebles CA. (2015) Engineering overexpression of orca3 and strictosidine glucosidase in *Catharanthus roseus* hairy roots increases alkaloid production. Protoplasma
94. Gongora-Castillo E, Childs KL, Fedewa G, Hamilton JP, Liscombe DK, Magallanes-Lundback M et al (2012) Development of transcriptomic resources for interrogating the biosynthesis of monoterpene indole alkaloids in medicinal plant species. PLoS One 7, e52506
95. Van Moerkercke A, Fabris M, Pollier J, Baart GJ, Rombauts S, Hasnain G et al (2013) Cathacyc, a metabolic pathway database built from *Catharanthus roseus* RNA-Seq data. Plant Cell Physiol 54:673–685
96. Xiao M, Zhang Y, Chen X, Lee EJ, Barber CJS, Chakrabarty R et al (2013) Transcriptome analysis based on next-generation sequencing of non-model plants producing specialized metabolites of biotechnological interest. J Biotechnol 166:122–134
97. Verma M, Ghangal R, Sharma R, Sinha AK, Jain M (2014) Transcriptome analysis of *Catharanthus roseus* for gene discovery and expression profiling. PLoS One 9, e103583
98. Liu LY, Tseng HI, Lin CP, Lin YY, Huang YH, Huang CK et al (2014) High-throughput transcriptome analysis of the leafy flower transition of *Catharanthus roseus* induced by peanut witches'-broom phytoplasma infection. Plant Cell Physiol 55:942–957
99. Duge de Bernonville T, Foureau E, Parage C, Lanoue A, Clastre M, Londono MA et al (2015) Characterization of a second secologanin synthase isoform producing both secologanin and secoxyloganin allows enhanced de novo assembly of a *Catharanthus roseus* transcriptome. BMC Genomics 16:619
100. Duge de Bernonville T, Clastre M, Besseau S, Oudin A, Burlat V, Glevarec G et al (2015) Phytochemical genomics of the *Madagascar periwinkle:* unravelling the last twists of the alkaloid engine. Phytochemistry 113:9–23
101. Kellner F, Kim J, Clavijo BJ, Hamilton JP, Childs KL, Vaillancourt B et al (2015) Genome-guided investigation of plant natural product biosynthesis. Plant J 82:680–692
102. Wilson SA, Roberts SC (2014) Metabolic Engineering approaches for production of biochemicals in food and medicinal plants. Curr Opin Biotechnol 26:174–182
103. Huang Y, Su CY, Kuo HJ, Chen YH, Huang PL, Lee KT (2013) A comparison of strategies for multiple-gene co-transformation via hairy root induction. Appl Microbiol Biotechnol 97:8637–8647

104. Untergasser A, Bijl GJ, Liu W, Bisseling T, Schaart JG, Geurts R (2012) One-step agrobacterium mediated transformation of eight genes essential for rhizobium symbiotic signaling using the novel binary vector system phuge. PLoS One 7, e47885
105. Runguphan W, O'Connor SE (2009) Metabolic reprogramming of periwinkle plant culture. Nat Chem Biol 5:151–153
106. Runguphan W, Maresh JJ, O'Connor SE (2009) Silencing of tryptamine biosynthesis for production of nonnatural alkaloids in plant culture. Proc Natl Acad Sci U S A 106:13673–13678
107. Minami H, Kim JS, Ikezawa N, Takemura T, Katayama T, Kumagai H et al (2008) Microbial production of plant benzylisoquinoline alkaloids. Proc Natl Acad Sci U S A 105:7393–7398
108. Santos CN, Koffas M, Stephanopoulos G (2011) Optimization of a heterologous pathway for the production of flavonoids from glucose. Metab Eng 13:392–400
109. Fossati E, Ekins A, Narcross L, Zhu Y, Falgueyret JP, Beaudoin GA et al (2014) Reconstitution of a 10-gene pathway for synthesis of the plant alkaloid dihydrosanguinarine in *Saccharomyces cerevisiae*. Nat Comm 5:3283
110. Brown S, Clastre M, Courdavault V, O'Connor SE (2015) De novo production of the plant-derived alkaloid strictosidine in yeast. Proc Natl Acad Sci U S A 112:3205–3210
111. Pan Q, Wang Q, Yuan F, Xing S, Zhao J, Choi YH et al (2012) Overexpression of ORCA3 and G10H in *Catharanthus roseus* plants regulated alkaloid biosynthesis and metabolism revealed by NMR-metabolomics. PLoS One 7, e43038
112. Owellen RJ, Hartke CA, Dickerson RM, Hains FO (1976) Inhibition of tubulin-microtubule polymerization by drugs of the Vinca alkaloid class. Cancer Res. 36(4):1499–1502
113. Burlat V, Oudin A, Courtois M, Rideau M, St-Pierre B (2004) Co-expression of three MEP pathway genes and geraniol 10-hydroxylase in internal phloem parenchyma of Catharanthus roseus implicates multicellular translocation of intermediates during the biosynthesis of monoterpene indole alkaloids and isoprenoid-derived primary metabolites. Plant J 38:131–141
114. Guirimand G, Guihur A, Phillips MA, Oudin A, Glévarec G, Melin C, Papon N, Clastre M, St-Pierre B, Rodríguez-Concepción M, Burlat V, Courdavault V (2012) A single gene encodes isopentenyl diphosphate isomerase isoforms targeted to plastids, mitochondria and peroxisomes in Catharanthus roseus. Plant Mol Biol 79:443–459
115. van der Fits L, Memelink J (2000) ORCA3, a jasmonate-reponsive transcriptional regulator of plant primary and secondary metabolism. Science 289(5477):295–297

Pathway Modulation of Medicinal and Aromatic Plants Through Metabolic Engineering Using *Agrobacterium tumefaciens*

17

Sana Khan and Laiq ur Rahman

Abstract

Plants are the most appropriate source of variety of compounds that are useful to mankind as food, fiber, medicines, natural products, industrial raw material, etc. A large number of metabolites (primary and secondary) have been utilized by mankind since many years ago. However, there are certain limitations like overharvesting the natural plant, low amount of metabolite/compound, etc., which have been associated with the availability of natural products/compounds. This common problem has become major hurdles for researchers these days. To address such problems, scientists have moved toward more efficient technique like genetic transformation methods. In today's era, *Agrobacterium tumefaciens*-mediated transformation (ATMT) strategies have shown to be an efficient and most sophisticated technique to understand about modifications, cloning, and diversification in biosynthetic pathways *in planta.* The existing knowledge and many successful achievements in biotechnology sector have facilitated the development of new methods like metabolic engineering to divert the target metabolic flux in transgenic plants.

Keywords

ATMT • *A. tumefaciens* • Biosynthetic pathway • Genetic engineering • Metabolic engineering • MAPs • Metabolic flux

S. Khan • L. ur Rahman (✉)
Plant Biotechnology Division, CSIR-Central Institute of Medicinal and Aromatic Plants, Lucknow, Uttar Pradesh, India
e-mail: sanakhan1905@gmail.com; faizslaiq@gmail.com

S. Jha (ed.), *Transgenesis and Secondary Metabolism*, Reference Series in Phytochemistry, DOI 10.1007/978-3-319-28669-3_15

Abbreviations

ATMT	*A. tumefaciens*-mediated genetic transformation
DW	Dry weight
MAPs	Medicinal and Aromatic Plants
ME	Metabolic engineering
PAL	Phenylalanine ammonia-lyase
T-DNA	Transfer DNA

Contents

1 Introduction

For millennia, significant progress has been made to use plant as an ultimate source of natural products. In highlighting strides to reengineer plant's genome over the past 20 years, genetic manipulation has shown much potential to benefit economies of scale in plant biotechnology sector. Metabolic/pathway engineering is basically an art to reengineer the already existing route in order to retrieve the target flux in enhanced concentration, and at times, accumulation of novel compounds has been reported so far. Plants in their native state comprise a phenomenal feat of metabolic cascades and innumerable diversion of biochemical pathways operated inside them. Although the conventional breeding programs have been proved fruitful that empower crop varieties to withstand biotic and abiotic stresses and to enhance the metabolite content, these breeding strategies are time consuming and laborious. However, metabolic engineering has significantly contributed many advances particularly in genetics/crop physiology and genomics by providing opportunities to alleviate the effect of these stresses and easily combat with factors involved in metabolism. Recent scientific discoveries in the area of ME have opened up new avenues which provide innumerable potential for plant improvement. Advancements in the ME leading to identification, annotation, and characterization of putative genes/enzymes in relation to particular aspect like primary or secondary metabolism have made this technology highly appreciable and feasible as well.

Strategies for diverting the metabolic flux include the engineering of an individual step in a particular pathway in order to enhance/reduce the content of target metabolite/product. In addition, to redirect the flux toward a more predictable aspect, sometimes suppression/arrest/blocking of competitive pathway has been also proved as an exciting tool. Moreover, reducing metabolic flux through competitive pathways overcome the rate-limiting steps in particular biosynthetic routes along with reduction in catabolic reactions, and the overexpression of regulatory genes also proved beneficial and key aspects of genetic engineering [1]. Recently, approaches have been applied to focus multiple genes/enzymatic steps in a particular pathway to control metabolite production. The parameters involved various strategies like the overexpression of enzymes and downregulation of consecutive genes/enzymes in a pathway while terminating those which are involved in competing pathways. This might also involve the various regulation transcription factors (TF) to recognize the multilocus genes which can control over a single and more consecutive pathways operating *in planta*. Some major technical limits, in context to how many numbers of genes can be transferred to a host plant, and new transformation techniques have been developed to transform a plant with multiple transgenes at a time and to express these transgenes in a coordinated manner [2]. The present scenario of paradigm in the biological science of complex cascades in plant system has diverted the current shift toward better, efficient transformation methods such as *Agrobacterium tumefaciens*-mediated genetic transformation systems (ATMT). The ATMT approach, an efficient method through which construct could be made per se, is used to investigate the type and the localization pattern (subcellular, intracellular) of gene of interest. Based on the studies, we here review the potential of metabolic engineering via *Agrobacterium tumefaciens* in the biosynthesis of novel compounds and discuss the prospects for establishing the technology to induce tailor-made synthesis of target metabolic flux. The above described aspects of genetic engineering have been made possible with approaches like biolistics, particle bombardment, but there are lacunas that still exist. The ATMT approach is highly feasible and efficient for these types of manipulation applied *in planta*.

2 Transformation Methods and Strategies

Recent progress has been made over last few years, which have led to the development of an efficient strategy that can be easily accessible with a broad range of agronomically important crops. There are approaches like biolistics (particle bombardment), electroporation, which considerably makes it easy to obtain transient expression at high levels (Table 1). The strategies have been used to check the efficiency of different gene constructs prior to transformation. While transient expression can be used to locate the presence of gene or study metabolic networks

Table 1 Common strategies used for transformation methods

Common methods used for introduction of foreign gene	Type of expression	Compatibility	Delivery system	Advantages	Disadvantages
Electroporation	Highly effective for transient expression	Compatible with every kind host range	High rate of DNA delivery	Very effective	Copy number is very high and therefore leads to gene co-suppression or silencing
Particle bombardment/ biolistic mechanism/gene gun	Highly effective for transient expression	Shows high compatibility with a range of host plant	High rate of DNA delivery	Highly effective	Copy number is very high and therefore leads to gene co-suppression or silencing
Agrobacterium-mediated transformation	Effective for both transient as well as stable expression but stable expression can be achieved in consecutive generations	Host range is limited due to hypersensitivity shown by plants	High rate of DNA delivery	Cheap, effective, and very simple. The technology may be useful in germ line transformation Copy number is often low and therefore high chances of getting transformants	Compatibility problem with a number of plant species especially leguminous species, monocot families, and tree species

in plant and cell tissue culture, it has been determined that the similar genes can also act differently in stable as well as in transiently transformed plants [3]. Transient expression is also used for analyzing the tissue-specific expression of target transgenes.

3 *Agrobacterium tumefaciens*: A Soil-Based Bioengineer in the Era of Advanced Biotechnology

Agrobacterium tumefaciens (scientifically named as *Rhizobium radiobacter*), a gram-negative, rod-shaped soil bacterium from family Rhizobiaceae, is responsible for crown gall disease in over 140 plus species of eudicots, dicotyledons, gymnosperms, and some monocots [4]. During the last years, the crown gall disease was acting as one of the major hurdle in horticultural biosector, as this disease leads to havoc loss in majority of economically important horticultural crops [5] like in grape [6], apple [7], cherry [8], and many more. However, problems like the mode of propagation through grafting of stems and the woody nature of these crops have been acted to be a common issue. The wounds on grafted tissue may serve as an infection channel and the crown gall disease takes place. In 1941 White and Braun observed that only a minute exposure of *A. tumefaciens* to wounded tissue may transform it permanently [9]. Soon after, tumorigenicity was proposed to be transferred through *A. tumefaciens* to host tissue [10, 11]. Further, during 1974, it was identified that the capacity to induce crown gall disease is inherently carried by Ti plasmid of *Agrobacterium* [12, 13]. However, in particular, the molecular characterization techniques like southern hybridization has led to the discovery that Ti plasmid transferred its DNA into the host plant cell and resulted in the origin of T-DNA from *A. tumefaciens* to the host plant cell [14–16]. The Ti plasmid consists of T-DNA which delimits by 25 bp direct repeats sequences a.k.a. T-DNA border sequences. This T-DNA from Ti plasmid gets transferred and integrated in the host plant where it encodes two enzymes, and results into the production of a massive amount of auxin and cytokinin lead to the uncontrolled cell proliferation called crown gall disease. In addition, the production of some low-molecular-weight compounds like amino acid and opines (sugar phosphate derivatives) could be metabolized specifically depending upon the *Agrobacterium* strain. Thus, *A. tumefaciens* strain is solely responsible for the production of different types of opines and ultimately creating an ecological niche which suits to a particular strain of *A. tumefaciens* [17, 18]. The formation of tumor in the host plant's cell is the resultant of T-DNA integration, and this factor has become an intrinsic interest of researchers/scientists for strategic engineering of the plant's genome. The same property of T-DNA has been used widely nowadays to insert a foreign gene of interest for target manipulation without showing any adverse effect in the host plant. The elucidation of the fact that a wild/normal type of T-DNA region can be easily replaced by desired sequence has made *A. tumefaciens* stand out of the queue of other delivery systems. Furthermore, the remarkable feature of stable integration of T-DNA inside the host plant cell has inspired the promise that *A. tumefaciens* may utilize as gene vector (either in homologous/heterologous system) in order to deliver target DNA sequence in the host plant cell.

In the early of 1980s, it was demonstrated that *A. tumefaciens* can be used as an efficient gene delivery system to produce a transgenic plant owing desired characteristics [19]. In addition, the transgenic plants have the ability to transmit the disarmed Ti plasmid to their progeny during the course of segregation, so less chances of a loss of desired gene. Apart from such attractive possibilities, the use of an antibiotic selection strategy makes the system efficient for the accurate selection of probable transgenic plants [20]. The eventual progress of using *A. tumefaciens* as a gene delivery vector for crop improvement has made it possible to use efficiently in the plant biotechnology sector, and the future began to look bright. During the 1990s, a monocot species, maize, was successfully transformed *by A. tumefaciens* [21]. In today's era, successful attempts have been made to transform horticultural cultivated flora and agronomically important species [17]. Recently, important plant species like soybean, canola, cotton, corn, potatoes, etc. have been transformed by *A. tumefaciens*, and improved varieties were released and successfully grown all over the world [22]. At present, apart from plant species, other organisms such as fungi, yeast, and mammalian cells were shown to be susceptible to ATMT [23]. However, nowadays the shift of plant molecular biology has been shifted toward more advanced experimentation which includes the elucidation of various aspects of molecular and biochemical mechanism/changes of T-DNA complex targeting the plant nucleus. Hence, based on the above presumed facts and present scenario, on how efficiently the *A. tumefaciens* could be used for genetic and metabolic manipulations, it can be considered as a bioengineer in advanced biotechnology.

4 An Overview on Expression of Transgene

For the perfect modification of target metabolite profile in respect to the utility of the particular plant, the expression of the gene of target enzymes or proteins has to be fine-tuned in a justified manner. For an appropriate expression of transgene in the host plant, the transcription factors need to be well identified. In context to the location and expression of metabolite, the role of promoters has been known to contribute significantly in engineering the specific host for a particular target. High level of transient expression can be obtained, but to get relatively stable transformed plants is not an easy task. Therefore, plant-specific promoters could be used to achieve a high-level as well as stable expression too. Promoters in metabolic engineering can be divided in three categories: constitutive promoters, organ-specific promoters, and inducible promoters.

The most commonly used constitutive promoter in plant genetic engineering is CaMV 35S promoter (cauliflower mosaic virus 35S) [24, 25]. In plant vectors the CaMV 35S promoter has been used extensively and is known to consist of more than one third of the full length sequence [26]. This promoter is of viral origin and has been thoroughly characterized in metabolic engineering of cell cultures and is able to drive a high level of transgene expression in plant tissues [27, 28]. It has been interpreted that partial duplication from −343 to −90 base pair region of promoter can amplify the expression of the target gene up to tenfold [29]. As, in context to the fact that, CaMV 35S is of viral origin so, could be able to drive a high level of transgene expression *in planta.* However, in contrary, sometimes it can also be resulted into deleterious effects like co-suppression of gene which ultimately lead to gene silencing [30].

On the basis on the type of expression, there are various types of promoters used to target the expression of the gene in a specific manner like organ/tissue/developmental stage. These kinds of tissue-/organ-specific promoters have been characterized at large level in plant engineering system. The respective application of these types of promoters will not hamper the normal plant growth and development. However, at times, it can also lead to the synthesis of desired or value-added novel compounds. The use of the inducible promoter's strategy [31] is highly efficient in order to examine the resultant effect of expression of the transgene in complex biosynthetic routes. For example, pristinamycin is a polyketide antibiotic which is not found in plants normally used to remove the bacterial infection/contamination in the bioengineering of cell cultures.

Likewise, safeners are known as "herbicide antidotes" and the exposure to the safener would lead to the activation of In 2–2 promoter in shoot and root tissues. Safener can be used in the treatment of seed to control the weeds during the planting of material. The use of safener would also be useful in seed bioengineering or in case the expression of transgene is required for a small period of time after germination. In addition, by selecting an appropriate promoter, it could be possible to redirect the transgene expression toward an appropriate need [32] (Table 2).

5 Engineering Primary Metabolism Using ATMT

5.1 Modulation in Primary Metabolites and Aspect of Engineering *In Planta*

The importance of modulating the primary metabolic pathway is a worthy strategy to provide strength at innate immunity level *in planta* to cope up with certain environmental imbalances and pathogen attack. Basically, the plants depend upon the innate immunity to protect themselves from pathogen attack. This innate immunity depends upon the induced defense responses [51]. The primary metabolic pathway undoubtedly plays an undeniable role in providing the cellular energy requirements which in turn help in strengthen the defense system in plants [52, 53]. The injury from ecological stress like pathogen attack, freezing, drought, and flooding often seems to be linked with the overproduction of free oxygen radicals. Superoxide dismutases (SODs) are a class of metalloproteins known to be actively involved in the detoxification of these free oxygen radicals, by breaking down into hydrogen peroxide and molecular oxygen. An overexpression of SODs was done in alfalfa to combat over winter stress [54]. Likewise, alfalfa is also found to be sensitive to aluminum

Table 2 Different plant-based promoters used in transformation experiments and engineering purposes *in planta*

Promoter	Name	Origin	Reference
Constitutive promoter	35S	Derived from viral origin, well characterized in plant metabolic engineering, highly expressed in plant vascular tissues with less expression in meristems tissues	[28]
	Ubiquitin	Plant origin promoter showing high constitutive expression but developmentally regulated	[33]
	Actin	Plant origin promoter showing high constitutive expression but developmentally regulated. Belongs to multigene family and highly expressed in between the tissues	[34–36]
Tissue/ Organ specific	StMCPI/patatin promoters	StMCPI and patatin promoters drives tuber-specific gene expression StMCPI is independent of hormonal and environmental fluctuations and the latter has high expression in sucrose treated leaves and in tubers	[37, 38]
	B-conglycinin promoter	The promoter regulates embryo-specific gene expression	[39, 40]
	OsGT1 promoter, ZmZ27 promoter, Opaque-2 promoter,	These are endosperm-specific promoters and expression is highly regulated by different developmental stages	[41, 42]
	Lhch3 promoter	A leaf-specific promoter derived from *Arabidopsis* and highly regulated by light	[43]
	Lat52 promoter	Lat52 promoter is a pollen-specific promoter and developmentally regulated with high expression during maturation of pollen	[44]
Inducible promoters	Apase promoter	Phosphatase-inducible expression in roots	[45]
	Safener inducible promoter, In 2–2 promoter	Maize In 2–2 promoter is stimulated by benzene sulfonamide herbicide safener	[46]

(*continued*)

Table 2 (continued)

Promoter	Name	Origin	Reference
	Pristinamycin responsive promoter	The promoter based on recombinant transcription factor fusion between VP16 and repressor of pristinamycin (Pip) transactivating domain of herpes simplex virus	[47]
	Ethanol-inducible promoter, glucocorticoid-inducible promoter, ecdysone-inducible promoter	The model is based upon the interaction between specifically designed transcription factor and inducer that can lead into activation of a synthetic promoter	[48–50]

(Al) toxicity, which leads into small roots and low yield. A multipurpose gene named *malate dehydrogenase* was incorporated into Alfalfa which resulted into the production of acetate, succinate, citrate, oxalate, and malate and with a concomitant increase in Al tolerance [55].

Although interminable progress has been made to enlighten the complex cascade steps underlying the developmental defense pathways, to date very little knowledge has been generated about the actual role and involvement of primary pathways in regulating the plant defense mechanisms. In addition, the linked pathways provide a connecting phase between secondary and primary metabolism as illustrated in Fig. 1. However, it has been well documented that the role of primary metabolism is to provide the elementary compounds and precursors for secondary metabolism routes. Based on the convincing experimental evidences only, we have chosen the modulated/engineered pathway studies involving essential primary metabolites like amino acid, carbohydrates, etc.

5.1.1 Essential Amino Acid Metabolism

Engineering the amino acid production pathway to enhance the accumulation of important amino acids like methionine, lysine, threonine, and tryptophan in fruits, food, and crops has been accomplished successfully [56]. Proline has been considered as osmotic protectant and plays a major role in both biotic and abiotic stresses in plants, so engineering the proline pathway has provided a way to cope up with these stresses *in planta* [57]. Transgenic rice with enhanced lysine content was developed by overexpression of seed stored in the β-phaseolin gene from *Phaseolus vulgaris* [58]. Another example includes the production of essential amino acid AmA1 (*Amaranth Albumin 1*) from *Amaranthus hypochondriacus* which leads to the increase in 2.5–4-fold higher content of methionine, tyrosine, and lysine transgenic potatoes [59].

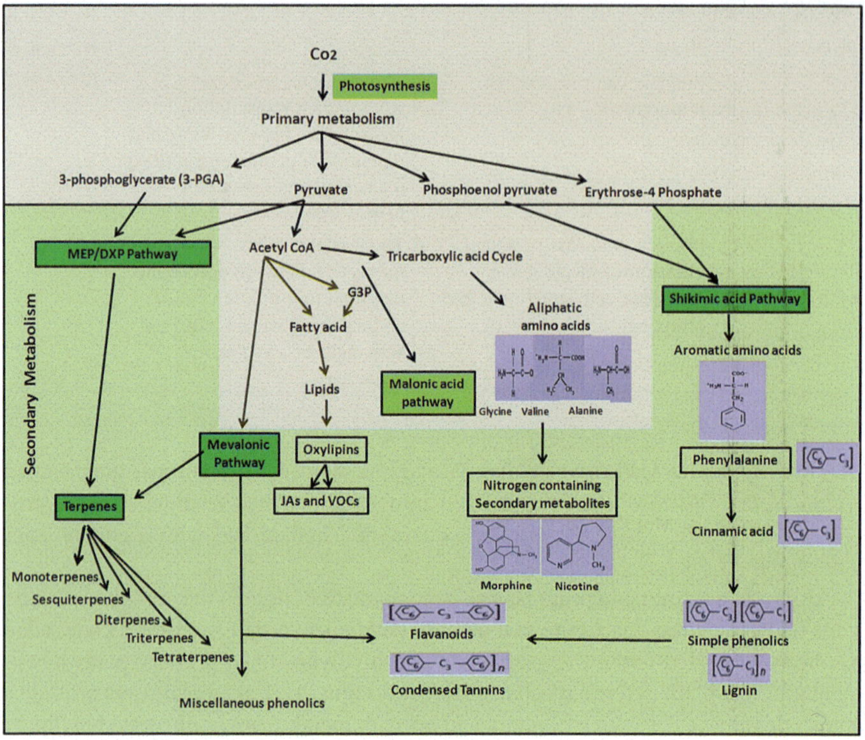

Fig. 1 Interconnection between primary and secondary metabolism *in planta*

In contrast, the lysine content in *Arabidopsis* was either increased by five times through knockout expression of lysine catabolism pathway or twelve times by modulation of bacterial feedback-insensitive dihydrodipicolinate synthase (DHPS) transgene [60].

5.1.2 Carbohydrate Metabolism

In plant kingdom, carbohydrate metabolism involves reversible sequential conversion of sugars into starch and cellulose as photosynthetic products. Starch is an example of stored carbohydrate which accumulates stably in roots, tubers, and seeds while transiently in aerial part like leaves. In the last few years, attention has been given to engineer the starch content and to modify the properties by coordinating relative proportion of amylose and amylopectin content. ME has been used in an attempt to produce novel starch using a heterologous expression of bacterial enzymes by changing the frequency and number of branching chains [61]. Engineering primary metabolite includes enhancing the adenylate pool to increase starch content along with the increase in yield of tubers in potato [62] and overexpression of potato sucrose transporter to enhance the sugar uptake in transgenic pea seeds [63]. Similar experiments have been carried out in rice and wheat species in order to modify the starch content along with seed biomass. A gene from maize was introduced for modification of ADP-glucose pyrophosphorylase (shrunken 2) which resulted in the production of 20% and 40% of higher seed biomass, respectively, than control plants [64, 65]. More recently, applying an antisense waxy (*Wx)* gene under *Wx* promoter was expressed in transgenic rice which led to production of low-level amylose in rice grain [66], and in contrast, applying the same approach to isoamylase-encoding gene has led to the increase in content of amylopectin in rice grain [67]. Strategies have been applied in sequential coordination of the enzymes/genes involved in the starch synthesis, along within the branches/skeleton in monomer unit. This stepwise engineering has successfully led to the production of high-amylopectin and high-amylose starch contents in transgenic potatoes [68]. A novel method for engineering of starch content has been explored, in which starch-binding domains were used to target recombinant enzymes during starch metabolism [69].

Apart from the engineering of starch, manipulations in cellulose biosynthetic pathway have been done. The cellulose is an essential source of fiber and pulp and will provide a way for value addition in industries. However, in addition, it serves as precursors for many commercially exploited products/polymers. Although the biosynthetic pathway for cellulose has not been fully understood to date, various enzymes/gene have been identified and characterized like Korrigan cellulases, sucrose synthase, etc. These studies led to the determination of consecutive aspects of metabolically progressive underlying mechanism operating *in planta.* Furthermore, the suppression of sucrose synthase belonging to Ces-A family of cellulose synthases has shown the simultaneous inhibition in the accumulation of fiber in transgenic cotton [70]. However, in *Arabidopsis*, dominant negative experiments and antisense approach including Ces-A genes have revealed the role of Ces-A proteins in the formation of primary and secondary cell wall [71, 72].

6 Engineering Secondary Metabolism Using ATMT

6.1 Modulation in Secondary Metabolites and Aspects of Engineering *In Planta*

Secondary metabolites are the outcome/products/compounds derived from secondary metabolism operated in plants. These metabolites are bioactive in nature and known to be actively involved in defense mechanism to overcome the fluctuations in the ecological environment and to protect itself from the pathogen attack. These compounds are produced in minute quantity/as per requirement depending upon the type of pathogen attack. However, being able to cope up with the disturbing ecological parameters and pathogen attack, these compounds are of intrinsic interest of scientists/researchers.

Plant secondary metabolites are categorized into three major groups: terpenes, phenolics, and nitrogen-containing compounds. In the modern era, scientists have often started to engineer the secondary metabolism for their intrinsic interests. Likewise, efforts have been made in engineering the pathways (terpenes, phenolics, and alkaloids) which proved out to be very beneficial for diverting the target metabolite flux as illustrated in Table 3.

Table 3 *Agrobacterium tumefaciens*-mediated genetic transformation (ATMT) resulting in significant alteration in primary and secondary metabolites in plant

Host plant	Gene	Isolated from	Inferences	Ref
Soybean (Glycine max L.)	*Cnx1 gene (GmCnx1)*	*Soybean*	Nitrate reductase (NR) and aldehyde oxidase (AO) were found approximately 2.6–3.9-folds higher than non-transgenic control plants	[90]
Tobacco	*PsnSuSy2 gene*	*Populus simonii × Populus nigra*	Increase in the cellulose content and fiber length and secondary cell wall was significantly thicker than normal plants	[91]
Withania somnifera	*Sterol glucosyltransferase (SGTL1) gene*	*Withania somnifera*	Increase in glycosylated withanolide and sterols	[92]
Artemisia annua	*HDR and ADS genes Artemisinic aldehyde Δ11 (13) reductase (DBR2) gene*	*A. annua*	Facilitates higher production of artemisinin content than non-transgenic lines	[93]
Artemisia annua L.	*HMG-Co A reductase gene (hmgr); amorpha-4,11-diene synthase (ads) gene*	*Catharanthus roseus (L) G. Don; A. annua*	Transgenic line was found to contain 7.65-fold higher (1.73 mg/gDW) artemisinin than the non-transgenic plant	[94]
Artemisia annua	*Cytochrome P450 monooxygenase (cyp71av1) and cytochrome P450 reductase (cpr) genes*	*A. annua*	38% higher accumulation of artemisinin content	[95]
Scutellaria baicalensis	*Phenylalanine ammonia-lyase isoforms (SbPAL1, SbPAL2, and SbAPL3) and one gene-encoding cinnamate 4-hydroxylase (SbC4H)*	*S. baicalensis*	The baicalin and baicalein contents in roots were 22 and 107 times higher than those in flowers	[96]

(*continued*)

Table 3 (continued)

Host plant	Gene	Isolated from	Inferences	Ref
Potato cultivar Désirée plants	*StAN11, a WD40-repeat gene*	*Chieftain (Solanum tuberosum L.).*	Dihydroflavonol reductase (DFR) was increased and StAN11 regulated anthocyanin biosynthesis	[97]
Barley cultivar Golden Promise	*Overexpression of cytokinin dehydrogenases (CKX)*	*Barley*	Cytokinin (CK) homeostasis mechanism in transgenic plants	[98]
Vicillin in cultivar Pusa 16	*γ-TMT gene*	*Perilla frutescens*	Elevated level of α-tocopherol	[99]
Panax ginseng	*Overexpression of CYP716A47 and CYP716A53v2 gene from CYP716A subfamily genes RNAi of CYP716A52v2 gene*	*Panax ginseng*	Overexpression of CYP716A52v2 greatly increased the content of oleanane-type ginsenoside (ginsenoside Ro), whereas RNA interference against CYP716A52v2 markedly reduced it	[100]
Atropa belladonna	*H6H gene*	*A. belladonna*	Increase in contents of hyoscyamine and scopolamine in roots, stems, leaves, and fruits of transgenic; content of scopolamine in transgenic line C8 was 2.17 mg $\times$ g^{-1} DW that was 4.2-folds of the non-transgenic ones (0.42 mg $\times$ g^{-1} DW)	[101]
Withania somnifera	*Squalene synthase*	*W. somnifera*	Fourfold significant increase in squalene synthase activity and 2.5-fold enhancement in withanolide; production of withaferin A was also observed	[102]

(*continued*)

Table 3 (continued)

Host plant	Gene	Isolated from	Inferences	Ref
Eucommia ulmoides Oliver.	*EuIPI gene*	*Eucommia ulmoides*	Three to fourfold increase in the total content of trans-polyisoprenes	[103]
Artemisia annua	*3-Hydroxy-3-methylglutaryl CoA reductase (HMGR) gene*	*Catharanthus roseus (L.) G. Don*	Increase of 22.5% artemisinin content	[104]
Cotton (G. hirsutum)	*Phytochrome B (PHYB) gene*	*Arabidopsis thaliana*	Highly increased in plant growth, with 35% more yield than normal plant. Moreover, decrease in apical dominance and increase in boll size were also seen to contribute in higher yield mass of transgenic cotton	[105]
Transgenic potato (Solanum tuberosum)	*Phytochrome B*	*Arabidopsis*	Greater biomass production, resulting in extended underground organs with significant increase in tuber yields	[106]
Coptis japonicus	*3'-Hydroxy-N-methylcoclaurine 4'-O-methyltransferase (4'OMT) gene*		Improvement of berberine yields (1.5-fold)	[107]
Chinese cabbage	*CYP79B2, CYP79B3, and CYP83B1*	*Arabidopsis*	Accumulation of higher levels of glucobrassicin, 4-hydroxy glucobrassicin, and 4-methoxy glucobrassicin	[108]
Eleutherococcus senticosus Rupr. and Maxim. plants	*PgSS1 gene (a squalene synthase gene*	*Panax ginseng*	Phytosterols (beta-sitosterol and stigmasterol) as well as triterpene levels was increased by 2–2.5-folds	[109]
Solanum tuberosum	*b-1,3-Glucanase class III (Glu-III)*	*Solanum tuberosum* cv. Igor	*In planta* protein production	[110]

(*continued*)

Table 3 (continued)

Host plant	Gene	Isolated from	Inferences	Ref
cv. Desiree and cv. Sante ´				
Arabidopsis thaliana	*Fatty Acyl Hydroxylase cDNA*	*Castor Bean (Ricinus communis L.)*	Accumulation of higher amount of ricinoleic, lesquerolic, and densipolic acids in seed	[111]
Arabidopsis thaliana	*SfN8DT-1 gene*	*Sophora flavescens*	Higher accumulation of prenylated apigenin, quercetin, kaempferol as well as 8-prenylnaringenin	[112]
Artemisia annua	*AaPYL9 gene*	*Artemisia annua*	Drought resistance and improved artemisinin content	[113]
Catharanthus roseus	*Deacetylvindoline-4-O-acetyltransferase (DAT) gene*	*Catharanthus roseus*	Increased yield of vindoline in transgenic plants	[114]
Atropa belladonna	*Hydroxylase gene*	*Hyoscyamus niger*	Improved alkaloid content	[115]
Rice	*Dammarenediol-II synthase*	*Panax ginseng*	Synthesis of dammarane-type sapogenin 20(S)-protopanaxadiol (PPD) and dammarane-type sapogenin 20(S)-protopanaxatriol (PPT)	[116]
Bacopa monnieri	*Cryptogein gene*	–	Accumulation of saponin in transgenic plants maximally up to 1.4–1.69%	[117]
Lycopersicum esculentum Mill.	*Stilbene synthase*	*Vitis vinifera*	Biosynthesis of two novel compounds named as trans-resveratrol and trans-resveratrol-glucopyranoside	[118]
Arabidopsis	*Myo-inositol oxygenase and an L-gulono-1,4-lactone oxidase*	*Arabidopsis*	Enhancement of Vitamin C content	[119]

(*continued*)

Table 3 (continued)

Host plant	Gene	Isolated from	Inferences	Ref
Panax notoginseng	*Squalene synthase (SS)*	*Panax notoginseng*	Enhancement in the biosynthesis of ginsenosides, saponins	[88]
Panax ginseng	*PgDDS. Dammarenediol-II synthase*	*Panax ginseng*	Production of a tetracyclic triterpenoid; dammarenediol-II	[120]
Panax notoginseng (Burk) F. H.	*PnFPS, PnSS, PnSE1, PnSE2, and PnDS*	*Panax notoginseng*	Biosynthesis of triterpene saponin	[87]
Kale (Brassica oleracea var. acephala)	*AtMYB12 transcription factor*	*Arabidopsis thaliana*	Several fold increase in both total phenolics content and flavonol accumulation	[121]
Lycopersicum esculentum	*S-linalool synthase (LIS) Gene*	*Clarkia breweri*	Enhancement in aroma and flavor compounds	[81]
Rose	*3',5'-Hydroxylase (F3'5'H), a key enzyme for delphinidin biosynthesis*	-	Accumulation of delphinidin; an anthocyanin results in blue color	[122]

6.1.1 Terpenes

The terpenoids family constitutes the major class of secondary metabolites. The family consists of more than 40,000 natural compounds, constituting both primary as well as secondary metabolites. Many efforts have been made to engineer the terpenoid metabolic pathway *in planta*. Many of the primary metabolite syntheses via terpenoid pathway consist of photosynthesis pigments, plant-derived phytohormones, and ubiquinones which are essential for respiration [73] (Fig. 1). Attempts have been made to produce monoterpenes in transgenic plants; this strategy is easier than to modify the terpenoids containing complexed longer chains [73]. To date, progress have been made to identify the formation, localization, stabilization, and subcellular compartmentalization of the target products as the plant system is highly complexed and consists of specialized structures for proper storage and transportation of volatile and hydrophobic compounds. The plant species which do not contain these specialized secretory structures made it difficult to introduce new pathways [74]. The strategy of overexpression of pathway enzyme in plant species proved to be fruitful in enhancing the desired substrate. Wu and coworkers provided opportunity to encourage the production of pharmaceutically or industrially relevant plants for commercialization at large scale [75, 76].

The biosynthetic pathway is started from acetyl-CoA or glycolytic intermediates. These are synthesized by the fusion of five-carbon isoprene units. Terpenes are produced by condensation of five-carbon elements that may have branched carbon skeleton of isopentane units. The terpenes are also known as isoprene units; at high temperature terpenes may decompose to give isoprene unit. Therefore, the terpenoid group is sometimes called as isoprenoids. The terpene group is categorized on the basis of number of five-carbon units, like monoterpenes (10-C), sesquiterpene (15-C), and diterpenes (20-C). In addition, triterpenes with 30 carbons, tetraterpenes with 40 carbons, and polyterpenoids with [C5]n carbons (where $n > 8$) have also been reported. There are two terpene biosynthetic pathways available *in planta*, i.e., mevalonic acid pathway, in which three molecules of acetyl-coA are condensed in stepwise manner to yield mevalonic acid which is a 6-C compound (the mevalonic acid is then pyrophosphorylated, decarboxylated, and followed by dehydration to produce isopentenyl diphosphate (IPP2)), and the MVA pathway which is known to yield triterpenes and sesquiterpenes which occur in the endoplasmic reticulum [73]. IPP is also discovered to be involved in the formation of intermediates in the carbon reduction cycle of photosynthesis through channeling individual set of chemical reactions known as methylerythritol phosphate (MEP) pathway which occur in the chloroplast/plastid [77, 78]. Larger units are formed via condensation of different sets of carbon ring like 10-carbon compound geranyl diphosphate (GPP), then 15-carbon compound farnesyl diphosphate (FPP), then 20-carbon compound geranylgeranyl diphosphate (GGPP), and finally 30-C compound triterpenes and the 40-C compound tetraterpenes, carotenoids [73]. These two leading pathways have been in focus for metabolic engineering so far [79].

To date, examples of metabolic engineering of monoterpenoid are like (S)-limonene synthase gene from *Clarkia breweri* in petunia resulting in the formation of linalool, which showed repulsion to aphids [80]. So far, the same gene under the control of a fruit-specific promoter encourages the accumulation of linalool in fruits of tomato [81]. Transgenic tomatoes expressing phytoene synthase resulted in the fourfold increment in the production of carotenoid than normal plants. Although, in the case of transgenic tomato expressing phytoene desaturase, no enhancement was observed in the overall carotenoid content, the content of β-carotenoid was threefold higher than the normal plants [82].

Attention has been given to carotenoid biosynthetic pathway, which led to the synthesis of pigments which takes part in photo protection system and light harvesting complex. The carotenoid pathway is started with the coupling reaction of two geranyl phosphate molecules resulted in the production of geranylgeranyl pyrophosphate (GGPP). This GGPP get converted into phytoene in the presence of phytoene synthase enzyme and eventually into Z-carotene in the presence of phytoene desaturase enzyme in a stepwise manner to yield lycopene and finally either into β-carotene (provitamin A) in the presence of β-cyclase enzyme or δ-carotene and then to α-carotene by enzymes ε-cyclase and β-cyclase, respectively

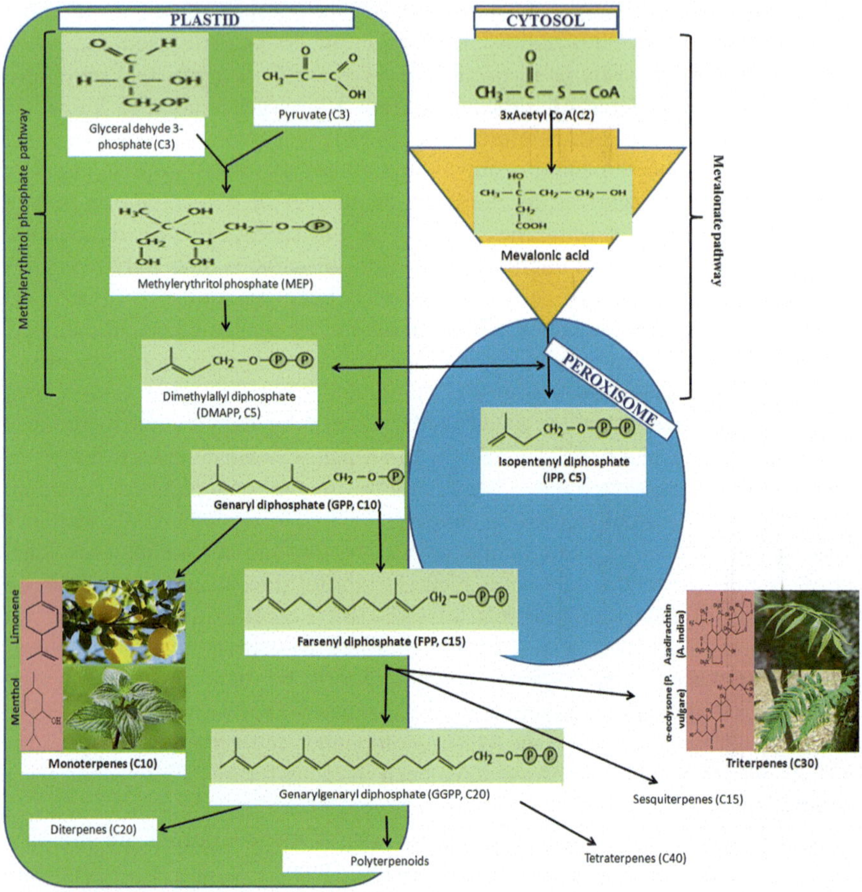

Fig. 2 Basic outline of terpene biosynthetic pathway *in planta*. Biosynthesis of terpenes proceeds via MVA and MEP pathways

[83]. Golden rice is supposed to be the best example which consists of two foreign genes: a bacterial phytoene desaturase and the daffodil phytoene synthase [84]. In addition, the overexpression of mono-/sequiterpene synthase FaNES1 from strawberry in *Arabidopsis* resulted in the production of nerolidol (CIS) derivatives named 4,8-dimethyl-1,3(E), 7-nonatriene [85].

Consistent with the notion, secondary metabolites, synthesis via terpenoid pathway such as monoterpene (C10), sesquiterpene (C15), diterpene (C20), and triterpene (C30), are actively involved in the ecological and physiological advantage to plants (Fig. 2). Basically, terpenes provide resistance to pathogen-related infection and are known to have antimicrobial, antifungal, and insecticidal properties. Terpenes are also having the ability to attract pollinators, repel pests, etc. In addition, apart from the use of terpenoids in flavor, fragrance, and medicines, they also contribute to agricultural sector. The benefit in manipulation in the inherent characteristic of plant in improvement of aromas and fragrance in horticultural sector and as a beneficial source of cosmetics/pharmaceutical has led the researcher to invest their time to engineer the terpenoid pathway [73]. The strategies for manipulation of terpenoid pathway have been reviewed by many workers [73, 74]. Studies like heterologous expression of genes were also done in order to upregulate/overexpress the desired gene. Likewise, a heterologous expression in tomato results in the reduction of a monoterpenoid which promotes the biosynthesis of undesirable flavor and color in mint species [74] and enhanced the production of aroma in ripening fruit [81]. The manipulation of ketocarotenoid biosynthetic pathway in medicinally important plants like tobacco and tomato species through the heterologous expression of bacterial gene also resulted in beneficial outcomes [86]. For example, tomatoes with β-carotene were further co-transformed with hydroxylase gene and β-carotene ketolase gene [86]. In addition, a crtW gene from bacteria was also introduced in tomato and tobacco leaves, and similarly, ketocarotenoid biosynthetic pathway was observed to be operated in leaves and nectary tissues of tobacco, respectively [86].

Metabolic engineering is not only meant to overexpress/downregulate the particular compound, but it could also provide a platform to enhance the yield of desired product up to various folds. Successful achievements regarding ME were discussed in Table 3. Pharmaceutically important terpenoids like diuretic glycyrrhizin, artemisinin (antimalarial drug from Artemisia), taxol (anticancerous; from *Taxus* species), and perilla alcohol have been isolated from plants. Although various bioactive compounds were derived from endangered species belonging to a threatened environment [76], the synthesis of these compounds by chemical methods is inefficient and also prohibitively costly [75]. Wu and workers have evaluated various factors that required for production of high amount of terpenoids in *Nicotiana* and were able to identify the various strategies for improving the yield up to 1000-fold [75]. Recently, Niu and coworkers have characterized PnFPS, PnSS, PnSE1, PnSE2, and PnDS genes, which are found to be responsible for the accumulation of triterpene saponins in *Panax notoginseng* [87]. However, it is noteworthy that the overexpression of squalene synthase (SS) in the same species leads to the production of ginsenosides [88]. Experiments have been set to manipulate the terpene pathway as a wish that the production of resultant/desired/novel product could be achieved via manipulation/engineering of the terpene synthase genes [89].

6.1.2 Alkaloids/Nitrogen-Containing Compounds

Alkaloid is one of the largest groups consisting of more than 12,000 low-molecular-weight compounds [123]. The nitrogen atom present in these compounds is a part of heterocyclic ring that consist of both nitrogen and carbon atom, which does not have nitrogen in peptide or amide group [124]. Alkaloids are derived from common amino acids, in particular tyrosine, lysine, and tryptophan, while some alkaloids are derived from ornithine, an intermediate in arginine biosynthesis also. Alkaloids are known for their striking pharmacological activities on vertebrate animals. The group consists of biogenically unrelated and structurally diverse molecules. Alkaloids which consist of similar structured compounds are the derivative of related biosynthetic pathways; however, many alkaloids are thought to be evolved from unique biosynthetic origins [125]. The substrate starting compound decides the type of structural class to which the alkaloid belongs [125]. Amino acids act as the starting precursor molecule for some alkaloid production; however, few are derived from purine derivatives also. Approximately 12,000 known alkaloids are known to have potent biological activities and can be exploited at commercial levels like stimulants, pharmaceuticals, poisons, and narcotics. The alkaloids are majorly known for their role in plant pathogen interaction, as they provide defense mechanism enabling the plants to cope with adverse ecological conditions by binding the cellular targets in antagonistic organism [126, 127]. Emphasis has been given to manipulate caffeine biosynthetic pathway in such a way that three *N*-methyltransferase genes were overexpressed, and the enhancement in caffeine content led to the increased tolerance from pests in tobacco plant [128, 129]. Additionally, in *Nicotiana*, the overexpression of methyltransferase genes from coffee results in 0.2–5 μg of caffeine/g fresh weight of leaf sample [129]. The various biosynthetic pathways/regulatory enzymes/genes involved in the production of alkaloid have been reviewed by Hashimoto and Yamada [130]. Genetic manipulations and characterization involving overexpression/suppression have been done with major emphasis on the key regulatory genes like PMT, ADC, and ODC which are involved in nicotine biosynthetic pathways [131, 132]. An antisense downregulation of berberine bridge enzyme in poppy leads to the reduction of benzophenanthridine alkaloids with concomitant increase in several amino acids [133]. In addition, the production of pharmaceutical important compounds derived from indole alkaloid biosynthetic pathway in *Catharanthus roseus* has always been in focus to overexpress the compounds (vinblastine and vincristine) having anticancer properties and may be applied as a new approach for manufacturing drugs [134]. Despite having various chemical diversities and metabolic complexities in alkaloid biosynthesis, efforts have been made which significantly contribute advancement in revealing the role and importance of alkaloid compounds. Table 3 represents some of the examples of significant metabolic manipulations in alkaloids biosynthetic pathway via ATMT approach.

6.1.3 Phenolics Compound

Large varieties of secondary products are produced in plants that consist of a phenol group. The phenolic compound consists of hydroxyl group as a functional aromatic ring. The group is chemically heterogeneous consisting of approximately 10,000 individual compounds. Phenolics are the derivatives of phenylpropanoid pathway, shikimate pathway, and pentose phosphate pathway via a phenylalanine as a precursor compound [135, 136]. This is an essential pathway in plants which provides the majority of carbon flux and the plant obtained approximately 20% of the total metabolites through phenylpropanoid pathway [137]. The pathway yields majority of product which includes flavonoids, condensed tannins, anthocyanin, lignins, lignans, and other miscellaneous phenolics. The shikimic acid pathway is responsible for the conversion of simple carbohydrate precursors to aromatic amino acids; these precursors are derived from the pentose phosphate and glycolysis pathway

[138]. The majority of secondary phenolic metabolites in plants are derived by the enzyme phenylalanine ammonia-lyase (PAL) which converts the phenylalanine to yield trans-cinnamic acid via the elimination of an ammonia molecule. PAL is perhaps the key enzyme present at a branch point between primary and secondary metabolism and yields majority of plant secondary metabolites. Flavonoids are compounds originated from phenylalanine and known to share a common pathway with lignan and lignin. 4-coumaroyl-Co-A is the first precursor for flavonoid biosynthesis which gets converted into chalcone in the presence of chalcone synthase enzyme. The stepwise conversions lead to the production of a variety of compounds which provide pigments like anthocyanin, phytoalexins (defense compounds), and many regulatory compounds [139]. During the early 1990s, the pathway has been focused for metabolic engineering to transfer multigene to target plants. Flavonoids are phenolic compounds with high antioxidant activity and therefore optioned as a favorite target for metabolic engineering. Manipulations like stacking of anthocyanidin synthase and dihydroflavonol 4-reductase transgenes in *Forsythia* by a series of sequential transformation events have been targeted to utilize the idea of sequential manipulation *in planta* [140].

Recently, strategies were applied to enhance the flower pigments that have contributed successfully to metabolic engineering [141]. In addition, nowadays, overexpression strategy has led to significant results like by overexpressing chalcone isomerase gene from *Petunia* which resulted in 20-fold increase in the flavonoid level in tomato paste and 80-fold increase in the flavonoid content in the tomato peel [142]. The flavonol synthase and chalcone synthase

Genes were also observed to significantly upregulate the flavonol biosynthesis in tomato fruits [143]. *Lycopersicon esculentum* (tomato) with a number of nutritional utilities (like flavonoid content, carotenoid content, ascorbic acid, anthocyanin pigments, etc.) has always been a favorite species for researchers/scientists to engineer these many qualities into quantities through ATMT approach. Some of the significant results were shown in Table 4.

6.1.4 Lignins

Like flavonoids, lignins are the derivatives of phenylalanine and have been classified as phenylpropenes. Lignins are highly complexed polymers which consist of ferulic acid, caffeic acid, and sinapic acid monomers joined to each other. Lignin is an essential component of the cell wall which provides rigidity to the plant architecture. Recently, ME have illustrated the utilization of multiple gene transfer in tree species to modify the lignin content and architecture of the desired plant [176]. Similarly, suppression/downregulation of multiple genes involved in the lignin biosynthetic pathway in tobacco leads to low lignin content without much change in the morphology of the transgenic plants [177, 178]. The transfer of tobacco LIM regulatory protein which is involved in lignification pathway and controls various enzymatic reactions in plant has also been reported [179]. In addition, two Myb-like transcription factors derived from *Eucalyptus gunnii* were expressed in xylem and determined the action sites of cinnamyl alcohol dehydrogenase (CAD) and hydroxycinnamoyl-CoA reductase (CCR) genes that are known to encode stepwise enzymatic reactions

Table 4 Improvement of organoleptic trait/quality tomato using genetic transformation

Character	Gene inserted	Phenotype of engineered fruit	References
Size	fw2.2	Size increased	[144]
Parthenocarpy	IAA9, Arf8	Induction of parthenocarpy	[145, 146]
Flavor	Thaumatin	Improved flavor	[147]
Firmness	β-galactosidase	Enhanced firmness	[148]
	EXPIA	Decreased firmness	[149]
	PME	Shelf life is reduced	[150]
	PG	Decreased softening	[151]
Flavor and aroma	LeAADC2, LeAADC1A	Enhanced/reduced 1-nitro-2-phenylethane, 2-phenylethanol, and phenylacetaldehyde	[152]
Nutritional quality			
Trait	**Targeted Gene**	**Fruit quality**	**References**
Ascorbic acid	ADCS and GCHI	Increase in fruit folate content	[153, 154]
	GME, GalLDH	reduction/enhancement in fruit ascorbic acid content	[155–157]
Soluble solid content	LCY-B, CYC-B	Enhancement in β-carotene and lycopene content	[158–160]
	Crt Y, Crt I, Crt B	Increase in carotenoid content	[82, 161]
	PSY-1	Increase in carotenoid content	[162]
	Dxs	Increase in carotenoid and phytoene content	[163]
	Lin5	Reduction in accumulation of sugars	[164]
Flavonoid	CHI, F3H, CHS, FLS	Flavonoids increased	[165]
	MYB12	Accumulation of flavonols	[166]
	Ros1, Del	High accumulation of anthocyanins	[167]
	STS, FNSII, CHS, CHI, CHR	Accumulation of deoxychalcones, resveratrol, and flavones	[168]
	CHI	Increase in fruit peel flavonol content	[142]
Carotenoid	FIBRILLIN	Increase in volatiles and carotenoid content	[169]
	Spermidine synthase	Increased in lycopene content	[170]
	CHY-B, LCY-B	Zeaxanthin and β-cryptoxanthin	[171]
	COP1LIKE, DET-1, CUL4	Enhanced flavonoid and carotenoid content	[172–174]
	CRY-2	Increase in carotene content	[175]

essential for the synthesis of p-coumaryl alcohol [180]. ME regarding lignin content in plants was beautifully reviewed by Boudet et al., Anterola and Lewis, and Boerjan et al. [180–182]. Moreover, Bell-Lelong et al. have observed that *Arabidopsis* C4H, a lignin-associated promoter, was much efficient than general promoter like

CaMV35S [183]. However, using lignin-specific promoter could prove more effective for manipulations or engineering purposes.

7 Conclusions

In today's era, ME strategies have been made focus to coordinate between a multipoint or multilocus gene to control the metabolic flux through cascading secondary metabolite pathways. The expression of target gene/loci and its localization is now starting to supersede using the specific promoters in aspect of particular compound. As discussed above, various points in a given biosynthetic pathway could be controlled either by using antisense or overexpressing genes/enzymes or via application of the transcriptional regulators to overrule the negative role of endogenous genes/enzymes. Recently, the era has moved forward in the area of ME by using various advanced approaches like proteomics, genomics, and metabolomics in order to expand the knowledge of pathways and routes *in planta*. The production of Golden rice has changed the global view and led the researchers/scientists to think and focus on using the *Agrobacterium* as a natural genetic manipulator so far. Hence, using ATMT strategy can help in multilocus genetic manipulation and remodeling the genome of target plants.

References

1. Verpoorte R, van der Heijden R, Memelink J (2000) Engineering the plant cell factory for secondary metabolite production. Transgenic Res 9:323–343
2. Naqvi S, Farre G, Sanahuja G, Capell T, Zhu C, Christou P (2009) When more is better: multigene engineering in plants. Trends Plant Sci 15(1):48–55
3. Ingelbrecht IL, Herman LM, Deckeyser RA, Van Montagu MC, Depicker AG (1989) Different 3′ end regions strongly influence the level of gene expression in plant cells. Plant Cell 1:671–680
4. De Cleene M, De Ley J (1976) The host range of crown gall. Bot Rev 42:389–466
5. Kennedy BW (1980) Estimates of U.S. crop losses to prokaryote plant pathogens. Plant Dis 64:674–676
6. Schroth MN, Mc-cain AH, Foott JH, Huisman OC (1988) Reduction in yield and vigor of grapevine caused by crown gall disease. Plant Dis 72:241–246
7. Ricker AJ, Berbee JG, Smalley EB (1959) Effects of crown gall and hairy root on the growth of apple trees. Phytopathology 49:88–90
8. Lopatin MI (1939) Influence of bacterial root center on the development of the cherry tree in the orchard. Plant Prot 18:167–173
9. White PR, Braun AC (1941) Crown gall production by bacteria-free tumor tissues. Science 94:239–241
10. Braun AC (1947) Thermal studies on the factors responsible for the tumor initiation in crown gall. Am J Bot 34:234–240
11. Braun AC, Mandle RJ (1948) Studies on the inactivation of the tumor-inducing principle in crown gall. Growth 12:255–269
12. Zaenen I, Van Larebeke N, Teuchy H, Van Montagu M, Schell J (1974) Supercoiled circular DNA in crown-gall inducing *Agrobacterium* strains. J Mol Biol 86:109–127

13. Van Larebeke N, Enbler G, Holsters M, Van Den Elsacker S, Zaenen I, Schilperoort RA, Schell J (1974) Large plasmid in *Agrobacterium tumefaciens* essential for crown gall-inducing ability. Nature 252:169–170
14. Depicker A, Van Montagu M, Schell J (1978) Homologous sequences in different Ti plasmids are essential for oncogenicity. Nature 275:150–152
15. Chilton MD, Drummond MH, Merlo DJ, Sciaky D, Montoya AL, Gordon MP, Nester EW (1977) Stable incorporation of plasmid DNA into higher plant cells: the molecular basis of crown gall tumorigenesis. Cell 11:263–271
16. Chilton MD, Drummond MH, Merlo DJ, Sciaky D (1978) Highly conserved DNA of Ti plasmids overlaps T-DNA maintained in plant tumors. Nature 275:147–149
17. Gelvin SB (2003) *Agrobacterium*-mediated plant transformation: the biology behind the "gene-jockeying" tool. Microbiol Mol Biol Rev 67:16–37
18. Escobar MA, Dandekar AM (2003) *Agrobacterium tumefaciens* as an agent of disease. Trends Plant Sci 8:380–386
19. Zambryski P, Joos PH, Genetello C, Leemans J, Van Montagu M, Schell J (1983) Ti plasmid vector for the introduction of DNA into plant cells without alteration of their normal regeneration capacity. EMBO J 2:2143–2150
20. Bevan M (1984) Binary *Agrobacterium* vectors for plant transformation. Nucleic Acids Res 12:8711–8721
21. Chilton MD (1993) *Agrobacterium* gene transfer: progress on a "poor man's vector" for maize. Proc Natl Acad Sci U S A 90:3119–3120
22. Vanlentine L (2003) *Agrobacterium tumefaciens* and the plant: the David and Goliath of modern genetics. Plant Physiol 133:948–955
23. Lacroix B, Tzfira T, Vainstein A, Citovsky V (2006) A case of promiscuity: *Agrobacterium's* endless hunt for new partners. Trends Genet 22:29–37
24. Guilley H, Dudley RK, Jonard G, Balazs E, Richards KE (1982) Transcription of cauliflower mosaic virus DNA: detection of promoter sequences and characterization of transcripts. Cell 30:763–773
25. Odell JT, Nagy F, Chua NH (1985) Identification of DNA sequences required for activity of the cauliflower mosaic virus 35S promoter. Nature 313:810–812
26. Fang RX, Nagy F, Sivasubramanian S, Chua NH (1989) Multiple cis regulatory elements for maximal expression of the cauliflower mosaic virus 35S promoter in transgenic plants. Plant Cell 1(1):141–150
27. Zarate R, Verpoorte R (2007) Strategies for the genetic modification of the medicinal plant *Catharanthus roseus* (L.) G. Don. Phytochem Rev 6:475–491
28. Benfey PN, Ren L, Chua NH (1990) Tissue specific expression from CaMV35S enhancer subdomains in early stages of plant development. EMBO J 9:1677–1684
29. Kay R, Chan A, Daly M, McPherson J (1987) Duplication of the CaMV 35S promoter sequences creates a strong enhancer for plant genes. Science 236:1299–1302
30. Koosha F, Muller RH, Davis SS (1989) Poly hydroxy butyrate as a drug carrier. Crit Rev Ther Drug Carrier Syst 6:117–130
31. Zuo J, Chua NH (2000) Chemical inducible systems for regulated expression of plant genes Curr. Opin Biotechnol 11:146–151
32. Lessard PA, Kulaveerasingam H, York GM, Strong A, Sinskey AJ (2002) Manipulating gene expression for the metabolic engineering of plants. Metab Eng 4:67–79
33. Plesse B, Criqui MC, Durr A, Parmentier Y, Fleck J, Genschik P (2001) Effects of the polyubiquitin gene Ubi U4 leader intron and first ubiquitin monomer on reporter gene expression in *Nicotiana tabacum*. Plant Mol Biol 45:655–667
34. An YQ, Huang S, McDowell JM, McKinney EC, Meagher RB (1996) Conserved expression of the *Arabidopsis* ACT1 and ACT 3 actin subclass in organ primordia and mature pollen. Plant Cell 8:15–30

35. An YQ, McDowell JM, Huang S, McKinney EC, Chambliss S, Meagher RB (1996) Strong, constitutive expression of the *Arabidopsis* ACT2/ACT8 actin subclass in vegetative tissues. Plant J 10:107–121
36. Becker D, Brettschneider R, Lorz H (1994) Fertile transgenic wheat from microprojectile bombardment of scutellar tissue. Plant J 5:299–307
37. Molnar A, Lovas A, Banfalvi Z, Lakatos L, Polgar Z, Horvath S (2001) Tissue-specific signal (s) activate the promoter of a metallo carboxy peptidase inhibitor gene family in potato tuber and berry. Plant Mol Biol 46:301–311
38. Liu XJ, Prat S, Willmitzer L, Frommer WB (1990) Cis regulatory elements directing tuber-specific and sucrose-inducible expression of a chimeric class I patatin promoter/GUS-gene fusion. Mol Gen Genet 223:401–406
39. Qinggele C, Mingchun L, Dongsheng W, Yi C, Laijun X (2007) Isolation and sequencing analysis on the seed-specific promoter from soybean. Front Agric China 1:17–23
40. Chen ZL, Schuler MA, Beachy RN (1986) Functional analysis of regulatory elements in a plant embryo-specific gene. Proc Natl Acad Sci U S A 83:8560–8564
41. Russell DA, Fromm ME (1997) Tissue-specific expression in transgenic maize of four endosperm promoters from maize and rice. Transgenic Res 6:157–168
42. Rossi V, Motto M, Pellegrini L (1997) Analysis of the methylation pattern of the maize opaque-2 (O2) promoter and in vitro binding studies indicate that the O2 B-Zip protein and other endosperm factors can bind to methylated target sequences. J Biol Chem 272:13758–13765
43. Ali S, Taylor WC (2001) The 3Œ non-coding region of a C4 photosynthesis gene increases transgene expression when combined with heterologous promoters. Plant Mol Biol 46:325–333
44. Bate N, Twell D (1998) Functional architecture of a late pollen promoter: pollen-specific transcription is developmentally regulated by multiple stage-specific and co-dependent activator elements. Plant Mol Biol 37:859–869
45. Haran S, Logendra S, Seskar M, Bratanova M, Raskin I (2000) Characterization of Arabidopsis acid phosphatase promoter and regulation of acid phosphatase expression. Plant Physiol 124:615–626
46. De Veylder L, Van Montagu M, Inze D (1997) Herbicide safener inducible gene expression in *Arabidopsis thaliana*. Plant Cell Physiol 38:568–577
47. Frey AD, Rimann M, Bailey JE, Kallio PT, Thompson CJ, Fussenegger M (2001) Novel pristinamycin-responsive expression systems for plant cells. Biotechnol Bioeng 74:154–163
48. Martinez A, Sparks C, Hart CA, Thompson J, Jepson I (1999) Ecdysone agonist inducible transcription in transgenic tobacco plants. Plant J 19:97–106
49. Aoyama T, Chua NH (1997) A glucocorticoid-mediated transcriptional induction system in transgenic plants. Plant J 11:605–612
50. Caddick MX, Greenland AJ, Jepson I, Krause KP, Qu N, Riddell KV, Salter MG, Schuch W, Sonnewald U, Tomsett AB (1998) An ethanol inducible gene switch for plants used to manipulate carbon metabolism. Nat Biotechnol 16:177–180
51. Mysore KS, Ryu CM (2004) Nonhost resistance: how much do we know? Trends Plant Sci 9:97–104
52. Bolton MD (2009) Primary metabolism and plant defense–fuel for the fire. Mol Plant Microbe Interact 22:487–497
53. Kangasjarvi S, Neukermans J, Li S, Aro EM, Noctor G (2012) Photosynthesis, photorespiration and light signalling in defence responses. J Exp Bot 63:1619–1636
54. McKersie BD, Bowley SR, Jones KS (1999) Winter survival of transgenic alfalfa overexpressing superoxide dismutase. Plant Physiol 119:839–847
55. Tesfaye M, Temple SJ, Allen DL, Vance CP, Samac DA (2001) Overexpression of malate dehydrogenase in transgenic alfalfa enhances organic acid synthesis and confers tolerance to aluminum. Plant Physiol 127:1836–1844

56. Galili G, Hofgen R (2002) Metabolic engineering of amino acids and storage proteins in plants. Metab Eng 4:3–11
57. Chen THH, Murata N (2002) Enhancement of tolerance of abiotic stress by metabolic engineering of betaines and other compatible solutes. Curr Opin Plant Biol 5:250–257
58. Zheng Z, Sumi K, Tanaka K, Murai N (1995) The bean seed storage protein p-Phaseolin 1s synthesized, processed, and accumulated in the vacuolar Type-II protein bodies of transgenic rice endosperm. Plant Physiol 109:777–786
59. Chakraborty S, Chakraborty N, Datta A (2000) Increased nutritive value of transgenic potato by expressing a nonallergenic seed albumin gene from *Amaranthus hypochondriacus*. Proc Natl Acad Sci U S A 97(7):3724–3729. doi:10.1073/pnas.050012697
60. Zhu XH, Galili G (2003) Increased lysine synthesis coupled with a knockout of its catabolism synergistically boosts lysine content and also transregulates the metabolism of other amino acids in *Arabidopsis* seeds. Plant Cell 15:845–853
61. Kok-Jacon GA, Ji Q, Vincken J-P, Visser RGF (2003) Towards a more versatile a-glucan synthesis in plants. J Plant Physiol 160:765–777
62. Regierer B, Fernie AR, Springer F, Perez-Melis A, Leisse A, Koehl K, Willmitzer L, Geigenberger P, Kossmann J (2002) Starch content and yield increase as a result of altering adenylate pools in transgenic plants. Nat Biotechnol 20:1256–1260
63. Rosche E, Blackmore D, Tegeder M, Richardson T, Schroeder H, Higgins TJV, Frommer WB, Offler CE, Patrick JW (2002) Seed-specific overexpression of the potato sucrose transporter increases sucrose uptake and growth rates of developing pea cotyledons. Plant J 30:165–175
64. Smidansky ED, Clancy M, Meyer FD, Lanning SP, Blake NK, Talbert LE, Giroux MJ (2002) Enhanced ADP-glucose pyrophosphorylase activity in wheat endosperm increases seed yield. Proc Natl Acad Sci U S A 99:1724–1729
65. Smidansky ED, Martin JM, Hannah LC, Fischer AM, Giroux MJ (2003) Seed yield and plant biomass increases in rice are conferred by deregulation of endosperm ADP-glucose pyrophosphorylase. Planta 216:656–664
66. Liu QQ, Wang ZY, Chen XH, Cai XL, Tang SZ, Yu HX, Zhang JL, Hong MM, Gu MH (2003) Stable inheritance of the antisense Waxy gene in transgenic rice with reduced amylose level and improved quality. Transgenic Res 12:71–82
67. Fujita N, Kubo A, Suh DS, Wong KS, Jane JL, Ozawa K, Takaiwa F, Inaba Y, Nakamura Y (2003) Antisense inhibition of isoamylase alters the structure of amylopectin and the physicochemical properties of starch in rice endosperm. Plant Cell Physiol 44:607–618
68. Andersson M, Melander M, Pojmark P, Larsson H, Leif B, Hofvander P (2006) Targeted gene suppression by RNA interference: an efficient method for production of high-amylose potato lines. J Biotechnol 123(2):137–148
69. Vincken QJJP, Suurs LCJM, Visser RGF (2003) Microbial starch-binding domains as a tool for targeting proteins to granules during starch biosynthesis. Plant Mol Biol 51:789–801
70. Ruan YL, Llewellyn DJ, Furbank RT (2003) Suppression of sucrose synthase gene expression represses cotton fiber cell initiation, elongation, and seed development. Plant Cell 15:952–964
71. Zhong RQ, Morrison WH, Freshour GD, Hahn MG, Ye ZH (2003) Expression of a mutant form of cellulose synthase AtCesA7 causes dominant negative effect on cellulose biosynthesis. Plant Physiol 132:786–795
72. Burn JE, Hocart CH, Birch RJ, Cork AC, Williamson RE (2002) Functional analysis of the cellulose synthase genes CesA1, CesA2, and CesA3 in *Arabidopsis*. Plant Physiol 129:797–807
73. Aharoni A, Jongsma MA, Bouwmeester HJ (2005) Volatile science? Metabolic engineering of terpenoids in plants. Trends Plant Sci 10:594–602
74. Mahmoud SS, Croteau RB (2002) Strategies for transgenic manipulation of monoterpene biosynthesis in plants. Trends Plant Sci 7(8):366–373
75. Wu M, Wu Y, Wang M (2006) Energy and emission benefits of alternative transportation liquid fuels derived from switchgrass: a fuel life cycle assessment. Biotechnol Prog 22(4):1012–1024

76. Bouwmeester HJ, Bertea CM, Kraker JW, Franssen MCR (2006) Research to improve artemisinin production for use in the preparation of anti-malarial drugs. In: Bogers RJ, Craker LE, Langer D (eds) Medicinal and aromatic plants. Springer, Amsterdam, pp 275–290
77. Lichtenthaler HK (1999) The 1-deoxy-D-xylulose-5-phosphate pathway of isoprenoid biosynthesis in plants. Annu Rev Plant Physiol Plant Mol Biol 50:47–65
78. Rodriguez-Concepcion M, Boronat A (2002) Elucidation of the methylerythritol phosphate pathway for isoprenoid biosynthesis in bacteria and plastids. A metabolic milestone achieved through genomics. Plant Physiol 103:1079–1089
79. Broun P, Somerville C (2001) Progress in plant metabolic engineering. Proc Natl Acad Sci U S A 98:8925–8927
80. Lucker J, Bouwmeester HJ, Schwab W, Blaas J, van der Plas LH, Verhoeven HA (2001) Expression of Clarkia S-linalool synthase in transgenic petunia plants results in the accumulation of S-linalyl-b-D-glucopyranoside. Plant J 27:315–324
81. Lewinsohn E, Schalechet F, Wilkinson J, Matsui K, Tadmor Y, Nam KH, Amar O, Lastochkin E, Larkov O, Ravid U et al (2001) Enhanced levels of the aroma and flavour compound S-linalool by metabolic engineering of the terpenoid pathway in tomato fruits. Plant Physiol 127:1256–1265
82. Fraser PD, Romer S, Shipton CA, Mills PB, Kiano JW, Misawa N, Drake RG, Schuch W, Bramley PM (2002) Evaluation of transgenic tomato plants expressing an additional phytoene synthase in a fruit-specific manner. Proc Natl Acad Sci U S A 99:1092–1097
83. Sandmann G (2001) Genetic manipulation of carotenoid biosynthesis: strategies, problems and achievements. Trends Plant Sci 6:14–17
84. Beyer P, Al-Babili S, Ye X, Lucca P, Schaub P, Welsch R, Potrykus I (2002) Golden rice: Introducing the b-carotene biosynthesis pathway into rice endosperm by genetic engineering to defeat vitamin A deficiency. J Nutr 132:506S–510S
85. Kappers IF, Aharoni A, van Herpen TW, Luckerhoff LL, Dicke M, Bouwmeester HJ (2005) Genetic engineering of terpenoid metabolism attracts bodyguards to *Arabidopsis*. Science 309:2070–2072
86. Ralley L, Enfissi EM, Misawa N, Schuch W, Bramley PM, Fraser PD (2004) Metabolic engineering of ketocarotenoid formation in higher plants. Plant J 39:477–486
87. Niu Y, Luo H, Sun C, Yang TJ, Dong L, Huang L, Chen S (2014) Expression profiling of the triterpene saponin biosynthesis genes FPS, SS, SE, and DS in the medicinal plant *Panax notoginseng*. Gene 533(1):295–303
88. Sun Y, Zhao HW, Ge F, Shi L, Liu DQ (2013) The construction of over-expression vector for *Panax notoginseng* SS gene and its transformation] [Article in Chinese] Yao Xue Xue Bao 48(1):138–143
89. Chappell J (2004) Valencene synthase-a biochemical magician and harbinger of transgenic aromas. Trends Plant Sci 9:266–269
90. Zhou Z, He H, Ma L, Yu X, Mi Q, Pang J, Tang G, Bao L (2015) Overexpression of a *GmCnx1* gene enhanced activity of nitrate reductase and aldehyde oxidase, and boosted mosaic virus resistance in soybean. PLoS One 10(4):e0124273
91. Wei Z, Qu Z, Zhang L, Zhao S, Bi Z, Ji X, Wang X, Wei H (2015) Overexpression of poplar xylem sucrose synthase in tobacco leads to a thickened cell wall and increased height. PLoS One 10(3):e0120669
92. Saema S, Rahman LU, Singh R, Niranjan A, Ahmad IZ, Misra P (2015) Ectopic overexpression of WsSGTL1, a sterol glucosyltransferase gene in *Withania somnifera*, promotes growth, enhances glycowithanolide and provides tolerance to abiotic and biotic stresses. Plant Cell Rep 35(1):195–211
93. Yuan Y, Liu W, Zhang Q, Xiang L, Liu X, Chen M, Lin Z, Wang Q, Liao Z (2014) Overexpression of artemisinic aldehyde 11(13) reductase gene–enhanced artemisinin and its relative metabolite biosynthesis in transgenic *Artemisia annua* L. Biotechnol Appl Biochem 62(1):17–23

94. Alam P, Abdin MZ (2011) Over-expression of HMG-CoA reductase and amorpha-4,11-diene synthase genes in *Artemisia annua* L. and its influence on artemisinin content. Plant Cell Rep 30(10):1919–1928
95. Shen Q, Chen YF, Wang T, Wu SY, Lu X, Zhang L, Zhang FY, Jiang WM, Wang GF, Tang KX (2012) Overexpression of the cytochrome P450 monooxygenase (cyp71av1) and cytochrome P450 reductase (cpr) genes increased artemisinin content in *Artemisia annua* (Asteraceae). Genet Mol Res 11(3):3298–3309
96. Xu H, Park NI, Li X, Kim YK, Lee SY, Park SU (2010) Molecular cloning and characterization of phenylalanine ammonia-lyase, cinnamate 4-hydroxylase and genes involved in flavone biosynthesis in *Scutellaria baicalensis*. Bioresour Technol 101(24):9715–9722
97. Li W, Wang B, Wang M, Chen M, Yin JM, Kaleri GM, Zhang RJ, Zuo TN, You X, Yang Q (2014) Cloning and characterization of a potato StAN11 gene involved in anthocyanin biosynthesis regulation. J Integr Plant Biol 56(4):364–372
98. Mrizova K, Jiskrova E, Vyroubalova S, Novak O, Ohnoutkova L, Pospisilova H, Frebort I, Harwood WA, Galuszka P (2013) Overexpression of cytokinin dehydrogenase genes in barley (*Hordeum vulgare* cv. Golden Promise) fundamentally affects morphology and fertility. PLoS One 8(11):e79029
99. Arun M, Subramanyam K, Theboral J, Sivanandhan G, Rajesh M, Kapil Dev G, Jaganath B, Manickavasagam M, Girija S, Ganapathi A (2014) Transfer and targeted overexpression of γ-tocopherol methyltransferase (γ-TMT) gene using seed-specific promoter improves tocopherol composition in Indian soybean cultivars. Appl Biochem Biotechnol 172(4):1763–1776
100. Han JY, Kim MJ, Ban YW, Hwang HS, Choi YE (2013) The involvement of β-amyrin 28-oxidase (CYP716A52v2) in oleanane-type ginsenoside biosynthesis in *Panax ginseng*. Plant Cell Physiol 54(12):2034–2046
101. Li JD, Qin BF, Yang CX, Lan XZ, Wu NB, Liao ZH (2013) Enhanced biosynthesis of scopolamine in transgenic *Atropa belladonna* by overexpression of h6h gene. Zhongguo Zhong Yao Za Zhi. [Article in Chinese] 38(11):1719–1724
102. Grover A, Samuel G, Bisaria VS, Sundar D (2013) Enhanced withanolide production by overexpression of squalene synthase in *Withania somnifera*. J Biosci Bioeng 115(6):680–685
103. Chen R, Harada Y, Bamba T, Nakazawa Y, Gyokusen K (2012) Overexpression of an isopentenyl diphosphate isomerase gene to enhance *trans*-polyisoprene production in *Eucommia ulmoides* Oliver. BMC Biotechnol 12:78
104. Aquil S, Husaini AM, Abdin MZ, Rather GM (2009) Overexpression of the HMG-CoA reductase gene leads to enhanced artemisinin biosynthesis in transgenic *Artemisia annua* plants. Planta Med 75(13):1453–1458
105. Rao AQ, Irfan M, Saleem Z, Nasir IA, Riazuddin S, Husnain T (2011) Overexpression of the phytochrome B gene from *Arabidopsis thaliana* increases plant growth and yield of cotton (Gossypium hirsutum). J Zhejiang Univ Sci B 12(4):326–334
106. Thiele A, Herold M, Lenk I, Quail PH, Gatz C (1999) Heterologous expression of *Arabidopsis* phytochrome B in transgenic potato influences photosynthetic performance and tuber development. Plant Physiol 120(1):73–82
107. Inui T, Kawano N, Shitan N, Yazaki K, Kiuchi F, Kawahara N, Sato F, Yoshimatsu K (2012) Improvement of benzylisoquinoline alkaloid productivity by overexpression of 3′-hydroxy-*N*-methylcoclaurine 4′-*O*-methyltransferase in transgenic *Coptis japonica* plants. Biol Pharm Bull 35(5):650–659
108. Zang YX, Kim JH, Park YD, Kim DH, Hong SB (2008) Metabolic engineering of aliphatic glucosinolates in Chinese cabbage plants expressing *Arabidopsis* MAM1, CYP79F1, and CYP83A1. BMB Rep 41:472–478
109. Seo JW, Jeong JH, Shin CG, Lo SC, Han SS, Yu KW, Harada E, Han JY, Choi YE (2005) Overexpression squalene synthase in *Eleutherococcus senticosus* increases phytosterol and triterpene accumulation. Phytochemistry 66:869–877

110. Dobnik D, Baebler S, Kogovšek P, Novak MP, Stebih D, Panter G, Janez N, Morisset D, Zel J, Gruden K (2013) β-1,3-glucanase class III promotes spread of PVY^{NTN} and improves *in planta* protein production. Plant Biotechnol Rep 7(4):547–555
111. Broun P, Somerville C (1997) Accumulation of ricinoleic, lesquerolic, and densipolic acids in seeds of transgenic *Arabidopsis* plants that express a fatty acyl hydroxylase cDNA from castor bean. Plant Physiol 113(3):933–942
112. Sasaki K, Mito K, Ohara K, Yamamoto H, Yazaki K (2008) Cloning and characterization of naringenin 8-prenyltransferase, a flavonoid-specific prenyltransferase of *Sophora flavescens*. Plant Physiol 146(3):1075–1084
113. Zhang F, Lu X, Lv Z, Zhang L, Zhu M, Jiang W, Wang G, Sun X, Tang K (2013) Overexpression of the Artemisia orthologue of ABA receptor, AaPYL9, enhances ABA sensitivity and improves artemisinin content in *Artemisia annua* L. PLoS One 8(2):e56697
114. Wang Q, Xing S, Pan Q, Yuan F, Zhao J, Tian Y, Chen Y, Wang G, Tang K (2012) Development of efficient *Catharanthus roseus* regeneration and transformation system using *Agrobacterium tumefaciens* and hypocotyls as explants. BMC Biotechnol 12:34
115. Yun DJ, Hashimoto T, Yamada Y (1992) Metabolic engineering of medicinal plants: transgenic *Atropa belladonna* with an improved alkaloid composition (scoolamlne/hyoscyamine 61-hydroxylase). Proc Natl Acad Sci U S A 89:11799–11803
116. Huang Z, Lin J, Cheng Z, Xu M, Huang X, Yang Z, Zheng J (2015) Production of dammarane-type sapogenins in rice by expressing the dammarenediol-II synthase gene from *Panax ginseng* C.A. Mey. Plant Sci 239:106–114
117. Majumdar S, Garai S, Jha S (2012) Use of cryptogein gene to stimulate the accumulation of Bacopa saponins in transgenic *Bacopa monnieri* plants. Plant Cell Rep 31(10):1899–1909
118. Giovinazzo G, D'Amico L, Paradiso A, Bollini R, Sparvoli F, DeGara L (2005) Antioxidant metabolite profiles in tomato fruit constitutively expressing the grapevine stilbene synthase gene. Plant Biotechnol J 3(1):57–69
119. Lisko KA, Torres R, Harris RS, Belisle M, Vaughan MM, Jullian B, Chevone BI, Mendes P, Nessler CL, Lorence A (2013) Elevating vitamin C content via overexpression of *myo*-inositol oxygenase and L-gulono-1,4-lactone oxidase in *Arabidopsis* leads to enhanced biomass and tolerance to abiotic stresses. In Vitro Cell Dev Biol Plant 49(6):643–655
120. Han JY, Wang HY, Choi YE (2014) Production of dammarenediol-II triterpene in a cell suspension culture of transgenic tobacco. Plant Cell Rep 33(2):225–233
121. Lannenpaa M (2014) Heterologous expression of AtMYB12 in kale (*Brassica oleracea* var. acephala) leads to high flavonol accumulation. Plant Cell Rep 33(8):1377–1388
122. Katsumoto Y, Mizutani MF, Fukui Y, Brugliera F, Holton TA, Karan M, Nakamura N, Sakakibara KY, Togami J, Pigeaire A, Tao GQ, Nehra NS, Lu CY, Dyson BK, Tsuda S, Ashikari T, Kusumi T, Mason JG, Tanaka Y (2007) Engineering of the rose flavonoid biosynthetic pathway successfully generated blue-hued flowers accumulating delphinidin. Plant Cell Physiol 48(11):1589–1600
123. Ziegler J, Facchini PJ (2008) Alkaloid biosynthesis: metabolism and trafficking. Annu Rev Plant Biol 59:735–769
124. Fischbach MA, Clardy J (2007) One pathway, many products. Nat Chem Biol 3:353–355
125. Facchini PJ (1999) Plant secondary metabolism: out of the evolutionary abyss. Trends Plant Sci 4:382–384
126. Facchini PJ (2001) Alkaloid biosynthesis in plants: biochemistry, cell biology, molecular regulation, and metabolic engineering applications. Ann Rev Plant Biol 52:29–66
127. Maplestone RA, Stone MJ, Williams DH (1992) The evolutionary role of secondary metabolites – a review. Gene 115:151–157
128. Kim YS, Sano H (2008) Pathogen resistance of transgenic tobacco plants producing caffeine. Phytochemistry 69:882–888
129. Kim YS, Uefuji H, Ogita S, Sano H (2006) Transgenic tobacco plants producing caffeine: a potential new strategy for insect pest control. Transgenic Res 15:667–672

130. Hashimoto T, Yamada Y (2003) New genes in alkaloid metabolism and transport. Curr Opin Biotechnol 14:163–168
131. Katoh A, Ohki H, Hashimoto T (2005) Molecular regulation of nicotine biosynthesis. Plant Biotechnol 22:389–392
132. De Boer KD, Dalton HL, Edward FJ, Ryan SM, Hamill JD (2013) RNAi-mediated down regulation of ornithine decarboxylase (ODC) impedes wound-stress stimulation of anabasine synthesis in *Nicotiana glauca*. Phytochemistry 86:21–28
133. Park SU, Yu M, Facchini PJ (2002) Antisense RNA-mediated suppression of benzophenanthridine alkaloid biosynthesis in transgenic cell cultures of California poppy. Plant Physiol 128:696–706
134. Verpoorte R, Memelink J (2002) Engineering secondary metabolite production in plants. Curr Opin Biotechnol 13:181–187
135. Dey PM, Harborne JB (1989) Methods in plant biochemistry: plant phenolics, vol 1. Academic, New York
136. Ryan D, Robards K, Prenzler P, Antolovich M (1999) Applications of mass spectrometry to plant phenols. TrAC Trends Anal Chem 18:362–372
137. Dixon RA, Paiva NL (1995) Stress-induced phenylpropanoid metabolism. Plant Cell 7:1085
138. Herrmann KM, Weaver LM (1999) The shikimate pathway. Annu Rev Plant Physiol Plant Mol Biol 50:473–503
139. Winkel-Shirley B (2001) Flavonoid biosynthesis, a colorful model for genetics, biochemistry, cell biology, and biotechnology. Plant Physiol 126:485–493. doi:10.1104/pp.126.2.485
140. Rosati C, Simoneau P, Treutter D, Poupard P, Cadot Y, Cadic A, Duron M (2003) Engineering of flower color in forsythia by expression of two independently-transformed dihydroflavonol 4-reductase and anthocyanidin synthase genes of flavonoid pathway. Mol Breed 12:197–208
141. Forkmann G, Martens S (2001) Metabolic engineering and application of flavonoids. Curr Opin Biotechnol 12:155–160
142. Muir SR, Collins GJ, Robinson S, Hughes S, Bovy A, De Vos CHR, Van Tunen AJ, Verhoeyen ME (2001) Overexpression of petunia chalcone isomerase in tomato results in fruits containing increased levels of flavonoids. Nat Biotechnol 19:470–474
143. Verhoeyen ME, Bovy A, Collins G, Muir S, Robinson S, de Vos CHR, Colliver S (2002) Increasing antioxidant levels in tomatoes through modification of the flavonoid biosynthetic pathway. J Exp Bot 53:2099–2106. doi:10.1093/jxb/erf044
144. Frary A, Nesbitt TC, Grandillo S, van der Knaap E, Cong B, Liu J, Meller J, Elber R, Alpert KB, Tanksley SD (2000) fw2.2: a quantitative trait locus key to the evolution of tomato fruit size. Science 289(5476):85–88
145. Wang H, Jones B, Li Z, Frasse P, Delalande C, Regad F, Chaabouni S, Latche A, Pech JC, Bouzayen M (2005) The tomato Aux/IAA transcription factor IAA9 is involved in fruit development and leaf morphogenesis. Plant Cell 17:2676–2692
146. Goetz M, Hooper LC, Johnson SD, Rodrigues JC, Vivian-Smith A, Koltunow AM (2007) Expression of aberrant forms of AUXIN RESPONSE FACTOR8 stimulates parthenocarpy in *Arabidopsis* and tomato. Plant Physiol 145:351–366
147. Bartoszewski G, Niedziela A, Szwacka M, Niemirowicz-Szczytt K (2003) Modification of tomato taste in transgenic plants carrying a thaumatin gene from *Thaumatococcus daniellii* Benth. Plant Breed 122(4):347–351
148. Smith DL, Abbott JA, Gross KC (2002) Down-regulation of tomato betagalactosidase 4 results in decreased fruit softening. Plant Physiol 129:1755–1762
149. Brummell DA, Harpster MH, Civello PM, Palys JM, Bennett AB, Dunsmuir P (1999) Modification of expansin protein abundance in tomato fruit alters softening and cell wall polymer metabolism during ripening. Plant Cell 11:2203–2216
150. Tieman DM, Handa AK (1994) Reduction in pectin methyltransferase activity modifies tissue integrity and cation levels in ripening tomato (*Lycopersicon esculentum* Mill.) fruits. Plant Physiol 106:429–436

151. Langley KR, Martin A, Stenning R, Murray AJ, Hobson GE, Schuch WW, Bird CR (1994) Mechanical and optical assessment of the ripening of tomato fruit with reduced polygalacturonase activity. J Sci Food Agric 66(4):547–554
152. Tieman DM, Zeigler M, Schmelz EA, Taylor MG, Bliss P, Kirst M, Klee HJ (2006) Identification of loci affecting flavour volatile emissions in tomato fruits. J Exp Bot 57:887–896
153. Diaz de la Garza RID, Quinlivan R, Klaus EP, Basset SM, Gregory GJ, Hanson AD (2004) Folate biofortification in tomatoes by engineering the pteridine branch of folate synthesis. Proc Natl Acad Sci U S A 101:13720–13725
154. Diaz de la Garza RID, Gregory JF, Hanson AD (2007) Folate biofortification of tomato fruit. Proc Natl Acad Sci U S A 104(10):4218–4222
155. Zhang C, Liu J, Zhang Y, Cai X, Gong P, Zhang J, Wang T, Li H, Ye Z (2011) Overexpression of SlGMEs leads to ascorbate accumulation with enhanced oxidative stress, cold, and salt tolerance in tomato. Plant Cell Rep 30:389–398
156. Gilbert L, Alhagdow M, Nunes-Nesi A, Quemener B, Guillon F, Bouchet B, Faurobert M, Gouble B, Page D, Garcia V, Petit J, Stevens R, Causse M, Fernie AR, Lahaye M, Rothan C, Baldet P (2009) GDP-D-mannose 3,5-epimerase (GME) plays a key role at the intersection of ascorbate and non-cellulosic cell-wall biosynthesis in tomato. Plant J 60:499–508
157. Garcia V, Stevens R, Gil L, Gilbert L, Gest N, Petit J, Faurobert M, Maucourt M, Deborde C, Moing A, Poessel JL, Jacob D, Bouchet JP, Giraudel JL, Gouble B, Page D, Alhagdow M, Massot C, Gautier H, Lemaire-Chamley M, Rolin D, Usadel B, Lahaye M, Causse M, Baldet P, Rothan C (2009) An integrative genomics approach for deciphering the complex interactions between ascorbate metabolism and fruit growth and composition in tomato. C R Biol 332:1007–1021
158. Ronen G, Carmel-Goren L, Zamir D, Hirschberg J (2000) An alternative pathway to β-carotene formation in plant chromoplasts discovered by map-based cloning of beta and old-gold color mutations in tomato. Proc Natl Acad Sci U S A 97:11102–11107
159. D'Ambrosio C, Giorio G, Marino I, Merendino A, Petrozza A, Salfi L, Stigliani AL, Cellini F (2004) Virtually complete conversion of lycopene into β-carotene in fruits of tomato plants transformed with the tomato lycopene β-cyclase (tlcy-b) cDNA. Plant Sci 166:207–214
160. Rosati C, Aquilani R, Dharmapuri S, Pallara P, Marusic C, Tavazza R, Bouvier F, Camara B, Giuliano G (2000) Metabolic engineering of β-carotene and lycopene content in tomato fruit. Plant J 24:413–420
161. Wurbs D, Ruf S, Bock B (2007) Contained metabolic engineering in tomatoes by expression of carotenoid biosynthesis genes from the plastid genome. Plant J 49:276–288
162. Romer S, Fraser PD, Kiano JW, Shipton CA, Mills PB, Drake R, Schuch W, Bramley PM (2000) Elevation of the provitamin A content of transgenic tomato plants. Nat Biotechnol 18:666–669
163. Enfissi EMA, Fraser PD, Lois LM, Boronat A, Schuch W, Bramley PM (2005) Metabolic engineering of the mevalonate and non-mevalonate isopentenyl diphosphate-forming pathways for the production of health-promoting isoprenoids. Plant Biotechnol J 3:17–27
164. Zanor MI, Osorio S, Nunes Nesi A, Carrari F, Lohse M, Usadel B, Kuhn C, Bleiss W, Giavalisco P, Willmitzer L, Sulpice R, Zhou YH, Fernie AR (2009) RNA interference of LIN5 in tomato confirms its role in controlling Brix content, uncovers the influence of sugars on the levels of fruit hormones, and demonstrates the importance of sucrose cleavage for normal fruit development and fertility. Plant Physiol 150:1204–1218
165. Colliver S, Bovy A, Collins G, Muir S, Robinson S, de Vos CHR, Verhoeyen ME (2002) Improving the nutritional content of tomatoes through reprogramming their flavonoid biosynthetic pathway. Phytochem Rev 1:113–123
166. Adato A, Mandel T, Mintz-Oron S, Venger I, Levy D, Yativ M, Dominguez E, Wang Z, De Vos RC, Jetter R, Schreiber L, Heredia A, Rogachev I, Aharoni A (2009) Fruit-surface flavonoid accumulation in tomato is controlled by a SlMYB12 regulated transcriptional network. PLoS Genet 5(12):e1000777

167. Butelli E, Titta L, Giorgio M, Mock HP, Matros A, Peterek S, Schijlen EG, Hall RD, Bovy AG, Luo J, Martin C (2008) Enrichment of tomato fruit with health promoting anthocyanins by expression of select transcription factors. Nat Biotechnol 26:1301–1308
168. Schijlen E, de Vos CHR, Jonker H, van den Broeck H, Molthoff J, van Tunen A, Martens S, Bovy A (2006) Pathway engineering for healthy phytochemicals leading to the production of novel flavonoids in tomato fruit. Plant Biotechnol J 4:433–444
169. Simkin AJ, Gaffe J, Alcaraz JP, Carde JP, Bramley PM, Fraser PD, Kuntz M (2007) Fibrillin influence on plastid ultrastructure and pigment content in tomato fruit. Phytochemistry 68(11):1545–1556
170. Neily MH, Matsukura C, Maucourt M, Bernillon S, Deborde C, Moing A, Yin YG, Saito T, Mori K, Asamizu E, Rolin D, Moriguchi T, Ezura H (2011) Enhanced polyamine accumulation alters carotenoid metabolism at the transcriptional level in tomato fruit over-expressing spermidine synthase. J Plant Physiol 168(3):242–252
171. Dharmapuri S, Rosati C, Pallara P, Aquilani R, Bouvier F, Camara B, Giuliano G (2002) Metabolic engineering of xanthophyll content in tomato fruits. FEBS Lett 519:30–34
172. Davuluri GR, van Tuinen A, Fraser PD, Manfredonia A, Newman R, Burgess D, Brummell DA, King SR, Palys J, Uhlig J, Bramley PM, Pennings HM, Bowler C (2005) Fruit-specific RNAi-mediated suppression of DET1 enhances carotenoid and flavonoid content in tomatoes. Nat Biotechnol 23:890–895
173. Wang S, Liu J, Feng Y, Niu X, Giovannoni J, Liu Y (2008) Altered plastid levels and potential for improved fruit nutrient content by downregulation of the tomato DDB1-interacting protein CUL4. Plant J 55(1):89–103
174. Liu Y, Roof S, Ye Z, Barry C, van Tuinen A, Vrebalov J, Bowler C, Giovannoni J (2004) Manipulation of light signal transduction as a means of modifying fruit nutritional quality in tomato. Proc Natl Acad Sci U S A 101:9897–9902
175. Giliberto L, Perrotta G, Pallara P, Weller JL, Fraser PD, Bramley PM, Fiore A, Tavazza M, Giuliano G (2005) Manipulation of the blue light photoreceptor cryptochrome 2 in tomato affects vegetative development, flowering time, and fruit antioxidant content. Plant Physiol 137:199–208
176. Li L, Zhou YH, Cheng XF, Sun JY, Marita JM, Ralph J, Chiang VL (2003) Combinatorial modification of multiple lignin traits in trees through multigene co-transformation. Proc Natl Acad Sci U S A 100:4939–4944
177. Abbott JC, Barakate A, Pincon G, Legrand M, Lapierre C, Mila I, Schuch W, Halpin C (2002) Simultaneous suppression of multiple genes by single transgenes. Down-regulation of three unrelated lignin biosynthetic genes in tobacco. Plant Physiol 128:844–853
178. Chabannes M, Barakate A, Lapierre C, Marita JM, Ralph J, Pean M, Danoun S, Halpin C, Grima-Pettenati J, Boudet AM (2001) Strong decrease in lignin content without significant alteration of plant development is induced by simultaneous down-regulation of cinnamoyl CoA reductase (CCR) and cinnamyl alcohol dehydrogenase (CAD) in tobacco plants. Plant J 28:257–270
179. Kawaoka A, Ebinuma H (2001) Transcriptional control of lignin biosynthesis by tobacco LIM protein. Phytochemistry 57:1149–1157
180. Boudet AM, Kajita S, Grima-Pettenati J, Goffner D (2003) Lignins and lignocellulosics: a better control of synthesis for new and improved uses. Trends Plant Sci 8:576–581
181. Anterola AM, Lewis NG (2002) Trends in lignin modification: a comprehensive analysis of the effects of genetic manipulations/mutations on lignification and vascular integrity. Phytochemistry 61:221–294
182. Boerjan W, Ralph J, Baucher M (2003) Lignin biosynthesis. Annu Rev Plant Biol 54:519–546
183. Bell-Lelong DA, Cusumano JC, Meyer K, Chapple C (1997) Cinnamate- 4-hydroxylase expression in Arabidopsis. Regulation in response to development and the environment. Plant Physiol 113:729–738

Metabolic Engineering for Improving Production of Taxol

18

Subrata Kundu, Sumita Jha, and Biswajit Ghosh

Abstract

Taxol (generic name paclitaxel), a complex diterpenoid, is an efficient antineoplastic drug extracted from the plant. It has been approved for the management of several cancers including lungs, breast, and ovary cancers. The bark of several *Taxus* species is the natural source of taxol, but its lower accumulation (0.01–0.04% dry weight) elevated the price of extraction. Its complex structure prohibits the complete chemical synthesis of the compounds in economical approach at the industrial level. Therefore, a plethora of approaches has been implemented by several researchers for alternative and economical production of taxol. The advent of recombinant DNA technologies has resulted in the commencement of metabolic engineering as an effective alternative for the production of pharmaceutically important plant natural products at industrial levels. Plants have emergence as a perfect system for metabolic engineering due to its relatively cheap price and easiness in growing. Plant cell factories provide an alternative source for the scale-up of the production of high added value secondary metabolites including the anticancer drug taxol that is biosynthesized in *Taxus* spp. in very tiny quantity. The demand for taxol and its derivatives has increased enormously owing to its unique antineoplastic activity, lack of the taxane ring in nature and complexity of chemical synthesis. Therefore, countless efforts have been executed in worldwide for the biotechnological production of taxol. In this

S. Kundu • B. Ghosh (✉)
Plant Biotechnology Laboratory, Department of Botany, Ramakrishna Mission Vivekananda Centenary College, Kolkata, India
e-mail: ghosh_b2000@yahoo.co.in

S. Jha
Department of Botany, Centre of Advanced Study, University of Calcutta, Kolkata, India

S. Jha (ed.), *Transgenesis and Secondary Metabolism*, Reference Series in Phytochemistry, DOI 10.1007/978-3-319-28669-3_29

chapter, we have discussed different features of metabolic engineering, including genetic manipulation of plants as well as microbes to increase production of taxol and its precursors.

Keywords
Paclitaxel • Taxol • *Taxus brevifolia* • Plant metabolic engineering • Anticancer drug

Abbreviations

10-DAB	10-Deacetylbaccatin III
BAPT	Baccatin III C-13-phenylpropanoyl-CoA transferase
BMS	Bristol-Myers Squibb
DBAT	10-Deacetylbaccatin III-10-O-acetyltransferase
DBTNBT	3′-N-debenzoyl-2′-deoxytaxol N-benzoyl transferase
DMAPP	Dimethylallyl pyrophosphate
FDA	Food and Drug Administration
FDP	Farnesyl diphosphate
GGPP	Geranylgeranyl diphosphate
GGPPS	Geranylgeranyl diphosphate synthase
IPI	Isopentenyl diphosphate isomerase
IPP	Isopentenyl pyrophosphate
MVA	Mevalonic acid
NCI	National Cancer Institute
PAM	Phenylalanine aminomutase
T10βH	Taxoid 10β-hydroxylase
TBT	Taxane 2α-O-benzoyl transferase
TDAT	Taxadiene-5α-ol-O-acetyltransferase
TXS	Taxadiene synthase
USDA	United States Department of Agriculture

Contents

1 Introduction

Plants as sessile organisms synthesize a wide range of chemical molecules collectively known as secondary metabolites. These versatile chemical compounds are not essential for basic growth and development but are primarily due to environmental adaptation against both biotic and abiotic stresses. Plant secondary metabolites are usually classified according to their biosynthetic pathways and have been categorized into three large molecules families including phenolics, terpenes, and steroids, and alkaloids [1–4]. Alkaloids are the oldest and structurally diverse group of nitrogen-containing secondary metabolites and associated pronounced pharmacological and medicinal importance [5]. Among different types of alkaloids, plants synthesize 20-carbon (C20) polycyclic isoprenoids collectively known as diterpenoids. They are generally confined to restricted groups and found to be the signature molecules [6]. The taxoids are diterpenoids compounds with a unique taxane (pentamethyl $[9.3.1.0]^{3,8}$ tricyclopentadecane) skeleton and generally found in *Taxus* (yew tree) species. There are about 400 taxoids whose structure and function have been characterized [7, 8]. The anticancer drug paclitaxel was found to be most important compared to other taxoids. The taxol was first extracted from the barks of *Taxus brevifolia* (Pacific yew tree), and the work was published in the year 1971 with over 4000 citations [9]. This important diterpene alkaloid is primarily found in the bark of different *Taxus* species, but its extraction cost is extremely high due to its minute quantity (0.01–0.04% dw) [9, 10]. The yew tree grows extremely slow and the mature trees are only suitable for the extraction of taxol. It has been reported that each mature tree could supply only about 2 kg of bark, and for the production of 500 mg taxol, 12 kg of bark was needed [9]. To extract sufficient amount of taxol at pharmaceutical industry level, destruction of the mature tree will also be harmful to nature. Subsequently, the growing demand for taxol greatly surpasses the supply, and alternative sources of the drug are needed. Due of its low yield from a natural source, intensive research has been accompanied to produce taxol more effectively in alternative source [11–13]. Though several groups of researchers have achieved the total chemical synthesis of taxol through sophisticated approaches, the high cost of this synthetic methodology forbids its commercial application [14–16].

Metabolic engineering was originally defined as "the improvement of cellular activities by manipulation of enzymatic, transport, and regulatory functions of the cell with the use of recombinant DNA technologies" [17]. The genetic engineering only focuses on a narrow range of phenotypic improvement through manipulation of target genes. In contrast to other cellular engineering strategies, the approaches of metabolic engineering include modulation of larger cellular metabolic networks. Metabolic engineering is considered as a powerful technique to manipulate the biosynthetic pathway and modulate the production of the natural product as well as novel molecules. The application of state-of-the-art molecular and genetic tool kit has enabled the modulation of enzyme activity either toward increase or decrease of metabolic flux toward desired metabolic pathway [17–22]. Nevertheless, modern expression and regulation system have allowed the refinement of metabolic pathway and creation of fusion pathways for the synthesis of novel compounds in the heterologous host [23–25]. The improvement in different features in metabolic engineering exposed the promises for large-scale production of important metabolites at industrial scale.

Plant metabolic engineering is enthusiastically gaining importance as an alternative to other biotechnological approaches for the production of novel compounds, for the industrial products, as well as for the cleaning of the environment through bioremediation. Plants are very attractive for biotechnological manipulation due to the cheap availability and simple growth nature. The engineering of rate-limiting step to induce or reduce metabolic flux toward target compounds, blocking of unwanted pathways, and introduction of shunting to divert metabolic flux in the desired direction are the major goals in plant metabolic engineering. In this chapter, we have emphasized the advancement in the production of taxol by metabolic engineering. The enormous efforts implemented by several researchers in the biotechnological production of taxol and its precursor, including cell culture, metabolic engineering in heterologous plants, as well as in microbial systems were highlighted.

2 Taxol: Most Potent Anticancer Drug from Nature

The classification of taxoid is based on the oxygenation pattern of the carbon skeleton as well as on existence of lateral chains. Among them, taxol and cephalomanine with C-13 lateral chain and baccatin III without lateral chain are most important toxoids. The chemical name of taxol (molecular formula $C_{47}H_{51}NO_{14}$) is 5β, 20-epoxy-1, 2α,4,7β,13α-hexahydroxytax-11-en-9one-4, 10-diacetate-2-benzoate 13 ester with molecular weight of 853.9 Da [26]. A, B, and C rings are present within the taxol. These rings also associated with functional groups including one benzoyl group, two hydroxyl groups, two acetyl groups, and an oxetane ring. The chemical structures of taxol, its important derivatives, and precursors are represented in Fig. 1. Among these Taxotere (semisynthetic analog of taxol) that is synthesized from the precursor 10-deacetylbaccatin III also found to be potent anticancer drug.

2.1 Historical Perspective of Taxol

In the year 1962, the bark of Pacific yew trees was first collected by the scientists of the US Department of Agriculture (USDA) in association with National Cancer Institute (NCI). The research groups of Dr. Monroe E. Wall and Dr. Mansukh Wani of Natural Product Laboratory at the Research Triangle Institute, Triangle Park, NC, discovered that extract of the bark has potent cytotoxic activity. This research group was engaged in identification and purification of the most active component of the extracts. In 1967, they were able to isolate and identified the active principal of the bark of *T. brevifolia* and named it taxol, according to its species of origin and the presence of hydroxyl groups [27, 28]. In 1977 nearly 1500 old yew tree of Pacific

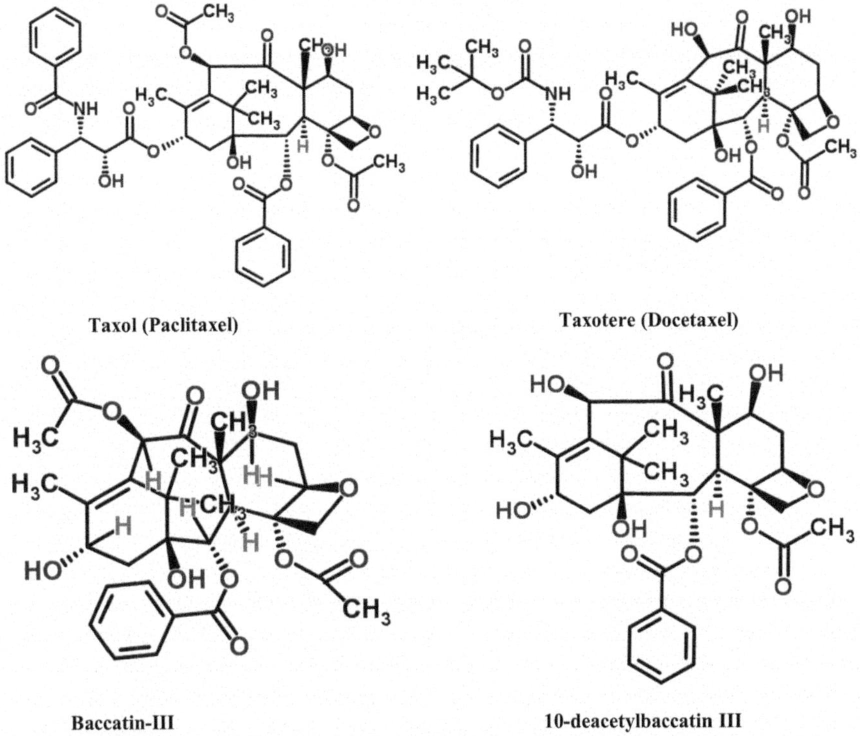

Fig. 1 The chemical structure of taxol, semisynthetic derivative Taxotere, and the two taxol precursors

Northwest forest sacrificed for collection of 700 pounds of bark which are used for taxol extraction. Later, the structure of taxol was published and it was incorporated in the NCI drug development program [9, 29]. Although taxol exhibited varied results in the preclinical trial against the tumor, but it represents efficacy against a subclass of mouse tumor models, including P388 leukemia [30]. Intensive clinical trials were hindered not only due to the shortage of the adequate amount of taxol but also for the inappropriate delivery system. Irrespective of the scarceness, clinical study on ovarian cancer was executed, and it was reported that 30% of patients with advanced ovarian cancer responded significantly [31]. Extensive demands of taxol ultimately resulted in severe depletion of *T. brevifolia* trees. In 1992, *T. brevifolia* was enlisted as endangered species to protect the trees [32]. Due to the limited accessibility of *T. brevifolia* tree for extraction of taxol, as well as its unique cytotoxic potential, a plethora of research group was involved in total chemical synthesis. However due to the complex chemical structure of taxol, about 40 reactions were found to be required for complete synthesis [11, 16]. Therefore, NCI decided to transfer taxol to a pharmaceutical company for its commercialization to obtain a large quantity of taxol. The Bristol-Myers Squibb (BMS) Company developed a cell culture technique to produce the drug. The bioreactor-based technology has been deployed by the BMS Company for large-scale production. They have trademarked the name "Taxol" and created the new generic name paclitaxel [33, 34]. The FDA approved taxol for the treatment for ovarian as well as for advanced breast cancer in 1990s. Till the date, it is the most lucrative and best-selling chemotherapeutic drug in clinical use identified by the plant screening program.

2.2 Sources of Taxol

Taxol was first obtained by extracting peeled bark of the Pacific yew tree. *Taxus* belongs to the class Pinopsida, the order Taxales, and the family Taxaceae. Different *Taxus* species can easily be separated geographically compared to morphologically. So far eight species of *Taxus* have been recognized including *T. baccata* (European yew), *T. brevifolia* (Pacific yew or Western yew), *T. canadensis* (Canadian yew), *T. chinensis* (Chinese yew), *T. cuspidata* (Japanese yew), *T. floridana* (Florida yew), *T. globosa* (Mexican yew), and *T. wallichiana* (Himalayan yew). The remarkable discovery of antitumor properties of the extract of the bark of *T. brevifolia*, taxol, has also been described from other species of the genus *Taxus* [35–38]. Although the mixture of taxoids produced is variable among different species and tissues of the same species [39, 40]. The discovery of 10-deacetylbaccatin III from needles and leaves of *T. brevifolia* as well as the European yew *T. baccata* provided the precursor for semisynthesis of taxol and its analog, Taxotere (docetaxel) [41–43]. The leaves and bark of *T. brevifolia*, *T. wallichiana*, and *T. baccata* have been extensively used for the extraction of taxol in medicinal purpose. Unfortunately, several *Taxus* species are disappearing at an alarming rate due to overexploitation.

The slow growth of *Taxus* species, low yield, and overwhelming pharmaceutical success lead to the search for alternative source of this important alkaloid. It was found *Taxomyces andreanae*, a novel endophytic fungus isolated from the phloem of the inner bark of Pacific yew, was capable of producing taxol growing in the semisynthetic liquid medium [44]. Another endophyte, *Pestalotiopsis guepinii* isolated from the inner bark of *T. wallichiana*, was also reported to produce taxol [45]. These findings created interest in the screening for alternate endophytes sources of taxol in other *Taxus* species as well as non-*Taxus* species, and several endophytes have been isolated [46–49].

2.3 Anticancer Properties and Mechanism of Action

The natural product taxol is considered as one of the most valuable anticancer drugs of plant origin [11, 50]. It is beneficial against ovarian and breast cancer and showed efficacy for other cancers including the head and neck, lung, bladder, and gastrointestinal [51, 52]. It has been reported that taxol displayed toxicity against 9 KB cancer cell culture, B16 murine melanoma, as well as against different leukemia cell lines [9, 30, 53–56]. Due to its versatility in antitumor activity, FDA in the United

States and other countries has approved taxol as first-line chemotherapeutic agent treatment for cancer [52].

The anticancer properties of taxol are quite unique and novel compared to other antitumor agents [57–59]. In spite of preventing the polymerization of tubulin into microtubule, taxol binds to assembled microtubules and blocks its depolymerization and ultimately inhibits mitosis. It has been reported that at 1–10 nM of concentration, taxol exerts its anticancerous properties through inhibiting with dynamics of the microtubule. The suppression of microtubule dynamics hinders the assemblage of mitotic spindle and separation of the chromosome during meiosis [60–62]. The arrests of the cell cycle are correlated with cytotoxicity that ultimately resulted in the induction of apoptosis through caspase-independent and caspase-dependent pathways particularly caspases 3, 8, and 10 [63–66]. The antitumor activity in taxol is due to side chain, A ring, C2 benzoyl group, and oxetane ring. It has been reported that the activity is sustained due to the C3′ amide-acyl group in the C13 chain and presence of hydroxyl group at C2′ improved its activity [41, 67]. The cytotoxicity and stabilization of microtubule are induced by the interaction of these components with β-tubulin [68].

2.4 Biosynthetic Pathway

In spite of their enormous biological and economic importance, taxol biosynthetic pathway and its regulation are not extensively known (Fig. 2). The enzyme taxadiene synthase (TXS) catalyzes the cyclization of geranylgeranyl diphosphate (GGPP) to the taxa-(4,5, 11,12)-diene. This is the first committed step of taxol biosynthesis pathway [69–71]. Then, oxygen and acyl groups are added to the taxane at different locations catalyzed by cytochrome P450 monooxygenases. The next step of taxol biosynthesis is the construction of taxa-4(20),11(12)-dien-5α-ol by the enzyme cytochrome P450 taxadiene-5α-hydroxylase (T5αH) [72, 73]. Then the enzyme taxadiene-5α-ol-O-acetyltransferase (TDAT) acrylates the taxa-4(20),11(12)-dien-5α-ol at C5 position to form taxa-4(20),11(12)-dien-5α-yl-acetate [74]. The product is then hydroxylated at C10 position by the enzyme taxoid 10β-hydroxylase (T10βH) to form taxadiene-5α-10β-diol-acetate [75]. It was followed by the hydroxylations, oxidation of C9, and epoxidation at the C4C5 double bond that resulted in the formation of 2-debenzolytaxane [40, 76]. Then taxane 2α-O-benzoyl transferase (TBT) catalyzes the alteration of 2-debenzolytaxane to 10-deacetylbaccatin III (10-DAB). It was followed by the hydroxylation at the C10 position of the 10-DAB to form a diterpene intermediate, baccatin III. The reaction is catalyzed by the enzyme 10-deacetylbaccatin III-10-O-acetyltransferase (DBAT). The enzyme phenylalanine aminomutase (PAM) transfers coenzyme A side chain to the amino acid β-phenyalanine that resulted in the formation of β-phenylalanyl-CoA [77]. The C13 hydroxyl group of baccatin III is esterified with β-phenylalanyl-CoA side chain by the enzyme baccatin III C-13-phenylpropanoyl-CoA transferase (BAPT). The product of this reaction, 3′-N-debenzoyl-2′-deoxytaxol (β-phenylalanyl baccatin III), is hydroxylated to form N-debenzoyltaxol by an unknown Cyt P450-dependent

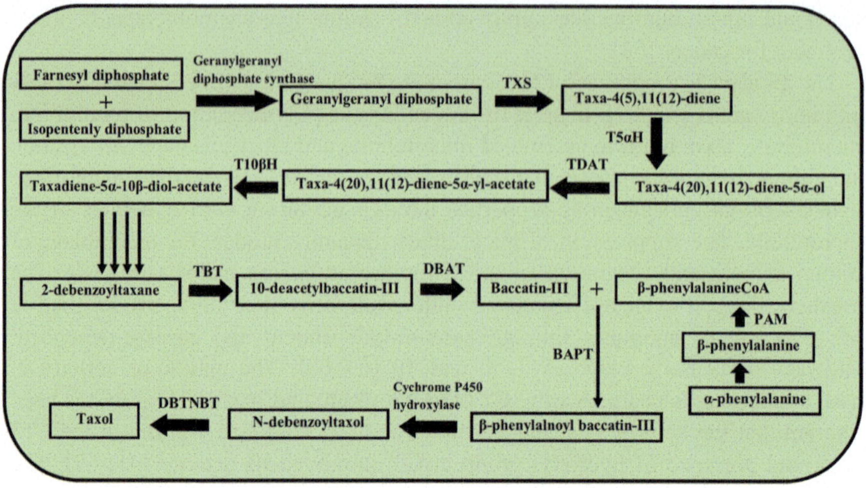

Fig. 2 Taxol biosynthetic pathway (*TXS* taxadiene synthase, *T5αH* cytochrome P450 taxadiene-5α-hydroxylase, *TDAT* taxadiene-5α-ol-O-acetyltransferase, *T10βH* taxoid 10β-hydroxylase, *TBT* taxane 2α-O-benzoyl transferase, *DBAT* 10-deacetylbaccatin III-10-O-acetyltransferase, *PAM* phenylalanine aminomutase, *BAPT* baccatin III C-13-phenylpropanoyl-CoA transferase)

hydroxylase enzyme. Finally, the enzyme 3′-N-debenzoyl-2′-deoxytaxol N-benzoyl transferase (DBTNBT) catalyzes the conjugation of benzoyl-CoA to 3′-N-debenzoyl-2′-deoxytaxol to form taxol. It is the final steps of 19 enzymes catalyzed biosynthetic pathway [77–79].

3 Metabolic Engineering

Metabolic engineering is the rewiring of cellular events through modulation of enzyme activities and regulatory functions of the cell with the implication of state-of-the-art recombinant DNA technologies [17, 22]. It includes leveraging of genetic and regulatory systems of individual cells to produce pharmaceutically or clinically important substances. The accumulation novel or desired compound could be achieved through modifying genes expression of cellular pathways to relocate the metabolic flux toward preferred pathway. The prospective applications of such modern technologies include the production of fuels, foods, and pharmaceuticals. The alteration of cells into effective factories is challenging due to the robust-regulated metabolic networks that prevent its modification. Nevertheless, the advancement in metabolic engineering has explored the potentiality in the production of complex biological molecules. Metabolic engineering can be executed in two approaches including direct and holistic approach. The direct approach is based on the introduction of genes into the host genome to control the outcome of biosynthetic pathway either by upregulation of an intermediate step or inhibit undesirable products (Fig. 3). In a holistic approach, the foreign gene codes transcription factor that

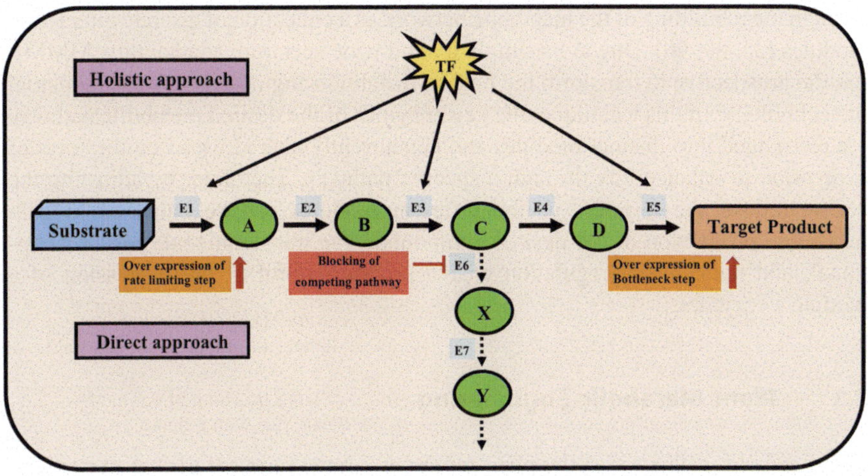

Fig. 3 Schematic representation of two approaches for metabolic engineering techniques to improve the biotechnological production of a target compound

regulates several biosynthetic pathways concurrently [80]. Endowed by the immense developments in molecular biology techniques in "post-genomic" era, the emphasis of metabolic engineering has progressively relocated from disturbing specific pathways toward modulation of the complete cellular activities that ultimately resulted in the generation of the concept of "systems metabolic engineering" [81–83]. The approaches have transformed the concept of metabolic engineering from conventional ideas of upregulation or inhibition of single gene expression toward the regulation and modification of complete cellular network [84–86].

Metabolic engineering is also applicable for transforming microorganisms into efficient cell factories for the synthesis of heterologous metabolites. The concept and techniques of system biology, synthetic biology, and evolutionary engineering have enabled a conceptual and technological framework to scale up the production of novel metabolic enzymes and biosynthetic pathways and modulation of preexisting pathway for optimizing desired product synthesis [87, 88]. The development of genetically modified microorganism capable of synthesizing various chemicals in industrial scale can be executed by systems metabolic engineering. Nevertheless, expansion of the metabolic capacity of the microbial system is a prerequisite for the large-scale production of a medicinally important metabolite. Identification of biochemical gap whose corresponding genes might not present in the host and artificial construction of metabolic pathways can efficiently be executed through the implication of different techniques of systems metabolic engineering including pathway prediction tools, incorporation of promiscuous enzymes that are compatible for of nonnatural reactions.

The pioneering approach in engineering secondary metabolism can also be achieved through multivariate modular metabolic engineering (MMME). The novelty of MMME approach includes assessment and exclusion of regulatory pathways through the rebuilding of the metabolic network as a collection of discrete interacting modules [25, 89, 90]. Due to its simplicity and wide spectrum adaptability, MMME has the prospective to transform the field of metabolic engineering toward industrial biotechnology. In this technique, the key enzymes of the desired metabolic pathway are rearranged into distinct modules and concurrently altered based on the level of expression to balance flux through a specific pathway. Therefore, by adjusting the concentrations the turnover of the different modules can be balanced to gain maximum production of the desired compound. The transcriptional, posttranscriptional, and translational regulation enable precise control of the expression of a module's enzymes.

3.1 Plant Metabolic Engineering

The metabolic pathways of the plant represent a huge resource of important compounds with pharmacological, biotechnological, and medicinal prominence. Generally, the plant metabolites were isolated from the natural sources or were semisynthesized from intermediates of metabolic pathways. Nevertheless, this process has been outshined due to small harvest and plentiful practical problems. The implementation of modern powerful techniques for the efficient transformation techniques along with the availability of complete genome sequence conveys plant metabolic engineering as a potential alternative to the traditional chemical synthesis of biologically important metabolites [91, 92]. Plant metabolic engineering includes the controlling of present metabolic pathways by either channelizing metabolic flux toward target metabolite or diverting flux from the undesirable compound and the synthesis of the innovative chemical compound through the integration of genes from heterologous organisms into its genome. Metabolic engineering in the plant is executed by the rerouting of the enzymatic pathways to modulate the production of medicinally important compounds that are normally synthesized in small quantities to industrial level [93–95]. The degradation of environmentally toxic compounds or conversion of the plant toward resistance to abiotic or biotic factors is also another important feature of plant metabolic engineering. Nevertheless, the principal objectives of utilizing plant as a green factory in the pharmaceutical industry or agriculture are the stimulation of the manufacture of final secondary metabolites and biosynthesis of precursor molecules of medicinally important compounds.

The engineering of microbial metabolism is established technology, but its position in plants delays due to the existence of large metabolic diversity with the complex metabolic network [96, 97]. Nevertheless, highly compartmentalized cellular structure particularly huge vacuoles and plastids make the plant more complicated compared to another organism. Therefore, in silico model-based methods have recently been implemented in plant engineering. The advancement in systems biology and bioinformatics has initiated to unravel the complication of plant metabolism and enable the construction of effective theoretical models of plant metabolism [98, 99]. The arrival of high-throughput skills of next-generation sequencing and phenotyping tools for phenomics has extended the study of the model as well as

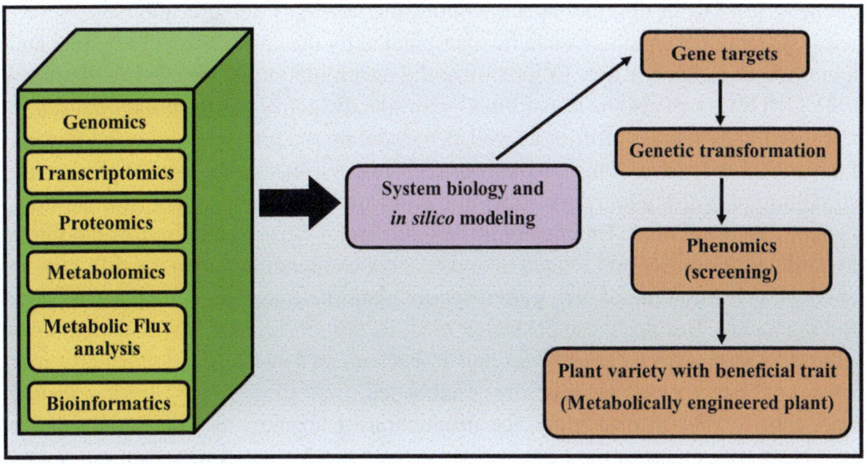

Fig. 4 Schematic representation of the application of "omics" technology for construction of metabolically engineered plants with beneficial traits

nonmodel plants through genome sequencing, transcriptome, proteomics, metabolome, and metabolic flux analysis which has subsequently provided fast, reliable, and dynamic evaluation of metabolically engineered crops with beneficial agricultural traits (Fig. 4).

4 Metabolic Engineering for Taxol Production

Taxol was first isolated from the bark of western yew, *Taxus brevifolia* [9, 54, 55]. It has been reported that the bark of yew is the primary source and it exists only in about 0.02% concentration. Additionally, the *Taxus* species are endangered, grow very slowly in areas of the Pacific Northwest, and peel off its bark which ultimately resulted in complete destruction of the plant. The endorsement of taxol for the treatment of ovarian cancer and due to its novel therapeutic application the demand of taxol is increasing worldwide [100, 101]. The economical way in the chemical synthesis of taxol is challenging due to its complex molecular architecture. Therefore, plethora of approaches including plant cell culture, metabolic engineering in heterologous plant, as well as in microbial system has been implemented from the scientists all over the world in the hunt for alternative approach to protect the limited resource of the yew plant as well as obtaining taxol in relatively inexpensive way to reduce the cost of therapy.

4.1 Plant Cell Culture

Cell culture provides an attractive way for the establishment of uniform conditions to optimize the production of medicinally important phytochemical. It provides the benefit of growth in minimal space without the necessity for greenhouses. Nevertheless, the culture conditions can be manipulated by the application of elicitors and other stresses in such a way to maximize the accumulation of targeted metabolites [102]. Cell lines can also be transformed with specific genes of metabolic pathway to modulate the metabolic profiles as well as to scale up the production. For harvesting of unusual metabolites from rare plant species, novel drugs or large-scale production cell culture provides a lucrative substitute compared to the plants that as impetuous to genetic manipulation. The metabolic engineering of *Taxus* cell culture to modulate the production of taxol and related toxoids is very challenging due to the difficulty in genetic transformation of the gymnosperm plantlike *Taxus* and due to its slow growing nature. Besides these obstacles, a wild strain of *Agrobacterium tumefaciens* was used to transform *T. brevifolia* and *T. baccata* in 1994 [103]. Later hairy root culture of *Taxus* was successfully established first time using *Agrobacterium rhizogenes* [104]. Unfortunately, the production of taxane was low, and therefore the system was inappropriate for industrial scale production. Hairy root culture of *Taxus* x *media* var. Hicksii was fed with precursors (L-phenylalanine and p-amino benzoic acid), or in combination with methyl jasmonate and accumulation of paclitaxel, baccatin III and 10-deacetylbaccatin III were monitored [105]. The highest amount of paclitaxel (221.8 μg L^{-1}) was accumulated in the medium supplemented with 100 μM of the precursors along with methyl jasmonate. The successful application of metabolic engineering toward the production of taxol using a direct approach was first successfully executed in 2005 [106]. Transgenic *T. marei* cell culture overexpressing 10-deacetylbaccatin III-10-O-acetyltransferase (DBAT) and taxadiene synthase (TXS) genes produce the highest level of taxol. *Taxus media* cell cultures with same genetic characteristics compared to the parent were established using *A. rhizogenes* LBA 9402 and the C58C1 strain of *A. tumefaciens* harboring the taxadiene synthase gene [107]. The highest taxane production (265% greater than control) was found in the transformed culture in optimal conditions and elicited with methyl jasmonate. 9-cis-epoxycarotenoid dioxygenase gene from *Taxus chinensis* was overexpressed in transgenic *T. chinensis* cells, and about 2.7-fold higher taxol production was recorded [108]. The hairy root culture of *Taxus* x *media* var. Hicksii was established by transformation with C58C1 strain of *A. tumefaciens* harboring the taxadiene synthase gene of *T. baccata* [109]. These transgenic root lines were elicited with different combinations of elicitors (sodium nitroprusside and methyl jasmonate), and 10.78 mg L^{-1} taxol was accumulated upon elicitation with methyl jasmonate and feeding with L-phenylalanine.

4.2 Engineering in Microbes Toward Production of Taxol and Its Precursor

Metabolic engineering within heterologous host such as microbes for the manufacture of plant metabolites includes improvement of regulatory processes that ultimately resulted in the increased yield of a specific compound through rechanneling the carbon flux. In recent year, the execution of recombinant expression systems to reconstruct natural product pathways has upgraded meaningfully. Among the different microorganism, *Escherichia coli* and *Saccharomyces cerevisiae* have usually been used to bypass complex technical issues associated with the metabolic engineering of plant cell cultures [24, 110, 111]. *E. coli* can be considered as superior host due to its simplicity in the central metabolic pathway and robust regulatory systems. The lacks of posttranslational modification and subcellular compartment and difficulty in the expression of complex protein are the major drawbacks in metabolic engineering of *E. coli*. On the other hand, yeast has ideal characters including larger cell size, better resistance against pH alteration, and unwanted products for final processing of end products. Additionally, mating allows improved cellular engineering that ultimately resulted in robust growth and increased adaptation to the adverse environment. In metabolic engineering, in order to improve product yields, the complete biosynthetic pathways are transferred from native hosts into heterologous organisms. Therefore, for improved synthesis of target product, the optimization of gene expression system, promoter strength precise control of the endogenous regulatory network is utmost required. The isoprenoids are also synthesized in the microbial system [112, 113]. In microbes, they are involved in different cellular activities including light absorption and protein modifications. Among the isoprenoids, biotechnological production of taxol and its precursors through the metabolically engineered microbial system has been enriched over the past few years [114].

The successful synthesis of taxadiene, the intermediate of taxol biosynthesis, was first reported in metabolically engineered *E. coli* in the year 2001 [115]. Overexpression of three genes encoding isopentenyl diphosphate isomerase (IPI), geranylgeranyl diphosphate synthase (GGPS), and taxadiene synthase (TXS) of terpenoid pathway was executed within genetically engineered *E. coli*. In the transgenic *E. coli*, taxadiene was found to be produced at the yield of 1.3 mg L^{-1} of cell culture. This success opened the window for the synthesis of taxoids in non-paclitaxel-producing organisms through metabolic engineering. In a subsequent study, the multivariate-modular approach was implemented to increase the titers of taxadiene up to 1 g L^{-1} [116]. In *E. coli*, methylerythritol-phosphate (MEP) pathway and mevalonic acid (MVA) pathway can produce isopentenyl pyrophosphate (IPP) and dimethylallyl pyrophosphate (DMAPP) (Fig. 5). The taxol can be synthesized from these two building blocks [117]. The researchers have separated the taxadiene biosynthetic pathway into two units: the MEP pathway that generates isopentenyl pyrophosphate (IPP) and heterologous downstream pathway that produces terpenoid. These modules were optimized to exploit the maximum production of taxadiene with reducing the accumulation of indole, the inhibitor of the pathway. In another study, in silico modeling was executed to optimize the biosynthesis of taxadiene in *E. coli* by comparing the maximum theoretical IPP yields and the thermodynamic properties of the 1-deoxy-D-xylulose 5-phosphate (DXP) pathway and MVA pathway using different hosts and carbon sources [118]. They have reported that genetic manipulation of the DXP pathway and chromosomal engineering were powerful tools for heterologous biosynthesis of taxadiene. The pedogenic DPX pathway was modulated by redesigning of codon usage and chromosomal

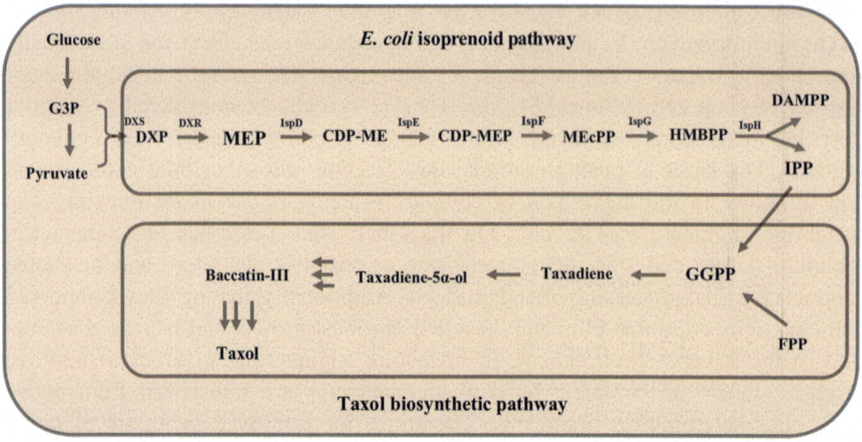

Fig. 5 Biosynthetic scheme for taxol production in *E. coli* (*DXS* DOXP synthase, *CDP-ME* methylerythritol cytidyl diphosphate, *CDP-MEP* 4-diphosphocytidyl-2-C-methyl-D-erythritol-2-phosphate, *HMBPP* 4-hydroxy-3-methyl-butenyl-1-diphosphate, *DXP* 1-deoxy-D-xylulose 5-phosphate, *MEP* 2-C-methylerythritol 4-phosphate, *MEcPP* 2-C-methyl-D-erythritol 2,4-cyclodiphosphate, *DAMPP* dimethylallyl pyrophosphate, *DXR DXP* reductoisomerase, *IspD* 2-C-methyl-D-erythritol 4-phosphate cytidylyltransferase, *IspE* 4-diphosphocytidyl-2-C-methyl-D-erythritol kinase, *IspF* 2-C-methyl-D-erythritol 2,4-cyclodiphosphate synthase, *IspG* 4-hydroxy-3-methylbut-2-en-1-yl diphosphate synthase, *IspH* 4-Hydroxy-3-methylbut-2-enyl diphosphate reductase)

engineering that ultimately gave yields of 876 ± 60 mg L^{-1} taxadiene. The K- and B-derived *E. coli* strains were used as a heterologous host for the production of taxadiene [119]. In this study, different inducible promoters (T7, Trc, and T5) and cellular backgrounds were varied during the biosynthesis of taxadiene. The T7 promoter system was examined for upstream MEP pathway, whereas T7, Trc, and T5 promoters were used for downstream MVA pathway. It was reported that K-derivative transgenic *E. coli* synthesized taxadiene about 2.5-fold higher compared to another one. The significant difference in pyruvate metabolism revealed through transcriptomics analysis established the discrepancy in taxadiene biosynthesis between these two strains. Different types of in vivo metabolic chemistries including oxygenation are vital for the production of complex secondary metabolites such as taxol. The expression of cytochrome P450 enzymes was modified to execute oxygenation chemistry in *E. coli*. The highest reported titer of oxygenated taxanes ($\sim 570 \pm 45$ mg L^{-1}) in *E. coli* was achieved through a series of optimizations including N-terminal modifications [120].

The yeast *Saccharomyces cerevisiae* can also be used for metabolic engineering for the scale-up of taxol biosynthesis. The study was executed to rebuild the initial steps in taxol biosynthetic pathway, and eight genes were expressed in yeast [121]. This initial study was a prerequisite for the construction of engineered yeast for the synthesis of taxol or its precursors. Heterologous genes encoding enzymes of early steps of taxol biosynthetic pathway and the modulation of controlling factor to prevent deviation toward other undesirable pathways were executed to increase the production of taxadiene within genetically modified *S. cerevisiae* [122]. The impact of these modulations in synthesis taxadiene was monitored. They have observed that only the expression of TXS gene of *T. chinensis* or coexpression of both TXS and GGPS was unable to rise taxadiene content significantly. They have reported that expression of modified 3-hydroxyl-3-methylglutaryl-CoA reductase (HMG-CoA reductase) isoenzyme 1 as well as UPC2-1 (mutant regulatory protein) in transgenic *S. cerevisiae* ultimately induced taxadiene content up to fifty percent. The substitution of *T. chinensis* GGPS gene with its complement part from *Sulfolobus acidocaldarius* and codon optimization resulted in high-level expression in taxadiene (8.7 ± 0.85 mg L^{-1}) that was found to be 40-fold higher compared to control. Later, protein modeling and in silico metabolomic study guided the establishment of transgenic *S. cerevisiae* that can efficiently modulate taxadiene content [123]. In this study, the biosynthesis pathway was reconstructed through transformation with TXS gene along with upregulation of erg20 and thmgr gene. The catalytic efficiency of GGPPS enzymes was anticipated using enzyme-substrate docking strategy. The in silico modeling was correlated with experimental outcomes, and it was found that transgenic yeast with *Taxus cuspidate* GGPPS gene caused the highest synthesis of taxadiene (72.8 mg L^{-1}).

4.3 Engineering in Heterologous Plant

There are a plethora of instances of successful metabolic engineering of taxol biosynthetic genes and enhanced production of taxol in microbial hosts. Nevertheless, complex plant-derived terpenoids including taxol are often problematic to produce within microbial host due to insolubility, membrane associations, as well as different types of posttranslational modifications. Plant cytochrome P450 enzymes are also utmost required for complete biosynthesis of the taxol. There are several obstacles in heterologous expression of terpenoids such as taxol due to the more complex plant transformation techniques for many species, time-consuming recovery of transgenic plants, and slow growth rates compared to microbial hosts. Therefore, plants are more preferable compared to a microbial system for metabolic engineering of natural plant products due to the existence of innate cytochrome P450-related biosynthetic genes. The unavailability of efficient transformation systems for *Taxus* species created a platform to screen alternative plants that can be genetically engineered to produce taxol or its precursors. Metabolic engineering of heterologous host plants for production of taxol has been reported by several research groups. The pioneer steps toward the genetically engineering of taxoids biosynthesis in angiosperms were published in 2004 [124]. The constitutive expression of recombinant *T. baccata* taxadiene synthase (TXS) gene without plastid targeting signature sequence and fusion of C-terminal histidine tagged (*Arabidopsis thaliana*) resulted in the production of taxadiene with associated growth hindrance in transgenic plants. The relatively lower accumulation of carotenoids also indicates relocation of geranylgeranyl diphosphate (GGDP) pool in transgenic plants. The redirection of GGDP pools toward the production of taxadiene was further supported by the successful establishment of transgenic system expressing a glucocorticoid-mediated system. In another study, the redirection of GGDP toward taxadiene production was reported in transgenic tomato plants. The expression of TXS gene in tomato line missing the capacity to utilize GGDP for carotenoid production ultimately resulted in the production of taxadiene about 160 mg kg^{-1} dried fruit [125]. Metabolic engineering of ginseng (*Panax ginseng* C.A. Meyer) roots with TXS gene of *T. brevifolia* was successfully executed without compromising in growth as well as phenotype [126]. In the transgenic TSS3-2 line accumulation of taxadiene, the precursor of taxol was found to be 9.1 μg g^{-1} dry weight. Elicitation by jasmonate also resulted in about 1.74-fold increment of the taxadiene content. The incorporation of TXS gene within the transcriptional regulation of CaMV 35S promoter in *Nicotiana benthamiana* resulted in the de novo production of taxadiene of 11–27 μg g^{-1} dry weight [127]. Upon elicitation with methyl jasmonate and shunting the terpenoid pathway with suppression of phytoene synthase gene also increased taxadiene content up to 1.9-fold in the transgenic line TSS8. The metabolic engineering of taxol biosynthetic pathway in *Artemisia annua* L plant was also successfully established [128]. The *Agrobacterium tumefaciens*-mediated transformation of TXS gene within pCAMBIA1304 promoter ensued in the accumulation of taxadiene up to 129.7 μg g^{-1} dry mass in the stem without affecting the artemisinin content as well as the growth of the transgenic plant.

5 Conclusions

Cancer is the most devastating disease compared to other life-threatening ailments, and it bears the majority of the burden for the people of low- and middle-income countries. A plethora of research work has been implicated to efficiently detect it in early stage and cost-effective treatment for the people below poverty line. Under such circumstance Taxol (generic name paclitaxel), isolated from the yew tree, was found to be one of the most effective anticancer drugs of plant origin. Unfortunately, due to the relatively lower amount of accumulation in native plant and unavailability of cost-effective chemical synthesis, the demand for the alternative synthesis of taxol is increasing enormously. The application of modern molecular biological techniques to metabolically engineered heterologous plant as well as microbes for the effective synthesis of taxol and its precursor has been successfully implemented. To scale up the biotechnological production of taxol in industrial level, metabolic engineering in a wide range of plant species along with efficient extraction and separation methods of taxol is in utmost priority among the researchers. The expression of taxol biosynthetic genes from the yew tree into diverse plant species through metabolic engineering is appreciated for the taxol biosynthesis, but the direct engineering in the yew species will be more sensible approach. The combinations of genetic modification techniques using *Agrobacterium* and particle bombardment, along with the established cell suspension technology for large-scale production, may lead to the higher productivities of taxol via metabolically engineered plant cells. More extensive research should be executed to completely unrevealed the regulation of taxol biosynthetic pathway as well as the key genes of rate-limiting steps. Therefore, the future perspective should be the implication of empirical as well as rational approaches of metabolic engineering toward the biotechnological production of taxol.

Acknowledgment Subrata Kundu gratefully acknowledges to Council of Scientific and Industrial Research (CSIR), India, for providing Research Associate (RA) fellowship [Sanction No. 08/525 (0004)/2014-EMR1].

References

1. Hartmann T (2007) From waste products to eco chemicals: fifty years research of plant secondary metabolism. Phytochemistry 68:2831–2846
2. Wink M (2008) Plant secondary metabolism: diversity, function, and its evolution. Nat Prod Commun 3:1205–1216
3. Harborne JB (1999) Classes and functions of secondary products. In: Walton NJ, Brown DE (eds) Chemicals from plants, perspectives on secondary plant products. Imperial College Press, London, pp 1–25
4. Bourgaud F, Gravot A, Milesi S, Gonntier E (2001) Production of plant secondary metabolites: a historical perspective. Plant Sci 161:839–851
5. Kutchan TM (1995) Alkaloid biosynthesis-the basis for metabolic engineering of medicinal plants. Plant Cell 7:1059–1070
6. Ashour M, Wink M, Gershenzon J (2010) Biochemistry of terpenoids: monoterpenes, sesquiterpenes and diterpenes. In: Wink M (ed) Annual plant reviews: biochemistry of plant secondary metabolism, vol 40, 2nd edn. Wiley, New York. doi:10.1002/9781444320503.ch5
7. Baloglu E, Kingston DG (1999) The taxane diterpenoids. J Nat Prod 62:1448–1472
8. Itokawa H (2003) Taxoids occurring in the genus *Taxus*. In: Itokawa H, Lee K-H (eds) Taxus – the genus Taxus. Taylor & Francis, London, pp 35–78
9. Wani MC, Taylor HL, Wall ME, Coggan P, McPhail AT (1971) Plant antitumor agents: VI. The isolation and structure of taxol, a novel antileukemic and antitumor agent from *Taxus brevifolia*. J Am Chem Soc 93:2325–2327
10. Goodman J, Walsh V (2001) The story of taxol. Cambridge University Press, Cambridge, p 282
11. Holton RA, Kim HB, Somoza C, Liang F, Biediger RJ, Boatman PD, Shindo M, Smith CC, Kim S (1994) First total synthesis of taxol. 2. Completion of the C and D rings. J Am Chem Soc 116:1599–1600
12. Expósito O, Bonfill M, Onrubia M, Jané A, Moyano E, Cusidó RM, Palazón J, Piñol MT (2008) Effect of taxol feeding on taxol and related taxane production in *Taxus baccata* suspension cultures. New Biotechnol 25:252–259
13. Kim JA, Baek KH, Son YM, Son SH, Shin H (2009) Hairy root cultures of *Taxus cuspidata* for enhanced production of paclitaxel. J Korean Soc Appl Biol Chem 52:144–150
14. Masters JJ, Link JT, Snyder LB, Young WB, Danishefsky SJ (1995) A total synthesis of taxol. Angew Chem Int Ed Eng 34:1723–1726
15. Nicolaou KC, Yang Z, Liu JJ, Ueno H, Nantermet PG, Guy RK, Claiborne CF, Renaud J, Couladouros EA, Paulvannan K et al (1994) Total synthesis of taxol. Nature 367:630–634
16. Nicolaou KC, Guy RK (1995) The conquest of taxol. Angew Chem Int Ed Eng 34:2079–2090
17. Bailey JE (1991) Toward a science of metabolic engineering. Science 252:1668–1675
18. Jacobsen JR, Khosla C (1998) New directions in metabolic engineering. Curr Opin Chem Biol 2:133–137

19. Grotewold E (2008) Transcription factors for predictive plant metabolic engineering: are we there yet? Curr Opin Biotechnol 19:138–144
20. Broun P (2004) Transcription factors as tools for metabolic engineering in plants. Curr Opin Plant Biol 7:202–209
21. Keasling JD (2010) Manufacturing molecules through metabolic engineering. Science 330:1355–1358
22. Staniek A, Bouwmeester H, Fraser PD, Kayser O, Martens S, Tissier A, van der Krol S, Wessjohann L, Warzecha H (2013) Natural products – modifying metabolite pathways in plants. Biotechnol J 8:1159–1171
23. Marienhagen J, Bott M (2013) Metabolic engineering of microorganisms for the synthesis of plant natural products. J Biotechnol 163:166–178
24. Xu P, Bhan N, Koffas MA (2013) Engineering plant metabolism into microbes: from systems biology to synthetic biology. Curr Opin Biotechnol 24:291–299
25. Yadav VG, De Mey M, Lim CG, Ajikumar PK, Stephanopoulos G (2012) The future of metabolic engineering and synthetic biology: towards a systematic practice. Metab Eng 14:233–241
26. Panchagnula R (1998) Pharmaceutical aspects of paclitaxel. Int J Pharm 72:1–15
27. Perdue RE Jr, Hartwell JL (1969) The search for plant sources of anticancer drugs. Morris Arboretum Bull 20:35–53
28. Wall ME, Wani MC (1995) Camptothecin and taxol: discovery to clinic – thirteenth Bruce F. Cain Memorial Award Lecture. Cancer Res 55:753–760
29. Wall ME, Wani MC (1977) Antineoplastic agents from plants. Annu Rev Pharmacol Toxicol 17:117–132
30. Fuchs DA, Johnson RK (1978) Cytologic evidence that taxol, an antineoplastic agent from *Taxus brevifolia*, acts as a mitotic spindle poison. Cancer Treat Rep 62: 1219–1222
31. McGuire WP, Rowinsky EK, Rosenshein NB, Grumbine FC, Ettinger DS, Armstrong DK, Donehower RC (1989) Taxol: a unique antineoplastic agent with significant activity in advanced ovarian epithelial neoplasms. Ann Intern Med 111:273–279
32. Walsh V, Goodman J (1999) Cancer chemotherapy, biodiversity, public and private property: the case of the anti-cancer drug taxol. Soc Sci Med 49:1215–1225
33. Walsh V, Goodman J (2002) From taxol to Taxol: the changing identities and ownership of an anti-cancer drug. Med Anthropol 21:307–336
34. Walsh V, Goodman J (2002) The billion dollar molecule: taxol in historical and theoretical perspective. Clio Med 66:245–267
35. Bala S, Uniyal GC, Chattopadhyay SK, Tripathi V, Sashidhara KV, Kulshrestha M et al (1999) Analysis of taxol and major taxoids in Himalayan yew, *Taxus wallichiana*. J Chromatogr A 858:239–244
36. Prasain JK, Stefanowicz P, Kiyota T, Habeichi F, Konishi Y (2001) Taxines from the needles of *Taxus wallichiana*. Phytochemistry 58:1167–1170
37. Jha S, Sanyal D, Ghosh B, Jha TB (1998) Improved taxol yield in cell suspension culture of *Taxus wallichiana* (Himalayan Yew). Planta Med 64:270–272
38. Mukherjee S, Ghosh B, Jha TB, Jha S (2002) Variation in content of taxol and related taxanes in Eastern Himalayan populations of *Taxus wallichiana*. Planta Med 68:757–759
39. Jha S, Jha TB (1995) A fast growing cell line of *Taxus baccata* L. (Himalayan yew) as a potential source of taxol precursor. Curr Sci 69:971–973
40. Croteau R, Ketchum REB, Long RM, Kaspera R, Wildung MR (2006) Taxol biosynthesis and molecular genetics. Phytochem Rev 5:75–97
41. Guenard D, Gueritte-Voegelein F, Poiter P (1993) Taxol and taxotere: discovery, chemistry and structure activity relationship. Acc Chem Res 26:160–167
42. Pronk LC, Stoter G, Verweij J (1995) Docetaxel (Taxotere): single agent activity, development of combination treatment and reducing side-effects. Cancer Treat Rev 21:463–478

43. Fumoleau P, Chevallier B, Kerbrat P, Dieras V, Azli N, Bayssas M, Glabbeke MV (1995) Current status of Taxotere® (docetaxel) as a new treatment in breast cancer. Breast Cancer Res Treat 33:39–46
44. Stierle A, Strobel G, Stierle D (1993) Taxol and taxane production by *Taxomyces andreanae*, an endophytic fungus of Pacific yew. Science 260:214–216
45. Strobel GA, Hess WM, Li JY (1997) Pestalotiopsis guepinii, a taxol producing endophyte of the Wollemi pine, Wollemia nobilis. Aust J Biotechnol 45:1073–1082
46. Wang J, Li G, Lu H, Zheng Z, Huang Y, Su W (2000) Taxol from Tuberfularia sp, strain TF5, an endophytic fungus of *Taxus mairei*. FEMS Microbiol Lett 193:249–253
47. Zhang P, Zhou PP, Jiang C, Yu H, Yu LJ (2008) Screening of Taxol – producing fungi based on PCR amplification from Taxus. Biotechnol Lett 30:2119–2123
48. Zhao K, Sun L, Ma X, Li X, Wang X, Ping W, Zhou D (2011) Improved Taxol production in *Nodulisporium sylviforme* derived from inactivated protoplast fusion. Afr J Biotechnol 10:4175–4182
49. Li JY, Strobel G, Sidhu R, Hess WM, Ford EJ (1996) Endophytic Taxol-producing fungi from bald cypress, *Taxodium distichum*. Microbiology 142:2223–2226
50. Schiff PB, Fant J, Horwitz SB (1979) Promotion of microtubule assembly *in vitro* by taxol. Nat Biotechnol 277:665–667
51. Rischin D, Smith J, Millward M, Lewis C, Boyer M, Richardson G, Toner G, Gurney H, McKendrick J (2000) A phase II trial of paclitaxel and epirubicin in advanced breast cancer. Br J Cancer 83:438–442
52. Fauzee NJS (2011) Taxanes: promising anti-cancer drugs. Asian Pac J Cancer Prev 12:837–851
53. Wall ME, Wani MC, Cook CE, Palmer CE, McPhail AT (1966) Plant antitumor agents: I. The isolation and structure of camptothecin, a novel alkaloidal leukemia and tumor inhibitor from *Camptotheca acuminata*. J Am Chem Soc 88:3888–3894
54. Wani MC, Kepler JA, Thompson JB, Wall ME, Levine SG (1970) Plant antitumour agents: alkaloids – synthesis of a pentacyclic camptothecin precursor. Chem Commun 7:404
55. Wall ME, Wani MC, Taylor HL (1976) Isolation and chemical characterization of antitumor agents from plants. Cancer Treat Rep 60:1011–1014
56. Patel RN (1998) Tour de paclitaxel: biocatalysis for semisynthesis. Annu Rev Microbiol 98:361–395
57. Horwitz SB (1992) Mechanism of action of taxol. Trends Pharmacol Sci 13:134–136
58. Schiff PB, Horwitz SB (1980) Taxol stabilizes microtubules in mouse fibroblast cells. Proc Natl Acad Sci U S A 77:1561–1565
59. Jordan MA, Toso RJ, Thrower D, Wilson L (1993) Mechanism of mitotic block and inhibition of cell proliferation by taxol at low concentrations. Proc Natl Acad Sci U S A 90:9552–9556
60. Jordan M, Wilson L (1998) Microtubules and actin filaments: dynamic targets for cancer chemotherapy. Curr Opin Cell Biol 10:123–130
61. Long B, Fairchild C (1994) Paclitaxel inhibits progression of mitotic cells to G1 phase by interference with spindle formation without affecting other microtubule functions during anaphase and telephase. Cancer Res 54:4355–4361
62. Donaldson K, Goolsby G, Wahl A (1994) Cytotoxicity of the anti-cancer agents cisplatin and taxol during cell proliferation and the cell cycle. Int J Cancer 57:847–855
63. Huisman C, Ferreira C, Broker L, Rodriguez J, Smit E, Postmus P, Kruyt F, Giaccone G (2002) Paclitaxel triggers cell death primarily via caspase-independent routes in the nonsmall cell lung cancer cell line NCI-H460. Clin Cancer Res 8:596–606
64. Oyaizu H, Adachi Y, Taketani S, Tokunaga R, Fukuhara S, Ikehara S (1999) A crucial role of caspase 3 and caspase 8 in paclitaxel-induced apoptosis. Mol Cell Biol Res Commun 2:36–41
65. Mielgo A, Torres V, Clair K, Barbero S, Stupack D (2009) Paclitaxel promotes a caspase 8-mediated apoptosis through death effector domain association with microtubules. Oncogene 28:3551–3562

66. Park S-J, Wu C-H, Gordon J, Zhong X, Emami A, Safa A (2004) Taxol induces caspase 10-dependent apoptosis. J Biol Chem 279:51057–51067
67. Kingston DGI (2000) Recent advances in the chemistry of taxol. J Nat Prod 63:726–734
68. He L, Jagtap GP, Kingston DGI, Shen HJ, Orr GA, Horwitz SB (2000) A common pharmacophore for taxol and the epothilones based on the biological activity of taxane molecule lacking a C-13 side chain. Biochemist 39:3972–3978
69. Hezari M, Ketchum REB, Gibson DM, Croteau R (1997) Taxol production and taxadiene synthase activity in *Taxus canadensis* cell suspension cultures. Arch Biochem Biophys 337:185–190
70. Koepp AE, Hezari M, Zajicek J, Stofer VB, LaFever RE, Lewis NG, Croteau RB (1995) Cyclization of geranylgeranyl diphosphate to taxa-4(5),11(12)-diene is the committed step of taxol biosynthesis in Pacific yew. J Biol Chem 270:8686–8690
71. Wildung MR, Croteau RA (1996) cDNA clone for taxadiene synthase, the diterpene cyclase that catalyzes the committed step of taxol biosynthesis. J Biol Chem 271:9201–9204
72. Jennewein S, Long RM, Williams RM, Croteau R (2004) Cytochrome p450 taxadiene 5alpha-hydroxylase, a mechanistically unusual monooxygenase catalyzing the first oxygenation step of taxol biosynthesis. Chem Biol 11:379–387
73. Hefner J, Rubenstein SM, Ketchum REB, Gibson DM, Williams RM, Croteau R (1996) Cytochrome p450-catalyzed hydroxylation of taxa-4(5),11(12)-dien to taxa-4(20),11(12)-dien-alpha-ol. The first oxygenation step in taxol biosynthesis. Chem Biol 3:479–489
74. Walker KD, Schoendorf A, Croteau R (2000) Molecular cloning of taxa-4(20),11(12)-dien-5alpha-ol-O-acetyl-transferase cDNA from Taxus and functional expresión in *Escherichia coli*. Arch Biochem Biophys 374:371–380
75. Schoendorf A, Rithner CD, Williams R, Croteau RB (2001) Molecular cloning of a cytochrome P450 taxane 10b-hydroxylase cDNA from Taxus and functional expresión in yeast. Proc Natl Acad Sci U S A 98:1501–1506
76. Vongpaseuth K, Roberts SC (2007) Advancements in the understanding of paclitaxel metabolism in tissue culture. Curr Pharm Biotechnol 8:219–236
77. Walker K, Klettke K, Akiyama T, Croteau RB (2004) Cloning, heterologous expression, and characterization of a phenylalanine aminomutase involved in taxol biosynthesis. J Biol Chem 279:53947–53954
78. Expósito O, Bonfill M, Onrubia M, Jané A, Moyano E, Cusidó RM et al (2009) Effect of taxol feeding on taxol and related taxane production in *Taxus baccata* suspension cultures. New Biotechnol 25:252–259
79. Guo BH, Kai GY, Jin HB, Tang KX (2006) Taxol synthesis. Afr J Biotechnol 5:15–20
80. Capell T, Christou P (2004) Progress in plant metabolic engineering. Curr Opin Biotechnol 15:148–154
81. Park JH, Lee SY (2008) Towards systems metabolic engineering of microorganisms for amino acid production. Curr Opin Biotechnol 19:454–460
82. Stafford DE, Stephanopoulos G (2001) Metabolic engineering as an integrating platform for strain development. Curr Opin Microbiol 4:336–340
83. Stephanopoulos G, Alper H, Moxley J (2004) Exploiting biological complexity for strain improvement through systems biology. Nat Biotechnol 22:1261–1267
84. Blazeck J, Alper H (2010) Systems metabolic engineering: genome-scale models and beyond. Biotechnol J 5:647–569
85. Tyo KE, Alper HS, Stephanopoulos GN (2007) Expanding the metabolic engineering toolbox: more options to engineer cells. Trends Biotechnol 25:132–137
86. Tyo KEJ, Kocharin K, Nielsen J (2010) Toward design-based engineering of industrial microbes. Curr Opin Microbiol 13:255–262
87. Lee JW, Na D, Park JM, Lee J, Choi S, Lee SY (2012) Systems metabolic engineering of microorganisms for natural and non-natural chemicals. Nat Chem Biol 8:536–546
88. Shin JH, Kim HU, Kim DI, Lee SY (2013) Production of bulk chemicals via novel metabolic pathways in microorganisms. Biotechnol Adv 31:925–935

89. Biggs BW, De Paepe B, Santos CN, De Mey M, Kumaran Ajikumar P (2014) Multivariate modular metabolic engineering for pathway and strain optimization. Curr Opin Biotechnol 29:156–162
90. Junjun W, Peiran L, Yongming F, Han B, Guocheng D, Jingwen Z, Chen J (2013) Multivariate modular metabolic engineering of *Escherichia coli* to produce resveratrol from l-tyrosine. J Biotechnol 167:404–411
91. DellaPenna D (2001) Plant metabolic engineering. Plant Physiol 125:160–163
92. Miralpeix B, Rischer H, Häkkinen ST, Ritala A, Seppänen-Laakso T, Oksman-Caldentey KM, Capell T, Christou P (2013) Metabolic engineering of plant secondary products: which way forward? Curr Pharm Des 19:5622–5639
93. Wilson SA, Roberts SC (2014) Metabolic engineering approaches for production of biochemicals in food and medicinal plants. Curr Opin Biotechnol 26:174–182
94. Ogita S (2015) Plant cell, tissue and organ culture: the most flexible foundations for plant metabolic engineering applications. Nat Prod Commun 10:815–820
95. Yuan L, Grotewold E (2015) Metabolic engineering to enhance the value of plants as green factories. Metab Eng 27:83–91
96. Fernie AR, Aharoni A, Willmitzer L, Stitt M, Tohge T, Kopka J, Carroll AJ, Saito K, Fraser PD, DeLuca V (2011) Recommendations for reporting metabolite data. Plant Cell 23:2477–2482
97. Morgan JA, Rhodes D (2002) Mathematical modeling of plant metabolic pathways. Metab Eng 4:80–89
98. Baghalian K, Hajirezaei MR, Schreiber F (2014) Plant metabolic modeling: achieving new insight into metabolism and metabolic engineering. Plant Cell 26:3847–3866
99. de Oliveira Dal'Molin CG, Nielsen LK (2013) Plant genome-scale metabolic reconstruction and modelling. Curr Opin Biotechnol 24:271–277
100. Bishop JF, Dewar J, Toner GC, Tattersall MH, Olver IN, Ackland S, Kennedy I, Goldstein D, Gurney H, Walpole E, Levi J, Stephenson J (1997) Paclitaxel as first-line treatment for metastatic breast cancer. The Taxol Investigational Trials Group, Australia and New Zealand. Oncology (Williston Park) 11:19–23
101. Gautam A, Koshkina N (2003) Paclitaxel (taxol) and taxoid derivates for lung cancer treatment: potential for aerosol delivery. Curr Cancer Drug Targets 3:287–296
102. Radman R, Saez T, Bucke C, Keshavarz T (2003) Elicitation of plants and microbial cell systems. Biotechnol Appl Biochem 37:91–102
103. Han KH, Fleming P, Walker K, Loper M, Chilton WS, Mocek U, Gordon MP, Floss HG (1994) Genetic transformation of mature Taxus: an approach to genetically control the in vitro production of the anticancer drug, taxol. Plant Sci 95:187–196
104. Furmanowa M, Syklowska-Baranek K (2000) Hairy root cultures of Taxus × media var. Hicksii Rehd. as a new source of paclitaxel and 10-deacetylbaccatin III. Biotechnol Lett 22:683–686
105. Syklowska-Baranek K, Pietrosiuk A, Kokoszka A, Furmanowa M (2009) Enhancement of taxane production in hairy root culture of Taxus × media var. Hicksii. J Plant Physiol 166:1950–1954
106. Ho CK, Chang SH, Lung J, Tsai CJ, Chen KP (2005) The strategies to increase taxol production by using *Taxus mairei* cells transformed with TS and DBAT genes. Int J Appl Sci Eng 3:179–185
107. Exposito O, Syklowska-Baranek K, Moyano E, Onrubia M, Bonfill M, Palazon J, Cusido RM (2010) Metabolic responses of Taxus media transformed cell cultures to the addition of methyl jasmonate. Biotechnol Prog 26:1145–1153
108. Li ST, Fu CH, Zhang M, Zhang Y, Xie S, Yu LJ (2012) Enhancing taxol biosynthesis by overexpressing a 9-cis-epoxycarotenoid dioxygenase gene in transgenic cell lines of *Taxus chinensis*. Plant Mol Biol Report 30:1125–1130
109. Sykłowska-Baranek K, Grech-Baran M, Naliwajski MR, Bonfill M, Pietrosiuk A (2015) Paclitaxel production and PAL activity in hairy root cultures of Taxus xmedia var. Hicksii carrying a taxadiene synthase transgene elicited with nitric oxide and methyl jasmonate. Acta Physiol Plant 37:218

110. Lim CG, Fowler ZL, Hueller T, Schaffer S, Koffas MA (2011) High-yield resveratrol production in engineered *Escherichia coli*. Appl Environ Microbiol 77:3451–3460
111. Zhao S, Jones JA, Lachance DM, Bhan N, Khalidi O, Venkataraman S, Wang Z, Koffas MA (2015) Improvement of catechin production in *Escherichia coli* through combinatorial metabolic engineering. Metab Eng 28:43–53
112. Gershenzon J, Dudareva N (2007) The function of terpene natural products in the natural world. Nat Chem Biol 3:408–414
113. Sacchettini JC, Poulter CD (1997) Creating isoprenoid diversity. Science 277:1788–1789
114. Chang MCY, Eachus RA, Trieu W, Ro DK, Keasling JD (2007) Engineering *Escherichia coli* for production of functionalized terpenoids using plant P450s. Nat Chem Biol 3:274–277
115. Huang Q, Roessner CA, Croteau R, Scott AI (2001) Engineering *Escherichia coli* for the synthesis of taxadiene, a key intermediate in the biosynthesis of taxol. Bioorg Med Chem 9:2237–2242
116. Ajikumar PK, Tyo K, Carlsen S, Mucha O, Phon TH, Stephanopoulos G (2008) Terpenoids: opportunities for biosynthesis of natural product drugs using engineered microorganisms. Mol Pharm 5:167–190
117. Ajikumar PK, Xiao WH, Tyo KE, Wang Y, Simeon F, Leonard E, Mucha O, Phon TH, Pfeifer B, Stephanopoulos G (2010) Isoprenoid pathway optimization for Taxol precursor overproduction in *Escherichia coli*. Science 330:70–74
118. Meng H, Wang Y, Hua Q, Zhang S, Wang X (2011) In silico analysis and experimental improvement of taxadiene heterologous biosynthesis in *Escherichia coli*. Biotechnol Bioprocess Eng 16:205–215
119. Boghigian BA, Salas D, Ajikumar PK, Stephanopoulos G, Pfeifer BA (2012) Analysis of heterologous taxadiene production in K- and B-derived *Escherichia coli*. Appl Microbiol Biotechnol 93:1651–1661
120. Biggs BW, Lim CG, Sagliani K, Shankar S, Stephanopoulos G, De Mey M, Ajikumar PK (2016) Overcoming heterologous protein interdependency to optimize P450-mediated Taxol precursor synthesis in *Escherichia coli*. Proc Natl Acad Sci U S A 113:3209–32114
121. Dejong JM, Liu Y, Bollon AP, Long RM, Jennewein S, Williams D, Croteau RB (2006) Genetic engineering of taxol biosynthetic genes in *Saccharomyces cerevisiae*. Biotechnol Bioeng 93:212–224
122. Engels B, Dahm P, Jennewein S (2008) Metabolic engineering of taxadiene biosynthesis in yeast as a first step towards Taxol (Paclitaxel) production. Metab Eng 10:201–206
123. Ding MZ, Yan HF, Li LF, Zhai F, Shang LQ, Yin Z, Yuan YJ (2014) Biosynthesis of Taxadiene in *Saccharomyces cerevisiae*: selection of geranylgeranyl diphosphate synthase directed by a computer-aided docking strategy. PLoS One 9:e109348
124. Besumbes O, Sauret-Güeto S, Phillips MA, Imperial S, Rodríguez-Concepción M, Boronat A (2004) Metabolic engineering of isoprenoid biosynthesis in Arabidopsis for the production of taxadiene, the first committed precursor of Taxol. Biotechnol Bioeng 88:168–175
125. Kovacs K, Zhang L, Linforth RS, Whittaker B, Hayes CJ, Fray RG (2007) Redirection of carotenoid metabolism for the efficient production of taxadiene [taxa-4(5),11(12)-diene] in transgenic tomato fruit. Transgenic Res 16:121–126
126. Cha M, Shim SH, Kim SH, Kim OT, Lee SW, Kwon SY, Baek KH (2012) Production of taxadiene from cultured ginseng roots transformed with taxadiene synthase gene. BMB Rep 45:589–594
127. Hasan MM, Kim HS, Jeon JH, Kim SH, Moon B, Song JY, Shim SH, Baek KH (2014) Metabolic engineering of *Nicotiana benthamiana* for the increased production of taxadiene. Plant Cell Rep 33:895–904
128. Li M, Jiang F, Yu X, Miao Z (2015) Engineering isoprenoid biosynthesis in *Artemisia annua* L. for the production of taxadiene: a key intermediate of taxol. Biomed Res Int 2015:504932

Metabolic Engineering of Phenylpropanoids in Plants

19

Satyabrata Nanda, Jatindra Nath Mohanty, Rukmini Mishra, and Raj Kumar Joshi

Abstract

Phenylpropanoids are diverse group of active secondary metabolites derived from the carbon backbone of amino acids phenylalanine and tyrosine. The phenylpropanoid pathway serves as the starting point for the biosynthesis of a wide range of organic compounds such as lignins, flavanols, isoflavanoids, anthocyanins, and stilbenes with an array of important functions including plant defense and structural support. Besides, they have major nutritional and pharmaceutical properties that find uses as food supplements, antioxidants, flavoring agents, insecticides, dyes, and pharmacological drugs. Major structural and regulatory genes of the phenylpropanoid pathway and associated branches have been isolated and characterized in the recent times. Consequently, the engineering of phenylpropanoid biosynthesis in plants and other ex-host systems have generated considerable scientific and economic importance to enhance their production. In this chapter, we summarize the recent advances in our knowledge of the phenylpropanoid biosynthesis. In addition, we discuss the recent strategies with respect to genetic and metabolic engineering of different phenylpropanoids in plants and microorganisms for their successful industrial production in the future.

Keywords

Phenylpropanoids • Biosynthetic pathway • Metabolic engineering • *E. coli* • *S. cereviceae* • Heterologous expression • Transcription factors

S. Nanda • J.N. Mohanty • R. Mishra • R.K. Joshi (✉)
Centre of Biotechnology, Siksha O Anusandhan University, Bhubaneswar, Odisha, India
e-mail: sbn.satyananda@gmail.com; jatindranathmohanty@gmail.com; rukmini.mishra@gmail.com; rajkumar.joshi@yahoo.co.in; rajjoshi@soauniversity.ac.in

S. Jha (ed.), *Transgenesis and Secondary Metabolism*, Reference Series in Phytochemistry, DOI 10.1007/978-3-319-28669-3_30

Abbreviations

4CL	4-Hydroxycinnamate CoA-ligase
ACC	Acetyl CoA carboxylase
ACS	Acetyl CoA synthase
AMP	Adenosine mono phosphate
ANS	Anthocyanidin synthase
C3′H	p-Coumaroyl-shikimate 3’ hydroxylate
C4H	Cinnamate 4-hydroxylase
CAD	Cinnamyl alcohol dehydrogenase
CCR	Cinnamoyl CoA reductase
CHI	Chalcone isolmerase
CHS	Chalcone synthase
CiED	Cipher of evolutionary design
COMT	Caffeoyl CoA 3-*O*-methyltransferase
CPR	Cytochrome P450
DFR	Dihydrofavonol reductase
DMAPP	Dimethyl allyl pyrophosphate
F3H	Flavonone 3-hydroxylase
FLS	Flavonone synthase
FPP	Farnesyl pyrophosphate
GPP	Geranyl pyrophosphate
H_2O_2	Hydrogen peroxide
HCT	Hydroxysinnamoyl-CoA shikimate:quinate hydroxylcinnamoyltrans-ferase
HID	2-Hydroxyisoflavanone dehydratase
IFS	Isoflavonone synthase
IPP	Isopentyl pyrophosphate
LAR	Leuco anthocyanidin reductase
PA	Proanthocyanidins
PAL	Phenylalanine ammonia lyase
PAP	Production of anthocyanin pigment
PLR	Pinoresinol/lariciresinol reductase

Contents

1 Introduction

Plants synthesize a variety of organic compounds which are mainly classified as primary and secondary metabolites. Primary metabolites including nucleotides, amino acids, organic acids, acyl lipids, and phytosterols have vital roles in essential biochemical processes like photosynthesis, respiration, growth, and development [1]. In contrast, secondary metabolites are phytochemicals that are specifically accumulated through particular biosynthetic pathways in a taxonomically restricted group of organisms [2]. Although the role of secondary metabolites in plant growth and physiology is not yet clear, they have a huge structural variation and accordingly exhibit a broad series of biological activities such as protection against pathogens and abiotic stresses [3, 4], acting as signalling molecules [5] and pheromones for insects and flies for pollination [6]. Besides, they are significantly used as major source for anticancer drugs, immunosuppressant, antibacterial and antifungal agents, herbicides, insecticides, flavoring agents, essential oils, dyes, and fibers [7–9].

Secondary metabolites are classified into one of a number of families, each having a particular structural characteristic arising from the way they have been built up. Terpenes and steroids are derived from dimethyl allyl pyrophosphate (DMAPP) and isopentyl pyrophosphate (IPP) precursors with five carbons [10]. These precursors' combines to yield geranyl pyrophosphate (GPP), leading to monoterpenes. Likewise, compounds derived from farnesyl pyrophosphate (FPP) lead to sesquiterpenes and two equivalents of FPP results in triterpenes. Polyketides and their derivatives are synthesized from acetate units derived from the acetyl coenzyme A [11].The acetate origin of these compounds leads to a predominance of even numbered carbon chains resulting in fatty acid structures. Polyketides often intersect with aromatic amino acid biosynthesis to generate phenylpropanoids [12]. They constitute a large variety of complex aromatic extractives with a phenolic backbone. Alkaloids constitute a broad range of secondary metabolites and characterized by the presence of a basic nitrogen-based heterocyclic ring within their molecules [13]. Besides, some carbohydrate and peptide molecules are also classified as secondary metabolites either forming core structures of other metabolites or act as independent modules [14, 15].

The phenylpropanoids are the most diverse group of physiologically active secondary metabolites derived from the carbon backbone of amino acids phenylalanine and tyrosine [16]. They are characterized by six carbon aromatic phenyl group and three carbon propene tail of cinnamic acid, the first product of the phenylpropanoids biosynthetic pathway [17]. They constitute a wide range of organic compounds such as lignins, flavanols, isoflavanoids, anthocyanins, and stilbenes [12] (Fig. 1). Although the phenylpropanoids are not directly required for growth and development, they are essential to plant survival by performing a wide array of important functions such as reinforcement of specialized cell walls [18], protection from ultraviolet radiation [19], protection from photo-oxidative effects [20], and symbiotic nitrogen fixation [21]. Besides, they have generated tremendous

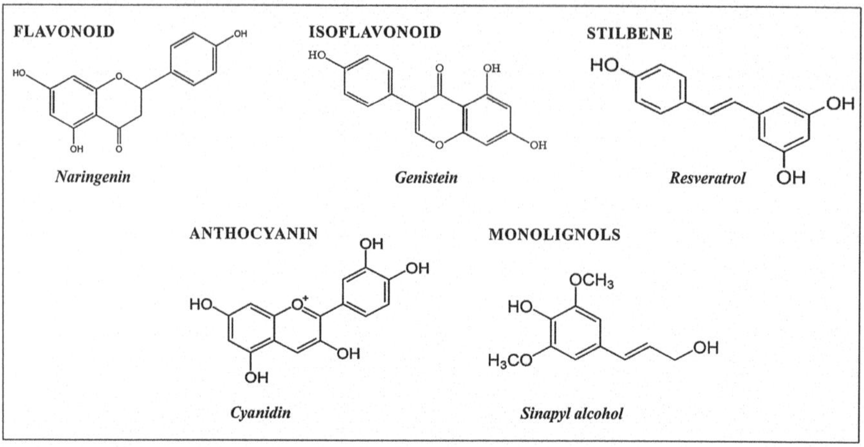

Fig. 1 Representative members of the major classes of organic compounds derived from the phenylpropanoid metabolism

commercial significance in the recent times due to their beneficial functions for human health including antimicrobial, immunomodulotary, and anticancerous attributes [22, 23]. Phenylpropanoids form an attractive system for genetic, enzymological, chemical, and molecular studies owing to their extensive availability, overwhelming diversity, and multiple functions. In recent years, the availability of large amount of information on phenylpropanoids biosynthesis and metabolism has provided the means to engineer their production in plants. Plants engineered with their phenylpropanoid profile could serve as a bridge for improving a large number of characteristics in different crop species such as improved pollination, superior commercial value of fruits and vegetables through better flavor and fragrance, enhanced resistance to diseases and abiotic stresses, and increased production of novel metabolites for medicinal and therapeutic use. Besides, alteration in the phenylpropanoid profile will also contribute to the fundamental aspects of their metabolism and regulation in plant system. Several attempts have been made to produce high levels of phenylpropanoids in transgenic plants of various species. For example, overexpression of *chalcone isolmerase* (CHI), a key enzyme of flavanoid biosynthesis, led to multiple fold increase on flavanoid levels in transgenic tomato [24]. Likewise, ectopic deposition of lignin was reported in *Populus tomentosa* through engineering the overexpression of *PtoMYB216*, a R2R3-MYB transcription factor associated with lignin biosynthesis [25].

In this chapter, we provide a concise outline of the advances in our understanding of the phenylpropanoid biosynthetic pathway in plants. We describe the recent work on the strategies of metabolic engineering of different phenylpropanoids to improve their production for different applications. Finally, we briefly overview the future prospects of engineering phenylpropanoids toward crop improvement and human health.

2 Biosynthesis of Phenylpropanoids

An intricate string of branching biochemical reactions constitutes the biochemical pathway of phenylpropanoids [26]. The plant shikimate pathway is the way in to the biosynthesis of phenylpropanoids through its end product, the aromatic amino acid phenylalanine. The difference in structures of plant phenylpropanoids is the result of various enzyme(s) complex actions involving methylation, acylation, hydroxylation, glycosylation, sulfation, aromatization, cyclization, and condensation reactions [27]. Phenylalanine ammonia lyase (PAL) catalyzes the oxidative deamination of phenylalanine to trans-cinnamic acid and initiates the carbon course from shikimate pathway to phenylpropanoids metabolism. Cinnamate 4-hydroxylase (C4H) mediates the hydroxylation of trans-cinnamate to p-coumaric acid which in turn is subjected to hydroxycinnamate CoA-ligase (4CL)-mediated thio-esterization to form p-coumaroyl-CoA. These three initial steps of the phenylpropanoids pathway catalyzed by PAL, C4H, and 4CL form the basis for all subsequent branches and resulting organic compounds (Fig. 2).

PAL is the first enzyme involved in the phenylpropanoid metabolism and has been extensively studied. PAL isoforms constitute a multigene family in various plant species such as four genes in *Arabidopsis thaliana*, five in *Populus trichocarpa*, and nine in *Oryza sativa* [28, 29]. Each of the individual PAL gene exhibits differential response to biotic and abiotic stresses, and their expression is highly variable across spatial regions and developmental stages [30]. In *Arabidopsis*, *PAL1*, *PAL2*, and *PAL4* are heterologously expressed at a higher level in the mature stems, while *PAL3* demonstrated a very low activity [31]. Considering the presence of specific promoter element in *PAL1* and *PAL2* genes, the enzyme encoded by them are considered to be the prominent one associated with *Arabidopsis* phenylpropanoid metabolism [16, 28, 29]. Usually PAL is a soluble enzyme residing in cytoplasm; however both cytoplasmic and membrane-bound localizations of several PAL isoforms have been reported [28, 32, 33]. Similarly, organ-specific expression analysis of poplar PAL genes revealed that three out of the five are active with the first one involved in lignin formation, the second one is specifically targeted for tannin formation, while the third one is associated with flowering [29, 34]. In contrast, as many as 20 *PAL* genes are associated with the phenylpropanoid biosynthesis in *Lycopersicon esculentum* although only one has been found to be constitutively expressed in all tissues [35]. The molecular mechanisms associated with the inactiveness of such a large number of PAL gene in tomato is still unclear. Additionally, a quadruple mutant *pal1pal1pal3pal4* in *Arabidopsis* demonstrated a marginal PAL activity suggesting that unidentified PAL-like genes are also linked to phenylpropanoid biosynthesis [36].

The subcellular association of PAL with the endoplasmic reticulum bound C4H has been hypothesized, enabling an effective conduit of metabolites through the phenylpropanoid pathway [37, 38]. C4H belongs to the oxidoreductase family of enzymes and one of the most studied plant cytochrome P450 monooxygenase involved in plant metabolism [39]. It is the only member of the *CYP73A* subfamily in *Arabidopsis* although it exists as multigene families in many species [40]. Heterologously expressed

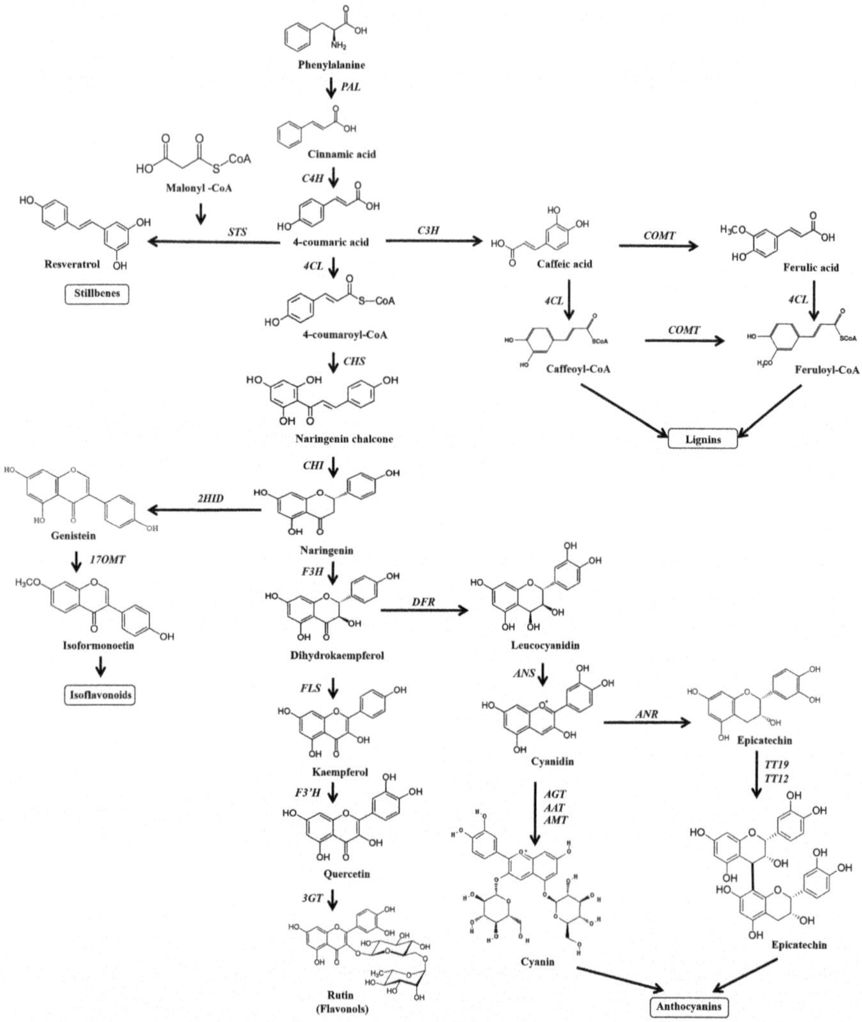

Fig. 2 Phenylpropanoid biosynthetic pathway with extended branching for various organic compounds. Enzymes are as follows: *PAL* phenylalanine ammonia lyase, *C4H* cinnamate-4 hydroxylase, *4CL* 4-hydroxycinnamate CoA-ligase, *CHS*- chalcone synthase, *CHI* chalcone isomerase, *F3H* flavonone 3-hydroxylase, *FLS* flavonone synthase, *3GT* flavonoid 3-*O*-glucosyltransferase, *HID* 2-hydroxy isoflavanone dehydratase, *I7OMT* isoflavone 7-*O*-methyltransferase, *DFR* dihydrofavonolreductase, *ANS* anthocyanidin synthase, *ANR* anthocyanidin reductase, *AGT* anthocyanin glycosyltransferase, *AAT* anthocyanin acetyltransferase, *AMT* anthocyanin methyltransferase, *TT19* transparent testa 19, *TT12* transparent testa 12, *STS* stilbene synthase, *C3H* p-coumaroyl-shikimate 3′ hydroxylate, *COMT* caffeoyl CoA 3-*O*-methyltransferase

C4H as well as the protein directly isolated from the plant exhibit high substrate specificity with t-cinnamic acid. Overexpression of C4H transcripts is often associated with high lignifications and enhanced production of phenolic compounds, while the downregulation of C4H results in altered lignin content [40, 41]. Allelic mutant of *Arabidopsis* C4H gene, *ref3*, has reported a reduced deposition of lignin and tannins, impaired water transport due to collapsed xylem, and reduced apical dominance and male sterility [42]. Taken together, these results suggest that C4H is a significant component of physiological processes that are critical to the survival of the plants besides phenylpropanoid metabolism. The subsequent step downstream of C4H is mediated by AMP forming small gene family of 4CL ligase resulting in the formation of the CoA thioester 4-coumaroyl CoA [43]. The importance of 4CL genes and proteins is evident from the fact that the activated 4-coumaroyl CoA represents the starting point of not only the phenylpropanoid biosynthesis, but the metabolism of a wide variety of secondary metabolites including flavonoids, isoflavonoids, and stilbenes [12, 16].

2.1 Biosynthesis of Monolignols, Lignin, and Lignan

A major downstream reaction in the phenylpropanoid pathway involves the biosynthesis of monolignols. These are the hydroxyl cinnamyl alcohol monomers that act as source materials for the biosynthesis of lignin and lignan. Following the synthesis of p-coumaroyl-CoA by 4CL, the p-coumaroyl group is transferred to shikimate by the enzyme hydroxylsinnamoyl-CoA shikimate to quinate hydroxylcinnamoyl-transferase (HCT) [44] (Fig. 2). The p-coumaroyl-shikimate is then hydroxylated at the 3′ end by p-coumaroyl-shikimate 3′ hydroxylate (C3′H) to form caffeoyl shikimate [45]. HCT undergoes a reverse reaction by transferring the caffeoyl part back into the CoA forming the caffeoyl CoA [44]. The first methyl transfer reaction is the phenylpropanoid metabolism catalyzed by caffeoyl CoA 3-*O*-methyltransferase (CCoAOMT) resulting in the formation of feruloyl CoA from caffeoyl CoA [12, 46]. The hydroxyl cimmamoyl CoA thioesters are subsequently reduced to hydroxyl cinnamaldehydes by the NADPH-dependent cinnamoyl CoA reductase (CCR) [47]. Consequently, cinnamyl alcohol dehydrogenase (CAD) catalyzes the next step of the reaction, reducing cinnamaldehydes to cinnamyl alcohols using NADPH [47]. CAD plays an extensive role in lignin biosynthesis and formation of other plant defense metabolites [48, 49].

Lignin is a prominent phenylpropanoid with vital role in plant structure and defense. It is the second most abundant organic complex after cellulose evenly deposited on the secondary walls of tracheary elements, sclereids cells, and to some extent in the middle lamella and primary cell walls of plants. Lignin is generated by oxidative polymerization of two hydroxy cinnamyl alcohols, namely, coniferyl alcohol and sinapyl alcohol to yield an intermediate radical [50] (Fig. 2). Constituents of lignin differ in angiosperm plants from that of the gymnosperms. In angiosperms, lignin consists of coniferyl and sinapyl alcohols [guaiacyl (G)-syringyl (S)], whereas in gymnosperms it consists mostly of coniferyl alcohol (G type) with small amounts of p-hydroxyphenyl (H)-type lignin [50]. Initial characterization suggested that the formation of lignin by monolignol polymerization was believed to be a random process mediated by H_2O_2-dependent peroxidases, H_2O_2-independent laccases, and polyphenol oxidases having phenoxy radicals as by-products

[50]. Conversely, later studies have shown that the dimerization of monolignols is a stereoselective to enantio-selective mechanism in which the bimolecular phenoxy radical coupling is guided by an auxiliary dirigent protein [51]. Further, lignin depositions at cell wall might be a result of the endwise polymerization of monolignols and its elongation via a template mechanism [52]. Additionally, other nonspecific oxidases are also involved in generating the free radical species from monolignols [53].

Lignans are phenylpropanoid dimeric metabolites where monomers are linked by the central carbon (C8) atoms [54]. Based upon the incorporation of oxygen into the carbon skeleton and subsequent cyclization pattern, lignans can be divided into eight subgroups such as furofuran, furan, dibenzylbutane, dibenzyl butyrolactone, aryltetralin, arylnaphthalene, dibenzo cyclooctadiene, and dibenzyl butyrolactol [54]. The level of oxidation of aromatic ring and the side chains vary along the subgroups of lignans. Lignan is formed by the enantio-selective dimerization of two coniferyl alcohol units along with dirigent protein to form pinoresinol or furofuran [55]. This reaction is catalyzed by the enzyme pinoresinol synthase (PS). Pinoresinol undergoes two consecutive stereo-specific reactions catalyzed by the DADPH-dependent pinoresinol-lariciresinol reductase (PLR) to synthesize (+)-lariciresinol (furan) which is further reduced to (+)-secoisolariciresinol. Consequent oxidation of the dibenzylbutane by secoisolariciresinol dehydrogenase (SDH) results in the production of matairesinol [56].

2.2 Biosynthesis of Flavonone, Isoflavonoids, and Anthocyanins

Flavonoids are a major group of plant secondary metabolites emerging out of the phenylpropanoid pathway. They are basically represented by the pigments that color most of the flowers, leaves, and fruits. Extensive research has been carried out on them, and this has resulted in the identification of more than 6000 flavonoids in a wide variety of plants [57]. They are broadly classified into seven major groups including flavones, flavonols, chalcones, flavondiols and anthocyanins, proanthocyanins, and aurones [57–59]. Legumes are characterized by the presence of specific flavonoid structures called as isoflavonoids [60, 61], and grapes, peanuts, and few unrelated species synthesize a chalcone derivative called as stilbenes [62], while few other species produce phlobaphenes [59]. A high diversity in their structure also accounts for myriads of biological functions including protection from phytopathogens and other physiological stresses, protection from photo-oxidative damages, nutrient retrieval, coloration of flowers, pollination, male sterility, nodular signaling, and hormonal transport [59, 63, 64]. 4-Coumaroyl CoA is transformed into various kinds of flavonoids by a series of downstream enzymatic activities [59] (Fig. 2). The first enzyme in the process is chalcone synthase (CHS), a type III polyketide synthase which catalyzes the claisen ester condensation concomitant with CO_2 liberation from p-coumaroyl-CoA to produce chalcone scaffolds from which all flavonoids are derived [65]. This is followed by subsequent isomerization and hydroxylation by chalcone isomerase (CHI) and flavonone 3-hydroxylase

(F3H), respectively, to produce dihydrokaempferol [66]. Dihydrokaempferol is further hydroxylated with F3H to synthesize dyhydroquercetin. Subsequently, flavonone synthase (FLS) catalyzes an oxidoreductase reaction by removing the dihydroxyl group from dihydrokaempferol and dihydroquercetin to synthesize kaempferol and quercetin, respectively [67]. Dihydrofavonols are consecutively catalyzed by dihydrofavonol reductase (DFR) and anthocyanidin synthase (ANS) to produce the anthocynidins. Subsequent glycosylation and acylation reactions convert the anthocyanidins into anthocyanin molecules [68, 69]. A multiple enzyme-based branching reaction emerges out from the central isoflavanone intermediate naringenin in the leguminous plants for the biosynthesis of isoflavonoids [60]. This first step in this process involves the migration of an aryl group from the two to three carbon position in the B ring of naringenin resulting in the formation of 2-hydroxyisoflavanone [70, 71]. This reaction is catalyzed by a CYP450 enzyme isoflavonone synthase (IFS). The 2-hydroxyisoflavanone is subsequently dehyadrated by 2-hydroxyisoflavanone dehydratase (HID) to form corresponding isoflavonoids [72].

Although the fundamental pathway for flavonoid biosynthesis is conserved, specific alteration in the basic skeleton of flavonoid metabolism could be seen amid plant species due to multiple steps driven by a wide range of enzymes such as isomerases, reductases, and dioxygenases [73]. Further, most of the flavonoid biosynthetic enzymes are present as enzyme complexes bound to the endoplasmic reticulum, while the synthesized pigments are located in the vacuoles or cell wall. However, recent reports also suggest the localization of some flavonoid biosynthetic enzymes in the nucleus as well as the tonoplast [74, 75].

3 Metabolic Engineering of Phenylpropanoids

Plant secondary metabolites constitute several groups and have multiple vital functions in plant life cycle. Besides, they are characterized by high medicinal and economic significance apart from natural in-plant functions [76]. In the attempt of producing high-yielding plants, often the productions of some specialized secondary metabolites are compromised [77]. Also, many plants producing important secondary metabolites grow slowly and are almost impossible for the in vitro cultures [2]. Therefore, it is equally important to maintain and/or improve the production of important secondary metabolites at the time of producing an elite cultivar. Metabolic engineering is a promising tool for the enhancement of production and quality of plant secondary metabolites in a highly selective and cost-effective approach (Fig. 3). Additionally, the engineering of metabolic pathways can lead to metabolite structural modifications, improvements, or even synthesis of new products [78]. Several plant secondary metabolites and their associated metabolic pathways have been improved by using metabolic engineering tools [79]. In some cases of pathway manipulation, the change of biosynthetic site also helps in improved yield of the particular secondary metabolite by overriding the plant regulatory mechanisms [80, 81]. Pathway engineering for enhanced production of secondary metabolites can

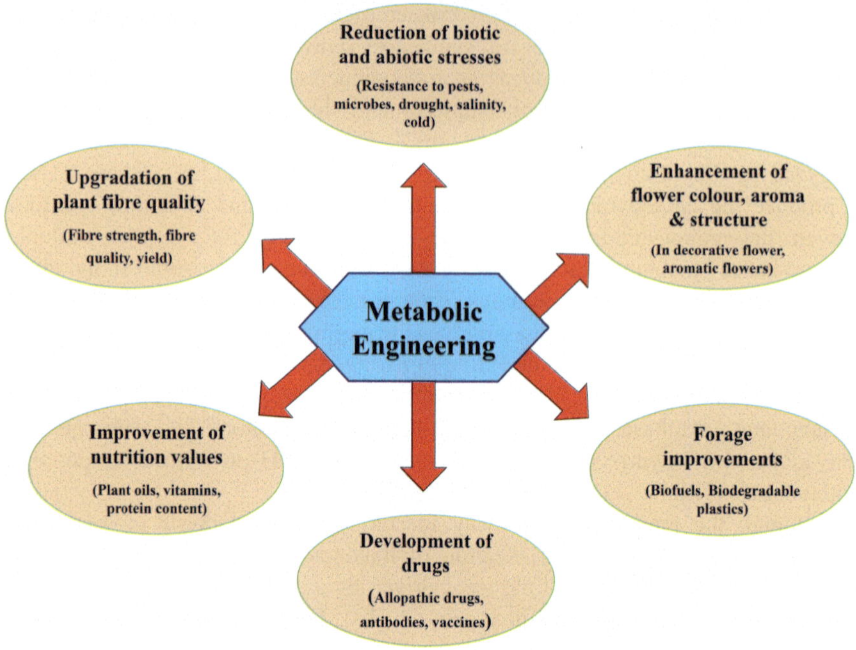

Fig. 3 Key applications of metabolic engineering in plants and microorganisms

be carried out in the native producer by the regulation of transcription factors and overexpression of biosynthetic genes [82]. Alternatively, the transfer of the metabolic pathways into heterologous hosts is a more acceptable strategy toward a successful metabolic engineering effort. Several heterologous hosts including *Streptomyces*, *Escherichia coli*, *Saccharomyces cerevisiae*, and *Nicotiana benthamiana* have been used to assemble large blocks of DNA for reconstruction of complex metabolic pathways for enhanced production of secondary metabolites [2].

Metabolic engineering of secondary metabolites can be achieved by bringing up alteration at many tiers [1]. The structural challenges in metabolic engineering should be addressed by the introduction of single or multiple rate-limiting enzymes encoding genes from the central metabolic pathway [83]. The overexpression of the rate-limiting enzymes in plants or their transformation into suitable microbes may lead to increased accumulation of the specific metabolite. Further, the biosynthesis of plant secondary metabolites is often governed at both transcriptional as well as translational stages. Therefore, the regulatory challenges have to be prevailed over by manipulation of transcriptional and translational apparatus [2]. Additionally, the generation of a particular metabolite is greatly affected by the metabolic pathways in-flux regulations. In such cases, the whole genomic or targeted transcriptomic and metabolomics analysis of a specific plant could be performed to overcome the flux-associated challenges. Lastly, the metabolic engineering should also result in high accumulation of the desired secondary metabolite, without any toxicity to the cells.

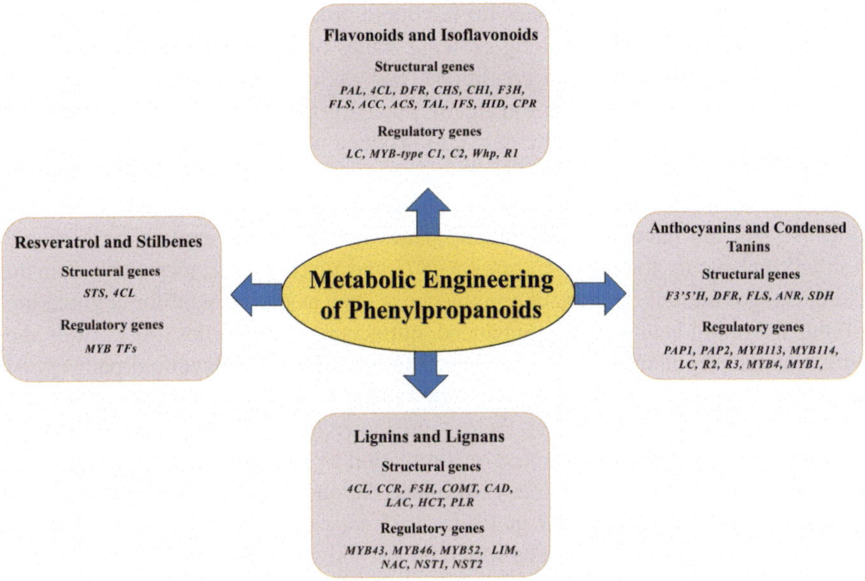

Fig. 4 Engineering of phenylpropanoid pathway through the manipulation of structural genes and regulatory factors. The names of the structural genes are mentioned in the legend of Fig. 2. The regulatory genes are as follows: *LC* leaf color, *MYB* myeloblastosis family of transcription factors, *NAC* NAM, ATAF and CUC family of transcription factors, *PAP* production of anthocyanin pigment

Therefore, specific transporters must be designated for specific secondary metabolites to ensure their movement to proper subcellular and/or extracellular location for sequestration, without resulting in any toxicity. Phenylpropanoid being an important group of secondary metabolites offers significant bioengineering potential for improvements [77]. Several key enzymes involved in both early and late stages of phenylpropanoid metabolism have been already cloned and characterized [84]. These enzymes and their sequence information serve as an asset for the metabolic engineering processes and improvement of the end products. A significant review of engineering strategies for the manipulation of structural and regulatory genes from the phenyl propanoid pathway is presented here (Fig. 4). We have discussed many specific examples on the engineering of different kinds of phenylpropanoids to point out crucial advancement in the field.

3.1 Engineering Major Flavonoid Metabolism

Flavanoids are the most significant group of commercial phenylpropanoids owing to multitude of functions in plant systems [85]. However, the flavonoid in plants is very low and largely depends upon developmental and environmental conditions. This limited availability is always a challenge for commercial production of flavonoids.

Besides, the alternative approaches for flavonoid production are highly challenging and cost expensive due to the long growing stages of the plants, poor culturability together with extreme reaction conditions, and toxicity of the chemical synthetic processes. Therefore, the engineering of phenylpropanoid pathway in plants or microbes could be a suitable alternative to produce novel flavonoid with important commercial and medicinal activities. Since the first report on heterologous expression of maize *dihydro quercetin 4-reductase* (*DFR*) gene resulting in orange *Petunia* flowers [86], the flavonoid biosynthetic pathways have been engineered in many horticultural plants for floral coloration [87] and disease resistance [88]. Tomato being an agriculturally important crop has been used as model plant for engineering of the flavonoid pathway for enhanced nutraceutical properties [89]. Both the structural and regulatory genes have been modified from the flavonoid pathway of tomato to increase the yield of the secondary metabolite. The overexpression of two flavonoid regulatory transcription factors, the maize *leaf color* (LC) and MYB type *C1* (*colorless*), resulted in enhanced accumulation of kaempferol in the flesh of the tomato fruits which normally doesn't produce flavonoids [90]. A previous heterologous expression of *LC* and *C1* genes in *Arabidopsis* also resulted in anthocyanin accumulation even in those tissues which do not show any pigmentation [91]. Interestingly, the tomato fruits did not accumulate any anthocyanin because of the low expression of the gene encoding flavanone-3,5-hydroxylase (F3,5-H) in tomato fruit together with strong preference of tomato dihydroflavonol reductase (DFR) enzyme to use F3,5-H-derived dihydromyricetin as substrate for the synthesis of anthocyanins [92]. Among the structural genes associated with flavonoid biosynthesis in tomato, *CHI* which converts naringenin chalcone to naringenin is the rate-limiting enzyme in the pathway leading to rutin. An ectopic overexpression of *CHI* gene of *Petunia* led to 70% enhancement in the accumulation of quercetin in the peels of tomato [24, 93]. Later, Colliver et al. [94] demonstrated that a concomitant expression of four *Petunia* genes *CHS*, *CHI*, *F3H*, and *FLS* enhanced the synthesis of quercetin and kaempferol in peel and flesh of tomato, respectively. In recent times, RNA interference has been widely used to downregulate the expression of specific flavonoid genes. RNAi inhibition of CHS resulted in drastic reduction of total flavonol together with significant phenotypic variations in tomato [95]. The majority of fruits were parthenocarpic in the RNAi lines suggesting that flavonoids have a significant role in fertilization and fruit and seed development. Engineering of flavonoid pathway has been reported in some plants for the regulation of pollen fertility. In maize, male sterile lines were produced through the mutation of *CHS* genes *C2* and *Whp* [87]. Antisense expressions of *CHS* and *STS* have also resulted in male sterility in *Petunia* and tobacco plants [96, 97].

E. coli has been the preferred ex-host system for metabolic engineering and heterologous production of flavonoids [98–100]. Hwang et al. [101] first reported the production of flavonoids in *E. coli* by engineering *PAL*, *CHS*, and *4CL* from *Streptomyces coelicolor*. However, the low levels of intracellular precursors greatly limited the flavonoid yield. Subsequently, Leonard et al. [102] introduced various strategies to reengineer the central metabolism in *E. coli* to enhance the production of intracellular precursor malonyl coenzyme A. Four acetyl CoA carboxylase (ACC) subunits from *Photorhabdus lumenescens* were introduced into *E. coli* to realize a 5.8-fold increase in flavonoid production [102]. Zha et al. [103] performed additional engineering in the internal metabolic pathway by simultaneous overexpression of ACC and acetyl CoA synthase (ACS) leading to 15-fold increase in cellular malonyl coenzyme A. Fowler et al. [98] developed a highly innovative cipher of evolutionary design (CiED) to identify genetic perturbations that could improve the host genotypes for better production of the metabolites. Employing CiED, it was revealed that the improved *E. coli* genotypes channelled the flux toward malonyl coenzyme A (CoA) more effectively to increase the flavonoid biosynthesis. The targeted deletion of native genes from the TCA cycle of *E. coli* as predicted by CiED and overexpression of genes required for synthesis of the plant-derived flavanones and coenzyme A biosynthetic pathway enhanced the production of flavonoids. The elevated yields in the modified *E. coli* cells were over 660% for naringenin (15–100 mg/l/optical density unit [OD]) and by over 420% for eriodictyol (13–55 mg/l/OD) [98]. Likewise, Santos et al. [100] reported the production of naringenin directly from glucose, without the supplement of any precursor amino acid. A complete pathway consisting of four enzymes, i.e., *CHS* and codon-optimized *TAL*, *4CL*, and *CHI*, was transformed into *E. coli* and its heterologous expression produced 29 mg/l naringenin from glucose. Besides *E. coli*, many flavonoid metabolic engineering have been carried out in *Saccharomyces cerevisiae* cultures as it provide a number of specific advantages of overexpression in other systems. Ro and Douglas [104] introduced *PAL* and *C4H* into *S. cerevisiae* which could effectively deaminate phenylalanine into cinnamic acid consequently producing 4-coumaryl CoA. A subsequent experiment involving the introduction of PAL, 4CL, and CHS into yeast cells produced significant amount of naringenin from phenylalanine [105]. Besides, several other studies have demonstrated that metabolic engineering in yeast could result in enhance production of natural as well as novel flavonoids using chalcone substrates.

3.2 Engineering Isoflavonoid Biosynthesis

Isoflavonoids are important derivative of central phenylpropanoid pathway and have direct role in plant defense and nitrogen fixation in leguminous plants [106]. Metabolic engineering has opened up new opportunities to alter or enhance the accumulation of specific or new isoflavonoids in a variety of species. Genetic modification of key structural genes including *IFS* and *HID* is directly associated with enhance accumulation of isoflavonoids in many plants. Silencing of *IFS* gene through RNAi resulted in reduced production of isoflavonoids and less resistance to microbial infection soybean [107]. In contrast, a constitutive expression of *M. truncatula IFS* (*MtIFS1*) led to significant production of genistein glucoside besides accumulating several other isoflavonones including formononetin, medicarpin, and daidzein in alfalfa leaves [108]. Similarly, RNAi silencing of *CHS6* genes caused a drastic decrease in the accumulation of daidzein, coumestrol, and genistein in transgenic soybean roots [109]. Likewise, the overexpression of *HID* also leads to substantial enhancement in the percentage of isoflavonoids daidzein and genistein in legumes as well as other plants [72, 110]. Additionally, the overexpression of regulatory transcription factor MYBs also plays an important role in the activation of isoflavonoid biosynthesis. Yu et al. [111] reported two to four folds increase in isoflavonoid levels of soybean seeds by ectopic expression of maize *C1* and *R1* TFs in conjunction with co-suppression of F3H. This has been replicated in various plant systems including maize, *Arabidopsis*, and tobacco [112, 113]. This suggests that activation/down-regulation of structural and regulatory genes is a viable strategy for the regulation of isoflavonoid biosynthesis in different plant species.

In addition to the native producers, heterologous hosts such as *E. coli* and *S. cereviceae* have also been used for the commercial production of isoflavonoids. However, the process is quite complex in bacteria owing to the absence of membrane-bound cytochrome P450 (CPR) enzymes and translational incompatibility due to the lack of endoplasmic reticulum. To overcome these problems, Leonard and Koffas [114] developed an artificial one component enzyme similar to bacterial P450BM3 that increases the turnover of bacterial P450 enzymes. These recombinant *E. coli* cells were found to have greater abilities for enhanced production of isoflavonoids genistein and daidzein. The heterologous expression of plant genes and the subsequent production of isoflavonoids in yeast are comparatively easy due to similarity in transcriptional and translational machinery. Co-expression of *CHI* and *IFS* in engineered yeast resulted in effective conversion of chalcone subtrates to flavones and subsequent isoflavones [115]. Recently, the co-expression of a three-enzyme system (IFS, HID, and CPRs) in *S. cereviceae* caused biotransformation of multiple natural and synthetic dihydroxy flavones into their corresponding isoflavones [116]. These results suggest that the identification of active enzymes and their engineering in heterologous hosts can be greatly used for mutasynthesis of isoflavonoids in commercial quantities.

3.3 Engineering the Production of Anthocyanin and Proanthocyanidins

Anthocyanins and proanthocyanidins (PA) (also known as condensed tannins) are major derivatives of phenylpropanoid pathway that occurs as focussed metabolites in various plants. Anthocyanins are glycosides of anthocyanidins, while proanthocyanidins are actually polymers of aglycones [117]. As they serve a variety of functions in plant systems, efforts have been taken to engineer their production for quality and quantity. Engineering of anthocyanin has been sought in many ornamental plants to generate novel floral colors. This includes the engineering of the anthocyanin and aurones biosynthetic pathway for the development of blue color in rose and *Petunia* or the introduction of red, orange, or yellow colors in other plants [118–120]. Engineering a functional *F3′5′H* gene to produce a delphidin-based anthocyanin together with the overexpression of a *Petunia DFR* gene led to blue coloration of the rose flower [118]. Likewise, the suppression of *F3′5′H* and overexpression of *Petunia DFR* genes in tobacco led to enhanced production of pelargonidin-based anthocyanin resulting in orange-red flowers [119]. A previous study had shown that the anthocyanin production could be further induced by downregulation of *FLS* gene [121]. Many new genes related to anthocyanin biosynthesis have been identified through genomic and transcriptomic analysis whose engineering has resulted in delicate modifications of floral color [122, 123]. In the recent times, scientists are more comfortable in engineering the regulatory TFs owing to the complexity of the anthocyanin biosynthetic pathway and the associated structural genes. As many as four MYB TFs, *PAP1*, *PAP2*, *MYB113*, and *MYB114*, have been identified as major regulators of anthocyanin biosynthesis, and their overexpression results in high accumulation of anthocyanin in leaf, stem, flowers, and roots of various plants [124–126]. Recently, the co-expression of *PAP1* and *Lc* genes led to significant elevation of anthocyanin levels along with other phenylpropanoids in *Saussurea involucrate* [127]. Besides, the modification of anthocyanin regulatory factors also contributes to the marketability of the fruits and vegetables. For example, the mutation of a R2-R3 MYB gene in cauliflower conferred anthocyanin accumulation resulting in intense purple color in the curds [128].

PA production is significantly induced during herbivory and microbial attack. Therefore, engineering the PA biosynthesis has been attempted to regulate plant defense response against pests and pathogens. Overexpression of a catechin biosynthetic gene *PtrLAR3* in *Populus trichocarpa* showed PA accumulation and demonstrated enhance resistance to the fungus *Marsonina brunnea* [129]. Higher PA content contributes to the bitterness and astringency in fruits and vegetables. Metabolic engineering is also used to suppress the PA biosynthetic genes to make the fruits and vegetables edible without artificial treatments. Co-suppression of catechin biosynthetic gene *ANR*, *F3′5′H*, and *shikimate dehydrogenase* in a persimmon (*Diospyrus kaki*) mutant resulted in a no astringent phenotype with reduced accumulation of PA in the fruit [130]. A MYB TF *DkMYB4* was found to be synchronously downregulated similar to the structural genes in persimmon transgenic lines. A *DkMYB4* knockdown line of *Diospyrus kaki* also revealed similar pattern of gene expression and phenotype in accordance with the mutant line [130]. PA modification also contributes to the quality of forage plants for better ruminal digestibility. Ectopic expression of a strawberry transcription factor *FaMYB1* inhibits the biosynthesis of PA in the leaves of the forage plant *Lotus corniculatus* [131]. Altogether, this suggests that metabolic engineering of anthocyanin and PA is necessary for qualitative and quantitative improvement of plant varieties.

3.4 Metabolic Engineering of Resveratrol

Resveratrol are the non-flavonoid monomeric units of dihydroxy stilbene which are produced by a limited number of plants species especially from the family Vitaceae and Fabaceae [132]. They have phytoalexin activity and form a significant part of the general plant defense mechanism. The trans-resveratrol has myriads of medicinal benefits such as anticancerous, antitumor, and antiaging besides being used in the treatment of neurological and cardiovascular disorders [62]. This has resulted in tremendous increase in the demand of trans-*R* and associated stilbenes in the recent times. As the native producers have a very low content of trans-*R*, metabolic engineering provides a suitable alternative for increasing the production of these natural phenolics compounds. Early application of trans-*R* engineering primarily focussed on increasing the production of phytoalexin for antimicrobial potential [133]. Since the first effort of transferring two grapevine *STS* genes *Vst1* and *Vst2* toward the development of tobacco resistance to *Botrytis cinerea* infection [134], the process has been reenacted in a variety of crop plants for protection against a wide range of phytopathogens (reviewed in Jeandet et al. [135]). However, the level of trans-*R* and its antimicrobial potential depended on the plant species, promoter used, and the ripening stage of the transgenic lines.

Current strategies for the bioproduction of trans-*R* largely involve the heterologous expression of either the entire phenylpropanoid pathway or selective genes in the baker yeast *S. cereviceae* [83, 99, 136]. Trantas et al. [99] introduced the entire pathway in *S. cereviceae* and produced 0.3 mg/l of trans-*R* from 10 mM phenylalanine as substrate. In a previous study, yeast strain transformed with *4CL* from tobacco and *STS* from grapevine resulted in higher quantity of trans-*R* only with p-coumaric acid as the substrate [137]. Further, Zhang et al. [138] constructed a 4CL::STS fusion to realize a tenfold increase in trans-*R* production. However, this yield is significantly lower as compared to the data obtained with bacterial host *E. coli* [139]. Recently, a new laboratory strain of yeast expressing *4CL1* from *A. thaliana* and *STS* from grapevine resulted in 262 mg/l of trans-*R* [136]. Additionally, the simultaneous expression of an *E. coli area E transporter* gene along with *4CL* and *STS* significantly increased the trans-R production in transformed yeast as compared to control cells [83]. These results suggest that a proper optimization of factors related to type of strains, precursors, and mode of transformation could result in higher level of trans-*R* production from microorganisms.

3.5 Metabolic Engineering of Lignans and Lignins

Lignin is a complex polymer generated by oxidative polymerization of cinnamyl alcohol derivatives called monolignols that are produced through the phenylpropanoid biosynthetic pathway [140]. They constitute a major part of the plant cell wall providing mechanical strength to plant stem and trunk. Lignin mostly occludes with plant cellulose preventing their enzymatic and microbial decomposition. This has tremendous impact on the usage of lignocellulosic biomass for their forage quality, pulping efficiency, and alternative source of fuels [141]. Metabolic engineering enables genetic alteration of the quality and quantity of lignin polymer in the lignocellulosic biomass. Downregulation of monolignol-specific pathway enzymes such as CCR, F5H, COMT, and CAD directly affects the lignin composition and increases the saccharification of the lignocellulosic biomass in many biomass crops (reviewed in Poovaiah et al. [142]). Besides, the modification of downstream laccases which helps in lignifications process also contributes to the decrease in lignin quantity. Barthet et al. [143] reported that silencing of two *A. thaliana* laccase genes *LAC4* and *LAC17* demonstrated low lignin levels and increased saccharification in the transgenic lines as compared to wild-type plants. Additionally, the manipulation of transcription factor and other regulatory elements associated with lignin biosynthesis also have dramatic effect on lignin content and composition in forage plants [142]. The recent discoveries of the biosynthetic enzymes for the alternative lignin monomers also provide a viable option for engineering lignin composition by replacing the traditional monolignols with alternative monomers [142]. However, these are preliminary considerations and a complete metabolic engineering is yet to be attempted for introducing alternative monomers into the lignin biosynthetic pathway.

Lignans are oligomers of dibenzylbutane that are produced by dimerization of cinnamyl alcohol derivatives by the help of dirigent proteins. Lignans and their glucoside derivatives exhibit a wide variety of bioactivities in plants and animals, and this has led to the molecular characterization and engineering of lignan biosynthetic genes to enhance their production to commercial levels. Current metabolic engineering strategies for ligan production mostly involve cell and organ culture of lignan-rich plants such as *Linum* and *Forsythia* species [144]. *Forsythia* cell line transfected with a pinoresinol-lariciresinol reductase (PLR)-RNA interference (RNAi) (PLR-RNAi) construct showed complete loss of matairesinol and 20-fold accumulation of pinoresinol [145]. Additionally, the co-expression of *PLR-RNAi* and a sesem in producing gene *CYP81Q1* exhibited production of sesem in as well as pinoresinol glucoside in the transgenic cell line. Likewise, RNAi metabolic engineering has been carried out in many *Linum* species for the production of important lignans including podophyllotoxin, justicidin, and pinoresinol glucoside [144, 146]. Together, this suggests that *Forsythia* and *Linum* could be used as a suitable medium for engineering the production of endogenous as well as exogenous lignans.

4 Conclusions and Future Perspectives

Engineering of secondary metabolic pathways primarily aims at increasing or decreasing the quantity of certain metabolic compounds. With the advances in our knowledge of the enzymes from the phenylpropanoid pathway and associated biosynthetic branches together with efficient transformation protocols, it is now possible to engineer the composition and content of phenylpropanoids in plants. Besides, the functional reconstruction of biosynthetic pathways in highly characterized heterologous host such as *E. coli* and *S. ceriviceae* provides greater opportunities for scaling the production of phenylpropanoids. However it is essential to understand and mitigate the structural, regulatory, and metabolic flux-associated challenges of the biosynthetic pathways to increase the efficiency of metabolic engineering in ex-host system. The advent of genome scale omics approaches such as whole genome/transcriptome sequencing, proteome analyses, molecular modelling, and metabolomics would be highly useful to iterate the complexities of engineering targets and substrate usage. Additionally, these tools will not only provide information about the new hosts and reconstructed targets but also will explore their conditions for the maintenance of engineered pathways and accumulation of exogenous phenolic compounds. These are interesting times to initiate more in-depth studies to develop and optimize new platforms of metabolic engineering for viable and large-scale production of phenylpropanoids as per the human requirements.

Acknowledgments SN and JNM are thankful for the award of Research Fellowship from the Dept. of Biotechnology, Govt. of India. We are thankful to Prof. Manoj Ranjan Nayak, President, Siksha 'O' Anusandhan University, Bhubaneswar, India, for his guidance and support. We also thank DST-FIST, Govt. of India, for the research infrastructure facilities provided to Centre of Biotechnology, Siksha O Anusandhan University.

References

1. Gandhi SG, Mahajan V, Bedi YS (2015) Changing trends in biotechnology of secondary metabolism in medicinal and aromatic plants. Planta 241:303–317. doi:10.1007/s00425-014-2232-x
2. O'Conner SE (2015) Engineering of secondary metabolism. Annu Rev Genet 49:5.1–5.24. doi:10.1146/annurev-genet-120213-092053
3. Naoumkina MA, He X, Dixon RA (2008) Elicitor-induced transcription factors for metabolic reprogramming of secondary metabolism in *Medicago truncatula*. BMC Plant Biol 8:132. doi:10.1186/1471-2229-8-132
4. Xu C, Sullivan JH, Garrett WM, Caperna TJ, Natarajan S (2008) Impact of solar ultraviolet-B on the proteome in soybean lines differing in flavonoid contents. Phytochemistry 69(1):38–48. doi:10.1016/j.phytochem.2007.06.010
5. Xu M, Dong J, Wang H, Huang L (2009) Complementary action of jasmonic acid on salicylic acid in mediating fungal elicitor induced flavonol glycoside accumulation of *Ginkgo biloba* cells. Plant Cell Environ 32:960–967. doi:10.1111/j.1365-3040.2009.01976.x
6. Gonzalez-Teuber M, Heil M (2009) Nectar chemistry is tailored for both attraction of mutualists and protection from exploiters. Plant Signal Behav 4:809–813. doi:10.4161/psb.4.9.9393
7. Croteau R, Ketchum REB, Long RM, Kaspera R, Wildung MR (2006) Taxol biosynthesis and molecular genetics. Phytochem Rev 5:75–97. doi:10.1007/s11101-005-3748-2
8. Dewick PM (2002) Medicinal natural products, a biosynthetic approach, 2nd edn. Wiley, Chichester, 507 p. ISBN 0-471-496413
9. Vaishnav P, Demain AL (2011) Unexpected applications of secondary metabolites. Biotechnol Adv 29:223–229. doi:10.1016/j.biotechadv.2010.11.006
10. Kitaoka N, Lu X, Yang B, Peters RJ (2015) The application of synthetic biology to elucidation of plant mono-, sesqui-, and diterpenoid metabolism. Mo Plant 8:6–16. doi:10.1016/j.molp.2014.12.002
11. McDaniel R, Welch M, Hutchinson CR (2005) Genetic approaches to polyketide antibiotics. Chem Rev 105:543–558. doi:10.1021/cr0301189
12. Vogt T (2010) Phenylpropanoid biosynthesis. Mol Plant 3:2–20. doi:10.1093/mp/ssp106
13. Ziegler J, Facchini PJ (2008) Alkaloid biosynthesis: metabolism and trafficking. Annu Rev Plant Biol 59:735–769. doi:10.1146/annurev.arplant.59.032607.092730
14. Patton GC, van der Donk WA (2005) New developments in lantibiotic biosynthesis and mode of action. Curr Opin Microbiol 8:543–551. doi:10.1016/j.mib.2005.08.008

15. McCranie EK, Bachmann BO (2014) Bioactive oligosaccharide natural products. Nat Prod Rep 31:1026–1042. doi:10.1039/c3np70128j
16. Fraser CM, Chapple C (2011) The phenylpropanoid pathway in *Arabidopsis*. Arabidopsis Book 9, e0152. doi:10.1199/tab.0152
17. Ververidis F, Trantas E, Douglas C, Vollmer G, Kretzschmar G, Panopoulos N (2007) Biotechnology of flavonoids and other phenylpropanoid-derived natural products. Part I: chemical diversity, impacts on plant biology and human health. Biotechnol J 2:1214–1234. doi:10.1002/biot.200700084
18. Chabannes M, Barakate A, Lapierre C, Marita JM, Ralph J, Pean M, Danoun S, Halpin C, Grima-Pettenati J, Boudet AM (2001) Strong decrease in lignin content without significant alteration of plant development is induced by simultaneous down-regulation of cinnamoyl CoA reductase (CCR) and cinnamyl alcohol dehydrogenase (CAD) in tobacco plants. Plant J 28(3):257–270. doi:10.1046/j.1365-313X.2001.01140.x
19. Xie Y, Xu D, Cui W, Shen W (2012) Mutation of *Arabidopsis HY1* causes UV-C hypersensitivity by impairing carotenoid and flavonoid biosynthesis and the down-regulation of antioxidant defence. J Expt Bot 63:3869–3883. doi:10.1093/jxb/ers078
20. Kataoka I, Beppu K (2004) UV irradiance increases development of red skin color and anthocyanins in 'Hakuho' peach. Hortscience 39(6):1234–1237
21. Parmar N, Dadarwal KR (1999) Stimulation of nitrogen fixation and induction of flavonoid like compounds by *Rhizobacteria*. J Appl Microbiol 86:36–44
22. D'Almeida RE, Alberto MR, Quispe C, Schmeda-Hirschmann G, Isla MI (2012) Antimicrobial phenylpropanoids from the Argentinean highland plant *Parastrephia* lucida (Meyen) Cabrera. J Ethno Pharmacol 142(2):407–414. doi:10.1016/j.jep.2012.05.010
23. Hemaiswarya S, Doble M (2013) Combination of phenylpropanoids with 5-fluorouracil as anti-cancer agents against human cervical cancer (HeLa) cell line. Phytomedicine 20(2):151–158. doi:10.1016/j.phymed.2012.10.009
24. Muir SR, Collins GJ, Robinson S, Hughes S, Bovy A, Ric De Vos CH, van Tunen AJ, Verhoeyen ME (2001) Overexpression of petunia *chalcone isomerase* in tomato results in fruit containing increased levels of flavonols. Nat Biotechnol 19(5):470–474. doi:10.1038/88150
25. Tian Q, Wang X, Li C, Lu W, Yang L, Jiang Y, Luo K (2013) Functional characterization of the poplar R2R3-MYB transcription factor *PtoMYB216* involved in the regulation of lignin biosynthesis during wood formation. PLoS One 8(10), e76369. doi:10.1371/journal.pone.0076369
26. Weisshaar B, Jenkins GI (1998) Phenylpropanoid biosynthesis and its regulation. Curr Opin Plant Biol 1(3):251–257
27. Noel JP, Austin MB, Bomati EK (2005) Structure-function relationships in plant phenylpropanoid biosynthesis. Curr Opin Plant Biol 8:249–253. doi:10.1016/j.pbi.2005.03.013
28. Raes J, Rohde A, Christensen JH, Van de Peer Y, Boerjan W (2003) Genome-wide characterization of the lignification toolbox in *Arabidopsis*. Plant Physiol 133:1051–1071. doi:10.1104/pp.103.026484
29. Hamberger B, Ellis M, Friedmann M, Souza CDA, Barbazuk B, Douglas CJ (2007) Genome-wide analyses of phenylpropanoid related genes in *Populus trichocarpa*, *Arabidopsis thaliana* and Oryza sativa: the *Populus* lignin toolbox and conservation and diversification of angiosperm gene families. Can J Bot 85:1182–1201. doi:10.1139/B07-098
30. Bhuiyan NH, Selvaraj G, Wei Y, King J (2009) Gene expression profiling and silencing reveal that monolignol biosynthesis plays a critical role in penetration defence in wheat against powdery mildew invasion. J Exp Bot 60(2):509–521. doi:10.1093/jxb/ern290
31. Cochrane FC, Davin LB, Lewis NG (2004) The *Arabidopsis phenylalanine ammonia lyase* gene family: kinetic characterization of the four PAL isoforms. Phytochemistry 65(11):1557–1564. doi:10.1016/j.phytochem.2004.05.006

32. Rasmussen S, Dixon RA (1999) Transgene-mediated and elicitor-induced perturbation of metabolic channeling at the entry point into the phenylpropanoid pathway. Plant Cell 8:1537–1552
33. Rohde A, Morreel K, Ralph J, Goeminne G, Hostyn V, De Rycke R, Kushnir S, Van Doorsselaere J et al (2004) Molecular phenotyping of the pal1 and pal2 mutants of *Arabidopsis thaliana* reveals far-reaching consequences on phenylpropanoid, amino acid, and carbohydrate metabolism. Plant Cell 16:2749–2771. doi:10.1105/tpc.104.023705
34. Kao YY, Harding SA, Tsai CJ (2002) Differential expression of two distinct phenylalanine ammonia-lyase genes in condensed tannin-accumulating and lignifying cells of quaking aspen. Plant Physiol 130:796–807. doi:10.1104/pp.006262
35. Chang A, Lim MH, Lee SW, Robb EJ, Nazar RN (2008) Tomato phenylalanine ammonia-lyase gene family, highly redundant but strongly underutilized. J Biol Chem 283:33591–33601. doi:10.1074/jbc.M804428200
36. Huang J, Gu M, Lai Z, Fan B, Shi K, Zhou YH, Yu JQ, Chen Z (2010) Functional analysis of the *Arabidopsis* PAL gene family in plant growth, development, and response to environmental stress. Plant Physiol 153:1526–1538. doi:10.1104/pp.110.157370
37. Achnine L, Blancaflor EB, Rasmussen S, Dixon RA (2004) Co localization of l-phenylalanine ammonia-lyase and cinnamate 4-hydroxylase for metabolic channeling in phenylpropanoid biosynthesis. Plant Cell 16:3098–3109. doi:10.1105/tpc.104.024406
38. Sato T, Takabe K, Fujita M (2004) Immuno-localization of phenylalanine ammonia-lyase and cinnamate-4-hydroxylase in differentiating xylem of poplar. Comptes Rendus Biol 327(9-10):827–836. doi:10.1016/j.crvi.2004.08.005
39. Petersen M, Simmonds MS (2003) Rosmarinic acid. Phytochemistry 62:121–125. doi:10.1016/S0031-9422(02)00513-7
40. Lu F, Ralph J, Akiyama T, Kim H, Schatz PF, Marita JM, Ralph SA, Reddy MSS, Chen F, Dixon RA (2006) Effects of coumarate 3-hydroxylase down-regulation on lignin structure. J Biol Chem 281:8843–8853. doi:10.1074/jbc.M511598200
41. Chen F, Reddy SM, Temple S, Jackson L, Shadle G, Dixon RA (2006) Multi-site genetic modulation of monolignol biosynthesis suggests new routes for formation of syringyl lignin and wall-bound ferulic acid in alfalfa (*Medicago sativa* L.). Plant J 48:113–124. doi:10.1111/j.1365-313X.2006.02857.x
42. Schilmiller AL, Stout J, Weng JK, Humphreys J, Ruegger MO, Chapple C (2009) Mutations in the *cinnamate4-hydroxylase* gene impact metabolism, growth and development in *Arabidopsis*. Plant J 60:771–782. doi:10.1111/j.1365-313X.2009.03996.x
43. Gulick AM, Starai VJ, Horswill AR, Homick KM, Escalante-Semerena JC (2003) The 1.75Å crystal structure of acetyl-CoA synthetase bound to adenosine-5′-propylphosphate and coenzyme A. Biochemistry 42:2866–2873. doi:10.1021/bi0271603
44. Hoffmann L, Maury S, Martz F, Geoffroy P, Legrand M (2003) Purification, cloning, and properties of an acyltransferase controlling shikimate and quinate ester intermediates in phenylpropanoid metabolism. J Biol Chem 278:95–103. doi:10.1074/jbc.M209362200
45. Franke R, Hemm MR, Denault JW, Ruegger MO, Humphreys JM, Chapple C (2002) Changes in secondary metabolism and deposition of an unusual lignin in the *ref8* mutant of *Arabidopsis*. Plant J 30:47–59. doi:10.1046/j.1365-313X.2002.01267.x
46. Do CT, Pollet B, Thévenin J, Sibout R, Denoue D, Barrière Y, Lapierre C, Jouanin L (2007) Both caffeoyl coenzyme A 3-*O*-methyltransferase 1 and caffeic acid *O*-methyltransferase 1 are involved in redundant functions for lignin, flavonoids and sinapoyl malate biosynthesis in *Arabidopsis*. Planta 226:1117–1129. doi:10.1007/s00425-007-0558-3
47. Petersen M, Strack D, Matern U (1999) Biosynthesis of phenylpropanoids and related compounds. Biochemistry of plant secondary metabolism. Annual Plan Rev 2:51–222. doi:10.1002/9781444320503.ch4
48. Anterola AM, Lewis NG (2002) Trends in lignin modification: a comprehensive analysis of the effect of genetic manipulations/mutations on lignification and vascular integrity. Phytochem 61:221–292. doi:10.1016/S0031-9422(02)00211-X

49. Boudet A, Hawkins S, Rochange S (2004) The polymorphism of the genes/enzymes involved in the last two reductive steps of monolignol synthesis: what is the functional significance? C R Biol 327:837–845. doi:10.1016/j.crvi.2004.04.007
50. Ibrahim RK (2001) Phenylpropanoid metabolism. In: Encyclopedia of life sciences. Wiley, Chichester. doi:10.1038/npg.els.0001912
51. Vanholme R, Demedts B, Morreel K, Ralph J, Boerjan W (2010) Lignin biosynthesis and structure. Plant Physiol 153:895–905. doi:10.1104/pp.110.155119
52. Davin LB, Lewis NG (2005) Lignin primary structures and dirigent sites. Curr Opin Biotechnol 16:407–441. doi:10.1016/j.copbio.2005.06.011
53. Davin B, Jourdes M, Patten AM, Kim KW, Vassao DG, Lewis NG (2008) Dissection of lignin macromolecular configuration and assembly: comparison to related biochemical processes in allyl/propenyl phenol and lignan biosynthesis. Nat Prod Rep 25(6):1015–1090. doi:10.1039/b510386j
54. Umezawa T (2003) Diversity in lignan biosynthesis. Phytochem Rev 2(3):371–390. doi:10.1023/B:PHYT.0000045487.02836.32
55. Umezawa T (2005) Biosynthesis of lignans, lignins, and norlignans. Kagaku Seibutsu 43:461–467
56. Suzuki S, Umezawa T (2007) Biosynthesis of lignans and norlignans. J Wood Sci 53(4):273–284. doi:10.1007/s10086-007-0892-x
57. Ferrer JL, Austin MB, Stewart C Jr, Noel JP (2008) Structure and function of enzymes involved in the biosynthesis of phenylpropanoids. Plant Physiol Biochem 46(3):356–370. doi:10.1016/j.plaphy.2007.12.009
58. Winkel-Shirley B (2006) The biosynthesis of flavonoids. In: Grotewold E (ed) The science of flavonoids. Springer, New York, pp 71–95
59. Ferreyra MLF, Ruis SP, Casati P (2012) Flavonoids: biosynthesis, biological functions, and biotechnological applications. Front Plant Sci 3:222. doi:10.2289/flps.2012.00222
60. Du H, Huang Y, Tang Y (2010) Genetic and metabolic engineering of isoflavonoid biosynthesis. Appl Microbiol Biotechnol 86(5):1293–1312. doi:10.1007/s00253-010-2512-8
61. Wang X (2011) Structure, function, and engineering of enzymes in isoflavonoid biosynthesis. Funct Integr Genom 11(1):13–22. doi:10.1007/s10142-010-0197-9
62. Chong J, Poutaraud A, Hugueney P (2009) Metabolism and roles of stilbenes in plants. Plant Sci 117:143–155. doi:10.1016/j.plantsci.2009.05.012
63. Winkel-Shirley B (2002) Biosynthesis of flavonoids and effects of stress. Curr Opin Plant Biol 5:218–223. doi:10.1016/S1369-5266(02)00256-X
64. Bradshaw HD, Schemske DW (2003) Allele substitution at a flower colour locus produces a pollinator shift in monkey flowers. Nature 426(6963):176–178. doi:10.1038/nature02106
65. Austin MB, Noel JP (2003) The chalcone synthase superfamily of type III polyketide synthases. Nat Prod Rep 20(1):79–110
66. Pelletier MK, Shirley BW (1996) Analysis of flavanone 3-hydroxylase in *Arabidopsis* seedlings. Plant Physiol 111:339–345. doi:10.1104/pp.111.1.339
67. Nakajima J, Tanaka I, Seo S, Yamazaki M, Saito S (2004) LC/PDA/ESI-MS profiling and radical scavenging activity of anthocyanins in various berries. J Biomed Biotech 5:241–247. doi:10.1155/S1110724304404045
68. Lipiniec L, Debeaujon I, Routaboul JM, Baudry A, Pourcel L, Nesi N, Caboche M (2006) Genetics and biochemistry of seed flavonoids. Annu Rev Plant Biol 57:405–430. doi:10.1146/annurev.arplant.57.032905.105252
69. Shi MZ, Xie DY (2014) Biosynthesis and metabolic engineering of anthocyanins in *Arabidopsis thaliana*. Recent Patents Biotechnol 8(1).47–60. doi:10.2174/1872208307666131218123538
70. Jung W, Yu O, Lau SMC, O'Keefe DPO, Odell J, Fader G, McGonigle B (2000) Identification and expression of isoflavone synthase, the key enzyme for biosynthesis of isoflavones in legumes. Nat Biotechnol 18(2):208–212. doi:10.1038/72671

71. Latunde-Dada A, Cabello-Hurtado F, Czittrich N, Didierjean L, Schopfer C, Werk-Reichhart-D, Ebel J (2001) Flavonoid 6-hydroxylase from soybean (Glycine max L.), anovel plant P-450 monoxygenase. J Biol Chem 276:1688–1695. doi:10.1074/jbc.M006277200
72. Akashi T, Aoki T, Ayabe SI (2005) Molecular and biochemical characterization of 2-hydroxyisoflavanone dehydratase involvement of carboxyl esterase-like proteins in leguminous isoflavone biosynthesis. Plant Physiol 137(3):882–891. doi:10.1104/pp.104.056747
73. Martens S, Preuss A, Matern U (2010) Multifunctional flavonoid dioxygenases: flavonols and anthocyanin biosynthesis in *Arabidopsis thaliana* L. Phytochemistry 71:1040–1049. doi:10.1016/j.phytochem.2010.04.016
74. Kuhn BM, Geisler M, Bigler L, Ringli C (2011) Flavonols accumulate asymmetrically and affect auxin transport in *Arabidopsis*. Plant Physiol 156:585–595. doi:10.1104/pp.111.175976
75. Toda K, Kuroiwa H, Senthil K, Shimada N, Aoki T, Ayabe S, Shimada S, Sakuta M et al (2012) The soybean F3′H protein is localized to the tonoplast in the seed coat hilum. Planta 236:79–89. doi:10.1007/s00425-012-1590-5
76. Newman DJ, Cragg GM (2012) Natural products as sources of new drugs over the 30 years from 1981 to 2010. J Nat Prod 75:311–335. doi:10.1021/np200906s
77. Peled-Zehavi H, Oliva M, Xie Q, Tzin V, Oren-Shamir M, Aharoni A, Galili G (2015) Metabolic engineering of the phenylpropanoid and its primary precursor pathway to enhance the flavor of fruits and the aroma of flowers. Bioengineering 2:204–212. doi:10.3390/bioengineering2040204
78. Cummings M, Breitling R, Takano E (2014) Steps towards the synthetic biology of polyketide biosynthesis. FEMS Microbiol Lett 351:116–125. doi:10.1111/1574-6968.12365
79. Verpoorte R, Memelink J (2002) Engineering secondary metabolite production in plants. Curr Opin Biotechnol 13:181–187. doi:10.1016/S0958-1669(02)00308-7
80. Wu S, Schalk M, Clark A, Miles RB, Coates R, Chappell J (2006) Redirection of cytosolic or plastidic isoprenoid precursors elevates terpene production in plants. Nat Biotechnol 24:1441–1447. doi:10.1038/nbt1251
81. Wu S, Jiang Z, Kempinski C, Eric Nybo S, Husodo S, Williams R, Chappell J (2012) Engineering triterpene metabolism in tobacco. Planta 236:867–877. doi:10.1007/s00425-012-1680-4
82. Bergmann S, Sch¨umann J, Scherlach K, Lange C, Brakhage A, Hertweck C (2007) Genomics-driven discovery of PKS-NRPS hybrid metabolites from *Aspergillus nidulans*. Nat Chem Biol 3:213–217. doi:10.1038/nchembio869
83. Wang H, Dong Q, Guan A, Meng C, Shi X, Guo Y (2011) Synergistic inhibition effect of 2-phenylethanol and ethanol on bioproduction of natural 2-phenylethanol by *Saccharomyces cerevisiae* and process enhancement. J Biosci Bioeng 112(1):26–31. doi:10.1016/j.jbiosc.2011.03.006
84. Zhao S, Tuan PA, Li X, Kim YB, Kim HR, Park CG et al (2013) Identification of phenylpropanoid biosynthetic genes and phenylpropanoid accumulation by transcriptome analysis of *Lycium chinense*. BMC Genomics 14:802. doi:10.1186/1471-2164-14-802
85. Yu O, Jez JM (2008) Nature's assembly line: biosynthesis of simple phenylpropanoids and polyketides. Plant J 54:750–762. doi:10.1111/j.1365-313X.2008.03436.x
86. Meyer P, Heidmann I, Forkmann G, Saedler H (1987) A new petunia flower colour generated by transformation of a mutant with a maize gene. Nature 330:677–678. doi:10.1038rgg/330677a0
87. Forkmann G, Martens S (2001) Metabolic engineering and applications of flavonoids. Curr Opin Biotechnol 12(2):155–160. doi:10.1016/S0958-1669(00)00192-0
88. Jeandet P, Douillet-Breuil AC, Bessis R, Debord S, Sbaghi M, Adrian M (2002) Phytoalexins from the Vitaceae: biosynthesis, phytoalexin gene expression in transgenic plants, antifungal activity, and metabolism. J Agric Food Chem 50:2731–274. doi:10.1021/jf011429s
89. Bovy A, Schijlen E, Hall RD (2007) Metabolic engineering of flavonoids in tomato (*Solanum lycopersicum*): the potential for metabolomics. Metabolomics 3:399–412. doi:10.1007/s11306-007-0074-2

90. Bovy A, de Vos R, Kemper M, Schijlen E, Almenar Pertejo M, Muir S, Collins G, Robinson S et al (2002) High-flavonol tomatoes resulting from heterologous expression of the maize transcription factor gene LC and C1. Plant Cell 14:2509–2526. doi:10.1105/tpc.004218
91. Lloyd AM, Walbot V, Davis RW (1992) *Arabidopsis* and *Nicotiana* anthocyanin production activated by maize regulators R and C1. Science 258(5089):1773–1775. doi:10.1126/science.1465611
92. Martens S, Forkmann G, Britsch L, Wellmann F, Matern U, Lukacin R (2003) Divergent evolution of flavonoid 2-oxoglutarate-dependent dioxygenases in parsley. FEBS Lett 544(1–3):93–98. doi:10.1016/S0014-5793(03)00479-4
93. Verhoeyen ME, Bovy A, Collins G, Muir S, Robinson S, de Vos CH, Colliver S (2002) Increasing antioxidant levels in tomatoes through modification of the flavonoid biosynthetic pathway. J Exp Bot 53(377):2099–2106. doi:10.1093/jxb/erf044
94. Colliver S, Bovy A, Collins G, Muir S, Robinson S, de Vos CHR, Verhoeyen ME (2002) Improving the nutritional content of tomatoes through reprogramming their flavonoid biosynthetic pathway. Phytochem Rev 1:113–123. doi:10.1023/A:1015848724102
95. Schijlen EG, de Vos CH, Martens S, Jonker HH, Rosin FM, Molthoff JW, Tikunov YM, Angenent GC et al (2007) RNA interference silencing of *chalcone synthase*, the first step in the flavonoid biosynthesis pathway, leads to parthenocarpic tomato fruits. Plant Physiol 144(3):1520–1530. doi:10.1104/pp.107.100305
96. van der Meer IM, Stam ME, van Tunen AJ, Mol JN, Stuitje AR (1992) Antisense inhibition of flavonoid biosynthesis in petunia anthers results in male sterility. Plant Cell 4(3):253–262. doi:10.1105/tpc.4.3.253
97. Fischer R, Budde I, Hain R (1997) Stilbene synthase gene expression causes changes in flower colour and male sterility in tobacco. Plant J 11:489–498. doi:10.1046/j.1365-313X.1997.11030489.x
98. Fowler ZL, Gikandi WW, Koffas MA (2009) Increased malonyl coenzyme a biosynthesis by tuning the *Escherichia coli* metabolic network and its application to flavanone production. Appl Environ Microbiol 75:5831–5839. doi:10.1128/AEM.00270-09
99. Trantas E, Panopoulos N, Ververidis F (2009) Metabolic engineering of the complete pathway leading to heterologous biosynthesis of various flavonoids and stilbenoids in *Saccharomyces cerevisiae*. Metab Eng 11:355–366. doi:10.1016/j.ymben.2009.07.004
100. Santos CN, Koffas M, Stephanopoulos G (2011) Optimization of a heterologous pathway for the production of flavonoids from glucose. Metab Eng 13:392–400. doi:10.1016/j.ymben.2011.02.002
101. Hwang EI, Kaneko M, Ohnishi Y, Horinouchi S (2003) Production of plant-specific flavanones by Escherichia coli containing an artificial gene cluster. Appl Environ Microbiol 69:2699–2706
102. Leonard E, Lim KH, Saw PN, Koffas MA (2007) Engineering central metabolic pathways for high-level flavonoid production in *Escherichia coli*. Appl Environ Microbiol 73:3877–3886. doi:10.1128/AEM.00200-07
103. Zha W, Rubin-Pitel SB, Shao Z, Zhao H (2009) Improving cellular malonyl-CoA level in Escherichia coli via metabolic engineering. Metab Eng 11:192–198
104. Ro DK, Douglas CJ (2004) Reconstitution of the entry point of plant phenylpropanoid metabolism in yeast (*Saccharomyces cerevisiae*): implications for control of metabolic flux into the phenylpropanoid pathway. J Biol Chem 279:2600–2607. doi:10.1074/jbc.M309951200
105. Jiang H, Wood KV, Morgan JA (2005) Metabolic engineering of the phenylpropanoid pathway in *Saccharomyces cerevisiae*. Appl Environ Microbiol 71:2962–2969. doi:10.1016/j.ymben.2005.11.001
106. Veitch NC (2007) Isoflavonoids of the leguminosae. Nat Prod Rep 24:417–464. doi:10.1039/c3np70024k
107. Subramanian S, Graham MY, Yu O, Graham TL (2005) RNA interference of soybean isoflavone synthase genes leads to silencing in tissues distal to the transformation site and to

enhanced susceptibility to *Phytophthora sojae*. Plant Physiol 137:1–9. doi:10.1104/pp.104.057257
108. Deavours BE, Dixon RA (2005) Metabolic engineering of isoflavonoid biosynthesis in alfalfa. Plant Physiol 138:2245–2259. doi:10.1104/pp.105.062539
109. Lozovaya VV, Lygin AV, Zernova OV, Ulanov AV, Li S, Hartman GL, Widholm JM (2007) Modification of phenolic metabolism in soybean hairy roots through down regulation of chalcone synthase or isoflavone synthase. Planta 225:665–679. doi:10.1007/s00425-006-0368-z
110. Shimamura M, Akashi T, Sakurai N, Suzuki H, Saito K, Shibata D, Ayabe S, Aoki T (2007) 2-Hydroxyisoflavanone dehydratase is a critical determinant of isoflavone productivity in hairy root cultures of Lotus japonicus. Plant Cell Physiol 48(11):1652–1657. doi:10.1093/pcp/pcm125
111. Yu O, Shi J, Hession AO, Maxwell AA, McGonigle B, Odell JT (2003) Metabolic engineering to increase isoflavone biosynthesis in soy-bean seeds. Phytochemistry 63:753–763. doi:10.1016/S0031-9422(03)00345-5
112. Sharma SB, Dixon RA (2005) Metabolic engineering of proanthocyanidins by ectopic expression of transcription factors in *Arabidopsis thaliana*. Plant J 44:62–75. doi:10.1111/j.1365-313X.2005.02510.x
113. Xie DY, Sharma SB, Wright E, Wang ZY, Dixon RA (2006) Metabolic engineering of proanthocyanidins through co-expression of anthocyanidin reductase and the *PAP1* MYB transcription factor. Plant J 45:895–907. doi:10.1111/j.1365-313X.2006.02655.x
114. Leonard E, Koffas MA (2007) Engineering of artificial cytochrome P450 enzymes for the synthesis of siflavones by *Escherichia coli*. Appl Environ Microbiol 73:7246–7251. doi:10.1128/AEM.01411-07
115. Ralston L, Subramanian S, Matsuno M, Yu O (2005) Partial reconstruction of flavonoid and isoflavonoid biosynthesis in yeast using soybean type I and type II chalcone isomerases. Plant Physiol 137:1375–1388. doi:10.1104/pp.104.054502
116. Chemler JA, Fowler ZL, McHugh KP, Koffas MA (2010) Improving NADPH availability for natural product biosynthesis in *Escherichia coli* by metabolic engineering. Metab Eng 12:96–104. doi:10.1016/j.ymben.2009.07.003
117. Dixon RA, Liu C, Jun JH (2013) Metabolic engineering of anthocyanins and condensed tannins in plants. Curr Opin Biotechnol 24(2):329–335. doi:10.1016/j.copbio.2012.07.004
118. Katsumoto Y, Fukuchi-Mizutani M, Fukui Y, Brugliera F, Holton TA, Karan M, Nakamura N, Yonekura-Sakakibara K et al (2007) Engineering of the rose flavonoid biosynthetic pathway successfully generated blue-hued flowers accumulating delphinidin. Plant Cell Physiol 48:1589–1600. doi:10.1093/pcp/pcm131
119. Nakatsuka T, Abe Y, Kakizaki Y, Yamamura S, Nishihara M (2007) Production of red-flowered plants by genetic engineering of multiple flavonoid biosynthetic genes. Plant Cell Rep 26:1951–1959. doi:10.1007/s00299-007-0401-010
120. Tanaka Y, Brugliera F, Kalc G, Senior M, Dyson B, Nakamura N, Katsumoto Y, Chandler S (2010) Flower color modification by engineering of the flavonoid biosynthetic pathway: practical perspectives. Biosci Biotechnol Biochem 74:1760–1769. doi:10.1271/bbb.100358
121. Davies KM, Schwinn KE, Deroles SC, Manson DG, Lewis DH, Bloor SJ, Bradley JM (2003) Enhancing anthocyanin production by altering competition for substrate between flavonol synthase and dihydroflavonol 4-reductase. Euphytica 131:259–268. doi:10.1023/A:1024018729349
122. Nakatsuka T, Mishiba K, Kubota A, Abe Y, Yamamura S, Nakamura N, Tanaka Y, Nishihara M (2010) Genetic engineering of novel flower colour by suppression of anthocyanin modification genes in gentian. J Plant Physiol 167:231–237. doi:10.1016/j.jplph.2009.08.00710
123. Yonekura-Sakakibara K, Fukushima A, Nakabayashi R et al (2012) Two glycosyl transferases involved in anthocyanin modification delineated by transcriptome independent component analysis in *Arabidopsis thaliana*. Plant J 69(1):154–167. doi:10.1111/j.1365-313X.2011.04779.x

124. Butelli E, Titta L, Giorgio M, Mock HP, Matros A, Peterek S, Schijlen EGMW, Hall RD, Bovy AG, Luo J, Martin C (2008) Enrichment of tomato fruit with health-promoting anthocyanins by expression of select transcription factors. Nat Biotechnol 26:1301–1308. doi:10.1038/nbt.1506
125. Gatica-Arias A, Farag MA, Stanke M, Matousek J, Wessjohann L, Weber G (2012) Flavonoid production in transgenic hop (*Humulus lupulus* L.) altered by PAP1/MYB75 from *Arabidopsis thaliana* L. Plant Cell Rep 31:111–119. doi:10.1007/s00299-011-1144-5
126. Zvi MM, Shklarman E, Masci T, Kalev H, Debener T, Shafir S, Ovadis M, Vainstein A (2012) PAP1 transcription factor enhances production of phenylpropanoid and terpenoid scent compounds in rose flowers. New Phytol 195:335–345. doi:10.1111/j.1469-8137.2012.04161.x
127. Qiu J, Gao F, Shen G, Li C, Han X, Zhao Q, Zhao D, Hua X, Pang Y (2013) Metabolic engineering of the phenylpropanoid pathway enhances the antioxidant capacity of saussurea involucrate. PLoS One 8(8), e70665. doi:10.1371/journal.pone.0070665
128. Chiu LW, Zhou X, Burke S, Wu X, Prior RL, Li L (2010) The purple cauliflower arises from activation of a MYB transcription factor. Plant Physiol 154(3):1470–1480. doi:10.1104/pp.110.164160
129. Yuan L, Wang L, Han Z, Jiang Y, Zhao L, Liu H, Yang L, Luo K (2012) Molecular cloning and characterization of *PtrLAR3*, a gene encoding leucoanthocyanidin reductase from *Populus trichocarpa*, and its constitutive expression enhances fungal resistance in transgenic plants. J Exp Bot 63(7):2513–2524. doi:10.1093/jxb/err425
130. Akagi T, Ikegami A, Tsujimoto T, Kobayashi S, Sato A, Kono A, Yonemori K (2009) DkMyb4 is a MYB transcription factor involved in proanthocyanidin biosynthesis in persimmon fruit. Plant Physiol 151(4):2028–2045. doi:10.1104/pp.109.146985
131. Paolocci F, Robbins MP, Madeo L, Arcioni S, Martens S, Damiani F (2011) Ectopic expression of a *basic helix-loop-helix* gene transactivates parallel pathways of proanthocyanidin biosynthesis. Structure, expression analysis, and genetic control of *leucoanthocyanidin 4-reductase* and *anthocyanidin reductase* genes in *lotuscorniculatus*. Plant Physiol 143(1):504–516. doi:10.1104/pp.106.090886
132. Morales M, Ros Barcelo A, Pedreno MA (2000) Plant stilbenes: recent advances in their chemistry and biology. Adv Plant Physiol 3:39–70
133. Halls C, Yu O (2008) Potential for metabolic engineering of resveratrol biosynthesis. Trends Biotechnol 26(2):77–81. doi:10.1016/j.tibtech.2007.11.002
134. Hain R, Reif HJ, Krause E, Langebartels R, Kindl H, Vornam B, Wiese W, Schmelzer E et al (1993) Disease resistance results from foreign phytoalexin expression in a novel plant. Nature 361(6408):153–156. doi:10.1038/361153a0
135. Jeandet P, Delaunois B, Aziz A, Donnez D, Vasserot Y, Cordelier S, Courot E (2012) Metabolic engineering of yeast and plants for the production of the biologically active hydroxy stilbene, resveratrol. J Biomed Biotechnol 2012:579089. doi:10.1155/2012/579089
136. Sydor T, Schaffer S, Boles E (2010) Considerable increase in resveratrol production by recombinant industrial yeast strains with use of rich medium. Appl Environ Microbiol 76(10):3361–3363. doi:10.1128/AEM.02796-09
137. Beekwilder J, Wolswinkel R, Jonker H, Hall R, de Vos CH, Bovy A (2006) Production of resveratrol in recombinant microorganisms. Appl Environ Microbiol 72:5670–5672. doi:10.1128/AEM.00609-06
138. Zhang Y, Li SZ, Li J, Pan X, Cahoon RE, Jaworski JG, Wang X, Jez JM, Chen F, Yu O (2006) Using unnatural protein fusions to engineer resveratrol biosynthesis in yeast and Mammalian cells. J Am Chem Soc 128:13030–13031. doi:10.1021/ja0622094
139. Katsuyama Y, Funa N, Miyahisa I, Horinouchi S (2007) Synthesis of unnatural flavonoids and stilbenes by exploiting the plant biosynthetic pathway in *Escherichia coli*. Chem Biol 14:613–621. doi:10.1016/j.chembiol.2007.05.004
140. Dixon RA, Chen F, Guo D, Parvathi K (2001) The biosynthesis of monolignols: a "metabolic grid", or independent pathways to guaiacyl and syringyl units? Phytochemistry 57(7):1069–1084. doi:10.1016/S0031-9422(01)00092-9

141. Vanholme R, Morreal K, Darrah C, Oyarce P, Grabber JH, Ralph J, Boerjan W (2012) Metabolic engineering of novel lignin in biomass crops. New Phytol 196:978–1000. doi:10.1111/j.1469.8137.2012.04337.x
142. Poovaiah CR, Rao NM, Soneji JR, Baxter HL, Stewart CN (2014) Altered lignin biosynthesis using biotechnology to improve lignocellulosic biofuel feedstocks. Plant Biotechnol J 12:1163–1173. doi:10.1111/pbi.12225
143. Berthet S, Caulet DN, Pollet B, Bidzinski P, Cezard L, Bris LP, Borrega N, Herve J, Blondet E, Balzergue S, Lapierre C, Jouanin L (2011) Disruption of LACCASE4 and 17 results in tissue-specific alterations to lignification of *Arabidopsis thaliana* stems. Plant Cell 23(3):1124–1137. doi:10.1105/tpc.110.082792
144. Satake H, Koyama T, Bahabadi SE, Matsumoto E, Ono E, Murata J (2015) Essences in metabolic engineering of lignan biosynthesis. Metabolites 5:270–290. doi:10.3390/metabo5020270
145. Kim HJ, Ono E, Morimoto K, Yamagaki T, Okazawa A, Kobayashi A, Satake H (2009) Metabolic engineering of lignan biosynthesis in Forsythia cell culture. Plant Cell Physiol 50:2200–2209. doi:10.1093/pcp/pcp156
146. Ionkova I, Antonova I, Momekov G, Fuss E (2010) Production of podophyllotoxin in *Linum* linearifolium in vitro cultures. Pharmacogn Mag 6(23):180–185. doi:10.4103/0973-1296.66932

Part IV

Metabolic Phytochemistry and Functional Analysis

Metabolic Phytochemistry-Based Approaches for Studying Secondary Metabolism Using Transformed Root Culture Systems

20

Adinpunya Mitra, Chiranjit Mukherjee, and Debabrata Sircar

Abstract

Though secondary metabolism is integrated in networks with primary metabolism, still the biosynthetic pathways for most of the structurally known plant secondary compounds are unknown. Transformed roots developed following integration and expression of genes from *Agrobacterium rhizogenes* have proved to be a valuable model system in studying secondary metabolic pathways since they generally reflect the *in planta* operation of these pathways both in their route and enzymology. Transformed root culture is an isolated organ-culture system and physiologically distinct from the whole-plant system. This is important on two counts. First, transformed root cultures offer a better and simpler system in which to relate enzyme activities and product formation, uncomplicated by potential transport and source-sink relationships. Secondly, there is scant information in literature concerning phenylpropanoid and terpenoid metabolisms in roots of any kind. In this chapter, we report our findings with transformed root

A. Mitra (✉)
Natural Product Biotechnology Group, Agricultural and Food Engineering Department, Indian Institute of Technology Kharagpur, Kharagpur, West Bengal, India
e-mail: adin@iitkgp.ac.in; adinpunya@gmail.com

C. Mukherjee
Natural Product Biotechnology Group, Agricultural and Food Engineering Department, Indian Institute of Technology Kharagpur, Kharagpur, West Bengal, India

Department of Botany, Chandernagore College, Chandernagore, West Bengal, India
e-mail: m.chiranjit84@gmail.com

D. Sircar
Natural Product Biotechnology Group, Agricultural and Food Engineering Department, Indian Institute of Technology Kharagpur, Kharagpur, West Bengal, India

Department of Biotechnology, Indian Institute of Technology Roorkee, Roorkee, Uttarakhand, India
e-mail: dsircar.iitkgp@gmail.com

S. Jha (ed.), *Transgenesis and Secondary Metabolism*, Reference Series in Phytochemistry, DOI 10.1007/978-3-319-28669-3_17

cultures of *Daucus carota* on a new biochemical pathway of hydroxybenzoic acid biosynthesis with an evidence of a phenylpropanoid chain cleavage enzyme for benzenoid formation and diversion of metabolic flows from phenylpropanoids to isoprenoids in green transformed roots upon photooxidative stress. Metabolite profiling using HPLC/ESI-MS/GC-MS, HPLC-based in vitro enzyme assays, inhibitor feeding experiments, and transcript analysis formed the basis of the above investigations.

Keywords
4-Hydroxybenzoic acid • *Daucus* • Green hairy roots • Isoprenoid • Methyl-erythritol 4-phosphate (MEP) • Phenylpropanoid • Volatile terpenoid

Abbreviations

4CL	4-Coumarate-CoA ligase
AOAA	Aminooxyacetic acid
APX	Ascorbate peroxidase
ATP	Adenosine triphosphate
BADH	Betaine aldehyde dehydrogenase
CAT	Catalase
DDC	Diethyldithiocarbamate
DTT	Dithiothreitol
DXR	1-Deoxy-D-xylulose 5-phosphate reductoisomerase
DXS	1-Deoxy-D-xylulose 5-phosphate synthase
EPSPS	5-Enolpyruvyl-shikimate-3-phosphate synthase
GPX	Guaiacol peroxidase
HBD	4-Hydroxybenzaldehyde dehydrogenase
HBS	4-Hydroxybenzaldehyde synthase
HPLC	High-performance liquid chromatography
MDCA	3 4-Methylenedioxycinnamic acid
MEP	Methyl-erythritol 4-phosphate
MVA	Mevalonic acid
NADPH	Nicotinamide adenine dinucleotide phosphate
PAL	Phenylalanine ammonia-lyase
PIP	Piperonylic acid
PK	Pyruvate kinase
SKDH	Shikimate dehydrogenase
SOD	Superoxide dismutase

Contents

1 Carrot Root Organs and Secondary Metabolism

Modified orange root of carrot (*Daucus carota* L.) is consumed worldwide due to its high levels of α-carotene and β-carotene [1]. Apart from carotenes, volatile terpenoids are also present in *D. carota* roots, which are mainly responsible for the typical aroma and flavor of carrots. Taken together, mono- and sesquiterpenes represent approximately 98% of the volatile compounds [2, 3]. Interestingly, *D. carota* is one of the few dicot species where accumulation of simple phenolic compounds even in the absence of pathogen attack was noted [4]. Several studies have demonstrated that carrot cell walls contain significant quantities of 4-hydroxybenzoic acid, which is presumed to be esterified to cell wall polymers [5]. 4-Hydroxybenzoic acid was associated predominantly with the branched pectic polysaccharides, in contrast to the 4-hydroxybenzaldehyde [6]. Among hydroxybenzoates, 4-hydroxybenzoic acid has received considerable attention because of its application in food [7], medicine [8, 9], cosmetics [10], and polymer industries [11]. This phenolic metabolite serves as biosynthetic building blocks for ubiquinone and naphthoquinone pigments such as shikonin [12]. Esters of 4--hydroxybenzoic acids are used in making antimicrobial food preservatives called parabens [6, 13]. In plants this phenolic acid plays an important role in defense response and also known to trigger the release of pathogenesis-related protein [14].

In addition, ferulic acid (mostly in dimer forms) and a range of other typical wall-bound phenolic aldehydes and cinnamic acid derivatives were also identified from *D. carota* roots [6, 15]. Thus, two major secondary metabolic pathways, viz., isoprenoid/carotenoid and phenylpropanoid/benzenoid, contribute to metabolite formation in *D. carota* root.

2 Understanding of 4-Hydroxybenzoic Acid Biosynthesis and Regulation Is Incomplete

Plant hydroxybenzoates are common mediators of plant responses of biotic and abiotic stress. Benzoate and hydroxybenzoate are simple small molecules. However, their evolving chemical and functional diversities and complexities are well reflected in their biosynthetic route [16]. Although much progress has been made in elucidating the biosynthetic route, still a part of the story remains elusive [17, 18]. Evidence is building to support the idea that hydroxybenzoic acid (or benzoic acid) could be synthesized either from phenylpropanoid pathway involving one of the three possible C_2-chain-shortening route or directly from shikimate/chorismate. There is a pressing need, however, to determine in what tissues and under which conditions these distinct pathways are utilized. Compared to lignin and flavonoid biosynthesis, relatively little is known about 4-hydroxybenzoate biosynthesis. However, till date no detailed information on the particular enzyme(s) and gene(s) involved in 4-hydroxybenzoic acid biosynthesis is available [5, 16, 19]. It is likely that a C_2-chain cleavage enzyme or enzyme system is involved in 4-hydroxybenzoic acid biosynthesis; however, the nature of this C_2-chain-shortening enzyme is poorly understood. This incomplete knowledge of biosynthetic pathway is a major reason for limited success of any metabolic engineering efforts for enhancement of 4-hydroxybenzoate production.

A metabolic pathway can be reconstructed by predicting intermediate reactions stepwise in the direction reverse to biosynthesis, from a given secondary compound to a starting basic metabolite. The chemical structures of natural compounds often give a clue to the selection of its precursor(s) and the reaction steps that have to take place to finish with the compound of interest [20, 21]. Therefore a complete understanding of the enzymes and genes involved in 4-hydroxybenzoic acid biosynthesis may provide new insights on our understanding about the regulation and function of this important metabolite. This will also help in designing proper strategies for pathway manipulation for enhancing 4-hydroxybenzoic acid formation.

3 Properties of Transformed Roots

Transformed roots (transformed roots) are obtained by genetic transformation of plant cells by the bacterium *Agrobacterium rhizogenes*. *A. rhizogenes* is a soil-living gram-negative bacterium which infects a wide range of dicotyledonous species and causes neoplastic "transformed root" disease. Upon infection, a copy of the T-DNA of the Ri plasmid is stably integrated into the plant genome. This event is mediated by the virulence (*vir*) genes located in the Ri plasmid; *vir* gene expression is induced by the phenolic compounds α-hydroxyacetosyringone and acetosyringone. Three genes, named *rolA*, *rolB*, and *rolC*, encoded in the TL portion of T-DNA of the Ri plasmid, have been shown to be involved in transformed root initiation and maintenance. The exact role of these genes is still unclear and may vary among plant species.

Typically, this disease syndrome is characterized by numerous fast-growing, highly branched adventitious roots at the site of infection, which can be grown continuously in vitro in hormone-free culture medium. Transformed roots (often referred to as "transformed roots" because of the profusion of root hairs commonly associated with them) are morphologically different from normal plant roots in that they are more highly branched and usually have faster overall growth rates than normal roots. Transformed roots are usually robust in culture and may be maintained for extended periods without detectable changes in their growth and biosynthetic capabilities and thus may be established as an experimental system for studying secondary metabolism.

In general, the secondary metabolism of transformed roots appears to reflect the *in planta* operation of the species from which they were developed. On the other hand, the absence of the aerial parts of the plant means that the normal transport of secondary metabolites from roots to the rest of the plant is disrupted and normally exported compounds may accumulate in the roots. Perhaps for these reasons, metabolites not normally prominent in roots of the normal plant may be present at detectable concentrations in transformed root cultures.

4 Transformed Roots of *D. carota* as a System to Investigate the Enzymatic Route to 4-Hydroxybenzoic Acid Formation

Major secondary metabolite constituents of *D. carota* include carotenes, anthocyanins, and phenolic acids. Carrot cell wall is known to accumulate a number of ester-linked phenolic acids such as 4-hydroxybenzoic acid, 4-hydroxybenzaldehyde, vanillic acid, vanillin, 4-coumaric acid, *cis*-ferulic acid, *trans*-ferulic acid, and 8-*O*-dihydroferulic acid; 4-hydroxybenzoic acid is the most abundant phenolics in *D. carota* cell wall [4, 6, 22], which appears to be ester-linked with lignin-like precursors [23] or to the more highly branched pectic polysaccharides [6].

In the past, elicited cell suspension culture of *D. carota* was explored for the production of 4-hydroxybenzoic acid [5]; however not much success was achieved in resolving 4-hydroxybenzoic acid biosynthesis at enzymatic level. The C_2-chain cleavage enzyme was found to be unstable in *D. carota* cell suspension cultures [5]. While examining these reports on 4-hydroxybenzoic acid accumulation in the cell wall of disorganized culture of *D. carota*, it was anticipated that perhaps an organized culture, in particular, transformed root culture, could be a suitable system to explore 4-hydroxybenzoic acid accumulation in carrot cell wall. Transformed roots are usually faster in growth rate as compared to other organ cultures because of the absence of source-sink metabolite transfer concept in transformed roots [24]. Therefore, these transformed root cultures often produce metabolites at higher levels than normal plant roots. Moreover, transformed roots are robust in culture and may be maintained for extended periods without detectable changes in their growth and biosynthetic capabilities [25]. It is further possible that although appropriate pathway structural genes for a desired natural product are present in both disorganized and organized cultures, they are frequently not expressed in disorganized cultures. This explains why some natural products are produced only in organ cultures [26]. Moreover, transformed root culture can be used to elucidate the intermediates and key enzymes involved in the biosynthesis of secondary metabolites [27, 28]. This is because, transformed roots generally reflect the *in planta* operation of the biochemical pathways operated in normal roots [29]. Transformed root cultures can also be treated with different biotic or abiotic elicitors, which is a standard approach to enhance or induce the expression of desired pathway genes and enzymes under investigations [30, 31]. Methyl jasmonate and chitosan are two widely used elicitors in plant system. Jasmonic acid and its related compounds such as methyl jasmonate have long been observed to be transducers of elicitor signals for the production of plant secondary metabolites [32]. Chitosan is a chemical inducer which stimulates the formation of a wide variety of secondary metabolites such as terpenoids, flavonoids, alkaloids, benzoates, phenylpropanoids, and many other categories of secondary metabolites in plants [33]. Chitosan (a polymer of ß-1,4-glucosamine residues) is an effective elicitor that is extensively used for stimulating plant secondary metabolite formation [19, 34–36].

The root disks of *Daucus carota* inoculated with *Agrobacterium rhizogenes* produced a number of transformed roots at the site of infection. The transformed root phenotype was characterized by fast hormone-independent growth, lack of geotropism, and profuse lateral branching. These *D. carota* transformed roots were used as a model system to study the enzymatic route of 4-hydroxybenzoic acid formation.

5 Profiling of Major Phenolic Compounds from Transformed Roots of *D. carota* by HPLC Analysis

Transformed roots of *D. carota* were analyzed by HPLC for the presence of soluble and wall-bound phenolic compounds. Soluble and wall-bound phenolics were separately analyzed and quantified by HPLC. Transformed roots growing in the log phase (15 days after subculture) were selected for the analysis of phenolic compounds. The identification of each phenolic compound was based on a combination of retention time, cochromatography, and spectral matching with authentic standard.

HPLC analysis of wall-bound phenolics from carrot transformed roots showed the presence of six phenolic compounds. They were 4-hydroxybenzoic acid, vanillic acid, 4-hydroxybenzaldehyde, vanillin, 4-coumaric acid, and ferulic acid. Phenolics were released by saponification, acidified and partitioned into organic solvent, and then analyzed by reverse phase HPLC. The identification of each compound was based on a combination of retention time, cochromatography, and spectral matching with authentic standard. The content of phenolic substances was determined by quantitative HPLC analysis. A typical HPLC chromatogram of wall-bound phenolics from *D. carota* transformed root has been presented in UV spectrum corresponding to each peak that was matched with spectrum of authentic reference compound. The most abundant phenolic compounds detected was 4-hydroxybenzoic acid (1.9 mg/g dry mass) followed by ferulic acid (0.62 mg/g dry mass) and 4-coumaric acid (0.45 mg/g dry mass). 4-Hydroxybenzaldehyde was present in scarce amount. No qualitative variation in phenolic composition was observed among different transformed root lines. However, quantitative variation of phenolic compounds was observed among different transformed root lines, particularly in the amount of 4-hydroxybenzoic acid.

HPLC analysis of soluble phenolics from transformed roots showed the presence of only 4-hydroxybenzoic acid and traces of caffeic acid. However, soluble 4-hydroxybenzoic acid was detected only in young transformed roots. The amount of soluble 4-hydroxybenzoic acid detected after 15 days of subculture was 0.12 mg/g dry mass. A gradual decrease in soluble 4-hydroxybenzoic acid content was observed with delay of subculture period. No free soluble 4-hydroxybenzoic acid was detectable after 28 days of subculture. Detection of a substantial high amount of 4-hydroxybenzoic acid in soluble fraction provided an opportunity to study the nature of occurrence of this phenolic acid. Hydroxybenzoates are known to occur in the cytosol usually in the form of conjugates (mostly as glycoside). Now, question arises if this accumulation of soluble 4-hydroxybenzoic acid in the cytosol occurred in conjugated form or in free form. To clarify this question, the aliquoted supernatant for soluble phenolic acid determination was treated separately with concentrated HCl and 2 M NaOH to remove the glycoside and ester bonds, respectively. HPLC analyses showed that all the three fractions (methanol-soluble fraction, HCl-treated methanol-soluble fraction, and NaOH-treated methanol-soluble fraction) had similar retention time as that of standard 4-hydroxybenzoic acid. Further, UV spectral analysis of all the three fractions matches with the corresponding spectra of standard 4-hydroxybenzoic acid. Similarity in HPLC retention time and UV spectral matching confirmed that 4-hydroxybenzoic acid occurred in cytosol as free acid instead of usual glycoside form. In order to get unambiguous confirmation for free nature of 4-hydroxybenzoic acid, HPLC elute corresponding to 4-hydroxybenzoic acid from methanol-soluble fraction was subjected to ESI-MS analyses. Spectral matching from ESI-MS analyses finally confirmed the non-conjugated (−free) nature of soluble 4-hydroxybenzoic acid (Fig. 1).

6 Elicitor-Induced Enhanced Accumulation of 4-Hydroxybenzoic Acid in Transformed Roots of *D. carota*

A range of chitosan concentration (50–300 mg/L) was also tested for inducing 4-hydroxybenzoic accumulation in transformed roots of *D. carota*. Transformed roots subcultured for 15 days (growing in liquid B5 medium) were used for elicitation experiments. Transformed root line L15 was chosen for these studies. Maximum 4-hydroxybenzoic acid accumulation was observed with 200 mg/L chitosan treatment. Higher chitosan concentration had less stimulatory effect on 4-hydroxybenzoic acid accumulation and also hampered root growth. Based on

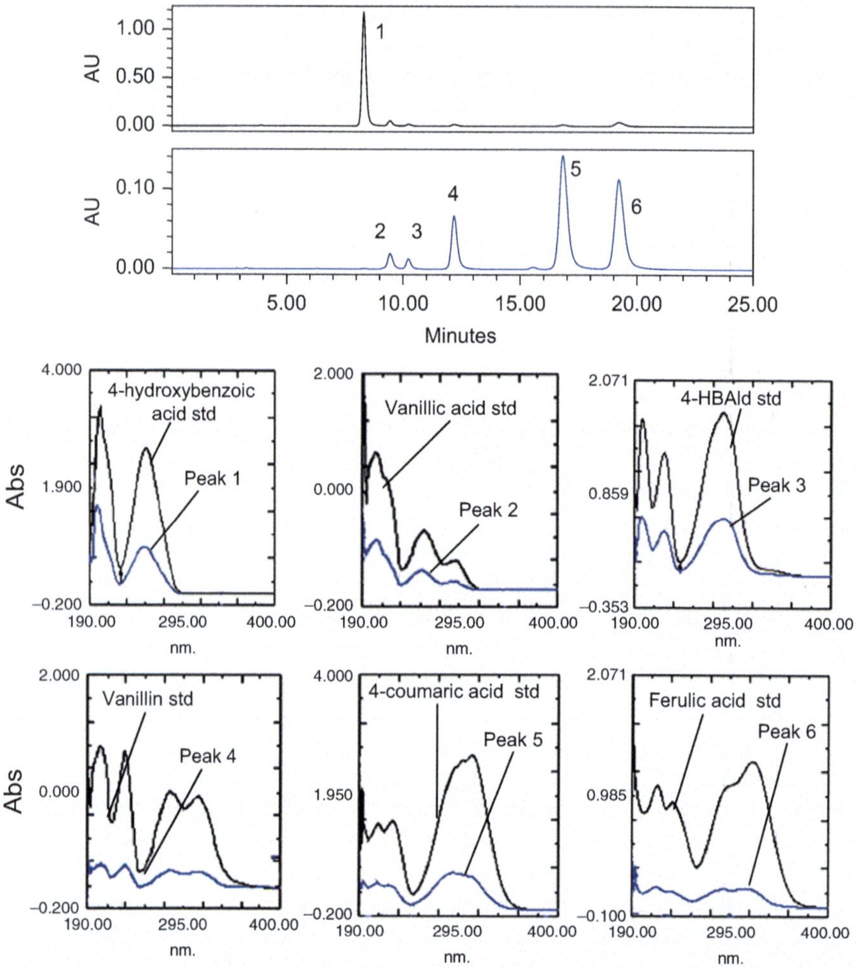

Fig. 1 Identification of wall-bound phenolic compounds on the basis of UV spectral matching. Upper spectrum of the stacked overly (*black color*) represented reference compound and lower spectrum (*blue color*) represented sample corresponding to HPLC eluted peak. Key to peak identity: *1*, 4-hydroxybenzoic acid; *2*, vanillic acid; *3*, 4-hydroxybenzaldehyde; *4*, vanillin; *5*, 4-coumaric acid; *6*, ferulic acid

dose-response study using 50–300 mg/L chitosan, a final concentration of 200 mg/L was chosen for all subsequent experiments that induced maximum 4-hydroxybenzoic acid accumulation. Maximum 4-hydroxybenzoic acid contents accumulated in soluble and wall-bound fractions were 0.5 ± 0.028 mg/g dry mass and 4.46 ± 0.24 mg/g dry mass, respectively, upon 36 h of chitosan treatment. As compared to non-elicited roots, these values of soluble and wall-bound phenolics were higher by 5.8-fold and 2.4-fold, respectively. Lignin content was found to be enhanced by twofold in elicited roots after 48 h of treatment. Non-elicited transformed root line showed a little or no change in 4-hydroxybenzoic acid and lignin content over the same time-course studied. Except 4-coumaric acid, no striking increase in other wall-bound phenolics was observed upon chitosan treatment.

7 Elucidating Enzymatic Pathway of 4-Hydroxybenzoic Acid Formation in Transformed Roots of *D. carota*

Cell-free extracts from chitosan-treated transformed roots were used to assay phenylalanine ammonia-lyase (PAL), cinnamate-4-hydroxylase (C4H), 4-hydroxybenzaldehyde synthase (HBS), and 4-hydroxybenzaldehyde dehydrogenase (HBD) enzymes.

PAL activity was detected by HPLC in chitosan-treated transformed roots. Chitosan-treated transformed roots were harvested at different time points after the onset of elicitation (3–48 h). Cell-free extracts were prepared and used for the determination of PAL activity. A maximum PAL activity of 4.5 nkat/mg protein was observed after 12 h of chitosan treatment, which was nearly 5.6-fold higher than the values obtained for nontreated controls. After 12 h elicitation, a rapid decline in the PAL activity was observed. At 48 h post-elicitation, PAL activity had nearly decreased to the basal level. The PAL activities remained more or less constant in non-elicited (control) roots (0.8 nkat/mg protein) over a similar time-course period.

A considerable C4H activity was detected in the microsomal fraction of elicited carrot transformed roots. The identity of product formation (4-coumaric acid) was ascertained by HPLC retention time, cochromatography, and UV spectral analysis with that of authentic reference. This enzymatic activity required NADPH as electron donor. Time-course analysis of C4H activity in chitosan-treated transformed roots showed maximum C4H activity at 18 h post-elicitation (0.15 nkat/mg proteins). C4H activity was found to be decreased rapidly thereafter and reached to basal level after 30 h. However, unlike PAL activity a very less basal activity was observed up to 12 h. Control roots showed little or no change in C4H activity over a similar time-course period. When PIP (200 μM) was added to the assay mixture (using cell-free extract from 18 h post-elicited roots that showed maximum C4H activity), a 60% reduction in product formation was observed. These results clearly indicated the PIP-mediated inhibition of C4H in vitro.

In order to detect and characterize the enzymes of 4-hydroxybenzoic acid biosynthesis, desalted cell-free extract from chitosan-treated 15-day-old transformed root tissue was used. Transformed roots respond to chitosan treatment by enhancing the accumulation of 4-hydroxybenzoic acid, mainly in the wall-bound form. Thus, chitosan-treated transformed roots were used to prepare cell-free extracts for the incubation with 4-coumaric acid, dithiothreitol (DTT), and NAD^+. Subsequent HPLC analysis demonstrated the formation of 4-hydroxybenzoic acid and scant amount of 4-hydroxybenzaldehyde. The identity of this enzymatic product was confirmed by cochromatography with a sample of authentic reference compound and UV spectroscopy. In addition, GC-MS analysis revealed that the mass spectrum of the enzymatic product agreed with that of the reference compound. Control assay with heat-denatured cell-free extract failed to produce any 4-hydroxybenzoic acid; however a small amount of 4-hydroxybenzaldehyde was detected even in control assay; this indicated a spontaneous nonenzymatic conversion of 4-coumaric acid to 4-hydroxybenzaldehyde in a slower rate. In the absence of NAD^+ in the reaction mixture, only 4-hydroxybenzaldehyde was formed instead of 4-hydroxybenzoic acid. The absence of DTT in the assay failed to produce 4-hydroxybenzoic acid in the assay. However, enzymatic conversion of 4-coumaric acid to 4-hydroxybenzoic acid was uplifted by the presence of thiol reagent, such as DTT.

To test whether in vitro conversion of 4-coumaric acid to 4-hydroxybenzoic acid was CoA dependent or not, cell-free extracts were incubated with CoA, ATP, and $MgCl_2$ in different combinations. When 4-coumaric acid was incubated with DTT (10 mM), ATP (2 mM), and $MgCl_2$ (1 mM), a substantial 4-hydroxybenzaldehyde formation was observed. Replacement of ATP with CoA (0.2 mM) decreased 4-hydroxybenzaldehyde formation. Furthermore, incubation of 4-coumaric acid with DTT (10 mM), ATP (2 mM), CoA (0.2 mM), $MgCl_2$ (1 mM), and NAD^+(2 mM) resulted in formation of 4-hydroxybenzoic acid (and trace amount of 4-hydroxybenzaldehyde), but it was always less than production with NAD^+(2 mM) alone. This suggests that this enzymatic reaction is independent of coumaroyl-CoA formation. Moreover, addition of MDCA (1 mM) in incubation mixture along with all other cofactors (DTT, ATP, CoA, NAD^+, and $MgCl_2$) led to the more production of 4-hydroxybenzoic acid. A similar observation was noticed when chitosan-treated (200 mg/L for 24 h) transformed roots were used for the preparation of cell-free extracts, the results of which were published elsewhere [37].

Formation of 4-hydroxybenzaldehyde as intermediate during in vitro conversion of 4-coumaric acid to 4-hydroxybenzoic acid by cell-free extracts of carrot clearly indicated the involvement of two separate enzyme activities: first, conversion of 4-coumaric acid to 4-hydroxybenzaldehyde followed by subsequent conversion of 4-hydroxybenzaldehyde to 4-hydroxybenzoic acid. These two enzymatic conversions were separately checked using desalted cell-free extracts.

To detect 4-hydroxybenzaldehyde synthase (HBS) activity that catalyzes the in vitro conversion of 4-coumaric acid into 4-hydroxybenzaldehyde, methyl jasmonate-treated transformed roots were used to prepare cell-free extract for the incubation with 4-coumaric acid and DTT. Subsequent HPLC analysis demonstrated the formation of 4-hydroxybenzaldehyde. The identity of this enzymatic product was confirmed by cochromatography with a sample of authentic reference compound, by UV spectroscopy, and finally by GC-MS analysis. GC-MS analysis revealed that the mass spectrum of the enzymatic product agreed with that of the reference compound. A small amount of nonenzymatic 4-hydroxybenzaldehyde formation was observed in control assays containing heat-denatured cell-free extract. Assay without enzyme also showed 4-hydroxybenzaldehyde formation in a low rate due to spontaneous conversion of 4-coumaric acid to 4-hydroxybenzaldehyde under the assay condition used. Spontaneous 4-hydroxybenzaldehyde formation was dependent on the concentration of 4-coumaric acid and DTT. However, no derailment product formation was observed with DTT. HBS activity was calculated by subtracting the amount of nonenzymatically formed 4-hydroxybenzaldehyde from the amount of HBS-catalyzed 4-hydroxybenzaldehyde formation. 4-Hydroxybenzaldehyde dehydrogenase (HBD) from *Daucus carota* transformed roots catalyzed the in vitro conversion of 4-hydroxybenzaldehyde into 4-hydroxybenzoic acid. HBD activity was tested in chitosan-treated transformed root cultures. Cell-free extracts prepared from chitosan-treated roots were incubated with 4-hydroxybenzaldehyde, NAD^{+}, and DTT. Enzymatic reaction was terminated by adding equal volume of methanol to acetic acid (9:1) and centrifuged. Subsequent HPLC analysis of supernatant showed the enzymatic formation of 4-hydroxybenzoic acid. The chemical identity of this enzymatic product was confirmed by cochromatography with a sample of authentic reference compound, by UV spectroscopy, and finally by GC-MS analysis. Mass spectrum of the enzymatic product agreed well with that of reference compound. No enzymatic 4-hydroxybenzoic acid formation was observed in control assays containing heat-denatured proteins (Fig. 2).

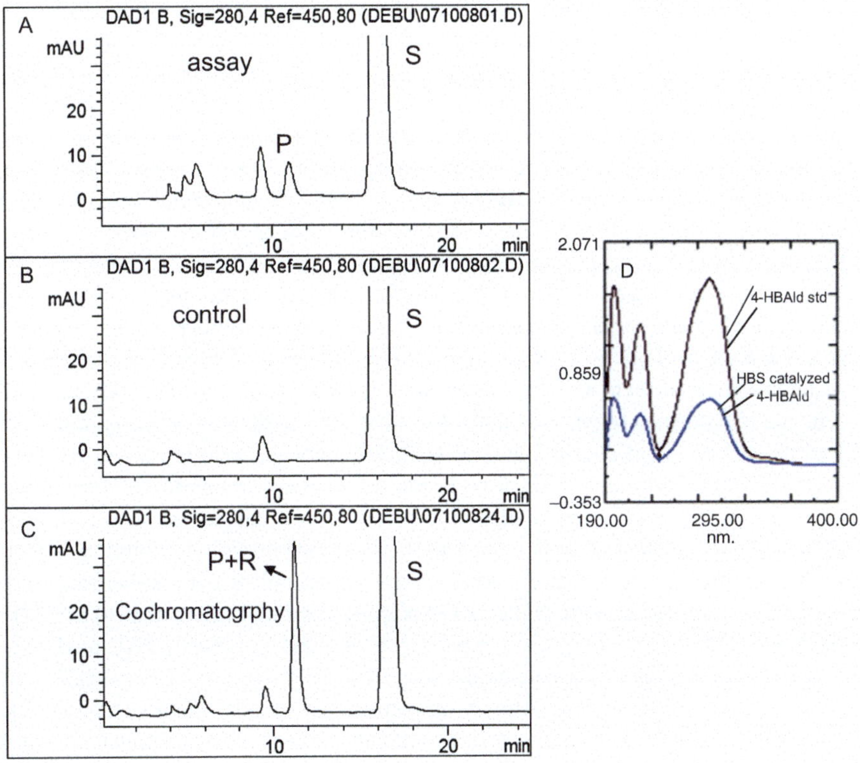

Fig. 2 HPLC analysis of 4-hydroxybenzaldehyde synthase (HBS). (**a**) Standard incubation, (**b**) control assay with denatured protein, (**c**) cochromatography of standard incubation and reference compound, (**d**) UV spectral overlay of standard 4-hydroxybenzaldehyde and product of HBS-catalyzed reaction. Key to peak identity: *S* substrate (4-coumaric acid), *P* enzymatic product (4-hydroxybenzaldehyde), *R* authentic reference compound (4-hydroxybenzaldehyde). Detection wavelength: 280 nm

8 Inhibitor Feeding Studies to Confirm Phenylpropanoid Pathway-Mediated Biosynthesis of 4-Hydroxybenzoic Acid

In order to ascertain whether 4-hydroxybenzoic acid biosynthesis involves phenyl ammonia-lyase (PAL), cinnamate-4-hydroxylase (C4H), and 4-coumaroyl-CoA ligase (4CL) enzyme activities, chitosan-elicited transformed roots (line L15) were fed with varying concentrations of aminooxyacetic acid (AOAA), piperonylic acid (PIP), and 3,4-methylenedioxycinnamic acid (MDCA), respectively, which were known to inhibit PAL, C4H, and 4CL enzyme, respectively. Although both chitosan and methyl jasmonate were found to be equally potent in eliciting 4-hydroxybenzoic acid formation, however, for all inhibitor feeding experiments, chitosan was used as elicitor because of three reasons: (i) chitosan treatment that led to higher accumulation of 4-hydroxybenzoic acid (4.46 mg/g dry mass), (ii) induction of lignin accumulation observed in transformed roots upon chitosan treatment, and (iii) low cost of elicitor to cope up with large number of experiments.

Effect of AOAA supplementation on 4-hydroxybenzoic acid production was investigated to test whether 4-hydroxybenzoic acid formation was dependent on PAL activity. AOAA is a potent inhibitor of L-phenylalanine ammonia-lyase (PAL). AOAA treatment inhibits PAL activity, which in turn prevents the formation of *trans*-cinnamic acid and all other downstream metabolites synthesizing from cinnamic acid. When chitosan-elicited transformed root cultures were supplemented with different concentrations of AOAA (0.2 – 1 mM), a concentration-dependent reduction of 4-hydroxybenzoic acid accumulation was observed. However, even with higher AOAA concentration, complete suppression of 4-hydroxybenzoic acid accumulation was never observed. Besides 4-hydroxybenzoic acid, AOAA also reduced 4-coumaric acid accumulation in a concentration-dependent manner. Elicited culture without AOAA supplementation showed maximum accumulation of 4-hydroxybenzoic acid and 4-coumaric acid at 36 h. When 1 mM AOAA was fed to elicited transformed roots, maximum 44% and 38% reductions in the contents of 4-hydroxybenzoic acid and 4-coumaric acid, respectively, were observed as compared to elicited roots alone. 4-Hydroxybenzoic acid and 4-coumaric acid content in elicited roots (without AOAA supplementation) were considered as maximum (100%). AOAA concentration beyond 1 mM had shown no further inhibitory effect on 4-hydroxybenzoic acid accumulation. Moreover, soluble 4-hydroxybenzoic acid completely disappeared with higher AOAA concentrations (more than 0.8 mM). Results of AOAA supplementation experiments suggest a primarily PAL-mediated biosynthesis of 4-hydroxybenzoic acid.

In order to confirm whether AOAA supplementation really inhibits endogenous PAL activity, experiments were carried out with cell-free extract prepared from AOAA-supplemented transformed roots. A reduced PAL activity was observed in AOAA (0.5–1 mM)-supplemented transformed roots. Chitosan-elicited transformed roots supplemented with 0.5–1 mM AOAA for 24 h were used to prepare cell-free extract for PAL assay. A 29–46% reduction in PAL activity was observed, when cell-free extract was prepared from 0.5 to 1 mM AOAA-supplemented transformed roots. Furthermore, when AOAA (0.05–0.5 mM) was added during in vitro PAL assay (from 24 h chitosan-elicited roots), a reduction in PAL activity was observed. A maximum of 68% inhibition was achieved with 0.5 mM AOAA, compared to assay without AOAA supplementation. This in vivo and in vitro inhibition of PAL activity indicated that AOAA supplementation inhibits endogenous PAL activity which in turn suppresses 4-hydroxybenzoic acid formation.

In elicited transformed roots, accumulation of 4-hydroxybenzoic acid was strongly inhibited by PIP treatment, a specific inhibitor of *trans*-cinnamate-4-hydroxylase (C4H) enzyme which occludes the formation of 4-coumaric acid and all other coumarate-derived downstream metabolites. Increasing concentration of PIP (0.2–1 mM) supplementation led to a subsequent decrease in the accumulation of 4-hydroxybenzoic acid and 4-coumaric acid. With 1 mM PIP, 4-hydroxybenzoic acid and 4-coumaric acid content decreased to a maximum 45–37%, respectively, as compared to elicited roots alone. 4-Hydroxybenzoic acid and 4-coumaric acid content in elicited roots were considered as 100%. Beyond 0.8 mM PIP concentration, soluble 4-hydroxybenzoic acid completely disappeared from the elicited roots. A simultaneous disappearance of 4-hydroxybenzoic acid and appearance of *trans*-cinnamic acid in soluble fraction was observed with higher PIP concentration more than 0.8 mM. Decrease in 4-hydroxybenzoic acid and 4-coumaric acid in PIP-supplemented root revealed that C4H enzyme activity is required for 4-hydroxybenzoic acid biosynthesis. Appearance of *trans*-cinnamic acid in soluble fraction unequivocally proved the PIP-mediated inhibition of C4H activity.

Being a methylenedioxy compound, MDCA was known to be an efficient inhibitor of 4-coumaroyl-CoA ligase (4CL) enzyme [38], which prevents the *in planta* conversion of 4-coumaric acid to 4-coumaroyl-CoA and thus the synthesis of all downstream metabolites derived from 4-coumaroyl-CoA. MDCA supplementation (0.2–1.25 mM) apparently had little or no inhibitory effect on 4-hydroxybenzoic acid accumulation in chitosan-elicited transformed roots of *D. carota*. However, with higher MDCA concentrations, a slight increase in *p*-hydroxybenzoic acid contents was observed, relative to the elicited roots (with DMSO added) alone. Soluble and wall-bound 4-hydroxybenzoic acid accumulation increased to a maximum 185–120%, respectively, in transformed roots supplemented with 1.25 mM of MDCA over transformed roots treated with chitosan alone. These results suggest that 4CL enzyme activity is not required for 4-hydroxybenzoic acid biosynthesis in *D. carota* transformed roots. Higher concentration of MDCA (1.25 mM) proved to be detrimental for the growth and viability of transformed roots.

MDCA supplementation inhibited 4CL activity, but considerably increased the 4-hydroxybenzoic acid accumulation. This apparently suggested that 4CL activity is not a prerequisite for 4-hydroxybenzoic acid biosynthesis in carrot transformed roots. Supplementation of MDCA (1 mM conc.) in elicited transformed roots resulted in a maximum of 35.5% decrease in 4CL activity (absolute value 214.75 U/mg protein) after 24 h of incubation as compared to 100% activity value of root lines treated with chitosan alone (absolute value 333.82 U/mg protein). In contrary to this, no such inhibitory effect of MDCA supplementation on 4-hydroxybenzoic acid accumulation was observed. Moreover, MDCA treatment to elicited roots led to an enhanced accumulation of both soluble and wall-bound 4-hydroxybenzoic acid. These results

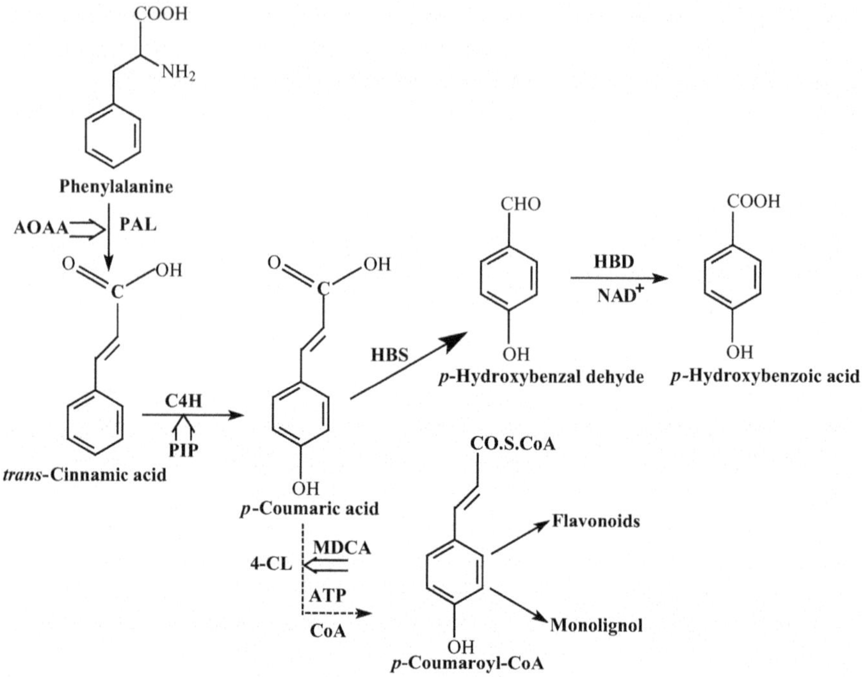

Fig. 3 Enzymatic pathway of 4-hydroxybenzoic acid biosynthesis in transformed roots of *D. carota*

clearly indicated that MDCA mediated blocking of 4CL activity had no inhibitory effect on 4-hydroxybenzoic acid biosynthesis. Since lignin biosynthesis is virtually not active in transformed roots grown under standard condition, chitosan-elicited transformed roots were used to study the in vivo effects of MDCA on 4CL activity and formation of lignin. Together with chitosan, MDCA at different concentrations were supplemented into transformed roots (grown for 15 days) and sampled after 36 h. When MDCA was added to the elicited transformed roots at a range of concentrations, significant decrease in lignin content was observed. With 1 mM MDCA, a maximum 31.6% decrease in lignin content was observed, compared to 100% lignin content of elicited controls (1.82 $\pm$ 0.12). Moreover, a maximum 16% decrease in total flavonoid content was also observed in MDCA-treated roots over the only elicitor-treated controls. This result was obvious because both lignin and flavonoid biosyntheses were known to occur via 4CL-mediated route (Fig. 3).

9 Greening of *D. carota* Transformed Roots Upon Cultivation in Continuous Light

The transformed roots used for these studies turned green, when incubated under continuous illumination (*ca.* 250 μmol.m^{-2}.s^{-1}.). These greening remained stable in transformed roots for more than 3 years, when maintained under continuous illumination with shaking and with a subculturing interval of 28 days. Anatomical examination has confirmed the presence of distinct chloroplasts in green transformed roots. Both green and normal transformed roots were studied for their growth in terms of biomass production for a period of 28 days at 7 days intervals. Initially biomass production was the same for both transformed roots, but after 14 days, green transformed roots showed more biomass production than normal transformed roots.

10 Physiological and Biochemical Changes in Green Transformed Roots Upon Cultivation in Continuous Light

Roots represent the belowground portion of the plant and therefore generally devoid of light exposure. But if roots are exposed to light even in a very low level (>10 μmol.m^{-2}.s^{-1}), it can be able to cause significant changes in root morphology and pigmentation [39]. Thus, greening of hairy roots under continuous light (250 μmol.m^{-2}.s^{-1}) appears to be a physiological process. This observation influenced us to study the physiological and biochemical changes, occurred during greening of transformed roots. Since, hairy roots are known to accumulate different phenolic compounds, the HPLC analysis of methanolic extracts of normal transformed roots revealed the formation of 4-hydroxybenzoic acid glucoside together with 4-hydroxybenzoic acid glucose ester and in much smaller amount of ferulic acid glucoside as soluble phenolics in normal transformed roots. These compounds were identified by diode-array UV spectral analysis in comparison with the spectra of authentic standards reported in the literature [24]. However, the reduced accumulations of these metabolites (about 50%) were observed in green transformed roots as compared to normal ones. Being the major wall-bound phenolic, the accumulation of 4-hydroxybenzoic acid was studied in both transformed roots. A gradual increase in the accumulation of wall-bound 4-hydroxybenzoic acid was found on time-course (7–28 days) analysis in both green and normal roots. In normal transformed roots, the 4-hydroxybenzoic acid contents recorded after 7 and 28 days were 1.2 ± 0.1 mg/g dry mass and 4.3 ± 0.19 mg/g dry mass, respectively, whereas in green transformed roots, these were 0.7 ± 0.12 mg/g dry mass and 3.2 ± 0.19 mg/g dry mass, respectively. At every time point of analysis, a reduced amount of 4-hydroxybenzoic acid content was observed in green transformed roots; in particular, a 48% reduction of 4-hydroxybenzoic acid content was observed in green transformed roots after 14 days of subculture, when compared with normal transformed roots.

Electrophoretic analysis of crude protein extract from both types of roots revealed the presence of a distinct new band in the crude protein profile of green transformed roots, but remained absent in the protein profile of normal transformed roots. This band was excised from gel and treated with trypsin to digest this protein. MALDI-TOF-MS/MS analysis of trypsin digested excised band of green transformed roots showed the large subunit of RuBisCO protein. MALDI-TOF-MS spectra were identified on the basis of molecular weight search score, percentage of sequence coverage with minimum five matched peptides. Continuous light has tendency to generate photooxidative stress on plant tissues. So, transformed hairy roots growing under continuous light are prone to photooxidative stress. Therefore, it was plausible to study the antioxidative enzyme activity to know photooxidative stress response in hairy root system. Superoxide dismutase (SOD) is a ubiquitous enzyme that constitutes the first line of defense against reactive oxygen species by catalyzing the conversion of hydrogen peroxide from reactive superoxide radical [40]. Increased activity of SOD in any tissue indicates response of this tissue against stress. In green transformed roots, the SOD activity was enhanced by 30% as compared to normal ones.

According to the metal cofactor present in the enzyme, SOD can be divided into three groups: iron SOD (Fe SOD), manganese SOD (Mn SOD), and copper-zinc SOD (Cu-Zn SOD). Three isozyme bands were present in both normal and green transformed root protein extracts, but the intensity of Cu-Zn SOD (band i) was more in green transformed roots, than normal transformed roots. However, the intensity of other bands (ii and iii) was more or less similar. These suggesting that specific activity of band i (Cu-Zn SOD) might contribute to an increase in total SOD activities in green transformed roots (Fig. 4).

In microorganism, plant, and animal peroxidase, a heme-containing glycoprotein catalyzes oxidoreduction between H_2O_2 and reductant to cleave the H_2O_2 [41]. Normal transformed roots showed around 60% higher guaiacol peroxidase activity as compared to green transformed roots. Ascorbate peroxidase catalyzes the reduction of H_2O_2 to water by preferential oxidation of ascorbate and therefore plays an important role in scavenging the H_2O_2, produced as a result of normal metabolism or environmental stresses [42, 43]. In green transformed roots, ascorbate peroxidase

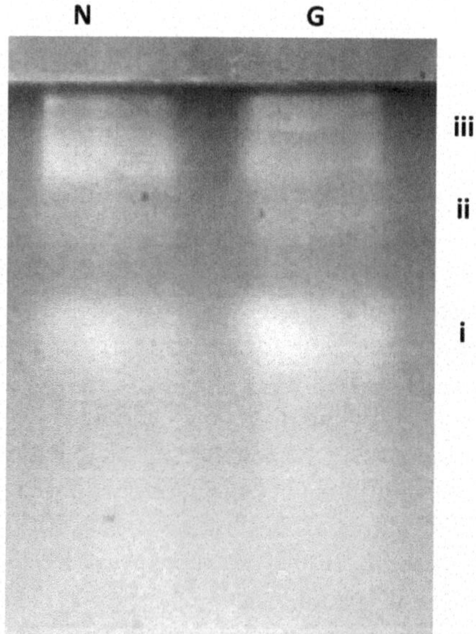

Fig. 4 In-gel SOD activity in total protein sample from normal (*N*) and green (*G*) hairy roots

activity was 10–15% higher than normal transformed roots. Catalase catalyzes the dismutation of two molecule of H_2O_2 to water and O_2 without consuming cellular reducing equivalents. Hence, catalase constitutes very energy-efficient antioxidative defense system in plant cells [44]. As catalase is a photosensitive antioxidative enzyme, green transformed roots grown under continuous illumination showed a 1.6-fold reduced catalase activity as compared to normal ones.

In plant cells, accumulation of hydrogen peroxide (H_2O_2) is usually observed as a product(s) of photosynthesis and photorespiration. Being a stable molecule, it plays a crucial role as signaling molecule to regulate several physiological processes. Hydrogen peroxide has a tendency to interact with thiol-containing proteins, which ultimately leads to the activation of different transcription factors that modulate the expression of different genes related to several important physiological processes. Plants under stress condition produced ample amount of H_2O_2 for expressing different genes related to antioxidative defense mechanism. In our study, green transformed roots kept under continuous light were prone to produce more reactive oxygen species as compared to normal transformed roots. As most of the antioxidative enzymes in green transformed roots showed enhance activities, it was necessary to measure the content of H_2O_2 in both green and normal transformed roots. Green transformed hairy roots accumulate higher amount of H_2O_2 than normal transformed hairy roots. It is plausible that the high level of H_2O_2 in green transformed roots downregulates the phenolic biosynthesis by some unknown mechanism. These findings raised questions as to a possible shift of the carbon flow toward other metabolic directions in green transformed roots. Glycine betaine is produced from toxic betaine aldehyde, by the action of enzyme betaine aldehyde dehydrogenase (BADH). Glycine betaine is known to stabilize photosystem II and RuBisCo during photosynthesis under stress condition [45]. Activity of BADH was found to be eightfold higher in green transformed roots as compared to normal ones.

11 Redirection of Secondary Metabolism from Hydroxybenzoates to Terpenoid Volatiles in Green Transformed Roots

In green transformed roots, a reduced accumulation of phenolic compounds was manifested along with enhanced activities of different antioxidative enzymes. It was anticipated that a photooxidative stress was exerted in transformed roots upon greening. Therefore, in order to combat this abiotic stress, a shift in carbon flows of secondary metabolism from phenolic biosynthesis to the production of volatile isoprenoids was predicted (Fig. 5). This hypothesis needs support from different biochemical experiments. Cell-free root extracts from both green and normal transformed roots were prepared in different protein extraction buffers and used to detect the in vitro activities of shikimate dehydrogenase (SKDH), phenylalanine ammonia-lyase (PAL), 4-hydroxybenzaldehyde dehydrogenase (HBD), pyruvate kinase (PK), and 1-deoxy-D-xylulose 5-phosphate reductoisomerase (DXR) enzymes. Except HBD, all other enzyme activities were determined on a time-course basis.

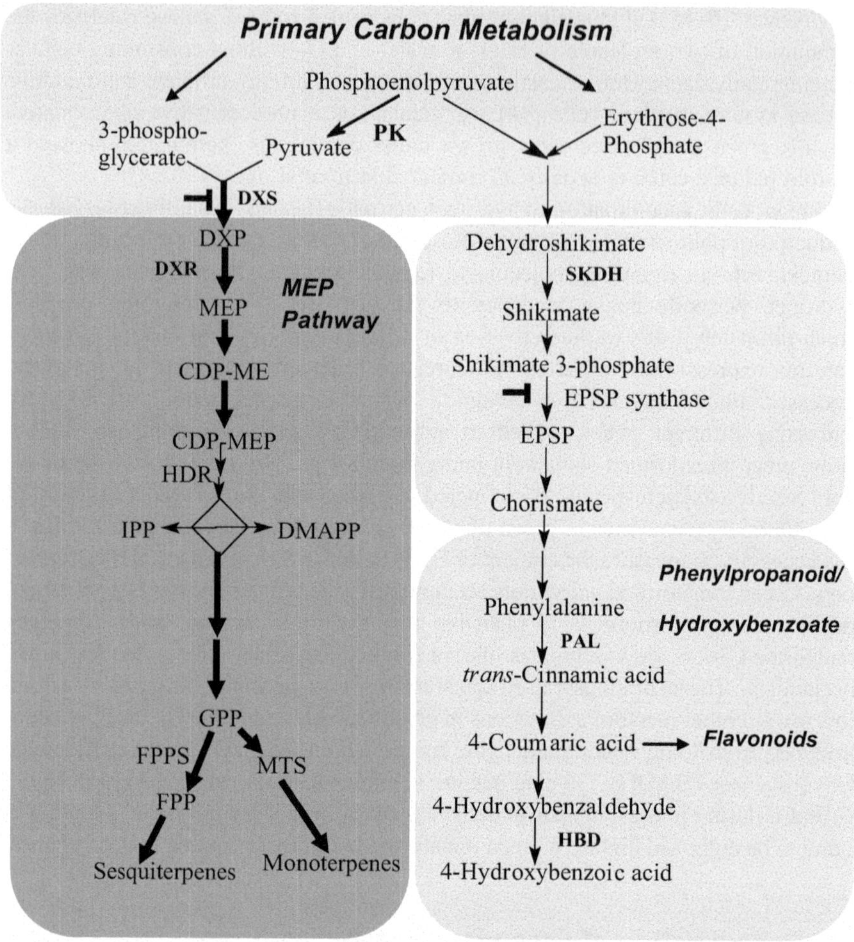

Fig. 5 Redirection of metabolic flow from phenolic pathway to terpenoid biosynthesis

HBD activities were determined from both green and normal transformed roots after 14 days of subculture. For determining time-course activities, 7–28-day-old roots were used. Every 7 days interval, the roots were harvested and used for enzyme assay. For DXR assay, 7-day-old cultures were not used due to generation of insufficient biomass. Therefore, for time-course DXR assay, 14–28-day-old transformed root cultures were used.

Shikimate dehydrogenase, one of the well-studied enzymes of the shikimate pathway is believe to regulate the synthesis of different phenolic compounds in plants [46]. In normal transformed roots, enzyme activity was highest (2 nkat/mg protein) in 14-day-old culture, whereas in green transformed roots after 7 days onward, activity started declining gradually and continued up to 28 days. A SKDH activity of 1.11 nkat/mg protein was detected in green transformed roots after 14 days of subculture – a 1.8-fold less than normal transformed roots.

While a PAL activity of 80 pkat/mg protein was observed in normal transformed roots after 14 days of subculture, in green transformed roots, the activity was found to be 54 pkat/mg protein, a 1.5-fold less than normal transformed roots. HBD is the ultimate enzyme of the 4-hydroxybenzoate pathway to produce 4-hydroxybenzoic acid from 4-hydroxybenzaldehyde. HPLC-based stopped assay method was used to determine the activities of HBD in both green and normal transformed roots of 14-day-old cultures. Normal transformed roots showed an activity of 6.0 pkat/mg protein, whereas, in green transformed roots, the HBD activity was found to be 0.4 pkat/mg protein, which was 15-fold less than normal root.

The MEP pathway, which is primarily responsible for the synthesis of different monoterpene compounds and the phytol chain of chlorophyll molecule, begins with joining of pyruvate and glyceraldehyde-3-phosphate – the products of primary metabolism. Greening of transformed roots indicated the formation of chlorophyll molecule in transformed root tissues. Chlorophyll biosynthesis required activation of methyl-erythritol 4-phosphate (MEP) pathway, which needs the supply of pyruvate for increasing the metabolic flux toward MEP pathway. Pyruvate kinase (PK) is one of the key enzymes in carbon metabolism that converts phosphoenolpyruvic acid into pyruvic acid. Pyruvate kinase activities can be determined through spectrophotometric method coupled with lactic acid dehydrogenase assay. The measurement of pyruvate kinase activity in both transformed roots tissues can be used to demonstrate the increment of MEP pathway in transformed root tissues. Time-course study revealed an enhanced PK activity in green transformed roots than the normal ones. After 14 days of subculture, the PK activity in green transformed roots was found to be 45 mU/mg protein, whereas in normal transformed roots, 29.3 mU/mg protein PK activity was evidenced, showing 1.5-fold lesser activity than the green ones..

DXR is one of the key enzymes of MEP pathway that plays a crucial role to control isoprenoid biosynthesis in plant [47]. The product of the DXR enzyme-catalyzed reaction is MEP. DXR activity was determined spectrophotometrically by measuring the oxidation of NADPH at 340 nm. Enzyme extracts were prepared from the plastids of both normal and green transformed roots. Time-course DXR activity was found to be increased up to 21 days then decreased in both types of transformed roots. At every time point, the DXR activity was higher in green transformed roots than normal ones.

Plant roots emit myriad of volatile organic compounds including terpenoids, fatty acid derivatives, and benzenoids in its surroundings [48]. Root volatiles play a major role in ecological interaction with belowground microorganisms, plants, and herbivores [49]. Volatile compounds of both types of transformed roots of *D. carota* were trapped by static headspace using an adsorbent Porapak Q. Compounds were separated in GC-MS and identified by comparing mass spectrum of the components to that of mass spectral library from NIST 05 (National Institute of Standards and Technology, Gaithersburg, USA) and Wiley 8.0 (Wiley, New York, USA). Retention index values of these compounds were calculated and compared from literatures [50–52].

Carrot root is known to emit several monoterpene compounds for its specific aroma [53]. In transformed roots different monoterpenes were detected and quantified. Emission of monoterpenes is critically regulated by different environmental cues specifically by light and temperature. A total of 31 compounds were identified from green transformed roots of which different monoterpenes and sesquiterpenes were predominant among the emitted volatiles. Monoterpenes constituted 42% of the total volatile mass. In monoterpenes *β*-myrcene, *p*-cymene, *γ*-terpinene, and limonene were major and common in both types of transformed roots. Sesquiterpenes were the most abundant terpenoids present in both green and normal transformed roots and constituted 54% of total volatile mass. Among the sesquiterpenes *β*-caryophyllene, trans-α-bergamotene, (*E*) β-farnesene, *β*-bisabolene, *β*-acoradiene, and (3*E*,6*E*) α-farnesene were detected. Only one benzenoid compound, methyl salicylate, was detected and quantified in both types of transformed roots. Comparative analysis revealed that green transformed roots growing in continuous light showed threefold increments in monoterpene content than normal.

Volatile compounds emitted from plants are dominated by sesquiterpene compounds and played a crucial role as attractant of pollinator, antagonist of herbivore, and substance for communication with other plants. The precursors of sesquiterpene biosynthesis generally come from cytosolic mevalonic acid (MVA) pathway. Emission of sesquiterpenes was largely influenced by several biotic and abiotic factors [54]. In several plant species, light plays a positive regulatory role in sesquiterpenes emission. In other plants, normal carrot roots are found to be an emitter of different arrays of sesquiterpenes. Transformed roots of *D. carota* were also found to emit different arrays of sesquiterpenes as in normal carrot roots. Different sesquiterpenes have been detected in both types of transformed roots. Comparative volatile analysis showed 1.7-fold increment of sesquiterpene content in green transformed roots than normal transformed roots.

Methyl salicylate this insect-induced plant volatiles have been detected in different plant species [55]. Methyl ester of plant hormone salicylic acid has been found to be released from different plant species including lima bean, cabbage, and *Arabidopsis* [56]. On herbivore-induced plant volatiles, methyl salicylate acts as attractant of carnivorous arthropod in plant [57–59]. Thus methyl salicylate mainly acts as defense molecule of plants [60] and formed through shikimic acid pathway [61]. In normal and green transformed roots of *D. carota*, considerable amounts of methyl salicylate have been detected as only benzenoid compound in transformed root volatiles. Comparative quantification of this particular benzenoid compound from both types of transformed roots showed 1.8-fold reductions in methyl salicylate content in green transformed roots than normal transformed roots.

Clomazone, a potent inhibitor of 1-deoxy-D-xylulose 5-phosphate synthase (DXS) was used for inhibitor feeding experiments using both the transformed root systems. In *Arabidopsis* treatment of clomazone prevents seedling establishment and photosynthetic development by inhibiting DXS enzyme [62]. In this study the application of clomazone inhibits chlorophyll biosynthesis completely in green transformed roots. A reduced biomass accumulation was also noticed in treated green transformed roots. Hence, plant monoterpene compounds are synthesized by MEP pathway; comparative volatile analyses of both clomazone-treated transformed roots and nontreated transformed roots showed that in clomazone-treated green transformed roots, no monoterpenes were detected as compared to nontreated green transformed roots. In treated normal transformed roots, monoterpene content was found to be reduced by 2.5-fold as compared to nontreated normal transformed roots. In clomazone-treated transformed roots, monoterpene contents were undetectable as compared to nontreated normal transformed roots.

Sesquiterpenes are synthesized through cytosolic mevalonate pathway in plant. Clomazone is found to be inhibiting MEP pathway not mevalonate pathway, but interestingly, in this study clomazone-treated green transformed roots showed no sesquiterpene formation as compared to nontreated green transformed roots. In treated normal transformed roots, sesquiterpene content was found to be reduced by threefold as compared to nontreated normal transformed roots. As green transformed roots had not produced any sesquiterpenes due to clomazone treatment, sesquiterpene content remained quite high in clomazone-treated normal transformed roots.

Methyl salicylate is only one benzenoid compound detected in volatile analysis of transformed roots. In clomazone-treated green transformed roots, methyl salicylate content was 1.3-fold more than green transformed roots, and in the case of clomazone-treated normal transformed roots, methyl salicylate content was also 1.4-fold more than nontreated normal transformed roots. It was observed that due to blocking of MEP pathway by clomazone, the amounts of methyl salicylate were increased in both normal and green transformed roots. However the content of methyl salicylate was increased by twofold in normal transformed roots as compared to green ones.

Glyphosate, a potent competitive inhibitor EPSP synthase enzyme is generally used to block shikimate pathway. In this study, glyphosate was also used to examine the effect on volatile terpenoid emission upon blocking the shikimate pathway. Feeding of glyphosate at a concentration of 0.05 mM was carried out according to a protocol used for tobacco [62]. An increased emission of volatile monoterpenes was observed in glyphosate-treated transformed roots; a 2.4-fold enhanced amount of volatile monoterpenes was noticed in glyphosate-treated green transformed roots as compared to treated normal transformed roots. However, the volatile sesquiterpene contents remained more or less unaffected in both types of transformed roots. Surprisingly, a huge increment of volatile methyl salicylate emission was observed in both green and normal transformed roots.

12 Conclusions

4-Hydroxybenzoic acid that usually originates from phenylpropanoid pathway is ubiquitously distributed in higher plants. Despite the importance of 4-hydroxybenzoic acid in plant defense response, the enzymatic route to this phenolic acid formation is not fully resolved. As elicitor-treated transformed root cultures of *D. carota* accumulate a high amount of 4-hydroxybenzoic acid, we used elicited transformed root cultures as a system for studying enzymatic route to of 4-hydroxybenzoic acid formation in *D. carota*. Biochemical investigation revealed the involvement of CoA-independent nonoxidative route of 4-hydroxybenzoic acid biosynthesis with phenylalanine as the starting point. Phenylalanine is converted into 4-hydroxybenzoic acid through four enzymatic steps. First, phenylalanine undergoes deamination to give rise to *trans*-cinnamic acid by a PAL-catalyzed reaction. This *trans*-cinnamate is subsequently converted into 4-coumaric acid by the action of cinnamate 4-hydroxylase, which serves as C_6-C_3 precursor for 4-hydroxybenzoic acid formation. 4-Coumaric acid was then converted nonoxidatively into 4-hydroxybenzaldehyde by a novel C_2-chain-shortening enzyme called 4-hydroxybenzaldehyde synthase (HBS). 4-Hydroxybenzaldehyde was subsequently converted into 4-hydroxybenzoic acid by a NAD^+-dependent dehydrogenase (HBD) enzyme.

Transformed roots of *D. carota* upon exposure to continuous light turned green, which appears to be a rare but physiological process. Surprisingly, a suppressed amount of 4-hydroxybenzoic acid accumulation was observed in these green transformed root cultures. This reduced 4-hydroxybenzoic acid contents in green roots apparently raised question of the possibility of metabolic shift to combat any stress, in particular photooxidative stress that might have occurred in green transformed roots [63]. In plants, stress conditions are associated with production of reactive oxygen species leading to cellular damage. Biochemical measurement of different marker antioxidative enzymes depicted the occurrence of photooxidative stress in green transformed roots. Volatile compound analysis from both types of transformed roots put some light in the question that arises due to reduction of phenolic content in green transformed roots as compared to normal one. Increased monoterpene and sesquiterpene contents in green transformed roots substantiated the activation of terpenoid biosynthesis in green transformed roots due to light irradiation. It is plausible that these volatile terpenoids perhaps protect the green transformed roots from photooxidative stress, thus justifying the metabolic expenses of volatile production. It is conceivable that, during greening, the primary metabolites that are normally channelized toward shikimate/phenylpropanoid pathways are used by the MEP pathway, thus limiting the carbon flow toward shikimic/phenylpropanoid pathways [64].

Acknowledgments The research findings described in this article were supported by Council of Scientific and Industrial Research (CSIR), India [grant no. 38(1201)/08/EMR-II to A. Mitra] and Department of Science and Technology (DST), India [grant no. SP/SR/BB-40/2002 to A. Mitra].

References

1. Fraser PD, Bramley PM (2004) The biosynthesis and nutritional uses of carotenoids. Prog Lipid Res 43:228–265
2. Alasalvar C, Grigor JM, Zhang D, Quantick PC, Shahidi F (2001) Composition of volatiles, phenolics, sugars, antioxidant vitamins and sensory quality of different coloured carrot varieties. J Agric Food Chem 49:1410–1416

3. Kjeldsen F, Christensen LP, Edelenbos M (2001) Quantitative analysis of aroma compounds in carrot (*Daucus carota* L.) cultivars by capillary gas chromatography using large-volume injection technique. J Agric Food Chem 49:4342–4348
4. Hartley RD, Harris PJ (1981) Phenolic constituents of cell walls of dicotyledonous. Biochem Syst Ecol 9:198–203
5. Schnitzler JP, Madlung J, Rose A, Seitz HU (1992) Biosynthesis of *p*-hydroxybenzoic acid in elicitor treated carrot cell cultures. Planta 188:594–600
6. Kang YH, Parker CC, Smith AC, Waldron KW (2008) Characterization and distribution of phenolics in carrot cell wall. J Agric Food Chem 56:8558–8564
7. Tomas-Barberan FA, Clifford MN (2000) Dietary hydroxybenzoic acid derivatives-nature, occurrence and dietary burden. J Sci Food Agric 80:1024–1032
8. Cho JY, Moon JH, Seong KY, Park KH (1998) Antimicrobial activity of 4- hydroxybenzoic acids isolated and identified from rice hull. Biosci Biotechol Biochem 62:2273–2276
9. Liu M, Liu S, Peterson SL, Miyake M, Liu KJ (2002) On the application of *p-hydroxybenzoic* acid as a trapping agent to study hydroxyl radical generation during cerebral ischemia and reperfusion. Mol Cell Biochem 234–235:379–385
10. Soni MG, Carabin IG, Burdock GA (2005) Safety assessment of esters of *p-hydroxybenzoic* acid (parabens). Food Chem Toxicol 43:985–1015
11. Tullo A (1999) Making a connection with liquid crystal polymers. Chem Mark Report 256:9
12. Löscher R, Heide L (1994) Biosynthesis of *p*-hydroxybenzoic acid from *p*-coumarate and *p*-coumaroyl-coenzyme A in cell free extract of *Lithospermum erythrorhizon* cell cultures. Plant Physiol 106:271–279
13. Mahuzier PE, Altria KD, Clark BJ (2001) Selective and quantitative analysis of 4-hydroxybenzoate preservatives by microemulsion electrokinetic chromatography. J Chromtogr A 924:465–470
14. Siebert M, Sommer S, Li S, Wang Z, Severin K, Heide L (1996) Genetic engineering of plant secondary metabolism – accumulation of 4-hydroxybenzoate glucosides as a result of the expression of the bacterial *ubiC* gene in tobacco. Plant Physiol 112:811–819
15. Sircar D, Roychowdhury A, Mitra A (2007) Accumulation of *p*-hydroxybenzoic acid in hairy roots of *Daucus carota*. J Plant Physiol 164:1358–1366
16. Wildermuth MC (2006) Variation on a theme: synthesis and modification of plant benzoic acids. Curr Opin Plant Biol 9:288–296
17. Qualley AV, Widhalm JR, Adebesin F, Kish CM, Dudareva N (2012) Completion of the core β-oxidative pathway of benzoic acid biosynthesis in plants. Proc Natl Acad Sci U S A 109:16383–16388
18. Gaid MM, Sircar D, Müller A, Beuerle T, Liu B, Ernst L, Hänsch R, Beerhues L (2012) Cinnamate: CoA ligase initiates the biosynthesis of a benzoate-derived xanthone phytoalexin in *Hypericum calycinum* cell cultures. Plant Physiol 160:1267–1280
19. Chakraborty D, Sircar D, Mitra A (2008) Phenylalanine ammonia-lyase-mediated biosynthesis of 2-hydroxy-4-methoxybenzaldehyde in roots of *Hemidesmus indicus*. J Plant Physiol 165:1033–1040
20. Oksman-Caldentey KM, Hakkinen ST, Rischer H (2007) Metabolic engineering of the alkaloid biosynthesis in plants: functional genomics approaches. In: Verpoorte R, Alfermann AW, Johnson TS (eds) Applications of plant metabolic engineering. Springer, Dordrecht, pp 109–127
21. Petersen M (2007) Current status of metabolic phytochemistry. Phytochemistry 68:2847–2860
22. Parr AJ, Ng A, Waldron KW (1997) Ester-linked phenolic components of carrot cell walls. J Agric Food Chem 45:2468–2471
23. Kadokura A, Yokota T, Yagi T, Sato T, Uchiyama T (2003) The nature of induced lignin in carrot (*Daucus carota* L.) callus cells elicited with hot water extracts from conidia of *Botrytis cinerea*. Plant Biotechnol 20:21–31
24. Mitra A, Mayer MJ, Mellon FA, Michael AJ, Narbad A, Parr AJ, Waldron K, Walton NJ (2002) 4-Hydroxycinnamoyl-CoA-hydratase/lyase, an enzyme of phenylpropanoid cleavage from

Pseudomonas, causes formation of C6-C1 glucose conjugates when expressed in hairy roots of *Datura stramonium*. Planta 15:79–89

25. Hamill JD, Parr AJ, Rhodes MJC, Robins RJ, Walton NJ (1987) New routes to plant secondary products. Biotechnology 5:800–804
26. Walton NJ, Alfermann AW, Rhodes MJC (1999) Production of secondary metabolites in cell and differentiated organ cultures. In: Wink M (ed) Function of plant secondary metabolites and their exploitation in biotechnology. Sheffield Academic Press, Sheffield, Annu Plant Rev 3: pp 311–346
27. Pletsch M, de Araujo BS, Charlood BV (1999) Novel biotechnological approaches in environmental remediation research. Biotechnol Adv 17:679–687
28. Hu ZB, Du M (2006) Hairy root and its application in plant genetic engineering. J Integr Plant Biol 48:121–127
29. Rhodes MJC, Parr AJ, Walton NJ (1997) Studies of metabolic pathways in transformed roots. In: Doran PM (ed) Hairy roots – culture and applications. Harwood Academic Publishers, Amsterdam, pp 31–42
30. Verpoorte R, Van der Heijden R, Hoge JHC, Ten Hoopen HJG (1994) Plant cell biotechnology for the production of secondary metabolites. Pure Appl Chem 66:2307–2310
31. Guillon S, Tremouillaus-Guiller J, Pati PK, Rideau M, Gantet P (2006) Hairy root research: recent scenario and exciting prospects commentary. Curr Opin Plant Biol 9:341–346
32. Farmer EE, Alméras E, Krishnamurthy V (2003) Jasmonates and related oxylipins in plant responses to pathogenesis and herbivory. Curr Opin Plant Biol 6:372–378
33. Zhao J, Davis LC, Verpoorte R (2005) Elicitor signal transduction leading to production of plant secondary metabolite. Biotechnol Adv 23:283–333
34. Jin JH, Shin JH, Kim JH, Chung IS, Lee HJ (1999) Effect of chitosan elicitation and media components on the production of anthraquinone colorants in madder (*Rubia akane* Nakai) cell culture. Biotechnol Bioprocess Eng 4:300–304
35. Pitta-Alvarez SA, Guilietti AM (1999) Influence of chitosan, acetic acid and citric acid on growth and tropane alkaloid production in transformed roots of *Brugmansia candida*: effect of medium, pH and growth phase. Plant Cell Tiss Org Cult 59:31–38
36. Komaraiah P, Naga Amrutha R, Kavi Kishor PB, Ramakrishna SV (2002) Elicitor enhanced production of plumbagin in suspension cultures of *Plumbago rosea* L. Enzym Microb Technol 31:634–639
37. Sircar D, Mitra A (2009) Accumulation of *p*-hydroxybenzoic acid in hairy roots of *Daucus carota* 2. Confirming biosynthetic steps through feeding of inhibitors and precursors. J Plant Physiol 166:1370–1380
38. Funk C, Brodelius PE (1990) Phenylpropanoid metabolism in suspension cultures of *Vanilla planifolia* Andr: 11. Effects of precursor feeding and metabolic inhibitors. Plant Physiol 94:95–101
39. Vollsnes AV, Melø TB, Futsaether CM (2012) Photomorphogenesis and pigment induction in lentil seedling roots exposed to low light conditions. Plant Biol 14:467–474
40. Alscher RG, Erturk N, Heath LS (2002) Role of superoxide dismutases (SODs) in controlling oxidative stress in plants. J Exp Bot 53:1331–1341
41. Hiraga S, Sasaki K, Ito H, Ohashi Y, Matsui H (2001) A large family of class III plant peroxidases. Plant Cell Physiol 42:462–468
42. Willekens H, Langebartels C, Tirě C, Montagu MV, Inzě D, Camp WV (1994) Differential expression of catalase genes in *Nicotiana plumbaginifolia* (L.). Proc Natl Acad Sci U S A 91:10450–10454
43. Rao MV, Paliyath G, Drmrod DP (1996) Ultraviolet-B and ozone induced biochemical changes in antioxidant enzymes of *Arabidopsis thaliana*. Plant Physiol 110:125–136
44. Scandalios JG (1994) Regulation and properties of plant catalases. In: Foyer CH, Mullineaux PM (eds) Causes of photooxidative stress and amelioration of defense systems in plants. CRC Press, Boca Raton, pp 275–315

45. Holmström KO, Somersalo S, Mandal A, Palva TE, Welin B (2000) Improvement tolerance to salinity and low temperature in transgenic tobacco producing glycine betaine. J Exp Bot 51:177–185
46. Herrmann KM (1995) The shikimate pathway: early steps in the biosynthesis of aromatic compounds. Plant Cell 7:907–919
47. Mahmoud SS, Croteau RB (2001) Metabolic engineering of essential oil yield and composition in mint by altering expression of deoxyxylulose phosphate reductoisomerase and menthofuran synthase. Proc Natl Acad Sci U S A 98:8915–8920
48. Loreto F, Schnitzler J (2010) Abiotic stress and induced BVOCs. Trends Plant Sci 15:156–166
49. Holopainen JK, Gershenzon J (2010) Multiple stress factors and the emission of plant VOCs. Trends Plant Sci 15:176–184
50. da Silva MHL, Andrade EHA, Zoghbi MGB, Luz AIR, da Silva JD, Maia JGS (1999) The essential oils of *Lantana camara* L. occurring in North Brazil. Flavour Fragance J 14:208–210
51. Daferera DL, Ziogas BN, Polissiou MG (2003) The effectiveness of plant essential oils on the growth of *Botrytis cinerea*, *Fusarium* sp. and *Clavibacter michiganensis* subsp. *michiganensis*. Crop Prot 22:39–44
52. Asuming WA, Beauchamp PS, Descalzo JT, Dev BC, Dev V, Frost S, Ma CW (2005) Essential oil composition of four *Lomatium* Raf. species and their chemotaxonomy. Biochem Syst Ecol 33:17–26
53. Kjeldsen F, Christensen LP, Edelenbos M (2003) Changes in volatile compounds of carrots (*Daucus carota* L.) during refrigerated and frozen storage. J Agri Food Chem 51:5400–5407
54. Duhl TR, Helmig D, Guenther A (2008) Sesquiterpene emissions from vegetation: a review. Biogeosci 5:761–777
55. Zhao N, Guan J, Ferrer JL, Engle N, Chern M, Ronald P, Tschaplinski TJ, Chen F (2010) Biosynthesis and emission of insect-induced methyl salicylate and methyl benzoate from rice. Plant Physiol Biochem 48:279–287
56. Snoeren TAL, Mumm R, Poelman EH, Yang Y, Pichersky E, Dicke M (2010) The Herbivore-Induced plant volatile methyl salicylate negatively affects attraction of the parasitoid *Diadegma semiclausum*. J Chem Ecol 36:479–489
57. DeBoer JG, Dicke M (2004) The role of methyl salicylate in prey searching behaviour of the predatory mite *Phytoseiulus persimilis*. J Chem Ecol 30:255–271
58. DeBoer JG, Posthumas MA, Dicke M (2004) Identification of volatiles that are used in discrimination between plants infested with prey or nonprey herbivores by a predatory mite. J Chem Ecol 30:2215–2230
59. Ishiwari H, Suzuki T, Maeda T (2007) Essential compounds in herbivore-induced plant volatiles that attract the predatory mite *Neoseiulus womersleyi*. J Chem Ecol 33:1670–1681
60. Paré PW, Tumlinson JH (1997) De novo biosynthesis of volatiles induced by insect herbivory in cotton plants. Plant Physiol 114:1161–1167
61. Flores-Pérez Ŭ, Pérez-Gil J, Closa M, Wright LP, Botella-Pavía P, Phillips MA, Ferrer A, Gershenzon J, Rodríguez-Concepción M (2010) Pleiotropic regulatory locus 1 (PRL1) integrates the regulation of sugar responses with isoprenoid metabolism in *Arabidopsis*. Mol Plant 3:101–112
62. Haderlie HC, Widholm JM, Slife FW (1977) Effect of glyphosate on carrot and tobacco cells. Plant Physiol 60:40–43
63. Mukherjee C, Sircar D, Chatterjee M, Das S, Mitra A. (2014) Combating photooxidative stress in green hairy roots of *Daucus carota* cultivated under light irradiation. J Plant Physiol 171:179–187.
64. Mukherjee C, Samanta T, Mitra A (2016) Redirection of metabolites biosynthesis from hydroxybenzoates to volatile terpenoids in green hairy roots of *Daucus carota*. Planta 243:305–320

Functional Analysis and the Role of Members of SGT Gene Family of *Withania somnifera*

21

Pratibha Misra, Gaurav Singh, Manoj K. Mishra, Vibha Pandey, and Syed Saema

Abstract

Sterol glycosyltransferases (SGTs) catalyze the attachment of a carbohydrate moiety to an aglycone sterol accepter molecule at different positions. SGTs are key enzymes for the biosynthesis of many precious natural plant products. SGTs of *Withania somnifera* (*Ws*SGTs) help in the glycosylation of withanolides, a pharmaceutically important C-28 phytochemical product and phytosterols, such as sitosterol and stigmasterol. SGTs of *W. somnifera* glycosylate the sterol backbone at C-3, C-17, and C-27 positions. Modified phytosterols and withanolides play an important role in maintaining metabolic plasticity during adaptive response. The expression of SGTs changed during different biotic and abiotic stresses indicating their role in maintaining the cellular disturbances. Overexpression of *WsSGTL1*, a gene member of *SGT* gene family and silencing of *SGT* members through RNAi and artificial miRNA technology, in homologous (*W. somnifera*) and heterologous (*Nicotiana tabacum* and *Arabidopsis thaliana*)

P. Misra (✉) • G. Singh
Genetics and Plant Molecular Biology Division, Council of Scientific and Industrial Research, National Botanical Research Institute, Lucknow, Uttar Pradesh, India
e-mail: pratibhaflora@gmail.com; pratibhamisra@nbri.res.in; g.singhbt@gmail.com

M.K. Mishra
Department of Botany, Lucknow University, Lucknow, Uttar Pradesh, India
e-mail: manoj30biotech@gmail.com

V. Pandey
Department of Plant Molecular Biology, University of Delhi South Campus, New Delhi, India
e-mail: vibhabiotech10@gmail.com

S. Saema
Department of Biotechnology, Integral University, Lucknow, Uttar Pradesh, India
e-mail: syeddssaema@gmail.com

S. Jha (ed.), *Transgenesis and Secondary Metabolism*, Reference Series in Phytochemistry, DOI 10.1007/978-3-319-28669-3_16

expression systems defines their role in growth and development of plants. The functional analysis of these genes has also been studied under abiotic (cold, heat, and salt) and biotic (SA, JA, *Alternaria alternata*, and *Spodoptera litura*) stress-providing tolerance to the plants. The chapter is concerned with the importance and application of SGTs in metabolic pathway engineering leading to biosynthesis of important bioactive compounds in *W. somnifera.*

Keywords
Glycosyltransferases • *Withania somnifera* • Sterol glycosyltransferases • Biotic and abiotic stresses • *Spodoptera litura* • *Alternaria alternata* • Artificial miRNA • Transgenic plants

Contents

1 Introduction

Plants are the major source of producing a huge diversity of low molecular weight natural products through primary and secondary metabolism. This diversity of compounds is produced by the modification of common backbone structures. Major modifications, such as methylation, acylation, hydroxylation, and glycosylation, play a significant role in maintaining the diversity of compounds in the plant. Glycosylation is one of the most extensive modifications in which monosaccharide molecules attach to the lipids or proteins. Carbohydrates maintain the structural and functional biology of plants, such as in the formation of glycoproteins, proteoglycans, glycolipids, polysaccharides, etc., and have important roles in cell growth, cell-cell interactions, immune defense, etc. [3, 14, 34, 54]. The modifications of protein or lipids by the stepwise assembly of carbohydrate molecules are made by enzymatic process of glycosyltransferases (GTs) which sequentially transfer the monosaccharide carbohydrate molecule from nucleotide donor to the required acceptor, such as lipids, and proteins resulting in the formation of a glycosidic bond [10, 17, 46]. The donor carbohydrates are typically limited to monosaccharides, such as fructose, galactose, glucose, arabinose, mannose, xylose, etc., linked to a nucleoside mono- or diphosphate, such as cytidine monophosphate (CMP), guanosine diphosphate (GDP), and uridine diphosphate (UDP) [37, 53]. Of these nucleoside sugar donor, more than 60% of all known GTs are reacted by UDP-glycosyltransferases (UGTs) [21]. Consequently, there are many known plant GTs with widely different specificities according to their donor and acceptors. Plant GTs are highly conserved in their sequence and mode of action [5, 15, 35]. They can recognize common features on a range of substrates including secondary metabolites, hormones, or xenobiotics. In *Withania somnifera*, a class of GTs called sterol glycosyltransferases (SGTs) help in the modification of withanolides and other free sterols, such as sitosterol and stigmasterol, for the conversion of its glycosylated forms [48, 51]. In this chapter, we have focused on the importance of SGTs in plants for their tolerance against different stress (biotic and abiotic) conditions.

In CSIR-NBRI, the first biochemical and kinetic properties of SGT were characterized by Madina et al. [30], Madina et al. [29], and Sharma et al. [50]. Chaturvedi et al. [6] described the expression of *sgtl3.1*, *sgtl3.2*, and *sgtl3.3* in biotic (salicylic acid, methyl jasmonate) and abiotic (heat and cold) stresses. Mishra et al. [32] overexpressed the *WsSGTL1* gene in *A. thaliana* and elucidated its role in abiotic stresses (heat, cold, and salt) by the analysis of transgenic lines. Pandey et al. [40] overexpressed *WsSGTL1* gene in *N. tabacum* transgenic lines and in *W. somnifera* for hairy root development and analyzed its role in the production of sterols, glycosylated sterols, and biotic stress (*Spodoptera litura*). Comparative interaction of WsSGTL1 and WsSGTL4 with withanolides and sterols was reported by Pandey et al. [41]. Effect of *WsSGTL1* gene silencing on the growth and glycosylation pattern by RNAi method was reported by Saema et al. [47]. Saema et al. [48] reported that overexpression of *WsSGTL1* in *W. somnifera* helps in growth, enhances glycowithanolides, and provides biotic and abiotic stress tolerance. Further, Singh et al. [51] reported that silencing of some members of *WsSGTs* by aMIR-VIGS system and silencing of genes modulate the withanolide biosynthesis and lines became susceptible to *Alternaria alternata* infection.

2 Evolutionary Conserved Nature of *WsSGTs*

Phylogenetic analysis performed in GTs explained the ancient origin during evolution from the time of divergence of prokaryotes and eukaryotes [11, 19]. On the basis of sequence similarity, GTs from prokaryotes to eukaryotes were grouped into four monophyletic superfamilies (named after identified four structural folds) GT-A, GT-B, GT-C, and GT-D. Among them, GT-A and GT-B were evolved as most diverse and ubiquitous group of GTs, as only GT-A includes *E. coli*, *Bacillus subtilis*, *Bos taurus*, *Oryctolagus cuniculus*, *Mus musculus*, *Neisseria meningitidis*, *Homo sapiens*, etc. [28]. According to CAZY online database (http://www.cazy.org/GT1.html), glycosyltransferases (GT) have more than 95 families in which SGTs have been grouped in family 1 because of having a distinct signature motif at the C-terminal, the PSPG domain (Fig. 1) [6, 32].

[FW]-x(2)-[LIVMYA]-[LIMV]-x(4-6)-[LVGAC]-[LVFYA]-[LIVMF]-[STAGCM]-[HNQ]-[STAGC]-G-x(2)-[STAG]-x(3)- [STAGL]-[LIVMFA]-x(4)-[PQR]-[LIVMT]-x(3)-[PA]-x(3)-[PA]- x(3)-[DES]-[QEHN]

Fig. 1 Highly conserved plant secondary product glycosyltransferases (PSPG) sequences [6]

Full-length characterization of amino acid sequence of *WsSGT* members (*sgtl3.1*-EU342379, *sgtl3.2*-EU342374, and *sgtl3.3*-EU342375) deduced their 45–67% open reading frame (ORF) homonology with known plant's SGTs [7]. Phylogenetic analysis of *WsSGTL1* and *WsSGTL4* proteins reveals the proximity with GTs of *S. lycopersicum* and *M. truncatula*, respectively [29, 41]. GENO3D online server is used for the structured modeling of *Ws*SGTL1 and *Ws*SGTL4 which gives us indications that both proteins belong to GT-B family glycosyltransferase because of having β-sheets in parallel orientation [41]. Studies revealed that almost all members of GT1 family (mostly plant UGTs) have GT-B domain [26, 57].

3 Docking of *Ws*SGTL1 and *Ws*SGTL4 with Sterols and Withanolides

Enzymes belonging to the families of glycosyltransferases (GTs) are responsible for the glycosylation of variety of metabolites. The reaction proceeds by the transfer of glycosyl moiety from activated nucleoside diphosphate sugar donor to an acceptor molecule [42]. Two identified SGT enzymes of *W. somnifera*, *Ws*SGTL1 and *Ws*SGTL4, have been compared by Pandey et al. [41] on the basis of their topological character and conserved nature and substrate specificity with different sterols. Differences in size as well as their affinity toward different substrates were observed through sequence alignment and docking experiments, respectively. The obtained results revealed that both WsSGTs show differences in their affinity towards all selected ligands and imply their broad substrate specificity. Among all the tested ligands, brassicasterol and withanolide A showed best interaction to form enzyme substrate complexes with *Ws*SGTL1 and *Ws*SGTL4, respectively [41].

4 In Vitro Enzymatic Activity of WsSGTs

The first study of biochemical and kinetic properties of a cytosolic SGT isolated from leaves of *W. somnifera* was performed by Madina et al. [30]. They described that UDP-glucose was sugar donor of that purified enzyme. It had broad glycosylation activity with several sterols and phytosterols with 3β-OH group and low activity with flavonoids and isoflavonoids [29]. The maximum Kcat/Km value of that enzyme was for 24-methylenecholesterol that resembles with sitoindosides VII and VIII of *W. somnifera*. After that Sharma et al. [50] biochemically characterized *WsSGTL1* in *E. coli*. They gave the evidence that the recombinant protein isolated from *E. coli* showed modification activity with different sterols but not with the salicylic acid [50]. Further, a 27β-hydroxy glucosyltransferase was purified from the cytosolic fraction of *W. somnifera* leaves to study its kinetic and biochemical properties [30]. After characterization, it was proved that purified enzyme showed broad sterol specificity by glycosylating a variety of withanolides/phytosterols with β-OH group at C-17, C-21, and C-27 positions. Singh et al. [51] further reported that silencing of *WsSGTL1*, *WsSGTL2*, and *WsSGTL4* together diminishes their activity in the silenced lines against stigmasterol and solanidine. This study indicated that these genes also participated in the glycosylation of stigmasterol and solanidine.

5 Strategies for Functional Analysis of *WsSGTs* Genes

5.1 Overexpression in Homologous and Heterologous Expression System

Expression of genes in heterologous organisms has contributed functional importance in animal and plant system. In animal research, *Xenopus* oocytes and cell cultures are major techniques for heterologous gene expression, whereas in plant domain, yeast has become the prevalent heterologous expression system [16, 58]. For the functional analysis of the *SGTL1* gene of *W. somnifera, Arabidopsis thaliana* and *Nicotiana tabacum* were used as heterologous expression systems [32, 40]. Leaves of *N. tabacum* were subjected to *A. tumefaciens*-mediated transformation to develop *WsSGTL1* expressing transgenic lines. Among homozygous T3 plants, three independent highly expressive transgenic lines were selected for further functional analysis. The *WsSGTL1* gene was transformed in *A. thaliana* through *Agrobacterium*-mediated transformation by floral dip method. The phenotypic and physiological parameters like seed weight, germination, root length, shoot weight, relative electrolyte conductivity, MDA content, SOD (and CAT) levels, relative electrolyte leakage, and chlorophyll measurements were compared between transgenic and wild-type *Arabidopsis/Nicotiana* plants. Transgenic *A. thaliana* was observed for tolerance under different abiotic stresses (salt, heat, and cold); however, *N. tabacum* was observed for tolerance to salt stress and biotic (insect) stress to confirm functional importance of *WsSGTL1*. An efficient transformation system of *W. somnifera* for homologus gene expression has also been devloped through transformation of *GUS* reporter gene [38]. Later, Saema et al. [48] introduced *WsSGTL1* gene in homologous system via *Agrobacterium tumefaciens*-mediated transformation [48]. The study proved the use of cotyledonary leaves and hypocotyls explants to get best transformation efficiency. Maximum regeneration and shoot organogenesis of explants were obtained in the presence of 2 mg/l BAP and 0.1 mg/l IAA. Biochemical analysis was done by HPLC-TLC/HPLC and radiolabeled enzyme assay, to confirm the glycosylation achieved by the activity of *Ws*SGTL1. Based on earlier studies, gene-specific primers were used to isolate the full-length open reading frame of *WsSGTL4* (EU342374) from cDNA. By using A4 strain of

A. rhizogenes, transgenic hairy root lines with overexpression of *WsSGTL4* gene have been developed [39].

5.2 Silencing of *WsSGT* Genes

Gene silencing is a universal word used to explain the regulation of gene expression. For the functional study of *WsSGT* gene members, siRNA and artificial microRNA (amiRNA) technologies have been applied using pFGC1008 and virus-induced gene silencing (VIGS) vectors, respectively [47, 51]. After siRNA-mediated gene silencing of *WsSGTL1* gene, silenced lines were subjected to metabolic profiling of important withanolides. For the functional analysis of a plant gene, virus-induced gene silencing has emerged as powerful tool for the reverse genetics. Silencing of *WsSGTL* members (*WsSGTL1*, *WsSGTL2, and WsSGTL4*) with the combination of artificial miRNA and VIGS (aMIR-VIGS) led to the enhanced accumulation of withanolides and phytosterols with decreased content of glycowithanolides in silenced lines. The variations in the concentrations of withanolide in silenced lines resulted in pathogen susceptibility. These results gave indications that a positive feedback regulation of withanolide biosynthesis occurred by silencing of *SGTLs* which resulted in reduced biotic tolerance.

6 Biological Roles of *WsSGTs*

6.1 Effects of *WsSGTs* on Growth, Development, and Physiology of Plants

Secondary metabolism produces some significant products that assist in the growth and development of plants but are not essential for the plant to survive. Glycosylation is an important phenomenon for the modification of essential secondary plant products during growth and development [2, 9]. This glycosylation can affect the solubility, transport, and biological activity of the hormones. In *W. somnifera*, overexpression of *WsSGTL1* gene positively affects the early and predominant growth [48]. One of the major effects of this overexpression was the increase of 18–38% in stomatal density and increased size of subsidiary cells in progeny of several transgenic lines. This included 42–56% increase in transpiration and stomatal conductance, but it did not affected photosynthetic rate, electron transport, quantum yield, and decreased in water use efficiency. During silencing of *WsSGT* gene members, plants grew with short height and less leaf area, which suggested that *SGT* activity might be affecting the growth hormone signaling pathway in the plants [51]. It has already been reported that the glycosylation of brassinosteroids and strigolactone, a sterol derivative hormone which helps in the growth of axillary bud and plays very important role in the development of plants [6, 22], has been carried out by the members of *SGT*s in plants. Mutation studies on fackel, *SMT1*, cyclopropylsterol isomerase (*cpi*), *hyd1*, *UGT80B1*, and cotyledon vascular pattern1 (*cvp1*) have revealed roles of sterol modifications in shoot and root patterning, embryo, cell expansion, vein, fertility, polarity, proliferation, gravitropism, hormone signaling, and cellulose level maintenance [4, 12]. An earlier report also revealed cold tolerance, reduced senescence, enhanced growth, and a substantial improvement in protoplasmic drought tolerance, suggesting that SDG8i glucosyltransferase activity might be affecting the strigolactone pathway in the overexpressing lines of *Arabidopsis* [22]. One member of UGTs was reported capable of affecting the biological activity of plant hormones, such as auxins, cytokinins, ABA, SA, jasmonic acid, and brassinosteroids via glycosylation [20, 23, 27, 44]. In heterologous system, expression of *WsSGTL1* reduced plant height and root length and fresh and dry weight of seedling stage of *WsSGTL1-Nt* lines of *Nicotiana tabacum* [40]. Overexpression of *WsSGTL1* gene in *A. thaliana* transgenics showed better seed germination and increased tolerance to salt, heat, and cold stress as compared to wild type [32]. Physiology of plant is dependent on several environmental, biochemical, and molecular changes. Chlorophyll fluorescence imaging (CFI) analysis of *WsSGTL1-Nt* lines revealed reduced fluorescence and absorbance as compared to control plants. Similar to seedlings, 4-month-old *WsSGTL1-Nt* plants also showed significantly decreased photochemical efficiency (Fv/Fm) along with lower photosynthetic rate, respiration rate, and stomatal conductance. Similarly, reduced level of chlorophyll-A, chlorophyll-B, total chlorophyll, and carotenoid were found reduced in leaves of 1-month-old and 3-month-old *WsSGTL1-Nt* plants. However, expression of *WsSGTL1* indirectly modulates the growth pattern as well as the physiological properties of *WsSGTL1-Nt* by influencing adaptive machinery of plants similar to reduced growth and photosynthetic rate of plants of severe climatic conditions [40].

6.2 Modulates the Glycosylation of Sterols

Cytoplasm and plastids are the compartments for sterol biosynthesis in plant cells by mevalonate pathway of isoprenogenesis through phenyl precursors (Fig. 2) [18]. As an alternative pathway in plants, bacteria, and protozoans, isoprenogenesis is also reported by 1-deoxy-D-xylulose-5-phosphate (DOXP) pathway, which is localized in the plastids [49]. These isoprene units direct the biosynthesis of 2,3-oxidosqualene, which serves as the common precursor of different forms of sterols. Sterol glycosyltransferases (SGTs) in plants catalyze the transfer of glycon moieties to the phytosterols and their related compounds to generate glycoconjugates or steryl glycosides (SGs). SGs are membrane associated sterols and comprised of a sugar moiety attached to C-3 hydroxyl (–OH) group of sterol [7]. For the modification of sterols by SGT enzymes, most preferred position is C-3OH followed by 27β OH present at the side chain of modified sterols [45, 56]. Enzymatic activity (glycosylating activity) of one member of *WsSGTs*, i.e., *WsSGTL1*, has been reported to have 3β-OH group [29, 31, 50]. Major sterols reported in plant system are sitosterol, campesterol, and stigmasterol having 3β-OH group. These sterols showed activity with

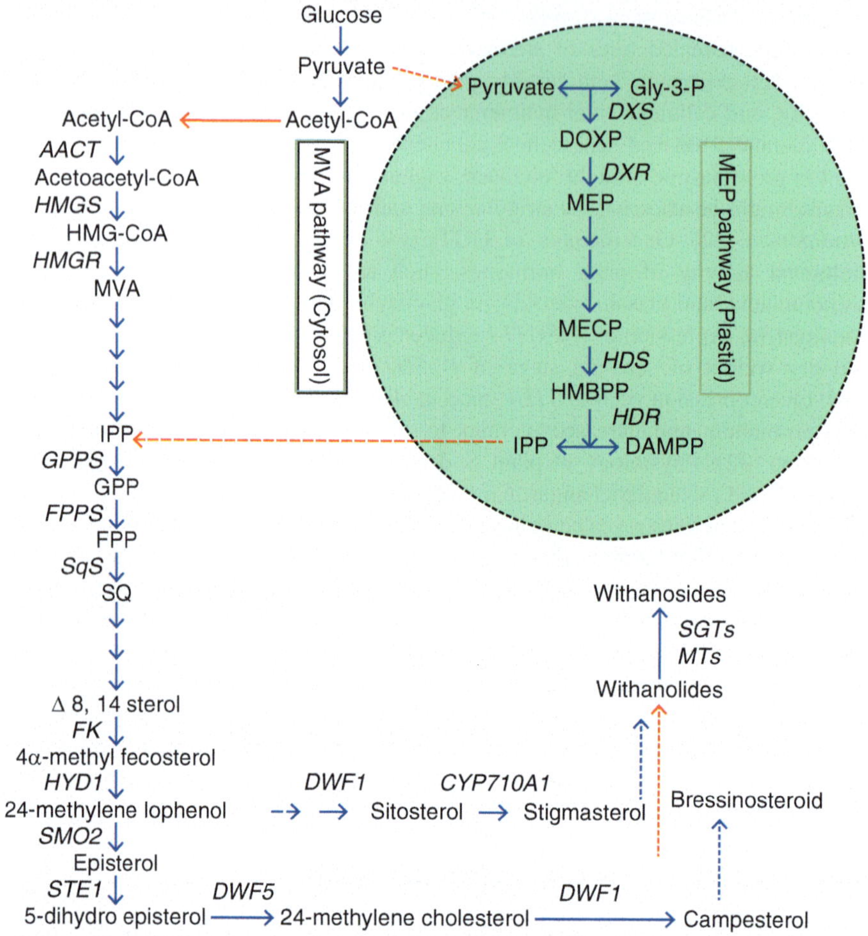

Fig. 2 Schematic diagram of mevalonate pathway (MVA) of withanoside (glycowithanolides) biosynthesis by the modifications action of SGTs on withanolides

WsSGTL1, while another important sterol, i.e., cholesterol, showed no activity with *WsSGTL1* [50]. In *W. somnifera*, SGTs help in the glycosylation of withanolides for the conversion to its glycoconjugates [47, 48, 51]. It was also reported that overexpression of *WsSGTL1* gene modulates the accumulation of glycosylated flavonoids which indicated the role of SGTs in glycosylation of rutin and quercetin [40]. The ratio of glycosylated versus nonglycosylated phytosterols (stigmasterol, sitosterol, and campesterol) also changes after changes in the expression of *WsSGT* in *Withania* or other systems [32, 40, 47, 48, 51]. Till date, hundreds of UDP-glycosyltransferase genes have been cloned and characterized from bacteria, fungus, model and non-model crops, and ornamental and medicinal plants. However, the target of specific *WsSGT* with their specific accepter molecules, withanolides, flavonoid, or phytosterol in *Withania* is still unknown due to their high sequence and functional similarity.

6.3 Role in Abiotic Stress

In plant membrane system, sterols along with lipids have been identified as major components and are necessary for organizing the important events such as protein targeting and signal transduction in plants, fungi, and animals. The expression of *SGTs* maintains the ratio of sterol and sterol glycosides in the plasma membrane by modulating its transcript level under different physiological conditions. Sterol glycosyltransferases (SGTs) are the intermediate enzymes involved in the modification of sterols and play significant role in maintaining the metabolic plasticity during adaptive responses. In *W. somnifera*, *SGTs* are of particular interest for their role in the biosynthesis of pharmacologically active substances [40]. Due to the presence of glycowithanolides in *Withania*, efforts have been made to identify and characterize *SGTs* from this important medicinal plant. Besides their medicinal value, *SGT* gene family members participate significantly different under-changed physiological conditions, growth and development by glycosylation of the growth hormones, and in abiotic and biotic stress conditions [32, 40, 48, 51]. Functional analysis of *WsSGTL1* (a member of *WsSGT* gene family) with its involvement in providing tolerance toward abiotic stresses has been extensively studied by overexpression in *A. thaliana*, *N. tabacum*, and *W. somnifera*. Initially, it was reported that the expression of *WsSGT* members (*WsSGTL1*, *WsSGTL3.1*, *WsSGTL3.2*, and *WsSGTL3.3*) under heat and cold stress indicated their role in abiotic stress tolerance [7]. Further, overexpression of *WsSGTL1* gene into *A. thaliana* showed better germination, salt tolerance, and heat and cold tolerance [32]. The expression level of *WsSGTL1* in the transgenic *A. thaliana* plants was elevated under heat, cold, and salt stress. Comparative study of *UGT80B1* mutant with their restored lines (*p35S: TTG15/UGT80B1*) of *A. thaliana* showed that this gene participated in the glycosylation of phytosterols. Due to low level of free and glycosylated sterols in the knockout mutant lines, physiology of plants was highly affected under cold and heat stress [31]. *WsSGTL1* overexpressed *N. tabacum* transgenic lines showed late germination, stunted growth, yellowish green leaves, and enhanced antioxidant system. These changes in the physiological parameters were due to the enhanced glycosylation by *WsSGTL1* gene. Further, overexpression of *WsSGTL1* gene in *W. somnifera* enhanced the ability to recover after cold stress, photosynthesis performance, chlorophyll, anthocyanin content, and quenching regulation of PSI and PSII. These results indicated that the *WsSGTL1* gene participated in the abiotic stress tolerance. Sterol glycosides play important role in membrane fluidity and permeability [58, 59]. Sterol glycosides and acylated sterol glycosides exchange gradually between the monolayer halves of membrane bilayer, in comparison to normal sterols, which serve to regulate sterol content composition and its distribution in the membrane [55, 58]. This modulation of sterols for maintaining the fluidity of biomembrane is the adaptation of plants toward different abiotic stresses [6, 52].

6.4 Enhanced Biotic Stress Tolerance

The accumulation of glycosylated sterols in the plant cells gives the tolerance against biotic stress also. There are very few reports available on antibacterial activity, in contrast to antifungal properties of saponins [13]. Silencing of a UDP-Glc-phenylpropanoid glycosyltransferase reduces the accumulation of scopoletin glycoside and weakens virus resistance by enhancing the oxidative stress in the cell [8]. Tomatine is a glycoalkaloid found in the stem and leaves of tomato plants that was reported to have antibacterial and antifungal effects on Gram-positive bacteria [24]. Avenacosides A and B are significant saponins, isolated from oat plants, and have sugar chains at C-3 and C-26 carbons. They lack antimicrobial activity, but by removing the C-26 sugar, it can be converted into biologically active forms. The C-3 position of the sterol moiety is the catalytic target of avenacosides, and these glycoalkaloids were active against the oat root pathogen *Gaeumannomyces graminis* [36]. It was reported that steroidal sapogenins glycosylated by *SAGT4* glycosyltransferase enzyme which catalyzes the 3-O-glycosylation are involved in the biosynthesis of saponin, such as diosgenin, nuatigenin, and tigogenin in *Solanum aculeatissimum* [25]. The involvement of SAGT4 enzyme in response to wounding stress indicates its role in plant defense system. An important potato SGT enzyme SAGT1 was identified by screening a wound-induced cDNA library in yeast suggesting the a possible role of SGT in biotic stress [33]. Brassinosteroids of *A. thaliana* and mycotoxin deoxynivalenol (DON) of *Fusarium* were glycosylated by *UGT73C5* gene which provides protection against the pathogen *Fusarium* by detoxifying DON [43]. The attachment of sugar chain to C-3 of the sterol is usually significant for both antifungal properties, and the permeability of membrane and elimination of these sugar molecules often result in the loss of biological activity of the membrane [1]. Some glycoalkaloid pairs that have common aglycon molecules but differ in the composition of their modified carbohydrate chains show synergism in their membranolytic and antifungal activity [6].

Overexpression of *WsSGTL1* in homologous system provides tolerance toward biotic stress. Transgenic lines prepared by the overexpression of *WsSGTL1* gene in *N. tabacum* and *W. somnifera* increase the insect resistance capacity of the plants, i.e., 100% mortality of *Spodoptera litura*. The larvae consumed a good amount of leaves of WT plants and grew normally, whereas negligible amount of transgenic leaves were fed by larvae. Feeding of *S. litura* larvae on transgenic leaves for 2–4 days resulted in 90–100% mortality. Heterologous system also caused the accumulation of steroidal saponins which increased the insect resistance property of the plants [40, 48]. In addition, in studies using artificial diets, it was proved that increased concentrations of tomatine caused growth retardation and delayed development of beetles feeding on Colorado potato, due to having tetrasaccharide moiety which has membranolytic action [6]. Recently, it was proved that the ratio of glycosylated and free sterols provides tolerance to the plants against the stress of *Alternaria alternata* [51]. SGT enzymes in plants are involved in modulation of sensitivity to stress hormones and changed tolerance to biotic and abiotic stresses [31, 32, 40, 48, 51]. SGTs also glycosylate steroidal hormones such as brassinosteroids, which has been participating as growth and development regulators in plants. In the field of metabolomics, it can be said that SGTs play significant role in plant metabolism and may offer future tools for crop improvement.

7 Conclusions

Sterol glycosyltransferases of *W. somnifera* participated into several physiological, molecular, and biochemical responses. Enzymatic and biochemical analysis of SGTs members indicated their role in modifications of important metabolites which have their own pharmacological value. Functional studies of these family members in homologous and heterologous system indicated their importance in plant system. The members of *SGT* gene family sustain the metabolic plasticity in the cell which maintains the homeostasis of the plants. From the previous research, we can conclude that SGTs maintain several biological processes in the cell, such as increasing the solubilization of compounds, helping in degradation of the xenobiotics, and maintaining the fluidity of membrane by glycosylation of phytosterols. But several important functions still need study in detail, such as how these genes participate in the growth and development and the specific sterol target of specific SGTs.

References

1. Armah CN, Mackie AR, Roy C, Price K, Osbourn AE, Bowyer P, Ladha S (1999) The membrane-permeabilizing effect of avenacin A-1 involves the reorganization of bilayer cholesterol. Biophys J 76:281–290
2. Bajguz A (2007) Metabolism of brassinosteroids in plants. Plant Physiol Biochem 45:95–107
3. Brandley BK, Schnaar RL (1986) Cell-surface carbohydrates in cell recognition and response. J Leukoc Biol 40:97–111
4. Carland F, Fujioka S, Nelson T (2010) The sterol methyltransferases *SMT1, SMT2,* and *SMT3* influence *Arabidopsis* development through nonbrassinosteroid products. Plant Physiol 153:741–756
5. Chang A, Singh S, Helmich KE, Goff RD, Bingman CA, Thorson JS, Phillips GN Jr (2011) Complete set of glycosyltransferase structures in the calicheamicin biosynthetic pathway reveals the origin of regiospecificity. Proc Natl Acad Sci U S A 108:17649–17654
6. Chaturvedi P, Misra P, Tuli R (2011) Sterol glycosyltransferases – the enzymes that modify sterols. Appl Biochem Biotechnol 165:47–68
7. Chaturvedi P, Mishra M, Akhtar N, Gupta P, Mishra P, Tuli R (2012) Sterol glycosyltransferases-identification of members of gene family and their role in stress in *Withania somnifera*. Mol Biol Rep 39:9755–9764
8. Chong J, Baltz R, Schmitt C, Beffa R, Fritig B, Saindrenan P (2002) Downregulation of a pathogen-responsive tobacco UDP-Glc:phenylpropanoid glucosyltransferase reduces scopoletin glucoside accumulation, enhances oxidative stress, and weakens virus resistance. Plant Cell 14:1093–1107
9. Clouse SD (1996) Molecular genetic studies confirm the role of brassinosteroids in plant growth and development. Plant J 10:1–8
10. Clutton-Brock J (1975) Letter: Design of anaesthesia apparatus. Lancet 1:1376
11. Coutinho PM, Deleury E, Davies GJ, Henrissat B (2003) An evolving hierarchical family classification for glycosyltransferases. J Mol Biol 328:307–317

12. DeBolt S, Scheible WR, Schrick K, Auer M, Beisson F, Bischoff V, Bouvier-Nave P, Carroll A, Hematy K, Li Y, Milne J, Nair M, Schaller H, Zemla M, Somerville C (2009) Mutations in UDP-glucose:sterol glucosyltransferase in *Arabidopsis* cause transparent testa phenotype and suberization defect in seeds. Plant Physiol 151:78–87
13. Didier A, Roselyne C, Jean-Marie L, Claudine P, Jean-Michel G, Roland P, Jean-Paul D, Daniel E (2003) Resistance to late blight and soft rot in six potato progenies and glycoalkaloid contents in the tubers. Am J Potato Res 80:125–134
14. Dilokpimol A, Geshi N (2014) *Arabidopsis thaliana* glucuronosyltransferase in family GT14. Plant Signal Behav 9:e28891
15. Egelund J, Petersen BL, Motawia MS, Damager I, Faik A, Olsen CE, Ishii T, Clausen H, Ulvskov P, Geshi N (2006) *Arabidopsis thaliana RGXT1* and *RGXT2* encode Golgi-localized (1,3)-alpha-D-xylosyltransferases involved in the synthesis of pectic rhamnogalacturonan-II. Plant Cell 18:2593–2607
16. Frommer WB, Ninnemann O (1995) Heterologous expression of genes in bacterial, fungal, animal, and plant cells. Annu Rev Plant Biol 46:419–444
17. Gloster TM (2014) Advances in understanding glycosyltransferases from a structural perspective. Curr Opin Struct Biol 28:131–141
18. Gupta P, Goel R, Pathak S, Srivastava A, Singh SP, Sangwan RS, Asif MH, Trivedi PK (2013) De novo assembly, functional annotation and comparative analysis of *Withania somnifera* leaf and root transcriptomes to identify putative genes involved in the withanolides biosynthesis. PLoS One 8:e62714
19. Hashimoto K, Madej T, Bryant SH, Panchenko AR (2010) Functional states of homooligomers: insights from the evolution of glycosyltransferases. J Mol Biol 399:196–206
20. Hou B, Lim EK, Higgins GS, Bowles DJ (2004) *N*-glucosylation of cytokinins by glycosyltransferases of *Arabidopsis thaliana*. J Biol Chem 279:47822–47832
21. Hu Y, Chen L, Ha S, Gross B, Falcone B, Walker D, Mokhtarzadeh M, Walker S (2003) Crystal structure of the MurG:UDP-GlcNAc complex reveals common structural principles of a superfamily of glycosyltransferases. Proc Natl Acad Sci U S A 100:845–849
22. Islam S, Griffiths CA, Blomstedt CK, Le TN, Gaff DF, Hamill JD, Neale AD (2013) Increased biomass, seed yield and stress tolerance is conferred in *Arabidopsis* by a novel enzyme from the resurrection grass *Sporobolus stapfianus* that glycosylates the strigolactone analogue GR24. PLoS One 8:e80035
23. Jackson RG, Lim EK, Li Y, Kowalczyk M, Sandberg G, Hoggett J, Ashford DA, Bowles DJ (2001) Identification and biochemical characterization of an *Arabidopsis* indole-3-acetic acid glucosyltransferase. J Biol Chem 276:4350–4356
24. Jadhav SJ, Sharma RP, Salunkhe DK (1981) Naturally occurring toxic alkaloids in foods. Crit Rev Toxicol 9:21–104
25. Kohara A, Nakajima C, Hashimoto K, Ikenaga T, Tanaka H, Shoyama Y, Yoshida S, Muranaka T (2005) A novel glucosyltransferase involved in steroid saponin biosynthesis in *Solanum aculeatissimum*. Plant Mol Biol 57:225–239
26. Lairson LL, Henrissat B, Davies GJ, Withers SG (2008) Glycosyltransferases: structures, functions, and mechanisms. Annu Rev Biochem 77:521–555
27. Lim EK, Doucet CJ, Li Y, Elias L, Worrall D, Spencer SP, Ross J, Bowles DJ (2002) The activity of *Arabidopsis* glycosyltransferases toward salicylic acid, 4-hydroxybenzoic acid, and other benzoates. J Biol Chem 277:586–592
28. Liu J, Mushegian A (2003) Three monophyletic superfamilies account for the majority of the known glycosyltransferases. Protein Sci 12:1418–1431
29. Madina BR, Sharma LK, Chaturvedi P, Sangwan RS, Tuli R (2007a) Purification and physicokinetic characterization of 3beta-hydroxy specific sterol glucosyltransferase from *Withania somnifera* (L) and its stress response. Biochim Biophys Acta 1774:392–402
30. Madina BR, Sharma LK, Chaturvedi P, Sangwan RS, Tuli R (2007b) Purification and characterization of a novel glucosyltransferase specific to 27beta-hydroxy steroidal lactones from *Withania somnifera* and its role in stress responses. Biochim Biophys Acta 1774:1199–1207

31. Mishra MK, Singh G, Tiwari S, Singh R, Kumari N, Misra P (2015) Characterization of *Arabidopsis* sterol glycosyltransferase *TTG15/UGT80B1* role during freeze and heat stress. Plant Signal Behav 10:e1075682
32. Mishra MK, Chaturvedi P, Singh R, Singh G, Sharma LK, Pandey V, Kumari N, Misra P (2013) Overexpression of *WsSGTL1* gene of *Withania somnifera* enhances salt tolerance, heat tolerance and cold acclimation ability in transgenic *Arabidopsis* plants. PLoS One 8:e63064
33. Moehs CP, Allen PV, Friedman M, Belknap WR (1997) Cloning and expression of solanidine UDP-glucose glucosyltransferase from potato. Plant J 11:227–236
34. Nieman DC (1998) Influence of carbohydrate on the immune response to intensive, prolonged exercise. Exerc Immunol Rev 4:64–76
35. Offen W, Martinez-Fleites C, Yang M, Kiat-Lim E, Davis BG, Tarling CA, Ford CM, Bowles DJ, Davies GJ (2006) Structure of a flavonoid glucosyltransferase reveals the basis for plant natural product modification. EMBO J 25:1396–1405
36. Osbourn AE (1996) Preformed antimicrobial compounds and plant defense against fungal attack. Plant Cell 8:1821–1831
37. Osmani SA, Bak S, Imberty A, Olsen CE, Moller BL (2008) Catalytic key amino acids and UDP-sugar donor specificity of a plant glucuronosyltransferase, *UGT94B1*: molecular modeling substantiated by site-specific mutagenesis and biochemical analyses. Plant Physiol 148:1295–1308
38. Pandey V, Misra P, Chaturvedi P, Mishra MK, Trivedi PK, Tuli R (2010) *Agrobacterium tumefaciens*-mediated transformation of *Withania somnifera* (L.) Dunal: an important medicinal plant. Plant Cell Rep 29:133–141
39. Pandey V, Srivastava R, Akhtar N, Mishra J, Mishra P, Verma P (2015) Expression of *Withania somnifera* steroidal glucosyltransferase gene enhances withanolide content in hairy roots. Plant Mol Bio Rep 34:1–9
40. Pandey V, Niranjan A, Atri N, Chandrashekhar K, Mishra MK, Trivedi PK, Misra P (2014) *WsSGTL1* gene from *Withania somnifera*, modulates glycosylation profile, antioxidant system and confers biotic and salt stress tolerance in transgenic tobacco. Planta 239:1217–1231
41. Pandey V, Dhar YV, Gupta P, Bag SK, Atri N, Asif MH, Trivedi PK, Misra P (2015) Comparative interactions of withanolides and sterols with two members of sterol glycosyltransferases from *Withania somnifera*. BMC Bioinformatics 16:120
42. Paquette SM, Jensen K, Bak S (2009) A web-based resource for the *Arabidopsis* P450, cytochromes b5, NADPH-cytochrome P450 reductases, and family 1 glycosyltransferases. Phytochemistry 70:1940–1947
43. Poppenberger B, Berthiller F, Lucyshyn D, Sieberer T, Schuhmacher R, Krska R, Kuchler K, Glossl J, Luschnig C, Adam G (2003) Detoxification of the *Fusarium* mycotoxin deoxynivalenol by a UDP-glucosyltransferase from *Arabidopsis thaliana*. J Biol Chem 278:47905–47914
44. Poppenberger B, Berthiller F, Lucyshyn D, Sieberer T, Schuhmacher R, Krska R, Kuchler K, Glossl J, Luschnig C, Adam G (2005) The *UGT73C5* of *Arabidopsis thaliana* glucosylates brassinosteroids. Proc Natl Acad Sci U S A 102:15253–15258
45. Potocka A, Zimowski J (2008) Metabolism of conjugated sterols in eggplant. Part 1. UDP-glucose: sterol glucosyltransferase. Acta Biochim Pol 55:127–134
46. Ritsema T, Smeekens SC (2003) Engineering fructan metabolism in plants. J Plant Physiol 160:811–820
47. Saema S, Rahman LU, Niranjan A, Ahmad IZ, Misra P (2015) RNAi-mediated gene silencing of *WsSGTL1* in *W. somnifera* affects growth and glycosylation pattern. Plant Signal Behav 10:e1078064
48. Saema S, Rahman LU, Singh R, Niranjan A, Ahmad IZ, Misra P (2016) Ectopic overexpression of *WsSGTL1*, a sterol glucosyltransferase gene in *Withania somnifera*, promotes growth, enhances glycowithanolide and provides tolerance to abiotic and biotic stresses. Plant Cell Rep 35:195–211

49. Schumacher K, Chory J (2000) Brassinosteroid signal transduction: still casting the actors. Curr Opin Plant Biol 3:79–84
50. Sharma LK, Madina BR, Chaturvedi P, Sangwan RS, Tuli R (2007) Molecular cloning and characterization of one member of 3beta-hydroxy sterol glucosyltransferase gene family in *Withania somnifera*. Arch Biochem Biophys 460:48–55
51. Singh G, Tiwari M, Singh SP, Singh S, Trivedi PK, Misra P (2016) Silencing of sterol glycosyltransferases modulates the withanolide biosynthesis and leads to compromised basal immunity of *Withania somnifera*. Sci Rep 6:25562
52. Steponkus PL, Uemura M, Joseph RA, Gilmour SJ, Thomashow MF (1998) Mode of action of the *COR15a* gene on the freezing tolerance of *Arabidopsis thaliana*. Proc Natl Acad Sci U S A 95:14570–14575
53. Thorsoe KS, Bak S, Olsen CE, Imberty A, Breton C, Lindberg Moller B (2005) Determination of catalytic key amino acids and UDP sugar donor specificity of the cyanohydrin glycosyltransferase *UGT85B1* from *Sorghum bicolor*. Molecular modeling substantiated by site-specific mutagenesis and biochemical analyses. Plant Physiol 139:664–673
54. Trouvelot S, Heloir MC, Poinssot B, Gauthier A, Paris F, Guillier C, Combier M, Trda L, Daire X, Adrian M (2014) Carbohydrates in plant immunity and plant protection: roles and potential application as foliar sprays. Front Plant Sci 5:592
55. Ullmann P, Bouvier-Nave P, Benveniste P (1987) Regulation by phospholipids and kinetic studies of plant membrane-bound UDP-glucose sterol beta-D-glucosyl transferase. Plant Physiol 85:51–55
56. Ullmann P, Ury A, Rimmele D, Benveniste P, Bouvier-Nave P (1993) UDP-glucose sterol beta-D-glucosyltransferase, a plasma membrane-bound enzyme of plants: enzymatic properties and lipid dependence. Biochimie 75:713–723
57. Wang X (2009) Structure, mechanism and engineering of plant natural product glycosyltransferases. FEBS Lett 583:3303–3309
58. Warnecke D, Erdmann R, Fahl A, Hube B, Muller F, Zank T, Zahringer U, Heinz E (1999) Cloning and functional expression of UGT genes encoding sterol glucosyltransferases from *Saccharomyces cerevisiae*, *Candida albicans, Pichia pastoris*, and *Dictyostelium discoideum*. J Biol Chem 274:13048–13059
59. Warnecke DC, Heinz E (1994) Purification of a membrane-bound UDP-glucose:sterol [beta]-D-glucosyltransferase based on its solubility in diethyl Ether. Plant Physiol 105:1067–1073

Phytochemicals of Nematode-Resistant Transgenic Plants

22

Phatu W. Mashela, Ashwell R. Ndhlala, Kgabo M. Pofu, and Zakheleni P. Dube

Abstract

In plant-nematode relations both plants and nematodes release biochemical compounds from the expressions of genes, which had been technically referred to as plant genes and gene products, respectively. Plant genes and gene products can have either a downregulation or an upregulation relationship, which might or might not result in successful plant-nematode partnerships. The uses of molecular technologies had enhanced the application of these biochemical compounds in transgenic plants for the management of nematodes. The chapter overviewed plant genes and gene products, along with their practical application in nematode management using nematode-resistant transgenic plants.

Keywords

Cyst nematode • Root-knot nematode • Gene products • Plant genes • Migratory phases • Sedentary phase • *Meloidogyne* spp. • *Heterodera* spp. • *Globodera* spp.

P.W. Mashela (✉) • Z.P. Dube (✉)
Green Technologies Research Centre, University of Limpopo, Sovenga, South Africa
e-mail: phatu.mashela@ul.ac.za; zakheleni_dube@yahoo.com

A.R. Ndhlala (✉)
Division Crop Science, Agricultural Research Council, Pretoria, South Africa

Green Technologies Research Centre, University of Limpopo, Sovenga, South Africa
e-mail: NdhlalaA@arc.agric.za; ashwelln@yahoo.com

K.M. Pofu (✉)
Green Technologies Research Centre, University of Limpopo, Sovenga, South Africa

Division Crop Protection, Agricultural Research Council, Pretoria, South Africa
e-mail: MarthaP@arc.agric.za

S. Jha (ed.), *Transgenesis and Secondary Metabolism*, Reference Series in Phytochemistry, DOI 10.1007/978-3-319-28669-3_26

Contents

1 Introduction

In plant-nematode interactions, nematodes attack plants through physical and biochemical strategies, whereas the plants mobilize defense responses which comprise both physical and biochemical strategies [1]. Generally, attacks from high nematode population densities overpower the plant defense responses, necessitating the need for human intervention to reduce the nematode numbers. The most effective nematode management strategy comprised the use of fumigant chemical nematicides, which had since 2005 been withdrawn from the agrochemical markets due to their being environment-unfriendly [2]. Three years prior to the cutoff withdrawal date, estimated global annual crop yield losses due to nematodes were at US$126 billion [3]. Three and eight years after the cutoff withdrawal date, the estimated yield losses were at US$157 [4] and US$173 [5] billions, respectively. These translated into the relative increase in yield losses of 25% and 37%, respectively.

Broadly, the damage nematodes induce depends on whether they are at the migratory or sedentary phases, along with the related strategies that they deploy. The nematode attacking strategies evolved in a manner that, in most cases, the plant defense responses are completely avoided. In exo-parasitic nematodes, for instance, nematodes spend all their life spans in soil solutions. The sting, (*Belonolaimus* sp.), the needle (*Longidorus* spp.), and the ring (*Criconemella* spp.) nematodes have long stylets that penetrate the cortical cells at specific feeding sites while the nematode bodies are out of reach to the plant defense responses. In contrast, those with shorter and weaker stylets like the pin (*Paratylenchus* spp.) nematodes feed on epidermal cells. In all cases, the exo-parasitic nematodes pierce the root surface and insert stylets into appropriate cells through physical force and then form the cushion-like structures between the distal part of the stoma and the root surfaces for support. The nematode releases hydrolytic chemical compounds into the cells, followed by the suction force to pull the hydrolyzed materials through the stylet canals, esophagi, and then into the intestines. After the feeding period, which ranges from a few hours to several days, the stylets are withdrawn, thereby avoiding the plant defense responses.

The endo-parasitic mobile nematodes spend most of their life spans inside the plant tissues, with various evolved strategies to counter the plant defense strategies. The nematodes restrict their movements to the apoplastic pathways [6, 7], from where they pierce the cell walls with their shorter but thicker and stronger stylets to suck up the cell contents, thereafter pull out and run away [7]. However, in apoplastic pathways there are extensive physical and biochemical barriers to nematode movements, which are reviewed later under sedentary nematodes. Additionally, mobile endo-parasitic nematodes feed on nonlignified cells at the nondifferentiated zones of the root systems. For one reason or the other, most uses of nematode resistance have focused much on sedentary nematodes [7].

Sedentary plant-parasitic nematodes like the cyst (*Globodera* spp., *Heterodera* spp.) and the root-knot (*Meloidogyne* spp.) nematodes are widely managed through nematode resistance [7]. Nematode resistance has a number of distinct advantages over other alternatives, which include: (a) complete prevention of nematode feeding, development, and reproduction, (b) does not require special application techniques and equipment, (c) cost-effective, (d) environment-friendly, and (e) sustainable [8]. However, high soil temperature [9], cyclic salinity [10], honeydew-inducing insects [11], and nematode races [12, 13] were shown to have deleterious effects on the efficacy of nematode resistance. Most of these drawbacks were attributed to imbalances of the osmoticum (Na, Cl, K) ions and osmotic organic compounds in nematode-infected plants [10, 14]. Increased efforts to manipulate nematode resistance through conventional plant breeding have had limited success due to the unavailability of nematode-resistant genotypes in certain cultigens, including their wild relatives. Alternatively, when nematode resistance was present in the wild relatives, uses of conventional breeding methods to integrate the relevant genes in nematode-susceptible cultigens failed due to various incompatibilities. For example, all commercial cultigens in the genus *Cucumis* are highly susceptible to root-knot nematodes, whereas their wild relatives, notably *Cucumis myriocarpus* and *Cucumis africanus*, are highly resistant [15]. However, the pollen grains of the *Cucumis* cultigens and their wild relatives are incompatible.

Molecular technological advances to explore gene expression in plant-nematode interactions had since provided unique information on biochemical compounds associated with the interactions. Such advances range from comparative protein profiling, differential screening of the cDNA library, and cDNA subtraction differential display of transcripts and promoter-β-glucuronidase (*GUS*) fusions, along with mRNA hybridization to reverse-transcription polymerase chain reaction of the selected genes [1, 7, 16]. The derived information distinguished among three types of biochemical compounds: (a) gene products (parasitones) – released by nematodes during their parasitic life cycles, (b) plant genes – released by plants to collaborate or to counter the gene products, and (c) secondary chemicals – released by cells around the wounded cells to prevent the spread of chemical damage to their own structures [7]. The second two groups of biochemical compounds constitute defense chemical compounds, which can be rapid and strong or late and weak [7]. The identification of the chemical compounds associated with rapid and strong plant defense responses through advanced technological approaches at the genetic level had since enabled the use of such information in the development of nematode-resistant transgenic plants. In this chapter we reviewed biochemical compounds associated with the migratory and sedentary phases of cyst and root-knot nematodes, with the view of singling out phytochemicals in nematode-resistant transgenic plants.

2 Gene Products in Sedentary Nematodes

Gene products are biochemical compounds (RNAs or proteins) resulting from expression of genes [16, 17]. A gene is a hereditary unit of DNA which is required to produce a functional product [17]. In plant-parasitic nematodes the subventral and dorsal gland cells sequentially produce different gene products (effectors) during migratory and sedentary phases, respectively. Cyst and root-knot nematodes were overviewed separately to highlight the different strategies the nematode species employ in plant-nematode interactions.

2.1 Cyst Nematodes

The cyst nematodes infect cool-season crops such as potato (*Solanum tuberosum*), soybean (*Glycine max*), and sugar beets (*Beta vulgaris*) which excel in regions with cold winters. The primary feature of female cyst nematodes is their ability to retain second-stage juveniles (J2) in their bodies and enter cryptobiosis for overwintering, with the bodies forming the protective cysts. Cyst nematodes are highly damaging to host crops [5], with yield reduction in sugar beets ranging from as high as 50% to complete crop failure [18]. Gene products associated with the migratory and sedentary phases of cyst nematodes are briefly overviewed.

2.1.1 Migratory Phase

During the migration phase, J2 crisscross the apoplastic and symplastic pathways, which form daunting barriers during movement to the feeding sites [7]. After penetration, cyst J2 migrate using the shortest distance through the apoplastic pathways using stylets and biochemical compounds to force their way to the feeding sites outside the vascular bundle [19–21]. The apoplastic pathways are cemented by complex barriers produced by the cell walls [22], which include celluloses, β-1,4-linked glucans, hemicelluloses (xyloglucan, glucomannan, xylan, and mixed-linkage glucans), and lignins [7, 23, 24]. These barriers are entrenched in an assortment of polysaccharide matrix of pectin, which comprises homogalacturonan and rhamnogalacturonan I and II [25]. Prior to the arrival at the feeding sites, cyst J2 have to traverse through the pericycle and the endodermis layers, which constitute the symplastic pathway [22]. Consequently, cyst J2 must contend with both apoplastic and symplastic pathways and physical and biochemical barriers, respectively, prior to their arrival at their feeding destinations. A large number of activated defense genes had been reported when cyst J2 interfere with structures in the apoplastic and the symplastic pathways, which include peroxidase, chitinase, lipoxygenase, extension and proteinase inhibitors [7]. Incidentally, cuticles around nematode bodies also contain lipids and chitins, which are sensitive to some of the listed activated plant genes. The upregulation of the plant genes due to recognizing the nematode molecular patterns are not limited to induced plant defense responses, but also induce secondary pathways to avoid damage spilling over all the cells by channeling some secondary biochemical compounds to the phytoalexin biosynthesis complex [26], along with the unusual biosynthesis of callose and lignin as physical barriers to seal off the damaged pathway [27]. In plants with rapid and strong defense responses, phytoalexins were previously viewed as the sole biochemical compounds responsible for nematode resistance [28]. These biochemicals result in necrosis in nematode-infected cells, which was previously viewed as being symbolic of hyperactive responses in nematode-resistant plants.

During the migratory phases, the J2 subventral gland cells are hyperactive [29, 30] and release copious gene products. The gene products, technically referred to as parasitones, facilitate the degradation of the cell wall networks [21, 30]. Some of the gene products used in degradation of the cell wall matrix by cyst nematodes include β-1,4 endoglucanase and pectate lyase [31–40]. The roles of these gene products in cell wall degradation were verified under in vitro trials through downregulation of targeted gene products, which resulted in cyst J2 failing to successfully migrate through the cortex to the vascular bundle [41].

Another gene product, the expansin from cysts J2 subventral gland cells of *Heterodera* species, was observed during the migratory phases [24, 41, 42]. Naturally, expansins do not have enzymatic activities, but have capabilities to relax noncovalent bonds in cellulose microfibrils and hemicelluloses [42], thereby reducing resistance against J2 migration to the feeding sites. Expansins also enhance passage of other gene products from the cyst J2 stylets into the polysaccharide matrix, thereby conferring synergistic effects that expand disruption of the physical barriers for easy passage of J2 to the feeding sites [43]. Once at the feeding sites, the subventral gland cells of the cyst J2 are atrophied, whereas the dorsal gland cells become hyperactive [42].

As the cyst J2 crisscross the apoplastic and symplastic pathways, the plant defense genes are activated to produce the plant genes in response to the inflicted physical and biochemical damages. The plant genes defense responses are elicited soon at the onset of biotic interactions [44, 45]. However, the outcome of the interactions depends on the rate of expression of the plant genes/gene products to execute such responses [7, 46, 47]. For instance, when the plant genes responses are too late and/or too weak, compatible interactions occur, whereas nematode resistance occurs when the responses are rapid and strong [1, 7].

2.1.2 Sedentary Phase

Upon arrival at the feeding sites, the cyst J2 establish the feeding sites parallel to the adjacent vascular bundle outside but parallel to the vascular bundle. The cell walls of a considerable number ($\geq$100 of cells) are dissolved and the protoplasts fused together to form elongated syncytium feeding sites [42]. Gene products secreted from the dorsal gland cells of sedentary cyst nematodes include chorismate mutase for altering auxin balance [33, 39], thioredoxin peroxidase for feeding cell formation [33, 34, 39], venom allergen-like protein for promoting early parasitism [46], and cytokinins for cell division [48–53]. Siddique et al. [49] demonstrated that cytokinin signaling is activated in both the syncytium cells and the neighboring cells and cytokinin-deficient conditions resulted in high susceptibility to nematode. As in the migratory phase, the roles of some of the gene products involved in the sedentary phase have been verified in situ and in vitro through downregulation techniques [7]. Gheysen and Fenoll [7] provided an extensive list of the plant genes upregulated at the nematode feeding sites of sedentary cyst nematodes, which include wound and defense responses and its functions along with analytical methods and assay plants response to cyst nematode species with appropriate references. In sedentary phases, for interactions to be compatible, most of the gene products from the dorsal gland cells of nematodes mimic plant genes which have essential roles in the physiology of the plant [34, 54]. The cytokinins [49–53] and auxins [33, 34, 39], for instance, produced by the dorsal gland cells in sedentary adult nematodes, have previously been known to be exclusively released by plants [22].

2.2 Root-Knot Nematodes

The genus *Meloidogyne* contains the largest number of nematode species which consists of over 63 species in the genus [55]. The distribution of *Meloidogyne* species spans in the major global zones where plants grow, with the widest hosts being in the tropical/subtropical and the temperate regions. Additionally, the genus has the widest plant hosts [56] and the largest number of biological races [57] that are morphologically identical nematode species but could be separated using differential hosts and molecular approaches.

2.2.1 Migratory Phase

After penetration of the root surface at the elongation zone [58], root-knot J2 move downward through the apoplastic pathways of the cortical to the meristematic cells at the root tips. The root-knot J2 then moves upward to enter the distal region of the vascular bundle and continues to move upward within the cylinder [7, 58, 59]. As described in the cyst J2, during migration from the penetration to the feeding sites, activities of roots are subjected to physical and chemical changes due to the stylets and gene products from the hyperactive subventral gland cells, respectively. Some of the gene products secreted and verified from the subventral gland cells of migratory phases of root-knot J2 include β-1,4 endoglucanase [31–34, 38–40], pectate lyase [34, 36, 37], and polygalacturonase [7], all for degradation of cell walls as described in cyst J2. Incidentally, the root-knot J2 gene products are more or less identical to those in cyst J2.

2.2.2 Sedentary Phase

Upon arrival of the root-knot J2 at the differentiation zones of the root system within the vascular bundles, the J2 insert the stylets through the cell wall of a selected cell, but may or may not be piercing through the plasmalemma [59, 60]. Secretions from the dorsal gland cells form a tube-like structure called the feeding tube, which goes within the cytoplasm of the feeding cell [60–63]. The gene products can be deposited either outside the plasmalemma of the feeding cell and interact with a membrane receptor or are injected into the cytoplasm of the feeding cell through a perforation in the plasmalemma at the stylet orifice to eventually interact with the multinucleate organelles [60]. The feeding cell and those around it respond by undergoing repeated mitosis without cytokinesis to create a multinucleate structure called the giant cell [62, 63]. The giant cell has a dense and metabolically hyperactive cytoplasm, with unusually thicker cell wall [65]. As the root-knot J2 develop through J3 and J4 to pear-shaped adult females, the adjacent root cells bulge to form a characteristic root gall.

During feeding the female lip regions rest on an amphidal secretion against the cell wall of the feeding cell [60]. The gene products observed in sedentary cyst nematodes, namely, the chorismate mutase, venom allergen-like protein, and cytokinin [64–66], it along with the gene product the calreticulin for enhancing early parasitism [64], were identified and verified from the dorsal gland cells of sedentary root-knot nematodes. All the gene products play indispensable roles in silencing the defense genes at the feeding sites [67]. Gheysen and Fenoll [7] also provided an extensive tabulated list of the plant genes upregulated at the nematode feeding sites of sedentary root-knot nematodes, which had both similarities and differences to those observed in sedentary cyst nematodes.

3 Gene Product and Plant Gene Interactions

Infection by cyst and root-knot nematodes invariably results in complex nematode-plant interactions. The feeding sites established by cyst and root-knot nematodes species are referred to as syncytium and giant cells, respectively. Infection by cyst nematodes, for example, cannot unduly result in the formation of giant cells and vice versa. The gene products secreted by J2 of the two nematode species during migration phases are more or less similar [68]. However, in sedentary adult stages, distinctly different gene products are released. For instance, thioredoxin peroxidase and calreticulin were exclusively observed in sedentary adult stages of *G. rostochiensis* [69] and *M. incognita* [64], respectively.

Nematode infections with subsequent releases of gene products initiate multiple changes in roots, triggering gene expressions that release plant genes [7, 66, 70, 71]. Subsequent to nematode infections and releases of various gene products by nematodes, a number of plant genes are either upregulated or downregulated [72]. The upregulated plant genes improve the compatibility of plant-nematode interactions that enhance the establishment of a suitable feeding site [67, 68, 73]. Puthoff et al. [74] carried out a GeneChip profiling of transcriptional responses to soybean cyst nematode colonization of soybean roots and observed that almost twice of the transcripts were upregulated by infection, whereas the downregulated set were smaller in number. In other similar studies [67–69], J2 of *Globodera*, *Heterodera*, and *Meloidogyne* species upregulated β-1,4 endoglucanase, pectate lyase, and/or polygalacturonase-like plant genes that encode for degradation of cell walls. In contrast, sedentary nematode stages of the three nematode species showed that upregulation of chorismate mutase, thioredoxin peroxidase, venom allergen-like proteins, calreticulin, pectin acetylesterase, and disaccharide carrier *At*SUC2 plant genes encoded for the successful establishment and maintenance of the feeding sites in respective nematode species [67–69].

An intriguing observation in the two nematode species was during the mobile or sedentary phases; where the gene products involved are chemical compounds that are historically produced by plants [33, 39, 49–53]. The sedentary nematode phases in both nematode species have developed advanced gene products that are released into the feeding sites, which trigger the plant to collaborate with nematode feeding without being detected as foreign entities by the defense responses. The feeding sites of the two sedentary adult nematodes comprise large, metabolically active and multinucleate cells, with numerous gene products and plant genes [75–81], which are distinctly similar and different. The latter could be ascribed to differences in the gene products produced by dorsal gland cells of cyst and root-knot nematodes. Due to the large number of gene products and plant genes involved in the formation of syncytium and giant cells in roots [75–81], it is increasingly difficult to pinpoint the champions. However, molecular approaches through upregulation and downregulation have shown that the initiation and maintenance of syncytium and giant cells involve an assortment of gene products and plant genes [7, 75–82]. The role of various chemical compounds from nematodes and plants in collaborative partnerships between nematodes and their feeding sites are increasingly being documented [82–84].

Plants are defined as being resistant to nematodes when they reduce the reproduction potential of nematodes [85, 86]. In plant-parasitic nematodes feeding is a prerequisite for development from J2 to the reproductive adult stages. Therefore, when syncytium and giant cells fail to develop, this could be viewed as nematode resistance. In such cases, the nematode gene products are being silenced by the plant genes. Alternatively, when the nematode gene products silence the plant genes, the plant could be viewed as being susceptible to nematodes. Plant genes with the capabilities to silence the gene products are collectively called the resistance (R) genes [16, 87, 88]. Pathogen recognition by the host plant is mediated by single R genes in the host plant and single genes in the pathogen referred to as avirulence (*Avr*) genes, resulting in a cascade of defense responses that silence the gene products [16, 88]. McDowell and Woffenden [16] provided an extensive review of R genes, their conserved motifs, durability, increasing the breadth of resistance, and potential prospects. Since then, some of the potential prospects have been successfully applied in the development of nematode-resistant strategies using information derived from the upregulation and downregulation concepts of the gene products and plant genes [89–91].

4 Biochemicals for Nematode Resistance in Transgenic Plants

Escobar and Fenoll [1] and others [17, 68, 92] provided a comprehensive review of antinematode genes, which are unique to nematode-resistant transgenic plants. Generally, these biochemicals are poorly expressed in nematode-susceptible plants. The R genes had since been transferred as transgenes through molecular approaches into many commercial cultigens that conventionally had no resistance genotypes to plant-parasitic nematodes [85] to develop/establish nematode-resistant transgenic plants. Incidentally, all the plant genes reviewed in the foregoing sections are candidates for use through molecular approaches in the development of nematode-resistant transgenic plants [7, 92].

4.1 Antigene Products Strategy

The nematode gene products are secreted sequentially by the subventral and dorsal gland cells through the nematode stylets during the parasitic life cycle of plant-parasitic nematodes [7, 93, 94]. Generally, as shown with the cytokinins in *H. schachtii* [60, 95], gene products involved in signaling serve multiple roles, including activation and maintenance of the nematode feeding site, nematode development, and mediation of the cell cycle activation that leads to the formation of specialized feeding sites [58, 60, 95].

The migration phases are literally the wounding phases, with the resultant activation of the plant defense genes that include peroxidase, chitinase, lipoxygenase, extension and proteinase inhibitors [7, 67–69]. However, the wounding phases are not exclusive to the migration phases since the expansion of cells and increase in female sizes during sedentary phases also induce cell wounding. The released nematode gene products are upregulated [7, 67–69] in order to induce the plant to produce appropriate plant genes that enhance migration and/or sedentary processes [7, 58, 94, 95]. In nematode-plant compatible interaction, both the nematode and the plant would produce the chemical compounds that would enhance partnerships in plant-nematode interactions [7, 58, 67–69]. In situations where the plant genes downregulate the nematode gene products, the interactions become hostile [7, 58, 67–69], with the nematode failing to feed, develop, and reproduce. Transgenic plants contain the desired R genes that downregulate the nematode gene products to stop the successful performance of the gene products, thereby establishing compatible relations [67–69, 88]. The differential expression of antinematode genes results in silencing of the nematode gene products in the transgenic plants during both migratory and sedentary phases.

4.2 Antiplant Gene Strategy

In this strategy, the host plant genes that help the nematode feeding and secretions, to allow for successful partnerships between nematode and plants, are silenced. In other words, phytotoxic chemical compounds that destroy the feeding structures, namely, syncytium and giant cells, are upregulated. Additionally, certain biochemical compounds that are released by plants to protect the nematode, such as chemicals could be suppressed, thereby leaving the bodies of nematodes unprotected [67–69]. The failure to develop and maintain the feeding structures arrest nematode development. The antiplant gene strategy in transgenic plants is being successfully used in cultigens.

4.3 RNA-Interference Strategy

RNA interference (RNAi) disrupts the nematode gene products through host-induced gene silencing (HIGS) approach [60, 67–69]. The RNAi genes have precise selectivity for the target organisms with minimal off-target effects [90]. Cathepsin L-like cysteine proteinases, produced in nematode-resistant transgenic plants, were shown to be an attractive group of candidate genes for RNAi-induced downregulation due to their high level of specificity to the target nematode gene products [90], thereby resulting in silencing of host-induced nematode gene products. Host-produced RNAi of Mi-cpl-1 gene confers resistance to *M. incognita* by inducing negative effects on nematode infection, development, and reproduction [90]. The RNA-interference strategy in transgenic plants is being successfully used in most cultigens [60, 96, 97]. The target nematode gene products had been identified for *Heterodera* spp., *Meloidogyne* spp., *Pratylenchus vulnus*, and *Globodera pallida* [30]. The observed phenotypic responses in nematodes ranged from interference with development of J2, reduced reproduction, and failure of formation of the feeding sites [30, 97].

5 Challenges and Future Possibilities

The identified three biochemical compounds in plant-nematode interactions in transgenic plants, namely, antigene products, antiplant gene, and RNAi, have emerged as a powerful technology for crop protection against nematodes in the twenty-first century. However, although most of the antigenes and RNAi in plants are highly sequence-specific, there are possibilities it might have detrimental effects on nontarget organisms [30]. Additionally, it had been shown that certain antiplant genes are highly sensitive to extremes in environmental factors such as high temperature [9], salinity [14], and attack by honeydew-sucking insects [11]. The future possibilities should focus on improving the resilience of the identified plant genes under diverse environments as well as preventing nematode penetration at the elongation zone [58].

Most of the plant genes required to confer nematode resistance in transgenic plants had been focusing exclusively on endo-parasitic nematodes. The feeding habits of exo-parasitic nematodes with long stylets make them not the likely candidates for being managed using the plant gene technologies. Identifying plant genes that can respond to stylet penetration by inducing the affected cells to release chemicals that could glue the stylets, thereby resulting in the nematodes losing their stylets when pulling off after feeding, could be one of the future possibilities.

6 Conclusions

During the migratory phases, gene products and plant genes in cyst and root-knot J2 are similar, with distinct differences during the sedentary phases. Interestingly, in all phases, regardless of the nematode species, the gene products from either the subventral or the dorsal gland cells were identical to biochemical compounds which had previously been observed in plants. The similarities in the biochemical compounds trigger the plant to collaborate with nematode feeding without being detected by the defense responses.

Acknowledgements We are grateful for financial support from the Land Bank Chair of Agriculture – University of Limpopo and the Agricultural Research Council-Universities Collaboration Centre.

References

1. Escobar C, Fenoll C (2015) Plant nematode interactions: a view on compatible interrelationships, vol 73, Advances in botanical research. Elsevier, New York. doi:10.3389/fpls.2016.00124
2. Mashela PW, Dube ZP, Pofu KM (2015) Managing the phytotoxicity and inconsistent nematode suppression in soil amended with phytonematicides. In: Meghvansi MK, Varma A (eds) Organic amendments and soil suppressiveness in slant disease management, 1st edn. Soil biology 46. Springer, Heidelberg, Switzerland, pp 147–173. doi: 10.1007/978-3-319-23075-7
3. Chitwood DJ (2003) Research on plant-parasitic nematode biology conducted by the United States Department of Agricultural Research Services. Pest Manag Sci 59:748–753. doi:10.1002/ps.684
4. Abad P, Gouz J, Aury JM, Castagnone-Sereno P, Danchin EG et al (2008) Genome sequence of the metazoan plant-parasitic nematode *Meloidogyne incognita*. Nat Biotechnol 26:909–915. doi:10.1038/nbt.1482
5. Elling AA (2013) Major emerging problems with minor *Meloidogyne* species. Phytopathology 103:1092–1102. doi:10.1094/PHYTO-01-13-0019-RVW
6. Wyss U (1997) Root parasitic nematodes: an overview. In: Fenoll C, Grundler FMW, Ohl SA (eds) Cellular and molecular aspects of plant-nematode interactions, vol 10 of the series developments in plant pathology. Springer, pp 5–22
7. Gheysen G, Fenoll C (2002) Gene expression in nematode feeding sites. Annu Rev Phytopathol 40:191–219. doi:10.1146/annurev.phyto.40.121201.093719
8. Stirling GR (2014) Biological control of plant-parasitic nematodes, 2nd edn. Biological Crop Protection Pty Ltd, Brisbane
9. Dropkin VH (1969) Cellular responses of plants to nematode infections. Ann Rev Phytopathol 7:101–122. doi:10.1146/annurev.py.07.090169.000533
10. Mashela PW, Duncan LW, McSorley R (1992) Salinity reduces resistance to *Tylenchulus semipenetrans* in citrus rootstocks. Nematropica. Springer, Heidelberg, Switzerland, 22:7–12

11. Pofu KM, Mashela PW, Mafeo TP (2013) Interaction of greenhouse whitefly (*Trialeurodes vaporariorum*) and root-knot nematode (*Meloidogyne javanica*) on nematode-resistance in wild watermelon. Acta Hort (ISHS) 1(1007):431–438, doi:10.17660/ActaHortic.2013.1007.49
12. Eddaoudi M, Ammati M, Rammah H (1997) Identification of resistance breaking populations of *Meloidogyne* on tomato in Morocco and their effect on new sources of resistance. Fundam Appl Nematol 20:285–289
13. Ornat C, Verdejo-Lucas S, Sorribas FJ (2001) A population of *Meloidogyne javanica* from Spain virulent to the *Mi* resistance gene in tomato. Plant Dis Rep 85:271–276. doi:10.1007/s10658-008-9413-z
14. Mashela PW, Nthangeni ME (2002) Osmolyte allocation in response to *Tylenchulus semipenetrans* infection, root pruning and stem girdling. J Nematol 34:273–277
15. Pofu KM, Mashela PW, Shimelis H (2012) Intergeneric grafting in watermelon for managing *Meloidogyne* species: a review. Sci Res Essay 7:107–113. doi:10.5897/SREX11.023
16. Williamson VM (1999) Plant nematode resistance genes. Curr Opin Plant Biol 2:327–331. doi:10.1016/S1369-5266(99)80057-0
17. Nussbaum R, McInnes RR, Willard H (2016) Thompson & Thompson genetics in medicine, 8th edn. Elsevier, Philadelphia
18. Mashela PW (2007) Undefeatable enemies: answering questions with questions. Inaugural Lecture, University of Limpopo Press, Sovenga
19. Wyss U, Zunke U (1986) Observations on the behavior of second stage juveniles of *Heterodera schachtii* inside host roots. Rev Nématol 9:153–165
20. Wyss U, Grundler FMW (1992) *Heterodera schachtii* and *Arabidopsis thaliana*, a model hos-parasite interaction. Nematologica 38:488–493. doi:10.1163/187529292X00450
21. Smant G, Stokkermans JP, Yan Y, De Boer JM, Baum TJ et al (1998) Endogenous cellulases in animals: isolation of β-1,4-endoglucanase genes from two species of plant-parasitic cyst nematodes. Proc Natl Acad Sci U S A 95:4906–4911
22. Reece JB (2015) Campbell biology, 10th edn. Benjamin Cummings, Redwood City
23. Li X, Chapple C (2010) Understanding lignification: challenges beyond monolignol biosynthesis. Plant Physiol 154:449–452. doi:10.1104/pp.110.162842
24. Scheller HV, Ulvskov P (2010) Hemicelluloses. Annu Rev Plant Biol 61:263–289. doi:10.1146/annurev-arplant-042809-112315
25. Harholt J, Suttangkakul A, Scheller H (2010) Biosynthesis of pectin. Plant Physiol 153:384–395. doi:10.1104/pp.110.156588
26. Kaplan DT, Keen NT, Thomason IJ (1980) Studies on the mode of action of glyceollin in soybean incompatibility to root-knot nematode, *Meloidogyne incognita*. Physiol Plant Pathol 16:319–325. doi:10.1590/S0102-33061987000100002
27. Balhadére P, Evans AAF (1995) Histo-pathogenesis of susceptible and resistant responses of wheat, barley and wild grasses to *Meloidogyne naasi*. Fundam Appl Nematol 18:531–538
28. McIntyre JL (1980) Defenses triggered by previous invaders: nematodes and insects. In: Horsfall JG, Cowling EB (eds) Plant disease, vol V, 1st edn. Academic, New York, pp 333–342
29. Wang X, Meyers D, Yan Y, Baum T, Smant G et al (1999) In planta localization of a β-1,4-endoglucanase secreted by *Heterodera glycines*. Mol Plant Microbe Interact 12:64–67. doi:10.1094/MPMI
30. Dutta TK, Banakar P, Rao U (2015) The status of RNAi-based transgenic research in plant nematology. Front Microbiol 5:1–7. doi:10.3389/fmicb.2014.00760
31. Bird AF, Loveys BR (1980) The involvement of cytokinins in a host-parasite relationship between the tomato (*Lycopersicon esculentum*) and a nematode *Meloidogyne javanica*. Parasitol 80:497–505. doi:10.1017/S0031182000000962
32. Doerner P, Jørgensen JE, You R, Steppuhn J, Lamb C (1996) Control of root growth and development by cyclin expression. Nature 380:520–523
33. Domingo C, Roberts K, Stacey NJ, Connerton I, Ruiz-Teran F, McCann MC (1998) A pectase lyase from *Zinnia elegans* is auxin inducible. Plant J 13:17–28

34. Duncan LH, Robertson WM, Kusel RJ, Phillips MS (1996) A putative nematode auxin binding protein from the potato cyst nematode *Globodera pallida*. Nematropica 26:259
35. Berg RH, Fester T, Taylor CG (2008) Development of the root-knot nematode feeding cell. In: Berg RH, Taylor CG (eds) Plant cell monographs. Springer, Berlin/Heidelberg, pp 115–152
36. Ehsanpou AA, Jones MGK (1996) Glucoronidase expression in transgenic tobacco roots with a Parasponia promotor on infection with *Meloidogyne javanica*. J Nematol 28:407–413
37. Elsasser S, Chi Y, Yang P, Campbell JL (1999) Phosphorylation controls timing of Cdc6p destruction: a biochemical analysis. Mol Biol Cell 10:3263–3277. doi:10.1371/journal.pgen.1003099
38. Hussey RS, Grundler FMW (1998) Nematode parasitism of plants. In: Perry RN (ed) The physiology and biochemistry of free-living and plant-parasitic nematodes. CBI Publication, Wallingford, pp 213–243
39. Huyangura P, Mathesius U, Jones MGK, Rolfe BG (1999) Auxins induction is a trigger for root gall formation caused by root-knot nematodes in white clover and is associated with the activation of the flavonoid pathway. Aust J Plant Physiol 26:221–231
40. Jasmer DP (1993) *Trichinella spiralis* infected skeletal muscles cells arrest in G_2/M and cease muscle gene expression. J Cell Biol 121:785–793
41. Chen Q, Rehman S, Smant G, Jones JT (2005) Functional analysis of pathogenicity proteins of the potato cyst nematode *Globodera rostochiensis* using RNAi. Mol Plant Microbe Interact 18:621–625. doi:10.1094/MPMI
42. Bohlmann H, Sobczak M (2014) The plant cell wall in the feeding sites of cyst nematodes. Front Plant Sci 5:1–10. doi:10.3389/fpls.2014.00089
43. Qin L, Kudla U, Roze EHA, Goverse A, Popeijus H et al (2004) Plant degradation: a nematode expansion acting on plants. Nature 427:30–30. doi:10.1093/jxb/erm209
44. Westwood JH, Yu X, Foy CL, Cramer CL (1998) Expression of defense-related 3-hydroxy-3-methylglutaryl CoA reductase gene in response to parasitisation of *Orobanche* spp. Mol Plant Biol 2:327–331. doi:10.1094/MPMI
45. Niehaus K, Kapp D, Pühler A (1993) Plant defense and delayed infection of alfalfa pseudonodules induced by an exopolysaccharide (EPS I)-deficient *Rhizobium meliloti* mutant. Planta 190:415–425
46. Siddique S, Matera C, Radokovic ZS, Shamin HM, Gutbrod P et al (2014) Parasitic worms stimulate host NADPH oxidases to produce reactive oxygen species that limit plant cell death and promote infection. Sci Signal 7(320):33. doi:10.1126/scisignal.2004777
47. Urwin PE, Lilley CJ, Atkinson HJ (2002) Ingestion of double-stranded RNA by preparasitic juvenile cyst nematodes leads to RNA interference. Mol Plant Microbe Interact 15(8):747–752. doi:10.1094/MPMI
48. Sukno SA, McCuiston J, Wong M, Wang X, Thon MR et al (2007) Quantitative detection of double-stranded RNA-mediated gene silencing of parasitism genes in *Heterodera glycines*. J Nematol 39(2):145–152
49. Siddique S, Radakovic ZS, De La Torre CM, Chronis D, Novák O et al (2015) A parasitic nematode releases cytokinin that controls cell division and orchestrates feeding site formation in host plants. Proc Natl Acad Sci U S A 112:12669–12674. doi:10.1073/pnas.1503657112
50. Vanstaden J, Dimalla GG (1977) Comparison of endogenous cytokinins in roots and xylem exudate of nematode resistant and susceptible tomato cultivars. J Exp Bot 107:1351–1356
51. Dimalla GG, Vanstaden J (1977) Cytokinins in root-knot nematode, *Meloidogyne incognita*. Plant Sci Lett 10:25–29
52. Shanks CM, Rice JH, Zubo Y, Schaller GE, Kieber JJ (2016) The role of cytokinin during infection of *Arabidopsis thaliana* by the cyst nematode *Heterodera schachtii*. Mol Plant Microbe Interact 29:57–68. doi:10.1094/MPMI-07-15-0156-R
53. Lohar DP, Schaff JE, Laskey JG, Kieber JJ, Bilyeu KD et al (2004) Cytokinins play opposite roles in lateral root formation and nematode and rhizobial symbioses. Plant J 38(2):203–214. doi:10.1111/j.1365-313X.2004.02038.x

54. Quentin M, Abad P, Favery B (2013) Plant parasitic nematode effectors target host defense and nuclear functions to establish feeding cells. Front Plant Sci 4:53. doi:10.3389/fpls.2013.00053
55. De Waele D, Elsen A (2007) Challenges in tropical plant nematology. Annu Rev Phytopathol 45:457–485. doi:10.1146/annurev.phyto.45.062806.094438
56. Roberts PA (1993) The future of nematology: integration of new and improved management strategies. J Nematol 25:383–394
57. Robertson L, Diez-Rojo MA (2008) New host races of *Meloidogyne arenaria*, *Meloidogyne incognita*, and *Meloidogyne javanica* from horticultural regions of Spain. Plant Dis 93:180–184. doi:10.1094/pdis-93-2-0180
58. Curtis RHC (2008) Plantnematode interactions: environmental signals detected by the nematode's chemosensory organs control changes in the surface cuticle and behaviour. Parasite 15:310–316. doi:10.1051/parasite/2008153310
59. Wyss U, Grundler F, Münch A (1992) The parasitic behaviour of second-stage juveniles of *Meloidogyne incognita* in roots of *Arabiodopsis thaliana*. Nematropica 38:98–111
60. Williamson VH, Hussey RS (1996) Nematode pathogenesis and resistance in plants. Plant Cell 8:1735–1745. doi:10.1105/tpc.8.10.1735
61. Hussey RS, Mims CW (1991) Ultrastructure of feeding tubes formed in giant-cells induced in plants by the root-knot nematode *Meloidogyne incognita*. Protoplasma 162:99–107
62. Van der Eycken W, de Almeida EJ, Inzé D, Van Montagu M, Gheysen G (1996) A molecular study of root-knot nematode-induced feeding sites. Plant J 9:45–54
63. Huang G, Gao B, Maier T, Allen R, Davis EL, Baum TJ (2003) A profile of putative parasitism genes expressed in the oesophageal gland cells of the root-knot nematode *Meloidogyne incognita*. Mol Plant Microbe Interact 16:376–381
64. Jaubert S, Laffaire JB, Abad P, Rosso MN (2002) A polygalacturonase of animal origin isolated from the root-knot nematode *Meloidogyne incognita*. FEBS Lett 522:109–112. doi:10.1016/S0014-5793(02)02906-X
65. Goellner M, Wang X, Davis EL (2001) Endo-β-1,4-glucanase expression in compatible plant-nematode interactions. Plant Cell 13:2241–2255. doi:10.1105/tpc.010219
66. Ellis J, Jones D (1998) Structure and function of proteins controlling strain-specific pathogen resistance in plants. Curr Opin Plant Biol 1:288–293
67. Hewezi T, Baum TJ (2015) Gene silencing in nematode feeding sites. In: Escobar C, Fenoll C (eds) Plant nematode interactions: a view on compatible interrelationships, vol 73, 1st edn, Advances in botanical research. Elsevier, New York, pp 221–239
68. Mentelin S, Thorpe P, Jones JT (2015) Suppression of plant defenses by plant-parasitic nematodes. In: Escobar C, Fenoll C (eds) Plant nematode interactions: a view on compatible interrelationships, vol 73, 1st edn, Advances in botanical research. Elsevier, New York, pp 325–337
69. Gardener M, Verma A, Mitchum MG (2015) Emerging roles of cyst nematode effectors in exploiting plant cellular processes. In: Escobar C, Fenoll C (eds) Plant nematode interactions: a view on compatible interrelationships, vol 73, 1st edn, Advances in botanical research. Elsevier, New York, pp 259–291
70. Williamson VM (1998) Root-knot nematode resistance genes in tomato and their potential for future use. Annu Rev Phytopathol 36:277–293. doi:10.1146/annurev.phyto.36.1.277
71. Lagudah E, Moullet O, Appels R (1997) Map-based cloning of a gene sequence encoding a nucleotide binding domain and a leucine-rich region at the Cre3 nematode resistance locus of wheat. Genome 40:659–665
72. Seah S, Sivasithamparam K, Karakousis A, Lagudah E (1998) Cloning and characterisation of a family of disease resistance gene analogs from wheat and barley. Theor Appl Genet 97:937–945
73. Jung C, Cai D, Kleine M (1998) Engineering nematode resistance in crop species. Trends Plant Sci 3:266–271
74. Puthoff DP, Ehrenfried ML, Vinyard BT, Tucker ML (2007) GeneChip profiling of transcriptional responses to soybean cyst nematode, Heterodera glycines, colonization of soybean roots. J Exp Bot 58:3407–3418. doi:10.1128/jmbe.v14i1.588

75. Veremis JC, Roberts PA (1996) Identification of resistance to *Meloidogyne javanica* in the *Lycopersicon peruvianum* complex. Theor Appl Genet 93:894–901
76. Yaghoobi J, Kaloshian I, Wen Y, Williamson VM (1995) Mapping a new nematode resistance locus in *Lycopersicon peruvianum*. Theor Appl Genet 91:457–464. doi:10.1007/BF00222973
77. Vercauteren I, De Almeida EJ, De Groodt R, Gheysen G (2002) An *Arabidopsis thaliana* pectin acetylesterase gene is upregulated in nematode feeding sites induced by root-knot and cyst nematodes. Mol Plant Microbe Interact 15:404–407. doi:10.1094/MPMI.2002.15.4.404
78. Fan DF, Maclachlan GA (1967) Massive synthesis of ribonucleic acid and cellulase in the pea epicotyl in response to indoleacetic acid, with and without concurrent cell division. Plant Physiol 42:1114–1122
79. Favery B, Lecomte P, Gil N, Bechtold N, Bouchez D et al (1998) RPE, a plant gene involved in early development steps of nematode feeding cells. EMBO J 17:6799–6811. doi:10.1093/emboj/17.23.6799/pdf
80. Fenoll C, Aristizábal FA, Sanz-Alférez S, de Campo FF (1997) Regulation of gene expression in feeding sites. In: Fenoll C, Grundler FMW, Ohl SA (eds) Cellular and molecular aspects of plant-nematode interactions, 1st edn. Kluwer, Dordrecht, pp 133–149
81. Gheysen G (1998) Chemical signals in the plant-nematode interaction: a complex system? 1st edn. In: Romeo JT, Downum KR, Verpoorte R (eds) Phytochemical signals and plant-microbe interactions. Springer US, pp 95–117. doi:10.10007/978-1-4615-5329-8_6
82. Wieczorek K, Gilecki B, Gerdes L, Heinen P, Szakasits D, Durachko DM et al (2006) Expansins are involved in the formation of nematode-induced syncytia in roots of *Arabidopsis thaliana*. Plant J 48:98–112. doi:10.1111/j.1365-313X.2006.02856.x
83. Wieczorek K, Hofmann J, Blochl A, Szakasits D, Bohlmann H, Grundler FM (2008) *Arabidopsis* endo-1,4-β-glucanases are involved in the formation of root syncytia induced by *Heterodera schachtii*. Plant J 53:336–351. doi:10.1111/j.1365-313X.2006.02856.x
84. Gheysen G, Van der Eycken W, Barthels N, Karimi M, Van Montagu M (1996) The exploitation of nematode-responsive plant genes in novel nematode control methods. Pestic Sci 47:95–101
85. Jones AM, Im KH, Savka MA, Wu MJ, DeWitt NG et al (1998) Auxin-dependent cell expansion mediated by overexpressed auxin-binding protein 1. Science 282:1114–1117. doi:10.1126/science.282.5391.1114
86. Jones MGK, Dropkin VH (1976) Scanning electron microscopy of nematode-induced giant transfer cells. Cytobios 15:149–161
87. Seinhorst JW (1965) The relationship between nematode density and damage to plants. Nematologica 11:137–154. doi:10.1163/187529265X00582
88. Trudgill DL (1992) Resistance to and tolerance of plant-parasitic nematodes in plants. Annu Rev Phytopathol 29:167–192
89. Hammond-Kosack KE, Jones DG (1996) Resistance gene-dependent plant defense responses. Plant Cell 8:1773–1791
90. McDowell JM, Woffenden BJ (2003) Plant disease resistance genes: recent insights and potential applications. Trends Biotechnol 21:178–183. doi:10.1016/S0167-7799(03)00053-2
91. Jung C (1998) A singular gene doubles up pest resistance. Nat Biotechnol 16:1315–1316. doi:10.1016/S1369-5266(99)80057-0
92. Dutta TK, Papolu PK, Banakar P, Choudhary D, Sirohi A, Rao U (2015) Tomato transgenic plants expressing hairpin construct of a nematode protease gene conferred enhanced resistance to root-knot nematodes. Front Microbiol 6:1–13. doi:10.3389/fmicb.2015.00260
93. Tripathi L, Babiryc A, Roderick H, Tripathi JN, Changa C et al (2015) Field resistance of transgenic plantain to nematodes has potential for future African food security. Sci Rep 5:1–10. doi:10.1038/srep08127
94. Burrows PR, De Waele D (1997) Engineering resistance against plant-parasitic nematodes using anti-nematode genes. In: Fenoll C, Grundler FMW, Ohl SA (eds) Cellular and molecular aspects of plant-nematode interactions, 1st edn. Kluwer, Dordrecht, pp 217–236

95. Siddique S, Radakovic ZS, De La Torre CM, Chronis D, Novák O et al (2014) Parasitic worms stimulate host NADPH oxidases to produce reactive oxygen species that limit plant cell death and promote infection. Sci Signal 7:ra33
96. Williamson VM, Lambert KN, Kaloshian I (1994) Molecular biology of nematode resistance in tomato. In: Lamberti F, De Giorgi C, Bird DMK (eds) Advances in molecular plant nematology, 1st edn. Plenum, New York, pp 211–219
97. Koch A, Kogel KH (2014) New wind in the sails: improving the agronomic value of crop plants through RNAi-mediated gene silencing. Plant Biotechnol J 12:821–831. doi:10.1111/pbi.12226

Part V

Applications and Future Prospects

Transgenesis and Plant Molecular Pharming

23

Sanjeewa Singhabahu, Kathleen Hefferon, and Abdullah Makhzoum

Abstract

The use of plants as efficient biopharmaceutical factories has significantly increased in the past two decades. This is mainly due to advancements in plant biotechnology which pave the way to high-yield production of biopharmaceuticals in plants, combined with efforts made to optimize yield through upstream, downstream, and preservation strategies of recombinant proteins. The FDA's approval to commercially release recombinant glucocerebrosidase enzyme produced in carrot cells by Protalix Biotherapeutics was the first plant-produced biopharmaceutical to be released for human consumption into the market. This is a major achievement in the field of molecular pharming. Although many other biopharmaceuticals produced in plants are in the pipeline for commercial release after undergoing various stages of clinical trials, there is room for improvement in enhancing recombinant protein yield in plants. These include exploration of innovative strategies involving genetics, genomics, epigenetics, in silico simulations and purification techniques. In this chapter, we discuss various approaches of plant biotechnology and plant genetic engineering that are being used in the molecular pharming of biopharmaceuticals.

Keywords

Molecular pharming • Bioproduction • Biopharmaceuticals • Vaccines • Recombinant proteins • Transgenic plants

S. Singhabahu
Human Genetics Unit, Faculty of Medicine, University of Colombo, Colombo, Sri Lanka
e-mail: sanjeewasinghabahu@gmail.com

K. Hefferon (✉)
Department of Food Sciences, Cornell University, Ithaca, NY, USA
e-mail: kathleen.hefferon@utoronto.ca; klh22@cornell.edu

A. Makhzoum (✉)
Department of Biology, The University of Western Ontario, London, ON, Canada
e-mail: abmakhzoum@gmail.com

S. Jha (ed.), *Transgenesis and Secondary Metabolism*, Reference Series in Phytochemistry, DOI 10.1007/978-3-319-28669-3_21

Abbreviations

ADA	Adenosine deaminase
AGPs	Arabinogalactan proteins
BY-2	Bright yellow-2
CaMV	Cauliflower mosaic virus
CPMV	Cowpea mosaic virus
DNA	Deoxyribonucleic acid
ER	Endoplasmic reticulum
ERT	Enzyme replacement therapy
FDA	Food and drug administration
GBP-Fc	Anti-green fluorescent protein antibody
GMOs	Genetically modified organisms
HAC1	Hemagglutinin antigen
HDEL	H-Histidine D-Aspartic acid, E-Glutamic acid, L-Leucine
hGM-CSF	Human Granulocyte-macrophage colony-stimulating factor
HIV	Human immunodeficiency virus
KDEL	K-Lysine D-Aspartic acid, E-Glutamic acid, L-Leucine
MERS	Middle East respiratory syndrome
mRNA	Messenger RNA
Pphas	β-phaseolin promoter
PpsbA	Photosystem II protein D1promoter
PSAD	Photosystem I subunit D1promoter
PVP	Polyvinylpyrrolidone
PVX	Potato Virus X
RAmy3D	Rice α-amylase 3D
RNA	Ribonucleic acid
SCID	Severe combined immune deficiency disorders
TMV	Tobacco mosaic virus
USDA	United States Department of Agriculture
UTRs	Untranslated regions
VHH-Fc	Nanobodies-crystallisable fragments
VLP	Virus-like particle
VNPs	Virus-based nanoparticles

Contents

1 Introduction

Plants and plant products have been used for centuries for the treatment of human diseases. Botanical gardens presented a wealth of *materia medica* for therapeutic use by the latter part of the sixteenth century. In the seventeenth century, a scientific approach and the discovery of new pharmacological remedies led to the identification of active biological compounds which could undergo purification for therapeutic use [1]. At present, about a quarter of prescription drugs contain medications which are of botanical origin. Current advances in biotechnology have led to renewed hopes in producing new medicines from botanical sources, including fully functional pharmacologically imperative recombinant human proteins [2]. Recombinant human insulin produced in bacteria was the first commercially produced biopharmaceutical to be made in 1982 by Lilly, USA, and coincides with the development of the first genetically modified plant [3]. The development of genetically modified plants was followed by the successful production of biopharmaceuticals in plants, such as human growth hormone fusion protein [4], interferon [5], monoclonal antibodies [6], and human serum albumin [7]. Since then, scores of biopharmaceuticals such as heterologous proteins, enzymes, and peptides have been produced in plant expression systems for treating ailments in both humans and animals (Table 1).

Bacteria are classified as convenient and cost-effective systems for producing smaller human proteins (e.g., human insulin). However, the absence of assembly steps and posttranslational modifications, which are essential for the biological activity of most therapeutic proteins, hinder their use for production of more complex proteins. In addition, effective removal of contaminating lipopolysaccharides from recombinant protein is extremely expensive in bacterial expression systems.

Plants, on the other hand, possess an efficient protein synthesis mechanism and, with established gene expression systems, can be used to produce large amounts of proteins (Fig. 1) [8, 9]. Plant expression systems, however, hold the potential to be alternative production systems for therapeutic proteins, due to the absence of mammalian pathogens present in mammalian cell expression systems. In addition, higher product safety, higher production scale, lower production costs, easy storage and distribution are some of the advantages of plant based production systems (Table 2) [2].

Increasing demand for therapeutic proteins has led to the development of plant production systems as an alternative for producing therapeutically imperative proteins. Successful investigations on biopharmaceutical production in plants have led to the FDA approval of the first plant-produced biopharmaceutical product, a poultry vaccine developed against Newcastle disease virus. This achievement by Dow AgroSciences is a significant breakthrough for plant-produced biopharmaceuticals. The Concert$^{(TM)}$ Plant-cell-produced, biocontained system used by Dow AgroSciences utilizes tobacco plant cells as an alternative to the expression of foreign proteins in whole plants [10].

This success in molecular pharming was also followed by the production of a human recombinant protein in carrot cell suspensions. The human

Table 1 Plant-produced therapeutic proteins published in the last 3 years

Therapeutic proteins	Host cells	Protein yields and signal used	Promoter	Localization	Reference
Isoform of human glutamic acid decarboxylase (GAD65) antigen	*N. benthamiana*	226.9 ± 42.1 μg/g (FLW)	*CaMV35S*	Leaves	[21]
HIV antigen P24	*Chlamydomonas reinhardtii*	0.25% of the TSP[a]	*PSAD promoter*	Chloroplast	[22]
HIV-neutralizing monoclonal antibody 2G12	*Oryza sativa cv.* Nipponbarre	46.4 μg/g	*Rice Glut1 (glutelin-1)*	Seed	[23]
Polio viral protein 1 (CTB-VP1)	*Tobacco (Nicotiana tabacum cv. Petit Havana) and Lettuce*	Up to 5% total leaf protein	*PpsbA, promoter*	Chloroplast	[24]
Human Adenosine deaminase	*N. tabacum* cv BY-2	16 mg L^{-1}(Ext, HP) 4 mg L^{-1}(PR1a) 1.5 mg L^{-1}(Ext)	*CaMV35S*	Secreted	[12]
Anti-HIV-1 monoclonal antibody 2G12	*Hordeum vulgare (Barley)*	1.2 g/kg dry weight (160 μg recombinant protein per g grain)	*GLO1 (GLOBULIN1)*	Seed	[25]
Avian influenza H5N1 antigen (M2)	*Nicotiana benthamiana*	125 to 205 mg Zera®M2e PB/kg fresh weight (FW)	*CaMV35S*	Intracellular	[26]
PVX and CPMV VNPs	*N. benthamiana*	~0.04 mg/g and ~0.05 mg/g FLW	*CaMV35S*	Leaves	[27]
Enterotoxin B (LTB) Cholera toxin B (CTB)	*Oryza sativa*	3.4 ng/μg of the TSP 21.3 ng/μg of the TSP	*Globulin (Glb)*	Seeds	[28]

H3N2 nucleoprotein (recombinant NP (rNP))	*Zea mays* (*Maize*)	70 μg/g of seed	*γ-zein promoter (P27γz)*	Seeds	[29]
Flg-4 M comprising flagellin of *Salmonella typhimurium* fused to four tandem copies of the M2e peptide of influenza A virus.	*Nicotiana benthamiana*	30% of TSP (about 1 mg/g of fresh leaf tissue)	*CaMV35S*	Leaves	[30]
Hemophilia B coagulation factor IX	*Lactuca sativa* (*Simpson Elite lettuce*)	~1 mg/g in lyophilized cells	*psbA – photosystem II protein D1*	Chloroplast	[31]
VHH-Fc antibodies GBP-Fc	*Nicotiana benthamiana*	GBP-Fc accumulated between 0.03 and 0.18 mg/g FW (1.2–5.3% TSP) VHH4-Fc reached 0.74 to 0.91 mg/gFW (26.5–27.7% of TSP)	*Pphas* *CaMV35S*	Leaves	[32]
	Arabidopsis thaliana	GBP-Fc accumulated up to 44.3 mg/g FW (13.1% TSP), which corresponds to 4.4% of seed weight	*Pphas* *CaMV35S*	Seeds	[32]
Antimicrobial peptides Cecropin A	*Oryza sativa*	1–3.3 μg/g seed	*Glutelin B1 (GluB1) or glutelin B4 (GluB4)*	Seeds	[33]

[a] *TSP* total soluble proteins

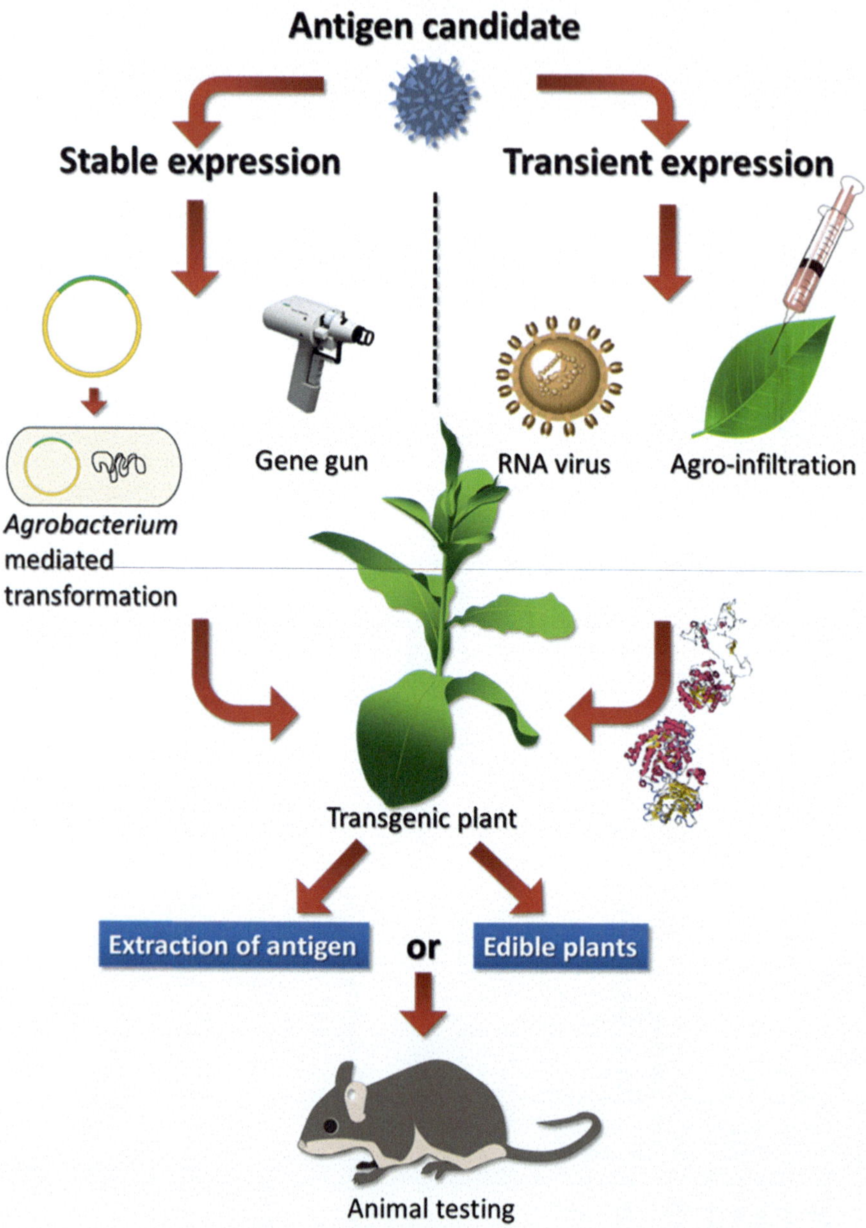

Fig. 1 Molecular pharming of vaccine antigens in transgenic plants (Figure produced by ChemBio Draw, Pain brush, and PowerPoint)

glucocerebrosidase (taliglucerase alfa), which is used in ERT to treat Gaucher's disease, has been produced by Protalix Biotherapeutics. This enzyme, which was approved by the US Food and Drug Administration in 2012, was the first

Table 2 Comparison of recombinant protein production in various production systems

	Transgenic plants	Plant cell cultures	Yeast	Bacteria	Mammalian cell cultures	Transgenic animals
Cost/storage	Cheap/RT	Cheap/−20 °C	Cheap/−20 °C	Cheap/−20 °C	Expensive	Expensive
Contamination risk	No	No	No	Yes	Yes	Yes
Distribution	Easy	Easy	Feasible	Feasible	Difficult	Difficult
Gene size	Not limited	Limited	Unknown	Unknown	Limited	Limited
Glycosylation	"Correct?"	"Correct?"	Incorrect	Absent	"Correct"	"Correct"
Multimeric protein assembly	Yes	No	No	No	No	Yes
Production cost	Low	Low	Medium	Medium	High	High
Product quality	High	High	Medium	Low	High	High
Production scale	Worldwide	Worldwide	Limited	Limited	Limited	Limited
Production vehicle	Yes	Yes	Yes	Yes	Yes	Yes
Propagation	Easy	Feasible	Easy	Easy	Hard	Feasible
Protein folding accuracy	High?	High?	Medium	Low	High	High
Protein homogeneity	High?	Medium	Medium	Low	Medium	High
Protein yield	High	Very high	High	Medium	Medium-high	High
Public perception of risk	High	High	Medium	Low	Medium	High
Safety	High	High	Unknown	Low	Medium	High
Scale-up costs	Low	Low	High[b]	High[b]	High[b]	High
Therapeutic risk[a]	Unknown	Unknown	Unknown	Yes	Yes	Yes
Time required	Medium	Low	Medium	Low	High	High

Table modified from Schillberg et al. [34] and Yao et al. [35]
[a]Residual viral sequences, oncogenes, endotoxins
[b]Large, expensive fermenters

commercialized plant-cell-produced human enzyme to be launched in the market (Table 3) [11].

Although there are a number of success stories in molecular farming, lower recombinant protein yield remains the major limitation in commercialization of this system [12]. Since the production of the first biopharmaceutical in plants [4], a substantial effort has been made to enhance pharmaceutical protein yield at the molecular, cell culture, upstream and downstream processing levels. Various molecular strategies have also been employed to enhance the yield of foreign proteins in plants. These include approaches such as enhancement of both gene transcription and translational efficiencies as well as the incorporation of novel protein-fusion and secretory pathway-targeting technologies.

The use of whole plant systems in molecular pharming has been revolutionized by the use of alternative plant systems such as suspension cultures, hairy roots, hydroponic culture, plant chloroplasts, and algae [12–18]. Furthermore, various plant species such as tobacco, potato, tomato, lettuce, wheat, and sorghum have been utilized to explore the potential uses in molecular pharming. These investigations have resulted in successful outcomes where, Sasou et al. [19], for example, have found that the rice endosperm can be used as a protective niche for heterologous recombinant proteins which can enhance the yield considerably. In another attempt to produce vaccines in carrot, Rosales-Mendoza et al. [20] and his coworkers were able to pioneer the production of prototype vaccines against ten different pathogens. Singhabahu et al. [12], on the other hand, was also able to produce human adenosine deaminase, an enzyme used in ERT of ADA-associated SCID, in tobacco BY-2 cell suspension cultures. This enzyme was successfully secreted into media through the use of signal peptide sequences.

The emergence of fatal viruses like Ebola and MERS has led to an increased demand for the rapid production of vaccines around the world, especially in underdeveloped countries. The potential of plants to act as green biofactories for the generation of biopharmaceuticals makes plants the ideal candidate for vaccine production. Among vaccines produced in plants, only a handful of plant-based human vaccines have reached clinical trials (Table 3). The anti-*Streptococcus* surface antigen I/II produced in tobacco against *Streptococcus mutans*-associated dental caries and the H5N1 influenza HA VLP antigen produced in tobacco against the H5N1 virus are currently in Phase II product development stage [36]. The commercial success of viral and bacterial subunit vaccines produced in plants destined for human consumption, however, has not yet been achieved. Various attempts have been made by scientists in recent years to produce vaccines in a number of plants such as tobacco, rice, maize, potato, alfalfa, lettuce, tomato, carrot, peanut, and soybean, either in whole plants, hairy roots, or cell suspension cultures. Takaiwa et al. [37] investigated the possibility of using rice seed as a vehicle for the delivery of vaccines to gut-associated lymphoid tissue (GALT), the biggest lymphoid organ in the body, and found that the delivery system not only plays an important role in exerting an immune response against invading pathogens but is also shown to maintain immune tolerance against nonpathogenic antigens such as foods.

Table 3 Plant-produced therapeutic proteins, antibodies, and vaccines in clinical development or on the market

Product	Plant host	Indication	Route of administration	Product development stage	Reference
Vaccines					
Newcastle disease virus HN	Tobacco cell suspension	Newcastle disease (poultry)	Subcutaneous	USDA approved (Marketed)	[39]
H1N1 influenza HAC1	*Nicotiana benthamiana*	H1N1 "swine" influenza	Intramuscular	Phase I (ongoing)	[40, 41]
Cholera CTB	Rice	Cholera	Oral	Phase I	[42, 43]
H5N1 influenza HAI1	*Nicotiana benthamiana*	H5N1 "avian" influenza	Intramuscular	Phase I (ongoing)	[36, 44]
H5N1 influenza HA VLP	*Nicotiana benthamiana*	H5N1 "avian" influenza	Intramuscular	Phase I (ongoing) Phase II (Health Canada approved and enrolling volunteers)	[36]
Personalized anti-idiotyped scFVs:	*Nicotiana benthamiana*	Non-Hodgkin's lymphoma	Subcutaneous	Phase I (ongoing)	[45]
E. coli LT-B	Maize Potato	Diarrhea	Oral	Phase I Phase I	[46] [47]
Rabies virus GP/NP	Spinach	Rabies	Oral	Phase I	[48, 49]
Norwalk virus CP	Potato	Diarrhea	Oral	Phase I	[50]
H7N9 influenza virus	*Nicotiana benthamiana*	H7N9 "Avian Influenza A"	Intramuscular	Phase I	[51]
Antibodies					
ZMapp™ a cocktail of highly purified monoclonal antibodies	Tobacco leaves	Ebola Virus	Intravenous	Concluded Prevail II clinical trials	[38]
Anti-*Streptococcus* surface antigen I/II (CaroRx antibody)	Tobacco	Dental caries	Topical	Phase II; EU approved	[52]
Anti-αCCR5	*Nicotiana benthamiana*	HIV	Topical	Preclinical	[53]

(*continued*)

Table 3 (continued)

Product	Plant host	Indication	Route of administration	Product development stage	Reference
Hepatitis B surface antigen	Potato Lettuce	Hepatitis B vaccine purification	Oral	Phase I	[54] [55]
Anti-CD20	Duckweed	Non-Hodgkin's lymphoma, rheumatoid arthritis	Intravenous	Preclinical	[56]
Norovirus Capsid protein	Potato	Norwalk virus (winter vomiting virus)	Oral	Phase I	[57]
Anti-HIV gp120	Maize *Nicotiana benthamiana*	HIV	Topical	Preclinical	[58]
Therapeutic and dietary proteins					
Intrinsic factor	*Arabidopsis thaliana*	Vitamin B12 deficiency	Oral	Phase III	[52]
Glucocerbrosidase	Carrot cell suspension	Gauchers disease	Intravenous	USFDA approved in 2012 (Marketed)	[59]
Insulin	Safflower	Diabetes	Subcutaneous	Phase I/II (marketing expected)	[60]
Gastric lipase	Maize	Cystic fibrosis, pancreatitis	Oral	Phase II (marketed as analytical reagent)	[61]
Lactoferrin	Maize	Gastrointestinal infections	Oral	Phase I (marketed as analytical reagent)	[62]

Table modified from Takeyama et al. [63], Yusibov et al. [64], and Paul and Ma [65]

Successful efforts have also been made recently to produce monoclonal antibodies (ZMapp™) in tobacco plants against Ebola virus. LeafBio, Inc., the commercial arm of Mapp Biopharmaceutical, Inc., has announced the conclusion of Prevail II clinical trial of the ZMapp™ therapy for the treatment of Ebola Virus Disease in February 2016. Although the license has not been acquired, the US Food and Drug Administration have encouraged Mapp Biopharmaceutical to continue to make ZMapp™ available to patients under an expanded access treatment protocol during the product's ongoing development [38] (Mapp Biopharmaceutical Press Releases, 2016).

2 Expression of Recombinant Proteins in Transgenic Plants

Expression of a foreign cDNA in plants requires a promoter to drive expression. Since its first use in the early 1980s, the most commonly used promoter to date is the Cauliflower Mosaic Virus (CaMV) 35S constitutive promoter [66]. Mason et al. [67] produced the first plant-derived vaccine in tobacco plants using a CaMV 35S promoter upstream of the hepatitis B surface antigen (HBsAg) coding region and a downstream nopaline synthatase (NOS) terminator. A wide variety of promoters and terminators are currently in use for the purpose of expressing foreign genes in plants. Apart from the CaMV 35S promoter, the *Ubiquitin* promoter [68, 69] and the hybrid (ocs) 3mas promoter [70] are two other constitutive promoters commonly used for foreign gene expression in plants.

The demand of production and the accumulation of therapeutic proteins in various plant tissues such as seeds, tubers, etc. has led to the prolific use of tissue-specific promoters in recent years. Barahimipour et al. [22] recently demonstrated the use of the PSAD promoter to produce HIV antigen P24 in *Chlamydomonas reinhardtii* chloroplasts. The glutclin1 promoter used in the expression of HIV-neutralizing monoclonal antibody 2G12 in *Oryza sativa cv.* Nipponbarre seeds [23] and the γ-zein promoter (P27γz) used in the expression of H3N2 nucleoprotein (recombinant NP (rNP)) in *Zea mays* seeds [29] are the seed-specific promoters commonly used in foreign gene expression in plants, especially monocotyledons. Inducible promoters, on the other hand, are used to regulate the expression of foreign genes and control the expression by external stimuli such as sucrose, salts, steroids, alcohol, or environmental stress factors [71]. The use of inducible promoters is advantageous in some cases as the process can maximize the protein yield [72]. A commonly used inducible rice α-amylase 3D (*RAmy3D*) promoter is governed by sucrose starvation and promotes high level expression of foreign genes. Rice α-amylase 3D (*RAmy3D*) promoter has been used to produce a range of human therapeutic proteins including growth hormone [73], α1-antitrypsin, serum albumin, lysozyme [71], and interlukin-12 [74].

The position of gene insertion in the plant genome [75] and the organization of gene integration [76] are known to affect transcription of foreign genes after plant transformation. It was reported that transgene silencing caused by DNA methylation can also affect transcription [76]. Tandem copies of T-DNA at a single locus is often a consequence of *Agrobacterium*-mediated transformation, and inverted T-DNA tandem repeats situated closer to the right border are also associated with promoter methylation and transgene silencing [77].

Hence, the choice of strong translational enhancer/regulation sequences is equally as important as the choice of a strong promoter in increasing recombinant protein yield. The incorporation of the 204-base 3′-untranslated region of Tobacco mosaic virus (TMV) to foreign mRNA increases gene expression by 100-fold, which resembles increased mRNA stability by the polyadenylation of mRNA [78]. Gallie [79] also demonstrated the ribosomal initiation factor (eIF4F) recruiting function of the TMV RNA 5′-leader sequence (omega-Ω) significantly enhances translation in both eukaryotes and prokaryotes. The 5′ UTR of the ribulose-1, 5-bisphosphate carboxylase/oxygenase (RUBISCO) enzyme is reported to enhance foreign gene expression in plants through recruiting translation factors as well [80]. RUBISCO is the most abundant enzyme found in the plant kingdom and catalyzes the primary stage in carbon fixation in plants. Patel et al.[81, 82] demonstrated the translational enhancement of the downstream transgenes by associated RUBISCO small subunit1 5′ untranslated region acquired from C4 *Amaranthus* and *Flaveria bidentis*.

RNA silencing is another factor that negatively affects foreign gene expression in plants. Saxena et al., [83] however, demonstrated the use of tomato bushy stunt virus (TBSV) P19 mutant (P19/R43W) to enhance heterologous gene expression through suppression of RNA silencing in tobacco plants.

Plastid transformation is another technique used in the overexpression of foreign genes in plants. The plastid, also known as chloroplast, is an important organelle in plant cells and eukaryotic algae. Chloroplasts carry out photosynthesis, carbon sequestration, oxygen evolution, amino acid, fatty acid and pigment biosynthesis, starch production, and metabolism of sulphur and nitrogen [84]. Plastids are known to have evolved overtime from free-living cyanobacterial progenitors and have given up the majority of their genes and function to accommodate energy transduction and metabolic machineries [85]. The presence of prokaryotic large circular chromosomes is a remnant primitive feature of the cynocobacterial progenitor. Most of these plastid chromosomes (50–290 kilobases) and encoded genes are similar across the entire plant kingdom, including algae [86], and have extremely high copy numbers [87].

Plastids are distributed throughout plant organs and tissues. Although all plastids contain identical copies of the plastid genome and are derived from undifferentiated proplastids, plastids in different organs and tissues differ in their morphology and function. Hence, photosynthetic chloroplasts are present in leaves and green tissues, pigmented chromoplasts are present in fruits and flowers, amyloplasts or elaioplasts are present in tubers and other storage organs, and leucoplasts are available in other nongreen tissues including roots [88].

Ever since the first chloroplast genetic transformation was performed two decades ago by Boynton and colleagues, many attempts have been made to express foreign genes in chloroplasts [89, 90]. Expression of *Bacillus thuringiensis* (Bt) Cry2Aa2 protein in the chloroplast led to accumulation of recombinant proteins up to 46.1% TSP in transgenic tobacco [91]. Furthermore, Chan and colleges were able to produce a cold chain and virus-free booster vaccine to confer immunity against different poliovirus serotypes in tobacco and lettuce chloroplast and accumulated up to 5% of the total leaf protein [24]. Barahimipour et al. [22] were also able to produce and accumulate HIV antigen P24 up to 0.25% of the TSP in *Chlamydomonas reinhardtii* chloroplast. As compared to nuclear transformation, chloroplast transformation can lead to significantly higher levels of foreign gene expression. Moreover, chloroplasts act as storage compartments for recombinant proteins and prevent them from proteolytic degradation [92]. Since the chloroplasts exhibit maternal inheritance the transgene flow in crops can also be contained [93]. Lack of transgene silencing in chloroplasts also leads to higher transgenic expression [94]. Another advantage is that the chromoplasts in nongreen tissues also can be used for foreign protein production through transformation [95]. Hence, chromoplasts or protoplast transformation of calli cells can be used in order to yield higher levels of recombinant proteins in plant cell suspensions. Chloroplasts also can carry out correct posttranslational modifications and folding [93, 96].

2.1 Strategies to Improve Protein Yield

While the use of various sequences such as the 35S CaMV promoter and the 5′ and/or 3′-untranslated regions of TMV can enhance foreign gene expression, the strategies involved in the accumulation of recombinant proteins are extremely important in achieving commercial success of plant-based production systems [97]. Recombinant proteins undergo extensive deterioration due to proteolysis and irreversible surface adsorption [97]. Hence, it is of prime importance to employ measures in order to reduce foreign protein degradation and loss. Anchoring recombinant proteins in the endoplasmic reticulum (ER) through C- terminal signal sequences such as KDEL and HDEL preserve foreign proteins from proteolytic degradation [98] due to the lack of endogenous protease-activity in plant cell ER, providing a protective environment for foreign proteins [99]. Petruccelli et al. [100] demonstrated the increase of monoclonal antibodies by 4–11-fold when retained in the ER as a fusion with the KDEL peptide. Furthermore, Hensel et al. [25] produced the anti-HIV1 monoclonal antibody 2G12 up to 160 ug per gram of grain using the GLOBULIN1 promoter along with the legumin B4 (LeB4) signal peptide and the endoplasmic reticulum (ER) retention signal (SE)KDEL sequences.

Knockout of the expression of protease encoding genes is another strategy employed to increase recombinant protein yield through reducing proteolytic degradation. Introduced by Kim et al. [73] the yield of hMG-CSF in rice cell suspensions was increased by suppression of cysteine protease gene Rep-1 by RNAi. In a similar attempt to enhance the accumulation of recombinant human IL-10 in tobacco leaves, Duwadi et al. [101] have used RNAi-mediated silencing of a cysteine protease-encoding gene, CysP6, which in turn enhances recombinant protein yield by 1.6-fold.

Coexpression of protease inhibitors is another strategy employed to enhance protein yield. Commonly expressed protease inhibitors for this purpose have specificity to cysteine, serine, or aspartic proteases [102]. The coexpression of trypsin and chymotrypsin inhibitor domains with recombinant human granulocyte-macrophage colony-stimulating factor (hGM-CSF) has been shown to increase the secreted hGM-CSF by twofold in rice suspension cultures [73]. Furthermore, Pillay et al. [103] showed the increased accumulation of the protein glutathione reductase in tobacco plants through the expression of the rice cysteine protease inhibitor oryzacystatin-I.

Prevention of recombinant proteins from proteolytic degradation through apoplast directing and subcellular targeting is another approach employed toward achieving higher protein yields. Located outside the cell membrane, the apoplast is a nonliving part of plant tissue consisting of intercellular space and the lumen of dead structures such as xylem cells. N-terminal signal peptide-tagged cell wall proteins are secreted into the apoplastic space. Signal sequences from proteins such as tobacco pathogenesis-related protein (PR1a) or extensin cell wall peptides, when used in conjunction with recombinant proteins, can direct foreign protein traffic into the apoplast [104, 105]. The N-terminal incorporation of PR1a signal peptide in endo-1, 4-β-D-glucanase of thermostable eubacterium increased the yield up to 26% of TSP in *Arabidopsis* leaves [106]. On the other hand, expression of bryodin 1 along with N-terminal extensin signal peptide increased the bryodin 1 yield up to 30 mg L^{-1} in tobacco cell cultures [105]. Singhabahu et al. [12] also expressed ADA in conjugation with PR1a and extensin signal sequences and reported a 4 mg L^{-1}(PR1a) and a1.5 mg L^{-1} (Ext) of ADA secreted into tobacco BY-2 suspension media, respectively.

Xu et al. [107] also demonstrated increased yields of recombinant interferon-α2 by 350–1,400-fold through incorporation of N-terminal extensin signal peptide and a C-terminal hydroxyproline (Hyp)-rich peptide. Production of recombinant proteins in conjugation with C-terminal Hyp-Ser residue leads to extensive *O*-glycosylation through the addition of arabino oligosaccharides. C-terminal *O*-glycosylation has been shown to stabilize recombinant proteins which resulted in increased yields [12, 107].

Another strategy employed to enhance recombinant protein yield is the supplementation of media with stabilizing agents. Recent trends in producing therapeutic proteins in plant cell suspension cultures enable the addition of protein-stabilizing agents to the culture medium. Wongsamuth and Doran [108] demonstrated increased yields of monoclonal IgG1 antibodies up to ninefold in the spent cultures of tobacco hairy roots by the addition of stabilizing agents such as gelatine or polyvinylpyrrolidone (PVP). Singhabahu et al. [12] also demonstrated increased ADA enzyme levels in the tobacco BY-2 suspension cultures that have been caused by supplementing media with 0.75 g L^{-1} of PVP (molecular weight 360,000) as described by LaCount et al. [109]. Furthermore, media supplemented with bovine serum albumin (BSA) was also shown to increase hGM-CSF levels in tobacco cell suspensions [110].

In silico and computational analysis will certainly help to compare mRNA and recombinant protein structures based on stability, efficiency, and thermodynamic properties [111, 112]. Consequently, this approach will lead to better optimization, modeling, and design of recombinant protein fusions in plant molecular pharming.

Indeed, Makhzoum et al. [111] showed computationally that adding a specific tag and a specific linker can lead to an improved overall structure and stability of a fusion recombinant protein SBA-ADA (soybean agglutinin-adenosine deaminase).

2.2 Glycosylation Patterns of Recombinant Proteins in Plants

Proteins undergo several posttranslational modifications such as glycosylation, phosphorylation, methylation, ADP-ribosylation, oxidation, and glycation during protein trafficking throughout plant cells [113]. Proteins can become glycosylated into two distinct forms: *N*-glycans (carbohydrates linked to the amide moiety of asparagine (Asn)) and *O*-glycans (carbohydrates linked to the hydroxyl group of threonine (Thr), serine (Ser), or hydroxyproline (Hyp)). Apart from nucleocytoplasmic *O*-GlcAcylation, both *N*- and *O*-glycosylation occur in the secretory pathway [114]. Glycosylation is extremely important in determining the biological activity of many therapeutic proteins. Hence, the production systems that can exert complex glycosylation structures are paramount in producing biologically active proteins. Plants as compared to bacteria or yeast-production systems have the capability to exert complex glycosylation patterns similar to that found in human cells.

The use of plant-made pharmaceuticals in human therapy, especially in vivo administration, is limited due to the inability of the plants to produce exact human-like glycosylation patterns onto recombinant proteins [115]. Immunogenic glycoepitopes composed of complex plant *N*-glycans are known to be immunogenic to mammals [116]. Attempts were made to redesign the protein *N*-glycan structures in plant expression systems. Various strategies such as retaining recombinant proteins in the ER, inhibition of host plant glycosyltransferases, and expression of heterologous glycosyltransferases for the addition of mammalian-like glycans have been employed over the years to reduce the adverse effects caused by plant glycosylation patterns in mammals [117].

Similar to plant *N*-glycans, *O*-glycans are also structurally different from human glycol-proteins. Mucin-type, *O*-GlcNAc, and a xylose residue linked to Ser and Thr residues are the most abundant *O*-glycans found in human cells. Extensins and various other arabinogalactan proteins (AGPs), on the other hand, are the most abundant *O*-glycosylated proteins that exist in plants. The galactose residues linked to most of the Serine (Ser) residues and arabinosyl residues linked to hydroxyproline (Hyp) residues are the most common *O*-glycosylation pattern of extensins [118]. The AGPs are the largest and the most extensively glycosylated proteins where Hyp residues are linked with very large arabinogalactan glycomodules or smaller arabino oligosaccharides [119]. Xu et al. [107] investigated the use of plant *O*-glycosylation to increase the stability of in vivo-administered recombinant proteins. The intravenous administration of AGP *O*-glycomodule fused human INFα2 into mice exhibited an increased serum half-life by 9 h with no adverse immune response [107]. Recombinant human growth hormone (hGH) fused to AGPs also exhibited no adverse immune response when injected into mice [120]. Xu et al. [107] also used an extensin signal peptide in conjunction with C-terminal hydroxyproline (Hyp)-rich peptide to produce a Hyp-rich glycoprotein. A C-terminal Hyp-Ser residue leads to extensive *O*-glycosylation through the addition of arabino oligosaccharides. Recombinant interferon-α2 produced in conjunction with the Hyp-Ser sequence increased recombinant protein yield by as much as 350–1,400-fold [107]. The findings of Singhabahu et al. [12] and Xu et al. [107] also confirmed the protein stabilizing effect of fused AGPs, where the incorporation of a signal for arabinogalactan in addition to ADA led to a 336-fold increase of recombinant protein yield in tobacco BY-2 suspension cultures.

3 Plant Virus Expression Vectors to Express Vaccine Proteins

Plant viruses have also been extensively utilized for the expression of pharmaceutical proteins in plants. The fact that heterologous proteins can be expressed transiently saves time spent for the generation of mature transgenic plants. This can facilitate the rapid response to produce vaccines against bioterrorism agents and diseases with pandemic potential such as influenza virus, Ebola virus, or other emergencies. Moreover, infection of plants by recombinant virus vectors allows for much higher yields of protein, thus enhancing the feasibility of acquiring the high doses of antigen that are required for oral immunization. Second-generation plant virus vectors, which will be discussed in more detail below, allow for biological containment and prevent the spread of heterogeneous sequences to weedy relatives. Since recombinant plant virus vectors are used to infect conventional nontransgenic plants, their usage circumvents public perception concerns regarding genetically modified organisms, or GMOs.

3.1 Construction of Plant Virus Expression Vectors

In general, plant viruses designed to express vaccines and other therapeutic proteins are based largely on the genomes of positive-sense RNA viruses or single-stranded DNA viruses. Infectious cDNA clones can be generated from RNA viruses, which facilitate their manipulation in the lab as well as on host plants. Review articles that cover the development of plant virus expression vectors for biopharmaceutical production include Hefferon [121], Yusibov et al. [122], and Salazar-González et al. [123].

The development of plant viruses as expression vectors can be presented in terms of first-generation versus second-generation vectors. First-generation vectors contain the entire virus genome, and the foreign gene of interest is expressed along with the coat protein either as part of a fusion protein or separately from an additional strong subgenomic promoter that is incorporated into the viral genome. Vaccine epitopes may also be presented on external portions of the virus capsid protein so that they are displayed on the surface of virus particles. The literature is full of examples of first-generation plant virus expression vectors and include Canizares et al. [124] and

Yusibov et al. [125]. A major drawback to using the full virus expression vector strategy to express vaccine proteins stems from the fact that the virus retains the ability to replicate, self assemble, and move from cell to cell, thus creating concerns with respect to the release of the expression vector to unintended plant species.

Second-generation virus vectors were constructed to address this and other issues, such as increased production levels, overcoming restrictions with respect to tissue specificity of virus infection to select plant tissues, and heterologous protein size limitations. These new "deconstructed vectors" are composed of the minimum virus components required for replication, thereby lacking both movement and coat protein genes. Deconstructed vectors are unable to infect plants on their own, instead they must be delivered to the host plant by alternative means, such as the vacuum infiltration of an agrobacterium suspension which harbors the vector into the leaves of young plants. This technique enables synchronous production in all plant tissues in a matter of days, thus increasing protein production in a reduced time period.

3.2 Examples of Plant Virus Expression Vectors Expressing Vaccine Proteins

Plant positive-sense RNA viruses include the rod-shaped tobamoviruses such as Tobacco mosaic virus (TMV). TMV is the first virus that has been engineered as a deconstructed vector system [126, 127]. Examples of articles that cover the use of TMV for pharmaceutical production include Noris et al. [128], Huang et al. [129], Musichek et al. [130], and Meti et al. [131]. Recent work by Petukhova et al. [132] involves the use of TMV as an epitope presentation system for different point mutation versions of the conserved M2e epitope from influenza virus.

TMV has more recently elicited a renewed interest due to the ability of the particle itself to act as an adjuvant. Banik et al. [133] found that TMV particles can act as both an adjuvant and an epitope display system for vaccine development against tularemia, caused by the facultative intracellular pathogen *Francisella tularensis*. *F. tularensis* is considered by the Centers for Disease Control to be a Category A biothreat agent due to its potential use as an agent of bioterrorism, as it can easily be transmitted by aerosol at a low dose of infection and is capable of causing extreme illness and even death within a short period of time. Several epitopes based upon protective antigens of the highly virulent SchuS4 strain of *F. tularensis*, OmpA-like protein (OmpA), chaperone protein DnaK, and lipoprotein Tul4, were conjugated so that they decorated the surface of TMV particles. These TMV-conjugate vaccine formulations were found to be safe when administered to mice in multiple doses and induced a strong humoral immune response which was successful in protecting mice against respiratory challenges with very high doses of *F. tularensis* LVS.

TMV has also been used by Jones et al. [134] to generate a malaria transmission blocking vaccine (TBV). TBVs which target proteins expressed in the mosquito midgut during sexual development of *Plasmodium falciparum*, such as the Pfs25 protein, are considered to be one avenue through which transmission of malaria can be reduced in regions where the disease is endemic. Mosquitoes which bite a vaccinated individual ingest antibodies which block parasite spread to the next human host. Vaccine proteins purified 6 days post-infiltration were shown to induce in both mice and rabbits transmission-blocking antibodies that persisted for up to 6 months postimmunization, indicating that new malaria vaccines could be developed on a large scale and inexpensively using this form of production line.

Cowpea mosaic virus (CPMV) is an icosahedral virus with a bipartite genome. CPMV has been utilized extensively for a number of functions including antigenic presentation and full-length protein expression [135, 136]. The deconstructed vector system pEAQ is based on CPMV RNA-2 and uniquely involves the expression of foreign proteins without the need for viral replication. Vectors have been used to generate vaccines against bluetongue virus, HIV, and influenza virus as well as the therapeutic protein human gastric lipase [137–141].

Potato Virus X (PVX), like TMV, is also a rod-shaped virus and has been used to express full-length proteins, fusion proteins in conjunction with the capsid protein, or epitopes that are displayed on the outer surface of the assembled virus particle. The use of PVX as an expression vector for vaccine production can be further illustrated by the work of Mardanova et al. [30]. This research group attempted to develop a universal influenza vaccine that is derived from the extracellular domain of the matrix protein 2 (M2e) of the virus. Since M2e is a weak immunogen, the researchers delivered it as part of a fusion protein with bacterial flagellin, a strong mucosal adjuvant. High levels of expression (1 mg/g fresh leaf tissue or 30% total soluble protein) were achieved in infected *Nicotiana benthamiana* plants using the PVX delivery vehicle. Mice who were inoculated intranasally with virus particles were protected from influenza lethal challenge, demonstrating the efficacy of this vaccine design and strategy.

Geminiviruses are single-stranded circular DNA viruses and possess a twinned capsid morphology. The fact that geminiviruses can replicate to a high copy number in a broad range of host plants make them attractive candidates for use as expression vectors. Several deconstructed versions of geminivirus have been developed and used to generate monoclonal antibodies to Ebola and West Nile viruses, respectively [129]. Other constructs based on BeYDV have been developed to generate vaccines against HIV and HPV [142].

4 Conclusions

Plants and plant materials have been used by humans throughout history and date back to the prehistoric era. However, recent advances in biotechnology led to the resurgence of using plants as biofactories to produce biopharmaceuticals (molecular pharming) through a completely different approach. To date, many plant varieties such as tobacco, potato, tomato, lettuce, wheat, and sorghum and also many plant systems such as suspension cultures, hairy roots, hydroponic culture, plant chloroplasts, and algae have been used in this regard. The lack of recombinant protein yield, however, is still an issue to be addressed. Ongoing efforts have been made to improve recombinant protein yield in transgenic plants. Strong promoters, strong 5′ and 3′ UTRs, suppression of promoter and mRNA silencing effects, etc. have been employed to enhance protein production. Also, subcellular and compartmental targeting, incorporating sequences to exert unique glycosylation patterns, and supplementing media with protein stabilizing and protease inhibitors can also preserve recombinant proteins. The use of viral vectors is another technique employed in enhancing recombinant protein yield.

The FDA approval of recombinant glucocerebrosidase enzyme produced in carrot cells is a major breakthrough and success story in molecular pharming. Many other plant-produced pharmaceuticals are in the pipeline to be released in the market shortly after completing rigorous clinical trials. However, investigations to improve recombinant protein yield are currently ongoing. Other improvements in the field of molecular pharming particularly at the level of functional genomics, epigenetics, and epigenomics will inevitably produce pharmagens more effectively and are likely to be achieved in the near future.

References

1. Winslow L, Kroll DJ (1998) Herbs as medicines. Arch Intern Med 158(20):2192–2199
2. Fischer R, Emans N (2000) Molecular farming of pharmaceutical proteins. Transgenic Res 9:279–299
3. Fraley RT, Rogers SG, Horsch RB, Sanders PR, Flick JS, Adams SP et al (1983) Expression of bacterial genes in plant cells. Proc Natl Acad Sci 80:4803–4807
4. Barta A, Sommergruber K, Thompson D, Hartmuth K, Matzke MA, Matzke AJM (1986) The expression of a nopaline synthase-human growth hormone chimaeric gene in transformed tobacco and sunflower callus tissue. Plant Mol Biol 6:347–357
5. De Zoeten GA, Penswick JR, Horisberger MA, Ahl P, Schultze M, Hohn T (1989) The expression, localization, and effect of a human interferon in plants. Virology 172:213–222
6. Hiatt A, Cafferkey R, Bowdish K (1989) The production of antibodies in transgenic plants. Nature 342:76–78
7. Sijmons PC, Dekker BMM, Schrammeijer B, Verwoerd TC, van den Elzen PJM, Hoekema A (1990) Production of correctly processed human serum albumin in transgenic plants. Nat Biotechnol 8(3):217–221
8. Makhzoum A, Benyammi R, Moustafa K, Trémouillaux-Guiller J (2014) Recent advances on host plants and expression cassettes' structure and function in plant molecular pharming. BioDrugs 28:145–159
9. Soria-Guerra RE, Moreno-Fierros L, Rosales-Mendoza S (2011) Two decades of plant-based candidate vaccines: a review of the chimeric protein approaches. Plant Cell Rep 30:1367–1382
10. Kaiser J (2008) Is the drought over for pharming? Science 320:473–475
11. Pfizer Press Releases (2012) Pfizer and Protalix BioTherapeutics Announce FDA Approval Of Elelyso™ (taliglucerase alfa) For The Treatment Of Gaucher Disease. http://press.pfizer.com/press-release/pfizer-and-protalix-biotherapeutics-announce-fda-approval-elelyso-taliglucerase-alfa-t. Accessed 04th July 2016
12. Singhabahu S, George J, Bringloe DH (2015) High yield production of apoplast-directed human adenosine deaminase in transgenic tobacco BY-2 cell suspensions. Biotechnol Appl Biochem 62(1):87–93
13. Makhzoum AB, Sharma P, Bernards MA, Trémouillaux-Guiller J (2013) Hairy roots: an ideal platform for transgenic plant production and other promising applications. In: Gang RD (ed) Phytochemicals, plant growth, and the environment. Springer, New York, pp 95–142

14. Oey M, Lohse M, Kreikemeyer B, Bock R (2009) Exhaustion of the chloroplast protein synthesis capacity by massive expression of a highly stable protein antibiotic. Plant J 57:436–445
15. Boyhan D, Daniell H (2011) Low-cost production of proinsulin in tobacco and lettuce chloroplasts for injectable or oral delivery of functional insulin and C-peptide. Plant Biotechnol J 9(5):585–598
16. Singhabahu S, George J, Bringloe DH (2013) Expression of functional human adenosine deaminase in tobacco plants. Transgenic Res 22(3):643–649
17. Skarjinskaia M, Ruby K, Araujo A, Taylor K, Gopalasamy-Raju V, Musiychuk K et al (2013) Hairy roots as a vaccine production and delivery system. In: Doran MP (ed) Biotechnology of hairy root systems. Springer, Berlin/Heidelberg, pp 115–134
18. Moustafa K, Makhzoum A, Trémouillaux-Guiller J (2016) Molecular farming on rescue of pharma industry for next generations. Crit Rev Biotechnol 36(5):840–850
19. Sasou A, Shigemitsu T, Saito Y, Tanaka M, Morita S, Masumura T (2016) Control of foreign polypeptide localization in specific layers of protein body type I in rice seed. Plant Cell Rep 35:1287–1295
20. Rosales-Mendoza S, Tello-Olea MA (2015) Carrot cells: a pioneering platform for biopharmaceuticals production. Mol Biotechnol 57(3):219–232
21. Merlin M, Gecchele E, Arcalis E, Remelli S, Brozzetti A, Pezzotti M, Avesani L (2016) Enhanced GAD65 production in plants using the MagnICON transient expression system: optimization of upstream production and downstream processing. Biotechnol J 11:542–553
22. Barahimipour R, Neupert J, Bock R (2016) Efficient expression of nuclear transgenes in the green alga *Chlamydomonas*: synthesis of an HIV antigen and development of a new selectable marker. Plant Mol Biol 90(4):403–418
23. Vamvaka E, Twyman RM, Murad AM, Melnik S, Teh AY-H, Arcalis E et al (2016) Rice endosperm produces an underglycosylated and potent form of the HIV-neutralizing monoclonal antibody 2G12. Plant Biotechnol J 14:97–108
24. Chan H-T, Xiao Y, Weldon WC, Oberste SM, Chumakov K, Daniell H (2016) Cold chain and virus-free chloroplast-made booster vaccine to confer immunity against different poliovirus serotypes. Plant Biotechnol J 11:2190–2200
25. Hensel G, Floss DM, Arcalis E, Sack M, Melnik S et al (2015) Transgenic production of an anti HIV antibody in the Barley endosperm. PLoS One 10(10), e0140476
26. Mbewana S, Mortimer E, Francisco P, Hitzeroth II, Edward R (2015) Production of H5N1 influenza virus matrix protein 2 ectodomain protein bodies in tobacco plants and in insect cells as a candidate universal influenza vaccine. Front Bioeng Biotechnol 3:197
27. Tinazzi E, Merlin M, Bason C, Beri R, Zampieri R, Lico C et al (2015) Plant-derived chimeric virus particles for the diagnosis of primary Sjögren syndrome. Front Plant Sci 6:1080
28. Soh HS, Chung HY, Lee HH, Ajjappala H, Jang K, Park J-H et al (2015) Expression and functional validation of heat-labile enterotoxin B (LTB) and cholera toxin B (CTB) subunits in transgenic rice (*Oryza sativa*). Springer Plus 4(1):1–14
29. Nahampun HN, Bosworth B, Cunnick J, Mogler M, Wang K (2015) Expression of H3N2 nucleoprotein in maize seeds and immunogenicity in mice. Plant Cell Rep 34(6):969–980
30. Mardanova ES, Kotlyarov RY, Kuprianov VV, Stepanova LA, Tsybalova LM, Lomonosoff GP et al (2015) Rapid high-yield expression of a candidate influenza vaccine based on the ectodomain of M2 protein linked to flagellin in plants using viral vectors. BMC Biotechnol 15(1):1–10
31. Su J, Zhub L, Shermanb A, Wangb X, Lina S, Kamesha A et al (2015) Low cost industrial production of coagulation factor IX bioencapsulated in lettuce cells for oral tolerance induction in hemophilia B. Biomaterials 70:84–93
32. De Myer T, Laukens B, Nolf J, Van Lerberge E, De Rycke R, De Beuckelaer A et al (2015) Comparison of VHH-Fc antibody production in *Arabidopsis thaliana*, *Nicotiana benthamiana* and *Pichia pastoris*. Plant Biotechnol J 13:938–947

33. Bundó M, Montesinos L, Izquierdo E, Campo S, Mieulet D, Guiderdoni E et al (2014) Production of cecropin A antimicrobial peptide in rice seed endosperm. BMC Plant Biol 14(1):1–13
34. Schillberg S, Fischer R, Emans N (2003) Molecular farming of recombinant antibodies in plants. Cell Mol Life Sci 60:433–445
35. Yao J, Weng Y, Dickey A, Wang KY (2015) Plants as factories for human pharmaceuticals: applications and challenges. Int J Mol Sci 16(12):28549–28565
36. Yusibov Y, Mett V, Musiychuk K (2012) Influenza vaccines, antigens, composition and methods. United States Patent 20120034253A1
37. Takaiwa F, Wakasa Y, Takagi H, Hiroi T (2015) Rice seed for delivery of vaccines to gut mucosal immune tissues. Plant Biotechnol J 13:1041–1055
38. Mapp Biopharmaceutical Press Releases (2016) LeafBio announces conclusion of ZMapp™ clinical trial; Therapy to treat Ebola shows promise. http://mappbio.com/leafbio-announces-conclusion-of-zmapp-clinical-trial/. Accessed 20th July 2016
39. Rybicki EP (2009) Plant-produced vaccines: promise and reality. Drug Discov Today 14:16–24. doi:10.1016/j.drudis.2008.10.002
40. Cummings J, Guerrero M, Moon J, Waterman P, Nielsen R, Jefferson S et al (2014) Safety and immunogenicity of a plant-produced recombinant monomer hemagglutinin-based influenza vaccine derived from influenza A (H1N1) pdm09 virus: a phase 1 dose-escalation study in healthy adults. Vaccine 32:2251–2259
41. Shoji Y, Farrance CE, Bautista J, Bi H, Musiychuk K, Horsey A et al (2012) A plant-based system for rapid production of influenza vaccine antigens. Influenza Other Respi Viruses 6(3):204–210
42. Yuki Y, Mejima M, Kurokawa S, Hiroiwa T, Takahashi Y, Tokuhara D et al (2013) Induction of toxin-specific neutralizing immunity by molecularly uniform rice-based oral cholera toxin B subunit vaccine without plant-associated sugar modification. Plant Biotechnol J 11:799–808
43. Nochi T, Yuki Y, Katakai Y, Shibata H, Tokuhara D, Mejima M et al (2009) A rice based oral cholera vaccine induces macaque-specific systemic neutralizing antibodies but does not influence pre-existing intestinal immunity. J Immunol 183:6538–6544
44. Chichester J, Jones R, Green B, Stow M, Miao F, Moonsammy G et al (2012) Safety and immunogenicity of a plant-produced recombinant hemagglutinin-based influenza vaccine (HAI-05) derived from A/Indonesia/05/2005 (H5N1) influenza virus: a phase 1 randomized, double-blind, placebo controlled, dose-escalation study in healthy adults. Viruses 4:3227–3244
45. Bendandi M, Marillonnet S, Kandzia R, Thieme F, Nickstadt A, Herz S et al (2010) Rapid, high-yield production in plants of individualized idiotype vaccines for non-Hodgkin's lymphoma. Ann Oncol 21(12):2420–2427
46. Tacket C, Pasetti M, Edelman R, Howard J, Streatfield S (2004) Immunogenicity of recombinant LT-B delivered orally to humans in transgenic corn. Vaccine 22:4385–4389
47. Mason HS, Haq TA, Clements JD, Arntzen CJ (1998) Edible vaccine protects mice against *Escherichia coli* heat-labile enterotoxin (LT): potatoes expressing a synthetic LT-B gene. Vaccine 16(13):1336–1343
48. Yusibov V, Hooper D, Spitsin S, Fleysh N, Kean R, Mikheeva T et al (2002) Expression in plants and immunogenicity of plant virus-based experimental rabies vaccine. Vaccine 20:3155–3164
49. Modelska A, Dietzschold B, Sleysh N, Fu ZF, Steplewski K, Hooper DC, Hilary K, Yusibov V (1998) Immunization against rabies with plant-derived antigen. Proc Natl Acad Sci 95(5):2481–2485
50. Mason HS, Ball JM, Shi JJ, Jiang X, Estes MK, Arntzen CJ (1996) Expression of Norwalk virus capsid protein in transgenic tobacco and potato and its oral immunogenicity in mice. Proc Natl Acad Sci 93(11):5335–5340
51. Influenza A virus vaccine H7N9 – Medicago http://adisinsight.springer.com/drugs/800039770

52. Williams SCP (2012) Drugmaker reaps what it sows with first plant-made biologic. Nat Med 18(1):5–5
53. Pogue GP, Vojdani F, Palmer KE, Hiatt E, Hume S, Phelps J et al (2010) Production of pharmaceutical-grade recombinant aprotinin and a monoclonal antibody product using plant-based transient expression systems. Plant Biotechnol J 8(5):638–654
54. Streatfield SJ (2006) Mucosal immunization using recombinant plant-based oral vaccines. Methods 38(2):150–157
55. Kapusta J, Modelska A, Figlerowicz M, Pniewski T, Letellier M, Lisowa O et al (1999) A plant-derived edible vaccine against hepatitis B virus. J Fed Am Soc Exp Biol 13(13):1796–1799
56. Cox KM, Sterling JD, Regan JT, Gasdaska JR, Frantz KK, Peele CG et al (2006) Glycan optimization of a human monoclonal antibody in the aquatic plant Lemna minor. Nat Biotechnol 24(12):1591–1597
57. Tacket C, Mason H, Losonsky G, Estes M, Levine M, Arntzen C (2000) Human immune responses to a novel norwalk virus vaccine delivered in transgenic potatoes. J Infect Dis 182:302–305
58. Yusibov V, Modelska A, Steplewski K, Agadjanyan M, Weiner D, Hooper DC, Koprowski H (1997) Antigens produced in plants by infection with chimeric plant viruses immunize against rabies virus and HIV-1. Proc Natl Acad Sci 94(11):5784–5788
59. Shaaltiel Y, Bartfeld D, Hashmueli S, Baum G, Brill-Almon E, Galili G et al (2007) Production of glucocerebrosidase with terminal mannose glycans for enzyme replacement therapy of Gaucher's disease using a plant cell system. Plant Biotechnol J 5(5):579–590
60. Markley N, Nykiforuk C, Boothe J, Moloney M (2006) Producing proteins using transgenic Oilbody-Oleosin technology. BioPharm Int 19(6):34–37
61. Zhong Q, Gu Z, Glatz CE (2006) Extraction of recombinant dog gastric lipase from transgenic corn seed. J Agric Food Chem 54(21):8086–8092
62. Samyn-Petit B, Gruber V, Flahaut C, Wajda-Dubos JP, Farrer S, Pons A et al (2001) *N*-glycosylation potential of maize: the human lactoferrin used as a model. Glycoconj J 18(7):519–527
63. Takeyama N, Kiyono H, Yuki Y (2015) Plant-based vaccines for animals and humans: recent advances in technology and clinical trials. Ther Adv Vaccines 3(5–6):139–154
64. Yusibov V, Streatfield SJ, Kushnir N (2011) Clinical development of plant-produced recombinant pharmaceuticals: vaccines, antibodies and beyond. Hum Vaccin 7(3):313–321
65. Paul M, Ma JKC (2011) Plant-made pharmaceuticals: leading products and production platforms. Biotechnol Appl Biochem 58(1):58–67
66. Odell JT, Nagy F, Chua NH (1985) Identification of DNA sequences required for activity of the cauliflower mosaic virus 35S promoter. Nature 313:810–812
67. Mason HS, Lam DM, Arntzen CJ (1992) Expression of hepatitis B surface antigen in transgenic plants. Proc Natl Acad Sci 89(24):11745–11749
68. Panahi M, Alli Z, Cheng X, Belbaraka L, Belgoudi J, Sardana R et al (2004) Recombinant protein expression plasmids optimized for industrial *E. coli* fermentation and plant systems produce biologically active human insulin-like growth factor-1 in transgenic rice and tobacco plants. Transgenic Res 13(3):245–259
69. Torres E, Vaquero C, Nicholson L, Sack M, Stöger E, Drossard J et al (1999) Rice cell culture as an alternative production system for functional diagnostic and therapeutic antibodies. Transgenic Res 8(6):441–449
70. Smith ML, Mason HS, Shuler ML (2002) Hepatitis B surface antigen (HBsAg) expression in plant cell culture: kinetics of antigen accumulation in batch culture and its intracellular form. Biotechnol Bioeng 80:812–822
71. Huang TK, McDonald KA (2009) Bioreactor engineering for recombinant protein production in plant cell suspension cultures. Biochem Eng J 45(3):168–184
72. Sharma AK, Sharma MK (2009) Plants as bioreactors: recent developments and emerging opportunities. Biotechnol Adv 2:811–832

73. Kim NS, Kim TG, Kim OH, Ko EM, Jang YS, Jung ES et al (2008) Improvement of recombinant hGM-CSF production by suppression of cysteine proteinase gene expression using RNA interference in a transgenic rice culture. Plant Mol Biol 68:263–275
74. Shin YJ, Lee NJ, Kima J, An XH, Yang MS, Kwon TH (2010) High-level production of bioactive heterodimeric protein human interleukin-12 in rice. Enzyme Microb Technol 46:347–351
75. Springer PS, McCombie WR, Sundaresan V, Martienssen RA (1995) Gene trap tagging of PROLIFERA, an essential MCM2-3-5-like gene in *Arabidopsis*. Science 268:877–880
76. Hepburn AG, Clarke LE, Pearson L, White J (1983) The role of cytosine methylation in the control of nopaline synthase gene expression in a plant tumor. J Mol Appl Genet 2:315–329
77. Stam M, De Bruijn R, Kenter S, Van Der Hoorn RAL, Van Blokland R, Mol JNM, Kooter JM (1997) Post-transcriptional silencing of chalcone synthase in Petunia by inverted transgene repeats. Plant J 12:63–82
78. Gallie DR, Walbot V (1990) RNA pseudoknot domain of tobacco mosaic virus can functionally substitute for a poly(A) tail in plant and animal cells. Genes Dev 4:1149–1157
79. Gallie DR (2002) The 5′-leader of tobacco mosaic virus promotes translation through enhanced recruitment of eIF4F. Nucleic Acids Res 30(15):3401–3411
80. Mazur BJ, Chui CF (1985) Sequence of a genomic DNA clone for the small subunit of ribulose bis-phosphate carboxylase-oxygenase from tobacco. Proc Natl Acad Sci 13:2373–2386
81. Patel M, Corey AC, Yin LP, Ali S, Taylor WC, Berry JO (2004) Untranslated regions from C4 *Amaranth* AhRbcS1 mRNAs confer translational enhancement and preferential bundle sheath cell expression in transgenic c4 *Flaveria bidentis*. Plant Physiol 136:3550–3561
82. Patel M, Siegel AJ, Berry JO (2006) Untranslated regions of FbRbcS1 mRNA mediate bundle sheath cell-specific gene expression in leaves of a C4 plant. J Biol Chem 281:25485–25491
83. Saxena P, Hsieh Y-C, Alvarado VY, Sainsbury F, Saunders K, Lomonossoff GP, Scholthof HB (2011) Improved foreign gene expression in plants using a virus-encoded suppressor of RNA silencing modified to be developmentally harmless. Plant Biotechnol J 9:703–712
84. Verma D, Daniell H (2007) Chloroplast vector systems for biotechnology applications. Plant Physiol 145(4):1129–1143
85. Moreira D, Guyader HL, Philippe H (2000) The origin of red algae and the evolution of chloroplasts. Nature 405(6782):69–72
86. Gray MW (1999) Evolution of organellar genomes. Curr Opin Genet Dev 9(6):678–687
87. Bendich AJ (1987) Why do chloroplasts and mitochondria contain so many copies of their genome? Bioessays 6:279–282
88. Heifetz PB, Lers A, Turpin DH, Gillham NW, Boynton JE, Osmond CB (1997) dr and spr/sr mutations of *Chlamydomonas reinhardtii* affecting D1 protein function and synthesis define two independent steps leading to chronic photoinhibition and confer differential fitness. Plant Cell Environ 20:1145–1157
89. Boynton JE, Gillaham NW, Harris EH, Hosler JP, Johnson AM, Jones AR et al (1988) Chloroplast transformation in *Chlamydomonas* with high velocity microprojectiles. Science 240(4858):1534–1538
90. Svab Z, Hajdukiewicz P, Maliga P (1990) Stable transformation of plastids in higher plants. Proc Natl Acad Sci U S A 87:8526–8530
91. De Cosa B, Moar W, Lee SB, Miller M, Daniell H (2001) Overexpression of the Bt cry2Aa2 operon in chloroplasts leads to formation of insecticidal crystals. Nat Biotechnol 19:71–74
92. Bogorad L (2000) Engineering chloroplasts: an alternative site for foreign genes, proteins, reactions and products. Trends Biotechnol 18(6):257–263
93. Daniell H, Muthukumar B, Lee SB (2001) Marker free transgenic plants: engineering the chloroplast genome without the use of antibiotic selection. Curr Genet 39:109–116
94. Lee SB, Kwon HB, Kwon SJ, Park SC, Jeong MJ, Han SE et al (2003) Accumulation of trehalose within transgenic chloroplasts confers drought tolerance. Mol Breed 11:1–13
95. Ruf S, Hermann M, Berger IJ, Carrer H, Bock R (2001) Stable genetic transformation of tomato plastids and expression of a foreign protein in fruit. Nat Biotechnol 19:870–875

96. Daniell H (2004) Medical molecular pharming: therapeutic recombinant antibodies, biopharmaceuticals, and edible vaccines in transgenic plants engineered via the chloroplast genome. In: Goodman RM (ed) Encyclopedia of plant and crop science. Marcel Decker, New York, pp 704–710
97. Doran PM (2006) Foreign protein degradation and instability in plants and plant tissue cultures. Trends Biotechnol 24(9):426–432
98. Conrad U, Fiedler U (1998) Compartment-specific accumulation of recombinant immunoglobulins in plant cells: an essential tool for antibody production and immunomodulation of physiological functions and pathogen activity. Plant Mol Biol 38:101–109
99. Nuttal J, Vine N, Hadlington JL, Drake P, Frigerio L, Ma JK (2002) ER-resident chaperone interactions with recombinant antibodies in transgenic plants. Eur J Biochem 269:6042–6051
100. Petruccelli S, Otegui MS, Lareu F, Tran Dinh O, Fitchette AC, Circosta A et al (2006) A KDEL-tagged monoclonal antibody is efficiently retained in the endoplasmic reticulum in leaves, but is both partially secreted and sorted to protein storage vacuoles in seeds. Plant Biotechnol J 4(5):511–527
101. Duwadi K, Chen L, Menassa R, Dhaubhadel S (2015) Identification, characterization and down-regulation of cysteine protease genes in tobacco for use in recombinant protein production. PLoS One 10, e0130556. doi:10.1371/journal.pone.0130556
102. Mandal MK, Ahvari H, Schillberg S, Schiermeyer A (2016) Tackling unwanted proteolysis in plant production hosts used for molecular farming. Front Plant Sci 7:267
103. Pillay P, Kibido T, Plessis M, Vyver C, Beyene G, Vorster BJ et al (2012) Use of transgenic oryzacystatin-I-expressing plants enhances recombinant protein production. Appl Biochem Biotechnol 168:1608–1620
104. Pen J, van Ooyen AJJ, van den Elzena PJM, Quaxb WJ, Hoekema A (1993) Efficient production of active industrial enzymes in plants. Ind Crops Prod 1:241–250
105. Francisco JA, Gawlak AL, Miller M, Bathe J, Russell D, Chace D (1997) Expression and characterization of Bryodin 1 and a Bryodin1-based single chain immunotoxin from tobacco cell culture. Bioconjug Chem 8:708–713
106. Ziegler TM, Thomas SR, Danna KJ (2000) Accumulation of a thermostable endo-1,4-β-D-glucanase in the apoplast of *Arabidopsis thaliana* leaves. Mol Breed 6:37–46
107. Xu J, Tan L, Goodrum KJ, Kieliszewski MJ (2007) High-yields and extended serum half-life of human interferon α2b expressed in tobacco cells as arabinogalactan-protein fusions. Biotechnol Bioeng 97(5):997–1008
108. Wongsamuth R, Doran PM (1997) Production of monoclonal antibodies by tobacco hairy roots. Biotechnol Bioeng 54:401–415
109. LaCount W, An G, Lee JM (1997) The effect of polyvinylpyrrolidone (PVP) on the heavy chain monoclonal antibody production from plant suspension cultures. Biotechnol Lett 19(1):93–96
110. James EA, Wang C, Wang Z, Reeves R, Shin JH, Magnuson NS et al (2000) Production and characterization of biologically active human GM-CSF secreted by genetically modified plant cells. Protein Expr Purif 19:131–138
111. Makhzoum A, Tahir S, Locke MEO, Trémouillaux-Guiller J, Hefferon K. (2014). An in silico overview on the usefulness of tags and linkers in plant molecular pharming. Plant Science Today 1:201–212
112. Zhang J, Yun J, Shang Z, Zhang X, Pan B (2009) Design and optimization of a linker for fusion protein construction. Prog Nat Sci 19:1197–1200
113. Gomord V, Faye L (2004) Post-translational modification of therapeutic proteins in plants. Curr Opin Plant Biol 7:171–181
114. Gomord V, Fitchette AC, Menu-Bouaouiche L, Saint-Jore-Dupas C, Plasson C, Michaud D, Faye L (2010) Plant-specific glycosylation patterns in the context of therapeutic protein production. Plant Biotechnol J 8(5):564–587

115. Faye L, Boulaflous A, Benchabane M, Gomord V, Michaud D (2005) Protein modifications in the plant secretory pathway: current status and practical implications in molecular pharming. Vaccine 23(15):1770–1778
116. Gomord V, Chamberlain P, Jefferis R, Faye L (2005) Biopharmaceutical production in plants: problems, solutions and opportunities. Trends Biotechnol 23:559–565
117. Saint-Jore-Dupas C, Faye L, Gomord V (2007) From planta to pharma with glycosylation in the toolbox. Trends Biotechnol 25:317–323
118. Showalter AM (2001) Arabinogalactan-proteins: structure, expression and function. Cell Mol Life Sci 58:1399–1417
119. Qi W, Fong C, Lamport DTA (1991) Gum Arabic glycoprotein is a twisted hairy rope: a new model based on *O*-galactosyl hydroxyproline as the polysaccharide attachment site. Plant Physiol 96:848–855
120. Xu J, Okada S, Tan L, Goodrum KJ, Kopchick JJ, Kieliszewski MJ (2010) Human growth hormone expressed in tobacco cells as an arabinogalactan-protein fusion glycoprotein has a prolonged serum life. Transgenic Res 19(5):849–867
121. Hefferon KL (2012) Plant virus expression vectors set the stage as production platforms for biopharmaceutical proteins. Virology 433(1):1–6
122. Yusibov V, Streatfield SJ, Kushnir N, Roy G, Padmanaban A (2013) Hybrid viral vectors for vaccine and antibody production in plants. Curr Pharm Des 19(31):5574–5586
123. Salazar-González JA, Bañuelos-Hernández B, Rosales-Mendoza S (2015) Current status of viral expression systems in plants and perspectives for oral vaccines development. Plant Mol Biol 87(3):203–217
124. Cañizares MC, Nicholson L, Lomonossoff GP (2005) Use of viral vectors for vaccine production in plants. Immunol Cell Biol 83(3):263–270
125. Yusibov V, Rabindran S, Commandeur U, Twyman RM, Fischer R (2006) The potential of plant virus vectors for vaccine production. Drugs R& D 7(4):203–217
126. Gleba Y, Marillonnet S, Klimyuk V (2004) Engineering viral expression vectors for plants: the 'full virus' and the 'deconstructed virus' strategies. Curr Opin Plant Biol 7:182–188
127. Gleba Y, Klimyuk V, Marillonnet S (2005) Magnifection – a new platform for expressing recombinant vaccines in plants. Vaccine 23(17–18):2042–2048
128. Noris E, Poli A, Cojoca R, Rittà M, Cavallo F, Vaglio S et al (2011) A human papillomavirus 8 E7 protein produced in plants is able to trigger the mouse immunsystem and delay the development of skin lesions. Arch Virol 156:587–591
129. Huang Z, Phoolcharoen W, Lai H, Piensook K, Cardineau G, Zeitlin L et al (2010) High-level rapid production of full-size monoclonal antibodies in plants by a single-vector DNA replicon system. Biotechnol Bioeng 106(1):9–17
130. Musiychuk K, Stephenson N, Bi H, Farrance CE, Orozovic G, Brodelius M et al (2007) A launch vector for the production of vaccine antigens in plants. Influenza Other Respi Viruses 1(1):19–25
131. Mett V, Musiychuk K, Bi H, Farrance CE, Horsey A, Ugulava N et al (2008) A plant-produced influenza subunit vaccine protects ferrets against virus challenge. Influenza Other Respi Viruses 2(1):33–40
132. Petukhova NV, Gasanova TV, Ivanov PA, Atabekov JG (2014) High-level systemic expression of conserved influenza epitope in plants on the surface of rod-shaped chimeric particles. Viruses 6(4):1789–1800
133. Banik S, Mansour AA, Suresh RV, Wykoff-Clary S, Malik M, McCormick AA, Bakshi CS (2015) Development of a multivalent subunit vaccine against tularemia using tobacco mosaic virus (TMV) based delivery system. PLoS One 10(6), e0130858
134. Jones RM, Chichester JA, Manceva S, Gibbs SK, Musiychuk K, Shamloul M et al (2015) A novel plant-produced Pfs25 fusion subunit vaccine induces long-lasting transmission blocking antibody responses. Human Vaccines Immunother 11(1):124–132

135. Sainsbury F, Lavoie P-O, D'Aoust M-A, Vézina L-P, Lomonossoff GP (2008) Expression of multiple proteins using full-length and deleted versions of cowpea mosaic virus RNA-2. Plant Biotechnol J 6:82–92
136. Sainsbury F, Thuenemann EC, Lomonossoff GP (2009) pEAQ: versatile expression vectors for easy and quick transient expression of heterologous proteins in plants. Plant Biotechnol J 7:682–693
137. Meshcheriakova YA, Saxena P, Lomonossoff GP (2014) Fine-tuning levels of heterologous gene expression in plants by orthogonal variation of the untranslated regions of a nonreplicating transient expression system. Plant Biotechnol J 12(6):718–727
138. Sainsbury F, Sack M, Stadlmann J, Quendler H, Fischer R, Lomonossoff GP (2010) Rapid transient production in plants by replicating and non-replicating vectors yields high quality functional anti-HIV antibody. PLoS One 5(11), e13976. doi:10.1371/journal.pone.0013976
139. Mardanova ES, Kotlyarov RY, Kuprianov VV, Stepanova LA, Tsybalova LM, Lomonossoff GP, Ravin NV (2016) High immunogenicity of plant-produced candidate influenza vaccine based on the M2e peptide fused to flagellin. Bioengineered 7(1):28–32
140. Vardakou M, Sainsbury F, Rigby N, Mulholland F, Lomonossoff GP (2012) Expression of active recombinant human gastric lipase in *Nicotiana benthamiana* using the CPMV-HT transient expression system. Protein Expr Purif 81(1):69–74
141. Thuenemann EC, Meyers AE, Verwey J, Rybicki EP, Lomonossoff GP (2013) A method for rapid production of heteromultimeric protein complexes in plants: assembly of protective bluetongue virus-like particles. Plant Biotechnol J 11(7):839–846
142. Regnard GL, Halley-Stott RP, Tanzer FL et al (2010) High level protein expression in plants through the use of a novel autonomously replicating geminivirus shuttle vector. Plant Biotechnol J 8(1):38–46

Biotechnological Approaches for Bioremediation: In Vitro Hairy Root Culture

24

Sonia Malik, Sara Adrián L. Andrade, Mohammad Hossein Mirjalili, Randolph R. J. Arroo, Mercedes Bonfill, and Paulo Mazzafera

Abstract

Environmental contamination has become a concerning issue worldwide due to number of risks it poses to human health and ecosystem functioning. Contaminants present in soils or waters can go up through the trophic chain via microbial or plant incorporation. Bio/Phytoremediation is an emerging technology for large-scale removal or detoxification of contaminants from the environment. It makes the use of plants and associated microbial communities to remove, transfer, or stabilize pollutants in an environmental friendly manner. This chapter discusses the biotechnological research including genetic engineering, hairy root culture, and identification of the genes or physiological processes in optimizing efficacy of plants as phytoremediators. Hairy root cultures constitute an important tool in phytoremediation research and may provide an ideal model

S. Malik (✉)
Graduate Program in Health Sciences, Biological and Health Sciences Center, Federal University of Maranhão, São Luís, MA, Brazil
e-mail: sonia.unicamp@gmail.com; 777soniamalik@gmail.com

S.A.L. Andrade (✉) • P. Mazzafera (✉)
Department of Plant Biology, Institute of Biology, State University of Campinas, Campinas, SP, Brazil
e-mail: sara.adrian@gmail.com; pmazza07@gmail.com

M.H. Mirjalili (✉)
Department of Agriculture, Medicinal Plants and Drugs Research Institute, Shahid Beheshti University, G.C., Tehran, Iran
e-mail: m-mirjalili@sbu.ac.ir

R.R.J. Arroo (✉)
Leicester School of Pharmacy, De Montfort University, Leicester, UK
e-mail: rrjarroo@dmu.ac.uk

M. Bonfill (✉)
Plant Physiology Laboratory, Faculty of Pharmacy, University of Barcelona, Barcelona, Spain
e-mail: mbonfill@ub.edu

S. Jha (ed.), *Transgenesis and Secondary Metabolism*, Reference Series in Phytochemistry, DOI 10.1007/978-3-319-28669-3_28

system to identify the role of plants in phytoremediation. Future studies on hairy roots system in relation to phytoremediation should focus on the engineering of target genes involved in this process and to extend the basic hairy root phytoremediation model to the environment.

Keywords
Agrobacterium rhizogenes • Detoxification • Environmental contamination • Phytoremediation • Phytovolatilization • Pollutants • Polycyclic aromatic hydrocarbons • In vitro culture • Rhizofiltration • Ri-plasmid

Contents

1 Introduction

Environmental contamination has become a concerning issue worldwide due to the risk it present to human health and ecosystem functioning. In developed and developing countries, the modern life style, with its increasing production of goods and the high consuming habits of people, is leading to an ever increasing number of areas modified by anthropogenic activities. These modifications include physical, chemical, and biological perturbances that may have detrimental effects on the living organisms. From an anthropocentric point of view, environmental contamination may affect soil productivity, socioeconomic well-being, and population's health. Environmental contamination may affect soil, water, and atmosphere with foreign organic or inorganic compounds accumulated as a result of misuse of the natural resources. As life on earth mainly concentrates above soil or water compartments, we will focus on contaminants affecting mainly soil ecosystems and in some cases water courses. Soil can be considered as the base for life on earth, where most plants grow and support many animal and microbes lives. Soil microbes are organisms of fundamental importance for geochemical and nutrient cycling, and the functioning of the soil ecosystem depends in a great manner on their activities allowing primary production. Thus, they have a central role in nutrient cycling and degradation of innumerous organic compounds.

Contaminants are present in soils or water courses can go up through the trophic chain via microbial or plant incorporation. As plants and microbes serve as energy source for other organisms, they too may become contaminated or harmed by the presence of contaminants. This fact may lead to the so-called biomagnification, defined as the increase in the concentration of a particular compound in living organisms due to the rate of absorption being higher than that of excretion. This process may lead to higher concentrations in organisms higher up in the trophic chain and may eventually adversely affect human health.

Among the organic pollutants in soils, herbicides and pesticides are very frequently found due to their excessive or inadequate use in agriculture. Examples of these class of compounds are atrazine, endosulphan, chlorpyrifos, 2,4-D, and cypermethrin, which are used on a large scale in agriculture and represent a real and constant threat for human health since its leaching and accumulation in soil directly affects soil functioning and thus threatens human and animal health. Agrotoxic residues in agricultural foods are relatively high for world population. It is estimated that in countries as Brazil, one of the most important producers of agricultural foods in the world, around 5 kg of agrotoxins are consumed annually per person. This has been related with the increasing incidence of diseases as cancer, neurological disorders, allergies, etc. [1–3].

Other important groups of organic pollutants are represented by so called polycyclic aromatic hydrocarbons (PAH), which include compounds such as naphthalene and pyrene; petroleum products as benzene, toluene, ethylbenzene, and xylene; and polychlorinated biphenyls (PCBs) [4]. Organic compounds related to pharmaceutical substances, viz., antibiotics, caffeine, or hormones, are emerging contaminants in environment with water courses as main receivers of these organics [5].

Degradation of organic pollutants has been traditionally accomplished by physicochemical technologies or by the use of bioremediation approaches. Recently, plant-engineering focuses the improvement of plant capabilities to degrade or transform organic compounds by overexpression of genes related to enzymes transforming xenobiotics [6]. Microbial genes involved in the degradation of specific organics can be introduced in plants in order to "in-planta remediation" or to diminish xenobiotic toxicity. Microbial endophytes inoculation with pollutant-degrading capabilities offers new functional resources for achieving organics remediation [7–9].

Among inorganic contaminants, metals and metalloids as lead, cadmium, copper, mercury, and arsenic are among the most toxic contaminants in soils and ground waters. Unlike organics, they cannot be degraded by physicochemical or biological agents becoming persistent in environment. In this case, controlling levels, mobility, and chemical specification is important in order to minimize the risk to their toxicity to biota and their entrance into food chain. The manipulation of metal-contaminated sites by changing in a controlled manner the physicochemical conditions or by introducing engineered systems can greatly diminish the mobility and bioavailability of inorganic contaminants, thus reducing toxicity levels to acceptable levels and allowing regeneration to proceed.

2 Biotechnological Approaches to Remove the Contaminants

Inorganics, petroleum, explosives, solvents, polycyclic aromatic hydrocarbons, and other organic compounds are common contaminants which are continually polluting soils and ground/surface waters [10]. The concentrations of the contaminants can vary from highly toxic concentrations from an accidental spill to barely detectable concentrations that, after long-term exposure, can be detrimental to human health [11]. Excavation, transport, soil washing, extraction, pumping, and treating of contaminated water, addition of reactants such as hydrogen peroxide or potassium permanganate, and incineration are prevalent engineering methods for the remediation of contaminated sites [7]. The cost of cleaning up with these methods is extremely high. A serious consequence of the high cost of remediation technologies is that polluted commercial properties are often abandoned rather than cleaned up. Using plants and their associated microbes to clean up toxic substances in the environment has been introduced from several decades ago [12, 13]. Phytoremediation takes advantage of the natural ability of plants to extract chemicals from water, soil, and air using energy from sunlight. The advantages of phytoremediation are its relatively low costs compared to mechanical methods, high public acceptance, little secondary waste generation, conservation of topsoil, and further it is passive and solar-driven [7].

It is well known that phytoremediation occur naturally in plants, with wide variation observed depending on plant genotype and environmental conditions. Adaptation to the particular soil-climatic zone productivity (fast-growing, large biomass formation), presence of the corresponding physiological (transpiration ability) and morphological (development of an appropriate leaf and root system) characteristics, adaptation to field conditions, and presence of the needed broad enzymatic systems are desirable characteristics for the plants which are interesting for phytoremediation [14]. In order to increase the ecological potential of plants, definite progress has already been achieved in the cloning of genes participating in metabolism, uptake, or transport of specific pollutants. In the last two decades, genetic engineering has substantially contributed on improving the ability of plants to remove environmental pollution [15, 16]. Genes from microbes, plants, and animals are being used successfully to enhance the ability of plants to tolerate, remove, and degrade soils and waters contaminants [7]. In addition to these transgenic approaches, the use of microbes that live within plants, termed endophytes, also led to improved tolerance to normally phytotoxic chemicals and increased removal of the contaminants [8, 17]. The first attempt to develop engineered plants for phytoremediation of organic pollutants targeted explosives and halogenated organic compounds in tobacco plants [18, 19]. The efficiency of transgenic plants to degrade chlorinated solvents, explosives, phenolics, etc. has been extensively acknowledged in the literature [7, 10, 20–23].

Radionuclides (e.g., Cs, P, U), metals/metalloids (e.g., As, Cd, Cu, Hg, Mn, Se, Zn), and plant fertilizers (e.g., nitrate, phosphate) are also known as common inorganic contaminants which occur in nature mainly as positively or negatively charged ions and depend on plant transporters for uptake and translocation. Inorganics can be altered (reduced/oxidized), moved into/inside plants, or in some cases volatilized (Hg, Se) but cannot be degraded. Thus, phytoremediation methods available for inorganics include immobilization (phytostabilization), sequestration in harvestable plant tissues (phytoextraction or rhizofiltration), and, in exceptional cases, phytovolatilization [7, 24]. Biotechnological approaches that have successfully altered the capacity of plants for phytoremediation of the contaminants especially inorganics have focused on both tolerance and accumulation. Genes targeted include metal transporter genes, as well as genes that facilitate chelator production. Also, in the case of elements that can be volatilized, genes that facilitate conversion to volatile forms were overexpressed [16]. The introduction of these genes can be readily achieved for many plant species using *Agrobacterium tumefaciens*-mediated plant transformation or direct DNA methods of gene transfer [7]. Phytoremediation of toxic metals has been successfully improved with transgenics [25–27]. Many of the genes involved in metal uptake, translocation, and sequestration have been identified using the model plant *Arabidopsis* or naturally hyperaccumulating plants. However, the phytoremediation capacity of these natural hyperaccumulators is limited by their small size, slow growth rates, and limited growth habitat [26]. Therefore, if the genes were transferred to plant species such as poplar and willow with their high biomass and extensive root systems, significant removal of the heavy metals should be achieved. Transgenic plants with improved metal uptake and/or sequestration have been developed for cadmium, zinc, mercury, arsenic, and selenium [28–30].

In vitro plant cell and tissue culture is also considered as an important biotechnological tool for fundamental studies that provide information about the plant-contaminant relationships, help to predict plant responses to environmental contaminants, and improve the design of plants with enhanced characteristics for phytoremediation [6]. Callus, cell suspensions, hairy roots, and shoot multiplication cultures are used to study the interactions between plants and contaminants under aseptic conditions [31, 32]. In vitro cultures offer a range of experimental advantages in studies aimed at examining the intrinsic metabolic capabilities of plant cells and their capacity for toxicity tolerance. The advantages of cell suspension culture system over intact plants are: (i) lack of interfering photochemical and microbial transformation of the pollutants, (ii) results may obtained more quickly, (iii) absence of chlorophyll and other plant pigments facilitates complete extraction and identification of metabolites, (iv) plant cells can survive at moderate concentrations of xenobiotics without any cellular damage, (v) plant cell suspensions can be easily cultured in scaled-up assays (up to 50 g fresh weight) and air-lift-fermenters (~500 g fresh weight), and (vi) in vitro culture of root and shoot cells allow indefinite propagation and experimentation using tissues derived from the same plant, avoiding the risks of variability among species [6, 33–35]. The ability to identify the contributions of plant cells to contaminant uptake and detoxification without interference from microorganisms is of particular significance in the search for fundamental knowledge about plants. However, if the ultimate goal of plant tissue culture experiments is the development of practical phytoremediation technology, the limitations inherent in the use of in vitro cultures as a representative of whole plants in the field must be recognized. The bioavailability of pollutants and the processes of contaminant uptake and metabolite distribution are likely to be substantially different in the two systems; this can lead to qualitative as well as quantitative differences in metabolic profiles and tolerance characteristics. Many plant species have an inherent ability to accumulate/metabolize a variety of contaminants, but they normally produce little biomass [36]. However, fast-growing plants such as poplar and willow trees are excellent candidates for phytoremediation because of their rapid growth, extensive root system, and high water uptake [7, 37].

Also, hairy roots, fine fibrous structures that are formed on plant tissues infected by *Agrobacterium rhizogenes* [38], can be grown in large mass in culture media in a controlled environment and can therefore be subjected to various physiological assays such as phytoremediation [6]. Also, these transformed roots are amenable to genetic manipulation and may facilitate the characterization of genes that influence the phytoremediation capacity of plants. Furthermore, the expression of suitable genes in root system enhances the rhizodegradation of highly recalcitrant compounds like PAHs, PCBs, etc. [15, 39]. Hairy root cultures could be also considered as initial screens for plants with enhanced capacity for phytoremediation [40]. Several advantages including the ability to grow rapidly in microbe-free conditions, providing a greater surface area of contact between contaminant and tissue, and genetic and metabolic stability in comparison to wild type have been offered for using hairy roots in phytoremediation [38, 41]. Another advantage of using hairy roots for studying phytoremediation is their ability to produce large quantities of exudates which are composed of enzymes and some metal chelating compounds that may detoxify or sequester harmful organic and inorganic contaminants [7, 41, 42].

3 Bio/Phytoremediation

3.1 What Is Phytoremediation?

Phytoremediation is an emerging technology for large-scale removal or detoxification of contaminants from soils, waters, sediments, and air [43, 44]. Phytoremediation uses plants and their associated microbial communities, to remove, transfer, or stabilize pollutants in an environmental friendly manner. Phytoremediation is frequently an in situ process and does not require the physical removal of the polluted soil which contributes to lowering costs and impacts on the ecosystem, preserving landscape that may reflect its acceptation by population and public actions.

The growing interest of phytoremediation may be that scientific studies in genetics, physiology, and biochemistry of plant tolerance to inorganic and organic contaminants have dramatically increased during the last decade. The choice of the plant-technique for restoring in specific level a determined site will depend on the chemical nature and concentration of the pollutant and also on the physicochemical and biological characteristics of the soil. However, there are still some barriers for the adoption of phytoremediation by private and public organisms that impairs successful applications of this remediation technique.

Phytoremediation has been subdivided in a number of phytoprocesses depending on the characteristic used by plants to remediate a polluted site, generating different phytotechnologies that includes the following:

- *Phytoextraction* and *phytoaccumulation* are probably the best known phytoremediation processes for soils contaminated with inorganics. In these techniques, metal-accumulating plants remove metals from soil concentrating them in plant parts that will be subsequently harvested. Metal-rich plant material obtained by these methods can be appropriately disposed or incinerated, or depending on the economical value of the metal and its concentration in plant tissues it can be phytomined and recovered from plant ashes [10, 32, 45]. In some cases, if the element extracted in plant material is a valuable nutrient for plants, the harvested metal-containing biomass can be used as fertilizer in deficient areas [46].
- *Phytostabilization* uses metal tolerant plants and its associated microbes to reduce pollutant mobility and bioavailability by means of its immobilization in the root zone. This process prevents metal leaching to ground water reservoirs and reduces the possible risks of offsite contamination through wind and rain erosion [47].
- *Phytovolatilization* uses plant mediated processes that favor the volatilization of pollutants. This process is possible only for some particular organic compounds and for inorganics such as mercury, in which after plant uptake, pollutant is released in a volatile form [48, 49];
- *Phyto*- or *rhizofiltration* is the use of plant roots to absorb and adsorb pollutants from waters; this process is mainly used to remove metals and inorganics from industrial and water treatment effluents. When seedlings are able to achieve this process, the technique is called blastofiltration [50].
- *Phytodegradation* is the degradation of pollutants by plant–microbe systems in which enzymatic activities can facilitate organics biodegradation [24]; and
- *Phytoscrubbing* is the plant removal of atmospheric pollutants [51], as in the case of organics such as benzene that can be efficiently removed from air by *Dracena sanderiana* plants [52]. Another recent example of phytoscrubbing is given by plants from the Ericaceae family which was screened to remove gaseous pollutants [53].

Biological techniques imply low energy, an input requirement that can translate in expressive cost advantages over other techniques. They also focus greater efforts in public, private, and scientific communities. However, other techniques as thermal, physical, and chemical strategies are successfully applied in some cases depending of the remediation problem. In some cases, uncontrolled biological activity can actually increase risks by remobilizing contaminants or by forming more harmful breakdown compounds.

3.2 Genes/Enzymes Involved in Detoxification of Pollutants

Biotechnology is the use of biological systems as tools to solve technological problems. A considerable part of biotechnological research is involved with optimizing the efficacy of the tools, e.g., optimizing efficacy of plants as phytoremediators by genetic engineering [54]. Another key part of the biotechnological process is the actual identification of the tools that are available for possible use, i.e., the identification of the genes or physiological processes that make some plants tolerate high levels of environmental contamination. The use of these tools make it possible for some plants to take up xenobiotics or high levels of heavy metals, and then survive or even thrive.

The rhizosphere is a complex and dynamic environment, where there is a constant interaction between plants and numerous microorganisms in the soil. For a full mechanistic understanding of phytoremediation, we need to be able to separate the contributions of microorganism and plants in the process. Yet, many studies have demonstrated that plant tissue cultures are an extremely valuable tool in phytoremediation research. For example, Doty et al. [55] demonstrated that plant cells of poplar are capable of metabolizing halogenated environmental pollutants in the absence of rhizospheric bacteria. Phytoremediation of benzophenone and bisphenol A using immobilized plant cells of *Nicotinia tabacum* has also been reported by Shimoda et al. [56]. Schmidt et al. introduced various P450 genes (CYP1A1, CYP2A2, CYP3A4) in tobacco cell cultures to study the degradation and detoxification of various insecticides, herbicides, and other xenobiotics [33, 34]. The results derived from tissue cultures can be used to predict the responses of plants to environmental contaminants and to improve the design and thus reduce the cost of subsequent conventional whole plant experiments. Hairy root cultures provide an ideal model system to identify the role of plants in phytoremediation and have been used frequently for this purpose [6]. They allow researchers to monitor and quantify the uptake of pollutants and follow the detoxification process in detail. Thus, hairy roots will help in gaining a greater understanding of the way plants deal with pollutants. In addition, hairy roots can be subjected to various physiological assays. This information will be valuable in choosing the best plant species for use in bioremediation [40]. Though, however valuable a tool they are, we will have to remain aware that aseptically grown hairy roots are a model system only. For example, in the rhizosphere microorganisms may directly affect phytoremediation by substantially altering the bioavailability of environmental contaminants or indirectly affect the process by affecting growth and development of roots and shoots [57].

Much of the research on phytoremediation has concentrated on the removal of polychlorinated biphenyls (PCBs) from contaminated water and soil. PCBs had found wide industrial application in the 1960s, due to their chemical stability, low flammability, and their electrical insulating properties. They were commonly used as coolants and insulating fluids or mixed in with plastics, lubricating oils, and flame retardants. In the 1970s, when the toxic effects of PCBs became widely known and their industrial use was severely restricted, the compounds had already widely spread into the environment through use and disposal. Their chemical stability now was a major drawback, and PCBs have become to this day persistent organic pollutants.

Plants respond to the presence of PCBs in their environment. In an effort to identify the genes that may play a role in phytoremediation, PCB-induced gene expression in *Arabidopsis thaliana* was monitored using a gene expression microarray [58]. A coordinated gene expression pattern was revealed; most PCB-responsive genes seemed to be involved in plant growth and development (e.g., ion transport, signal transduction, transcriptional regulation) and various metabolic pathways. Plants are known to exude oxidoreductive enzymes (e.g., peroxidases and laccases), which have been considered as possible metabolic tools transform organic contaminants [59]. Whereas the main function of the oxidoreductive enzymes seems to be the oxidation of various phenolic compounds, hairy root cultures of *Solanum nigrum* have been shown to also convert PCBs into hydroxyl-PCBs and further conjugates [60, 61].

Species of bacteria living in symbiosis with plant roots [62] actively degrade PCBs into chlorobenzoic acids (CBAs). These degradation products are taken up from aqueous solutions by plant roots, through a mix of passive processes (e.g., sorption to plant tissue) and active metabolism. The common osier, *Salix viminalis*, could remove up to 90% of 4-CBA from hydroponic solutions [63].

Attempts have been made to genetically engineer plants so that they can degrade PCBs into less environmentally damaging products [15]. To this aim, plants have been transformed with genes coding for mammalian oxidoreductive enzymes [64] or bacterial biphenyl dioxygenases [65, 66]. In another example, a bacterial gene encoding γ-glutamylcysteine synthetase, the synthesis of which is the rate-limiting factor in glutathione biosynthesis has been introduced in some poplar transgenic clones characterized by γ-glutamylcysteine supersynthesis. The transgenic poplars grown on soil containing the chloracetinyl herbicides metolachlor and acetochlor, accumulated biomass much more intensively than the wild poplar hybrids. The transgenic clones were also distinguished by increased activities of γ-glutamylcysteine synthetase and glutathione-S-transferase (GST) in the leaves, indicating the close relationship of these biosynthetic processes. The mechanism of uptake of cadmium and zinc by *Thlaspi caerulescens* has also been studied and involves a highly expressed metal transporter [67]. The transporter gene, *ZNT1*, encodes a high-affinity zinc/low-affinity cadmium transporter, as demonstrated in yeast. The zinc/cadmium pumping ATPase was recently purified directly from *T. caerulescens* and was shown to transport both zinc and cadmium [68]. Constructed *Arabidopsis* plants with the γ-*ECS* gene and the arsenate reductase C (*ArsC*) gene to control both the mobility and the sequestration of arsenic have been also reported by Dhankher et al. [69]. The transgenic plants coexpressing these two genes grew substantially better, with healthy shoots, on medium containing 200 μM arsenate compared with the wild-type controls which were stunted and chlorotic. The double transgenic plants cultured on this medium showed sixfold greater biomass than the wild-type plants after 3 weeks. These findings could be useful as an important first step towards giving another plant species the characteristics of a hyperaccumulator. Mercury is another highly toxic pollutant with expensive clean up. Ruiz and Daniell

[70] have reported that by expressing some bacterial genes, namely, the *merA* gene encoding mercuric ion reductase and the *merB* gene encoding organomercurial lyase, transgenic plants had improved resistance to the toxic effects of mercury. They confirmed that resistance was further improved when *merB* was expressed in the endoplasmic reticulum. Since the chloroplast is the primary target for mercury poisoning, their results helped lead the research in the direction of chloroplast genome engineering. Greatly improved resistance to mercury was achieved with high expression levels of *merA* and *merB* in the chloroplast. Engineering of the plastid genome has many advantages over the traditional engineering of the nuclear genome [8]. Bacterial genes, such as *merA* and *merB*, are expressed more efficiently in plastid genomes that are believed to be of bacterial origin [8, 71]. Very high expression levels can be achieved through plastid engineering without the issue of gene silencing. Furthermore, since the plastid genomes are maternally inherited, there is no transgenic escape via pollen. For example, by expressing *merA* and *merB* in the chloroplast genome, transgenic tobacco plants exhibited much higher resistance to phenyl mercuric acetate and were able to efficiently transport organic mercury and then volatilize elemental mercury [8]. The chelation of the metals by a ligand has also been introduced as a general mechanism for detoxification of them in plants. Peptide ligands, phytochelatins, have been identified in a wide variety of plant species including angiosperms, gymnosperms, algae, fungi, and marine diatoms [72–74]. Phytochelatins are enzymatically synthesized from glutathione and are rapidly induced in cells and tissues exposed to a range of metal ions such as Cd, Zn, Ag, Cu, Ni, Hg, and Pb and anions such as arsenate and selenite [75]. Several studies have suggested that phytochelatin synthase activity is regulated at the level of enzyme activation by metal ions. It has been shown that phytochelatin synthase activity has an important role in Cd tolerance. For example, in transgenic plants of Indian mustard plants, increases in the expression of enzymes of glutathione biosynthesis pathway led to an increase in phytochelatin biosynthesis and Cd-tolerance. Plant cell and hairy root cultures have provided an excellent model for investigating the phytotoxicity and metabolic fate of xenobiotics since test conditions can be closely controlled and extraction of metabolic products is relatively easy [15]. Several in vitro assays have been developed to monitor the toxicity of post-removal solutions, e.g., the AMPHITOX bioassay which measures the effect of pollutants on amphibian embryo development [54, 76], the mitotic index of meristematic cells of *Allium cepa* [77], or an assay measuring the germination and development of *Lactuca sativa* seeds [78].

4 Transport of Chemicals Inside the Root Cells

A major research effort has gone into understanding the mechanism of hyperaccumulation in plants. Hyperaccumulation involves the active uptake of large amounts of heavy metals from the soil, followed by translocation to the shoot and subsequent accumulation in leaves, without showing symptoms of phytotoxicity [79]. Hairy roots have been instrumental in elucidating the process of heavy metal uptake by roots. In response to high levels of heavy metals, roots produce glutathione oligomers called phytochelatin, and related peptides [80–82]. In addition, hairy roots of the hyperaccumulators, *Thlaspi caerulescens* and *Alyssum bertolonii* (both Brassicaceae), contain high constitutive levels of citric, malic, and malonic acids, which indicates that organic acid complexation also plays a role in heavy metal uptake [83].

After uptake, heavy metals are usually transported to the shoots. However, hyperaccumulation does not necessarily depend on the presence of shoots or root-shoot translocation. Hairy root cultures of the hyperaccumulator *A. bertolonii* remained healthy in appearance and continued to grow in the presence high levels of nickel, whereas hairy roots of *Nicotiana tabacum* turned dark brown and growth was negligible [84]. This observation may be explained by different mechanisms of transport of heavy metals; when hairy roots of the hyperaccumulator *T. caerulescens* were cultured at high Cd concentrations, they initially stored metal in the wall fraction, whereas in *N. tabacum* roots most metal was transported directly into the symplasm. The delay in transmembrane uptake may represent an important defensive strategy against metal poisoning [32].

Much work on metal transport in plants is done with the hyperaccumulator *T. calerulescens*. Several ecotypes of this plant species exist; some are adapted to growth in metalliferous soil. Molecular genetics comparison of different ecotypes has resulted in the identification of genes involved in transmembrane metal transport and hyperaccumulation [85–87], a key role is played by genes belonging to the ZIP, HMA, MATE, YSL, and MTP families [79].

4.1 Induction and Establishment of Hairy Root Culture System

Hairy root disease is caused by a naturally occurring gram negative soil bacterium *Agrobacterium rhizogenes* that contains root inducing (Ri) plasmid (more than 200 kb). T-DNA on Ri-plasmid is transferred and integrated into the nuclear genome of the host plant upon infection with *A. rhizogenes* [88–90]. Opines are produced as a result of infection, and based on the opines produced, *A. rhizogenes* strains are classified into octopine, agropine, nopaline, mannopine, and cucumopine type [91]. Choice of proper explant, an appropriate *A. rhizogenes* strain, culture medium, as well as selection of antibiotic to remove the bacteria after cocultivation are important factors for establishing hairy root culture [92]. Different explants like root, stem, leaf, hypocotyl, cotyledons, nodal segments, and embryo axis can be used for initiation of hairy roots. Out of all the strains, agropine is the most commonly used due to its strong induction ability. Hairy roots are characterized by a large number of highly branched fast growing adventitious roots at the wound site which are capable of growing on hormone free medium. These are plagiotrophic in nature and are found to be genetically as well as biochemically stable over a long period of time [90, 93].

4.2 Advantages of Hairy Root Cultures in Phytoremediation

Hairy roots provide several benefits compared to dedifferentiated cells and tissues including callus and cell suspension cultures. Hairy roots present phenotypic, genetic, as well as biochemical stable system for a long period compared to dedifferentiated cells that are generally found to show somaclonal variations and product instability [94, 95]. Moreover, plant roots are in direct contact with soil or water hence pollutants present therein.

Hairy roots provide a large surface area due to fast growth and highly branched nature and hence contact between the contaminants and tissue in comparison to naturally growing roots thus providing reliable and reproducible experimental system to study the pollutants and their response to toxic substances [40, 96]. Introduction of foreign genes and their resultant proteins to metabolize environmental pollution in transformed hairy roots can be expressed for a long term due to genetic stability [97]. Another advantage associated with the hairy roots is absence of shoots that help in understanding the mechanisms present only in roots for removal of contaminants without the translocation effects [98]. These roots can also be used to understand the enzymatic processes involved in bioconversion of toxic pollutants to nontoxic compounds [99]. Literature studies showed the number of reports wherein hairy roots have been successfully used to study phytoremediation. Some of the examples of hairy root cultures from different plant species employed to uptake and degrade the various pollutants are presented in Table 1.

5 Limitations of Phytoremediation in Land Plants

The term phytoremediation comprises plant-based technologies used to clean up toxic compounds in the environment and it has been extensively used in studies with toxic metals. The term is usually misused and confounded with phytoextraction, which is the technique related to the removal of toxic metals from contaminated soils using plants [112]. For decontamination of aquatic environments, the process is called rhizo- or phytofiltration [113]. Phytostabilization and phytovolatilization are also techniques to clean the environment, but in the first case, plants are used to stabilize the toxic metal in the soil, and in the second, plants are used to extract certain metals from soil and further released to atmosphere through volatilization [112].

Some plants with the capacity to absorb and tolerate high concentrations of toxic metals in their tissues are called hyperaccumulators, many from the Cucurbitaceae family, then being herbaceous and having annual cycle [111, 114]. Hyperaccumulators should be differentiated from excluders, which nearly do not absorb metals as a strategy to avoid intoxication [115]. On the other hand, hyperaccumulators do absorb metals and accumulate large amounts in their tissues without visible metabolic damage.

Phyto- or rhizoextraction may be simply defined as the removal of a toxic metal from the contaminated area to be accumulated in a plant part. Clearly the dimension

Table 1 Hairy root cultures of different plant species utilized for phytoremediation of various environmental pollutants

Pollutant	Hairy root culture of plant species	Reference
Cadmium	*Beta vulgaris, Nicotiana tabacum*	[100]
	Solanum nigrum	[101]
	Thlaspi caerulescens	[32]
	Adenophora lobophylla, A. potaninii	[82]
Copper	*Rubia tinctorum*	[102]
	Hyptis capitata	[32]
Uranium	*Chenopodium amaranticolor, Brassica juncea*	[96]
Nickel	*Alyssum* sp.	[84]
	Alyssum murale	[103]
Phenol	*Brassica juncea*	[104]
	B. napus	[77, 105]
	Lycopersicon esculentum	[106, 107]
	Armoracia lapathifolia	[108]
	Daucus carota	[109]
RDX and HMX	*Catharanthus roseus*	[31]
Tetracycline, Oxytetracycline	*Helianthus annuus*	[41]
DDT	*Brassica juncea, Cichorium intybus*	[110]
TCE	*Atropa belladonna*	[97]
PCBs	*Solanum nigrum*	[111]

DDT dichloro-diphenyl-trichloroethane, *HMX* oxtahydro-1,3,5,7-tetranitro-1,3,5,7-tetrazocine, *PCBs* polychlorinated biphenyls, *RDX* hexahydro-1,3-5-trinitro-1,3-5-triazine, *TCE* Trichloroethylene

of the root system plays an important role in the efficiency of the process. Contamination of soil deep layers may restrict the efficiency of land plant accumulators [116].

As a process, the practical use of phytoremediation depends on in which plant part metals accumulate. In aquatic plants, this is secondary since the whole plant can be removed from the contaminated place, but in land plants, the accumulation of toxic metals in roots implies the necessity of soil movement or removing large amounts of soil, making the process costly. Unfortunately, in a number of plant species, the main accumulation site is the root system [117].

Soil particles may also tightly hold toxic metals, becoming unavailable for root absorption. Pb is strongly retained in nearly all types of soil, thus being poorly absorbed by plants [117]. When absorbed, Pb is poorly transported to the shoot [118–121]. In such cases, application of chelators to the soil could be an alternative, by increasing the availability of the metal for plant absorption [122–125]. However, lack of detailed studies on the persistence of the metal-chelating agent in the soil and the potential risk of metals reach the groundwater has raised serious concerns to adopt chelators in contaminated areas [126].

Other plant species accumulate toxic metals in the shoot [114] but low growth rates and low biomass production may limit their use for phytoremediation

[127]. Biomass production may also be low due to adverse local climatic conditions or nonadaptation of the hyperaccumulator to the local environment because of different geographical origin [127]. Altogether, this implies the need of selection of an appropriate species for a given metal and for a given contaminated site, not excluding adoption of breeding strategies to select genotypes with desirable phytoremediation characteristics [128].

If biomass production is not a problem, the next step is to decide its destination (and the metal contained therein). At this stage, some questions must be asked: Can the metal be recovered from the plant matrix? Is the metal recovering process economically and/or technically viable? Is the recovered metal sufficiently pure or has enough quality to attend a specific demand or to be marketed? Does the metal price afford the whole recovery process? Can the recovered metal be a component of another product or process, which will not pollute or damage the environment? Thus, plant-based phytoremediation is viable if plants efficiently absorb the metal, accumulation occurs mainly in the shoot, produce large amount of biomass, possess ability to grow in different environments, and biomass destination removes the metal from the environment or it is used/applied in other safe process or products [129, 130].

A practical problem in using phytoremediation is the time the hyperaccumulator plant has to be harvested from the contaminated site. Studies investigating the accumulation of Cd in rice grain showed a dynamic metal movement between various plant organs during development [131, 132]. Thus, a compromise between metal concentration in the tissues and biomass production may define the right time for plant shoot harvesting.

Trees might be an alternative to herbaceous and short-cycle hyperaccumulator plants. The metal tolerance of trees has been reported in some research studies, but the information is restricted to a few species [133]. Moreover, genetic variability in trees is high and the definition of what could be tolerance is variable, since it has been observed that within a population there is great variation in the ability to tolerate a specific metal. To understand tree metal tolerance, some studies have used tissue culture or cell suspension cultures, but most have assessed seedlings since the slow growth of trees imposes time limitation [134, 135]. However, this strategy has been questioned, since it cannot reflect the behavior of mature trees [136].

Studies with adult trees demand long periods of time and consequently there is little literature on this topic. However, the slow growth may be an advantage since it would also slow the metal absorption avoiding toxicity development [133]. Association of slow growth and slow absorption, preventing high metal accumulation, has been proposed as an explanation for the capability trees have to acclimate to contaminated soils [133]. Thus, although slow growth imposes a limitation, over a long period of time phytoremediation should be viewed positively and the total amount of metal absorbed by the whole plant should be considered rather than only concentration per unit mass [137].

The metal compartmentalization in trees has been little explored and few species have been investigated. Among them, *Salix* and *Populus* are those that have deserved more attention in North America and Europe [138]. According to Pulford and Watson [133], trees accumulate metals preferentially in zones of active growth, such as shoots and young leaves, but significant part of the studies show that depending on the metal, accumulation also occurs in roots. Most of the metal nonretained in the roots was transported to the stem in *Acer pseudoplatanus* (sycamore) growing in soil contaminated with Pb [136]. Analysis of shoots of willow plants growing in contaminated soil showed a higher concentration of Pb in stems, along with Cr and Cu, while Zn, Cd, and Ni accumulated more in leaves [139]. Comprehensive reviews on metal tolerance in trees can be found in the literature [133, 140–142].

As for herbaceous plants, some aspects are important for the choice of tree species used for phytoremediation as: (1) high biomass production, (2) a deep root system, (3) high growth rate, (4) ability to grow in soil with low nutrient availability, and (5) high capacity to allocate metals in the stem [140, 143]. These features could allow for high-density planting, exploiting fast-growing trees such as *Eucalyptus*, *Pinus*, *Populus*, and *Salix* [143].

The use of trees for phytoremediation allows the possibility of exploiting a true immobilization of the metal in the timber, which could then be used for several purposes. For example, timber could be employed in house and building structures and furniture covered by sealants or any protective material (laminates for example) avoiding human contact. Such timber would be traceable and producers would receive a certificate.

6 Conclusions and Future Directions

Phytoremediation is a promising tool for cleaning up pollutants from the environment. This plant-based technology offers different solutions for the removal of contaminants, including phytodegradation, phytoextraction, phytostabilization, phytostimulation, and phytovolatilization. Although the main role in the phytoremediation process is taken by the whole plant, specific plant constituents, such as cells and hairy roots, can be used for the same purpose, mainly to study how the process works.

Amongst all the toxic compounds (inorganics, explosives, polychlorinated biphenyls, and other organics) that pollute water and soil, polychlorinated biphenyls and inorganics, including heavy metals and radionuclides, have been the main targets in phytoremediation research.

Hairy root cultures have been shown to convert polychlorinated biphenyls into hydroxyl derivates and other conjugates. Genetic manipulation of these roots with suitable genes has enabled hairy roots to degrade these toxic compounds (phytodegradation) and enhanced their rhizo/phytoremediation capacity.

Since heavy metals and radionuclides cannot be degraded, other plant-based technologies included in phytoremediation such as phytoextraction, phytostabilization, and phytovolatilization are carried out in which the plant neutralizes these pollutants. Hairy roots also play an important role here, above all in the studies of heavy metal uptake by roots and the production of chelating compounds.

Other advantages of using hairy roots in phytoremediation are their high biomass production resulting in a large contact surface area between contaminant and tissue; their high growth rate, which enables studies to be completed more quickly; genetic and metabolic stability in the engineered hairy roots; their constant size regardless of the size of the whole plant, and their usefulness as a tool to screen hyperaccumulator plants.

There is no doubt that hairy root cultures constitute an important tool in phytoremediation research but they are a model system only, so it is necessary to remain aware of their limitations in the development of practical phytoremediation technology.

Future studies on hairy roots in phytoremediation should remain focused on the engineering of target genes involved in this process, and new studies are required to extend the basic hairy root phytoremediation model to the environment.

Acknowledgments SM would like to acknowledge the Foundation for the Support of Research, Scientific, and Technological Development of the state of Maranhão – FAPEMA for financial support.

References

1. Chrisman JR, Koifman S, de Novaes Sarcinelli P, Moreira JC, Koifman RJ, Meyer A (2009) Pesticide sales and adult male cancer mortality in Brazil. Int J Hyg Environ Health 212:310–321
2. LaFiura KM, Bielawski DM, Posecion NC, Ostrea EM, Matherly LH, Taub JW, Ge Y (2007) Association between prenatal pesticide exposures and the generation of leukemia-associated T (8; 21). Pediatr Blood Cancer 49:624–628
3. Thrasher JD, Heuser G, Broughton A (2002) Immunological abnormalities in humans chronically exposed to chlorpyrifos. Arch Environ Health 57:181–187
4. Hooper SW, Pettigrew CA, Sayler GS (2009) Ecological fate, effects and prospects for the elimination of environmental polychlorinated biphenyls (PCBs). Environ Toxicol Chem 9:655–667
5. Zhang DQ, Tan SK, Gersberg RM, Sadreddini S, Zhu J, Tuan NA (2012) Removal of pharmaceutical compounds in tropical constructed wetlands. Ecol Eng 37:460–464
6. Doran PM (2009) Application of plant tissue cultures in phytoremediation research: incentives and limitations. Biotechnol Bioeng 103:60–76
7. Doty SL (2008) Enhancing phytoremediation through the use of transgenics and endophytes. New Phytol 179:318–333
8. Dowling DN, Doty SL (2009) Improving phytoremediation through biotechnology. Curr Opin Biotechnol 20:204–206
9. Weyens N, van der Lelie D, Taghavi S, Newman L, Vangronsveld J (2009) Exploiting plant-microbe partnerships to improve biomass production and remediation. Trends Biotechnol 27:591–598
10. Meagher RB (2000) Phytoremediation of toxic elemental and organic pollutants. Curr Opin Plant Biol 3:153–162

11. Alexander M (1999) Biodegradation and bioremediation. Academic, Ithaca
12. Barber DA, Lee RB (1974) The effect of micro-organisms on the absorption of manganese by plants. New Phytol 73:97–106
13. Raskin I, Kumar PB, Dushenkov S, Salt D (1994) Bioconcentration of heavy metals by plants. Curr Opin Biotechnol 5:285–290
14. Kvesitadze G, Khatisashvili G, Sadunishvili T, Ramsden JJ (2006) Biochemical mechanisms of detoxification in higher plants: basis of phytoremediation. Springer, Berlin
15. Abhilash PC, Jamil S, Singh N (2009) Transgenic plants for enhanced biodegradation and phytoremediation of organic xenobiotics. Biotechnol Adv 27:474–488
16. Dhankher OM, Pilon-Smith EAH, Meagher RB, Doty S (2012) Biotechnological approaches for phytoremediation. In: Altman A, Hasegawa PM (eds) Plant biotechnology and agriculture, prospects for the 21st century. Academic, Oxford
17. Barac T, Taghavi S, Borremans B, Provoost A, Loeyen JV, Colpaert J, Vangronsveld J, van der Lelie D (2004) Engineered endophytic bacteria improve phytoremediation of water-soluble volatile organic pollutants. Nat Biotechnol 22:583–588
18. French CE, Rosser SJ, Davies GJ, Nicklin S, Bruce NC (1999) Biodegradation of explosives by transgenic plants expressing pentaerythritol tetranitrate reductase. Nat Biotechnol 17:491–494
19. Doty SL, Shang QT, Wilson AM, Tangen J, Westergreen A, Newman LA et al (2000) Enhanced metabolism of halogenated hydrocarbons in transgenic plants containing mammalian P450 2E1. Proc Natl Acad Sci U S A 97:6287–6291
20. Eapen S, Singh S, D'Souza SF (2007) Advances in development of transgenic plants for remediation of xenobiotic pollutants. Biotechnol Adv 25:442–451
21. Macek T, Kotrba P, Svatos A, Novakova M, Demnerova K, Mackova M (2008) Novel roles for genetically modified plants in environmental protection. Trends Biotechnol 26:146–152
22. Mackova M, Barriault D, Francova K, Sylvestre M, Möder M, Vrchotova B, Lovecka P, Najmanová J, Demnerova K, Novakova M, Rezek J, Macek T (2006) Phytoremediation of polychlorinated biphenyls. In: Mackova M, Dowling D, Macek T (eds) Phytoremediation and rhizoremediation. Theoretical background, Focus on biotechnology. Springer, Dordrecht
23. McCutcheon SC, Schnoor JL (2003) Phytoremediation: transformation and control of contaminant. Wiley, New York
24. Pilon-Smits E (2005) Phytoremediation. Annu Rev Plant Biol 56:15–39
25. Kraemer U, Chardonnens AN (2001) The use of transgenic plants in the bioremediation of soil contaminated with trace elements. Appl Microbiol Biotechnol 55:661–672
26. Meagher RB, Heaton ACP (2005) Strategies for the engineered phytoremediation of toxic element pollution: mercury and arsenic. J Indust Microbiol Biotechnol 32:502–513
27. Rugh CL (2004) Genetically engineered phytoremediation: one man's trash is another man's transgene. Trends Biotechnol 22:496–498
28. Ellis DR, Sors TG, Brunk DG, Albrecht C, Orser C, Lahner B et al (2004) Production of Se-methylselenocysteine in transgenic plants expressing selenocysteine methyltransferase. BMC Plant Biol 4:1–11
29. Gasic K, Korban SS (2007) Transgenic Indian mustard (*Brassica juncea*) plants expressing an *Arabidopsis* phytochelatin synthase (AtPCS1) exhibit enhanced As and Cd tolerance. Plant Mol Biol 64:361–369
30. Heaton AC, Rugh CC, Kim T, Meagher RB (2003) Toward detoxifying mercury-polluted aquatic sediments with rice genetically engineered for mercury resistance. Environ Toxicol Chem 22:2940–2947
31. Bhadra R, Wayment DG, Williams RK, Barman SN, Stone MB, Hughes JB, Shanks JV (2001) Studies on plant-mediated fate of the explosives RDX and HMX. Chemosphere 44:1259–1264
32. Nedelkoska TV, Doran PM (2000) Hyperaccumulation of cadmium by hairy roots of *Thlaspi caerulescens*. Biotechnol Bioeng 67:607–615
33. Schmidt B, Faymonville T, Gembé E, Joußen N, Schuphan I (2006) Comparison of the biotransformation of the 14C-labelled insecticide carbaryl by non-transformed and human

CYP1A1, CYP1A2-, and CYP3A4-transgenic cell cultures of *Nicotiana tabacum*. Chem Biodivers 3:878–896
34. Schmidt B, Joußen N, Bode M, Schuphan I (2006) Oxidative metabolic profiling of xenobiotics by human P450s expressed in tobacco cell suspension cultures. Biochem Soc Trans 34:1241–1245
35. Suresh B, Ravishankar GA (2004) Phytoremediation – a novel and promising approach for environmental clean-up. Crit Rev Biotechnol 24:97–124
36. Ma LQ, Komar KM, Tu C, Zhang WH, Cai Y, Kennelley ED (2001) A fern that hyperaccumulates arsenic: a hardy versatile fast-growing plant helps to remove arsenic from contaminated soils. Nature 409:579
37. Han KH, Meilan R, Ma C, Strauss SH (2000) An *Agrobacterium tumefaciens* transformation protocol effective on a variety of cottonwood hybrids (genus *Populus*). Plant Cell Rep 19:315–320
38. Georgiev MI, Pavlov AI, Bley T (2007) Hairy root type plant *in vitro* systems as sources of bioactive substances. Appl Microbiol Biotechnol 74:1175–1185
39. Gerhardt KE, Huang XD, Glick BR, Greenberg BM (2009) Phytoremediation and rhizoremediation of organic soil contaminants: potential and challenges. Plant Sci 176:20–30
40. Suza W, Harris RS, Lorence A (2008) Hairy roots: from high-value metabolite production to phytoremediation. Electron J Integr Biosci 3(1):57–65
41. Gujarathi NP, Haney BJ, Park HJ, Wickramasinghe SR, Linden JC (2005) Hairy roots of *Helianthus annuus*: a model system to study phytoremediation of tetracycline and oxytetracycline. Biotechnol Prog 21:775–780
42. Bais HP, Weir TL, Perry LG, Gilroy S, Vivanco JM (2006) The role of root exudates in rhizosphere interactions with plants and other organisms. Annu Rev Plant Biol 57:233–266
43. Brooks R (1998) Geobotany and hyperaccumulators. In: Plants that hyperaccumulate heavy metals. CAB International, New York, pp 55–94
44. Chaudhry T, Hayes W, Khan A, Khoo C (1998) Phytoremediation – focusing on accumulator plants that remediate metal-contaminated soils. Aust J Ecotoxicol 4:37–51
45. Sheoran V, Sheoran A, Poonia P (2009) Phytomining: a review. Miner Eng 22:1007–1019
46. Robinson B, Green S, Chancerel B, Mills T, Clothier B (2007) Poplar for the phytomanagement of boron contaminated sites. Environ Pollut 150:225–233
47. Salt DE, Prince RC, Pickering IJ, Raskin I (1995) Mechanisms of cadmium mobility and accumulation in Indian mustard. Plant Physiol 109:1427–1433
48. Schnoor JL, Light LA, McCutcheon SC, Wolfe NL, Carreia LH (1995) Phytoremediation of organic and nutrient contaminants. Environ Sci Technol 29:318–323
49. Terry G, Stokes N, Hewitt C, Mansfield T (1995) Exposure to isoprene promotes flowering in plants. J Exp Bot 46:1629–1631
50. Elless MP, Poynton CY, Willms CA, Doyle MP, Lopez AC, Sokkary DA, Ferguson BW, Blaylock MJ (2005) Pilot-scale demonstration of phytofiltration for treatment of arsenic in New Mexico drinking water. Water Res 39:3863–3872
51. Gadd G (2010) Metals, minerals and microbes: geomicrobiology and bioremediation. Microbiology 156:609
52. Treesubsuntorn C, Thiravetyan P (2012) Removal of benzene from indoor air by *Dracaena sanderiana*: effect of wax and stomata. Atmos Environ 57:317–321
53. Yang H, Liu YJ, Shao W (2012) Screening houseplants to remove ammonia from indoor air through fumigation. Adv Mater Res 378:397–400
54. Talano MA, Busso DC, Paisio CE, Gonzalez PS, Purro SA, Medina MI, Agostini E (2012) Phytoremediation of 2,4-dichlorophenol using wild type and transgenic tobacco plants. Environ Sci Pollut Res 19(6):2202–2211
55. Doty SL, Dosher MR, Singleton GL, Moore AL, van Aken B, Stettler RF, Strand SE, Gordon MP (2005) Identification of an endophytic *Rhizobium* in stems of *Populus*. Symbiosis 39:27–36

56. Shimoda K, Hamada H, Hamada H (2009) Phytoremediation of benzophenone and bisphenol A by glycosylation with immobilized plant cells. Environ Health Insights 3:19–25
57. Rajkumar M, Sandhya S, Prasad MNV, Freitas H (2012) Perspectives of plant-associated microbes in heavy metal phytoremediation. Biotechnol Adv 30:1562–1574
58. Jin XF, Shuai JJ, Peng RH, Zhu B, Fu XY, Tian YS, Zhao W, Han HJ, Chen C, Xu J, Yao QH, Qu SC, Xiong AS (2011) Identification of candidate genes involved in responses of *Arabidopsis* to polychlorinated biphenyls based on microarray analysis. Plant Growth Regul 65(1):127–135
59. Rodríguez-Dorantes A, Guerrero-Zúñiga LA (2012) Phenoloxidases activity in root system and their importance in the phytoremediation of organic contaminants. J Environ Chem Ecotoxicol 4(3):35–40
60. Rezek J, Macek T, Mackova M, Triska J (2007) Plant metabolites of polychlorinated biphenyls in hairy root culture of black nightshade *Solanum nigrum* SNC-90. Chemosphere 69(8):1221–1227
61. Rezek J, Macek T, Doubsky J, Mackova M (2012) Metabolites of 2,2′-dichlorobiphenyl and 2,6-dichlorobiphenyl in hairy root culture of black nightshade *Solanum nigrum* SNC-90. Chemosphere 89(4):383–388
62. Boldt TS, Sorensen J, Karlson U, Molin S, Ramos C (2004) Combined use of different Gfp reporters for monitoring single-cell activities of a genetically modified PCB degrader in the rhizosphere of alfalfa. FEMS Microbiol Ecol 48(2):139–148
63. Deavers K, Macek T, Karlson UG, Trapp S (2010) Removal of 4-chlorobenzoic acid from spiked hydroponic solution by willow trees (*Salix viminalis*). Environ Sci Pollut Res 17(7):1355–1361
64. Shimazu S, Inui H, Ohkawa H (2011) Phytomonitoring and phytoremediation of agrochemicals and related compounds based on recombinant cytochrome P450s and aryl hydrocarbon receptors (AhRs). J Agric Food Chem 59(7):2870–2875
65. Mohammadi M, Chalavi V, Novakova-Sura M, Laliberte JF, Sylvestre M (2007) Expression of bacterial biphenyl-chlorobiphenyl dioxygenase genes in tobacco plants. Biotechnol Bioeng 97(3):496–505
66. Novakova M, Mackova M, Chrastilova Z, Viktorova J, Szekeres M, Demnerova K, Macek T (2009) Cloning the bacterial bphC gene into *Nicotiana tabacum* to improve the efficiency of PCB-phytoremediation. Biotechnol Bioeng 102(1):29–37
67. Pence NS, Larsen PB, Ebbs SD, Letham DL, Lasat MM, Garvin DF, Eide D, Kochian LV (2000) The molecular physiology of heavy metal transport in the Zn/Cd hyperaccumulator *Thlaspi caerulescens*. Proc Natl Acad Sci U S A 97:4956–4960
68. Parameswaran A, Leitenmaier B, Yang M, Kroneck PM, Welte W, Lutz G, Papoyan A, Kochian LV, Kupper H (2007) A native Zn/Cd pumping P(1B) ATPase from natural overexpression in a hyperaccumulator plant. Biochem Biophys Res Commun 363:51–56
69. Dhankher OP, Li Y, Rosen BP, Shi J, Salt D, Senecoff JF, Sashti NA, Meagher RB (2002) Engineering tolerance and hyperaccumulation of arsenic in plants by combining arsenate reductase and gamma-glutamylcysteine synthetase expression. Nat Biotechnol 20:1140–1145
70. Ruiz ON, Daniell H (2009) Genetic engineering to enhance mercury phytoremediation. Curr Opin Biotechnol 20:213–219
71. Rugh CC, Summers AO, Meagher RB (1996) Mercuric ion reductase and resistance in transgenic *Arabidopsis thaliana* expressing modified bacterial merA gene. Proc Natl Acad Sci U S A 93:3182–3187
72. Gekeler W, Grill E, Winnacker E-L, Zenk MH (1989) Survey of the plant kingdom for the ability to bind heavy metals through phytochelatins. Z Naturforsch C 44:361–369
73. Kneer R, Kutchan TM, Hochberger A, Zenk MH (1992) *Saccharomyces cerevisiae* and *Neurospora crassa* contain heavy metal sequestering phytochelatin. Arch Microbiol 157:305–310

74. Morelli E, Pratesi E (1997) Production of phytochelatins in the marine diatom *Phaeodactylum tricornutum* in response to copper and cadmium exposure. Bull Environ Contam Toxicol 59:657–664
75. Rauser WE (1995) Phytochelatins and related peptides. Plant Physiol 109:1411–1419
76. Paisio CE, Gonzalez PS, Gerbaudo A, Bertuzzi ML, Agostini E (2010) Toxicity of phenol solutions treated with rapeseed and tomato hairy roots. Desalination 263(1-3):23–28
77. Gonzalez PS, Maglione GA, Giordana M, Paisio CE, Talano MA, Agostini E (2012) Evaluation of phenol detoxification by *Brassica napus* hairy roots, using *Allium cepa* test. Environ Sci Pollut Res 19(2):482–491
78. Angelini VA, Orejas J, Medina MI, Agostini E (2011) Scale up of 2,4-dichlorophenol removal from aqueous solutions using *Brassica napus* hairy roots. J Hazard Mater 185(1):269–274
79. Rascio N, Navari-Izzo F (2011) Heavy metal hyperaccumulating plants: how and why do they do it? And what makes them so interesting? Plant Sci 180(2):169–181
80. Kubota H, Sato K, Yamada T, Maitani T (2000) Phytochelatin homologs induced in hairy roots of horseradish. Phytochemistry 53(2):239–245
81. Son KH, Kim DY, Koo N, Kim KR, Kim JG, Owens G (2012) Detoxification through phytochelatin synthesis in *Oenothera odorata* exposed to Cd solutions. Environ Exp Bot 75:9–15
82. Wu SX, Zu YG, Wu M (2001) Cadmium response of the hairy root culture of the endangered species *Adenophora lobophylla*. Plant Sci 160(3):551–562
83. Boominathan R, Doran PM (2003) Organic acid complexation, heavy metal distribution and the effect of ATPase inhibition in hairy roots of hyperaccumulator plant species. J Biotechnol 101(2):131–146
84. Nedelkoska TV, Doran PM (2001) Hyperaccumulation of nickel by hairy roots of *Alyssum* species: comparison with whole regenerated plants. Biotechnol Prog 17(4):752–759
85. Plaza S, Tearall KL, Zhao FJ, Buchner P, McGrath SP, Hawkesford MJ (2007) Expression and functional analysis of metal transporter genes in two contrasting ecotypes of the hyperaccumulator *Thlaspi caerulescens*. J Exp Bot 58(7):1717–1728
86. Plessl M, Rigola D, Hassinen VH, Tervahauta A, Karenlampi S, Schat H, Aarts MGM, Ernst D (2010) Comparison of two ecotypes of the metal hyperaccumulator *Thlaspi caerulescens* (J. & C. PRESL) at the transcriptional level. Protoplasma 239(1-4):81–93
87. Visioli G, Pirondini A, Malcevschi A, Marmiroli N (2010) Comparison of protein variations in *Thlaspi caerulescens* populations from metalliferous and non-metalliferous soils. Int J Phytoremediation 12(8):805–819
88. Nilsson O, Olsson O (1997) Getting to the root: the role of the *Agrobacterium rhizogenes rol* genes in the formation of hairy roots. Physiol Plant 100:463–473
89. Riva GA, González-Cabrera J, Vázquez-Padrón R, Ayra-Pardo C (1998) *Agrobacterium tumefaciens*: a natural tool for plant transformation. Electron J Biotechnol 3:1–16
90. Sevón N, Oksman-Caldentey KM (2002) *Agrobacterium rhizogenes*-mediated transformation: root cultures as a source of alkaloids. Planta Med 68:859–868
91. Zhou L, Wang J, Yang C (1998) Progress on plant hairy root culture and its chemistry. 1. Induction and culture of plant hairy roots. Nat Product Res Dev 10:87–95
92. Hu ZB, Du M (2006) Hairy root and its application in plant genetic engineering. J Integr Plant Biol 48(2):121–127
93. Chilton MD, Tepfer DA, Petit A, David C, Casse-Delbart F, Tempé J (1982) Agrobacterium rhizogenes inserts T-DNA into the genomes of the host plant root cells. Nature 295:432–434
94. Giri A, Narasu ML (2000) Transgenic hairy roots. Recent trends and applications. Biotechnol Adv 18:1–22
95. Benyammi R, Paris C, Khelifi-Slaoui M, Zaoui D, Belabbassi O, Bakiri N, Aci MM, Harfi B, Malik S, Makhzoum A, Desobry S, Khelifi L (2016) Screening and kinetic studies of catharanthine and ajmalicine accumulation and their correlation with growth biomass in *Catharanthus roseus* hairy roots. Pharm Biol 17:1–11

96. Eapen S, Suseelan KN, Tivarekar S, Kotwal SA, Mitra R (2003) Potential for rhizofiltration of uranium using hairy root cultures of *Brassica juncea* and *Chenopodium amaranticolor*. Environ Res 91:127–133
97. Bernejee S, Shang TQ, Wilson AM, Moore AL, Strand SE, Gordon MP, Doty SL (2002) Expression of functional mammalian P450 2E1 in hairy root cultures. Biotechnol Bioeng 77(4):462–466
98. Majumder A, Jha S (2012) Hairy roots: a promising tool for phytoremediation. In: Satyanarayana T et al (eds) Microorganisms in environmental management: microbes and environment. Springer, Dordrecht, pp 607–629
99. Macek T, Mackova M, Kas J (2000) Exploitation of plants for the removal of organics in environmental remediation. Biotechnol Adv 18:23–24
100. Metzger L, Fouchault I, Glad C, Prost R, Tepfer D (1992) Estimation of cadmium availability using transformed roots. Plant Soil 143:249–257
101. Macek T, Kotbra P, Suchova M, Skacel F, Demnerova K, Ruml T (1994) Accumulation of cadmium by hairy-root cultures of *Solanum nigrum*. Biotechnol Lett 16:621–624
102. Maitani T, Kubota H, Sato K, Takeda M, Yoshikira K (1996) Induction of phytochelatin (class III metallothionein) and incorporation of copper in transformed hairy roots of *Rubia tinctorum* exposed to cadmium. J Plant Physiol 147:743–748
103. Vinterhalter B, Savić J, Platiša J, Raspor M, Ninković S, Mitić N, Vinterhalter D (2008) Nickel tolerance and hyperaccumulation in shoot cultures regenerated from hairy root cultures of *Alyssum murale* Waldst et Kit. Plant Cell Tiss Org Cult 94:299–303
104. Singh S, Melo JS, Eapen S, D' Souza SF (2006) Phenol removal using *Brassica juncea* hairy roots: role of inherent peroxidase and H_2O_2. J Biotechnol 123:43–49
105. Coniglio MS, Busto VD, González PS, Medina MI, Milrad S, Agostini E (2008) Application of *Brassica napus* hairy root cultures for phenol removal from aqueous solutions. Chemosphere 72:1035–1042
106. Wevar-Oller AL, Agostini E, Talano MA, Capozucca C, Milrad SR, Tigier HA, Medina MI (2005) Overexpression of a basic peroxidase in transgenic tomato (*Lycopersicon esculentum* Mill. cv. Pera) hairy roots increases phytoremediation of phenol. Plant Sci 169:1102–1111
107. Gonzalez PS, Capozucca CE, Tigierm HA, Milrad SR, Agostini E (2006) Phytoremediation of phenol from waste water, by peroxidases of tomato hairy root cultures. Enzyme Microb Technol 39:647–653
108. Flocco CG, Giulietti AM (1998) In vitro hairy root cultures as a tool for phytoremediation research. In: Willey N (ed) Methods in biotechnology, vol. 23: phytoremediation: methods and reviews. Humana Press, Totowa
109. Santos D, Charlwood BV, Pletsch M (2002) Tolerance and metabolism of phenol and chloroderivatives by hairy root cultures of *Daucus carota* L. Environ Pollut 117:329–335
110. Suresh B, Sherkhane PD, Kale S, Eapen S, Ravishankar GA (2005) Uptake and degradation of DDT by hairy root cultures of *Cichorium intybus* and *Brassica juncea*. Chemosphere 61:1288–1292
111. Mackova M, Macek T, Kucˇerova P, Burkhard J, Pazlarova J, Demnerova K (1997) Degradation of polychlorinated biphenyls by hairy root culture of *Solanum nigrum*. Biotechnol Lett 8:787–790
112. Prasad MNV, Freitas HMO (2003) Metal hyperaccumulation in plants – biodiversity prospecting for phytoremediation technology. Electron J Biotechnol 6(3):285–321
113. Dushenkov S, Kapulnik Y (2000) Phytofilitration of metals. In: Raskin I, Ensley BD (eds) Phytoremediation of toxic metals – using plants to clean-up the environment. Wiley, New York, pp 89–106
114. Gratão PL, Prasad MNV, Cardoso PF, Lea PJ, Azevedo RA (2005) Phytoremediation: green technology for the clean up of toxic metals in the environment. Braz J Plant Physiol 17:53–64
115. Seregin I, Kozhevnikova A (2008) Roles of root and shoot tissues in transport and accumulation of cadmium, lead, nickel, and strontium. Russ J Plant Physiol 55:1–22

116. Cunningham SD, Berti WR, Huang JW (1995) Phytoremediation of contaminated soils. Trends Biotechnol 13(9):393–397
117. Kabata-Pendias A, Pendias H (2001) Trace elements in soils and plants, 3rd edn. CRC Press, Boca Raton, p 432
118. Boonyapookana BP, Parkplan S, Techapinyawat RD, DeLaune RD, Jugsujinda A (2005) Phytoaccumulation of lead by sunflower (*Helianthus annuus*), tobacco (*Nicotiana tabacum*), and vetiver (*Vetiveria zizanioides*). J Environ Sci Health 40:117–137
119. Cunningham SD, Berti WR (2000) Phytoextraction and phytostabilization: technical, economic, and regulatory considerations of the soil-lead issue. In: Terry N, Banuelos G (eds) Phytoremediation of contaminated soils and water. CRC Press, Boca Raton, pp 365–369
120. Odjegba VJ, Fasidi IO (2004) Accumulation of trace elements by *Pistia stratiotes*: implications for phytoremediation. Ecotoxicology 13:637–646
121. Wei SH, Zhou QX, Wang X, Cao W, Ren LP, Song YF (2004) Potential of weed species applied to remediation of soils contaminated with heavy metals. J Environ Sci 16:868–873
122. Karczewska A, Orlow K, Kabala C, Szopka K, Galka B (2011) Effects of chelating compounds on mobilization and phytoextraction of copper and lead in contaminated soils. Commun Soil Sci Plant Anal 42:1379–1389
123. Seth CS, Misra V, Singh RR, Zolla L (2011) EDTA-enhanced lead phytoremediation in sunflower (*Helianthus annuus* L.) hydroponic culture. Plant and Soil 347:231–242
124. Shahid ME, Pinelli B, Pourrut J, Silvestre DC (2011) Lead-induced genotoxicity to *Vicia faba* L. roots in relation with metal cell uptake and initial speciation. Ecotoxicol Environ Saf 74:78–84
125. Zhao H, Lin L, Yan Q, Yang Y, Zhu X, Shao J (2011) Effects of EDTA and DTPA on lead and zinc accumulation of ryegrass. J Environ Prot 2:932–939
126. Lombi E, Zhao FJ, Dunham SJ, MacGrath SP (2001) Phytoremediation of heavy metal-contaminated soils: natural hyperaccumulation versus chemically enhanced phytoextraction. J Environ Qual 30:1919–1926
127. Kamnev AA, van der Lelie D (2000) Chemical and biological parameters as tools to evaluate and improve heavy metal phytoremediation. Biosci Rep 20:239–258
128. Alkorta I, Hernández-Allica J, Becerril JM, Amezaga I, Albizu I, Garbisu C (2004) Recent findings on the phytoremediation of soils contaminated with environmentally toxic heavy metals and metalloids such as zinc, cadmium, lead, and arsenic. Rev Environ Sci Biotechnol 3:71–90
129. Karenlampi S, Schat H, Vangronsveld J, Verkleij JAC, van der Lelie D, Mergeay M, Tervahauta AI (2000) Genetic engineering in the improvement of plants for phytoremediation of metal polluted soils. Environ Pollut 107:225–231
130. Pilon-Smits E, Pilon M (2002) Phytoremediation of metals using transgenic plants. Crit Rev Plant Sci 21:439–456
131. Kensuke TK, Fujimaki S, Fujiwara T, Yoneyama T (2007) Quantitative estimation of the contribution of the phloem in cadmium transport to grains in rice plants (*Oryza sativa* L.). Soil Sci Plant Nutr 53:72–77
132. Uraguchi S, Mori S, Kuramata M, Kawasaki A, Arao T, Ishikawa S (2009) Root- to-shoot Cd translocation via the xylem is the major process determining shoot and grain cadmium accumulation in rice. J Exp Bot 60:2677–2688
133. Pulford ID, Watson C (2003) Phytoremediation of heavy metal-contaminated land by trees – a review. Environ Int 29:529–540
134. Dickinson NM, Turner AP, Watmough SA, Lepp NW (1992) Acclimation of trees to pollution stress: cellular metal tolerance traits. Ann Bot 70:569–572
135. Watmough SA, Dickinson NM (1995) Dispersal and mobility of heavy metals n relation to tree survival in an aerially contaminated woodland soil. Envrion Pollut 90:135–142
136. Turner AP, Dickinson NM (1993) Survival of *Acer pseudoplatanus* L. (sycamore) seedlings on metalliferous soils. New Phytol 123:509–521

137. Landberg T, Greger M (1994) Can heavy metal tolerant clones of *Salix* be used as vegetation filters on heavy metal contaminated land? In: Aronsson P, Perttu P (eds) Willow vegetation filters for municipal wastewaters and sludges. A biological purification system. Swedish University of Agricultural Sciences, Uppsala, pp 133–144
138. Unterbrunner R, Puschenreiter M, Sommer P, Wieshammer G, Tlustos P, Zupan M, Wenzel WW (2007) Heavy metal accumulation in trees growing on contaminated sites in Central Europe. Environ Pollut 148:107–114
139. Hasselgren K (1999) Utilisation of sewage sludge in short-rotation energy forestry: a pilot study. Waste Manage Res 17:251–262
140. Almeida AAF, Valle RR, Mielke MS, Gomes FP (2007) Tolerance and prospection of phytoremediator woody species of Cd, Pb, Cu and Cr. Braz J Plant Physiol 19:83–98
141. Marmiroli M, Pietrini F, Maestri E, Zacchini M, Marmiroli N, Massacci A (2011) Growth, physiological and molecular traits in Salicaceae trees investigated for phytoremediation of heavy metals and organics. Tree Physiol 31:1319–1334
142. Pulford ID, Watson C, McGregor SD (2001) Uptake of chromium by trees: prospects for phytoremediation. Environ Geochem Health 23:307–311
143. Shukla OP, Juwarkar AA, Singh SK, Khan S, Rai UN (2011) Growth and metal accumulation capabilities of woody plants during the phytoremediation of tannery sludge. Waste Manag 31:115–123

Genetic Manipulation and Its Contribution to Pharmaceuticals: Past and Future Perspectives

25

Ashwell R. Ndhlala, Bhekumthetho Ncube, Phatu W. Mashela, Ian du Plooy, Hafiz A. Abdelgadir, and Johannes van Staden

Abstract

Genetic manipulation or metabolic engineering in plants has resulted in the ability to control plants with the aim to synthesize desired plant secondary metabolites for the pharmaceutical and nutraceutical industry. The evolution of "omics" has further provided advanced resources that allow for elucidation of uncharacterized biosynthetic pathways, enabling novel pharming approaches. Enhanced production levels of alkaloids with anticancer activities in *C. roseus*, vinblastine and vincristine through genetic manipulation, have been achieved. Various natural antioxidants in the form of phenolic compounds including flavonoids and proantocyanidins have not been spared. This chapter discusses the exciting topic of metabolites in transgenic plants and the investigation into the large-

A.R. Ndhlala (✉)
Division Crop Science, Agricultural Research Council, Pretoria, South Africa

Green Technologies Research Centre, University of Limpopo, Sovenga, South Africa
e-mail: NdhlalaA@arc.agric.za; ashwelln@yahoo.com

B. Ncube (✉) • J. van Staden (✉)
Research Centre for Plant Growth and Development, School of Life Sciences, University of KwaZulu-Natal Pietermaritzburg, Scottsville, South Africa
e-mail: Ncubeb@ukzn.ac.za; bbkncube@yahoo.com; rcpgd@ukzn.ac.za

P.W. Mashela (✉)
Green Technologies Research Centre, University of Limpopo, Sovenga, South Africa
e-mail: Phatu.Mashela@ul.ac.za

I. du Plooy (✉)
Agricultural Research Council, Vegetable and Ornamental Plants (VOP), Pretoria, South Africa
e-mail: IduPlooy@arc.agric.za

H.A. Abdelgadir (✉)
Agricultural Research Council, Vegetable and Ornamental Plants (VOP), Pretoria, South Africa

University of Limpopo, Sovenga, South Africa
e-mail: AbdelgadirH@arc.agric.za

S. Jha (ed.), *Transgenesis and Secondary Metabolism*, Reference Series in Phytochemistry, DOI 10.1007/978-3-319-28669-3_27

scale production of pharmaceutical ingredients in a number of heterologous systems including plant cell and organ cultures. The chapter also reveals the unknown and unexpected results in metabolic pathway engineering which still need to be researched and understood including the behavior of transgenes in the environment. In an attempt to map the future of transgenic plants, next generation technologies are put on the spotlight.

Keywords
Genetic engineering • Hairy root cultures • Pharming • Pharmaceuticals

Contents

1 Introduction

Genetic engineering in pharmaceuticals, also known as "pharming", involves the use of genetically modified organisms including plants (transgenic plants) as factories for pharmaceutical production. The purpose of genetic manipulation or metabolic engineering in plants is to improve by increasing the production or expression of an existing compound, reduce the production of a specific compound, or introduce the production of novel compounds through the modulation of a group or single sets of enzymatic reactions in a biosynthetic pathway [1]. Pharming is achieved by the use of cells, organs, and/or organisms to manipulate the genetic make of a plant so as to hijack the biochemical pathways to diverge metabolic fluxes towards the upregulation or downregulation of desired compounds [1]. To complement this, the fast evolving "omics" era provides advanced resources that allow for elucidation of uncharacterized biosynthetic pathways, enabling novel pharming approaches [2].

The ability presented by the "omics" to genetically manipulate plants has led to changes in the course of plant metabolism and the potential for enhancing the content and nature of plant secondary metabolites of commercial value. This makes plants potential factories for the production of a variety of useful compounds [1]. By definition, secondary metabolites are compounds produced by specialized biosynthetic pathways in plants, often counseling their functions as defenses against invasion by pathogens, pests and herbivores while regarded as nonessential for normal growth and development. More often than not, secondary metabolites confer beneficiary effects for the survival and/or behavior of plants, microbes, insects, mammals, and most importantly as useful pharmacological agents in humans and animals. Usually the pharmacological compounds can only be obtained as extracts from medicinal plants which in most cases grow slowly and are difficult to cultivate. Once pharmacological agents are identified, chemical synthesis, which is often impractical or uneconomical due to the complexity of their molecular structures, is used for industrial production. This opens up a whole new and exciting field of metabolites in transgenic plants and the investigation into the large-scale production of pharmaceutical ingredients in a number of heterologous systems including microbes, plant cell/organ cultures, and intact plants. In this chapter, the biosynthetic platforms for pharmacological compounds in transgenic plants are discussed. We highlight bottlenecks that remain to be overcome for a successful next generation metabolic bioengineering for the benefit of humans.

2 Gene Manipulation and Plant Transformation

Under natural circumstances, secondary metabolic pathways result in the production of tens of thousands of compounds and intermediary components involved in various biological responding processes, under stimuli of specific external environmental stress elicitors as well as signal molecules of normal growth and development [3, 4]. However, in some instances, depending on compound demand, enhancing production of specific compounds using different methods is called for. Since the successful introduction and expression of diverse foreign genes in tobacco during the early 1980s, gene manipulation and plant transformation became the core technological tool in plant biology and a practical tool for cultivar improvement and enhancement of specific compounds for both plant basic research and for agricultural biotechnology applications [5, 6]. The successes of the gene manipulation tools have now been extended to include most major economic crops, vegetables, ornamental, fruit, tree, pasture plants [5], and most recently medicinal plants. However, the rapid and simultaneous developments in the evolution of technology and information technology make tabulations of transformed species quickly out of date; thus it is advisable to always locate current transformation methods for species of interest. The fast evolution of gene manipulation process is aimed for higher diversification and refinement of transformation techniques for greater convenience, higher efficiency, broader genotype range, and desired molecular characteristics of transformants.

2.1 Secondary Metabolites and Pharmaceuticals

Because of their photoautotrophic nature, plants have developed an unlimited capacity for growth and development which is driven by a relatively plastic metabolism that allows them to rapidly adapt to changing environmental factors, pathogens, and other biotic and abiotic challenges [1]. This phenomenon is reflected in the wide variety of secondary metabolites accumulated by plants in their organs. Be that as it may, the biosynthetic origin and roles of the secondary metabolites in the plant is still poorly understood, yet they are of considerable interest because of their potential industrial, pharmacological, and medicinal value [1].

Humans are in the process of mastering the use of natural products derived from plants growing wild in nature or cultivated as crops as the prime ingredients in many drugs, beverages, flavors, and cosmetics. Pharmanutraceuticals are positioned between food and pharmaceuticals (functional foods). The nutraceuticals comprise foods and substances and/or combinations of substances that consist of molecules or elements found in nature for the purpose of maintaining or improving health and treating or preventing diseases. Major food and pharmaceutical companies are currently investing heavily in research and development (R&D) and marketing budgets to secure market share and promote further growth and development. In most cases, the R&D is aimed at enhancing secondary metabolites with pharmanutraceutical potentials.

Gene manipulation has achieved significant progress in understanding and enhancing the biosynthesis of a number of pharmacologically useful secondary metabolites. The mostly studied and often used as an example of successful gene manipulation to produce transgenic plants with enhanced biosynthetic pathway and gene regulation research is that of *Catharanthus roseus* (L.) G.Don (Apocynaceae) production of anticancer compounds, vinblastine and vincristine (Fig. 1). Due to the low contents of these two compounds in naturally growing plants, it becomes

Catharanthus roseus (Madagascar periwinkle)

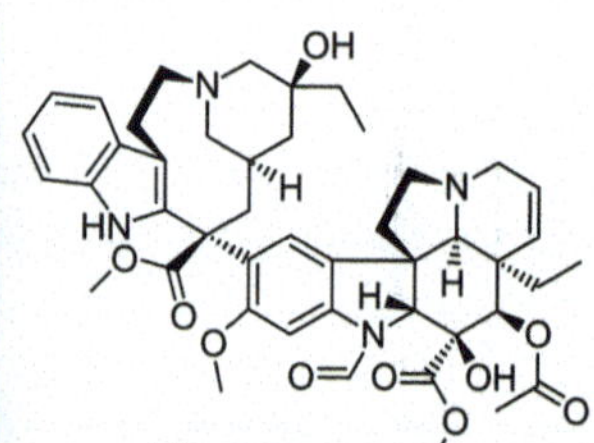

Vincristine

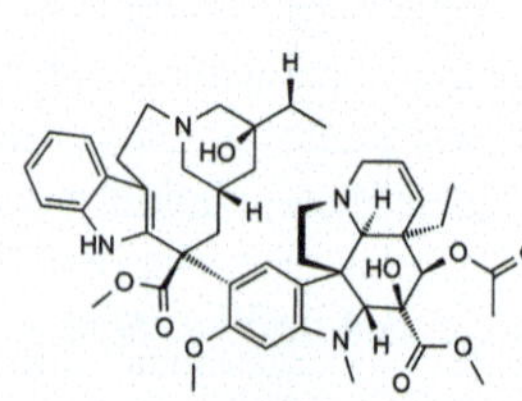

Vinblastine

Fig. 1 *Catharanthus roseus* (L.) G.Don (Apocynaceae) and the two dimeric alkaloids, vinblastine and vincristine. The two compounds are used in the treatment of acute leukemia since the 1960s. The recent advances in the "omics" have enabled researchers to characterize the key genes and their resulting enzymes in the production of vinblastine and vincristine

important to clone and regulate expression of key genes (DXR, SLS, G10H, STR) involved in the pathway, in order to achieve high yields in in vitro culture systems. This was made possible because of the in-depth understanding of the biosynthetic pathways of the terpenoid indole alkaloids in *C. roseus* [7].

3 Gene Regulatory Networks for Secondary Metabolites

Most of the plant secondary metabolites, including phenolic compounds, flavonoids, alkaloids and theanine, are important components of plant chemistry and are closely related to the health benefits exhibited by most plants. Secondary metabolite biosynthesis in plants is differentially regulated in different networks and tissues during growth and development. However, little is still known about the expression patterns of genes involved in secondary metabolic pathways or their regulatory mechanisms.

3.1 Broadening the Horizons of Cancer Therapy: A Case of Vinblastine and Vincristine from Transgenic *C. roseus*

Catharanthus roseus (Madagascar periwinkle), presented in Fig. 1, accumulates more than 100 indole alkaloids including the monomeric alkaloids such as ajmalicine, serpentine, vindoline, and catharanthine. Ajmalicine is used to treat circulatory diseases such as hypertension, while vindoline and catharanthine are precursors of dimeric alkaloids such as vinblastine and vincristine (Fig. 1) which are used in the treatment of acute leukemia since the 1960s. Vinblastine and vincristine accumulates in the areal parts of *C. roseus* in very low amounts, making them very costly for commercial use.

A whole panthera of information has been generated in the direction of understanding the growth of cell suspensions, regulation of indole alkaloid production and the signals that trigger their biosynthesis. Monomeric alkaloids are easily, though noneconomic, produced by suspension-cultured cells. Transgenic *C. roseus* presents a good procedure to improve the alkaloid production and even to create new pathways for the synthesis of new pharmaceuticals. Alkaloids are mainly derived from the decarboxylation of amino acids precursors like ornithine, lysine, tryptophan, and histidine to give respective amines. Some are derived from anthranilic acid or nicotinic acid. The remarkable ability of plants to piece together amines and different chemical partners presents a number of chemical backbones for bioactive pharmanutraceuticals [8]. Characterization of genes responsible for the generation of these backbone and central chemical structures and their development into bioactive components have revolutionized the manipulation of biosynthesis of exciting molecules.

As mentioned before, the current advances in the "omics" have enabled researchers to characterize the key genes and their resulting enzymes and in some cases clone them. *Strictosidine synthase1* (*Str1*), a key enzyme in the biosynthesis of terpenoid indole alkaloids by catalyzing the formation of strictosidine from tryptamine and the monoterpenoid secologanin in *C. roseus* has now been cloned [8, 9]. Strictosidine and secologanin are central backbones in the synthesis of monoterpenoid indole alkaloids like isoquinoline (e.g., berberine – fungitoxic and antibacterial), acridine (e.g., rutacridone – fungitoxic and antibacterial), and pyrrolizidine (e.g., scenecionine – hepatotoxic, insect pheromone precursor and antileukemic) [8, 10].

Vinblastine and vincristine are derived from the piecing together, through dimerization of catharathine and vindoline with some late stages involving cytochrome p450 monooxygenase activity. Tabersonine 16-hydroxylase (T16H) is a cytochrome p450 that has now been cloned and has added to the much anticipated manipulation of the biosynthesis of vinblastine and vincristine [11].

Perhaps in the future, functional genomics and proteomic analysis of alkaloid biosynthesis using expression-based analyses and computational modeling systems will be used to accelerate the comprehensive understanding of specialized and compartmentalization in the production of vinblastine and vincristine. For example, a glimpse into the future suggests that cell-specific compartmentation of alkaloid biosynthesis in *C. roseus* occurs in the epidermal, idioblast, and laticifer cells [12]. This is valuable information which could be used to enrich these pathways and allow growth of specific cells, in this case epidermal, idioblast, and laticifer, in a laboratory for the manufacturing of specific alkaloid as factories.

3.2 Enhancing Production of Antioxidants Including Vitamin A, C, and E in Plants

Most organisms including animals, microbes, and plants derive their energy for use in other processes from the oxidation of foodstuffs in the cells. However, there is a need to strike a balance between the oxygen consumption and metabolism as this process can lead to the production of excess aggressive forms of oxygen, reactive oxygen species (ROS), which are capable of serious damage to cell constituents, including membranes and DNA in humans and plants. Plant cells have evolved mechanisms of mopping up ROS, preventing uncontrolled oxidation, regulation of electron transport processes, and control of enzymatic reactions through natural antioxidants including flavonoids, phenols, tannic acid, glutathione, ascorbic acid, carotenoids, and enzymatic antioxidants such as superoxide dismutase (SOD). The production of compounds in plants with antioxidative capacity has generated interest among human health researchers and has directed human nutrition towards the use of enriched and biofortified foods with potential to prevent and decrease incidence of several diseases. With recent advances in genetic engineering to enhance the production of useful secondary metabolites in crop plants, there has been a renewed interest among scientists especially human nutritionists to enhance well-known antioxidants such as vitamins (Fig. 2) required for metabolic functions, i.e., vitamin C, vitamin E, vitamin B, and phytochemicals such as phenolic compounds (Fig. 3) including catechins, carotenoids, β-carotene, lycopene, zeaxanthin, diterpene, curcumin, and anthocyanins [13].

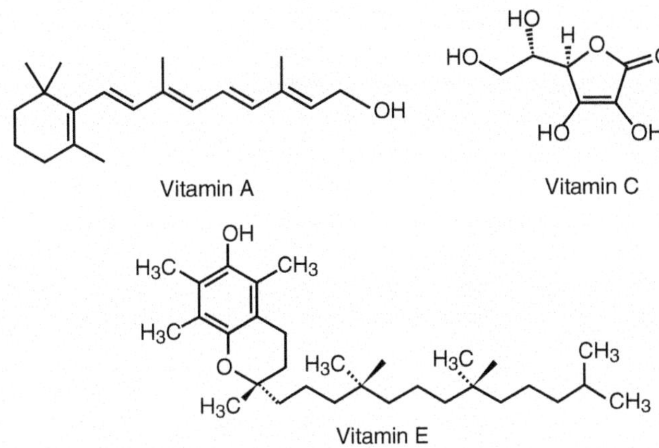

Fig. 2 Chemical structures of vitamins A, C, and E

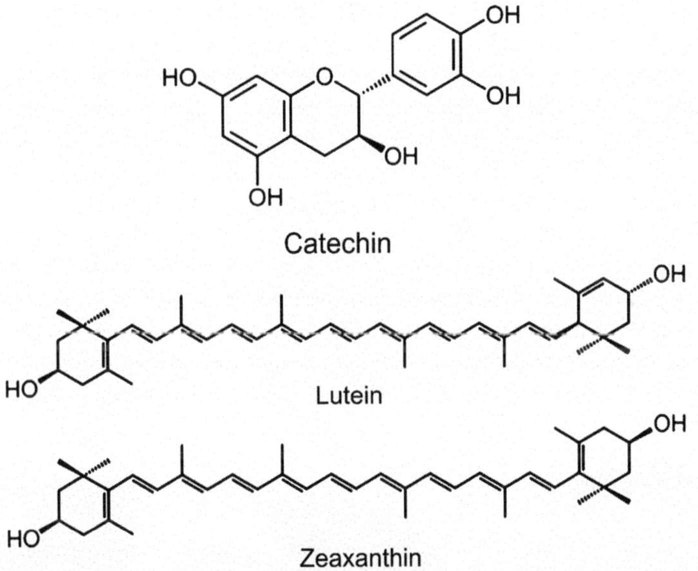

Fig. 3 Chemical structures of catechin, lutein, and zeaxanthin whose production can be enhanced through genetic manipulation of specific plants

3.2.1 Targeting Vitamin A

Dietary shortages of vitamin A (Vitamin A Deficiency – VAD) were estimated to cause fatalities to about 190 million children and 19 million pregnant women, in 122 countries in 2005 [14]. VAD is also responsible for more than 500,000 cases of irreversible blindness and millions of cases of xerophthalmia annually. The most vulnerable group is children under the age of five and pregnant women. Perhaps genetic engineering of plants could offer the antidote to VAD among children and pregnant women [15]. In many communities where the diet is deficient in vitamin A, the much needed vitamin is supplemented orally and by injection. Be that as it may, several agricultural strategies coupled with genetic engineering have been employed across the world to combat undernutrition and most particularly VAD. In a classical example, a nutritious orange fleshed sweet potato variety is growing in popularity and becoming an important strategy to improve VAD across Africa.

In South Africa, the Agricultural Research Council, Institute of Vegetable and Ornamental Plants (ARC-VOP) is at the helm of promoting the vitamin A rich orange-fleshed sweet potato (OFSP) [*Ipomoea batatas* (L.) Lam – Convolvulaceae]. Funded by the South African Government's Department of Rural Development and Land Reform (DRDLR), agricultural extension researchers work closely with farmers, groups, and other parties to ensure widespread OFSP availability and sustainability. Elsewhere in Africa, OFSP is being disseminated with support from USAID under Feed the Future, the US Government's global hunger and food security initiative.

The R2R3-type protein IbMYB1 is a key regulator of some secondary metabolite biosynthesis in the storage roots of OFSP. Future transgenic work will likely look into introducing other necessary pharmaceutical ingredients into OFSP (Fig. 4). Recent studies on R2R3-type protein IbMYB1 have provided insight into further biofortification of OFSP by introducing genes that will overexpress anthocyanin pigmentation together with a provitamin A molecule, β-carotene (Park et al. [16]). Consequently, all IbMYB1 transgenic plants will have much higher antioxidant activities compared to the normal OFSP [16].

In another initiative, the ARC-VOP in collaboration with the Council for Scientific and Industrial Research (CSIR) which is South Africa's central and premier scientific research and development organization and Nestle (South Africa) is involved in the production of a biofortified household product noodles with a provitamin A rich *Amarathus* sp., dubbed *morogo* (indigenous vegetables) 2 min noodles (Fig. 4). Presently, the *Amarathus* sp. used is not transgenic, but elicitation

Fig. 4 β-Carotene biofortified foods: *Golden* rice with the potential of producing 23-fold increased β-carotene. *Orange*-fleshed sweet potato with the R2R3-type protein IbMYB1 set to further increase β-carotene in the edible parts of the crop, *Morogo* (indigenous vegetables) 2 min noodles biofortified with a provitamin A rich *Amarathus* sp.

methods of boosting the production of provitamin A β-carotene are being used to maximize the vitamin A benefit. However, future trend will focus on transgenic *Amarathus* sp. for the purpose.

In another classical example, genetic engineering has given rise to golden rice, a variety of rice (*Oryza sativa* L. – Poaceae) produced to biosynthesize large amounts of β-carotene, a provitamin A molecule, in the edible parts of rice. The transgenic golden rice differs from its parental strain by the addition of three β-carotene biosynthesis genes. The parental rice plant can naturally produce β-carotene in its leaves, where it is utilized during the photosynthesis processes. The plant does not normally produce the pigment in the endosperm, where photosynthesis does not occur. With the evolution of transgenic tools, a second generation variety of golden rice, 2, was announced, which produces elevated β-carotene up to 23-fold to the original golden rice. Issues on bioavailability of the β-carotene from golden rice have been tested and confirmed. This will make both the OFSP and golden rice effective sources of vitamin A for humans and will offer solutions for a range of natural-source β-carotene formulations that offer ease of use, superior color stability, and guaranteed minimum color intensity for the pharmaceutical industry.

3.2.2 Enhancing Vitamin C

Vitamin C or L-ascorbic acid has been classified as an essential nutrient for humans and many other animal species. It is also one of the well-researched plant secondary metabolites, possessing important pharmacological properties including being an antioxidant, enzyme cofactor, electron donor, and acceptor in the electron transport system. Apart from these, Vitamin C also participates in several physiological processes, among others, immune stimulation, synthesis of collagen, hormones, neurotransmitters, and iron absorption [17]. Its official IUPAC name is 2-oxo-L-threo-hexono-1,4-lactone-2,3-enediol. Deficiencies in vitamin C are widespread, affecting both developing and developed countries, with the latter actually trying to overcome this lack through dietary supplements and food fortification. It is therefore apparent that new strategies aimed to increase vitamin C production in plants would be of interest to benefit human health absorption [17].

All higher plants, some higher animals and a number of yeasts are able to synthesize ascorbic acid. However, some animals, including humans, are unable to synthesize the molecules due to the nonfunctional L-gulono-1,4-lactone oxidase gene which acts in the last step of the molecule's synthesis. The major pathway of ascorbic acid synthesis is the Smirnoff-Wheeler pathway where the molecule is synthesized from D-glucose via a complex 10-step pathway involving phosphorylated sugar intermediates and sugar nucleotides [18]. Interestingly, the pathway can be enhanced to provide a living bioreactor for vitamin C production in optimal growing conditions. The enhancement can be done through genotype selection by genetic engineering, classical breeding, and changes of the agronomic conditions, on the basis of the emerging concepts that plants can enhance vitamin C synthesis as part of defense responses [17].

As mentioned earlier, in developed countries vitamin C supplementation is largely adopted especially for preventing/reducing cold-related diseases. It has been confirmed that the industrial production of vitamin C for the pharmaceutical industry represents a low efficient and expensive technology [17, 19]. In the last decade, an interesting phenomenon has also been demonstrated which shows that vitamin C from plant-derived food is more bioavailable than the chemically synthesized or purified molecule used in supplementation in the pharmaceutical scenario [17, 20]. This is because the molecule works in cohort with other molecules like vitamin E, polyphenols, and flavonoids in their pharmacological actions.

3.2.3 Enhancing Vitamin E

The term vitamin E is generic to a group of lipid-soluble antioxidant compounds, the tocochromanols, more specifically tocopherols and tocotrienols. Tocotrienols form the primary form of vitamin E in most plants. They contribute to the nutritive and pharmaceutical value of most plants as potent antioxidants in human diets, protecting polyunsaturated fatty acids against lipid peroxidation [21]. The initial stages of vitamin E synthesis involve the production of homogentisate from hydroxyphenylpyruvate (HPP) in a complex enzymatic reaction involving hydroxyphenylpyruvate dioxygenase (HPPD). The dioxygenase enzyme, HPPD has a key location in the tocochromanol pathway and is an important activity regulating tocochromanol fluxes in plants. However, overexpression of HPPD in Arabidopsis [*Arabidopsis thaliana* (L.) Heynh – Brassicaceae] and tobacco (*Nicotiana tabacum* L. – Solanaceae) resulted in the increase of tocopherols [22]. This is due to the fact HPP production is regulated by feedback inhibition of arogenate dehydrogenase by its product Tyr [22]. In the efforts to enhance the production of vitamin E, researchers succeeded in bypassing this feedback inhibition by expressing in tobacco a yeast (*Saccharomyces cerevisiae*) prephenate dehydrogenase (PDH) that catalyzes HPP production directly from prephenate [23]. This was even enhanced by the coexpression of PDH and Arabidopsis HPPD in tobacco which resulted in up to eightfold increase in vitamin E production in the tobacco leaves. Surprising this resultant eightfold increase in vitamin E production was mainly due to the increase in tocotrienols, which are normally produced in tobacco seeds but not in tobacco leaves. This presents a classical breakthrough in enhancing the production of vitamin E using transgenic plants.

3.3 Modulating Production of Functional Phenolic Compounds: A Case of Flavonoids

Among the major groups of bioactive compounds in plants, phenolic acids are the most abundant and provide the most beneficiary properties for human health. These are molecules containing a phenolic ring and an organic carboxylic acid function that function predominantly in electron exchange system by donating or accepting electrons which can delay or inhibit the oxidation of biomolecules such as DNA, proteins, and lipids [24]. Apart from this, the bioactive properties of phenolic acids are numerous. This has resulted in an increasing interest in production of plants with enhanced content of phenolic acids. Increasing the content in phenolic acids of plants can be achieved by a variety of means, including development of improved cultivars, use of specific cultivation conditions, and application of postharvest treatments as well as genetic engineering to produce new cultivars [24].

Flavonoids are a ubiquitous group of plant secondary metabolites with numerous pharmacological functions ranging from antioxidants, anti-inflammatory, anticancer, antimicrobial and inhibition of coagulation, thrombus formation or platelet aggregation in blood, reducing the risks of cardiovascular diseases. They consist of a family of phenolic molecules with variable structures naturally occurring in vegetables, fruits, grains, flowers, and beverages. In the last two decades, the flavonoid biosynthetic pathway in plants together with the numerous enzymes required for the production of different structures and their regulation have been well studied and have been characterized [25]. Using this knowledge, enhancing flavonoid biosynthesis in plants may provide unlimited ingredients that have the potential for use in novel pharmanutraceuticals designed to benefit human health. Several plant models have been used including maize, petunia tomato, and Arabidopsis. With understanding of the biosynthesis pathways, genetic manipulation has made it possible to generate several tomato lines with altered flavonoid content. Of most notable success was the ectopic expression of a biosynthetic enzyme, chalcone isomerase which resulted in up to 78-fold increase in total fruit flavonols [25].

4 Potential of the Hairy Root for Enhancing Active Metabolite Production

During the last two decades, there has been much excitement about the ability to genetically engineer plants using gene isolation and insertion techniques. These techniques allow for the construction of transgenic plants that contain and express a single, well-defined gene from any source including microbes, animals, or other plant species. Hairy root cultures are obtained from the root transformation with *Agrobacterium rhizogenes* that inserts transfer-DNA into the genomes of the infected root cells, unbalancing their hormone physiology. The genetically transformed root cultures can produce and accumulate high levels of secondary metabolites comparable to that of intact plants. Hairy root cultures are currently used for high throughput production and enhanced bioactivity of some secondary metabolites used as pharmanutraceuticals, pigments, and flavors from many plants.

Rosmarinic acid has been found to accumulate in the hairy root cultures of *Salvia miltiorrhiza* Bunge (Lamiaceae, Danshen). Rosmarinic acid is a potent antioxidant with potential for Alzheimer's disease treatment [26]. *Arachis hypogaea* L. (Fabaceae, peanuts) hairy root cultures were reported to secrete pharmacologically important stilbenoids including 30-fold increase in resveratrol, arachidin-1,m and arachidin-3 [27]. Enhancement of many other pharmacologically important compounds has been reported in the literature, these include the anti-inflammatory compound phenylethanoid verbascoside from hairy root cultures of *Harpagophytum procumbens* (Burch.) DC. ex Meisn. (Pedaliaceae, Devils claw) [28]; antioxidant compounds caftaric, caffeic, and chlorogenic acids in *Echinacea purpurea* (L.)

Moench (Asteraceae, coneflower) hairy root cultures [29]; and the antimalarial drug artemisinin in hairy root cultures of *Artemisia annua* L (Asteraceae) [30].

5 Finding Nemo: Exploring Transgenic Organisms of the Aqua World for Important Secondary Metabolites

Recent trends in drug research from natural sources have shown that organisms from the aqua world, more especially algae, are promising organisms to furnish novel biochemically active compounds [31]. Just like plants, freshwater and marine algae have developed defense strategies for competitive survival, resulting in production of numerous structural and chemical diverse secondary metabolites from different metabolic pathways [32]. Currently, around 18 million tonnes (wet weight) of aquatic plants and organisms are cultivated/harvested annually with an estimated value of 4.4 billion US dollars [33]. Exploration of algae for pharmaceutical purposes has resulted in numerous prototypes for the discovery of new agents. This stimulates the use of sophisticated physical techniques and new syntheses of compounds with biomedical application. One such technique will be transgenesis for the enhancement of specific secondary metabolites.

However, transgenesis in algae is complex requiring selectable transformation techniques, and other genetic tools [34]. Be that as it may, transgenesis methods for some algae are already available with about 25 species already accessible to genetic transformation [34]. Recently, sequencing projects have been completed for several of the important species, providing the vast amount of genomic data on a number of algae, dramatically enlarging the algae's molecular toolbox [34]. Already, genetically modified algae are being used as bioreactors for the production of biofuel, recombinant antibodies, vaccines, and insecticidal proteins [35]. This unlocks promises and potentials on a much broader field of application, metabolic pharming, with emphasis on production of proteins or metabolites that are valuable to medicine. Perhaps the genetic modifications aimed at enhancing the physiological properties of algal strains and optimization of algal production systems will further improve the potential of this auspicious technology in the future.

6 The Curse of the Transgenic Plant: Biosafety Issues Surrounding Transgenic Plants

Because the metabolic pathways in plants and other organisms are intertwined to one another to form a complex system, it is expected that perturbation of a single step in the network usually have extensive effects on metabolic flux [36]. Metabolic engineering usually focuses on production of only one metabolite or a single metabolic gene and normally generates unexpected metabolic consequences [37]. In other scenarios, can promiscuous transgenic plants spread genes to other plants? Transgenic plants, like wild plants, are expected to breed with closely related species to produce hybrids, a phenomenon called out-crossing. Similarly, the possibility of transgenic algae cultured near natural surface waters raises questions as raised above about transgenic plants and their potential to become invasive [35]. In worst-case scenarios, escaped transgenic algae might persist and produce toxins or might become so abundant that they create harmful algal blooms. If it is possible for free-living GE algae to become more invasive, more toxic, or more tolerant of extreme abiotic conditions than their wild counterparts are, this would be cause for concern [35]. So far, too little attention has been paid to these ecological questions. Research to support environmental risk assessment of novel transgenic plants and algae should be prioritized.

7 The Next-Generation Metabolic Engineering

One possible approach for improving metabolic engineering lies in the ability to predict variant effects on the product function and stability of the products of transgenic plants [38]. With the emergence of next-generation sequencing technologies, genetic variation in transgenic plants can now be determined on an unprecedented scale and resolution by resequencing thousands of plants systematically. To complement this, constraint-based reconstruction and analysis (COBRA) methods have gained popularity, tools for next-generation metabolic engineering [39]. The technique uses a genome-scale *in silico* representation of the metabolic network of a host organism to predict optimal genetic modifications that improve the rate and yield of secondary metabolite synthesis and production. A new generation of COBRA models and methods is under review to encompass a host of biological processes and simulation strategies to enable new types of predictions [39]. This will result in more efficient and precise production of desired secondary metabolites from transgenic plants. COBRA methods can also be used to predict some biosafety issues associated with transgenic plants. However, these methods are highly dependent on technology and database tools to predict variant metabolites.

8 Conclusions

Advanced genomics and biotechnology has resulted in the production of transgenic plants with the ability to synthesize desired plant secondary metabolites for the pharmaceutical and nutraceutical industry, in a phenomenon called pharming. A classic example is the production of golden rice and OFSP, with the ability to produce 23-fold β-carotene, a provitamin A molecule. Similarly, enhanced levels of alkaloids with anticancer activities in *C. roseus*, vinblastine and vincristine, have been achieved. Various natural antioxidants in the form of phenolic compounds including flavonoids and proantocyanidins have not been spared. Several techniques including hairy root cultures have been employed to enhance production of these. However, the unknown and unexpected results in metabolic pathway engineering still need to be researched and understood. The behavior of these transgenes in the environment still leaves a lot to be desired. While a decade ago production of transgenic plants was expensive, time consuming, and had limitations on the host plants, next generation technologies are bringing efficient and precise methods enabling researchers to quickly edit genomes in a precise and accurate fashion at the single base-pair resolution level at multiple loci simultaneously. Perhaps it is what now lies in the future of multi-omics which could bring a concerted action between modelers, geneticists, microbiologists, and bioinformaticians to allow the full achievement of the predicted changes and novel metabolic capabilities.

Acknowledgments The work was partially supported by the National Research Foundation, Pretoria (Grants Nos. 93185 and 95662).

References

1. Teli NP, Timko MP (2004) Recent developments in the use of transgenic plants for the production of human therapeutics and biopharmaceuticals. Plant Cell Tiss Org Cult 79:125–145
2. Wilson SA, Roberts SC (2014) Metabolic engineering approaches for production of biochemicals in food and medicinal plants. Curr Opin Biotechnol 26:174–182
3. Zhao J, Davis LC, Verpoorte R (2005) Elicitor signal transduction leading to production of plant secondary metabolites. Biotechnol Adv 23:283–333
4. Lv Q, Cheng R, Shi T (2014) Regulatory network rewiring for secondary metabolism in *Arabidopsis* thaliana under various conditions. BMC Plant Biol 14:180
5. Birch RG (1997) Plant transformation: problems and strategies for practical application. Annu Rev Plant Physiol Plant Mol Biol 48:297–326
6. Gelvin SB (2010) Manipulating plant histone genes to improve *Agrobacterium*- and direct DNA-mediated plant genetic transformation. ISB News report 1–4
7. Zárate R, Verpoorte R (2007) Strategies for the genetic modification of the medicinal plant *Catharanthus roseus* (L.) G. Don. Phytochem Rev 6:475–491
8. De Luca V, Laflamme P (2001) The expanding universe of alkaloid biosynthesis. Curr Opin Plant Biol 4:225–233
9. Pasquali G, Erven AS, Ouwerkerk PB, Menke FL, Memelink J (1999) The promoter of the strictosidine synthase gene from periwinkle confers elicitor-inducible expression in transgenic tobacco and binds nuclear factors GT-1 and GBF. Plant Mol Biol 39:1299–1310
10. St-Pierre B, Vázquez-Flota FA, De Luca V (1999) Multicellular compartmentation of *Catharanthus roseus* alkaloid biosynthesis predicts intercellular translocation of a pathway intermediate. Plant Cell 11:887–900
11. De Luca V, St-Pierre B (2000) The cell and developmental biology of alkaloid biosynthesis. Trends Plant Sci 5:168–173
12. Qu Y, Easson MALE, Froese J, Simionescu R, Hudlicky T, De Luca V (2015) Completion of the seven-step pathway from tabersonine to the anticancer drug precursor vindoline and its assembly in yeast. Proc Natl Acad Sci U S A 112:6224–6229
13. Sanghera GS, Malhotra PK, Sidhu GS, Sharma VK, Sharma BB, Karan R (2013) Genetic engineering of crop plants for enhanced antioxidants activity. Int J Adv Res Technol 2:428–458
14. WHO (2009) Global prevalence of vitamin A deficiency in populations at risk 1995–2005. WHO global database on vitamin a deficiency. World Health Organization, Geneva, Retrieved 12 Mar 2016
15. Humphrey JH, West KP Jr, Sommer A (1992) Vitamin a deficiency and attributable mortality in under-5-year-olds. Bull WHO 70:225–232
16. Park SC, Kim YH, Kim SH, Jeong YJ, Kim CY, Lee JS, Bae JY, Ahn MJ, Jeong JC, Lee HS, Kwak SS (2015) Overexpression of the IbMYB1 gene in an orange-fleshed sweet potato

cultivar produces a dual-pigmented transgenic sweet potato with improved antioxidant activity. Plant Physiol 153:525–537
17. Locato V, Cimini S, De Gara L (2013) Strategies to increase vitamin C in plants: from plant defense perspective to food biofortification. Front Plant Sci 4:1–12
18. Smirnoff N, Wheeler GL (2000) Ascorbic acid in plants: biosynthesis and function. Crit Rev Biochem Mol Biol 35:291–314
19. Hancock RD, Viola R (2002) Biotechnological approaches for L-ascorbic acid production. Trends Biotechnol 20:299–305
20. Vissers MC, Bozonet SM, Pearson JF, Braithwaite LJ (2011) Dietary ascorbate intake affects steady state tissue concentrations in vitamin C-deficient mice: tissue deficiency after suboptimal intake and superior bioavailability from a food source (kiwifruit). Am J Clin Nutr 93:292–301
21. Cahoon EB, Hall SE, Ripp KG, Ganzke KE, Hitz WD, Coughlan SJ (2003) Metabolic redesign of vitamin E biosynthesis in plants for tocotrienol production and increased antioxidant content. Nat Biotechnol 21:1082–1087
22. Matringe M, Ksas B, Rey R, Havaux M (2008) Tocotrienols, the unsaturated forms of vitamin E, can function as antioxidants and lipid protectors in tobacco leaves. Plant Physiol 147:764–778
23. Rippert P, Scimemi C, Dubald M, Matringe M (2004) Engineering plant shikimate pathway for production of tocotrienol and improving herbicide resistance. Plant Physiol 134:92–100
24. Kaushik P, Andújar I, Vilanova S, Plazas M, Gramazio P, Herraiz FJ, Brar NS, Prohens J (2015) Breeding vegetables with increased content in bioactive phenolic acids. Molecules 20:18464–18481
25. Verhoeyen ME, Bovy A, Collins G, Muir S, Robinson S, de Vos CHR, Colliver S (2002) Increasing antioxidant levels in tomatoes through modification of the flavonoid biosynthetic pathway. J Exp Bot 53:2099–2106
26. Tan RH, Zhang JJ, Zhao SJ (2014) Optimization of induction and culture conditions for hairy roots of Salvia miltiorrhiza. China J Chin Mater Med 39:3048–3053
27. Abbott JA, Medina-Bolivar F, Martin EM, Engelberth AS, Villagarcia H, Clausen EC, Carrier DJ (2010) Purification of resveratrol, arachidin-1, and arachidin-3 from hairy root cultures of peanut (*Arachis hypogaea*) and determination of their antioxidant activity and cytotoxicity. Biotechnol Prog 26:1344–1351
28. Homova V, Weber J, Schulze J, Alipieva K, Bley T, Georgiev M (2010) Devil's claw hairy root culture in flasks and in a 3-L bioreactor: bioactive metabolite accumulation and flow cytometry. Z Naturforsch C 65:472–478
29. Liu CZ, Abbasi BH, Gao M, Murch SJ, Saxena PK (2006) Caffeic acid derivatives production by hairy root cultures of *Echinacea purpurea*. J Agric Food Chem 54:8456–8460
30. Liu C, Wang Y, Guo C, Ouyang F, Ye H, Li G (1998) Enhanced production of artemisinin by *Artemisia annua* L hairy root cultures in a modified inner-loop airlift bioreactor. Bioprocess Eng 19:389–392
31. Cardozo KHM, Guaratini T, Barros MP, Falcão VR, Tonon AP, Lopes NP, Campos S, Torres MA, Souza AO, Colepicolo P, Pinto C (2007) Metabolites from algae with economical impact. Comp Biochem Physiol C 146:60–78
32. Barros MP, Pinto E, Sigaud-Kutner TCS, Cardozo KHM, Colepicolo P (2005) Rhythmicity and oxidative/nitrosative stress in algae. Biol Rhythm Res 36:67–82
33. Kannan RRR, Aderogba MA, Ndhlala AR, Stirk WA, Van Staden J (2013) Acetylcholinesterase inhibitory activity of phlorotannins isolated from the brown alga, *Ecklonia maxima* (Osbeck) Papenfuss. Food Res Int 54:1250–1254
34. Hallmann A (2007) Algal transgenics and biotechnology. Transgenic Plant J 1:81–98
35. Snow AA, Smith VH (2012) Genetically engineered algae for biofuels: a key role for ecologists. BioScience 62:765–768
36. Nielsen J, Olsson L (2002) An expanded role for microbial physiology in metabolic engineering and functional genomics: moving towards systems biology. FEMS Yeast Res 2:175–181

37. Sweetlove LJ, Last RL, Fernie AR (2003) Predictive metabolic engineering: a goal for systems biology. Plant Physiol 132:420–425
38. Cardoso JGR, Andersen MR, Herrgård MJ, Sonnenschein N (2015) Analysis of genetic variation and potential applications in genome-scale metabolic modeling. Front Bioeng Biotechnol 3:1–12
39. King ZA, Lloyd CJ, Feist AM, Palsson BO (2015) Next-generation genome-scale models for metabolic engineering. Curr Opin Biotechnol 35:23–29

Index

S. Jha (ed.), *Transgenesis and Secondary Metabolism*, Reference Series in Phytochemistry, DOI 10.1007/978-3-319-28669-3

The manufacturer's authorised representative in the EU is Springer
Nature Customer Service Centre GmbH, Europaplatz 3, 69115 Heidelberg,
Germany. If you have any concerns regarding our products, please
contact ProductSafety@springernature.com

Printed and bound by CPI Group (UK) Ltd, Croydon, CR0 4YY
05/10/2026
02228129-0008